Mucosal Delivery of Biopharmaceuticals

José das Neves • Bruno Sarmento
Editors

Mucosal Delivery of Biopharmaceuticals

Biology, Challenges and Strategies

Springer

Editors
José das Neves
IINFACTS – Instituto de Investigação e Formação Avançada em Ciências e Tecnologias da Saúde
Instituto Superior de Ciências da Saúde-Norte, CESPU
Gandra, Portugal

Bruno Sarmento
NEWTherapies Group
INEB – Instituto de Engenharia Biomédica
Porto, Portugal

ISBN 978-1-4614-9523-9 ISBN 978-1-4614-9524-6 (eBook)
DOI 10.1007/978-1-4614-9524-6
Springer New York Heidelberg Dordrecht London

Library of Congress Control Number: 2013958086

Printed on acid-free paper

Springer is part of Springer Science+Business Media (www.springer.com)

Foreword

We have witnessed great progress in the delivery of biopharmaceuticals in the past few decades, empowered by multidisciplinary interactions in biomaterial science, polymer science, protein and peptide chemistry, recombinant biotechnology, isolation/purification and processing technology, formulation technology, nanotechnology including aerosol and micro/nanoparticle research, and many more. Delivery science involves transdermal, oral, intravenous/arterial/subcutaneous, pulmonal/tracheal, nasal, ocular, vaginal, vesical, cavitary and enteric routes, and more directly target-oriented arterial infusion of nanomedicines warrants both better delivery and marked clinical efficacy, and lesser adverse effect. Transmucosal delivery plays a very important role in biopharmaceutical delivery and this work covers the state-of-the art of the science and technology involved.

Nasal, buccal, pulmonary, ocular, vaginal and enteric routes have become more popular recently. For any of these routes of biopharmaceutical drug administration, multi-disciplinary factors or components need to be well considered. As drug action becomes more direct, adverse effects could be so as well. For instance, the presence of a surface-active agent or permeability enhancer that might facilitate the interaction with dendritic cells or immune cells more effectively might result in different outcome than without detergent. There was an incidence of an allergic reaction in a soap preparation, in which a portion of hydrolyzed wheat protein was used as foam stabilizer [1]. This protein fragment effectively penetrated epidermis in the presence of detergents, and it became the cause of allergic reaction in some users with fatal accidents.

Progress in nanotechnology—including more diverse requirements in nanomedicine—such as sustained release, stealth character, long plasma half-life, with receptor binding probes, or biocompatibility and yet efficient cell uptake, need to be achieved to fulfill the needs of pharmacological properties that would ultimately benefit patients. In cases of cancer and inflammation, we can take advantage of the enhanced vascular permeability or extravation property of nanoparticles (by EPR effect) at the diseased site. This effect can be further enhanced by modulating vascular mediators such as nitric oxide releasing agents or by elevating blood pressure. Conventional low molecular weight (MW) anticancer drugs rarely exhibit intratumoral drug accumulation more than 2-fold of plasma level. Thus, systemic toxicity is unavoidable at the therapeutic dose of low MW drugs. These inherent drawbacks of low MW drugs could be solved by using nanomedicines. This pathophysiological uniqueness of the targeted lesion may be utilized in many clinical settings.

In considering many of these strategies, chemical and physical properties are important components to make optimization for appropriate biopharmaceutical formulation. Chemistry can offer polymers, macromolecular biomaterials or bioadhesives that fit the objectives for a given biopharmaceutical delivery *in vivo*. In this book, a wide range of these issues are discussed, and many state-of-the-art science and technology in mucosal biopharmaceutical drug delivery can be found as a useful sources of reference for present and future investigations.

Reference

1. Chinuki Y, Morita E. Wheat-dependent exercise-induced anaphylaxis sensitized with hydrolyzed wheat protein in soap. Allergol Int. 2012;61(4):529–37.

August 2013

Hiroshi Maeda, Ph.D., M.D.
Sojo University
Institute of Drug Delivery System Science,
Kumamoto, Japan

Preface

Recent years have seen the rise of biopharmaceuticals as promising tools in the treatment, prophylaxis and diagnosis of multiple diseases. With more than 300 molecules approved worldwide (while others are on late stages of development) and an estimated global market of over US$ 166 billion by 2017 [1], biopharmaceutical medicines are now an important part of the armamentarium of modern therapeutics [2–3]. The term "biopharmaceuticals" (often used interchangeably with "biologics" or "biological products") is widely used, but its definition has been often neglected and a topic of discussion [4]. For the purpose of this book, biopharmaceuticals are broadly defined as molecules with inherently biological origin and/or manufactured using biotechnological techniques that usually comprise the use of living organisms, cells or their components. This class of pharmaceuticals is fairly heterogeneous and includes different molecular entities such as protein- and peptide-based molecules (antibodies, hormones, toxins, enzymes, growth factors, among others), and genetic material (plasmid DNA, small interference RNA, ribozymes, aptamers, among others).

Even if parenteral routes are typically considered for their administration, the mucosal delivery of biopharmaceuticals may present important advantages that make it preferential, namely by providing direct access to target sites, abbreviating patient compliance issues, mimicking physiological processes, enhancing safety, and allowing taking advantage of the distinctive characteristics of the mucosal immune system. However, frequent unfavorable physical-chemical properties of these active biomolecules lead to reduced stability in different biological fluids and poor permeability. This poses an important hurdle to their mucosal administration and the attainment of significant bioavailability values. In particular, challenges in developing adequate materials and delivery systems that allow the use of biopharmaceuticals in daily life are huge [5].

Accumulated knowledge and achievements in developing successful biopharmaceutical delivery systems that may explore the mucosal pathway to exert local effects or enter the bloodstream are emerging. This book aims at providing a concise and up-to-date overview of the biological features justifying the use of different human mucosa as delivery routes for biopharmaceuticals, the technological strategies that have been adopted so far regarding the optimization of mucosal potentialities, as well as the challenges that arise with the advent of new biopharmaceutical drugs and alternative means of administration. These exciting and innovative topics are addressed in their different perspectives by some of the most important academic authorities and industrial experts in the field. The work is divided into four parts. The first section of this book addresses general aspects of the biology of mucosal tissues and their unique aspects towards beneficial or deleterious interaction with biopharmaceuticals and their delivery systems. The second section is dedicated to the different delivery strategies that have recently been investigated for different mucosal sites. The third section describes the development and clinical applications, either factual or potential, of particular pharmaceutical delivery systems/products enclosing biopharmaceuticals for mucosal delivery. Special focus is set on the most successful case studies of recent years, namely by some of those directly engaged in developing such solutions in a concise and practical way. The last section briefly centers on pertinent aspects about the regulatory, toxicological and market issues of biopharmaceuticals intended for mucosal administration.

We hope that scientists and researchers in the fields of drug delivery, materials and biomedical sciences and bioengineering, as well as professionals in the pharmaceutical, biotechnology and health-care industries will find in this work an important compendium of fundamental concepts and practical tools for their daily research and activities. In particular, extensive emphasis on case studies of successfully developed and some already marketed systems/products for mucosal delivery of biopharmaceuticals was pursued. Also, focus on regulatory issues makes this book a valuable tool for decision-makers in the pharmaceutical industry and in regulatory bodies worldwide.

January 2014

José das Neves
Bruno Sarmento

References

1. International Market Analysis Research and Consulting Group. Global Biopharmaceutical Market Report & Forecast (2012–2017). 2012.
2. Walsh G. Biopharmaceutical benchmarks 2010. Nat Biotechnol. 2010;28(9):917–24.
3. Walsh G. New biopharmaceuticals—a review of new biologic drug approvals over the years, featuring highlights from 2010 and 2011. BioPharm Int. 2012;25(6):34–38.
4. Rader RA. (Re)defining biopharmaceutical. Nat Biotechnol. 2008;26(7):743–51.
5. Jorgensen L, Nielson HM, editors. Delivery technologies for biopharmaceuticals: peptides, proteins, nucleic acids and vaccines. Chichester: Wiley; 2009.

Contents

Contributors

Kazunari Akiyoshi Department of Polymer Chemistry, Graduate School of Engineering, Kyoto University, Kyoto, Japan

ERATO Bio-nanotransporter Project, Japan Science and Technology Agency, Kyoto, Japan

António J. Almeida iMed.UL, Faculty of Pharmacy, University of Lisbon, Lisbon, Portugal

Patrick V. Almeida Division of Pharmaceutical Technology, Faculty of Pharmacy, University of Helsinki, Helsinki, Finland

Mansoor Amiji Department of Pharmaceutical Sciences, School of Pharmacy, Northeastern University, Boston, MA, USA

Fernanda Andrade Laboratory of Pharmaceutical Technology, Faculty of Pharmacy, University of Porto, Rua de Jorge Viterbo Ferreira, Porto, Portugal

Nanoprobes & Nanoswitches Group, Institute for Bioengineering of Catalonia (IBEC), Barcelona, Spain

Gavin P. Andrews School of Pharmacy, The Queen's University of Belfast, Medical Biology Centre, Belfast, Northern Ireland, UK

Yong Bai School of Public Health, University of California, Berkeley, CA, USA

Meena Bansal Mehr Chand Polytechnic College, Jalandhar, Panjab, India

Sanjay Bansal Mehr Chand Polytechnic College, Jalandhar, Panjab, India

Barbara C. Baudner Vaccine Research, Novartis Vaccines and Diagnostics Srl, Siena, Italy

David B. Bennett Pharmaceutical Consultants, Mountain View, CA, USA

Isabel Buettel Paul-Ehrlich-Institut, Federal Institute for Vaccines and Biomedicines, Langen, Germany

Rachael Burchfield Program in Comparative Biochemistry, University of California, Berkeley, CA, USA

Ana I. Camacho Department of Microbiology, University of Navarra, Pamplona, Spain

Merve Cansız Department of Pharmaceutical Technology, Faculty of Pharmacy, Hacettepe University, Ankara, Turkey

Philippe Daull Novagali Pharma, Evry, France

Katrin Féchir Paul-Ehrlich-Institut, Federal Institute for Vaccines and Biomedicines, Langen, Germany

Simona Gallorini Vaccine Research, Novartis Vaccines and Diagnostics Srl, Siena, Italy

Carlos Gamazo Department of Microbiology, University of Navarra, Pamplona, Spain

Jean-Sébastien Garrigue Novagali Pharma, Evry, France

Karen Brigitta Goetz Paul-Ehrlich-Institut, Federal Institute for Vaccines and Biomedicines, Langen, Germany

Ana Grenha CBME—Centre for Molecular and Structural Biomedicine/IBB—Institute for Biotechnology and Bioengineering, Faculty of Sciences and Technology, University of Algarve, Faro, Portugal

Michael Hinchcliffe Paracelsis Ltd, Nottingham, UK

Juan M. Irache Department of Pharmacy and Pharmaceutical Technology, University of Navarra, Pamplona, Spain

Shardool Jain Department of Pharmaceutical Sciences, School of Pharmacy, Northeastern University, Boston, MA, USA

David S. Jones School of Pharmacy, The Queen's University of Belfast, Medical Biology Centre, Belfast, Northern Ireland, UK

Takanori Kanazawa Laboratory of Pharmaceutics and Drug Delivery, Department of Pharmaceutical Science, School of Pharmacy, Tokyo University of Pharmacy and Life Sciences, Tokyo, Japan

Joseph A. Katakowski Department of Microbiology and Immunology, Albert Einstein College of Medicine, Bronx, NY, USA

Hiroshi Kiyono Division of Mucosal Immunology, Institute of Medical Science, The University of Tokyo, Tokyo, Japan

International Research and Development Center for Mucosal Vaccines, Institute of Medical Science, The University of Tokyo, Tokyo, Japan

Core Research for Evolutional Science and Technology, Japan Science and Technology Agency, Tokyo, Japan

Kenji Kono Department of Applied Chemistry, Graduate School of Engineering, Osaka Prefecture University, Sakai, Osaka, Japan

Awie F. Kotzé Unit for Drug Research and Development, North-West University, Potchefstroom, South Africa

Evelyne Kretzschmar Paul-Ehrlich-Institut, Federal Institute for Vaccines and Biomedicines, Langen, Germany

Rachna Kumria Swift School of Pharmacy, Rajpura, Panjab, India

Frédéric Lallemand Novagali Pharma, Evry, France

Fenyong Liu School of Public Health, University of California, Berkeley, CA, USA

Program in Comparative Biochemistry, University of California, Berkeley, CA, USA

Holly Lorentz Chemical Engineering, McMaster University, Hamilton, ON, Canada

Sangwei Lu School of Public Health, University of California, Berkeley, CA, USA

Program in Comparative Biochemistry, University of California, Berkeley, CA, USA

Catarina Moura Faculty of Engineering, University of Porto, Porto, Portugal

José das Neves IINFACTS – Instituto de Investigação e Formação Avançada em Ciências e Tecnologias da Saúde, Instituto Superior de Ciências da Saúde-Norte, CESPU, Gandra, Portugal

Faculty of Pharmacy, University of Porto, Porto, Portugal

Hanne Mørck Nielsen Department of Pharmacy, University of Copenhagen, Copenhagen, Denmark

Tomonori Nochi Laboratory of Functional Morphology, Graduate School of Agricultural Science, Tohoku University, Sendai, Miyagi, Japan

Derek T. O'Hagan Vaccine Research, Novartis Vaccines and Diagnostics Srl, Siena, Italy

Hiroaki Okada Laboratory of Pharmaceutics and Drug Delivery, Department of Pharmaceutical Science, School of Pharmacy, Tokyo University of Pharmacy and Life Sciences, Tokyo, Japan

Deborah Palliser Department of Microbiology and Immunology, Albert Einstein College of Medicine, Bronx, NY, USA

Willi Paul Biomedical Technology Wing, Sree Chitra Tirunal Institute for Medical Sciences and Technology, Poojappura, Thiruvananthapuram, India

Catarina Pinto Reis CBIOS—Laboratory of Nanoscience and Biomedical Nanotechnology, Universidade Lusófona, Lisbon, Portugal

Lissinda H. du Plessis Unit for Drug Research and Development, North-West University, Potchefstroom, South Africa

Michael J. Rathbone School of Pharmacy, International Medical University, Kuala Lumpur, Malaysia

Hélder A. Santos Division of Pharmaceutical Technology, Faculty of Pharmacy, University of Helsinki, Helsinki, Finland

Bruno Sarmento NEW Therapies Group, INEB – Instituto de Engenharia Biomédica, Porto, Portugal

IINFACTS – Instituto de Investigação e Formação Avançada em Ciências e Tecnologias da Saúde, Instituto Superior de Ciências da Saúde-Norte, CESPU, Gandra, Portugal

Thomas M. Scherer Genentech, Late Stage Pharmaceutical Development, San Francisco, CA, USA

Sevda Şenel Department of Pharmaceutical Technology, Faculty of Pharmacy, Hacettepe University, Ankara, Turkey

Mohammed-Ali Shahbazi Division of Pharmaceutical Technology, Faculty of Pharmacy, University of Helsinki, Helsinki, Finland

Chandra P. Sharma Biomedical Technology Wing, Sree Chitra Tirunal Institute for Medical Sciences and Technology, Poojappura, Thiruvananthapuram, India

Heather Sheardown Chemical Engineering, McMaster University, Hamilton, ON, Canada

Steven J. Shire Genentech, Late Stage Pharmaceutical Development, San Francisco, CA, USA

Catarina Oliveira Silva CBIOS—Laboratory of Nanoscience and Biomedical Nanotechnology, Universidade Lusófona, Lisbon, Portugal

Alan Smith Archimedes Development Ltd, Albert Einstein Centre, Nottingham, UK

Cynthia L. Stevenson Pharmaceutical Consultants, Mountain View, CA, USA

Yuansheng Sun Paul-Ehrlich-Institut, Federal Institute for Vaccines and Biomedicines, Langen, Germany

Sunita Prem Victor Biomedical Technology Wing, Sree Chitra Tirunal Institute for Medical Sciences and Technology, Poojappura, Thiruvananthapuram, India

Peter Watt Archimedes Development Ltd, Albert Einstein Centre, Nottingham, UK

Tao Yu School of Pharmacy, The Queen's University of Belfast, Medical Biology Centre, Belfast, Northern Ireland, UK

Eiji Yuba Department of Applied Chemistry, Graduate School of Engineering, Osaka Prefecture University, Sakai, Osaka, Japan

Yoshikazu Yuki Division of Mucosal Immunology, Institute of Medical Science, The University of Tokyo, Tokyo, Japan

International Research and Development Center for Mucosal Vaccines, Institute of Medical Science, The University of Tokyo, Tokyo, Japan

Core Research for Evolutional Science and Technology, Japan Science and Technology Agency, Tokyo, Japan

Hongbo Zhang Division of Pharmaceutical Technology, Faculty of Pharmacy, University of Helsinki, Helsinki, Finland

About the Editors

José das Neves is a researcher at Instituto Superior de Ciências da Saúde-Norte, CESPU, Gandra, and in the Faculty of Pharmacy, University of Porto, Portugal, where he earned a Ph.D. in Pharmaceutical Sciences. His previous work has spanned multiple aspects of the development of vaginal drug delivery systems, and his current research interests include the development of nanotechnology-based solutions for the development of anti-HIV microbicides and mucosal delivery of biopharmaceuticals.

Bruno Sarmento is an affiliated researcher at INEB—Instituto de Engenharia Biomédica, University of Porto, Portugal. He is also an assistant professor of pharmaceutical and biopharmaceutical technology at Instituto Superior de Ciências da Saúde-Norte, CESPU, Gandra, Portugal. He earned a Ph.D. in Pharmaceutical Technology at the University of Porto. He has extensive work in the development of nanocarriers for the oral delivery of biopharmaceuticals—namely insulin—and the establishment of novel *in vitro* intestinal permeability models. His current research focuses on nanomedicines and their applications in the pharmaceutical and biomedical fields.

Part I
Biology of Mucosal Sites

Chapter 1
Concepts in Mucosal Immunity and Mucosal Vaccines

Simona Gallorini, Derek T. O'Hagan and Barbara C. Baudner

1.1 Introduction

Mucosal surfaces of the digestive, respiratory, and reproductive systems, with a combined surface area of about 400 m^2, are the primary site for transmission of numerous viral and bacterial diseases. Therefore, mucosal tissues are in a constant state of alert, but also adapted to the presence of foreign microorganisms and their products. Most foreign antigens in the intestine are derived from food and the commensal microbial flora; both generally do not trigger defensive immune responses in spite of the fact that such antigens regularly enter the mucosa. This is because mucosal antigen-presenting cells (APCs), lymphocytes, and even the epithelium itself play important but poorly understood roles in modulating immune responses to incoming antigens. Indeed, a major role of the mucosal immune system is the downregulation or suppression of immune responses to food antigens and commensal bacteria. The exact sites or mechanisms of this oral tolerance are still controversial and have been reviewed elsewhere [1–2]. As a result, vaccines that would produce vigorous immune responses if injected into a sterile environment, such as muscle, might be "ignored" when given mucosally where the tissue is constantly exposed to microorganisms.

Overall, mucosal respiratory and gastrointestinal infections kill five million children under age five in developing countries and cause more than ten billion disease episodes each year. This has a tremendous negative impact on global health and overall economic development [3]. Similarly, there is a great need for vaccines that can protect against human immunodeficiency virus (HIV) and other sexually transmitted infections that affect millions of adults and adolescents. It is highly probable that an infection with mucosal pathogens by and inter-person transmission can be effectively controlled by mucosal vaccines, provided these vaccines are rationally designed and formulated to be administered through an appropriate route. However, the nature of the pathogen and of the target mucosal tissue will determine whether

S. Gallorini (✉) · D. T. O'Hagan · B. C. Baudner
Vaccine Research, Novartis Vaccines and Diagnostics Srl,
Via Fiorentina 1, 53100 Siena, Italy
e-mail: simona.gallorini@novartis.com

J. das Neves, B. Sarmento (eds.), *Mucosal Delivery of Biopharmaceuticals,* DOI 10.1007/978-1-4614-9524-6_1,

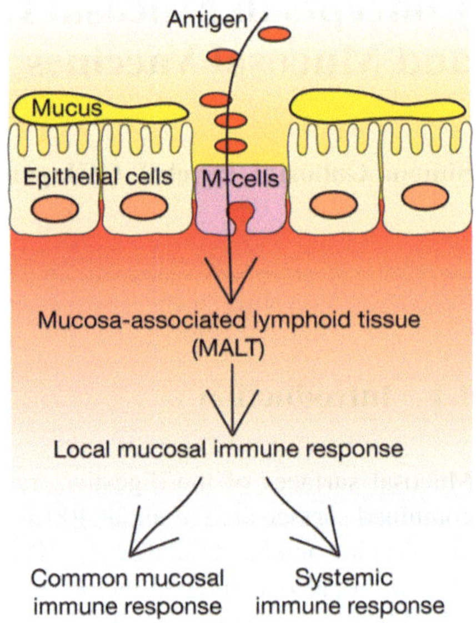

Fig. 1.1 Mucosal immunization induces systemic and mucosal immune responses. Mucosal immunization at one mucosal site can induce specific responses at distant sites through an immunological intranet—common mucosal immune system

the vaccine should be given mucosally or parenterally to be efficacious. Parenteral vaccines induce good systemic immune responses but only limited mucosal immune responses [4–5]. Mucosal vaccination more efficiently elicits mucosal immune responses at the most common sites of infectious agents entry and additionally elicits systemic immune responses (Figs. 1.1 and 1.2), thereby resulting in two layers of host protection [6–7].

A mucosal vaccination route seems to be critical for protection against non invasive infections at mucosal surfaces and infections that involve pathogens which remain on the apical (luminal) side of mucosal epithelia, i.e., at sites that are poorly accessible to antibodies transudating from blood, and where blood-derived monomeric immunoglobulin G (IgG) or immunoglobulin A (IgA) are insufficiently concentrated on the apical cell surface (due to the lack of receptor-mediated transport) or are unstable to function in the external mucosal environment. Cholera and noninvasive enterotoxigenic *Escherichia coli* (ETEC) are typical examples of infections in which vaccine-induced protection appears to be mediated mainly, if not exclusively, by locally produced secretory IgA (S-IgA) antibodies and is associated with immunological memory [7].

On the other hand, when infections occur at mucosal surfaces, like the respiratory and urogenital tract, which are more permeable to transudation by serum antibodies, or when mucosal pathogens are able to quickly enter the blood for systemic spread, a parenteral route of vaccination most likely will be effective [8]. Compared to the parenteral route, needle-free vaccine administration has many advantages, as for example the potential to improve safety for the vaccinator, vaccinee, and community. A primary safety concern is the risk of transmission of infectious diseases

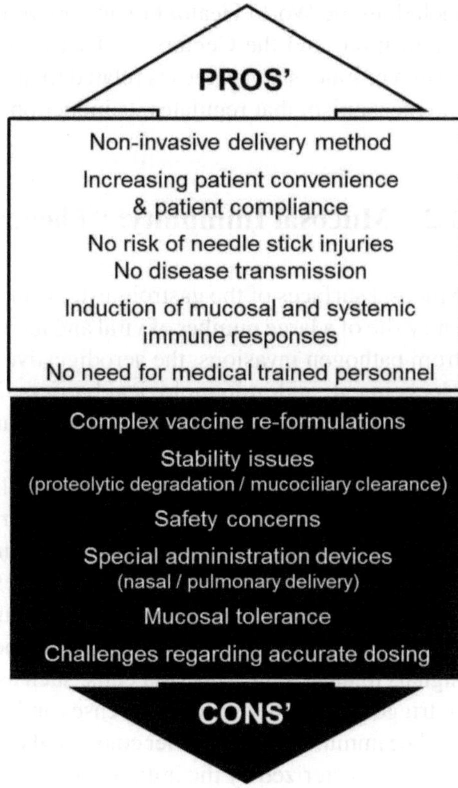

Fig. 1.2 Advantages and drawbacks of mucosal vaccination

between patients or between patients and healthcare providers. In both the developed and developing world, the administration of vaccines poses an occupational risk, through needle-stick injuries. Another advantage of needle-free vaccine delivery is an expected increase in compliance with recommended vaccination schedules. Poor compliance with schedules is often due to parental concern regarding the number of vaccine injections administered to children and to "needle phobia," which is common in both adults and children. Several recent studies have addressed fear of injections and methods to minimize pain associated with vaccines. The use of certain methods of needle-free vaccine delivery is expected to decrease pain and suffering, including actual injection site pain, anticipatory and perceived pain, and local side effects from injections. Needle-free vaccine delivery may increase the ease of vaccine delivery. Administering vaccine without the use of a needle and syringe means less healthcare training needed to give vaccines. Increasing the speed of vaccine delivery, while not compromising on safety, has obvious advantages. Decreasing the time required for each individual to be vaccinated means less healthcare worker person-time is required to complete vaccination of the same number of individuals (Fig. 1.2). Generally, needle-free vaccines are more stable in storage and have the potential to avoid cold chain, thus reducing cost for storage. For these reasons, needle-free vaccine delivery is supported by many prominent public health organizations, including the World Health Organization, the Global Alliance for Vaccines and Immunization, and the Centers for Disease Control and Prevention. Importantly, the design of mucosal vaccines is related to the understanding of mucosal immunity and the mechanism that regulates its induction.

1.2 Mucosal Immunity: "Theory-Examples"

Mucosal surfaces of the gastrointestinal and respiratory tracts represent the principal entry site of a large number of viral and microbial pathogens. To protect mucosal sites from pathogen invasions, the aerodigestive tract is equipped with multiple physical, biochemical, and immunological barriers. Mucosal surfaces of the respiratory, gastrointestinal, and urogenital tracts are separated from the outside world by delicate epithelial barriers. Epithelia and their associated glands (such as the salivary glands) produce nonspecific or innate defenses including mucins and antimicrobial proteins [9]. Nevertheless, foreign antigens and microorganisms frequently breach the epithelial barrier, and mucosal tissues are sites of intense immunological activity. Epithelial cells are active participants in mucosal defense. They function as sensors that detect dangerous microbial components through pattern-recognition receptors such as Toll-like receptors (TLRs) [10]. They respond by sending cytokine and chemokine signals to underlying mucosal cells, such as dendritic cells (DCs) and macrophages, to trigger innate, nonspecific defenses and promote adaptive immune responses [11].

The immunological barrier consists of both innate and acquired immunity, with the latter characterized by the initiation of antigen-specific immune response in mucosa-associated lymphoid tissues (MALTs) including the gut-associated lymphoid tissue (GALT), the nasopharynx-associated lymphoid tissue (NALT), and the bronchus-associated lymphoid tissue (BALT) [12–14]. In particular, Peyer's patches (PPs) and NALT are thought to be representative MALT in the gastrointestinal and respiratory tract, respectively (Fig. 1.3).

Additionally, isolated lymphoid follicles (ILFs), which are located throughout the intestine, have been identified and characterized as an additional organized lymphoid tissue in the digestive tract [15]. These tissues contain an interfollicular area that is abundant in T lymphocytes and in high endothelial venules (HEVs), as well as a germinal center (GC), characterized by a dense network of follicular DCs, providing a source of antigen-primed IgA-committed B cells. They also are overlaid by a follicle-associated epithelium (FAE) that is specialized for uptake of antigens and microbes from the lumen, and this effectively localizes such uptake to sites where incoming antigens and pathogens can be efficiently processed and presented for induction of appropriate immune responses [16–18]. FAE contains antigen-sampling microfold (M) cells (Fig. 1.4), allowing selective transport of antigen from the lumen to underlying APCs such as DCs and macrophages [19].

In addition to M cells, DCs in the lamina propria extend their dendrites into the lumen and sample antigens (Fig. 1.5) [10, 20–23]. It is not clear whether this mechanism is constitutively active or is induced in response to signals from epithelial

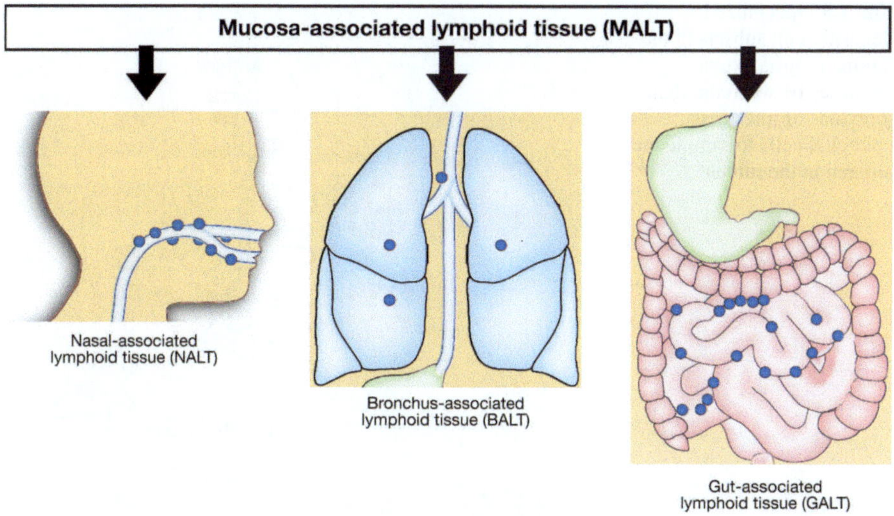

Fig. 1.3 Antigen-specific immune responses are initiated in organized mucosa-associated lymphoid tissues (MALTs) and include nasopharynx-associated lymphoid tissue (NALT), bronchus-associated lymphoid tissue (BALT), and gut-associated lymphoid tissue (GALT)

cells that have been in contact with pathogens or high numbers of nonpathogenic bacteria in the lumen. Based on these anatomical and histological characteristics of MALT, it has been generally considered that MALTs act as inductive tissues for the generation and priming of antigen-specific T- and B-cell responses, and that they communicate with effector tissues (e.g., intestinal lamina propria and nasal passages) via an immunological intranet known as the common mucosal immune system (CMIS) [24–25].

Fig. 1.4 Specialized intestinal epithelial microfold (M) cells overlie Peyer's patches and lymphoid follicles to facilitate luminal sampling. M cells have modified apical and basolateral surfaces compared to epithelial cells secreting mucin and antimicrobial peptides

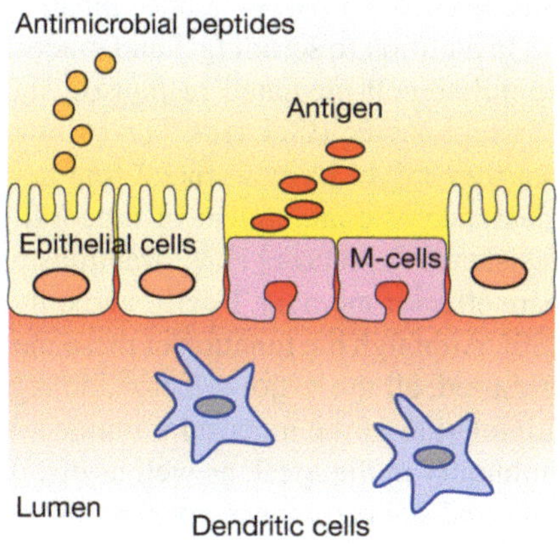

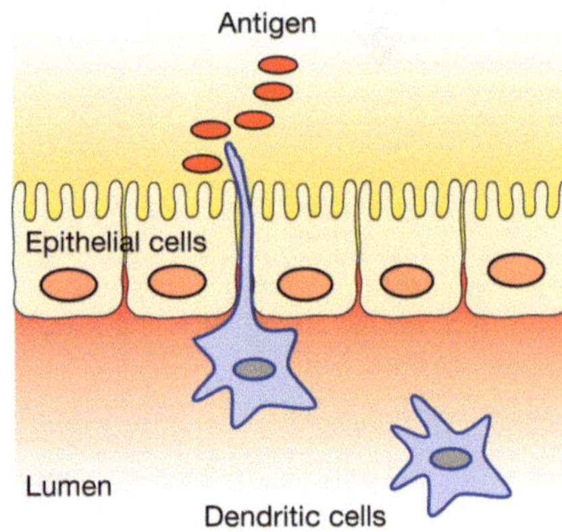

Fig. 1.5 Specialized dendritic-cell subsets in the lamina propria extend dendrites between the tight junctions of intestinal epithelial cells to sample the antigen in the lumen

An important characteristic of the mucosal adaptive immune response is the local production and secretion of dimeric or multimeric IgA antibodies that, unlike other antibody isotypes, are resistant to degradation in the protease-rich external environments of mucosal surfaces. The epithelial polymeric immunoglobulin receptor (pIgR) mediates transport of dimeric IgA across epithelial cells to the lumen (Fig. 1.6) [26]. S-IgA has multiple roles in mucosal defense [27]; e.g., it promotes the entrapment of antigens or microorganisms in the mucus, preventing direct contact of pathogens with the mucosal surface, a mechanism that is known as "immune exclusion." Alternatively, S-IgA of the appropriate specificity might block or sterically impede the microbial surface molecules that mediate epithelial attachment [28]. Local IgG synthesis also can occur in the mucosal tissues following the administration of antigen or vaccine to mucosal surfaces [29]. This IgG, as well as S-IgA, could play a significant role in blocking infection. Intact IgG in mucosal tissues, whether locally produced or from serum, can potentially neutralize pathogens that enter the mucosa and prevent systemic spread.

In addition to serum IgG and mucosal IgA antibodies, mucosal immunization can stimulate cell-mediated responses, including CD4$^+$ Th cells and CD8$^+$ cytotoxic T lymphocytes (CTLs). CTLs in mucosal tissues cannot prevent pathogen entry, but they might have a crucial role to eliminate intracellular pathogens [30] and in clearance or containment of mucosal viral infections as demonstrated in mice for resistance to mucosal HIV viral transmission [31]. Most T cells in the lamina propriaare effector memory T cells, and only low numbers of naive T cells are found there [30]. Although the function of these memory T cells in mucosal tissues is not fully understood, all the major effector and regulatory CD4$^+$ T-cell subsets are present. The stimulation of the mucosal immune system at one mucosal site can lead to mucosal immunity in the local, as well as distal, mucosal surfaces. The immunization at one mucosal site can induce specific responses at distant sites because of the expression of mucosa-specific homing receptors (site-specific integrins) by mucosally primed

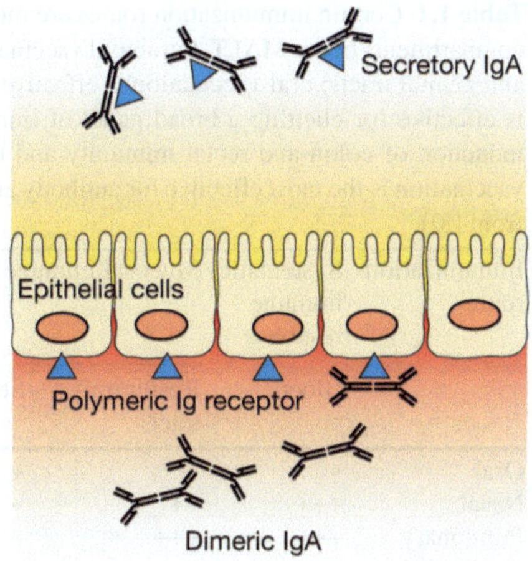

Fig. 1.6 Dimeric forms of IgA become secretory IgA by binding to polymeric Ig receptors that are displayed on the monolayer of epithelial cells lining the mucosa. Secretory IgA is then released into the nasal passage and intestinal tract

lymphocytes and complementary mucosal-tissue specific receptors (addressins) on the vascular endothelial cells [7]. For example, nasal vaccination is effective at inducing systemic and mucosal immunity in the respiratory and genital tracts [32–33]. There are studies in which antigen stimulation of the PP in the gastrointestinal tract (GIT) produced S-IgA-producing B cells not only in the intestine, but also in the bronchi and genitourinary tract [34–35].

This interconnected network is important because protective immunity (for instance, against sexually transmitted diseases) could be induced in segregated mucosal sites in a practical way, such as by sublingual (s.l.) or intranasal (i.n.) immunization, and without hampering cultural or religious barriers. At the same time, because chemokines, integrins, and cytokines are differentially expressed among mucosal tissues, within the mucosal immune system, a degree of compartmentalization is still present linking specific mucosal inductive sites with particular effector sites (e.g., the gut with the mammary glands and the nose with the respiratory and genital mucosae) (Table 1.1).

1.3 History of Mucosal Vaccines

First, reports of a mucosal vaccination practice date back to the fifteenth century in China, when healthy people acquired immunity to smallpox either by sniffing powdered smallpox pustules or by inserting them into small cuts in the skin (a technique called variolation), or finally by the oral administration of fleas from cows with cowpox [36–38]. In Europe, the scientific era of mucosal vaccinology started in the early eighteenth century with the introduction of the skin inoculation of cowpox pus to prevent smallpox, and the first clinical investigations were conducted in 1796 by the English Edward Jenner [39]. Two centuries later, in the early 1960s, the Sabin

Table 1.1 Certain immunization routes are more effective at stimulating immunity within specific compartments of the MALT. Intranasal vaccination is preferred for targeting the respiratory, gastric, and genital tracts; oral vaccination is effective for immunity in the gut; the sublingual vaccination is effective for eliciting a broad panel of immune response; rectal immunization is best for the induction of colon and rectal immunity and to some extent genital tract immunity; intravaginal vaccination is the most effective for antibody and T-cell immunity in the genital tract. (Data adapted from [8])

Immunization route	Systematic immune response	Mucosal immune response					
	Blood	Respiratory tract	Stomach	Small intestine	Colon	Rectum	Reproductive tract
Oral	+	–	+	+++	++	+/–	–
Nasal	+++	++	–	–	–	–	++
Pulmonary	+++	+++	–	–	–	–	++
Sublingual	+++	+++	+	+++	NA	NA	+++
Vaginal	+/–	–	–	–	–	–	++
Rectal	+/–	–	–	–	++	+++	–
Transdermal	+++	+++	NA	+	+	NA	NA

NA not applicable

oral polio vaccine (OPV) became available and brought mucosal immunization to prominence (Table 1.2). OPV proved to be superior in administration, eliminating the need for sterile syringes and making the vaccine more suitable for mass vaccination campaigns, by playing an essential role in the global eradication of polio. The first oral rotavirus vaccine, Wyeth's RotaShield, was approved in August 1998 in the USA; however, after several cases of intussusception, the vaccine was withdrawn from the US market only a year later in October 1999. After this first drawback, two rotavirus vaccines, "RotaTeq" (Merck & Co) and "Rotarix" (GlaxoSmithKline), were approved in 2006. Additional oral vaccines are available against typhoid fever (Ty21a) "Vivotif" (Crucell), Cholera "Dukoral" (Crucell 1992) and *Vibrio cholerae* "Orochol" (Crucell 1994).

In 2000, an i.n., virosomal influenza vaccine was launched in Europe (however, withdrawn in 2002 due to association with facial paralysis). Since 2003, FluMist [40], an i.n. life cold-adapted influenza vaccine, is the first approved in the USA. In the last 10 years, many studies were done in the field of improved vaccine delivery and led to the approval of Instanza in 2009, the first intradermal microinjection influenza vaccine (Table 1.2). When compared with most licensed injectable vaccines, it is interesting to note that currently there are no pure subunit vaccines formulated and licensed for mucosal administration. The majorities of marketed mucosal vaccines are either attenuated or inactivated microorganisms which can survive intestinal degradation by virtue of having, for example, digestion-resistant bacterial cell walls. The most successful products include the OPV, the two rotavirus vaccines, as well as vaccines against typhoid fever and cholera (Table 1.3).

Table 1.2 History of mucosal vaccination

Year	Vaccine
16th century	First attempts of vaccination against smallpox through skin puncturing or oral administration
1796	Edward Jenner demonstrates smallpox vaccination through cutaneous administration
1961	Sabin oral poliovirus (OPV) reaches the market
1992 1994	Oral vaccines against *Salmonella Typhi*, *Vibrio cholerae* and Cholera (all Crucell) are approved
1998	"RotaShield " (Wyeth's) first oral rotavirus vaccine was approved - withdrawal in 1999
2000	First intranasal Influenza vaccine "Nasalflu " (Berna Biotech) is launched - withdrawal in 2002
2003	"FluMist" (MedImmune) first intranasal seasonal influenza vaccine approved in the US
2006	Oral rotavirus vaccines "RotaTeq" (Merck) and "Rotarix " (GlaxoSmithKline) were approved
2009	"Instanza" (Sanofi Pasteur) the first intradermal microinjection influenza vaccine is approved in Europe
2010	"NasoVac "(Serum Institute of India Ltd) nasal-spray Influenza vaccine (swine flu) launched in India

TIME

Table 1.3 Licensed mucosal vaccine

Pathogen	Vaccine - Composition	Immunization Route	Trade Name (Company)
Poliovirus	Live attenuated vaccine (OPV)	Oral	Various
Cholera	CT-B/Killed whole-cell cholera vaccine	Oral	Dukoral (Crucell)
Vibrio cholera	CVD 103-HgR, live attenuated *V. Cholerae* 01 strain	Oral	Orochol (Crucell)
Salmonella typhi	Ty21a live attenuated vaccine	Oral	Vivotif (Crucell)
Rotavirus	Live attenuated monovalent human rotavirus strain	Oral	RotaTeq (Merck), Rotarix (GSK)
Influenza type A and B virus	Live attenuated cold-adapted influenza virus reassortant strain	Intranasal	Flu-Mist (MedImmune)

Attenuated live vaccines mimic natural infection. The ability of a live mucosal vaccine to propagate and colonize the mucosa of vaccinees enables it to persist for a relatively long period of time, thus allowing ample opportunity for immune stimulation. Further advantages of live-attenuated vaccines include the expression of a broad cocktail of antigens (proteins, polysaccharides, glycolipids) and immunomodulating nucleic acid sequences, including antigens which are only produced under in vivo conditions, the expression of native antigens, their correct post-translational modification, and their long-term expression. While these vaccines are stable and efficacious, there is an inherent safety risk, particularly for older products that were developed by passaging the vaccine organism in culture until it lost its pathogenicity. The example of the Sabin polio vaccine has shown that live-attenuated pathogens can occasionally mutate back into pathogenic forms able to cause disease [41]. While vaccine design for injected formulations has moved on to safe and efficacious split- and subunit vaccines over recent decades, these strategies are difficult to apply for mucosal administrations since the new subunit vaccines based on highly purified recombinant proteins are poorly immunogenic and mobilize insufficient immune responses for protective immunity. Moreover, the effectiveness of the subunit vaccines is troubled due to several physiological and immunological barriers like low pH or proteolitic enzymes. As a result, many companies are currently investigating technologies to protect antigens from digestive degradation and to increase their immunogenicity.

Adjuvants might be an answer to these needs. Adjuvants are components added to vaccines to increase the immunogenicity to the target antigen. In particular, delivery systems can help to overcome mucosal barriers, by protecting the vaccine from proteolytic enzymes and the harsh local mucosal environment. Adjuvants often possess intrinsic immunopotentiating activity and/or can be customized towards a given immunological profile by the appropriate combination of delivery systems with immunopotentiating compounds that specifically activate cells of the immune system [42]. Advanced approaches consist of antigen delivery within a stable "capsule," which can contain various encapsulation materials such as poly-lactic-*co*-glycolic acid (PLGA), polystyrene, carboxymethylcellulose, polyethylene glycol (PEG), polydimethylaminoethyl methacrylate, or liposomes [36]. Other approaches for mucosal vaccine delivery include starch microparticles, virus-like particles, or microspheres with pH-dependent antigen release.

Nevertheless, only few adjuvants are currently approved for human use and none of them for mucosal vaccine delivery, thus there is the need for the development of effective and safe adjuvants that in addition to humoral immunity can stimulate cellular or mucosal immunity, or combinations thereof, depending on the requirements for protection against the specific disease.

Recent data from humans and experimental models have shown that the choice of adjuvant can dramatically affect not only the immediate immune response but also the long-term protective effect of a vaccine. Also, the quality of the immune response—especially the development of high-affinity B-cell clones, long-lived memory B cells and plasma cells—can be influenced by the choice of adjuvant [43].

1.4 Potential Routes for Mucosal Vaccine Delivery: Challenges and Strategies

The compartmentalization of mucosal immune responses imposes constraints on the selection of vaccine administration route. Traditional routes of mucosal immunization include oral and nasal routes. Other routes recently investigated are the s.l., vaginal, and rectal routes (Fig. 1.7).

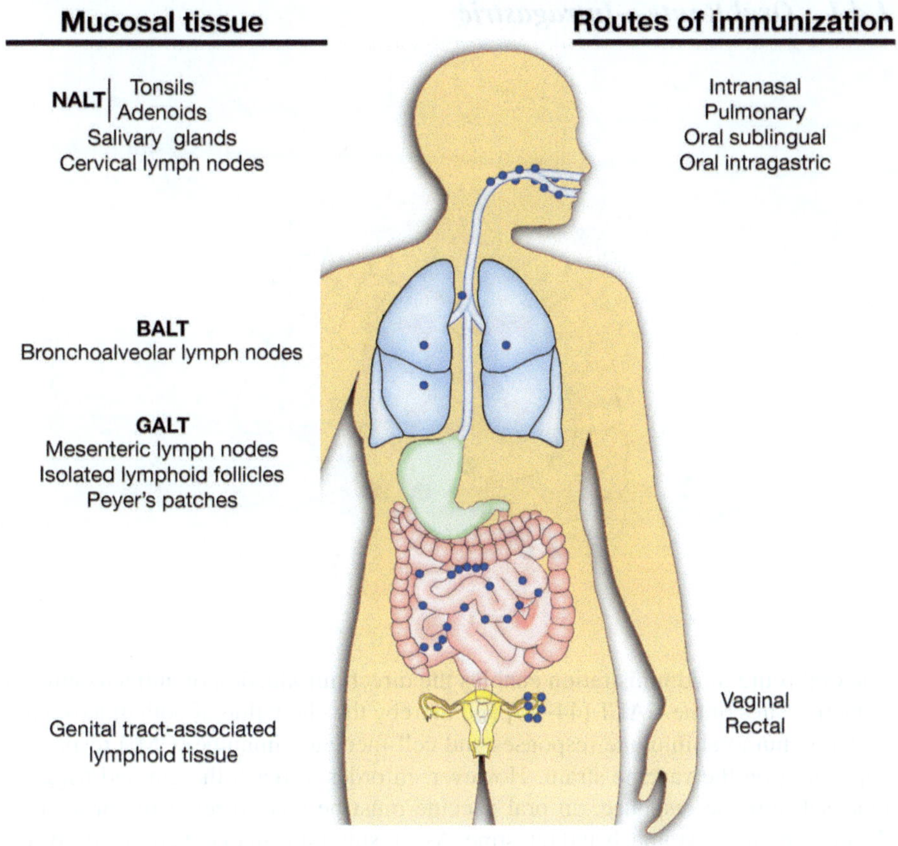

Fig. 1.7 Routes of immunization and correlated mucosal tissue

1.4.1 Oral Route—Intragastric

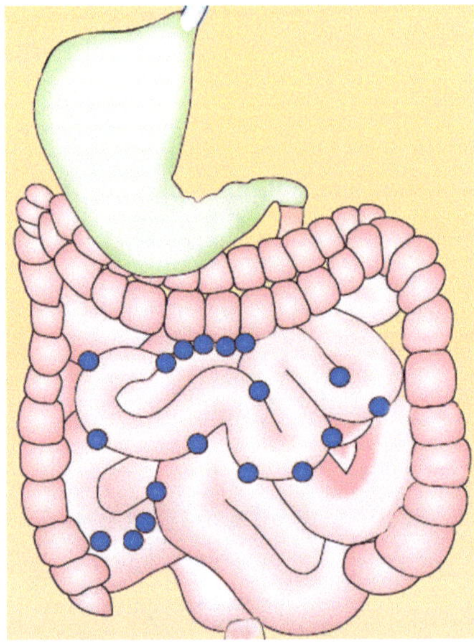

The oral route of administration enables the direct introduction of antigens into the inductive sites of the GALT [44–45] and thereby the elicitation of both mucosal and systemic humoral immune responses and cell-mediated immune (CMI) responses, depending on the vaccine strain. However, in order to reach the gut and trigger a mucosal immune response, an oral vaccine must be able to resist the proteolytic degradation in the stomach and intestine. As a result, most marketed orally delivered vaccines are live-attenuated microorganisms which can survive intestinal degradation by virtue of having digestion-resistant bacterial cell walls and are primarily directed against intestinal or respiratory pathogens, which enter the body through mucosal surfaces. The most successful products include the oral Sabin polio vaccine, as well as the two rotavirus vaccines RotaTeq (Merck & Co) and Rotarix (GlaxoSmithKline) and oral vaccines against *V. cholerae* Dukoral and Orochol (both Crucell) and *Salmonella typhi* Vivotif (Crucell).

These live-attenuated pathogen-based vaccines are associated with a risk of reconverting into their pathogenic forms, therefore able to cause disease. The example of the Sabin polio vaccine has shown that live-attenuated pathogens can occasionally mutate back into pathogenic forms causing a polio outbreak in Haiti and the Dominican Republic in 2000, which in some cases resulted in paralysis [36]. Another example is the rotavirus vaccine, RotaShield, for which several cases of intussusception, a situation in which one part of the intestine prolapses into another intestinal section, were reported, which led to the withdrawal of RotaShield from the US market in 1999. Importantly, this kind of safety risk has decreased substantially since live-attenuated strains can be developed by targeted genetic engineering instead of laboratory passaging.

A different approach to reduce the safety risks associated with live-attenuated vaccines is the use of entirely inactivated pathogens. While these do not share the risk of reactivation, their efficacy has always been doubted by experts [36]. Nevertheless, several oral vaccines based on inactivated pathogens have been developed and are marketed locally in various countries. These include, among others, a killed, whole-cell cholera toxin (CT) recombinant B subunit vaccine developed in Sweden (WC/rBS); a simpler version of the cholera vaccine without the recombinant B subunit, manufactured in Vietnam (biv-WC); as well as Russian tableted cholera bivalent vaccine with Ogawa and Inaba antigens (Microb, Saratov) [36].

Also, the use of attenuated viral or bacterial vectors, which have been genetically modified to express recombinant antigens either through insertion of the antigen into a plasmid or its integration into the host chromosome, was explored for vaccine antigen delivery. The vector protects the antigen from degradation in the stomach and intestine and facilitates delivery to the APCs in the GALT. Various species have been studied as vectors for orally delivered vaccines, including *Salmonella spp., Shigella flexneri*, *Listeria monocytogenes*, *V. cholerae*, *Yersinia enterocolitica, Bordetella pertussis*, and Bacille Calmette–Guerin [46], as well as several viral vectors like adenovirus-5 vector. While attenuated pathogen vectors provide an efficacious trafficking system of antigens to the GALT, their development is associated with significant challenges. The use of live-attenuated viruses or bacteria is associated with potential safety risks, and furthermore, in preclinical development of some vectors, such as *Salmonella,* no suitable animal model exists for these exclusively human pathogens [47]. In addition, the immunogenicity of vector-based vaccines is sometimes suboptimal, requiring a high dose of bacteria to penetrate the host cells effectively and stimulate a sufficient immune response. A further challenge is the genetic design of the antigen expression system.

Finally, particularly for viral vectors, stability can be an issue. As of 2009, the two most advanced candidates for viral and bacterial vectors in oral vaccine delivery are adenovirus and *Salmonella enterica serovar typhi,* with various companies including Emergent Biosolutions, Vaxart, and Barr developing oral vaccines based on both approaches. Vaxart has developed an oral delivery system based on a nonreplicating chimeric adenovirus-5 vector, engineered to express various antigens, and a TLR3 ligand as a vaccine adjuvant. The viral vector is then administered in an enterically coated formulation to withstand degradation in the stomach.

Challenges associated with oral vaccine delivery are the poor transport of antigens across the intestinal epithelium to reach the underlying GALT and the induction of oral tolerance [48] instead of protective immunity by the GALT. Moreover, protein antigens not only have to survive the low gastric pH and degradation by proteolytic enzymes present in the GIT, they often have to circumvent the interference by the lactogenic immunity, such as neutralizing antibodies and milk factors. For this reason, the oral route for vaccine delivery is the most challenging and the most difficult to achieve, and progress in oral vaccine development has been rather slow.

More recent developments in the sector have focused on subunit vaccines, which are delivered orally by means of encapsulation and often contain targeting molecules or adjuvants in order to guarantee sufficient immunity. The encapsulation of vaccine antigens in biodegradable particulate delivery systems can protect antigens from digestive enzymes and the maternal immunity. Uptake of these particulate delivery systems by the epithelium is rather poor and although modification of their size, surface charge, or hydrophobicity can increase the efficiency of epithelial uptake, surface decoration of the antigen-loaded particulates with targeting ligands, specific for epithelial receptors, could further enhance the uptake and transepithelial transport of antigens [49–50]. Moreover, this could potentially overcome the induction of tolerance since receptor-mediated endocytosis mostly induces antigen-specific mucosal immune responses. In addition, the incorporation of mucosal adjuvants in particulate delivery systems could lead to a more potent activation of the innate and adaptive immune system.

A better understanding of the intestinal mucosa and its role in the overall immune system and of the molecular and cellular pathways will, therefore, be crucial for the future development of improved oral vaccines. For instance, it is already known that *S. typhi* bacteria pass through M cellsto cause a systemic disease [51]; human immunodeficiency virus-1 (HIV-1) is endocytosed and transcytosed by M cells of mice and rabbits [52]. Owen et al. inoculated *V. cholerae* into the intestinal lumen and observed by transmission electron microscopy that they were phagocytosed by M cells into vesicles which were released from the basolateral membrane to the underlying lymphocytes and macrophages of the PPs [53]. Cationized ferritin (CF) has been used to investigate uptake and transport by M cells in comparison with absorptive enterocytes [54]; and even inert particles have been shown to be taken up from the intestinal lumen specifically by M cells [55]. Since such inert substances as latex microparticles and CF are taken up by M cells, this indicates that specific receptor binding is not required for uptake. This means that antigens adsorbed onto microparticles might mimic the route by which many intestinal pathogens naturally infected the body.

1.4.2 Intranasal Route

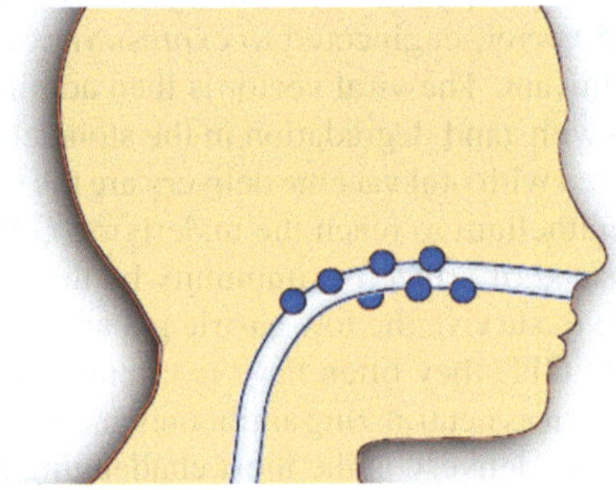

Nasal vaccination has several interesting advantages; the nose is easily accessible and the nasal cavity is equipped with a high density of DCs that can mediate strong systemic and local immune responses against pathogens that invade the human body through the respiratory tract [9, 56]. Local immunity in the upper airways, as well as systemic immunity, is mainly mediated by the lymphoid tissue referred to as NALT. The uptake of nasally administered vaccines is probably mediated by M cells, which can transport particulate antigens to the NALT by transcytos is [57]. Alternatively, live-attenuated vaccines trigger an immune response by mimicking natural infection. Furthermore, the enzymatic activity in the nose is relatively low [58], which is favorable for antigen stability at the administration site.

In general, i.n. vaccination is an attractive approach, as much lower antigen and adjuvant doses are required compared with oral vaccination. Thus, nasal immunization is an effective method for stimulating both mucosal and systemic immunity. As a consequence, many companies have embarked on the development of nasally delivered vaccines, primarily against respiratory pathogens which naturally infect the body through the upper respiratory tract, including influenza, respiratory syncytial virus, or parainfluenza virus. However, although research and development of nasal vaccines has gained momentum over the last few years, only one nasally delivered vaccine, AstraZeneca's FluMist, is currently approved in humans, reflecting the substantial challenges for i.n. vaccine delivery. Intranasal administration of a live-attenuated influenza virus vaccine (FluMist; MedImmune) has proven effective at protecting against seasonal infection, and it even provides cross protection against drifted influenza virus strains. A promising strategy in HIV-1 vaccine development has been suggested by a study in which rhesus macaques were intranasally vaccinated with a virosome-coupled trimeric gp41 protein, which elicited strong protective IgA antibody responses in the genital tract and also prevented the transmission of infection [59]. Most candidates in the sector are still based on live-attenuated pathogens, an approach that is associated with safety risks, particularly in immunocompromised populations.

The most important obstacle for nasal vaccines is the limited time a vaccine persists in the nose before it is evicted, a process referred to as mucociliary clearance. Consequently, a vaccine has to be taken up very rapidly by the nasal mucosa in order to be efficacious. This uptake can be inefficient, particularly for split- or subunit antigens.

One solution to these issues is the development of vaccine formulations that contain mucoadhesive structures in order to prolong the nasal residence time as well as strong adjuvants, molecular delivery, and targeting systems to increase uptake and immunogenicity of the vaccine. Mucoadhesives, usually polymers like chitosan, with enhanced permeabilizing properties to facilitate contact and retention of vaccine antigens in the epithelium are likely to become a major milestone for the future emergence of needle-free vaccines [42, 60–64]. Uptake of antigens through the mucosal epithelium can be increased by incorporation into particles [65]. For instance, i.n. administration in mice of antigens incorporated in nanoparticles composed of poly lactide-*co*-glycolide (PLGA), a biodegradable polymer, led to an increased antibody response in comparison with aqueous solution of protein antigens [66–67]. Because M cells are extremely efficient in the uptake of luminal antigens, it is an effective strategy to target antigens to these cells. For nasal vaccination, several studies pointed to small (nano)particles being more rapidly absorbed by nasal M cells [65, 68–71], but no boundaries have been determined. Fujimura et al. [72] showed that particles coated with the cationic polymers chitosan or poly-l-lysine were taken up by the NALT with an increased uptake of smaller particles.

Ligands that selectively target M cells include isolectin B_4 and *Maackia amurensis I* lectin [73], which recognize α-(1–3)-linked galactose and sialic acid, respectively [74]. Besides lectin binding domains, several other receptors have recently been identified as potential M cell-targeting ligands, especially β_1-integrin [75].

Nevertheless, safety concerns were reported with some i.n. vaccines because antigens and/or adjuvants might be redirected to the central nervous system (CNS) through the olfactory epithelium [76–78]. The first intranasally applied vaccine to reach the market was Berna Biotech's virosomal flu vaccine Nasalflu, which was launched in Switzerland and Germany in late 2000. However, the vaccine was withdrawn from both markets for further clinical studies in September 2001, in order to investigate possible links between its use and incidents of Bell's Palsy, a temporary facial paralysis, in vaccine recipients. In June 2002, the company concluded that a possible association could not be excluded based on preliminary results of the clinical studies. Experts now believe that the association with Bell's Palsy most likely resulted from the adsorption of heat-labile toxin (LT), a known mucosal adjuvant which was present in the vaccine, to facial nerve fibers followed by retrograde transport and subsequent neuronal damage [79–80].

Intranasal vaccines, unlike other formulations such as orally delivered products, cannot be administered directly but require special delivery devices. This increases the costs of vaccine delivery and requires partnerships between device manufacturers and vaccine developers. Vaccine delivery via aerosol spray and droplets is an attractive possibility owing to the development of new delivery devices [81]. Looking forward, advances made in the development of adjuvant and molecular delivery systems have the potential to shift nasal vaccine development towards safe and efficacious subunit vaccines. However, the combination of various technologies will be needed to succeed in the sector.

1.4.3 Pulmonary Route

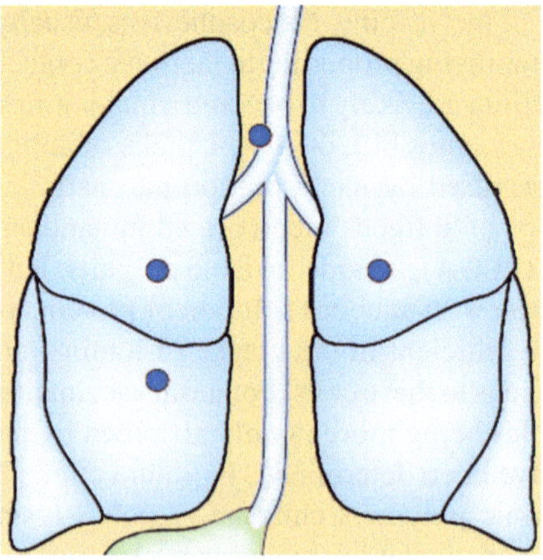

Pulmonary delivery of vaccines mimics the natural pathway of infection for many pathogens and is another promising approach. Immunization through the lungs may provide an excellent first barrier to prevent disease and appears very promising since the lungs contain a highly responsive immune system [82]; and airway mucosal DCs, the most frequent efficient APCs in the larger airways, may enhance immune responses against pathogens [83]. The favorable physiological environment of the deep lungs, relative to other mucosal tissues such as nose or gut, may prevent common problems for other mucosal delivery routes, such as mucociliary clearance, proteolytic degradation, or antigenic tolerance, which can develop in tissues frequently exposed to common environmental substances [84]. There is evidence that mice genetically lacking spleen, lymph nodes, and PP can generate strong primary B- and T-cell responses to inhaled influenza. These responses appear to be initiated at sites of the induced BALT, which functions as an inducible secondary lymphoid tissue for respiratory immune responses [85].

Many aerosol exposure methods have been used to vaccinate animals and human by delivery to the lungs. Aerosol vaccine delivery involves creating small particles, usually generated by a nebulizer, that reach the lungs [86]. The exposure of the lungs to various aerosol formulations designed to protect against influenza virus showed to be more effective than i.n. administration or parenteral injections, indicating that a local response was generated in the respiratory tract [87]. Also, intratracheal instillation and insufflation allow direct delivery of liquids and powders to the lungs. When targeting specific lung compartments, the particle size (defined as aerodynamic diameter), size distribution, particle shape, and density of the antigen are important factors determining deposition within the respiratory tract and vaccine efficiency [88]. Interestingly, pulmonary vaccination was first used against Newcastle disease in 1952 in chickens via inhalation of a live vaccine. Since then, pulmonary vaccination has been used worldwide to immunize poultry against Newcastle disease [89], and there have been numerous successful aerosol immunization trials of fowl and pigs against a number of diseases, including fowl pox, hog cholera, erysipelas, pseudorabies, gastroenteritis, pasteurellosis, and mycoplasmosis [90–91]. However, in humans, besides small-scale vaccination trials in the Soviet Union, the measles vaccine is the only successful use of pulmonary immunization on a large scale [92].

One of the main challenges regarding pulmonary immunization is the potential to worsen respiratory diseases, such as bronchitis, pneumonia, and allergic asthma. The excipients in aerosol formulations may be allergenic and irritating, inducing unanticipated and undesirable inflammation [93].

Although the field of pulmonary vaccine delivery is still in its infancy and some challenges need to be met before use can be made of successful new vaccination protocols, future strategies for vaccination using the pulmonary route are promising. Pulmonary vaccination may provide a mean to rapidly immunize a large population, either in a bioterrorism setting or in a mass vaccination program in developing countries [86].

1.4.4 Oral Route—Sublingual

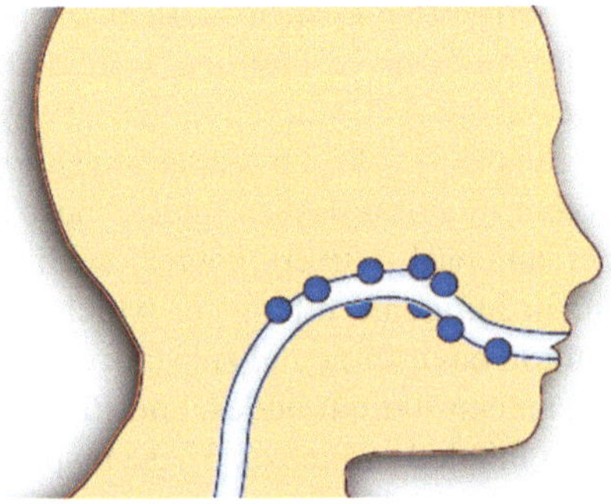

Oral mucosae, including buccal, s.l. (underside of the tongue), and gingival mucosa, have recently received much attention as novel delivery sites because they do not subject proteins and/or peptides to the degradation associated with gastrointestinal administration. Since compounds administered sublingually can elicit a local response, while additionally being absorbed rapidly through the oral mucosa and into the systemic compartments, the s.l. route is commonly used for immunotherapeutic treatment of allergies [94–95]. On the basis of these findings, International Vaccine Institute researchers assumed that the s.l. route might be promising for delivery of vaccines targeting infectious diseases. Recent in vivo data suggest that s.l. immunization uses the same cellular trafficking system as i.n. immunization. However, in contrast to i.n. delivery, s.l. administration of inactivated influenza virus with a mucosal adjuvant was not associated with migration to the CNS [96–97]. Furthermore, s.l. administration of viral antigen elicited antigen-specific immune responses in the respiratory tract and the oral/nasal cavity of mice that were comparable to those elicited by i.n. immunization. Sublingual administration of a variety of soluble and particulate antigens, including live and killed bacteria and viruses, subunit vaccines, and virus-like particles can evoke secretory and systemic antibody responses and also mucosal and systemic CTL responses. Sublingual responses have been far superior in magnitude and duration while requiring significantly (10–50-fold) lower amounts of antigens compared to responses induced by the intragastric route. Importantly, s.l. vaccination evoked broadly disseminated immune responses, including genital immunity. Thus, s.l. immunization may offer an advantageous alternative to oral immunization for vaccine administration [96, 98–99]. The above studies indicate that s.l. vaccination can induce broadly disseminated humoral and cell-mediated immune responses and may thus overcome the compartmentalization of mucosal immune responses observed when vaccines are administered by the more traditional orogastric and i.n. routes (Table 1.3).

Vaccine antigen is taken up by intraepithelial CD11c-positive DCs present in the s.l. mucosa and transported to the draining lymph nodes for antigen presentation and priming of T and B cells. The s.l. route has been explored for administration of vaccines against a range of bacterial and viral diseases, and various mucosal adjuvants have been tested for s.l. use [96]. The mechanism and outcome of s.l. vaccination with a soluble protein antigen plus mucosal adjuvant (i.e., CT) shows that the CCR7 + CCL19/CCL21 pathway on CD11c + DCs is responsible for efficient antigen-specific systemic and mucosal immune responses (including T and B cells) by the s.l. route [100]. However, even though s.l. administration induced qualitatively similar responses to the i.n. route, the magnitude of response was lower after s.l. administration [96, 101].

This may be due to enzymatic degradation by salivary enzymes, deglutition, or differences in the antigen uptake and processing mechanisms between the two routes. Supporting the latter explanation, the NALT has shown distinct phenotypical properties compared with other mucosal sites and could be a superior mucosal site for inducing effective immune responses following vaccination. Furthermore, M cells are apparently lacking in the s.l. epithelium, and unlike the nasal mucosa, the s.l. mucosa is devoid of any organized MALT. Targeting antigens in the s.l. mucosa to the DCs that imprint adequate adaptive T-cell responses will require novel mucosal vaccine strategies, including effective adjuvant and immunomodulatory molecules. Prototype vaccines have successfully targeted DCs by using antibodies specific for cell-surface receptors such as DEC205 and DC-SIGN, or by using the natural ligands of these receptors, such as mannan and mannosylated liposomes [102–105]. TLRs are important signaling molecules which DCs use to sense danger. It is, therefore, a logical approach to use either purified or synthetic TLR ligands as adjuvants to activate DCs [106–108]. Bacterial ADP-ribosylating exotoxins possess a high degree of adjuvanticity and are, therefore, the adjuvants that are most often used preclinically for mucosal immunizations. Among them, CT and *E. coli* LT are the ones most intensively studied [109–110]. It was demonstrated that the application of CT as adjuvant under the tongue increases the recruitment of DCs in the s.l. epithelium [96]. The understanding of the functional specialization of DC subsets might allow modulating the immune system by targeted delivery of antigen and adjuvant predominantly to one of these DC subsets.

1.4.5 Vaginal and Rectal Route

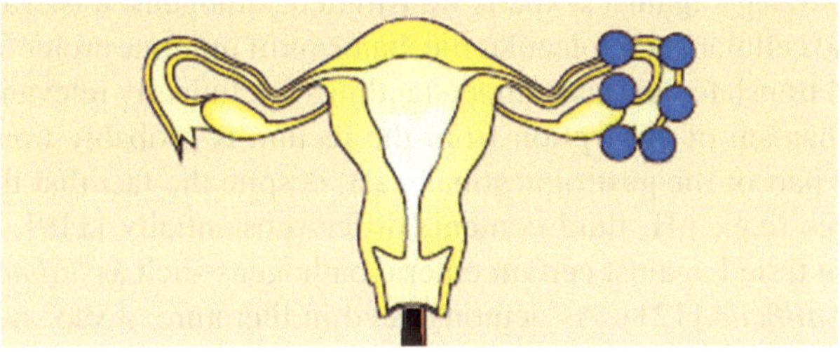

Vaginal immunization, especially during the midfollicular phase of the menstrual cycle, similarly induces strong local mucosal immune responses without producing notable distal immune responses [29, 111–114].

Compared to the monolayer epithelia in the intestine and in the lung, the vaginal tract is covered with stratified epithelia. In addition, the vaginal mucosa differs from other mucosae with respect to mucus composition, microbiota, and innate and adaptive immune mechanisms. At steady state, vaginal epithelial layer and the submucosa are surveyed by innate leukocytes and lymphocytes, but the recruitment of antigen-specific T and B cells to the vagina is restricted. Once infected, both epithelial cells and innate leukocytes produce type I interferons (IFNs), inflammatory cytokines and induce chemokines that recruit *natural killer* (NK) cells, monocytes, plasmacytoid DCs (pDCs), and neutrophils. Virions and viral antigens are taken up and processed by migrant submucosal DCs or by LN-resident DCs and presented to T cells. Activated effector T cells are recruited to the vagina and can persist for a long period [115]. Vaginal epithelial cells lack pIgR for transport of S-IgA. Instead, virus-specific IgG is transcytosed by neonatal Fc receptor for IgG (FcRn) into the vaginal lumen, and provides protection. Recent studies demonstrate that adaptive immunity in the vaginal mucosa is uniquely regulated compared to other mucosal organs. In particular, development of virus-specific $CD4^+$ and $CD8^+$ T cells is critically important for antiviral defense in vagina. Despite a great success in prophylactic systemic HPV vaccine [116], neither therapeutic vaccine has been made against any sexually transmitted viruses, nor is there an efficacious preventive vaccine against HIV-1 and HSV infection. There is evidence that CTLs can control AIDS virus replication in the absence of antibodies. The first indications that CTLs could suppress HIV-1 replication in vivo were observations that the reduction in viremia in acute infection was temporally associated with the appearance of HIV-1-specific CTLs [117]. Unfortunately, the most promising approach for inducing CTL responses tested clinically to date, an Ad5-based vaccine regimen, has recently failed in human efficacy trials. The vaccine's failure to control HIV-1 replication may have been due to the Ad5 vector, the choice of HIV-1 transgenes,or a combination of these two factors. It is possible that a replication-defective Ad5 vector is simply unable to stimulate CMI responses of sufficient breadth to control HIV-1 infection. Furthermore, many people have been infected with Ad5 and, therefore, have immunity to this virus. Preexisting Ad5-specific $CD8^+$ T-cell responses could also potentially reduce the potency and breadth of vaccine-induced HIV-1-specific $CD8^+$ T-cell responses [118]. While much has been learned from infection models in other mucosal tissues and skin, for a better vaccination strategy against sexually transmitted pathogens, it is critically important to understand cellular and molecular mechanisms of immune protection in the genital mucosa, and translate our basic understandings to clinically relevant outcome.

The mechanism of absorption from the rectum is probably no different to that in the upper part of the gastrointestinal tract, despite the fact that the physiological circumstances (e.g., pH, fluid content) differ substantially [119]. Rectal vaccination has been tested against certain enteric pathogens such as *Salmonella* [120] and *Clostridium difficile* [121]. As demonstrated in literature, a vaccine that stimulates mucosal immunity in the gut should be an appropriate line of defense against respective pathogens. Drawbacks of rectal drug administration include the interruption of absorption by defecation and lack of patient acceptability mainly due to cultural or religious barriers.

1.4.6 Skin Delivery to Induce Mucosal Immunity

An alternative route to induce robust systemic and mucosal immune responses against pathogens is the transdermal delivery [122–125]. The skin represents the boundary between the body and external environment and includes three primary layers with different features and functions: epidermis, dermis, and hypodermis [126]. The skin is more than a passive barrier protecting the host against physical or chemical damage. Both the innate and adaptive arms of the immune system are represented in the skin. Noninflamed skin is an immunologically active site that contains numerous cell populations of immune-responsive cells. The presence and function of these cells determines the response to antigens that permeate layers of epidermis, dermis, and the main cell types involved in immune surveillance, antigen uptake, and initiation of immune responses (Fig. 1.8). Contributors to the cutaneous immune response include keratinocytes, epidermal and dermal DCs (DDCs), T lymphocytes, NK-T cells, mast cells, and macrophages, among others [127]. APCs in the skin perform an essential role in processing incoming antigens [128]. For these reasons, it is possible that delivery of vaccines to the epidermis or dermis may result in superior immune responses compared to other anatomical compartments [129]. Alternatively, the skin delivery has a potential for dose sparing, meaning that an equivalent immune response could be stimulated by delivery of a smaller quantity of vaccine to the skin. A prerequisite for successful cutaneous delivery of vaccines is that the vaccine antigens can reach the skin DCs, as these cells are essential to initiate the immunization. The DCs in the epidermis are called Langherans cells (LCs) [130]. For many years, LCs were designated as the major APCs in the skin. Now, it is clear that the DDCs are also important and some reports suggest that DDCs are more important than LCs in immunity [131].

Induction of antigen-specific antibody responses in mucosal tissues after transdermal has been studied in animal model. Antigen-specific IgA and IgG antibodies have been observed in the gastrointestinal, respiratory, and genitourinary tracts [122, 132–133]. The mechanisms involved in these immune responses are not well understood, but recent studies have documented the migration of DCs activated in the skin to the gut mucosa [134]. The transdermal delivery is able to induce not only humoral immune response but also antigen-specific $CD8^+$ T-cell responses [135–136]. Following transdermal delivery of the vaccine, antigen-specific $CD8^+$ CTLs were observed in the PP of the small intestine and in the spleen [137]. The advantages and safety profiles of transdermal immunization predicted from animal studies have stimulated to initiate a number of clinical trials. The safety of transdermal route has been demonstrated in several clinical trials [123, 138–141]. To increase the systemic and mucosal immune response after transdermal immunization, many rational approaches might be used. Because the cornified layer and thin junctions limit the penetration of molecules larger than 500–600 Da, vaccines cannot simply be applied onto the skin. Both barriers need to be disrupted to enable vaccine antigens to enter the skin. Disruption of the skin barrier increases the transcutaneous permeation of antigen and makes it more readily available for sampling by APCs. Moreover, it is

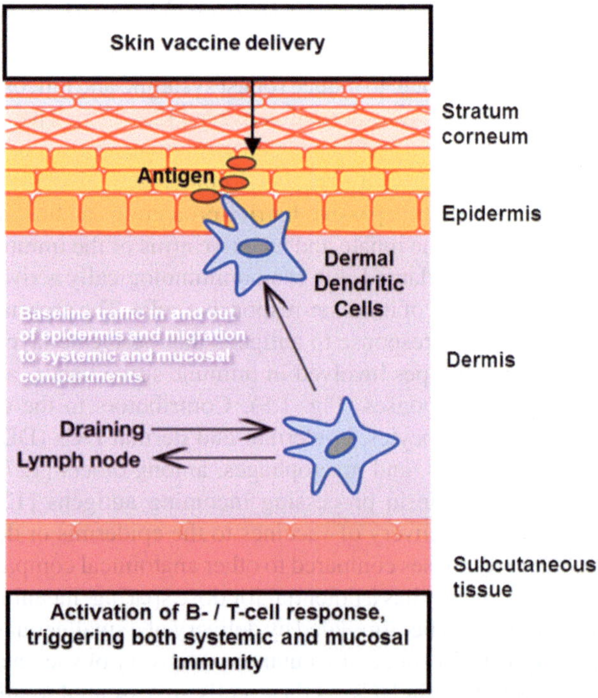

Fig. 1.8 Antigen delivered in the epidermis or dermis is taken up by antigen-presenting cells that most likely migrate to mucosal tissues inducing systemic and mucosal immune response

known that skin barrier disruption can activate the immune system, inducing the secretion of proinflammatory cytokines by keratinocytes and resulting in DC activation [142–143].

This makes it attractive to develop physical methods to overcome the skin barrier. Different devices have been used during the years to pierce the skin and thereby deposit vaccine in the epidermal/dermal space. The most recent devices include techniques such as microneedles [144–145] and tattooing [146]. Among microneedle technologies, three major approaches are under investigation: (1) Hollow microneedles through which liquid vaccines can be injected; one example is the licensed seasonal influenza vaccine "Intanza" from Sanofi Pasteur, which is delivered through a prefilled microinjection system from Becton Dickinson's, approved in Europe in February 2009 [147]. (2) Solid microneedles, which are coated with the antigen in the form of a powder or film, deposit the antigen into the skin upon administration. A key challenge to this approach is the dosing efficiency, as it is often necessary to coat the microneedles with an excessive amount of antigen to guarantee a sufficient immune response [148]. (3) Dissolvable microneedles, where the antigen is formulated into a dissolvable matrix [149]. For example dissolvable microneedles designed by Georgia Institute of Technology ("GA Tech") where the microneedle patch is hard and sharp when dry (comparable to the other two technologies), but as soon as it is applied to the skin, the body's own fluids begin to dissolve it and allow the vaccine antigen to diffuse into the skin. One of the main challenges is that vaccines need to be reformulated to be suitable for microneedle technologies. The most promising systems combine barrier disruption with the addition of an adjuvant to the vaccine formulation, particularly if subunit antigens are used [150].

1.5 Conclusions and Future Perspectives

Mucosal vaccine delivery is likely to progress over the next decade as the currently limited knowledge of the molecular mechanisms of mucosal immunity is expanding. A key challenge will be the design of efficacious and safe vaccines, with stability being the most important bottleneck. Oral and nasal vaccines have already both demonstrated their feasibility and will continue to gain importance, whereas pulmonary vaccine delivery is associated with significant drawbacks compared with the other two main mucosal delivery routes.

An advantage of oral delivery is that it is probably safer, most likely, there are less stringent production conditions, and it is the most convenient way of delivery, "you just give someone a tablet and they take it themselves." However, it is by far the most challenging route as it is extremely difficult to get vaccines to work orally; in order to be efficacious, oral vaccines have to overcome simultaneously a series of challenges: the acidity and enzymes present in the stomach and in the intestine, the dilution effect because of the volume, the mucus layer that have to be crossed, the peristalsis that has to be avoided, and the epithelial cells that have to be bound to and got across. The s.l. delivery combined with the design of a fast releasing vaccine formulation that could provide mucoadhesive properties might avoid some of these challenges. While live-attenuated viruses and viral vectors are still the prevailing approaches for antigen design, several companies are moving towards subunit vaccines, which are delivered by means of encapsulation and novel technologies such as adjuvants, molecular delivery vehicles, viral and bacterial vectors in order to guarantee sufficient immunity. However, the development of commercial products is still hampered by a challenging regulatory environment.

A significant number of companies are currently developing i.n. vaccines, with the majority of clinical programs targeting respiratory pathogens such as influenza, respiratory syncytial virus, and parainfluenza virus. While these are the obvious candidates, nasal delivery could potentially also be suitable for a wider range of indications. For instance, some experts draw attention to a potential link between nasal immunization and the generation of mucosal immunity in the vagina, making nasal delivery potentially attractive for vaccines directed against some sexually transmitted diseases and other vaginal pathogens.

A strong argument for transdermal vaccine delivery is the potential to elicit both systemic and mucosal immune responses at multiple mucosal sites, a mechanism that still remains to be fully understood. Furthermore, the skin is easily accessible and patches benefit from the ease and flexibility of administration of vaccines, making them a very promising option for quick mass immunization; e.g., influenza is a promising target due to the large population receiving the vaccine on an annual basis.

The combination of a broad panel of routes, adjuvants, and delivery technologies holds tremendous promise for effective, safe, needle-free vaccines, and the delivery landscape is set for rapid change over the next decade. However, the most efficient way to induce a potent mucosal immunity still needs to be found, which will require joint efforts from immunologists, vaccinologists, and pharmaceutical scientists. Additionally, a lot of the research behind administration technologies is happening in small companies and academic institutions, therefore partnerships between vaccine developers and key delivery technology companies and respective academic institutions are essential. Only then, needle-free immunization can be further improved and will essentially revolutionize the current vaccination practice.

Acknowledgments The authors are grateful to Giorgio Corsi for the artwork.

References

1. Mowat AM, Millington OR, Chirdo FG. Anatomical and cellular basis of immunity and tolerance in the intestine. J Pediatr Gastr Nutr. 2004;39(3):723–4.
2. Mowat AM, Parker LA, Beacock-Sharp H, Millington OR, Chirdo F. Oral tolerance: overview and historical perspectives. Ann NY Acad Sci. 2004;1029:1–8.
3. Lycke N. Recent progress in mucosal vaccine development: potential and limitations. Nat Rev Immunol. 2012;12(8):592–605.
4. Cox RJ, Haaheim LR, Ericsson JC, Madhun AS, Brokstad KA. The humoral and cellular responses induced locally and systemically after parenteral influenza vaccination in man. Vaccine. 2006;24(44–46):6577–80.
5. Atmar RL, Keitel WA, Cate TR, Munoz FM, Ruben F, Couch RB. A dose-response evaluation of inactivated influenza vaccine given intranasally and intramuscularly to healthy young adults. Vaccine. 2007;25(29):5367–73.
6. Yuki Y, Kiyono H. New generation of mucosal adjuvants for the induction of protective immunity. Rev Med Virol. 2003;13(5):293–310.
7. Holmgren J, Czerkinsky C. Mucosal immunity and vaccines. Nat Med. 2005;11(4):45–53.
8. Czerkinsky C, Holmgren J. Mucosal delivery routes for optimal immunization: targeting immunity to the right tissues. Curr Top Microbiol Immunol. 2012;354:1–18.
9. Neutra MR, Mantis NJ, Kraehenbuhl JP. Collaboration of epithelial cells with organized mucosal lymphoid tissues. Nat Immunol. 2001;2(11):1004–9.
10. Chieppa M, Rescigno M, Huang AY, Germain RN. Dynamic imaging of dendritic cell extension into the small bowel lumen in response to epithelial cell TLR engagement. J Exp Med. 2006;203(13):2841–52.
11. Kagnoff MF, Eckmann L, Epithelial cells as sensors for microbial infection. J Clin Invest. 1997;100(1):6–10.
12. Hamada H, Hiroi T, Nishiyama Y, Takahashi H, Masunaga Y, Hachimura S, Kaminogawa S, Takahashi-Iwanaga H, Iwanaga T, Kiyono H, Yamamoto H, Ishikawa H. Identification of multiple isolated lymphoid follicles on the antimesenteric wall of the mouse small intestine. J Immunol 2002;168(1):57–64.
13. Kiyono H, Fukuyama S. NALT- versus Peyer's-patch-mediated mucosal immunity, Nature reviews. Immunology. 2004;4(9):699–710.

14. Kunisawa J, Fukuyama S, Kiyono H. Mucosa-associated lymphoid tissues in the aerodigestive tract: their shared and divergent traits and their importance to the orchestration of the mucosal immune system. Curr Mol Med. 2005;5(6):557–72.
15. Pearson C, Uhlig HH, Powrie F. Lymphoid microenvironments and innate lymphoid cells in the gut. Trends Immunol. 2012;33(6):289–96.
16. Kraehenbuhl JP, Neutra MR. Molecular and cellular basis of immune protection of mucosal surfaces. Physiol Rev. 1992;72(4):853–79.
17. Neutra MR, Frey A, Kraehenbuhl JP. Epithelial M cells: gateways for mucosal infection and immunization. Cell. 1996;86(3):345–48.
18. Neutra MR, Pringault E, Kraehenbuhl JP. Antigen sampling across epithelial barriers and induction of mucosal immune responses. Ann Rev Immunol. 1996;14:275–300.
19. Styers ML, Kowalczyk AP, Faundez V. Intermediate filaments and vesicular membrane traffic: the odd couple's first dance? Traffic. 2005;6(5):359–65.
20. Rescigno M, Rotta G, Valzasina B, Ricciardi-Castagnoli P. Dendritic cells shuttle microbes across gut epithelial monolayers. Immunobiol. 2001;204(5):572–81.
21. Rescigno M, Urbano M, Valzasina B, Francolini M, Rotta G, Bonasio R, Granucci F, Kraehenbuhl JP, Ricciardi-Castagnoli P. Dendritic cells express tight junction proteins and penetrate gut epithelial monolayers to sample bacteria. Nat Immunol. 2001;2(4):361–7.
22. Niess JH, Brand S, Gu X, Landsman L, Jung S, McCormick BA, Vyas JM, Boes M, Ploegh HL, Fox JG, Littman DR, Reinecker HC. CX3CR1-mediated dendritic cell access to the intestinal lumen and bacterial clearance. Science. 2005;307(5707):254–8.
23. Niess JH, Reinecker HC. Lamina propria dendritic cells in the physiology and pathology of the gastrointestinal tract. Curr Opin Gastroen. 2005;21(6):687–91.
24. McGhee JR, Kiyono H, Michalek SM, Mestecky J. Enteric immunization reveals a T cell network for IgA responses and suggests that humans possess a common mucosal immune system. A Van Leeuw. 1987;53(6):537–43.
25. McGhee JR, Xu-Amano J, Miller CJ, Jackson RJ, Fujihashi K, Staats HF, Kiyono H. The common mucosal immune system: from basic principles to enteric vaccines with relevance for the female reproductive tract. Reprod Fert Develop. 1994;6(3):369–79.
26. Kaetzel CS, Robinson JK, Chintalacharuvu KR, Vaerman JP, Lamm ME. The polymeric immunoglobulin receptor (secretory component) mediates transport of immune complexes across epithelial cells: a local defense function for IgA. Proc Natl Acad Sci U S A. 1991;88(19):8796–800.
27. Lamm ME. Interaction of antigens and antibodies at mucosal surfaces. Annu Rev Microbiol. 1997;51:311–40.
28. Hutchings AB, Helander A, Silvey KJ, Chandran K, Lucas WT, Nibert ML, Neutra MR. Secretory immunoglobulin A antibodies against the sigma1 outer capsid protein of reovirus type 1 Lang prevent infection of mouse Peyer's patches. J Virol. 2004;78(2):947–57.
29. Kozlowski PA, Williams SB, Lynch RM, Flanigan TP, Patterson RR, Cu-Uvin S, Neutra MR. Differential induction of mucosal and systemic antibody responses in women after nasal, rectal, or vaginal immunization: influence of the menstrual cycle. J Immunol. 2002;169(1):566–74.
30. Sheridan BS, Lefrancois L. Regional and mucosal memory T cells. Nat Immunol. 2011;12(6):485–91.
31. Belyakov IM, Ahlers JD, Brandwein BY, Earl P, Kelsall BL, Moss B, Strober W, Berzofsky JA. The importance of local mucosal HIV-specific CD8($^{+}$) cytotoxic T lymphocytes for resistance to mucosal viral transmission in mice and enhancement of resistance by local administration of IL-12. J Clin Invest. 1998;102(12):2072–81.
32. Brandtzaeg P. Potential of nasopharynx-associated lymphoid tissue for vaccine responses in the airways. Am J Resp Crit Care. 2011;183(12):1595–604.
33. Mestecky J, Alexander RC, Wei Q, Moldoveanu Z. Methods for evaluation of humoral immune responses in human genital tract secretions. Am J Reprod Immunol. 2011;65(3):361–67.
34. Mestecky J, Michalek SM, Moldoveanu Z, Russell MW. Routes of immunization and antigen delivery systems for optimal mucosal immune responses in humans. Behring Inst Mitt. 1997;(98):33–43.

35. Nugent J, Po AL, Scott EM. Design and delivery of non-parenteral vaccines. J Clin Pharm Ther. 1998;23(4):257–85.
36. Silin DS, Lyubomska OV, Jirathitikal V, Bourinbaiar AS. Oral vaccination: where we are? Expert Opin Drug Deliv. 2007;4(4):323–40.
37. Fulginiti VA, Papier A, Lane JM, Neff JM, Henderson DA. Smallpox vaccination: a review, part II. Adverse events. Clin Infect Dis. 2003;37(2):251–71.
38. Fulginiti VA, Papier A, Lane JM, Neff JM, Henderson DA. Smallpox vaccination: a review, part I. Background, vaccination technique, normal vaccination and revaccination, and expected normal reactions. Clin Infect Dis. 2003;37(2):241–50.
39. Hilleman MR. Vaccines in historic evolution and perspective: a narrative of vaccine discoveries. Vaccine. 2000;18(15):1436–47.
40. No authors. FluMist: an intranasal live influenza vaccine. The medical letter on drugs and therapeutics. 2003;45(1163):65–6.
41. Pliaka V, Kyriakopoulou Z, Markoulatos P, Risks associated with the use of live-attenuated vaccine poliovirus strains and the strategies for control and eradication of paralytic poliomyelitis. Expert Rev Vaccines. 2012;11(5):609–28.
42. Vajdy M, Singh M. The role of adjuvants in the development of mucosal vaccines. Expert Opin Biol Ther. 2005;5(7):953–65.
43. Galli G, Hancock K, Hoschler K, DeVos J, Praus M, Bardelli M, Malzone C, Castellino F, Gentile C, McNally T, G. Del Giudice, Banzhoff A, Brauer V, Montomoli E, Zambon M, Katz J, Nicholson K, Stephenson I. Fast rise of broadly cross-reactive antibodies after boosting long-lived human memory B cells primed by an MF59 adjuvanted prepandemic vaccine. Proc Natl Acad U S A. 2009;106(19):7962–967.
44. Brandtzaeg P, Baekkevold ES, Farstad IN, Jahnsen FL, Johansen FE, Nilsen EM, Yamanaka T. Regional specialization in the mucosal immune system: what happens in the microcompartments? Immunol Today. 1999;20(3):141–51.
45. Neutra MR, Kozlowski PA. Mucosal vaccines: the promise and the challenge. Nat Rev Immunol. 2006;6(2):148–58.
46. Garmory HS, Griffin KF, Brown KA, Titball RW. Oral immunisation with live aroA attenuated Salmonella enterica serovar Typhimurium expressing the Yersinia pestis V antigen protects mice against plague. Vaccine. 2003;21(21–22):3051–57.
47. Pasetti MF, Levine MM, Sztein MB. Animal models paving the way for clinical trials of attenuated Salmonella enterica serovar Typhi live oral vaccines and live vectors, Vaccine. 2003;21(5–6):401–18.
48. Mowat AM. Anatomical basis of tolerance and immunity to intestinal antigens. Nat Rev Immunol. 2003;3(4):331–41.
49. Chadwick S, Kriegel C, Amiji M. Nanotechnology solutions for mucosal immunization. Adv Drug Deliv Rev. 2010;62(4–5):394–407.
50. Rice-Ficht AC, Arenas-Gamboa AM, Kahl-McDonagh MM, Ficht TA. Polymeric particles in vaccine delivery. Curr Opin Microbiol. 2010;13(1):106–12.
51. Clark MA, Jepson MA, Simmons NL, Hirst BH. Preferential interaction of Salmonella typhimurium with mouse Peyer's patch M cells. Res Microbiol. 1994;145(7):543–52.
52. Amerongen HM, Weltzin R, Farnet CM, Michetti P, Haseltine WA, Neutra MR. Transepithelial transport of HIV-1 by intestinal M cells: a mechanism for transmission of AIDS. J Acquir Immune Defic Syndr. 1991;4(8):760–65.
53. Owen RL, Pierce NF, Apple RT, Cray WC Jr. M cell transport of Vibrio cholerae from the intestinal lumen into Peyer's patches: a mechanism for antigen sampling and for microbial transepithelial migration. J Infect Dis. 1986;153(6):1108–18.
54. Neutra MR, Phillips TL, Mayer EL, Fishkind DJ. Transport of membrane-bound macromolecules by M cells in follicle-associated epithelium of rabbit Peyer's patch. Cell Tissue Res. 1987;247(3):537–46.
55. Honda K, Nakano H, Yoshida H, Nishikawa S, Rennert P, Ikuta K, Tamechika M, Yamaguchi K, Fukumoto T, Chiba T, Nishikawa SI. Molecular basis for hematopoietic/mesenchymal interaction during initiation of Peyer's patch organogenesis. J Exp Med. 2001;193(5):621–30.

56. van der Ven I, Sminia T. The development and structure of mouse nasal-associated lymphoid tissue: an immuno- and enzyme-histochemical study. Region Immunol. 1993;5(2):69–75.
57. Brooking J, Davis SS, Illum L. Transport of nanoparticles across the rat nasal mucosa. J Drug Target. 2001;9(4):267–79.
58. Sarkar MA. Drug metabolism in the nasal mucosa. Pharm Res. 1992;9(1):1–9.
59. Bomsel M, Tudor D, Drillet AS, Alfsen A, Ganor Y, Roger MG, Mouz N, Amacker M, Chalifour A, Diomede L, Devillier G, Cong Z, Wei Q, Gao H, Qin C, Yang GB, Zurbriggen R, Lopalco L, Fleury S. Immunization with HIV-1 gp41 subunit virosomes induces mucosal antibodies protecting nonhuman primates against vaginal SHIV challenges. Immunity. 2011;34(2):269–80.
60. Czerkinsky C, Cuburu N, Kweon MN, Anjuere F, Holmgren J. Sublingual vaccination. Hum Vaccin. 2011;7(1):110–4.
61. Baudner BC, Verhoef JC, Giuliani MM, Peppoloni S, Rappuoli R, Del Giudice G, Junginger HE. Protective immune responses to meningococcal C conjugate vaccine after intranasal immunization of mice with the LTK63 mutant plus chitosan or trimethyl chitosan chloride as novel delivery platform. J Drug Target. 2005;13(8–9):89–498.
62. Baudner BC, Morandi M, Giuliani MM, Verhoef JC, Junginger HE, Costantino P, Rappuoli R, Del Giudice G Modulation of immune response to group C meningococcal conjugate vaccine given intranasally to mice together with the LTK63 mucosal adjuvant and the trimethyl chitosan delivery system. J Infect Dis. 2004;189(5):828–32.
63. Baudner BC, Giuliani MM, Verhoef JC, Rappuoli R, Junginger HE, Giudice GD. The concomitant use of the LTK63 mucosal adjuvant and of chitosan-based delivery system enhances the immunogenicity and efficacy of intranasally administered vaccines. Vaccine. 2003;21(25–26):3837–44.
64. O'Hagan AH, Irvine AD, Allen GE, Walsh M. Pseudoporphyria induced by mefenamic acid. Brit J Dermatol. 1998;139(6):1131–32.
65. Koping-Hoggard M, Sanchez A, Alonso MJ. Nanoparticles as carriers for nasal vaccine delivery. Expert Rev Vaccines. 2005;4(2):185–96.
66. Moore A, McGuirk P, Adams S, Jones WC, McGee JP, O'Hagan DT, Mills KH. Immunization with a soluble recombinant HIV protein entrapped in biodegradable microparticles induces HIV-specific $CD8^+$ cytotoxic T lymphocytes and $CD4^+$ Thl cells. Vaccine. 1995;13(18):1741–9.
67. Shephard MJ, Todd D, Adair BM, Po AL, Mackie DP, Scott EM. Immunogenicity of bovine parainfluenza type 3 virus proteins encapsulated in nanoparticle vaccines, following intranasal administration to mice. Res Vet Sci. 2003;74(2):187–90.
68. Jung T, Kamm W, Breitenbach A, Hungerer KD, Hundt E, Kissel T. Tetanus toxoid loaded nanoparticles from sulfobutylated poly(vinyl alcohol)-graft-poly(lactide-co-glycolide): evaluation of antibody response after oral and nasal application in mice. Pharm Res. 2001;18(3):352–60.
69. Nagamoto T, Hattori Y, Takayama K, Maitani Y. Novel chitosan particles and chitosan-coated emulsions inducing immune response via intranasal vaccine delivery. Pharm Res. 2004;21(4):671–4.
70. Vila A, Sanchez A, Evora C, Soriano I, McCallion O, Alonso MJ. PLA-PEG particles as nasal protein carriers: the influence of the particle size. Int J Pharm. 2005;292(1–2):43–52.
71. Slütter B, Bal S, Keijzer C, Mallants R, Hagenaars N, Que I, Kaijzel E, van Eden W, Augustijns P, Löwik C, Bouwstra J, Broere F, Jiskoot W. Nasal vaccination with N-trimethyl chitosan and PLGA based nanoparticles: nanoparticle characteristics determine quality and strength of the antibody response in mice against the encapsulated antigen. Vaccine. 2010;28(38):6282–91.
72. Fujimura Y, Akisada T, Harada T, Haruma K. Uptake of microparticles into the epithelium of human nasopharyngeal lymphoid tissue. Med Mol Morphol. 2006;39(4):181–6.
73. Takata S, Ohtani O, Watanabe Y. Lectin binding patterns in rat nasal-associated lymphoid tissue (NALT) and the influence of various types of lectin on particle uptake in NALT. Arch Histol Cytol. 2000;63(4):305–12.

74. Giannasca PJ, Boden JA, Monath TP. Targeted delivery of antigen to hamster nasal lymphoid tissue with M-cell-directed lectins. Infect Immun. 1997;65(10):4288–98.
75. Tyrer P, Foxwell AR, Cripps AW, Apicella MA, Kyd JM. Microbial pattern recognition receptors mediate M-cell uptake of a gram-negative bacterium, Infect Immun. 2006:74(1):625–31.
76. Fujihashi K, Koga T, van Ginkel FW, Hagiwara Y, McGhee JR. A dilemma for mucosal vaccination: efficacy versus toxicity using enterotoxin-based adjuvants. Vaccine. 2002;20(19–20):2431–8.
77. Lemiale F, Kong WP, Akyurek LM, Ling X, Huang Y, Chakrabarti BK, Eckhaus M, Nabel GJ. Enhanced mucosal immunoglobulin A response of intranasal adenoviral vector human immunodeficiency virus vaccine and localization in the central nervous system. J Virol. 2003;77(18):10078–87.
78. Armstrong ME, Lavelle EC, Loscher CE, Lynch MA, Mills. Proinflammatory responses in the murine brain after intranasal delivery of cholera toxin: implications for the use of AB toxins as adjuvants in intranasal vaccines. J Infect Dis. 2005;192(9):1628–33.
79. Mutsch M, Zhou W, Rhodes P, Bopp M, Chen RT, Linder T, Spyr C, Steffen R. Use of the inactivated intranasal influenza vaccine and the risk of Bell's palsy in Switzerland. New Engl J Med. 2004;350(9):896–903.
80. Lewis DJ, Huo Z, Barnett S, Kromann I, Giemza R, Galiza E, Woodrow M, Thierry-Carstensen B, Andersen P, Novicki D, Del Giudice G, Rappuoli R. Transient facial nerve paralysis (Bell's palsy) following intranasal delivery of a genetically detoxified mutant of Escherichia coli heat labile toxin, PloS One. 2009;4(9):e6999.
81. Hanif SN, Garcia-Contreras L. Pharmaceutical aerosols for the treatment and prevention of Tuberculosis. Front Cell Infect Microbiol. 2012;2:118.
82. Lu D, Hickey AJ. Pulmonary vaccine delivery. Expert Rev Vaccines. 2007;6(2):213–26.
83. Blank F, Stumbles P, von Garnier C. Opportunities and challenges of the pulmonary route for vaccination. Expert Opin Drug Deliv. 2011;8(5):547–63.
84. Moyle PM, McGeary RP, Blanchfield JT, Toth I. Mucosal immunisation: adjuvants and delivery systems. Curr Drug Deliv. 2004;1(4):385–96.
85. Moyron-Quiroz JE, Rangel-Moreno J, Kusser K, Hartson L, Sprague F, Goodrich S, Woodland DL, Lund FE, Randall TD. Role of inducible bronchus associated lymphoid tissue (iBALT) in respiratory immunity. Nat Med. 2004;10(9):927–34.
86. Roth Y, Chapnik JS, Cole P. Feasibility of aerosol vaccination in humans. Ann Otol Rhinol Laryngol. 2003;112(3):264–70.
87. Smith DJ, Bot S, Dellamary L, Bot A. Evaluation of novel aerosol formulations designed for mucosal vaccination against influenza virus. Vaccine. 2003;21(21–22):2805–12.
88. Pilcer G, Amighi K. Formulation strategy and use of excipients in pulmonary drug delivery. Int J Pharm. 2010;392(1–2):1–19.
89. Rautenschlein S, Sharma JM, Winslow BJ, McMillen J, Junker D, Cochran M. Embryo vaccination of turkeys against Newcastle disease infection with recombinant fowlpox virus constructs containing interferons as adjuvants. Vaccine. 1999;18(5–6):426–33.
90. Murphy D, Van Alstine WG, Clark LK, Albregts S, Knox K. Aerosol vaccination of pigs against Mycoplasma hyopneumoniae infection. Am J Vet Res. 1993;54(11):1874–80.
91. Deuter A, Southee DJ, Mockett AP. Fowlpox virus: pathogenicity and vaccination of day-old chickens via the aerosol route. Res Vet Sci. 1991;50(3):362–64.
92. Moss WJ, Griffin DE. Measles, Lancet. 2012;379(9811):153–64.
93. Rottem M, Shoenfeld Y. Vaccination and allergy. Curr Opin Otolaryngol Head Neck Surg. 2004;12(3):223–31.
94. Brimnes J, Kildsgaard J, Jacobi H, Lund K. Sublingual immunotherapy reduces allergic symptoms in a mouse model of rhinitis. Clin Exp Allergy. 2007;37(4):488–97.
95. Kildsgaard J, Brimnes J, Jacobi H, Lund K. Sublingual immunotherapy in sensitized mice. Ann Allergy Asthma Immunol. 2007;98(4):366–72.
96. Cuburu N, Kweon MN, Song JH, Hervouet C, Luci C, Sun JB, Hofman P, Holmgren J, Anjuere F, Czerkinsky C. Sublingual immunization induces broad-based systemic and mucosal immune responses in mice. Vaccine. 2007;25(51):8598–610.

97. Song JH, Nguyen HH, Cuburu N, Horimoto T, Ko SY, Park SH, Czerkinsky C, Kweon MN. Sublingual vaccination with influenza virus protects mice against lethal viral infection. Proc Natl Acad U S A. 2008;105(5):1644–49.
98. Cuburu N, Kweon MN, Hervouet C, Cha HR, Pang YY, Holmgren J, Stadler K, Schiller JT, Anjuere F, Czerkinsky C. Sublingual immunization with nonreplicating antigens induces antibody-forming cells and cytotoxic T cells in the female genital tract mucosa and protects against genital papillomavirus infection. J Immunol. 2009;183(12):7851–9.
99. Hervouet C, Luci C, Cuburu N, Cremel M, Bekri S, Vimeux L, Maranon C, Czerkinsky C, Hosmalin A, Anjuere F. Sublingual immunization with an HIV subunit vaccine induces antibodies and cytotoxic T cells in the mouse female genital tract. Vaccine. 2010; 28(34):5582–90.
100. Song JH, Kim JI, Kwon HJ, Shim DH, Parajuli N, Cuburu N, Czerkinsky C, Kweon MN. CCR7-CCL19/CCL21-regulated dendritic cells are responsible for effectiveness of sublingual vaccination. J Immunol. 2009;182(11):6851–60.
101. Pedersen GK, Ebensen T, Gjeraker IH, Svindland S, Bredholt G, Guzman CA, Cox RJ. Evaluation of the sublingual route for administration of influenza H5N1 virosomes in combination with the bacterial second messenger c-di-GMP. PloS One. 2011;6(11):e26973.
102. Feinberg H, Tso CK, Taylor ME, Drickamer K, Weis WI. Segmented helical structure of the neck region of the glycan-binding receptor DC-SIGNR. J Mol Biol. 2009;394(4):613–20.
103. Shortman K, Lahoud MH, Caminschi I. Improving vaccines by targeting antigens to dendritic cells. Exp Mol Med. 2009;41(2):61–66.
104. Romani N, Thurnher M, Idoyaga J, Steinman RM, Flacher V. Targeting of antigens to skin dendritic cells: possibilities to enhance vaccine efficacy. Immunol Cell Biol. 2010; 88(4):424–30.
105. Flamar AL, Zurawski S, Scholz F, Gayet I, Ni L, Li XH, Klechevsky E, Quinn J, Oh S, Kaplan DH, Banchereau J, Zurawski G. Noncovalent assembly of anti-dendritic cell antibodies and antigens for evoking immune responses in vitro and in vivo. J Immunol. 2012;189(5):2645–55.
106. Peiser M, Koeck J, Kirschning CJ, Wittig B, Wanner R. Human Langerhans cells selectively activated via Toll-like receptor 2 agonists acquire migratory and CD4 + T cell stimulatory capacity. J Leukoc Biol. 2008;83(5):1118–27.
107. Oh JZ, Kurche JS, Burchill MA, Kedl RM. TLR7 enables cross-presentation by multiple dendritic cell subsets through a type I IFN-dependent pathway. Blood. 2011;118(11):3028–38.
108. Romani N, Brunner PM, Stingl G. Changing views of the role of Langerhans cells. J Invest Dermatol. 2012;132(3 Pt 2):872–81.
109. Freytag IC, Clements JD. Bacterial toxins as mucosal adjuvants. Curr Top Microbiol Immunol. 1999;236:215–36.
110. Pizza M, Giuliani MM, Fontana MR, Monaci E, Douce G, Dougan G, Mills KH, Rappuoli R, Del Giudice G. Mucosal vaccines: non toxic derivatives of LT and CT as mucosal adjuvants. Vaccine. 2001;19(17–19):2534–41.
111. Kozlowski PA, Cu-Uvin S, Neutra MR, Flanigan TP. Comparison of the oral, rectal, and vaginal immunization routes for induction of antibodies in rectal and genital tract secretions of women. Infect Immun. 1997;65(4):1387–94.
112. Johansson EL, Wassen L, Holmgren J, Jertborn M, Rudin A. Nasal and vaginal vaccinations have differential effects on antibody responses in vaginal and cervical secretions in humans. Infect Immun. 2001;69(12):7481–86.
113. Wassen L, Schon K, Holmgren J, Jertborn M, Lycke N. Local intravaginal vaccination of the female genital tract. Scand J Immunol. 1996;44(4):408–14.
114. Nardelli-Haefliger D, Wirthner D, Schiller JT, Lowy DR, Hildesheim A, Ponci F, De Grandi P. Specific antibody levels at the cervix during the menstrual cycle of women vaccinated with human papillomavirus 16 virus-like particles. J Natl Cancer Inst. 2003;95(15):1128–37.
115. Iwasaki A. Antiviral immune responses in the genital tract: clues for vaccines, Nat Rev Immunol. 2010;10(10):699–711.
116. Schiller JT, Castellsague X, Villa LL, Hildesheim A. An update of prophylactic human papillomavirus L1 virus-like particle vaccine clinical trial results. Vaccine. 2008;26(Suppl 10):K53–61.

117. Borrow P, Lewicki H, Hahn BH, Shaw GM, Oldstone MB. Virus-specific CD8$^+$ cytotoxic T-lymphocyte activity associated with control of viremia in primary human immunodeficiency virus type 1 infection. J Virol. 1994;68(9):6103–10.
118. Watkins DI, Burton DR, EGKallas, Moore JP, Koff WC. Nonhuman primate models and the failure of the Merck HIV-1 vaccine in humans. Nat Med. 2008;14(6):617–21.
119. de Boer AG, Moolenaar F, de Leede LG, Breimer DD. Rectal drug administration: clinical pharmacokinetic considerations. Clin Pharmacokinet. 1982;7(4):285–311.
120. Kantele A, Hakkinen M, Moldoveanu Z, Lu A, Savilahti E, Alvarez RD, Michalek S, Mestecky J. Differences in immune responses induced by oral and rectal immunizations with Salmonella typhi Ty21a: evidence for compartmentalization within the common mucosal immune system in humans. Infect Immun. 1998;66(12):5630–5.
121. Pechine S, Deneve C, Le Monnier A, Hoys S, Janoir C, Collignon A. Immunization of hamsters against Clostridium difficile infection using the Cwp84 protease as an antigen. FEMS Immunol Med Microbiol. 2011;63(1):73–81.
122. Yu J, Cassels F, Scharton-Kersten T, Hammond SA, Hartman A, Angov E, Corthesy B, Alving C, Glenn G. Transcutaneous immunization using colonization factor and heat-labile enterotoxin induces correlates of protective immunity for enterotoxigenic Escherichia coli. Infect Immun. 2002;70(3):1056–68.
123. Glenn GM, Flyer DC, Ellingsworth LR, Frech SA, Frerichs DM, Seid RC, Yu J. Transcutaneous immunization with heat-labile enterotoxin: development of a needle-free vaccine patch. Expert Rev Vaccines. 2007;6(5):809–19.
124. Uddowla S, Freytag LC, Clements JD. Effect of adjuvants and route of immunizations on the immune response to recombinant plague antigens. Vaccine. 2007;25(47):7984–93.
125. Vogt A, Mahe B, Costagliola D, Bonduelle O, Hadam S, Schaefer G, Schaefer H, Katlama C, Sterry W, Autran B, Blume-Peytavi U, Combadiere B. Transcutaneous anti-influenza vaccination promotes both CD4 and CD8 T cell immune responses in humans. J Immunol. 2008;180(3):1482–89.
126. Proksch E, Brandner JM, Jensen JM. The skin: an indispensable barrier. Exp Dermatol. 2008;17(12):63–1072.
127. Lawson LB, Clements JD, Freytag LC. Mucosal immune responses induced by transcutaneous vaccines. Curr Top Microbiol Immunol. 2012;354:19–37.
128. Nicolas JF, Guy B. Intradermal, epidermal and transcutaneous vaccination: from immunology to clinical practice. Expert Rev Vaccines. 2008;7(8):1201–14.
129. Glenn GM, Kenney RT. Mass vaccination: solutions in the skin. Curr Top Microbiol Immunol. 2006;304:247–68.
130. Romani N, Clausen BE, Stoitzner P. Langerhans cells and more: langerin-expressing dendritic cell subsets in the skin. Immunol Rev. 2010;234(1):120–41.
131. Teunissen MB, Haniffa M, Collin MP. Insight into the immunobiology of human skin and functional specialization of skin dendritic cell subsets to innovate intradermal vaccination design. Curr Top Microbiol Immunol. 2012;351:25–76.
132. Gockel CM, Bao S, Beagley KW. Transcutaneous immunization induces mucosal and systemic immunity: a potent method for targeting immunity to the female reproductive tract. Mol Immunol. 2000;37(9):537–44.
133. Naito S, Maeyama J, Mizukami T, Takahashi M, Hamaguchi I, Yamaguchi K. Transcutaneous immunization by merely prolonging the duration of antigen presence on the skin of mice induces a potent antigen-specific antibody response even in the absence of an adjuvant. Vaccine. 2007;25(52):8762–70.
134. Novak N, Bieber T. 2. Dendritic cells as regulators of immunity and tolerance. J Allergy Clin Immunol. 2008;121(2 Suppl):S370–4; quiz S413.
135. Rechtsteiner G, Warger T, Osterloh P, Schild H, Radsak MP. Cutting edge: priming of CTL by transcutaneous peptide immunization with imiquimod. J Immunol. 2005;174(5):2476–80.
136. He Y, Zhang J, Donahue C, Falo LD Jr. Skin-derived dendritic cells induce potent CD8($^+$) T cell immunity in recombinant lentivector-mediated genetic immunization. Immunity. 2006;24(5):643–56.

137. Belyakov IM, Hammond SA, Ahlers JD, Glenn GM, Berzofsky JA. Transcutaneous immunization induces mucosal CTLs and protective immunity by migration of primed skin dendritic cells. J Clin Invest. 2004;113(7);998–1007.
138. Glenn GM, Taylor DN, Li X, Frankel S, Montemarano A, Alving CR. Transcutaneous immunization: a human vaccine delivery strategy using a patch. Nat Med. 2000;6(12):1403–6.
139. Etchart N, Hennino A, Friede M, Dahel K, Dupouy M, Goujon-Henry C, Nicolas JF, Kaiserlian D. Safety and efficacy of transcutaneous vaccination using a patch with the live-attenuated measles vaccine in humans. Vaccine. 2007;25(39–40):6891–9.
140. Glenn GM, Villar CP, Flyer DC, Bourgeois AL, McKenzie R, Lavker RM, Frech SA. Safety and immunogenicity of an enterotoxigenic Escherichia coli vaccine patch containing heat-labile toxin: use of skin pretreatment to disrupt the stratum corneum. Infect Immun. 2007;75(5):2163–70.
141. Frech SA, Dupont HL, Bourgeois AL, McKenzie R, Belkind-Gerson J, Figueroa JF, Okhuysen PC, Guerrero NH, Martinez-Sandoval FG, Melendez-Romero JH, Jiang ZD, Asturias EJ, Halpern J, Torres OR, Hoffman AS, Villar CP, Kassem RN, Flyer DC, Andersen BH, Kazempour K, Breisch SA, Glenn GM. Use of a patch containing heat-labile toxin from Escherichia coli against travellers' diarrhoea: a phase II, randomised, double-blind, placebo-controlled field trial. Lancet. 2008;371(9629):2019–25.
142. Wood LC, Jackson SM, Elias PM, Grunfeld C, Feingold KR. Cutaneous barrier perturbation stimulates cytokine production in the epidermis of mice. J Clin Invest. 1992;90(2):482–7.
143. Cumberbatch M, Dearman RJ, Kimber I. Langerhans cells require signals from both tumour necrosis factor-alpha and interleukin-1 beta for migration. Immunology. 1997;92(3):388–95.
144. Kim YC, Jarrahian C, Zehrung D, Mitragotri S, Prausnitz MR. Delivery systems for intradermal vaccination. Curr Top Microbiol Immunol. 2012;351:77–112.
145. Kommareddy S, Baudner BC, Oh S, Kwon SY, Singh M, O'Hagan DT. Dissolvable microneedle patches for the delivery of cell-culture-derived influenza vaccine antigens. J Pharm Sci. 2012;101(3):1021–7.
146. Verstrepen BE, Bins AD, Rollier CS, Mooij P, Koopman G, Sheppard NC, Sattentau Q, Wagner R, Wolf H, Schumacher TN, Heeney JL, Haanen JB. Improved HIV-1 specific T-cell responses by short-interval DNA tattooing as compared to intramuscular immunization in non-human primates. Vaccine. 2008;26(26):3346–51.
147. Leroux-Roels I, Weber F. Intanza (®) 9 μg intradermal seasonal influenza vaccine for adults 18 to 59 years of age. Hum Vaccin Immunother. 2013;9(1):115–21.
148. Kommareddy S, Baudner BC, Bonificio A, Gallorini S, Palladino G, Determan AS, Dohmeier DM, Kroells KD, Sternjohn JR, Singh M, Dormitzer PR, Hansen KJ, O'Hagan DT. Influenza subunit vaccine coated microneedle patches elicit comparable immune responses to intramuscular injection in guinea pigs. Vaccine. 2013;31(34):3435-41.
149. Kim YC, Park JH, Prausnitz MR. Microneedles for drug and vaccine delivery. Adv Drug Deliv Rev. 2012;64(14):1547–68.
150. Weldon WC, Zarnitsyn VG, Esser ES, Taherbhai MT, Koutsonanos DG, Vassilieva EV, Skountzou I, Prausnitz MR, Compans RW. Effect of adjuvants on responses to skin immunization by microneedles coated with influenza subunit vaccine. PloS One. 2012;7(7):e41501.

Chapter 2
Mucoadhesion and Characterization of Mucoadhesive Properties

Tao Yu, Gavin P. Andrews and David S. Jones

2.1 Introduction

An adhesive is a material that attaches to another substrate surface and resists separation [1]. Adhesion involves the formation of attractive bonds between two substrates that resist separation. Bioadhesion is a specific case of adhesion in which at least one of the two substrates involves a biological tissue [2]. Furthermore, if the adherent substrate surface is a mucosal surface, e.g., a mucosal membrane, bioadhesion is specifically referred to as mucoadhesion [3–5].

The use of mucoadhesive materials for the enhanced delivery of therapeutic agents has been of interest for several years owing to several important advantages concerning the in vitro and in vivo performance of dosage forms. Mucoadhesive formulations are capable of providing localized drug release in desirable regions such as nasal cavity, eye, mouth, stomach, intestine, and vagina to enhance their clinical efficacy. The employment of mucoadhesive materials in formulations may modify the permeability of mucosal tissue or membranes and hence facilitate the adsorption of macromolecules, e.g., peptides. Furthermore, the interaction between mucoadhesive formulations and mucosal surface offers potential to prolong the residence time of the dosage form at the site of application, thereby reducing the dosing frequency and increasing patient compliance [3, 5–8].

Since the first report of the first application of mucoadhesive systems by Scrivener and Schantz [9], there have been many publications regarding the design, development, and testing of bioadhesive and mucoadhesive platforms. Examples of reported mucoadhesive drug delivery dosage forms include tablets, films, gels, creams, ointments, viscous solutions, micro- and nanoparticulate suspensions, and sprays [8, 10]. Commercially, one of the earliest mucoadhesive products was Orabase®, which

D. S. Jones (✉) · T. Yu · G. P. Andrews
School of Pharmacy, The Queen's University of Belfast,
Medical Biology Centre, 97 Lisburn Road, BT9 7BL,
Belfast, Northern Ireland, UK
e-mail: d.jones@qub.ac.uk

J. das Neves, B. Sarmento (eds.), *Mucosal Delivery of Biopharmaceuticals*, DOI 10.1007/978-1-4614-9524-6_2,

consists of natural gums to facilitate mucoadhesion. So far, a number of mucoadhesive products have been commercialized, e.g., Replens®, Zidoval® gels, for vaginal therapies.

2.2 Structure, Composition, and Functions of Mucosa, Mucus, and Mucin

Mucoadhesive formulations are designed to form specific interactions with mucin-coated mucosal membranes which, typically, are composed of specialized epithelium, lamina propria, and glands (depending upon their type and location) [11–13]. The mucosal membrane covers the epithelium and facilitates the exchange of gases and nutrients between the underlying epithelium and the external environment. In addition, such membranes inherently lubricate cavities and passages and form a barrier to protect the epithelium from damage associated with pathogens and noxious substances [11, 14–15]. Mucus which is secreted by goblet cells within mucosal membranes is the most crucial component responsible for mucosal protective functions as well as adsorption and exchange of other components, notably drugs [15, 16]. Structurally, mucus is a complex viscous gel that is primarily composed of water (circa 95 %) and mucin (a glycoprotein), electrolytes, fatty acids, phospholipids, cholesterol, proteins, and other various species in smaller proportions [5, 11]. The presence of mucin on the epithelia of many cavities including the mouth, nose, eyes, vagina, rectum, and the stomach has been confirmed. Furthermore, it has been shown that there are several types of mucins in vaginal fluid [17–18], saliva [19], tears [20], and within the gastrointestinal tract [21]. The mucin glycoproteins exhibit a highly entangled network of macromolecules that associate with one another through non-covalent bonds. Such molecular association is central to the structure of mucus and is responsible for its rheological properties [5, 22–27]. Mucin can be considered as an anionic polyelectrolyte at neutral pH owing to the presence of pendant sialic acid and sulfate groups located on the glycoprotein molecules through covalent bonds [27]. These acidic groups exhibit pKa values from 1.0 to 2.6 resulting in their complete ionization under physiological conditions [8, 28]. The negative charge of mucin has been reported to be important in partitioning and complex formation with pharmaceutical preparations [27, 29]. Mucin is also fundamental to cytoprotection; the endothelial and leukocyte classes of mucins being adhesion molecules that are involved in lymphocyte homing and lymphocyte activation or are part of the adhesion cascade that plays a role in the initiation of inflammation [30–31]. A thorough understanding of mucin is fundamental to the potential pharmaceutical applications of mucoadhesive dosage forms.

Mucins are macromolecules with the molecular weight range from 0.5 up to 20 MDa. The basic composition of a typical mucin includes about 80 % carbohydrates, namely *N*-acetylgalactosamine, *N*-acetylglucosamine, fucose, galactose, sialic acid, and traces of mannose and sulfate, and 20 % protein core [8, 15]. The central protein segments are composed of a large number of tandem repeats rich in serine, threonine, and proline that are all linked by the interspersed regions possessing little *O*-glycosylation, a few *N*-glycosylation sites, and a high proportion of cysteine domains ($> 10\,\%$) [32]. In addition, it has been shown that large carbohydrate side chains link to the hydroxyl side chains of serine and threonines of the protein core via *O*-glycosidic covalent bonds [15]. The terminal regions of an entire mucin chain where little glycosylation occurs are referred to as "naked protein regions" [5, 33].

Although mucin has been reported to be difficult to characterize due to the large molecular weight, polydispersity, and high degree of glycosylation, the significant interest and progress in research have been focused on identifying and distinguishing mucin genes. There have been at least 19 mucin sequenced by cDNA cloning, and three of them have been totally sequenced, namely MUC1, MUC2, and MUC5B [15, 34]. Other techniques employed to characterize mucin include light scattering [35–39], nuclear magnetic resonance (NMR) [40–43], transmission electron micrographic (TEM) [30, 44], and atomic force microscopy (AFM) [45–48].

2.3 Mucoadhesion Theories of Polymer Attachment

Mucoadhesion is a complex process and has not yet been fully understood [8, 26]. Several theories have been proposed to explain mucoadhesion, notably: (1) the wetting theory [49–52], (2) the mechanical interlocking theory [1], (3) the electronic transfer theory [1], (4) the diffusion-interpenetration theory [50], (5) the adsorption theory [53], and (6) the fracture theory [53]. They are briefly detailed below.

2.3.1 Wetting Theory

The wetting theory attributes the bonding between the formulation and the surface tissue to intermolecular interaction and interfacial tension. This theory is usually applied for liquid or low viscosity mucoadhesive systems and is essentially a measure of the "spreadability" of a drug delivery system across the biological substrate [5, 54]. The spreadability of the system is indicative of interactions and can be measured by the liquid–solid contact angle. Adhesive forces between a liquid and solid enable a liquid drop to spread across the surface, whereas, cohesive forces within the liquid cause the drop to ball up and avoid contact with the surface. Generally, contact angles less than 90° indicate that the wetting of the surface is favorable, and the liquid tends to spread out to a large area. A contact angle greater than 90° indicates the wetting of the surface is unfavorable; the interaction among liquid molecules maintains the shape of the droplet and minimizes its contact area to the solid surface [55].

The contact angle may be experimentally measured from which interfacial tension (γ) may be derived using the Young equation [49, 56]:

$$\gamma_{SG} = \gamma_{SL} + \gamma_{LG} \cos\theta \tag{2.1}$$

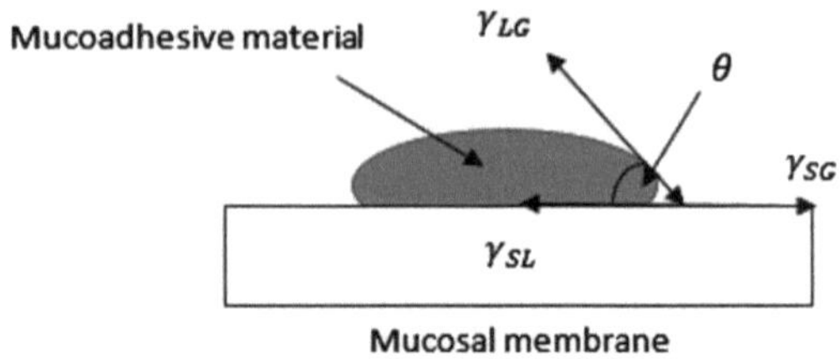

Fig. 2.1 Contact angle measurement between a droplet and solid surface. (Modified from [5])

Where γ_{SG} is the interfacial tension between solid and gas; γ_{SL} is the interfacial tension between solid and liquid; γ_{LG} is the interfacial tension; and θ is the contact angle between solid and liquid interface (Fig. 2.1).

The interfacial tension associated with contact angle θexhibits the degree of wetting. When the contact angle θ is 0°, wetting is complete, the liquid having fully spread across the surface of the substrate. In contrast, a contact angle of 180° is indicative of nonwettability. Wetting between the liquid (formulation) and the substrate (e.g., mucus) substance occurs whenever the contact angle ranges between 0° and 180° [54, 57].

2.3.2 *Mechanical Interlocking Theory*

The mechanical interlocking theory only considers the adhesion between liquid and a rough surface or a surface rich in pores [58–60] and essentially proposes that the adhesion between the two substrates is due to mechanical interlocking of the adhesive into the irregularities of the substrate surface [1]. Adhesion between the mucoadhesive system and the rough surface typically occurs within a diverse biological environment and accordingly this theory does not fully explain the adhesive properties in vivo [1, 60–61].

2.3.3 *Electronic Transfer Theory*

In the electronic transfer theory, mucoadhesion occurs as the result of the transfer of electrons between mucus and the mucoadhesive platform. The electronic transfer between two different layers results in the formation of a double-layered electronic charge at the interface. This theory suggests that the electrostatic forces are critical in generating bond adhesions rather than high joint strength [1, 60, 62–65].

2.3.4 *Adsorption Theory*

There are various surface interactions that result in adhesion, including primary bond and secondary bond formation. In the former situation, primary bonds (e.g., ionic and covalent bonds) are undesirable because they form a strong energy barrier, which may result in permanent interactions with mucus or tissue layer [1]. In contrast, secondary (weaker) bonds such as van der Waals forces, hydrogen bonding, electrostatic attraction, and hydrophobic interactions are more desirable, resulting in semipermanent interactions (an important criterion for drug delivery systems) [1, 5, 53, 66].

2.3.5 *Fracture Theory*

According to the fracture theory, the adhesive strength, also known as the fracture strength, is related to the force required to separate the platform and the mucus surfaces. Previous publications have suggested the use of Young's modulus of elasticity (E), the fracture energy, (ε) and the critical crack length (L) to determine the fracture strength (σ). The equation is shown below [52]:

$$\sigma = \left(\frac{E \times \varepsilon}{L}\right)^{1/2} \tag{2.2}$$

The fracture energy can be obtained from the reversible adhesive work:

$$\varepsilon = \varepsilon_{\mathrm{r}} + \varepsilon_{\mathrm{d}} \tag{2.3}$$

where ε_{r} is the required energy for producing new fractured surfaces and ε_{d} is the work of plastic deformation provoked by the removal of a proof tip until the disruption of the adhesive bond.

Depending on the position of their occurrences, fractures may be divided into platform fracture, platform-mucus fracture, and mucus fracture. As a result, the fracture theory is not only used to measure adhesive strength between the platform surface and the mucus surface, but can also be used to evaluate the strength of intermolecular interactions within the platform.

2.3.6 *Diffusion-Interpenetration Theory*

The diffusion-interpenetration theory is a commonly employed theory to describe mucoadhesion and involves the interpenetration and entanglement between the polymer chains and the mucus chains [67]. The first step in this process involves the creation of an initial contact between the bioadhesive polymer chains and the mucus chains. In this step, weak physical forces, e.g., attraction and electronic force, dominate the mobility of the polymer chains. The second step involves the interpenetration of polymer chains from the delivery system into mucus layer to achieve mucoadhesion via more substantial bond formation. The profile of the

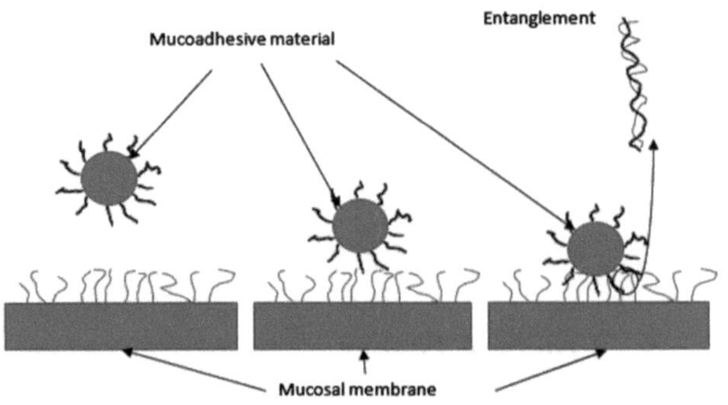

Fig. 2.2 Schematic profile of interpenetration steps. (1) Polymer chain approaching the mucus layer. (2) Interpenetration of polymer into mucus chains. (3) Polymer chains and mucus layer contact by physical-chemical forces

diffusion-interpenetration theory is shown in Fig. 2.2. The depth of interpenetration is dependent upon the diffusion coefficient of both polymer and substrate, time of contact, and the adhesive strength of the bioadhesive polymer [4, 5, 60, 61, 68, 69]. Mikos and Peppas introduced the relationship between interpenetration depth and characteristic time:

$$\tau = \mathrm{l}^2/\mathrm{D_b} \tag{2.4}$$

Where l is the interpenetration depth and D_b the bioadhesive diffusion coefficient through mucus [50].

For significant interpenetration to occur, diffusion of the polymer chains of the dosage form into the mucin layer (and vice versa) must occur. Furthermore, the two components should have similar chemical structure to obtain the strongest mucoadhesive interaction [64].

Although the theories that have been described in this section may be helpful in describing the mucoadhesive behavior between polymer and mucus theoretically, the real situation is more complex to be explained or modeled using a single theory. Thus, a combination of two or more theories is always employed to characterize the complex phenomenon [4, 60].

2.4 Mucoadhesive Polymers

The possibility of mucoadhesion and the interaction strength can be influenced by the polymer structural and functional groups [5]. At present, the most commonly used mucoadhesive polymers are composed of polar chemical functional groups such as hydroxyl (—OH), carboxyl (—COOH), amide (—NH2), and sulfate (—SO4H) groups that are able to interact with the mucin glycoproteins [5, 60, 70]. The interactions between polymers and mucin include physical entanglements and secondary interactions notably hydrogen bonds [71]. The contributions from such forces facilitate the formation of a strengthened cross-linked network and hence achieve mucoadhesion [27].

Table 2.1 Commonly used adhesive polymers. (Modified from [118])

Type	Common polymers
Anionic polymers	Carbopol®
	Polycarbophil®
	Sodium alginate
	Sodium carboxymethylcellulose
Cationic polymers	Chitosan
Nonionic polymers	Hydroxypropylmethylcellulose
	Hydroxypropylcellulose
	Methylcellulose
	Polyethylene glycol
	Polyvinylpyrrolidone
	Hydroxyethylcellulose
Stimuli-sensitive polymers	Poloxamer

Mucoadhesive polymers can be classified according to the chemical characterization of the polar functional groups as shown in Table 2.1. The applications of these polymers as mucoadhesive drug delivery platforms is described in other chapters in this book and will not be addressed in this chapter.

2.5 Techniques Utilized for the Assessment of Mucoadhesive Strength

Despite the accumulation of numerous studies concerning the in vitro and in vivo performance of mucoadhesive drug delivery systems, surprisingly, there has not been a standard technique designed for mucoadhesive measurement or any analytical method that can be employed to qualify mucoadhesive strength. The lack of a uniform method to assess retention at the site of application may compromise the selection of formulations for clinical examination. At present, researchers have developed several approaches to rank the mucoadhesive properties of polymers and formulations and to understand their adhesive behavior. The developed techniques can be categorized as in vitro or in vivo methods.

2.5.1 In Vitro Techniques Used for Mucoadhesion Characterization

In vitro tests are the most common and convenient methods to assess the mucoadhesive properties of candidate formulations [72]. These techniques typically assess mucoadhesion using tensile force measurements that assess forces of attachment and detachment, and/or flowing techniques that evaluate the influence of shear stress, and measurement of the residence time of a mucoadhesive formulation on mucosal membrane. These methods have evolved from simple analytical techniques to more sophisticated and comprehensive procedures.

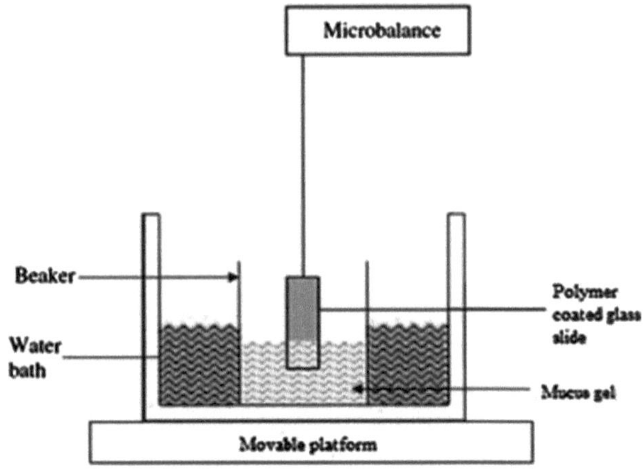

Fig. 2.3 The apparatus for the modified Wilhelmy plate technique. (Modified from [5])

Tensile Force Measurements There have been several reports of methods that have been used to characterize mucoadhesion based on the measurement of tensile force to break the interaction between the mucoadhesive drug delivery platform and the test substrate. Examples of these are detailed below.

Tensile measurement using microbalance (the modified Wilhelmy plate technique) This method was first employed to determine the detachment force by Smart and coworkers in 1984 [73]. The method is based on the Wilhelmy plate method and consists of a glass plate and a microforce balance (Fig. 2.3). In their work, the glass plate was coated by dipping into a 1 % solution of the test material, and subsequently drying the moisture at 60 °C in an oven to constant weight. The glass plate was hung on to a microforce balance and in contact with 1 ml homogenized mucus contained in a 5 ml glass vial which was placed on a vertical movable platform. The platform was raised until the plate had penetrated the mucus gel or model material to touch the base of the container. The platform was lowered subsequently following 7 min contacting of mucus gel and sample material while the maximum force for detachment was recorded by the microforce balance. To make a standard experiment, the detachment force between mucus gel and a clean glass plate was tested, and the coated plate force was then expressed as a percentage of the clean plate force. In their work, several materials have been tested to obtain the mucoadhesiveness. Those materials were ranked according to the strength of mucoadhesion and the rank order agreed with that published by Chen and Cyr [74]. Furthermore, the effects of contact time, molecular weight, and pH value on adhesion were studied. The mucoadhesive forces of several materials are listed in Table 2.2.

This modified Wilhelmy plate technique was probably the first method employed to screen the potential of polymers as mucoadhesive platforms for use as buccal films [75]. It is a simple and efficient method that provides information about possible mucoadhesive strength; however, it may be limited due to the dissolution of polymer candidates in mucus gel and the absence of biological tissue [50, 75].

Table 2.2 Rank order of mucoadhesive force. (Modified from [74])

Coating material	Mean adhesive force (%)	Standard deviation
75P SCMC	192.5	12.0
Carbopol 937	185.0	10.3
Tragacanth	154.4	7.5
Gantrez AN	147.7	9.7
Sodium alginate (H.V.)	126.2	12.0
Hypromellose (M.V.)	125.2	4.8
Gelatin	115.8	5.6
Pectin	100.0	2.4
P.V.P	97.6	3.9
Acacia	97.6	5.9
PEG 6000	96.0	7.6

H.V. high viscosity, *M.V.* medium viscosity

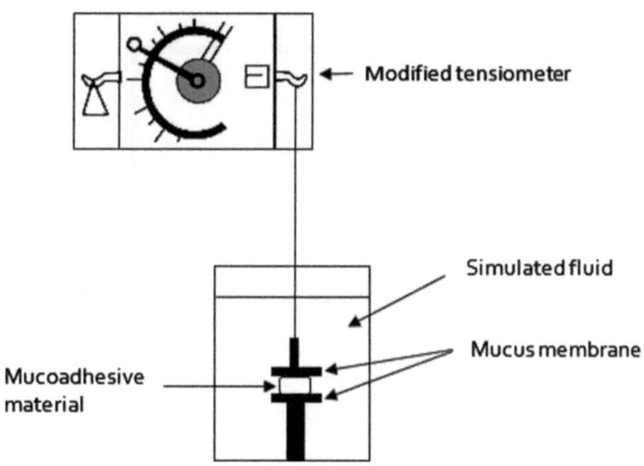

Fig. 2.4 A modified surface tensiometer for measurement of detachment force of mucoadhesion. (Modified from [75])

Techniques based on modified tensiometers and advanced dual tensiometers These techniques are modifications of the Wilhelmy plate technique [76]. As shown in Fig. 2.4, a mucoadhesive material is placed between two tissues, with the upper tissue suspended from a tensiometer spring to record forces, and the lower tissue fixed on a weighed glass vial in a beaker containing simulated fluid. The upper tissue is subsequently raised following a period of contact between the mucoadhesive material and tissues. A detachment force is recorded by the tensiometers to present the maximum loading that the mucoadhesive interaction could withstand prior to separation [77].

More recently, Abruzzo and coauthors [78] presented an alternative means to characterize chitosan/gelatin films. In their study, the in vitro and in vivo mucoadhesion characterizations were conducted using the modified surface tensiometer method in five healthy volunteers aged 25–40 years, respectively. They demonstrated that the in vivo residence time of the film in the buccal cavity was related to the in vitro mucoadhesive strength, notably detachment force.

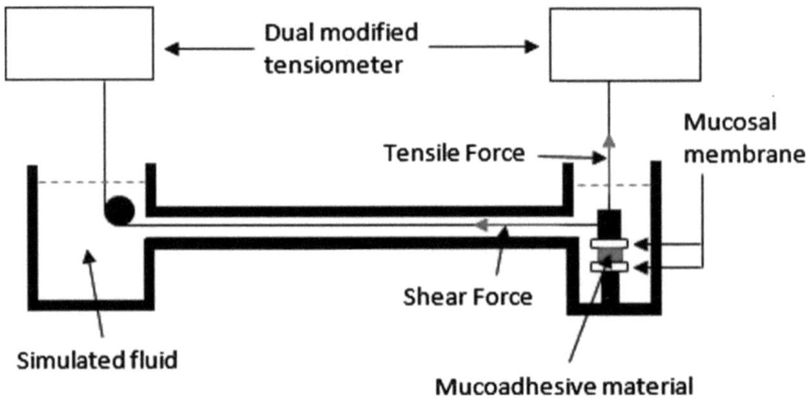

Fig. 2.5 Dual tensiometer apparatus. (Modified from [3])

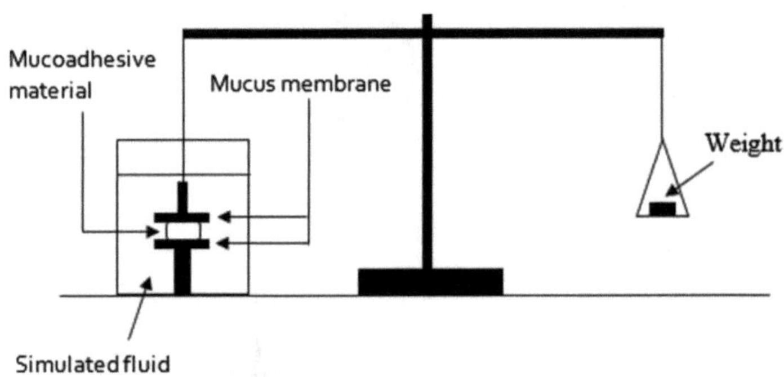

Fig. 2.6 Modified physical balance used to measure mucoadhesive strength. (Modified from [79])

In a further modification, the dual tensiometer method has been designed to characterize the effect of shear stress on mucoadhesion. The apparatus was constructed using two tensiometers that measured tensile stress and shear stress, respectively. The second tensiometer was connected to a standard single tensiometer apparatus to stretch the upper tissue from the left or right side (Fig. 2.5). A shear stress is recorded on the second tensiometer to indicate the strength of mucoadhesion in the horizontal direction.

Tensile force measurement using a balance A series of methods to assess the mucoadhesive properties of materials using a modified physical balance have been described by Gupta [79]. As shown in Fig. 2.6, this consists of a balance with two arms on which weights are suspended, the weight corresponding to the detachment force being determined.

More specifically, in this method the formulation is located between the two tissue layers in a glass beaker containing a defined amount of fluid. Weights are gradually added to one arm of the balance; the fracture of the mucus/sample interface being

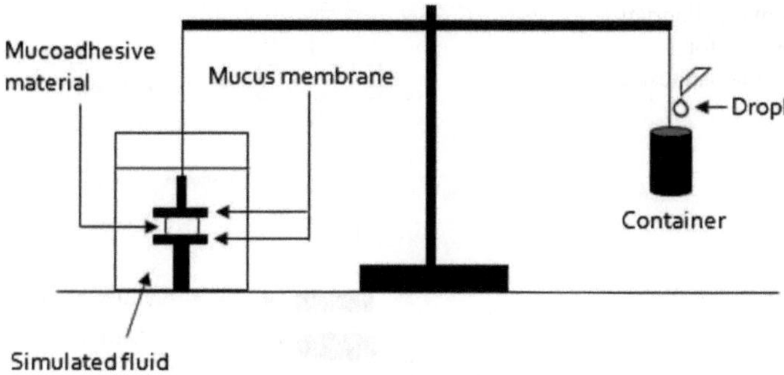

Fig. 2.7 Mucoadhesive force-measuring device using water droplet. (Modified from [81])

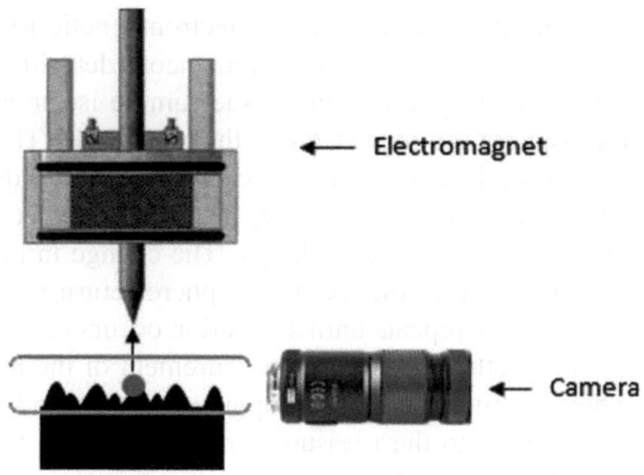

Fig. 2.8 The working elements of EMFT. (Modified from [83])

visually observed, and the detachment force obtained according to the magnitude of the loaded weight. This balance technique has been used in several published articles, and some modifications on the construction of apparatus have been performed [80]. In a related study, a modified two-arm balance method has been reported by Qi and coworkers. As seen in Fig. 2.7, instead of weight, the newer apparatus included a dropping bottle which dropped water into a glass vial with a constant flow rate. The weight of the water in the glass vial increased until the gel and the (corneal) tissue were detached [81]. This method may offer improved control of the applied force and hence offers an accurate measurement of the mucoadhesive bond strength.

A further modification of the physical balance method which does not involve the use of any simulated fluid has been reported to be suitable for the measurement of the detachment force between mucus and a semisolid material [82].

Mucoadhesive measurement using electromagnetic force transducer (EMFT) The electromagnetic force transducer (EMFT) technique was initially reported by Hertzog and Mathiowitz as a method for the assessment of mucoadhesion of microspheres (< 300 μm) [83]. Typically, the apparatus consists of an electromagnet element, tissue stage, and a camera for observation as shown in Fig. 2.8. Instead of applying

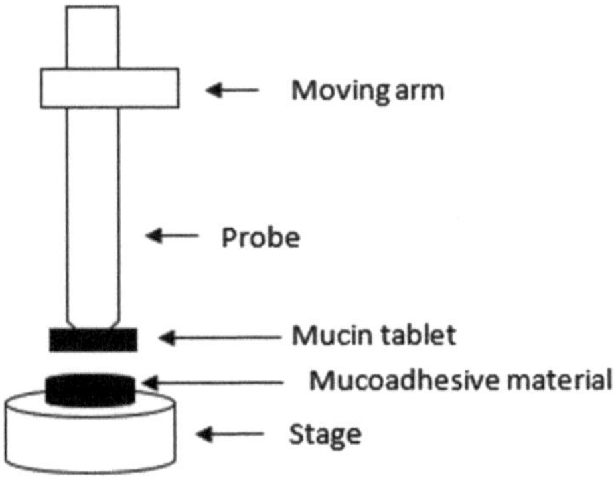

Fig. 2.9 Schematic diagram of texture analyzer for assessment of mucoadhesive tensile stress

a weight, the EMFT loads an electromagnetic force to a magnetic-loaded polymer microsphere and senses the simultaneous detachment of the microsphere from tissue sample. During a test, the tissue sample is mounted in a special chamber with a microsphere set directly under the magnet tip. The tissue chamber is slowly moved away from the magnet tip while the camera records the behavior of the microsphere. When the sphere moves away from the magnet tip, the control system increases the magnet current accordingly. The change in magnetic field strength generates a force that makes the magnetic sphere return to its original position. The process continuously repeats until separation occurs on sphere and tissue interface [10, 83].

This method allows the measurement of the mucoadhesive properties of microspheres, which have been implanted in vivo and then excised (along with the host tissue) prior to the measurement. In addition, this technique can also be used to evaluate the bioadhesion of polymers to specific cell types and hence may be used to develop mucoadhesive drug delivery system to target-specific tissues [67, 83].

Texture profile analyzer (TPA) In a common adaptation, the solid formulation is attached to the end of the probe of the texture profile analyzer (TPA), and the mucosal substrate fixed on a platform (Fig. 2.9). The moving arm is lowered until it reaches the mucosa and the polymer allowed interacting with mucosa by the application of a downward force for a predetermined period. The moving arm is then raised and the force to break the interaction between the formulation and mucosa determined from the force-time relationship.

TPA is a widely employed technique to assess mucoadhesive properties developed by Tobyn et al. [84]. Subsequently, the characterization of the adhesive properties of gels and semisolids using such tensile measurements was originally described by Jones et al. [85–86]. The author reported that the mechanical properties including hardness, adhesiveness, elasticity, and compressibility of formulations were significantly influenced by the concentration of polymers. Increasing concentration of polymers (polyvinylpyrrolidone (PVP), hydroxyethylcellulose (HEC), and polycarbophil) resulted in significant increase in adhesiveness [85, 86]. In a later research by Jones [87], the mechanical/textural, viscoelastic, and mucoadhesive properties

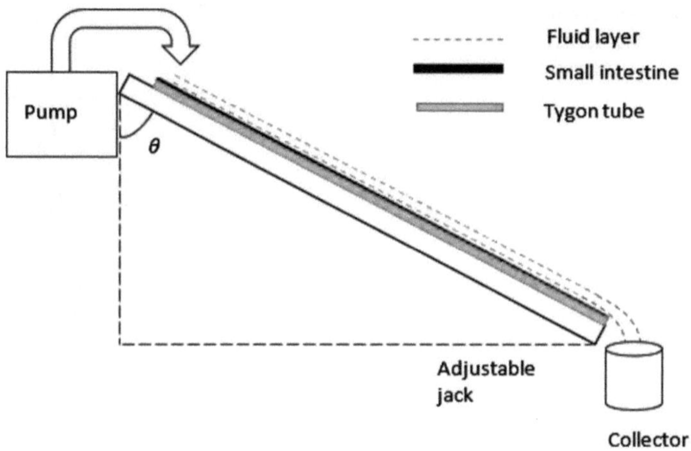

Fig. 2.10 Schematic diagram of the falling liquid film perfusion system. (Modified from [88])

of a range of aqueous gels composed of either HEC or sodium carboxymethylcellulose (NaCMC) were examined. The mucoadhesive properties of aqueous gels were dependent upon both concentration of polymers and contact duration with mucin [87].

2.5.2 *Methods Based on Flow Properties*

These methods have been developed to principally examine the behavior of a mucoadhesive platform within a simulated environment in which dilution with the biological fluid may occur. Examples of these are described below.

The falling liquid film Teng and Ho reported the earliest flowing technique for the quantification of the interaction between material and mucosa [88]. As exhibited in Fig. 2.10 the falling liquid film system generally consists of four parts: (1) a single drive syringe pump, (2) a supporting platform for the intestinal segment, (3) an adjustable jack, and (4) a collector for the liquid effluent. In this method, the retention of particles following the steady-state flow of a dilute suspension from an infinite reservoir over the mucosal surface has been quantitatively studied [10]. Initially, the concentration of particles in the suspension is determined and subsequently related to the concentration of particles that leaves the intestinal segment, thereby enabling the steady-state fraction of particles retain to be determined [89].

This modified method was reported by Rao and Buri in 1989 [89]. As shown in Fig. 2.11, the apparatus consisted of: (1) reservoir containing the washing solution, (2) peristaltic pump, (3) plastic support, (4) tissue, (5) pin, and (6) receiver for collecting the washings. In a similar fashion to the previous description, a determined number of uncoated and/or polymer-coated glass spheres were uniformly placed on

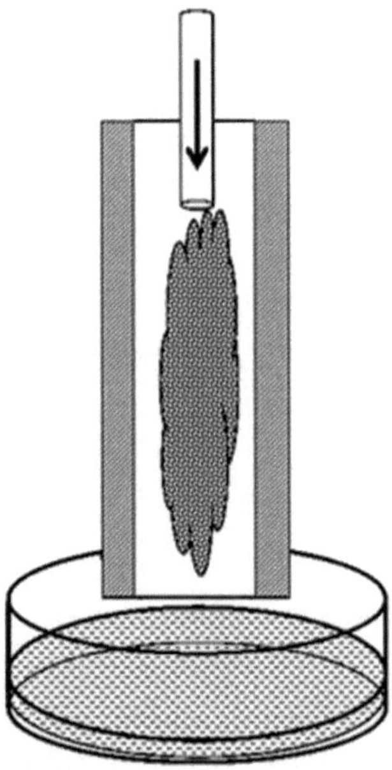

Fig. 2.11 Top view schematic diagram of the glass sphere flowing apparatus. (Modified from [89])

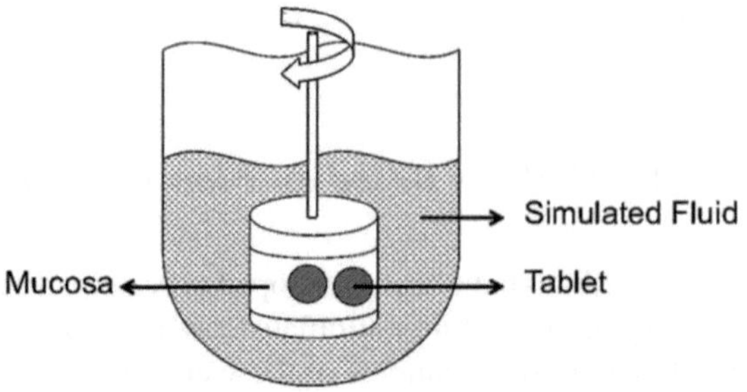

Fig. 2.12 Schematic presentation of the test system. (Modified from [7])

mucosa of intestine (4.5 × 1 cm), which had been fixed on a polyethylene support. This tissue was then placed in a desiccator and maintained at > 80 % relative humidity and 20 ± 1 °C for 20 min to allow the polymer to hydrate and to interact with the glycoprotein. After this conditioning, the polyethylene support was located into a plastic tube and washed with predetermined fluid for 5 min. A further drying was conducted in a hot air oven at 70 °C for all collected washings to enable the calculation of the percentage of beads that were washed away, and this quantifies mucoadhesion in terms of retention.

This technique was successfully applied by the authors to rank a series of polymers with respect to their mucoadhesive properties [76, 89].

Rotational cylinder method The evaluation of mucoadhesion using rotation cylinder was initially reported by Bernkop-Schnürch and Steininger [7]. It is a simple apparatus that consists of a vessel containing fluid and a stainless steel cylinder coated with freshly excised intestinal porcine mucosa (Fig. 2.12). A mucoadhesive tablet

Table 2.3 Residence time of polymers on the agar plate. (Data from [94])

Sample	Residence time (min)
XG	41
TG	14
HPC	13
PVA	4

or may be other dosage forms are thereby attached to the mucosal membrane. The cylinder is then rotated at a predetermined speed and the subsequent detachment, disintegration, and/or erosion of test formulation determined [7, 90].

This test specifically allowed the effect of shear stress on mucoadhesion to be determined under conditions that may be experienced in vivo.In Bernkop-Schnürch and Steininger's study, the total work of adhesion (TWA) of thiomers was evaluated using a tensiometer and compared with those determined using the rotation cylinder. The mucoadhesive properties of polycarbophil (a strongly mucoadhesive polymer) were compared to those of both unmodified NaCMC and NaCMC thiomers at pH of 5. The authors reported that the tablets based on the polycarbophil-cysteine conjugate pH 5 remained attached to the mucosa even after 10 h. In contrast, the corresponding control detached from the mucosa within half of this time [7]. In subsequent studies, the same apparatus was used to assess mucoadhesion of chitosan thiomers [24, 91–93].

Agar plate method This method was first reported by Nakamura and coworkers to investigate mucoadhesion of selected polymers (hydroxypropylcellulose (HPC), xanthan gum (XG), tamarind gum (TG), and polyvinyl alcohol (PVA) to nasal mucosa [94]. The apparatus includes an agar plate of 7 cm in diameter and the BP disintegration test apparatus. The agar was spray coated with 5 mg of the mucoadhesive polymer sample in Finntip® and was subsequently moved up and down vertically within pH 7.2 phosphate buffer at body temperature. The disintegration time was then measured and used as a measure of the residence time of the sample on the agar plate.

In Nakamura and coworkers' investigation, the mucoadhesion of several polymers was ranked using the agar plate method and compared to results of in vivo on rabbits. Although the observed residence time according to the in vitro technique did not exactly match with in vivo results, the rank of mucoadhesive strength for the polymers could be properly evaluated. Both in vitro (Table 2.3) and in vivo data exhibited a rank of residence time that was XG > TG > HPC > PVA. It can be seen that XG showed outstanding residence time in both disintegration (41 min) and in vivo (retention after 6 h) tests.

A more recent article using the agar plate technique was reported by Bachhav and Patravale [95]. In their work, a commercial product (Candid-V® gel) was evaluated, and the retention time was observed to be circa 24 min. This was then compared to the retention properties of a formulated microemulsion product composed of fluconazole, capryol 90, cremophor EL, benzyl alcohol, chlorocresol, and water. The authors reported a significant enhancement of retention (twofold) associated

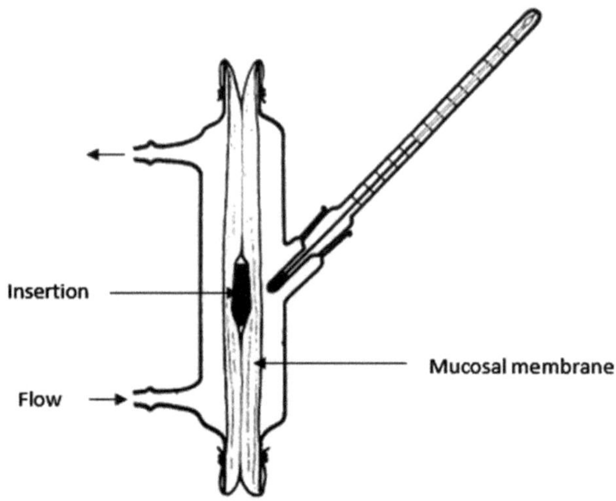

Fig. 2.13 Schematic diagram of Setnikar and Fantelli apparatus. (Modified from [96])

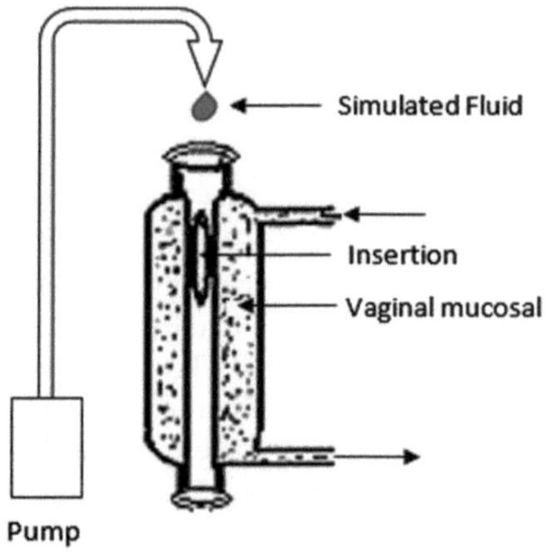

Fig. 2.14 Apparatus for ex vivo mucoadhesion experiment. (Modified from [97])

with the formulated product which might be indicative of prolonged in vivo residence time.

Modified Setnikar and Fantelli apparatus The original Setnikar and Fantelli apparatus was introduced in 1962 for measuring the liquefaction time of rectal suppositories as shown in Fig. 2.13 [96]. The apparatus includes a cellophane tube with cellulose dialysis tubing placed in the channel of the tube. The temperature is maintained by water circulation and monitored using a thermometer. In an experiment, a suppository was inserted into the dialysis bag and the liquefaction time measured.

In a later publication from Alam and coworkers, this technique was modified to measure retention of vaginal tablets (Fig. 2.14) [97]. In their experiment, excised and cleaned buffalo vagina was vertically suspended in the glass cell. The ends of the tube were averted on the upper and lower ends of the glass cell and crimped using rubber bands. Simulated vaginal fluid was allowed to fall drop-wise into the temperature-controlled vertically suspended vaginal tube and the retention was recorded as the time required for the tablet to leak from the lower end of the cell [97]. In their work, vaginal tablets showed considerable retention (> 24 h) in comparison to that of a commercial product Infa-V which started leaking after 5–8 h. It was also reported that a preferred tablet weight was 1.2 g to minimize the effects of the gravity-induced leakage from the vagina [97]. This modified technique was also utilized to measure retention of vaginal gels [98].

2.5.3 Other In Vitro Techniques used for the Assessment of Mucoadhesion

In addition to the tensile and the flow methods, several other techniques, namely the mucin-gold staining method, fluorescent, and rheological techniques have also been suggested for the indirect measurement of mucoadhesive interactions between mucosa and polymers.

Mucin gold staining Given that mucin plays an important role in stabilizing colloidal gold particles and that the solution concentration of the stabilized colloidal gold particles can be easily measured from the absorbance at a visible wavelength of 525 nm, Park reported the use of the mucin-gold staining method for the determination of mucoadhesion [6]. Mucin-gold conjugates were prepared by dissolving the predetermined weight of mucin in deionized distilled water and, following dialysis, 0.1 ml of this mucin solution was added to 1 ml of colloidal gold particles. Subsequently, 1 ml of 10 % NaCl was then added to the solution and rapidly mixed after 30 min equilibration. The minimum amount of mucin which prevented aggregation and a color change from red to light blue was identified and used to generate the mucin-gold conjugates. These were then separated and diluted as required and used in the staining experiment. In this, 2.5 ml of mucin-gold solution was gradually added to a polyacrylate cuvette containing a vertically orientated test material. At predetermined time intervals, the polymer sample was removed and gently rinsed in a buffer solution. The intensity of red color, known as the colloidal gold density, on the polymer sample was quantified by measuring absorbance at 525 nm and compared to a transparent control polymer sample and used as a measure of mucoadhesion. This method has also been used in a subsequent study to study the interaction between the polymer and the mucin [99].

Aside from mucin-gold staining, rheological techniques, notably flow and oscillation methods, have been used to investigate the mucoadhesive properties of formulations [27, 100]. Viscosity and viscoelasticity that are measured by flow and oscillation, respectively, will vary following interactions between mucin and polymer. As the most widely used technique, flow rheology provides information concerning the deformation of the material within a broad range of applied shearing rates. Subsequently, the rheological properties of mucoadhesive formulations in the presence and absence of mucin may be modeled using flow models, namely Power Law, Cross, and Herschel–Bulkley [101–102], which may aid the understanding of formulation structure and the possible evaluation of in vivo performance [103–104]. Of these models the Power Law equation is the simplest equation known to measure the flow properties of pharmaceutical system [105].

$$\sigma = K * \dot{\gamma}^n \tag{2.5}$$

Where σ is the applied shear stress, unit Pa; σ_0 is the yield stress, unit Pa; K is consistency, unit Pa.s^n; $\dot{\gamma}$ is the shear rate, unit s^{-1}; n is the flow index.

An advanced equation based on the Power Law equation is the Herschel–Bulkley flow model [106]:

$$\sigma - \sigma_0 = K * \dot{\gamma}^n \tag{2.6}$$

where σ is the applied shear stress, unit Pa; σ_0 is the yield stress, unit Pa; K is consistency, unit Pa.s^n; $\dot{\gamma}$ is the shear rate, unit s^{-1}; n is the flow index.

For both the Power Law and Herschel–Bulkley equations, n is used to identify the fluid type. For a Newtonian fluid and plastic fluid, $n = 1$ indicating the initial viscosity equals to the consistency. For pseudoplastic and dilatant fluids, $n < 1$ and $n > 1$, respectively.

The Cross equation [101] calculates the zero-rate viscosity and the infinity rate viscosity directly based on experimental results. The Cross equation is expressed as:

$$\eta = \eta_\infty + \frac{\eta_0 - \eta_\infty}{1 + (C \cdot \dot{\gamma})^m} \tag{2.7}$$

Where η is the viscosity at any shear rate; η_0 and η_0 are the initial rate viscosity and the infinity rate viscosity, respectively; C is the Cross Time Constant (or the Consistency); $\dot{\gamma}$ is the shear rate; m is the Cross Rate Constant [101, 107]. In all of these models, the effect of mucin on the flow properties of the mucoadhesive formulations are examined, with an observed and increasing synergy being recorded for mucoadhesive formulations.

In a similar fashion, oscillatory analysis has been used to quantify mucoadhesion in terms of the changes in the viscoelastic properties of the system in the absence and presence of mucin. From these measurements, the interaction parameter is normally calculated as follows (modified from [108]):

$$\Delta G' = G'_{\text{mucin-polymer}} - G'_{\text{polymer}} \tag{2.8}$$

In the above equation, the rheological parameter under investigation is the storage modulus but other parameters, e.g., the dynamic viscosity or the loss modulus could be used. As in the case of flow measurements, the synergy term ($\Delta G'$, in the above equation) increases as the interaction between the mucoadhesive formulation and mucin increases, thereby enabling an accurate ranking of the mucoadhesive properties of various materials.

Another method employed by Park and Robinson to evaluate mucoadhesion involves the use of fluorescent probes [2]. In this, the lipid bilayer of cultured human conjunctiva cells is labeled with the fluorescent probe pyrene. The adhesion of polymers to these cells caused a change in fluorescence due to surface compression when compared to control cells. This degree of change in fluorescence is proportional to the amount of polymer binding and hence mucoadhesion [5].

2.6 In Vivo Techniques used for Mucoadhesion Characterization

The main purpose of using mucoadhesive polymers is to prolong the retention of dosage forms. Although in vivo techniques do not provide information concerning the strength of mucoadhesion, detachment force, effect of shear force, and dilution directly, they are the most useful and reliable methods to specifically define the retention of the dosage form within the biological environment. Examples of techniques used to investigate retention in vivo include, gamma scintigraphy, magnetic resonance imaging, and visual method in association with dyes [10, 109–117].

2.7 Conclusions and Future Perspectives

Mucoadhesion has been widely used to affect the retention of dosage forms at the site of application. Although several theories have been suggested to explain the phenomenon of mucoadhesion, the mechanisms of interaction between polymers and mucosa are complex, and still insufficiently understood. There are several methods that may be used to characterize mucoadhesion, defined primarily by the nature of the dosage form; however, there remains the need for a universal test for mucoadhesion. These tests do, however, provide an insight into the process of mucoadhesion and have allowed research groups to identify formulations that may show promise for increased retention within the biological environment.

References

1. Kinloch A. The science of adhesion. J Mater Sci. 1982;17(3):617–51.
2. Park K, Robinson JR. Bioadhesive polymers as platforms for oral-controlled drug delivery: method to study bioadhesion. Int J Pharm. 1984;19(2):107–27.
3. Leung S-HS, Robinson JR. The contribution of anionic polymer structural features to mucoadhesion. J Control Release. 1988;5(3):223–31.
4. Smart JD. The basics and underlying mechanisms of mucoadhesion. Adv Drug Deliv Rev. 2005;57(11):1556–68.
5. Andrews GP, Laverty TP, Jones DS. Mucoadhesive polymeric platforms for controlled drug delivery. Eur J Pharm Biopharm. 2009;71(3):505–18.
6. Park K. A new approach to study mucoadhesion: colloidal gold staining. Int J Pharm. 1989;53(3):209–17.
7. Bernkop-Schnürch A, Steininger S. Synthesis and characterisation of mucoadhesive thiolated polymers. Int J Pharm. 2000;194(2):239–47.

8. Khutoryanskiy VV. Advances in mucoadhesion and mucoadhesive polymers. Macromol Biosci. 2011;11(6):748–64.
9. Scrivener CA, Schantz CW. Penicillin; new methods for its use in dentistry. J Am Dent Assoc. 1947;35(9):644–7.
10. Baloglu E, Senyigit ZA, Karavana SY, Bernkop-Schnürch A. Strategies to prolong the intravaginal residence time of drug delivery systems. J Pharm Pharm Sci. 2009;12(3):312–36.
11. Allen A. Structure and function of gastrointestinal mucus. In: Johnson L, editor. Physiology of the gastroenterology tract. New York: Raven; 1981. pp. 617–39.
12. Moran DT, Rowley JC 3rd, Jafek BW, Lovell MA. The fine structure of the olfactory mucosa in man. J Neurocytol. 1982;11(5):721–46.
13. Burkitt HG, Young B, Heath JW. Histologia Funcional. 3rd ed. Rio de Janeiro: Guanabara Koogan; 1994.
14. Neutra M, Forstner J. Gastrointestinal mucus: synthesis, secretion, and function. In: Johnson L, editor. Physiology of the gastrointestinal tract. New York: Raven; 1987. pp. 975–1009.
15. Bansil R, Turner BS. Mucin structure, aggregation, physiological functions and biomedical applications. Curr Opin Colloid Interface Sci. 2006;11(2):164–70.
16. Stefan R-I, Draghici I, Baiulescu G-E. Determination of urinary oxalate using oxalate-selective membrane electrodes. Sens Actuators B Chem. 2000;65(1):250–2.
17. Gipson IK, Ho SB, Spurr-Michaud SJ, Tisdale AS, Zhan Q, Torlakovic E, Pudney J, Anderson DJ, Toribara NW, Hill J. Mucin genes expressed by human female reproductive tract epithelia. Biol Reprod. 1997;56(4):999–1011.
18. Dasari S, Pereira L, Reddy AP, Michaels JE, Lu X, Jacob T, Thomas A, Rodland M, Roberts CT Jr, Gravett MG, Nagalla SR. Comprehensive proteomic analysis of human cervical-vaginal fluid. J Proteome Res. 2007;6(4):1258–68.
19. Chiappin S, Antonelli G, Gatti R, De Palo EF. Saliva specimen: a new laboratory tool for diagnostic and basic investigation. Clin Chim Acta. 2007;383(1–2):30–40.
20. Spurr-Michaud S, Argueso P, Gipson I. Assay of mucins in human tear fluid. Exp Eye Res. 2007;84(5):939–50.
21. Toribara NW, Roberton AM, Ho SB, Kuo WL, Gum E, Hicks JW, Gum JR Jr, Byrd JC, Siddiki B, Kim YS. Human gastric mucin. Identification of a unique species by expression cloning. J Biol Chem. 1993;268(8):5879–85.
22. Marriot C, Gregory N. Mucus physiology and pathology. In: Lenaerts V, Gurny R, editor. Bioadhesive drug delivery systems. Boca Raton: CRC; 1990. pp. 1–24.
23. Mortazavi S, Carpenter B, Smart J. A comparative study on the role played by mucus glycoproteins in the rheological behaviour of the mucoadhesive/mucosal interface. Int J Pharm. 1993;94(1):195–201.
24. Mortazavi SA, Smart JD. An investigation into the role of water movement and mucus gel dehydration in mucoadhesion. J Control Release. 1993;25(3):197–203.
25. Rossi S, Bonferoni M, Ferrari F, Bertoni M, Caramella C. Characterization of mucin interaction with three viscosity grades of sodium carboxymethylcellulose. Comparison between rheological and tensile testing. Eur J Pharm Sci. 1996;4(3):189–96.
26. Edsman K, Hägerström H. Pharmaceutical applications of mucoadhesion for the non-oral routes. J Pharm Pharmacol. 2005;57(1):3–22.
27. Capra RH, Baruzzi AM, Quinzani LM, Strumia MC. Rheological, dielectric and diffusion analysis of mucin/carbopol matrices used in amperometric biosensors. Sens Actuators B Chem. 2007;124(2):466–76.
28. Yang X, Robinson JR. Bioadhesion in mucosal drug delivery. In: Okano T, editor. Biorelated polymers and gels: controlled release and applications in biomedical engineering. San Diego: Academic;1998. pp. 135–192.
29. Riley RG, Smart JD, Tsibouklis J, Dettmar PW, Hampson F, Davis JA, Kelly G, Wilber WR. An investigation of mucus/polymer rheological synergism using synthesised and characterised poly(acrylic acid)s. Int J Pharm. 2001;217(1–2):87–100.
30. Fiebrig I, Harding SE, Rowe AJ, Hyman SC, Davis SS. Transmission electron microscopy studies on pig gastric mucin and its interactions with chitosan. Carbohydr Polym. 1995;28(3):239–44.

31. Van Klinken BJ, Dekker J, Büller HA, Einerhand AW. Mucin gene structure and expression: protection vs. adhesion. Am J Physiol. 1995;269(5 Pt 1):G613–27.
32. Bell S, Xu G, Khatri I, Wang R, Rahman S, Forstner J. N-linked. oligosaccharides play a role in disulphide-dependent dimerization of intestinal mucin Muc2. Biochem J. 2003;373:893–900.
33. Davies JM, Viney C. Water-mucin phases: conditions for mucus liquid crystallinity. Thermochim Acta. 1998;315(1):39–49.
34. Perez-Vilar J, Hill RL. Mucin family of glycoproteins. In: Lennarz WJ, Lane MD, editors. Encyclopedia of biological chemistry. Oxford: Academic/Elsevier; 2004. pp. 758–764.
35. Bettelheim FA, Hashimoto Y, Pigman W. Light-scattering studies of bovine submaxillary mucin. Biochim Biophys Acta. 1962;63:235–42.
36. Bettelheim F, Scheinthal B. Light scattering of mucins in concentrated solutions. J Polym Sci C Polym Symp. 1970;30:117–24
37. Harding SE. The macrostructure of mucus glycoproteins in solution. Adv Carbohydr Chem Biochem. 1989;47:345–81.
38. Bansil R, Stanley E, LaMont JT. Mucin biophysics. Annu Rev Physiol. 1995;57:635–57.
39. Bastardo L, Claesson P, Brown W. Interactions between mucin and alkyl sodium sulfates in solution. A light scattering study. Langmuir. 2002;18(10):3848–53.
40. Dua VK, Rao BN, Wu SS, Dube VE, Bush CA. Characterization of the oligosaccharide alditols from ovarian cyst mucin glycoproteins of blood group A using high pressure liquid chromatography (HPLC) and high field 1H NMR spectroscopy. J Biol Chem. 1986;261(4):1599–1608.
41. Naganagowda G, Gururaja T, Satyanarayana J, Levine M. NMR analysis of human salivary mucin (MUC7) derived O-linked model glycopeptides: comparison of structural features and carbohydrate-peptide interactions. J Peptide Res. 1999;54(4):290–310.
42. Thomsson KA, Prakobphol A, Leffler H, Reddy MS, Levine MJ, Fisher SJ, Hansson GC. The salivary mucin MG1 (MUC5B) carries a repertoire of unique oligosaccharides that is large and diverse. Glycobiology. 2002;12(1):1–14.
43. Kinarsky L, Suryanarayanan G, Prakash O, Paulsen H, Clausen H, Hanisch FG, Hollingsworth MA, Sherman S. Conformational studies on the MUC1 tandem repeat glycopeptides. implication for the enzymatic O-glycosylation of the mucin protein core. Glycobiology. 2003;13(12):929–39.
44. Paz HB, Tisdale AS, Danjo Y, Spurr-Michaud SJ, Argueso P, Gipson IK. The role of calcium in mucin packaging within goblet cells. Exp Eye Res. 2003;77(1):69–75.
45. McMaster TJ, Berry M, Corfield AP, Miles MJ. Atomic force microscopy of the submolecular architecture of hydrated ocular mucins. Biophys J. 1999;77(1):533–41.
46. Hong Z, Chasan B, Bansil R, Turner BS, Bhaskar KR, Afdhal NH. Atomic force microscopy reveals aggregation of gastric mucin at low pH. Biomacromolecules. 2005;6(6):3458–66.
47. Round AN, McMaster TJ, Miles MJ, Corfield AP, Berry M. The isolated MUC5AC gene product from human ocular mucin displays intramolecular conformational heterogeneity. Glycobiology. 2007;17(6):578–85.
48. Haugstad KE, Gerken TA, Stokke BT, Dam TK, Brewer CF, Sletmoen M. Enhanced self-association of mucins possessing the T and Tn carbohydrate cancer antigens at the single-molecule level. Biomacromolecules. 2012;13(5):1400–9.
49. Young T. An essay on the cohesion of fluids. Philos Trans R Soc Lond. 1805;95:65–87.
50. Mikos A, Peppas N. Systems for controlled release of drugs. V: bioadhesive systems. STP Pharma Sci. 1986;2(19):705–15.
51. Gandhi R, Robinson JR. Bioadhesion in drug delivery. Indian J Pharm Sci. 1988;50(3): 145–52.
52. Gu JM, Robinson JR, Leung SH. Binding of acrylic polymers to mucin/epithelial surfaces: structure-property relationships. Crit Rev Ther Drug Carrier Syst. 1988;5(1):21–67.
53. Jiménez-Castellanos MR, Zia H, Rhodes C. Mucoadhesive drug delivery systems. Drug Dev Ind Pharm. 1993;19(1–2):143–94.
54. Shaikh R, Raj Singh TR, Garland MJ, Woolfson AD, Donnelly RF. Mucoadhesive drug delivery systems. J Pharm Bioallied Sci. 2011;3(1):89–100.

55. Shafrin EG, Zisman WA. Constitutive relations in the wetting of low energy surfaces and the theory of the retraction method of preparing monolayers1. J Phys Chem. 1960;64(5):519–24.
56. Pritchard WH. The role of hydrogen bonding in adhesion. In: Alder D, editor. Aspects of adhesion. London: London University Press; 1970. pp. 11–23.
57. Krishnakumar P. Wetting and spreading phenomena, physics 563 Phase Transitions and the Renormalization Group. Urbana-Champaign: University of Illinois; 2010.
58. Peppas NA, Sahlin JJ. Hydrogels as mucoadhesive and bioadhesive materials: a review. Biomaterials. 1996;17(16):1553–61.
59. Packham DE. The mechanical theory of adhesion—a seventy year perspective and its current status. In: Van Ooij WJ, Anderson JHR, editors. First international congress on adhesion science and technology. The Netherlands: VSP BV; 1998. pp. 81–108.
60. Carvalho FC, Bruschi ML, Evangelista RC, Gremião MPD. Mucoadhesive drug delivery systems. Braz J Pharm Sci. 2010;46(1):1–17.
61. Lee JW, Park JH, Robinson JR. Bioadhesive-based dosage forms: the next generation. J Pharm Sci. 2000;89(7):850–66.
62. Derjaguin B, Aleinikova I, Toporov YP. On the role of electrostatic forces in the adhesion of polymer particles to solid surfaces. Powder Technol. 1969;2(3):154–8.
63. Derjaguin B, Toporov YP, Muller V, Aleinikova I. On the relationship between the electrostatic and the molecular component of the adhesion of elastic particles to a solid surface. J Colloid Interface Sci. 1977;58(3):528–33.
64. Chickering DE 3rd, Mathiowitz E. Definitions, mechanisms, and theories of bioadhesion. In: Mathiowitz E, Chickering DEIII, Lehr C-M, editors. Bioadhesive drug delivery systems: fundamentals, novel approaches, and development. New York: Dekker; 1999. pp. 1–10.
65. Dodou D, Breedveld P, Wieringa PA. Mucoadhesives in the gastrointestinal tract: revisiting the literature for novel applications. Eur J Pharm Biopharm. 2005;60(1):1–16.
66. Ahagon A, Gent A. Effect of interfacial bonding on the strength of adhesion. J Polym Sci Polym Phys Ed. 1975;13(7):1285–300.
67. Vasir JK, Tambwekar K, Garg S. Bioadhesive microspheres as a controlled drug delivery system. Int J Pharm. 2003;255(1–2):13–32.
68. Huang Y, Leobandung W, Foss A, Peppas NA. Molecular aspects of muco- and bioadhesion: tethered structures and site-specific surfaces. J Control Release. 2000;65(1–2):63–71.
69. Hägerström H, Edsman K, Strømme M. Low-frequency dielectric spectroscopy as a tool for studying the compatibility between pharmaceutical gels and mucous tissue. J Pharm Sci. 2003;92(9):1869–81.
70. Madsen F, Eberth K, Smart JD. A rheological assessment of the nature of interactions between mucoadhesive polymers and a homogenised mucus gel. Biomaterials. 1998;19(11–12):1083–92.
71. Hagesaether E, Sande SA. In vitro measurements of mucoadhesive properties of six types of pectin. Drug Dev Ind Pharm. 2007;33(4):417–25.
72. Accili D, Menghi G, Bonacucina G, Martino PD, Palmieri GF. Mucoadhesion dependence of pharmaceutical polymers on mucosa characteristics. Eur J Pharm Sci. 2004;22(4):225–34.
73. Smart JD, Kellaway IW, Worthington HE. An in-vitro investigation of mucosa-adhesive materials for use in controlled drug delivery. J Pharm Pharmacol. 1984;36(5):295–9.
74. Chen J, Cyr GN. Compositions producing adhesion through hydration. In: Manly RS, editor. Adhesion in biological systems. New York: Academic;1970. pp. 163–81.
75. Nair AB, Kumria R, Harsha S, Attimarad M, Al-Dhubiab BE, Alhaider IA. In vitro techniques to evaluate buccal films. J Control Release. 2013;166(1):10–21.
76. Ch'ng HS, Park H, Kelly P, Robinson JR. Bioadhesive polymers as platforms for oral controlled drug delivery II: synthesis and evaluation of some swelling, water-insoluble bioadhesive polymers. J Pharm Sci. 1985;74(4):399–405.
77. Gandhi RB, Robinson JR. Oral cavity as a site for bioadhesive drug delivery. Adv Drug Deliv Rev. 1994;13(1):43–74.
78. Abruzzo A, Bigucci F, Cerchiara T, Cruciani F, Vitali B, Luppi B. Mucoadhesive chitosan/gelatin films for buccal delivery of propranolol hydrochloride. Carbohydr Polym. 2012;87(1):581–8.

79. Gupta A, Garg S, Khar RK. Measurement of bioadhesion strength of mucoadhesive buccal tablet design of an in vitro assembly. Indian Drugs. 1992;30:152–5.
80. Pendekal MS, Tegginamat PK. Formulation and evaluation of a bioadhesive patch for buccal delivery of tizanidine. Acta Pharm Sin B. 2012;2(3):318–24.
81. Qi H, Chen W, Huang C, Li L, Chen C, Li W, Wu C. Development of a poloxamer analogs/carbopol-based in situ gelling and mucoadhesive ophthalmic delivery system for puerarin. Int J Pharm. 2007;337(1–2):178–87.
82. Choi H-G, Jung J-H, Ryu J-M, Yoon S-J, Oh Y-K, Kim C-K. Development of in situ-gelling and mucoadhesive acetaminophen liquid suppository. Int J Pharm. 1998;165(1):33–44.
83. Hertzog BA, Mathiowitz E. Novel magnetic technique to measure bioadhesion. In: Mathiowitz E, Chickering DE III, Lehr C-M, editor. Bioadhesive drug delivery systems: fundamentals, novel approaches, and development. Boca Raton: CRC Press; 1999. pp. 147–174.
84. Tobyn MJ, Johnson JR, Dettmar PW. Factors affecting in vitro gastric mucoadhesion. I: test conditions and instrumental parameters. Eur J Pharm Biopharm. 1995;41(4):235–41.
85. Jones DS, Woolfson AD, Djokic J. Texture profile analysis of bioadhesive polymeric semisolids mechanical characterization and investigation of interactions between formulation components. J Appl Polym Sci. 1996;61(12):2229–34.
86. Jones DS, Woolfson AD, Djokic J, Coulter WA. Development and mechanical characterization of bioadhesive semi-solid, polymeric systems containing tetracycline for the treatment of periodontal diseases. Pharm Res. 1996;13(11):1734–8.
87. Jones DS, Woolfson AD, Brown AF. Textural, viscoelastic and mucoadhesive properties of pharmaceutical gels composed of cellulose polymers. Int J Pharm. 1997;151(2):223–33.
88. Teng C, Ho N. Mechanistic studies in the simultaneous flow and adsorption of polymer-coated latex particles on intestinal mucus I: methods and physical model development. J Control Release. 1987;6(1):133–49.
89. Rao K, Buri P. A novel in situ method to test polymers and coated microparticles for bioadhesion. Int J Pharm. 1989;52(3):265–70.
90. Grabovac V, Guggi D, Bernkop-Schnürch A. Comparison of the mucoadhesive properties of various polymers. Adv Drug Deliv Rev. 2005;57(11):1713–23.
91. Duchê ne D, Ponchel G. Principle and investigation of the bioadhesion mechanism of solid dosage forms. Biomaterials. 1992;13(10):709–14.
92. Kast CE, Bernkop-Schnürch A. Thiolated polymers-thiomers: development and in vitro evaluation of chitosan-thioglycolic acid conjugates. Biomaterials. 2001;22(17):2345–52.
93. Kast CE, Valenta C, Leopold M, Bernkop-Schnürch A. Design and in vitro evaluation of a novel bioadhesive vaginal drug delivery system for clotrimazole. J Control Release. 2002;81(3):347–54.
94. Nakamura F, Ohta R, Machida Y, Nagai T. In vitro and in vivo nasal mucoadhesion of some water-soluble polymers. Int J Pharm. 1996;134(1):173–81.
95. Bachhav YG, Patravale VB. Microemulsion based vaginal gel of fluconazole: formulation in vitro and in vivo evaluation. Int J Pharm. 2009;365(1–2):175–9.
96. Setnikar I, Fantelli S. Liquefaction time of rectal suppositories. J Pharm Sci. 1962;51:566–71.
97. Alam MA, Ahmad FJ, Khan ZI, Khar RK, Ali M. Development and evaluation of acid-buffering bioadhesive vaginal tablet for mixed vaginal infections. AAPS PharmSciTech. 2007;8(4):E109.
98. Ahmad FJ, Alam MA, Khan ZI, Khar RK, Ali M. Development and in vitro evaluation of an acid buffering bioadhesive vaginal gel for mixed vaginal infections. Acta Pharm. 2008;58(4):407–19.
99. Mahrag Tur K, Ch'ng H-S. Evaluation of possible mechanism(s) of bioadhesion. Int J Pharm. 1998;160(1):61–74.
100. Yu T, Malcolm K, Woolfson D, Jones DS, Andrews GP. Vaginal gel drug delivery systems: understanding rheological characteristics and performance. Expert Opin Drug Deliv. 2011;8(10):1309–22.
101. Cross MM. Rheology of non-Newtonian fluids: a new flow equation for pseudoplastic systems. J Colloid Sci. 1965;20(5):417–37.

102. Bird RB, Dai G, Yarusso BJ. The rheology and flow of viscoplastic materials. Rev Chem Eng. 1983;1(1):1–70.
103. Banerjee R, Bellare JR, Puniyani R. Effect of phospholipid mixtures and surfactant formulations on rheology of polymeric gels, simulating mucus, at shear rates experienced in the tracheobronchial tree. Biochem Eng J. 2001;7(3):195–200.
104. Liu H-H, Li L, Birkholzer J. Unsaturated properties for non-Darcian water flow in clay. J Hydrol. 2012;430:173–8.
105. Barnes HA, Hutton JF, Walters K. An Introduction to Rheology. Amsterdam:Elsevier; 1989.
106. Deem DE. Rheology of dispersed systems. In: Lieberman HA, Rieger MM, Banker GS, editors. Pharmaceutical dosage forms: disperse systems. New York:Marcel Dekker; 1988.
107. Giboreau A, Cuvelier G, Launay B. Rheological behaviour of three biopolymer/water systems, with emphasis on yield stress and viscoelastic properties. J Texture Stud. 1994;25(2):119–38.
108. Jones DS, Bruschi ML, de Freitas O, Gremião MP, Lara EH, Andrews GP. Rheological, mechanical and mucoadhesive properties of thermoresponsive, bioadhesive binary mixtures composed of poloxamer 407 and carbopol 974P designed as platforms for implantable drug delivery systems for use in the oral cavity. Int J Pharm. 2009;372(1–2):49–58.
109. Richardson JL, Whetstone J, Fisher AN, Watts P, Farraj NF, Hinchcliffe M, Benedetti L, Illum L. Gamma-scintigraphy as a novel method to study the distribution and retention of a bioadhesive vaginal delivery system in sheep. J Control Release. 1996;42(2):133–42.
110. Brown J, Hooper G, Kenyon CJ, Haines S, Burt J, Humphries JM, Newman SP, Davis SS, Sparrow RA, Wilding IR. Spreading and retention of vaginal formulations in post-menopausal women as assessed by gamma scintigraphy. Pharm Res. 1997;14(8):1073–8.
111. Witter FR, Barditch-Crovo P, Rocco L, Trapnell CB. Duration of vaginal retention and potential duration of antiviral activity for five nonoxynol-9 containing intravaginal contraceptives. Int J Gynaecol Obstet. 1999;65(2):165–70.
112. Vermani K, Garg S, Zaneveld LJ. Assemblies for in vitro measurement of bioadhesive strength and retention characteristics in simulated vaginal environment. Drug Dev Ind Pharm. 2002;28(9):1133–46.
113. Chatterton BE, Penglis S, Kovacs JC, Presnell B, Hunt B. Retention and distribution of two 99mTc-DTPA labelled vaginal dosage forms. Int J Pharm. 2004;271(1–2):137–43.
114. Albrecht K, Greindl M, Kremser C, Wolf C, Debbage P, Bernkop-Schnürch A. Comparative in vivo mucoadhesion studies of thiomer formulations using magnetic resonance imaging and fluorescence detection. J Control Release. 2006;115(1):78–84.
115. Braga PC, Dal Sasso M, Spallino A, Sturla C, Culici M. Vaginal gel adsorption and retention by human vaginal cells. Visual analysis by means of inorganic and organic markers. Int J Pharm. 2009;373(1–2):10–5.
116. Poelvoorde N, Verstraelen H, Verhelst R, Saerens B, De Backer E, dos Santos SGL, Vervaet C, Vaneechoutte M. De Boeck F, Van Bortel L, Temmerman M, Remon JP. In vivo evaluation of the vaginal distribution and retention of a multi-particulate pellet formulation. Eur J Pharm Biopharm. 2009;73(2):280–4.
117. Mehta S, Verstraelen H, Peremans K, Villeirs G, Vermeire S, De Vos F, Mehuys E, Remon JP, Vervaet C. Vaginal distribution and retention of a multiparticulate drug delivery system, assessed by gamma scintigraphy and magnetic resonance imaging. Int J Pharm. 2012;426 (1–2):44–53.
118. das Neves J, Bahia MF. Gels as vaginal drug delivery systems. Int J Pharm. 2006;318(1–2): 1–14.

Chapter 3
Mucus as a Barrier for Biopharmaceuticals and Drug Delivery Systems

Hongbo Zhang, Mohammed-Ali Shahbazi, Patrick V. Almeida and Hélder A. Santos

3.1 Introduction

The development of advanced drug delivery technologies to improve drug pharmacokinetics and facilitate localized delivery to target tissues or cells can enhance tremendously the efficacy of various therapies [1, 2]. In order to be efficient, the nanoparticulate systems should have a number of suitable characteristics for therapy, such as locally sustained and controlled drug release [3, 4], deep tissue penetration [5–8], and protection of cargo therapeutics at both the extracellular and intracellular levels [9–11].

Mucosal membranes are moist surfaces lining on the walls of various body cavities such as respiratory, gastrointestinal, and reproductive tracts as well as the nostrils, the eyes, and the mouth. Mucus plays an important role in protecting the cellular epithelia from chemical and mechanical damage. The mucosal membranes also provide lubrication and wettability of the cell epithelial surface and regulate its moisture content [12]. In order to improve the therapeutic efficacy of many drugs across the mucus layers, the drug formulations should have mucoadhesive properties. Mucoadhesion is defined as an attractive interaction at the interface between a pharmaceutical dosage form and a mucosal membrane [13].

In the 1980s, Nagai and coworkers were one of the pioneers showing the potential of mucoadhesion in drug delivery by demonstrating the applicability of viscous gel ointments and mucoadhesive tablets for drug administration in the oral cavity [14, 15]. The great advantages associated with the use of mucoadhesive materials in drug delivery applications include [13, 16]: increased dosage form's residence time, improved drug bioavailability, reduced administration frequency, simplified administration of a dosage form, and termination of a therapy as well as the possibility of targeting particular body sites and tissues.

H. A. Santos (✉) · H. Zhang · M.-A. Shahbazi · P. V. Almeida
Division of Pharmaceutical Technology, Faculty of Pharmacy,
University of Helsinki, Helsinki FI-00014, Finland
e-mail: helder.santos@helsinki.fi

J. das Neves, B. Sarmento (eds.), *Mucosal Delivery of Biopharmaceuticals,* DOI 10.1007/978-1-4614-9524-6_3,

The majority of the mucoadhesive drug delivery systems (MDDS) are formulated as tablets, solid inserts, wafers, films, gels, viscous solutions, particulate suspensions (micro and nano), in situ gelling systems, and sprays. In addition, these MDDS often incorporate polymeric excipients, which play a major role in their mucoadhesivity [17–21]. The mucoadhesive polymers not only increase the dosage form's residence time at the site of administration, but also enhance the drug permeability through the epithelium by altering the tight junctions between the cells [22–25].

In the last few decades, there has been an increase in interest in the research worldwide for the development of controlled release systems for drug and gene delivery to mucosal surfaces such as, for example, the lung airways, the gastrointestinal tract (GIT), eyes, and the female reproductive tract [26–52]. However, the mucus layer that lines all mucosal tissues acts as a protective barrier in the body, trapping and removing the foreign particles and hydrophobic molecules [53]. As a consequence, there is a very limited permeability of drug delivery particles and hydrophobic drugs that can cross the mucus barrier and, in most of the cases, they are rapidly cleared from the delivery site [54].

In order to reach their targets, drug nanodelivery systems have to cross at least the outermost layers of the mucus barrier rapidly to avoid a fast mucus clearance [12, 55–57]. It is well recognized that to penetrate the mucus, the nanoparticles must avoid adhesion to mucin fibers and be small enough to avoid significant steric inhibition by the dense fiber mesh [54]. Lai and coworkers have shown that nanoparticles of 500 nm coated with a muco-inert polymer cross the physiological human mucus with diffusivities of only four-fold less compared to their rates in pure water [57].

The proper design of mucus-penetrating particles (MPPs) can therefore enhance the sustained drug release at the mucosal surfaces, enhancing the efficacy of the therapeutics and reducing possible side effects of the drugs. However, and despite several decades of research, mucoadhesion is still not fully understood. This is due to the large complexity of the interactions between various polymer-based mucoadhesive dosage forms and the biopolymer net of the mucus gel present on the surface of the mucosal membranes [13, 16].

In this chapter, we start by briefly introducing some of the important properties of mucus and mucosal membranes that need to be overcome in drug delivery applications. We then address some of the roles of mucus in blocking nanoparticulate drug delivery systems. We further highlight the mucoadhesive properties of particulates, the design and development of MDDS to avoid rapid mucus clearance and to provide targeted or sustained drug delivery for localized therapies in mucosal tissues (e.g., buccal, nasal, ocular, gastro, vaginal, and rectal). Next, we also present an example of MPPs used to target a disease state mucosa. Finally, we conclude the chapter with a brief overview of our visions of the future of MDDS and their potential to overcome the mucus limitations in drug delivery.

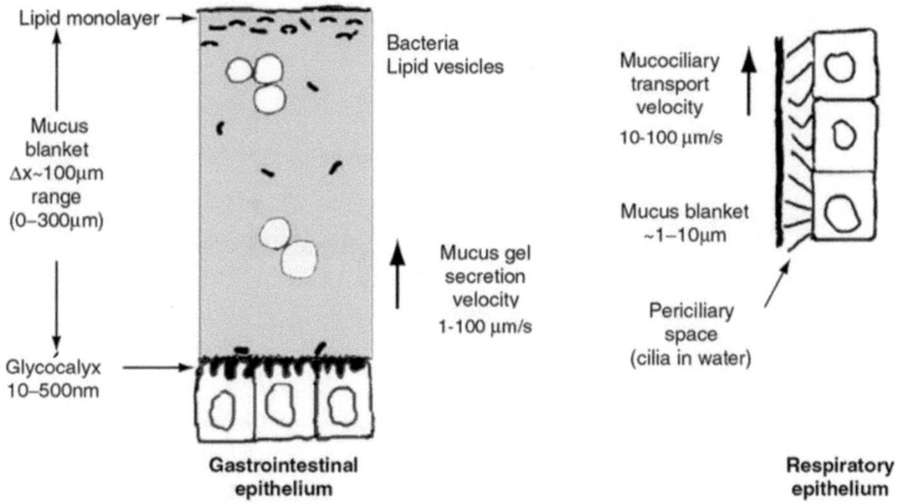

Fig. 3.1 Schematic representation of the characteristics of the mucus of the gastrointestinal (*left*) and respiratory (*right*) epithelium. (Reprinted with permission from [17])

3.2 Mucus as a Critical Barrier for Penetration: Structure, Functions and Role of the Mucosal Membranes

Mucus is a viscoelastic, adhesive gel that lines on the epithelial surfaces in humans (Fig. 3.1). Mucus serves to protect the epithelium by lubricating, trapping, and removing possible infectious agents. Due to the special biological role, mucus significantly limits the drug and gene delivery across biological barriers [54]. The development of MPPs can dramatically improve the effectiveness of diagnostic and therapy of numerous diseases [26].

As a tenacious, viscoelastic gel layer mucus can stick to most particles preventing their penetration to the epithelial surface [54]. Mucus is typically 10^3–10^4 times more viscous than water. The epithelial cells of the mucosal tissues are coated by two types of mucins: the membrane-bound and the secreted (soluble) biomacromolecules forming the fully-hydrated gel layer. Mucus is composed of cross-linked and entangled mucin fibers. These mucins are typically high-molecular weight glycoproteins (0.5–40 MDa), which are composed of 0.3–0.5 MDa size mucin monomers and coated with highly diverse array of proteoglycans [12], as well as by subunits linked together by peptide linkages and intramolecular cysteine–cysteine disulfide bridges [12, 13].

Mucins are mainly divided into two families: secreted mucins and cell-associated mucins [58]. The secreted mucin monomers with 0.2–0.6 μm in length are linked together end to end by disulfide bonds and form the several-micrometer long, secreted mucins. About 90 % of SH groups in secreted mucins form disulfide bonds. In mucin granules, the long mucin fibers are tightly packed together and form a network [12].

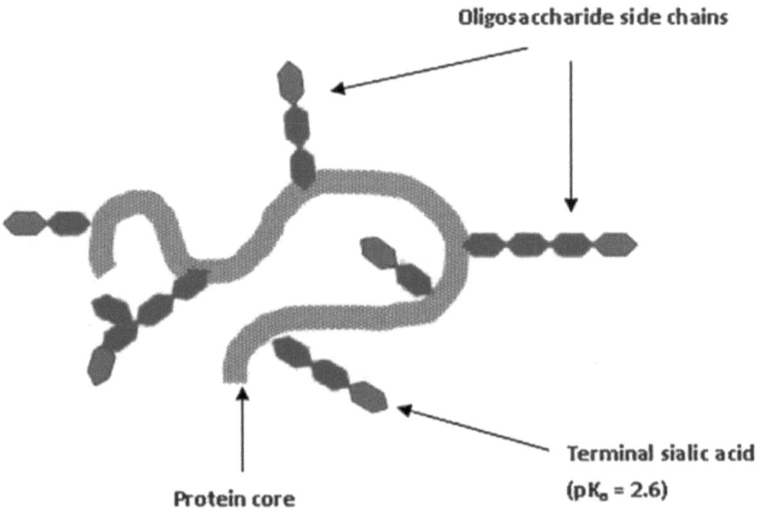

Fig. 3.2 Schematic representation of the structure of mucin subunits. (Reprinted with permission from [33])

The mucus gels are formed by 2–5 wt% of mucin content and 90–98 wt% of water content [59–64]. Cell-associated mucins are 100–500 nm in length, anchored to the cell surface by a transmembrane domain, and form the glycocalyx to protect the cell surfaces. Cell-associated mucins have a unique SEA (sea-urchin sperm protein, enterokinase, and agrin) domain that undergoes auto-proteolysis. There are two hydrophobic patches on the external surface of the SEA domain, which are potential interaction sites for proteins and other hydrophobic molecules [65]. These properties are important for the development of MPPs. In addition to mucins, there are also DNA, lipids, salts, proteins, cells, and cellular debris in the mucus gels [58].

Mucins are usually negatively charged due to the presence of carboxylate groups (sialic acid) and ester sulfates at the terminus of some sugar units (Fig. 3.2). The pK_a of these acidic groups is ~ 1.0–2.6. The pH of the mucus can vary dramatically depending on the local environment [54]. Typical pH values for the lung and nasal mucus are neutral or slightly acidic (pH = 5.5–6.5), for the eye mucus is ~ 7.8 and the mouth 6.2–7.4, whereas the typical pH values for the GIT mucus range from 1 to 8 and the vaginal mucus from 3.5 to 4.5.

The thickness of the mucus is another important property to consider for the development of MPPs. The mucus thickness depends on the rate of secretion and rate of degradation and shedding. Typically, foreign compounds have stimulation effect on mucus secretion. Moreover, the thickness of the mucus layers varies significantly for different mucosal surfaces, ranging from 5 μm in ocular up to 170 μm in gastric mucosa [54].

Mucus is secreted in a continuous manner, with nearly 10 L of mucus secreted into the GIT per day. When foreign particles enter the body, mucus generally converts

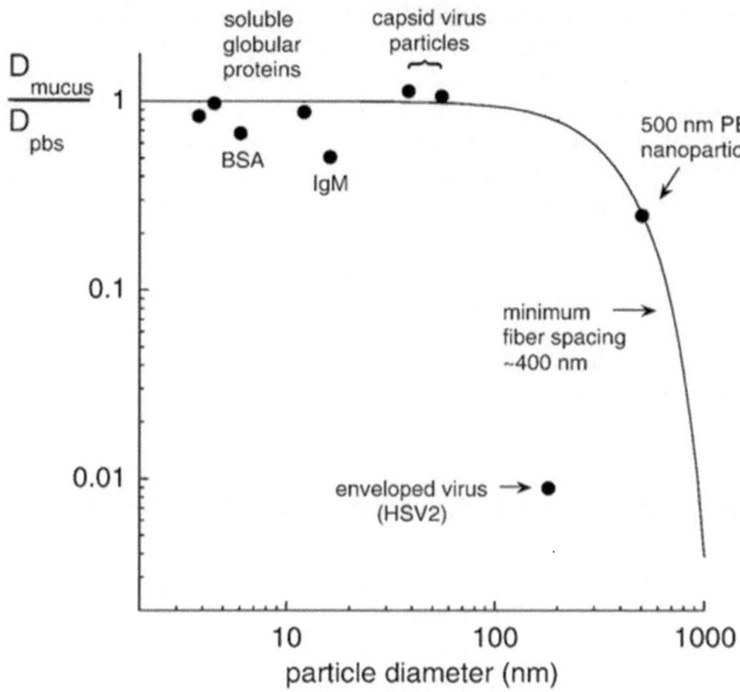

Fig. 3.3 Ratio of the diffusion constants in mucus and water (D_{mucus}/D_{pbs}) as a function of the hydrodynamic diameter of several diffusing particles. (Reprinted with permission from [12])

them into mucus-covered "slugs" independently of the size, density, or composition. This is a general property of all kinds of mucus, including the extremely thin mucus of the eyes [54]. However, mucus often has short lifetime ranging from few minutes to hours. Thinner mucus layer has faster turnover. For example, in the nasal tract, the mucus layer is renewed every 20 min, which leads to very fast clearance of inhaled particulates [66], whereas in the GIT, the mucus layer renovation time takes several hours [67].

It has been recognized that the diffusion through mucus gels is possible for small molecules and that larger molecules, such as globular proteins, are too large to penetrate the intestinal mucus [12]. Interestingly, recent reports demonstrate that nanoparticles as large as 500 nm can also diffuse through mucus gels [55, 57, 66, 68]. Figure 3.3 shows the speed at which particles of various sizes diffuse through the human cervico-vaginal mucus (CVM) compared with the speed at which they diffuse in water [12]. In this example, the ratio of the diffusion constant in mucus, D_{mucus}, is divided by the diffusion constant in water, D_{water}, and plotted as a function of particle size (Stokes diameter) for several globular proteins including bovine serum albumin, human immunoglobulin M (IgM), virus-like particles (Norwalk and human papilloma virus), and 500 nm PEGylated nanoparticles [39, 66, 69].

It is observed that almost every soluble globular protein can diffuse freely through mucus ($D_{mucus}/D_{pbs} \sim 1$), whereas secreted antibodies such as IgM are slightly slowed by mucus ($D_{mucus}/D_{pbs} \sim 0.4$). The diffusion of antibodies through the mucus has been attributed to the weak, low affinity bonds formed with mucin fibers allowing their accumulation on the surfaces of the pathogens to which they bind tightly and specifically [70]. In addition, small molecules that partition into oil, or other nonpolar solvents, diffuses more slowly through mucus than through water, and their diffusion constants in mucus decrease in proportion to their nonpolar/polar partition coefficient [71, 72]. Contrarily, small cationic molecules can bind tightly, and polyvalently, to the negatively charged glycan domains, such as large positively charged nanoparticles coated with chitosan, and bind tightly to mucus gels [73]. In very high concentrations, these molecules can lead to the collapse of the mucus gel, forming large channels that may provide access to epithelial surfaces for other molecules [74].

3.3 Biochemical Factors Governing the Properties of Mucus

As discussed above, mucus is a complex biological fluid composed by a wide variety of macromolecules and biological elements, namely mucins, proteins, lipids, salts, DNA, cells and cellular fragments, and water [74, 75]. The individual regulation of all these biochemical factors in vivo has a strict impact on the fluid dynamics and viscoelastic properties of mucus, influencing its functions such as clearance, physical barrier properties, and adhesiveness to mucosal epithelial cells.

The rheological properties of mucus are mainly influenced by the mucin content, composition, and glycosylation degree, as well as by the mucus hydration. However, the mucus viscoelasticity appears to be influenced by variations in mucin concentration. For example, CVM experiences a 100-fold reduction in its viscosity during ovulation, owing to a decrease of 2–4 times of mucin concentration, strictly related to a greater hydration [76]. Regarding the airway mucus, the viscoelasticity of the tracheal mucus is also different from that found in small airways, possibly due to different transepithelial water-diffusion patterns, thus resulting in distinct mucin concentrations [77].

Some proteins, particularly antibodies, are believed to interact with mucin glycoproteins via low-affinity bonds established between them and antibodies' ferrocene (Fc) moieties. Olmsted and coworkers studied the transportation rates of IgM and small aggregates of immunoglobulin A (IgA) through mucus and have found that they were considerably slower when compared to their diffusion in water [66]. The same effect was not observed for larger virus-like particles, neither for proteins without Fc domain. These remarkable findings suggest that individual protein expression and biochemistry are crucial factors affecting the physical properties of mucus.

The lipid fraction in mucus composition is typically around 1–2 wt%. However, in some diseases, such as cystic fibrosis (CF), the lipid content of mucus secretions is substantially higher, resulting in an increase in the mucus viscoelasticity [78]. Therefore, rheological properties of mucus seem to be also highly influenced by its lipid content.

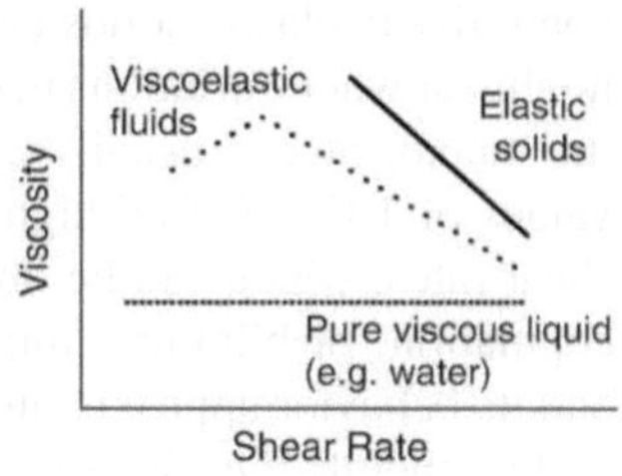

Fig. 3.4 Illustration of the steady state viscosity of elastic solids, pure viscous liquids, and viscoelastic fluids, as a function of the shear rate. (Reprinted with permission from [58])

The mucus viscoelasticity is also influenced by variations in ionic strength. In fact, mucus secretions are subject to shrinking and swelling phenomena, depending on the ions' concentration [58]. In general, an increase in salt concentration is reflected with an increase in mucus viscosity [79]. Viscoelasticity of mucus was also found to be increased by harsh acidic conditions, as for example those found in the gastric environment. This effect can be explained by the protonation of carboxyl groups of sialic acid residues present in the glycosylated parts of mucus [80].

Mucus is also constituted by nucleic acids, particularly DNA, which usually represents 0.02 wt% of human mucus [58]. Lethem and coworkers demonstrated that most of the DNA constituting mucus secretions has its origin in fragments of epithelial cells, and its accumulation generally leads to an increase on mucus viscosity [81]. In this study, an increase of 30 % in the viscoelasticity of the mucus collected from CF patients was observed.

3.3.1 Macro- and Microrheology of Mucus

The rheological features of complex biological fluids, including mucus, can be mainly described as a function of two physical properties: viscosity or loss modulus (G″), which is a measure of the resistance of fluids to flow; and elasticity or storage modulus (G′), established as the resistance to deformation and capacity of restoring the original state, when an external force is applied to fluids. These two parameters of fluids depend not only on their composition, but also on the frequency, amplitude and rate of the applied shear stress. Another parameter—phase angle or loss tangent value, δ—relates to the aforementioned physical properties by representing the inverse tangent of G″/G′, and can similarly be used for characterizing fluids ($\delta = 0°$ for an elastic solid; $\delta = 90°$ for a viscous liquid; $\delta < 45°$ for a viscoelastic solid; and $\delta > 45°$ for a viscoelastic liquid) [58].

3.3.1.1 Macrorheology of Mucus

Regarding the macrorheology of the mucus (i.e., the physical behavior as a bulk fluid), it is considered to be a viscoelastic gel, presenting viscosity properties between those of a pure viscous liquid and an elastic solid. In fact, the physical behavior of mucus is complex, and so it is described as a non-Newtonian fluid. As shown in Fig. 3.4, contrarily to elastic solids and viscous liquids, the viscosity of viscoelastic fluids is nonlinear when variations in shear rate are taken into account [58]. At low shear rates, the human mucus usually exhibits a strong resistance to flow, reaching viscosity values of 10^4–10^6-fold higher than that of water [82]. On the other hand, at high shear rates, which can be observed in physiological phenomena such as coughing, copulation, or blinking, mucus' resistance to deformation decreases significantly, and its behavior approximates the one of viscous liquids.

In general, the viscosity of human mucus varies within 10^3Pa·s–10^2Pa·s, depending mainly on the considered anatomical site and physiopathological condition. Concerning the respiratory system, the rheological profile of mucus differs between upper and lower airways, as well as between small and large ducts inside lungs. Thus, the viscoelasticity appears to significantly increase from nasal cavities toward lungs, owing to the lower solid content of tracheobronchial mucus [83, 84].

Furthermore, an increase in the viscoelasticity of mucus can be usually found in patients with pulmonary disorders, such as CF, chronic obstructive pulmonary disorder, and asthma, resulting from an increased fraction of mucin glycoproteins and a lower hydration [85, 86]. In contrast, patients suffering from rhinitis or bronchitis typically present a considerable decrease in nasal mucus' viscoelasticity [84, 87]. In the case of the GIT, mucus secretions obtained from different segments (i.e., stomach, intestine, and colon) revealed to possess similar rheological properties [88]. Although exhibiting viscosity values within the aforementioned range of human mucus, CVM normally becomes less viscous during ovulation periods, owing in part to the increased water content, playing a key role in fertilization, since this change in rheological properties of mucus turns it more penetrable to sperm [89]. Apart from ovulatory mucus, some mucus secretions (e.g., saliva and tears) exhibit distinctly lower viscosity, ranging from 10^2 to 10^3 times higher to approximately the same viscosity of water, at low and high shear rates, respectively [82].

Notwithstanding the importance of bulk rheological properties of mucus for comprehending some of its essential functions, such as mucus clearance and lubrication, the study of microrheological properties of mucus is distinctly needed for understanding its barrier properties to pathogens, toxins, and taken up by particles, including MPPs.

3.3.1.2 Microrheology of Mucus

Similar to other complex biological fluids, mucus consists of a biopolymer network, which closely interacts with its surrounding and interpenetrating fluid at nanoscopic and microscopic levels.

The microrheology of the mucus refers to the detailed characterization of its viscoelastic properties at high-resolution scale, influenced by both the biopolymer network itself and the interactions between this and its local environment. Regarding this, microrheology is strictly correlated with the fluid dynamics of micro- and nanoscale entities interpenetrating the pores in the heterogeneous mucus mesh, therefore being influenced by its length scale. As the length scale increases toward pore size of the mucus network, the capacity of these entities to diffuse gets reduced as a consequence of steric interactions, apparently resulting in a higher viscosity [74]. For example, considering macromolecules, such as proteins (< 10 nm), or even small capsid viruses (< 55 nm), it is observed that the resistance to diffusion through mucus (i.e., microviscosity) does not diverge from that of water [66]. Lai and coworkers corroborated this theory, although increasing the low-viscosity length scale from $\sim$55 to $\sim$500 nm [57]. In this case, a reduction on the effective diffusivity of 200 nm and 500 nm polymeric nanoparticles, through human mucus of only six-fold and four-fold, respectively, was observed when compared to water. Lai and coworkers also showed that the microviscosity of the mucus is expected to increase when oversized particles are taken into account [56]. For example, muco-inert- coated particles around 1 μm in size experienced a significant reduction on the diffusion rates when compared to water. Therefore, an understanding of the viscosity and elasticity of the biological barrier at the applicable length scale is highly required when developing MPP drug delivery systems.

3.4 In Vitro and Ex Vivo Models for Drug Transport Studies Across the Mucosa

3.4.1 In Vitro Models

Since the transport of therapeutic molecules across the mucus and intestinal epithelium is the most important factor determining in vivo bioavailability of orally administered drugs, different in vitro models have been recently developed to assess drug absorption via the GIT route [90].

One of the main reasons for the application of in vitro models is the less laborious and more cost-effective benefits in comparison to the in vivo models [91]; however, their success in the prediction of drug absorption depends on their efficiency for mimicking intestine properties and functionalities [92]. Although there are different in vitro models to screen drug permeability through the intestinal cell models [90], there is no perfect model to mimic all characteristics of the epithelial cells. For example, despite a wide range of studies with Caco-2 cell line as an intestinal absorption model, these cells suffer from the lack of mucus layer, high transepithelial resistance, and are a poor representation of the paracellular route and easy access to microvilli. In addition, other cell culture models such as T84, 2/4/A1, and IEC-18 cells represent high transepithelial electrical resistance, lack of transporters, and low cloning efficiency, respectively [90].

Therefore, according to the aim of study and also characteristics of the in vitro models, various types of cell cultures can be used in the initial steps of oral drug delivery research; however, researchers are trying to simultaneously find new cell models for specific studies and also to develop new in vitro models that possess very close properties to an in vivo condition. For example, HT-29-based in vitro

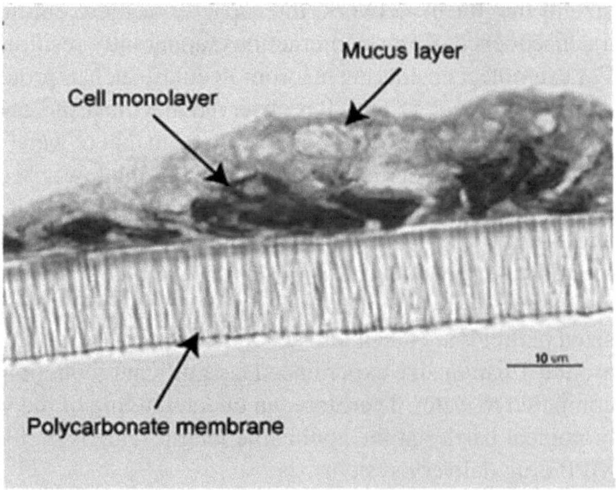

Fig. 3.5 Coculture of HT29-MTX (10 %) and Caco-2 (90 %) cells on a polycarbonate membrane showing the mucus layer with the thickness of 2–10 μm. (Reprinted with permission from [97])

cell cultures (a human adenocarcinoma cell line) were developed to compensate the absence of goblet cells and mucus layer in the cellular Caco-2 monolayer [91, 93]. This cell line has attracted a lot of attention for the study of mucus role in drug permeability.

Currently, different subtypes of HT-29-intestinal model exist for some specific studies. For example, HT-29GlucH was developed as a multilayer model with high proportion of goblet cells and large amount of mucin secretion, allowing the study of the impact of the mucus layer on drug transport [94]. HT-29MTXE12 is another highly reliable subclone of cells with desirable mucus thickness and presence of tight junctions in its structure, providing a very good in vivo correlation and easy prediction of mucus effect in intestinal absorption studies [95].

In contrast to the Caco-2 cells, HT-29 cells allow hydrophilic drugs to cross more easily through the paracellular pathway; therefore, a coculture of Caco-2 and HT-29 cells, compared to the Caco-2 and HT-29 cells alone, can provide very rational biophysical in vitro model with higher resemblance to human intestinal cells [96, 97]. Because these models also produce mucus, a decrease in the overall tightness resistance values (lower P-gp activity) is observed and can be used to determine the extent of the nanoparticles uptake by both carrier-mediated mechanisms and passive paracellular pathways [90]. Figure 3.5 shows the mucus production in a coculture of HT29-MTX (10 %) and Caco-2 (90 %) cells.

Since the mucus layer thickness varies in different parts of the GIT, it is very difficult to create a proper model in vitro [54]. Nevertheless, researchers have observed that different subtypes of HT-29 show varying depths of the mucus layer, therefore, they can be applied to study drug absorption in different parts of the GIT. For example, HT29-MTX- E12 makes mucus layers of ~ 142 μm while HT29-D1 results in mucus layers of ~ 53 μm [98].

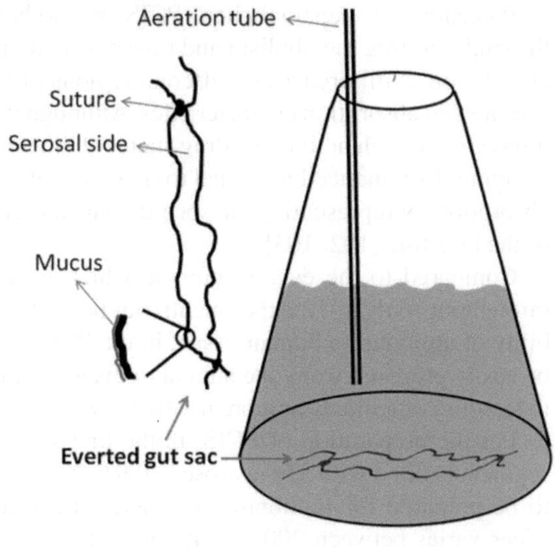

Fig. 3.6 A representative schematic for everted intestinal gut sac model

3.4.2 Ex Vivo Models

As it is now well recognized that the intestinal permeability of drugs cannot be precisely predicted by a single in vitro method, various ex vivo models have been developed to establish a more acceptable in vitro and in vivo correlation.

To this end, everted gut sac and precision-cut intestinal slices (PCIS) are two valuable ex vivo models that have attracted a lot of interest in recent years. The everted gut sac model was developed in 1954 for the first time and improved over time to be used as an efficient tool for drug permeability investigations and to study the mechanisms involved in the absorption, drug metabolism or prodrug conversion in GIT, drug interactions, multidrug resistance, efflux transport, and the influence of efflux transport modulators on the drug absorption [99, 100].

The main benefits of this model compared to the other available ones include relatively large surface areas as well as the presence of mucus layer. However, this model suffers from some limiting factors that cause controversies for its application. These drawbacks include short tissue viability and metabolic activity (around 2 h) [100, 101] as well as the presence of the muscularis mucosa that might elicit an underestimation for drug transportation. Although there are different reports in the literature about everted gut sac model developed from different animals such as sheep, frog, rabbit, catfish, pigs, etc., the most commonly used is the everted rat intestinal sac [100, 101].

A schematic illustration of the everted gut sac model is shown in Fig. 3.6. It is worth pointing out the impact of several factors on the outcome and functionality of everted gut sac studies. These factors include: (1) animal factors such as age, sex, and disease state; (2) intestinal segments, such as ileum, jejunum, duodenum, or colon, used for the study; and (3) experimental factors such as temperature, pH, and drug concentration [100].

Precision-cut intestinal slice (PCIS) is another ex vivo model usually used for the study of drug metabolism and transport in all parts of the intestine, allowing the elucidation of differences in different regions of the intestine in terms of metabolic activity and absorption characteristics. Although this model is viable for only 8–24 h, it has shown high activity of drug-metabolizing enzymes, a promising potential for studying drug-induced intestinal toxicity and also the effect of transporters on drug absorption by representing the specific physiological and anatomical characteristics of the intestine [102, 103].

Compared to the everted gut sac which is relatively labor intensive with low throughput (only 1–12 experiments can be performed per animal) and the impossibility of applying to human tissue, in the PCIS model, a large number of slices can be easily prepared from one animal or even one piece of human tissue, contributing to less loss of animals or more trustworthy results by using human tissue [104, 105].

For the preparation of PCIS, in the first step, ice-cold buffer washed intestinal segment is filled with 3 % agarose in 0.9 % NaCl, and then, inserted it in 3 % agarose to be prepared for sectioning by a Krumdieck tissue slicer. The thickness of the slices varies between 200 and 400 μm. The protocol for human small intestine is very similar. The only difference is that the human intestine is used after stripping off the muscle while the tissue is used unstrapped in the rat [106].

All the in vitro and ex vivo models described above allow a better understanding of the transport of molecules or particulates across the mucus membranes of the GIT, and are also important to pinpoint the effects of the mucus on such transport.

3.5 Mucoadhesive Particulate Systems

Mucoadhesive particulates are designed to adhere to mucosal membranes in the human body. They interact with the mucosal epithelial surface and/or mucin molecules and enable prolonged drug retention at the site of application. The ability of mucoadhesion depends on the structure of mucosal membranes, the properties of mucus gels, and the physicochemical properties of the mucoadhesive polymers [107]. The MDDS have been developed for buccal, oral, nasal, ocular, rectal, and vaginal drug delivery. In the next sections, we summarize the common theories used to describe the mucoadhesion of particulates to the mucosal membranes.

3.5.1 Theories of Mucoadhesion

Mucoadhesion is a rather complex phenomenon. There are several theories that have been described in the literature to explain the mucoadhesion phenomena, and they are summarized in Table 3.1.

Table 3.1 Theories and mechanisms of bioadhesion

Theory	Mechanism	Type of polymer	Reference
Wetting	Ability of bioadhesive polymer to spread on mucus layer	Liquid form	[107]
Electronic	Electron transfer between the polymer and mucus layer	Charged polymer	[109]
Diffusion	Polymers penetrate into mucus gel at depth of 0.2–0.5 mm	Good mutual solubility	[108]
Adsorption	Polymer interacts with mucus with ionic, covalent, and metallic bonding or van der Waals forces, hydrophobic interactions, and hydrogen bonding	Polymer with multiple functional groups	[110]
Fracture	Difficulty of separating after polymer mucus adhesion	Solid and/or rigid	[111]
Mechanical	Surface roughness increases the surface contact, thus enhancing adhesion	Rough and/or porous	[13]

3.5.1.1 Wetting Theory

The wetting theory describes the spreading ability of mucoadhesive polymers on the mucus layer, which is mainly applied for a liquid form mucoadhesive system [108]. The mucoadhesion can be predicted by contact angle measurement. Lower contact angle presents reduction of surface and interfacial energies and higher mucoadhesion [107].

3.5.1.2 Electronic Theory

The mucoadhesion occurs by electron transfer between the mucus layer and the mucoadhesive system [109]. This theory is applicable when the mucoadhesive system has different electronic characteristics or is oppositely charged from the mucus layer.

3.5.1.3 Diffusion Theory

The diffusion theory describes that mucoadhesive polymers are merged into mucus gel and form semipermanent adhesive bonds with the mucin chains. This theory applies when the mucoadhesive materials have good mutual solubility and have similar chemical structures as the mucus. The depth of diffusion depends on several parameters such as the gradient concentrations, molecular weight, flexibility, hydrodynamic size, and mobility of the mucoadhesive macromolecules. It is also affected by the

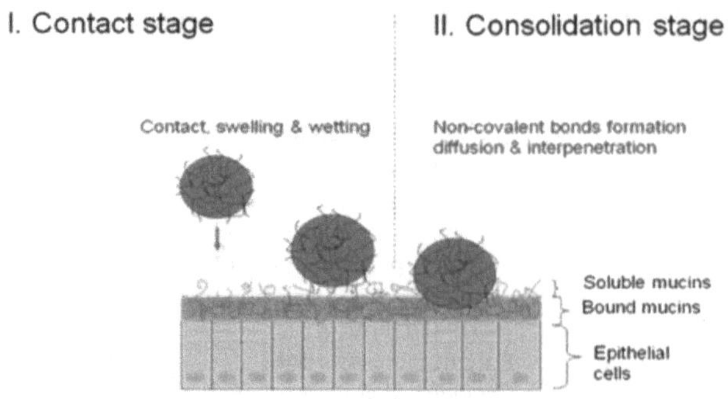

Fig. 3.7 Stages for the mucosal adhesion: **I** contact and **II** consolidation. (Reprinted with permission from [13])

diffusion coefficients and the time of contact of the mucoadhesive compounds. Usually, it is considered that the mucoadhesion is efficient when the interpenetration layer reaches 0.2–0.5 mm [108].

3.5.1.4 Adsorption Theory

The adsorption theory considers that the adhesion between mucus and mucoadhesive materials is because of the various surface interactions. For example, the strong interaction due to ionic, covalent, and metallic bonding or weak interactions due to van der Waals forces, hydrophobic interactions, and hydrogen bonding [110].

3.5.1.5 Fracture Theory

The fracture theory relates the difficulty of separating the two surfaces after adhesion is established. This theory is considered to be appropriate for solid and rigid mucoadhesive materials, when the polymer chains do not penetrate into the mucus layer [111].

3.5.1.6 Mechanical Theory

Surface roughness will favor the adhesion due to increased contact surface area, thus strengthening the mucoadhesion. This theory is applicable for rough and porous materials [13].

3.5.1.7 Summary

During the mucoadhesion process, different theories are complementary to each other rather than independent of each other. There might be several theories applied at the same stage or at different stages, and even with the aid of these theories, the mucoadhesion process is still not fully understood [112]. As depicted in Fig. 3.7, it is generally assumed that first, the dosage form wets and swells (wetting theory), after which noncovalent (physical) bonds are created within the mucus–polymer interface (electronic and adsorption theories), followed by the polymer and protein chains interpenetration (diffusion theory) and entanglement to form further noncovalent (physical) and covalent (chemical) bonds (electronic and adsorption theories) [108].

Regarding the several methods that have been described in the literature to measure and study mucoadhesion, the readers are directed to the works reported elsewhere [13, 17–19, 38, 108, 113–115].

3.5.2 Factors Affecting Mucoadhesion

Mucoadhesion may be affected by numerous factors, including the molecular weight, flexibility, cross-linking, swelling, spatial conformation, concentration, surface charge, hydrogen bonding capacity of the polymer, and the pH of the mucoadhesion interface [107, 110]. Below, several of these factors are discussed.

The molecular weight effect on the mucoadhesion depends on the type of polymer used [116]. For example, it has been found that the mucoadhesion of polyoxyethylene polymers is significantly increased by increasing the molecular weight of the polymer from 200 kDa to 4,000 kDa [117].

Based on the diffusion theory, increased chain interpenetration will strengthen the mucoadhesion. The structural flexibility of polymers will affect the diffusion coefficients, thus affecting the mucoadhesion. Polymers with higher structural flexibility will have greater mucoadhesion [107, 111].

Higher density of cross-linking reduces the flexibility of polymer and the degree of swelling, thus decreasing the mucoadhesion [111]. However, too high swelling degree will reverse the effect due to the slippy mucilage that can be easily removed [118]. Therefore, light but reasonable cross-linked polymers are often favorable for mucoadhesion.

Spatial conformation is also an important factor that affects mucoadhesion. For example, dextrans with molecular weight of 19.5 MDa have similar adhesive strength as poly(ethylene glycol) (PEG) with molecular weight of 200 kDa. PEG polymers have linear conformation and the active groups for mucoadhesion are exposed, while the dextrans have helical conformation and a great part of the active groups are shielded [110].

This factor is on the basis of the development of strong adhesive bond between the polymer chain and the mucus. The number of functional polymer chains for mucus interaction increases with the concentration of the polymer. Therefore, more concentrated polymers would benefit mucoadhesion. However, if the concentration is too high, the polymers will form coiled structures by themselves. As a result, the concentration effect will be saturated or even reduced [107]. It has been reported that polymeric films based on polyvinylpyrrolidone or poly(vinyl alcohol) (PVA) have optimal mucoadhesion at concentrations of 2–10 wt%. Further increase of the polymer concentration did not enhance or decrease the mucoadhesion [119].

Some cationic polymers can also produce superior mucoadhesion. For example, chitosan-based systems have been applied in various mucoadhesion formulations, and the extent of mucus adsorption has had good correlation to the absolute values of the positive zeta potential of chitosan microspheres [120].

Hydrogen bonding is also an important factor in mucoadhesion. The mucoadhesion of PVA, hydroxylated methacrylate, poly(methacrylic acid) as well as all their copolymers are based on their good hydrogen bonding capacity and flexibility for hydrogen bonding [121].

Finally, the pH at the adhesion interface can dramatically influence the mucoadhesion of polymers. The pH value not only affects the protonation of polymer but also alters the ability to form hydrogen bonds. The poly(acrylic acid) (PAA) family of polymers have pK_a of 4–5, and they achieve maximum mucoadhesion at pH 4–5, probably because of protonated carboxyl groups, rather than the ionized carboxyl groups that react with mucin molecules through hydrogen bonds [110].

3.5.3 Examples of Mucoadhesive Materials

In order to improve the mucoadhesivity of dosage forms, polymer (hydrophilic)-based excipients are usually used to stick to mucosal membranes. The most common polymers used are those possessing charged groups or nonionic functional groups capable of forming hydrogen bonds with the mucosal surfaces. Basically, polymers are good candidates for mucoadhesion if they comprise in their structure the following properties [13]: (1) strong hydrogen bonding groups (e.g., carboxyl, hydroxyl, amino, and sulfate groups); (2) strong anionic or cationic charges; (3) high molecular weight; (4) chain flexibility; and (5) surface energy properties favoring spreading onto mucus.

Numerous hydrophilic groups are an important feature of mucoadhesive polymers. The mucoadhesive polymers can be divided into different classes according to their physicochemical properties as described below [13]. The weakly anionic carboxyl-containing polymers have often been related to mucoadhesion because of the hydrogen bonds between the carboxyl group and the oligosaccharide chains of mucins. The hydrogen bond between PAA and the glycoprotein component of mucus has been shown to play a significant role on mucoadhesion [122]. Hydrogen bonds between the carboxyl groups of the bovine submaxillary mucin and poly(acrylic acid-block-methyl methacrylate) (PAA-b-PMMA) copolymer have been confirmed by infrared spectroscopic ellipsometry [123]. Chitosan and some synthetic poly(methacrylate) are good examples of mucoadhesive cationic polymers. The mucoadhesive properties of chitosan and its derivatives have been widely exploited for drug, protein, and gene delivery to the mucosal tissue.

In addition to anionic and cationic polymers, thiomers are polymers containing thiol-bearing functional group. These polymers are capable of forming disulfide bridges with the mucus glycoproteins [127, 128]. Thiolate-modified PAA was found to have 3–20-fold prolonged disintegration time than the correspondent unmodified polymer [129]. The chitosan/N-acetyl-L-cysteine (NAC) conjugate has been shown to have 50-fold longer residence time on the mucosa than the correspondent unmodified chitosan [130]. For example, thiolated microspheres have been successfully applied to deliver insulin via nasal route [131].

Aminated gelatin microspheres have also stronger interaction with mucin than gelatin [132]. Mono-N-carboxymethyl chitosan was demonstrated to enhance the absorption of anionic macromolecules (low-molecular weight heparin) both in vitro and in vivo [133]. In addition to this, the bacterial pili also exhibit intestinal mucus-binding capacity. The binding mechanism of the bacterial pili from Gram-positive bacterium *Lactobacillus rhamnosus* GG (LGG) has been demonstrated to be dependent on the protein SpaC of LGG pili, which has multifunctional adhesive properties [134]. The pili formed a zipper-like adhesion with SpaC molecules distributed along the pili. The flexibility of pili enabled it to bend and resist to high force strengthening adhesion as well as to withstand shear stresses in the natural environment.

Many other polymers have been described in the literature as potential mucoadhesive materials. For example, poly(amidoamine) dendrimers with multiple functional groups have been applied to deliver pilocarpine nitrate and tropicamide through the ocular route [135]. Copolymers of N-acryloyl-m-aminophenylboronic acid with N, N-dimethylacrylamide were also shown to interact with mucin from porcine stomach and form insoluble complexes at pH 9.0 [136].

3.5.4 Routes for Mucoadhesive Drug Delivery Systems (MDDS)

MDDS have been extensively investigated in the past decade, which can be applied in various drug delivery routes, including buccal, nasal, ocular, gastrointestinal, vaginal, and rectal as shown in Table 3.2. Examples of several commercially available mucoadhesive products are also presented in Table 3.3. In the next sections, we present some examples of the application of mucoadhesive systems for drug delivery application by the different routes of administration.

The oral cavity has been used as a site for local and systemic drug delivery. Adhesive tablets are the most commonly described dosage forms for buccal drug delivery such as, for example, special tablets that attach to the mucosa in mouth without disturbing drinking, eating, and speaking [137]. Mucoadhesive patches have also been employed for buccal drug delivery [138]. For example, the mucoadhesive patch containing Toluidine blue O (TBO) was used for photodynamic therapy of fungal infections of the mouth [139]. Moreover, a liquid aerosol formulation Oralin (Generex Biotechnology) was developed to deliver insulin through the oral mucosa route [140].

Nasal mucosa has highly dense vascular network, relatively permeable membrane structure, and 150 cm^2 area, which renders it a good route for drug delivery applications [141]. For example, a lyophilized nasal insert formulation has shown extended nasal residence time and has been applied for insulin delivery [142]. Furthermore,

Table 3.2 Examples of routes of administration of mucoadhesive drug delivery systems

Routes of administration	Example	Reference
Buccal	Toluidine blue O (TBO) containing patch for fungal infection	[139]
	Oralin for insulin delivery	[140]
Nasal	Lyophilized nasal insert for insulin delivery	[142]
	Starch and cross-linked PAA mixture for inactivated influenza vaccine	[143]
Ocular	Polycarbophil microspheres for sulfacetamide sodium delivery	[144]
	Hyaluronan–chitosan nanoparticles for gene delivery	[145]
Gastrointestinal	Carbohydrate polymers for riboflavin delivery to stomach	[147]
	Gut bacterial adhesion model for drug delivery to intestine	[148]
	Chitosan and thiolated-chitosan-coated sub-100 μm Ca^{2+} alginate microcapsules for probiotic bacteria delivery to colon	[149]
Rectal	Mucoadhesive hydrogels for diclofenac delivery	[150]
Vaginal	Mucoadhesive thermo responsive systems for vulvo vaginal candidiasis treatment	[152]

Table 3.3 Bioadhesive drug formulation products already in the market. (Batchelor 2004)

Product name	Polymer	Routes of administration	Form	Company
Aci-jel	Tragacanth and Acacia	Vaginal	Gel	Janssen-Cilag
Buccastem	PVP, Xanthum gum, Locust Bean gum	Buccal	Tablet	Reckitt Benckiser
Corlan pellet	Acacia gum	Oromucosal	Tablet	EllTech
Corsodyl gel	HPMC	Oromucosal	Gel	GlaxoSmithKline
Crinone	Carbomer	Vaginal	Gel	Serono
Gaviscon Liquid	Solium alginate	Gastrointestinal	Liquid	Reckitt Benckiser
Gyol-II	SCMC and PVP	Vaginal	Gel	Janseen-Cilag
Nyogel	Carbomer and PVA	Ocular	Gel	Novartis
Pilogel	Carbomer	Ocular	Gel	Alcon
Suscard	HPMC	Buccal	Tablet	Forest
Timoptol-LA	Gellan gum	Ocular	Gel solution	Merck, Sharp and Dohme
Zidoval	Carbomer	Vaginal	Gel	3-M

spray-dried mixtures of starch and cross-linked PAA have been also applied for intranasal deliver of inactivated influenza vaccine [143].

Ocular drug delivery is often limited by the protection systems in eyes (e.g., tear production, tear flow, and blinking). Mucoadhesive systems have also been developed to prolong the drug resistance in eyes. For example, bioadhesive sulfacetamide sodium-loaded polycarbophil microspheres have been shown to be highly effective in the treatment of ocular keratitis [144]. Moreover, the bioadhesive hyaluronan–chitosan nanoparticles have been shown as an alternative strategy for ophthalmic gene therapy [145].

The GIT delivery is undoubtedly the most favored route for drug administration. A variety of mucoadhesion systems have been investigated for enhanced drug delivery in GIT, for stomach, small intestine, and colon [146]. Carbohydrate polymers significantly increased the bioavailability and retention time of riboflavin in stomach [147]. A novel gut bacterial adhesion model has been developed for probiotic and other delivery applications [148]. In addition, chitosan and thiolated chitosan-coated sub-100 micro Ca^{2+} alginate microcapsules enhanced the colonic delivery of probiotic bacteria [149].

Rectal is also a common route for mucoadhesion drug delivery. Mucoadhesive hydrogels containing diclofenac sodium–chitosan microspheres have shown a controllable drug release pattern and prolonged drug action [150].

Bioadhesive vaginal drug delivery systems have also been used for local treatment of diseases such as candidiasis, vaginal dryness, and others [151]. Mucoadhesive thermo-responsive systems have been developed to treat vulvo vaginal candidiasis [152].

3.6 Mucus-Penetrating Particles (MPPs)

For the development of MPPs, the nanoparticles must be small and smooth enough. Since mucus rapidly clears the foreign particles, the particles must be fast enough to traverse at least the outermost layers of the mucus barrier (Fig. 3.8) [26, 54]. MPPs readily penetrate the luminal mucus layer (LML) and enter the underlying adherent mucus layer (AML). Conventional mucoadhesive particles (CPs) are largely immobilized in the LML. MPPs are able to enter the AML staying in close proximity to the cells, and thus, exposing cells to a greater dose of drugs when compared to drug released from CPs. As the LML layer is cleared, CPs are removed along with the LML, whereas MPPs in the AML are retained, leading to prolonged residence time of MPPs at the mucosal surface. Therefore, at long period of times, there is almost no drug dosing to cells using CPs. On the other hand, because MPPs are retained longer, they can continue to release drugs to cells. Since MPPs can penetrate both the LML and AML, a fraction of the particles may reach and bind to the underlying epithelia and further improve drug delivery.

Although Fig. 3.8 reflects the mucosal physiology of the GI and cervico-vaginal tracts, the same behavior is expected for the respiratory airways. In this case, CPs are mostly immobilized in the luminal stirred mucus gel layer, whereas MPPs penetrate the mucus gel and enter the underlying periciliary layer. Upon mucociliary clearance, a significant fraction of MPPs remain in the periciliary layer, resulting in prolonged retention [54].

When administered to mucosal tissues, nanoparticles are likely to be trapped by the mucus and cleared thereafter. To overcome the short transit times, mucoadhesion

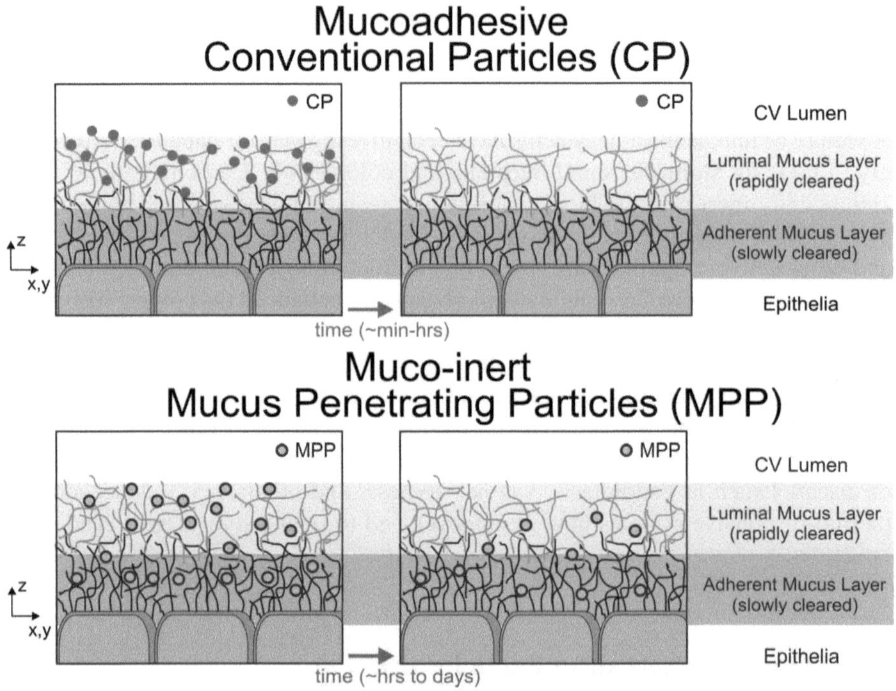

Fig. 3.8 Schematic illustration of the fate of MPPs and conventional CP administered to a mucosal surface. This reflects mainly the mucosal physiology of the GIT and cervico-vaginal tracts and does not depict the glycocalyx adjacent to the epithelial surface, which may contribute as an additional steric barrier to cellular entry of MPP. (Reprinted with permission from [54])

systems have been developed, particularly for oral administration. The mucoadhesion properties of the particles can have the same transit times of the mucus renewal.

Currently, mucoadhesive systems are the predominant approaches in mucosal drug delivery; however, the efficiency of these systems is determined by the physiological turnover time of the different mucus layers. Moreover, if certain polymers have mucoadhesive properties, they are incapable of penetrating the mucus layer, thus being highly unsuitable for the delivery of drug and gene molecules that require intracellular delivery [54].

To overcome these problems and increase the particle's transit time, MPPs have been developed as depicted in Fig. 3.9. The diffusion of the nanoparticles or other macromolecules in mucus can be estimated by measuring the permeation rates through a thin layer of mucus in a diffusion chamber [153]. The diffusion of variously sized polystyrene particles has been investigated through diffusion chambers with a reconstituted porcine gastric mucin gel in between [154]. Later on, a modified thin layer wicking technique was developed and validated to allow the determination of absolute surface hydrophobicity of intact microparticles [155]. Another method for tracking MPPs is to measure the dynamic transport of the nanoparticles using

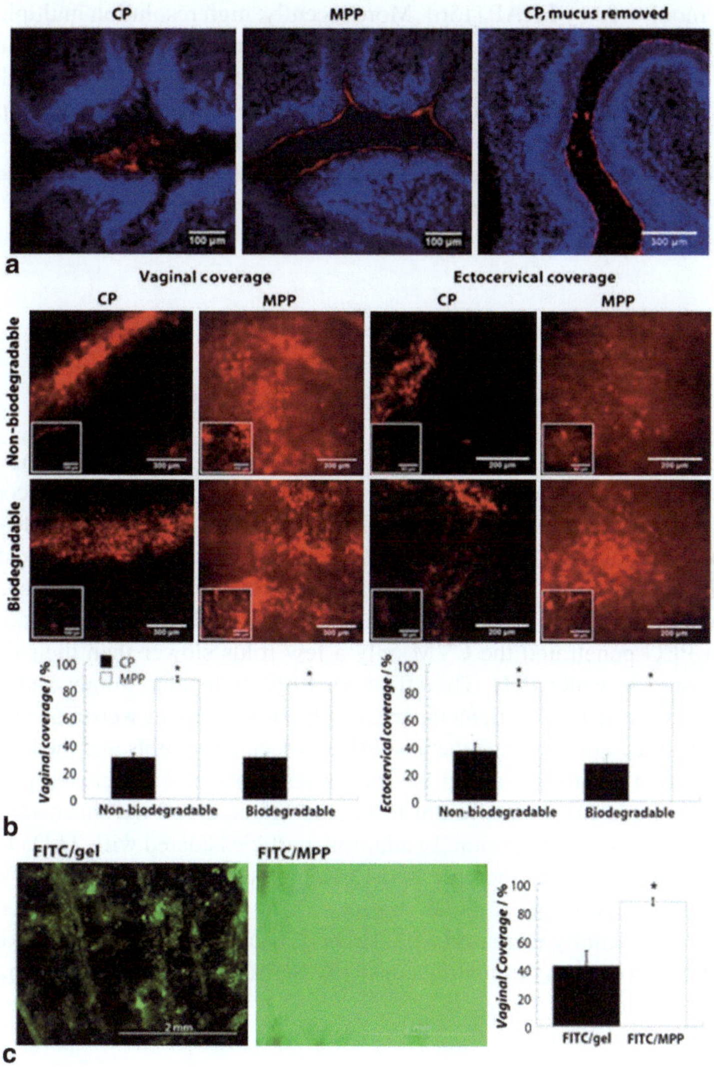

Fig. 3.9 **a** Distribution of red fluorescent nonbiodegradable CPs and MPPs in transverse cryosections of mouse vaginal tissue with an intact mucus layer or mucus removed by lavage and swabbing (mucus removed). **b** Distribution of nonbiodegradable and biodegradable CPs and MPPs on flattened mouse vaginal and ectocervical tissue (*insets* are images of higher magnification). **c** Distribution and retention of a model fluorescent dye, FITC, in the mouse vagina delivered in gel form or encapsulated in biodegradable MPPs. Fluorescent images of flattened mouse vaginal tissue after 24 h. Student's t-test set of $*p < 0.05$ to compare CP or FITC/gel. (Reprinted with permission from [162])

fluorescent microscopy techniques, such as fluorescence recovery after photobleaching (FRAP). The diffusion of plasmid DNAs of various sizes (2.7–8.3 kb) in mucus has been monitored by FRAP [156]. More recently, high resolution multiple particle tracking (MPT) techniques have been developed for monitoring MPPs. The diffusion coefficients of hundreds of individual amine-modified and carboxylate polystyrene particles (diameter 100–500 nm) embedded in fresh human sputum obtained from patients with CF were determined by MPT [74]. MPT was also applied to quantify the transport rates of individual polymeric particles of various sizes and surface chemistries in fresh human CVM [57].

3.6.1 Engineering Particles to Cross Mucus Barriers

3.6.1.1 The Importance of the Surface Characteristics

With the development of the nanomedicine field, nanoparticles have shown many desirable properties on mucosal drug delivery, and many types of mucoadhesion particles have been developed. However, those particles are also rapidly cleared from the human body due to the dynamics of the mucus layers. Thus, MPPs are needed to deliver the therapeutic agents to achieve a slower clearance from the mucus layer or epithelial surface [26]. For example, it has been found that a dense surface coating of 2 kDa PEG penetrated the CVM only a few folds slower than their theoretical diffusion rates in water [57]. The effects of PEG molecular weight and degree of surface coverage on the rate of nanoparticle diffusion in mucus were also investigated [65]. The results indicated that the low-PEG molecular weights and the high density of PEG coating clearly enhanced the mucus penetration of the coated particles. In contrast, the high-molecular-weight PEG coating increased the mucoadhesion and decreased the mucus penetration. In addition, particles coated with 2 kDa PEG were shown to have 1,000-fold increase in the mean-square-displacement in CVM (time scale 1 s) compared to the particles coated with 10 kDa PEG. The transport rates in CVM also significantly increased with 2 kDa PEG coating compared to 10 kDa PEG coating [65]. Moreover, the coating density was also found to be an important factor for mucus penetration.

3.6.1.2 Transport of Viruses in Mucus

Another example to evaluate the mucus-penetrating properties of particles is studying the virus particles. The mesh spacing between mucin fibers is large enough (20–200 nm) for small viruses to diffuse through mucus [66]. In this respect, polystyrene nanoparticles covalently modified with carboxyl groups were found to be completely immobilized in human cervical mucus at size of 59 nm [66]. Many viruses are capable of penetrating the mucus barrier at high transportation rate. Mimicking the essential surface properties of viruses may produce highly efficient MPPs for drug delivery [54].

Many mucus-penetrating viruses are coated with an equal density of positively and negatively charged groups at high density and have a net neutral surface. This feature may facilitate the efficiency of mucus transportation by avoiding electrostatic adhesive interactions [54]. Moreover, it has been hypothesized that some viruses cross the mucus barriers by hitchhiking on bacteria or sperm cells, which can transport them across the mucosal layers [157].

3.6.1.3 MPPs for Drug and Gene Delivery

PEG is an uncharged hydrophilic polymer routinely used in nanomedicine. As described above, the size of the PEG molecule attached to nanoparticles can affect the mucus penetration properties of the particles [65]. Recently, PEGylated solid lipid nanoparticles (SLNs) have been demonstrated to enhance the drug bioavailability by mucus penetration [158]. In addition, it has also been reported that copolymers of poly(sebacic acid) and PEG (PSA-PEG) also enabled mucus penetration [159]. Table 3.4 shows the summary of PEG and other polymer-based MPPs.

The treatment of mucus with mucolytic agents may also improve the nanoparticle penetration through mucus layers [58]. Pulmozyme® is a human DNAase (rhDNAase), which is commonly used as mucolytic agent in CF. It has been shown that the rhDNAase treatment dramatically narrows the distribution of individual particle diffusion rates [74]. Mucinex® (NAC) is another mucolytic agent. It has been reported that the NAC enhanced the gene transfer efficiency for gene therapy of CF in an ex vivo model of sheep tracheal epithelium [160]. Other mucolytic agents, such as Nacystelyn, Gelsolin, and thymosin β4 have also shown varying effects after mucus penetration, which suggests that the use of mucolytic agents must be carefully selected and evaluated when the aim is to enhance the mucus penetration of compounds/particles [54].

MPPs can also be applied to deliver numerous therapeutic molecules, including peptides and proteins. Poly(lactic-co-glycolic acid) (PLGA)-PEG, PSA-PEG copolymers, and vitamin E-PEG conjugate coated with PLGA have been shown to rapidly diffuse or penetrate the human CVM [26, 161]. Moreover, both polystyrene-based and biodegradable MPPs, generally recognized as safe ingredients, were able to penetrate chronic rhino sinusitis mucus [57]. In addition, MPPs have provided uniform distribution over the vaginal epithelium and have improved the vaginal drug distribution and retention [162].

In addition to drug delivery across the mucus, gene delivery through mucosal membranes has also been studied. For example, CF gene therapy is of great interest for MPPs. DNA nanoparticles in sputum pretreated with NAC have shown improved diffusion on an ex vivo mouse tracheal tissue and have mediated improved airway gene delivery [163]. Highly compacted DNA nanoparticles with low-molecular weight PEG coatings have also exhibited high gene transfer rate to lung airways following inhalation in BALB/c mice [164].

All the above-mentioned examples show the importance of the development of MPPs in order to enhance the drug and gene delivery across the mucus, which are described in more detail in the other chapters of this book.

Table 3.4 MPPs based on PEG and other polymers

Polymer	Function	Reference
Dense surface coating of 2 kDa PEG	Penetrated CVM	[57, 65]
59 nm polystyrene nanoparticles covalently modified with carboxyl groups	Immobilized in human cervical mucus	[66]
PEGylated SLNs	Enhanced drug bioavailability	[158]
Copolymer of PSA-PEG	Enabled mucus penetration	[159]
Pulmozyme® treatment	Narrows the distribution of individual particle diffusion rates	[74]
Mucinex® (NAC)	Enhanced efficiency for gene therapy of CF	[160]
PLGA-PEG and PSA-PEG copolymers	Rapidly diffused through cervico-vaginal mucus	[26]
VitaminE-PEG-PLGA	Rapidly penetrating in human cervico-vaginal mucus	[161]
Polystyrene-based and biodegradable MPPs with Generally Regarded as Safe (GRAS) ingredients	Rapidly penetrating in chronic rhinosinusitis mucus	[57]
Nonbiodegradable and biodegradable MPPs	Improved the vaginal drug distribution and retention	[162]
DNA nanoparticles in sputum pretreated with NAC	Mediated improved airway gene delivery	[163]
Highly compacted DNA nanoparticles with low-molecular weight PEG coatings	High gene transfer rate to lung airways in BALB/c mice	[164]

3.7 Particulate Drug Carriers for Targeting of the Inflamed Intestinal Mucosa

The transport of nutrients, drugs, and other molecules across mucosal membranes can be greatly affected by disease state of mucosal membranes. For example, daily high doses of anti-inflammatory drugs or immune-suppressant drugs are needed for the conventional treatment of inflammatory bowel disease (IBD). Due to the unspecific drug targeting, long-term IBD treatment often causes serious adverse effects. Thus, nano- and microparticulates that can deliver the drug specifically to the inflamed intestinal regions and prolong the drug release at the site would dramatically enhance the therapeutic efficiency of IBD [165].

Currently, the pathogenesis of IBD is still not fully understood. However, there have been several therapeutic targets which entered the clinical routine. The most prominent way of inflamed intestine targeting is the use of tumor necrosis factor-alpha (TNF-α) antibodies. Another strategy is to target the protein subunit p40. There are already several p40 antibodies in clinical trials for patients with Crohn's disease (CD). Natalizumab blocks the unregulated adhesion molecules during inflammation. The use of natalizumab for CD treatment has been approved in the USA [166].

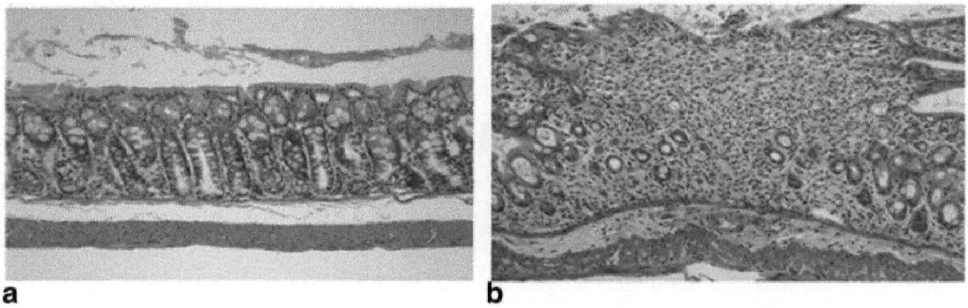

Fig. 3.10 Histological sections of the mouse intestinal colonic mucosa. Healthy control (**a**) and dextran sodium sulfate (DSS) model colitis (**b**): severe, segmental, chronic, suppurative-ulcerous-nectrotisizing colitis with oedema of the intestinal wall, segmentally retained fibrosis, and moderately suppurative serositis. (Reprinted with permission from [165])

3.7.1 Inflammatory Bowel Disease (IBD) and Therapy

IBD is a group of inflammatory conditions in large and small bowel (Fig. 3.10). IBD has two major types: the CD and ulcerative colitis (UC). IBD belongs to autoimmune diseases, in which the disordered immune system attacks the digestive system in intestine and colon [167].

IBD can limit the quality of life of the patients due to the pain, vomiting, and diarrhea. It is also reported that patients with long-term IBD have an increased risk of colorectal cancer. It is estimated that as many as about 2 million people suffer from IBD in North America and 2.2 million in Europe [168].

IBD has been well known for several decades, but the pathogenesis is still not fully clear. Analyses of the genes and genetic loci implicated in IBD show several pathways that are critical for intestinal homeostasis. These pathways include epithelial barrier, epithelial restitution, solution transport, microbial defense, innate mucosal defense, reactive oxygen species generations, endoplasmic reticulum stress, immune toleration, and metabolic pathways associated with cellular homeostasis (Fig. 3.11) [169].

Currently, there is no permanent cure for IBD. The current therapy of IBD is mainly focused on the induction and maintenance of remission [165]. The most commonly prescript IBD drug for mild-to-moderate CD or UC is mesalamine (5-ASA). Mesalamine has poor systemic bioavailability due to extensive metabolism in the liver [170]. Thus, local delivery of mesalamine to the inflamed intestine and colon will significantly improve the therapeutic efficiency. Glucocorticosteroid drugs have also been applied for IBD, singularly or in combination with mesalamine. However, this type of drugs may cause severe side effects, such as risk of infections due to immunosuppression, osteoporosis, diabetes mellitus, and Cushing's syndrome [165].

The particulate (micro and nano) drug delivery systems have significantly enhanced the therapeutic efficiency of IBD due to the more specific targeted drug delivery and the preferential uptake by antigen-presenting cells (Fig. 3.12) [165].

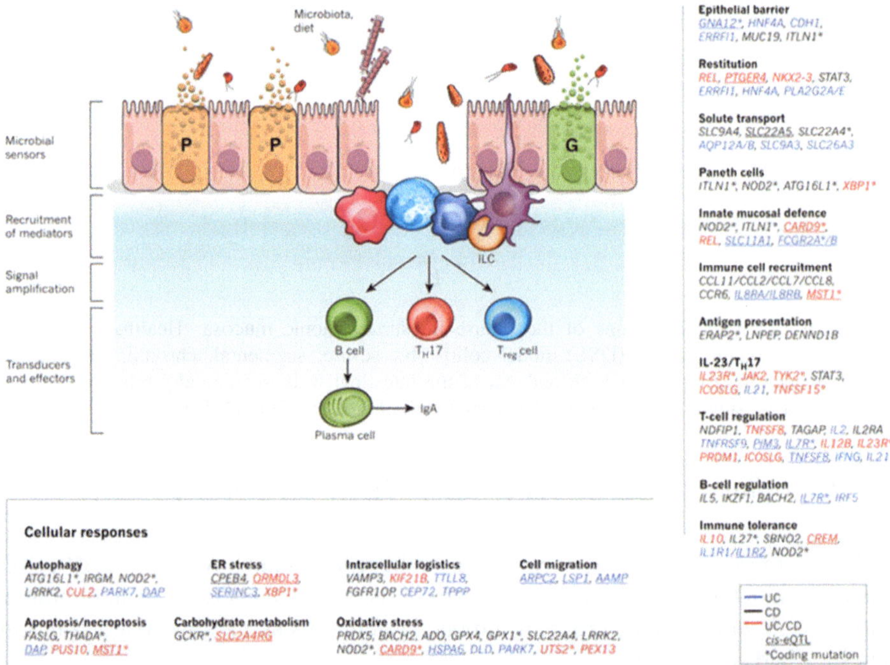

Fig. 3.11 Intestinal homeostasis involves the coordinated actions of epithelial, innate, and adaptive immune cells. Barrier permeability permits microbial incursion, which is detected by the innate immune system, which then orchestrates appropriate tolerogenic, inflammatory, and restitutive responses in part by releasing extracellular mediators that recruit other cellular components, including adaptive immune cells. Genes in linkage disequilibrium with IBD-associated single nucleotide polymorphisms (SNPs) are classified according to their function(s) in the context of intestinal homeostasis and immunity. Text color indicates whether the genes are linked to risk loci associated with CD (*black*), UC (*blue*), or both (*red*). *Asterisk* denotes corresponding coding mutations; cis-eQTL effects are *underlined*. *G* goblet cell, *P* Paneth cell. (Reprinted with permission from [169])

Numerous particulate-based systems have been developed for the treatment of IBD as shown in Table 3.5.

For example, dexamethasone-loaded PLA microspheres, which target immune-regulating cells, facilitate mucosal repair in dextran sodium sulphate (DSS) mouse colitis mode [171]. Chitosan-Ca-alginate microparticles are used to specifically deliver 5-ASA to colon after oral administration [172]. Eudragit P-4135F polymer has been used to prepare tacrolimus microparticles for colonic delivery [173]. Rolipram has shown therapeutic potential for various TNF-α-dependent diseases. Different nanoscaled delivery systems were developed for the delivery of rolipram [165]. PCL nanoparticles loaded with rolipram prepared by pressure homogenization-emulsification method has been shown to control and sustain the drug release [174]. Tacrolimus (FK506)-loaded PLGA nanoparticles have also been encapsulated into pH-sensitive microspheres to achieve colon-specific drug delivery when administered orally [175].

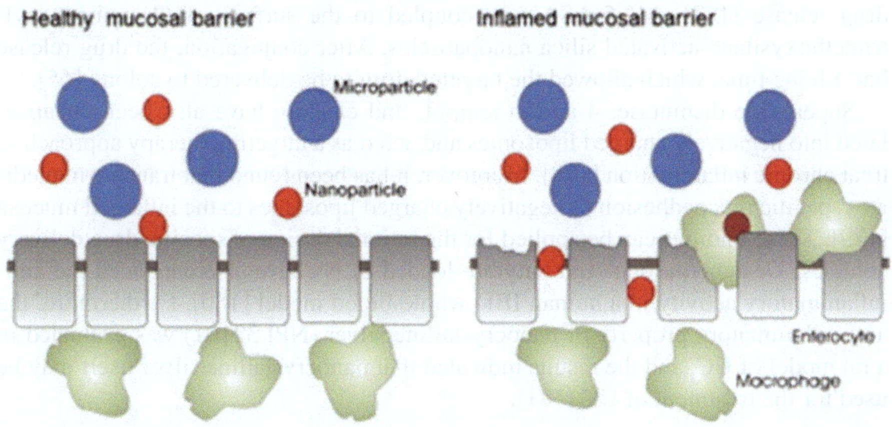

Fig. 3.12 Proposed mechanism of the size-dependent accumulation of particles in the inflamed intestinal mucosa. Small enough particles may accumulate in the gaps between epithelial cells or be taken up by invading immune cells. (Reprinted with permission from [165])

Table 3.5 Nano- and microparticulate drug delivery systems for IBD therapy

Polymer or particle	Loaded drug	System	Reference
PLA microspheres	Dexamethasone	Microparticulate	[171]
Chitosan-Ca-alginate	5-ASA	Microparticulate	[172]
Eudragit P-4135F polymer	Tacrolimus	Microparticulate pH-sensitive	[173]
Poly(epsilon-caprolactone) (PCL)	Rolipram	Nanoparticulate	[174]
PLGA encapsulated into pH-sensitive microspheres	Tacrolimus	pH-sensitive	[175]
Polymeric mixture of PLGA and a pH-sensitive methacrylate copolymer	Budesonide	pH-sensitive	[176]
PCL, covalent conjugation	5-ASA	Sustained release	[177]
(3-aminopropyl)-trimethoxysilane activated silica, covalent conjugation	Me-5-ASA	Sustained release	[165]
Negatively charged liposomes	Superoxidedismutase, 4-amino tempol and catalase	Liposomes	[178]
SLN	Dexamethasone and butyrate	SLN	[180]
Nanocrystalline silver (NPI 32101)		Nanocrystals	[181]

The polymeric mixture of a pH-sensitive methacrylate copolymer was also designed to deliver budesonide to the ulcerated and inflamed mucosal tissue of the rat colon [176]. 5-ASA has been covalently bound to PCL to achieve a sustained drug release [177]. Me-5-ASA was coupled to the surface of (3-aminopropyl) trimethoxysilane-activated silica nanoparticles. After conjugation, the drug release had 8 h lag time, which allowed the targeted drug to be delivered to colon [165].

Superoxide dismutase, 4-amino tempol, and catalase have also been encapsulated into negatively charged liposomes and acted as a targeting therapy approach to treat chronic inflammation [178]. Moreover, it has been found that transferrin mediates specific mucoadhesion of negatively charged liposomes to the inflamed mucosa [179]. These findings can be applied for the rational design of specific drug delivery vehicles. Dexamethasone- and butyrate-loaded SLNs have also enhanced the anti-inflammatory activity in a human IBD whole-blood model [180]. Furthermore, the anti-inflammatory property of nanocrystalline silver (NPI 32101) was evaluated in a rat model of UC, and the results indicated that nanocrystalline silver itself may be used for the treatment of UC [181].

3.7.2 Disease State and Changes in Mucus Layer Functionality

One of the important issues for oral nanoparticulate drug delivery is the possibility of changes in the protective characteristic of the mucus in disease states. Most of the time, mucus is confronted with different materials such as fluids, nutrients, and even toxic substances, bacteria, and viruses. These compounds represent a wide variety of physicochemical properties and, therefore, mucus layer should work as an intelligent barrier to maintain homeostasis despite very large surface area and exposure to more than 1,000 different types of bacteria [182].

As there are only few studies related to the changes in mucus layer in disease state, there is not a lot of information in this regard. However, it is very well known that some histochemical alterations occur in many GIT diseases [183]. For example, it has been shown that in inflammatory disorders the changes in glycosylation patterns of mucus occur, while no alteration in polymorphism or MUC gene expression takes place [184]. Shirazi and coworker have showed that in IBD, the length of the oligosaccharide chain and the degree of sulfation and sialylation become different compared to the normal mucus layer [185]. They also demonstrated that these changes may eventually affect the viscoelastic characteristics and resistance to pathogens.

Moreover, it has been demonstrated that UC can reduce the obstructive characteristics of the mucus in different manners, including reduction in the thickness of mucus layer in the colorectum, reduction in the expression of the protective trefoil peptides, and also alteration in the lipid composition of the mucosa [182]. Recently, the role of lipid composition change compared to the other destructive mechanisms in mucus change in UC has been demonstrated [186].

Rieux and coworkers have reported that intestinal inflammatory diseases can improve nanoparticle's permeability via destructive changes in the epithelium structure. Likewise, they have showed upregulation of particle transport via Peyer's patches after bacterial invasion [187]. In recent years, some studies have also displayed the influence of tumor presence on the mucus layer. For example, Wang and coworkers have reported that enhanced secretion of mucins by tumor cells and irregular glycosylation may delay drug uptake and cause resistance to the cancer therapy [188].

It is also worth to mention that despite some preliminary and basic studies concerning mucus change in disease state, this field is in its infancy and needs to be taken into account in upcoming research works, improving our knowledge for a more rational design of nanoparticulate drug delivery systems in a disease state.

3.8 Conclusions and Future Perspectives

In this chapter, we attempted to present a brief overview on the mucus as a barrier for biopharmaceuticals and drug delivery systems. We started by briefly introducing some of the important properties of mucus and mucosal membranes that need to be overcome in drug delivery applications. We then addressed some of the roles of mucus in blocking nanoparticulate drug delivery systems. We further highlighted the mucoadhesive properties of particulates, the design and development of mucus-penetrating delivery systems to avoid rapid mucus clearance and to provide targeted or sustained drug delivery for localized therapies in mucosal tissues (e.g., buccal, nasal, ocular, gastro, vaginal, and rectal). Next, we also presented an example of MPPs used to target a disease state mucosa. Finally, we described some examples of MPPs and provided some examples of particulate systems used to target a disease state mucosa.

As a result of the complex structure and functionality of the mucus of the various mucosal membranes, different strategies have been employed to overcome these barriers, for example, by developing novel mucoadhesive polymers and novel dosage forms for the various routes of drug administration (e.g., buccal, oral, nasal, ocular, and vaginal). There is currently a huge effort in this area in order to design mucoadhesive materials to improve the performance of the biopharmaceuticals. It is well recognized that the design of nanoparticles for controlled drug or gene delivery at the mucosal sites can lead to more effective treatments of diseases. However, conventional therapeutic particulate systems are still far from being ideal as they often cannot penetrate the human mucosal barriers, which rapidly clears them from the body. The development of MPPs is envisaged as of great potential to overcome the mucus barrier. In addition, MPPs offer the prospect of sustained drug delivery with great potential to treat diseases in the mucosal tissues.

Despite the various routes of administration, oral delivery is still the most commonly used and readily accepted form of drug administration. Since many orally administered drugs suffer from poor water solubility, stability, and/or bioavailability, the development of carriers for these drugs is of utmost importance. However, as aforementioned, these particles have to overcome the mucus barrier of the GIT in order to allow an efficient delivery of the therapeutics. The unique rheological and adhesive properties of mucus protect the epithelium from both mechanical forces and foreign pathogens and particles, leading to a rapid mucus secretion and clearance rates, which limit the residence time of orally administered nanoparticles. Therefore, MPP particles can potentially improve oral drug delivery by quickly traversing and penetrating the mucus layer, avoiding a rapid mucus clearance and increasing the GIT residence time and the distribution over the epithelium, thus leading to more effective treatments.

Acknowledgments Hélder A. Santos acknowledges the Academy of Finland (projects numbers 252215 and 256394), the University of Helsinki, and the European Research Council under the European Union's Seventh Framework Programme (FP7/2007–2013)/ERC Grant agreement number 310892 for financial support.

References

1. Shmulewitz A, Langer R. The ascendance of combination products. Nat Biotechnol. 2006;24(3):277–80.
2. Santos HA, Bimbo LM, Lehto VP, Airaksinen AJ, Salonen J, Hirvonen J. Multifunctional porous silicon for therapeutic drug delivery and imaging. Curr Drug Discov Technol. 2011;8(3)228–49.
3. Langer R. Drug delivery and targeting. Nature. 1998;392(6679 Suppl):5–10.
4. Farokhzad OC, Langer R. Nanomedicine: developing smarter therapeutic and diagnostic modalities. Adv Drug Deliv Rev. 2006;58(14):1456–9.
5. Wong C, Stylianopoulos T, Cui J, Martin J, Chauhan VP, Jiang W, Popovic Z, Jain RK, Bawendi MG, Fukumura D. Multistage nanoparticle delivery system for deep penetration into tumor tissue. Proc Natl Acad U S A. 2011;108(6):2426–31.
6. Yuan F, Leunig M, Huang SK, Berk DA, Papahadjopoulos D, Jain RK. Microvascular permeability and interstitial penetration of sterically stabilized (stealth) liposomes in a human tumor xenograft. Cancer Res. 1994;54(13):3352–6.
7. Jiang X-M, Wang L-M, Chen C-Y. Cellular uptake, intracellular trafficking and biological responses of gold nanoparticles. J Chinese Chem Soc. 2011;58(3):273–81.
8. Chiu Y-L, Ho Y-C, Chen Y-M, Peng S-F, Ke C-J, Chen K-J, Mi F-L, Sung H-W. The characteristics, cellular uptake and intracellular trafficking of nanoparticles made of hydrophobically-modified chitosan. J Control Release. 2010;146(1):152–9.
9. Tarn D, Xue M, Zink JI. pH-responsive dual cargo delivery from mesoporous silica nanoparticles with a metal-latched nanogate. Inorg Chem. 2013;52(4):2044–9.
10. Agostini A, Mondragon L, Bernardos A, Martinez-Manez R, Marcos MD, Sancenon F, Soto J, Costero A, Manguan-Garcia C, Perona R, Moreno-Torres M, Aparicio-Sanchis R, Murguia JR. Targeted cargo delivery in senescent cells using capped mesoporous silica nanoparticles. Angew Chem Int Ed Engl. 2012;51(42):10556–60.
11. Aznar E, Mondragon L, Ros-Lis JV, Sancenon F, Marcos MD, Martinez-Manez R, Soto J, Perez-Paya E, Amoros P. Finely tuned temperature-controlled cargo release using paraffin-capped mesoporous silica nanoparticles. Angew Chem Int Ed Engl. 2011;50(47):11172–5.
12. Cone RA. Barrier properties of mucus. Adv Drug Deliv Rev. 2009;61(2):75–85.
13. Khutoryanskiy VV. Advances in mucoadhesion and mucoadhesive polymers. Macromol Biosci. 2011;11(6):748–64.
14. Ishida M, Machida Y, Nambu N, Nagai T. New mucosal dosage form of insulin. Chem Pharm Bull. 1981;29(3):810–6.
15. Nagai T. Adhesive topical drug delivery system. J Control Release. 1985;2:121–34.
16. Andrews GP, Laverty TP, Jones DS. Mucoadhesive polymeric platforms for controlled drug delivery. Eur J Pharm Biopharm. 2009;71(3):505–18.

17. Bagan J, Paderni C, Termine N, Campisi G, Lo Russo L, Compilato D, Di Fede O. Mucoadhesive polymers for oral transmucosal drug delivery: a review. Curr Pharm Des. 2012;18(34):5497–514.
18. de Araujo Pereira RR, Bruschi ML. Vaginal mucoadhesive drug delivery systems. Drug Dev Ind Pharm. 2012;38(6)643–52.
19. Shinkar DM, Dhake AS, Setty CM. Drug delivery from the oral cavity: a focus on mucoadhesive buccal drug delivery systems. PDA J Pharm Sci Technol/PDA. 2012;66(5):466–500.
20. Singh RM, Kumar A, Pathak K. Mucoadhesive in situ nasal gelling drug delivery systems for modulated drug delivery. Expert Opin Drug Deliv. 2013;10(1):115–30.
21. Swain S, Behera A, Beg S, Patra CN, Dinda SC, Sruti J, Rao ME. Modified alginate beads for mucoadhesive drug delivery system: an updated review of patents. Recent Pat Drug Deliv Formul. 2012;6(3):259–77.
22. Sandri G, Rossi S, Ferrari F, Bonferoni MC, Zerrouk N, Caramella C. Mucoadhesive and penetration enhancement properties of three grades of hyaluronic acid using porcine buccal and vaginal tissue, Caco-2 cell lines, and rat jejunum. J Pharm Pharmacol. 2004;56(9): 1083–90.
23. Bonferoni MC, Chetoni P, Giunchedi P, Rossi S, Ferrari F, Burgalassi S, Caramella C. Carrageenan-gelatin mucoadhesive systems for ion-exchange based ophthalmic delivery: in vitro and preliminary in vivo studies. Eur J Pharm Biopharm 2004;57(3):465–72.
24. Sandri G, Rossi S, Ferrari F, Bonferoni MC, Muzzarelli C, Caramella C. Assessment of chitosan derivatives as buccal and vaginal penetration enhancers. Eur J Pharm Sci. 2004; 21(2–3):351–59.
25. Sandri G, Bonferoni MC, Rossi S, Ferrari F, Boselli C, Caramella C. Insulin-loaded nanoparticles based on N-trimethyl chitosan: in vitro (Caco-2 model) and ex vivo (excised rat jejunum, duodenum, and ileum) evaluation of penetration enhancement properties. AAPS PharmSciTech. 2010;11(1):362–71.
26. Ensign LM, Schneider C, Suk JS, Cone R, Hanes J. Mucus penetrating nanoparticles: biophysical tool and method of drug and gene delivery. Adv Mater. 2012;24(28):3887–94.
27. Paderni C, Compilato D, Giannola LI, Campisi G. Oral local drug delivery and new perspectives in oral drug formulation. Oral Surg Oral Med Oral Pathol Oral Radiol. 2012;114(3):e25–34.
28. Carvalho FC, Rocha e Silva H, da Luz GM, Barbi Mda S, Landgraf DS, Chiavacci LA, Sarmento VH, Gremiao MP. Rheological, mechanical and adhesive properties of surfactant-containing systems designed as a potential platform for topical drug delivery. J Biomed Nanotechnol. 2012;8(2):280–9.
29. Gee CM, Nicolazzo JA, Watkinson AC, Finnin BC. Assessment of the lateral diffusion and penetration of topically applied drugs in humans using a novel concentric tape stripping design. Pharm Res. 2012;29(8):2035–46.
30. Fulgencio Gde O, Viana FA, Ribeiro RR, Yoshida MI, Faraco AG, Cunha-Junior Ada S. New mucoadhesive chitosan film for ophthalmic drug delivery of timolol maleate: in vivo evaluation. J Ocul Pharmacol Ther. 2012;28(4):350–8.
31. Liu J, Wang Z, Liu C, Xi H, Li C, Chen Y, Sun L, Mu L, Fang L. Silicone adhesive, a better matrix for tolterodine patches-a research based on in vitro/in vivo studies. Drug Dev Ind Pharm. 2012;38(8):1008–14.
32. Mahmoud AA, El-Feky GS, Kamel R, Awad GE. Chitosan/sulfobutylether-beta-cyclodextrin nanoparticles as a potential approach for ocular drug delivery. Int J Pharm. 2011;413(1–2):229–36.
33. Movassaghian S, Barzegar-Jalali M, Alaeddini M, Hamedyazdan S, Afzalifar R, Zakeri-Milani P, Mohammadi G, Adibkia K. Development of amitriptyline buccoadhesive tablets for management of pain in dental procedures. Drug Dev Ind Pharm. 2011;37(7):849–54.
34. Al-Hezaimi K, Al-Askar M, Selamhe Z, Fu JH, Alsarra IA, Wang HL. Evaluation of novel adhesive film containing ketorolac for post-surgery pain control: a safety and efficacy study. J Periodontol. 2011;82(7):963–8.

35. Morrow DI, McCarron PA, Woolfson AD, Juzenas P, Juzeniene A, Iani V, Moan J, Donnelly RF. Novel patch-based systems for the localised delivery of ALA-esters. J Photochem Photobiol B. 2010;101(1):59–69.
36. Gullick DR, Pugh WJ, Ingram MJ, Cox PA, Moss GP. Formulation and characterization of a captopril ethyl ester drug-in-adhesive-type patch for percutaneous absorption. Drug Dev Ind Pharm. 2010;36(8):926–32.
37. Zhang J, Deng L, Zhao H, Liu M, Jin H, Li J, Dong A. Pressure-sensitive adhesive properties of poly(N-vinyl pyrrolidone)/D, L-lactic acid oligomer/glycerol/water blends for TDDS. J Biomater Sci Polym Ed. 2010;21(1):1–15.
38. Jones DS, Bruschi ML, de Freitas O, Gremiao MP, Lara EH, Andrews GP. Rheological, mechanical and mucoadhesive properties of thermoresponsive, bioadhesive binary mixtures composed of poloxamer 407 and carbopol 974P designed as platforms for implantable drug delivery systems for use in the oral cavity. Int J Pharm. 2009;372(1–2):49–58.
39. Hung CF, Lin YK, Huang ZR, Fang JY. Delivery of resveratrol, a red wine polyphenol, from solutions and hydrogels via the skin. Biol Pharm Bull. 2008;31(5):955–62.
40. Martin MD, Sherman J, van der Ven P, Burgess J. A controlled trial of a dissolving oral patch concerning glycyrrhiza (licorice) herbal extract for the treatment of aphthous ulcers. Gen Dent. 2008;56(2):206–10; quiz 211–202, 224.
41. Chandrashekar NS, Hiremath SR. Transdermal delivery of 5-fluorouracil for induced ehrlich ascites carcinoma tumor in BALB/c mice and pharmacokinetic study. Recent Pat Anticancer Drug Discov. 2007;2(3):235–9.
42. Abdulmajed K, Heard CM. Topical delivery of retinyl ascorbate. 3. Influence of follicle sealing and skin stretching. Skin Pharmacol Physiol. 2008;21(1):46–9.
43. Jain AK, Chalasani KB, Khar RK, Ahmed FJ, Diwan PV. Muco-adhesive multivesicular liposomes as an effective carrier for transmucosal insulin delivery. J Drug Target. 2007;15(6):417–27.
44. Valtcheva-Sarker RV, O'Reilly JD, Sarker DK. Administration of drug and nutritional components in nano-engineered form to increase delivery ratio and reduce current inefficient practice. Recent Pat Drug Deliv Formul. 2007;1(2):147–59.
45. Donnelly RF, McCarron PA, Zawislak AA, Woolfson AD. Design and physicochemical characterisation of a bioadhesive patch for dose-controlled topical delivery of imiquimod. Int J Pharm. 2006;307(2):318–25.
46. Barnhart K. Vaginal drug delivery. IDrugs. 1999;2(8): 756–9.
47. Jones DS, Lawlor MS, Woolfson AD. Rheological and mucoadhesive characterization of polymeric systems composed of poly(methylvinylether-co-maleic anhydride) and poly(vinylpyrrolidone), designed as platforms for topical drug delivery. J Pharm Sci. 2003;92(5):995–1007.
48. Bian S, Doh HJ, Zheng J, Kim JS, Lee CH, Kim DD. In vitro evaluation of patch formulations for topical delivery of gentisic acid in rats. Eur J Pharm Sci. 2003;18(2):141–7.
49. Baeyens V, Felt-Baeyens O, Rougier S, Pheulpin S, Boisrame B, Gurny R. Clinical evaluation of bioadhesive ophthalmic drug inserts (BODI) for the treatment of external ocular infections in dogs. J Control Release. 2002;85(1–3):163–8.
50. Artusi M, Santi P, Colombo P, Junginger HE. Buccal delivery of thiocolchicoside: in vitro and in vivo permeation studies. Int J Pharm. 2003;250(1):203–13.
51. Taware CP, Mazumdar S, Pendharkar M, Adani MH, Devarajan PV. A bioadhesive delivery system as an alternative to infiltration anesthesia. Oral Surg Oral Med Oral Pathol Oral Radiol Endod. 1997;84(6):609–15.
52. Genta I, Conti B, Perugini P, Pavanetto F, Spadaro A, Puglisi S. Bioadhesive microspheres for ophthalmic administration of acyclovir. J Pharm Pharmacol. 1997;49(8):737–42.
53. Knowles MR, Boucher RC. Mucus clearance as a primary innate defense mechanism for mammalian airways. J Clin Invest. 2002;109(5):571–7.
54. Lai SK, Wang YY, Hanes J. Mucus-penetrating nanoparticles for drug and gene delivery to mucosal tissues. Adv Drug Deliv Rev. 2009;61(2):158–71.

55. Lai SK, Wang YY, Hida K, Cone R, Hanes J. Nanoparticles reveal that human cervicovaginal mucus is riddled with pores larger than viruses. Proc Natl Acad Sci U S A. 2010;107(2):598–603.
56. Lai SK, Wang YY, Cone R, Wirtz D, Hanes J. Altering mucus rheology to "solidify" human mucus at the nanoscale. PloS One. 2009;4(1):e4294.
57. Lai SK, O'Hanlon DE, Harrold S, Man ST, Wang YY, Cone R, Hanes J. Rapid transport of large polymeric nanoparticles in fresh undiluted human mucus. Proc Natl Acad Sci U S A. 2007;104(5):1482–7.
58. Lai SK, Wang YY, Wirtz D, Hanes J. Micro- and macrorheology of mucus. Adv Drug Deliv Rev. 2009;61(2):86–100.
59. Samet JM, Cheng PW. The role of airway mucus in pulmonary toxicology. Environ Health Perspect. 1994;102(Suppl 2):89–103.
60. Quraishi MS, Jones NS, Mason J. The rheology of nasal mucus: a review. Clin Otolaryngol Allied Sci. 1998;23(5):403–13.
61. Allen A, Flemstrom G, Garner A, Kivilaakso E. Gastroduodenal mucosal protection. Physiol Rev. 1993;73(4):823–57.
62. Carlstedt I, Lindgren H, Sheehan JK, Ulmsten U, Wingerup L. Isolation and characterization of human cervical-mucus glycoproteins. Biochem J. 1983;211(1):13–22.
63. Chao CC, Butala SM, Herp A. Studies on the isolation and composition of human ocular mucin. Exp Eye Res. 1988;47(2):185–96.
64. Engel E, Guth PH, Nishizaki Y, Kaunitz JD. Barrier function of the gastric mucus gel. Am J Physiol. 1995;269(6 Pt 1):G994–9.
65. Wang YY, Lai SK, Suk JS, Pace A, Cone R, Hanes J. Addressing the PEG mucoadhesivity paradox to engineer nanoparticles that "slip" through the human mucus barrier. Angew Chem Int Ed Engl. 2008;47(50):9726–9.
66. Olmsted SS, Padgett JL, Yudin AI, Whaley KJ, Moench TR, Cone RA. Diffusion of macromolecules and virus-like particles in human cervical mucus. Biophys J. 2001;81(4):1930–7.
67. Crater JS, Carrier RL. Barrier properties of gastrointestinal mucus to nanoparticle transport. Macromol Biosci. 2010;10(12):1473–83.
68. Yoncheva K, Gomez S, Campanero MA, Gamazo C, Irache JM. Bioadhesive properties of pegylated nanoparticles. Expert Opin Drug Deliv. 2005;2(2):205–18.
69. Saltzman WM, Radomsky ML, Whaley KJ, Cone RA. Antibody diffusion in human cervical mucus. Biophys J. 1994;66(2 Pt 1):508–15.
70. Corthesy B, Kraehenbuhl JP. Antibody-mediated protection of mucosal surfaces. Curr Top Microbiol Immunol. 1999;236:93–111.
71. Larhed AW, Artursson P, Grasjo J, Bjork E. Diffusion of drugs in native and purified gastrointestinal mucus. J Pharm Sci. 1997;86(6):660–5.
72. Matthes I, Nimmerfall F, Vonderscher J, Sucker H. Mucus models for investigation of intestinal absorption mechanisms. 4. Comparison of mucus models with absorption models in vivo and in situ for prediction of intestinal drug absorption. Pharmazie. 1992;47(10):787–91.
73. Kas HS. Chitosan: properties, preparations and application to microparticulate systems. J Microencapsul. 1997;14(6):689–711.
74. Dawson M, Wirtz D, Hanes J. Enhanced viscoelasticity of human cystic fibrotic sputum correlates with increasing microheterogeneity in particle transport. J Biol Chem. 2003;278(50):50393–401.
75. Thornton DJ, Sheehan JK. From mucins to mucus: toward a more coherent understanding of this essential barrier. Proc Am Thorac Soc. 2004;1(1):54–61.
76. Wolf DP, Blasco L, Khan MA, Litt M. Human cervical mucus. I. Rheologic characteristics. Fertil Steril. 1977;28(1):41–6.
77. Boucher RC, Stutts MJ, Bromberg PA, Gatzy JT. Regional differences in airway surface liquid composition. J Appl Physiol. 1981;50(3):613–20.
78. Girod S, Galabert C, Lecuire A, Zahm JM, Puchelle E. Phospholipid composition and surface-active properties of tracheobronchial secretions from patients with cystic fibrosis and chronic obstructive pulmonary diseases. Pediatr Pulmonol. 1992;13(1):22–7.

79. Yeates DB, Besseris GJ, Wong LB. Physicochemical properties of mucus and its propulsion. In: Crystal RG, et al. editors. The lung: scientific foundations. Philadelphia: Lippincott-Raven; 1997. pp. 487–503.
80. Lamont JT. Mucus: the front line of intestinal mucosal defense. Ann N Y Acad Sci. 1992;664:190–201.
81. Lethem MI, James SL, Marriott C. The role of mucous glycoproteins in the rheologic properties of cystic fibrosis sputum. Am Rev Respir Dis. 1990;142(5):1053–8.
82. Cone R. Mucus, In: Ogra PL, et al. editors. Mucosal immunology. San Diego: Academic; 1999. pp. 43–64.
83. App EM, Zayas JG, King M. Rheology of mucus and transepithelial potential difference: small airways versus trachea. Eur Respir J. 1993;6(1):67–75.
84. Rubin BK, Druce H, Ramirez OE, Palmer R. Effect of clarithromycin on nasal mucus properties in healthy subjects and in patients with purulent rhinitis. Am J Respir Crit Care Med. 1997;155(6):2018–23.
85. Rubin BK. Mucus structure and properties in cystic fibrosis. Paediatric Respir Rev. 2007;8(1):4–7.
86. Voynow JA, Gendler SJ, Rose MC. Regulation of mucin genes in chronic inflammatory airway diseases. Am J Respir Cell Mol Biol. 2006;34(6):661–5.
87. Hattori M, Majima Y, Ukai K, Sakakura Y. Effects of nasal allergen challenge on dynamic viscoelasticity of nasal mucus. Ann Otol Rhinol Laryngol. 1993;102(4 Pt 1):314–7.
88. Allen A, Cunliffe WJ, Pearson JP, Sellers LA, Ward R. Studies on gastrointestinal mucus. Scand J Gastroenterol. 1984;Supplement 93:101–13.
89. Clift AF. Early studies on the rheology of cervical mucus. Am J Obstetr Gynecol. 1979;134(7):829–32.
90. Shahbazi MA, Santos HA. Improving oral absorption via drug-loaded nanocarriers: absorption mechanisms, intestinal models and rational fabrication. Curr Drug Metab. 2013;14(1): 28–56.
91. Wood KM, Stone GM, Peppas NA. The effect of complexation hydrogels on insulin transport in intestinal epithelial cell models. Acta Biomaterialia. 2010;6(1):48–56.
92. Balimane PV, Chong S, Morrison RA. Current methodologies used for evaluation of intestinal permeability and absorption. J Pharmacol Toxicol Meth. 2000;44(1):301–12.
93. Antunes F, Andrade F, Araujo F, Ferreira D, Sarmento B. Establishment of a triple co-culture in vitro cell models to study intestinal absorption of peptide drugs. Euro J Pharm Biopharm. 2013;83(3):427–35.
94. Wikman A, Karlsson J, Carlstedt I, Artursson P. A drug absorption model based on the mucus layer producing human intestinal goblet cell line HT29-H. Pharm Res. 1993;10(6): 843–52.
95. Keely S, Rullay A, Wilson C, Carmichael A, Carrington S, Corfield A, Haddleton DM, Brayden DJ. In vitro and ex vivo intestinal tissue models to measure mucoadhesion of poly (methacrylate) and N-trimethylated chitosan polymers. Pharm Res. 2005;22(1):38–49.
96. Nollevaux G, Deville C, El Moualij B, Zorzi W, Deloyer P, Schneider YJ, Peulen O, Dandrifosse G. Development of a serum-free co-culture of human intestinal epithelium cell-lines (Caco-2/HT29–5M21). BMC Cell Biol. 2006;7:20.
97. Mahler GJ, Shuler ML, Glahn RP. Characterization of Caco-2 and HT29-MTX cocultures in an in vitro digestion/cell culture model used to predict iron bioavailability. J Nutr Biochem. 2009;20(7):494–502.
98. Gamboa JM, Leong KW. In vitro and in vivo models for the study of oral delivery of nanoparticles. Adv Drug Deliv Rev. 2013;65:800–10.
99. Barthe L, Woodley JF, Kenworthy S, Houin G. An improved everted gut sac as a simple and accurate technique to measure paracellular transport across the small intestine. Eur J Drug Metab Pharmacokinet. 1998;23:313–23.
100. Alam MA, Al-Jenoobi FI, Al-Mohizea AM. Everted gut sac model as a tool in pharmaceutical research: limitations and applications. J Pharm Pharmacol. 2012;64(3):326–36.

101. Carreno-Gomez B, Duncan R. Everted rat intestinal sacs: a new model for the quantitation of P-glycoprotein mediated-efflux of anticancer agents. Anticancer Res. 2000;20(5A):3157–61.
102. van de Kerkhof EG, de Graaf IA, Ungell AL, Groothuis GM. Induction of metabolism and transport in human intestine: validation of precision-cut slices as a tool to study induction of drug metabolism in human intestine in vitro. Drug Metab Dispos. 2008;36(3):604–13.
103. van de Kerkhof EG, de Graaf IA, Groothuis GM. In vitro methods to study intestinal drug metabolism. Curr Drug Metab. 2007;8(7):658–75.
104. Groothuis GM, de Graaf IA. Precision-cut intestinal slices as in vitro tool for studies on drug metabolism. Curr Drug Metab. 2013;14(1):112–9.
105. van Midwoud PM, Merema MT, Verpoorte E, Groothuis GM. A microfluidic approach for in vitro assessment of interorgan interactions in drug metabolism using intestinal and liver slices. Lab Chip. 2010;10(20):2778–86.
106. de Kanter R, Tuin A, van de Kerkhof E, Martignoni M, Draaisma AL, de Jager MH, de Graaf IA, Meijer DK, Groothuis GM. A new technique for preparing precision-cut slices from small intestine and colon for drug biotransformation studies. J Pharmacol Toxicol Meth. 2005;51(1):65–72.
107. Boddupalli BM, Mohammed ZN, Nath RA, Banji D. Mucoadhesive drug delivery system: an overview. J Adv Pharm Technol Res. 2010;1(4):381–7.
108. Smart JD. The basics and underlying mechanisms of mucoadhesion. Adv Drug Deliv Rev. 2005;57(11):1556–68.
109. Derjaguin BV, Aleinikova IN, Toporov YP. On the role of electrostatic forces in the adhesion of polymer particles to solid surfaces. Progr Surf Sci. 1994;45(1–4):119–23.
110. Shaikh R, Raj Singh TR, Garland MJ, Woolfson AD, Donnelly RF. Mucoadhesive drug delivery systems. J Pharm Bioallied Sci. 2011;3(1):89–100.
111. Gu JM, Robinson JR, Leung SH. Binding of acrylic polymers to mucin/epithelial surfaces: structure-property relationships. Crit Rev Ther Drug Carrier Syst. 1988;5(1):21–67.
112. Dodou D, Breedveld P, Wieringa PA. Mucoadhesives in the gastrointestinal tract: revisiting the literature for novel applications. Eur J Pharm Biopharm. 2005;60(1):1–16.
113. das Neves J, Amiji M, Sarmento B. Mucoadhesive nanosystems for vaginal microbicide development: friend or foe? Wiley Interdiscip Rev Nanomed Nanobiotechnol. 2011;3(4):389–99.
114. das Neves J, Bahia MF, Amiji MM, Sarmento B. Mucoadhesiveicines: characterization and modulation of mucoadhesion at the nanoscale. Expert Opin Drug Deliv. 2011;8(8):1085–104.
115. Andrews GP, Donnelly L, Jones DS, Curran RM, Morrow RJ, Woolfson AD, Malcolm RK. Characterization of the rheological, mucoadhesive, and drug release properties of highly structured gel platforms for intravaginal drug delivery. Biomacromolecules. 2009;10(9):2427–35.
116. Gurny R, Meyer JM, Peppas NA. Bioadhesive intraoral release systems: design, testing and analysis. Biomaterials. 1984;5(6):336–40.
117. Tiwari D, Goldman D, Sause R, Madan PL. Evaluation of polyoxyethylene homopolymers for buccal bioadhesive drug delivery device formulations. AAPS PharmSci. 1999;1(3):E13.
118. McCarron PA, Woolfson AD, Donnelly RF, Andrews GP, Zawislak A, Price JH. Influence of plasticizer type and storage conditions on properties of poly(methyl vinyl ether-co-maleic anhydride) bioadhesive films. J Appl Polym Sci. 2004;91(3):1576–89.
119. Solomonidou D, Cremer K, Krumme M, Kreuter J. Effect of carbomer concentration and degree of neutralization on the mucoadhesive properties of polymer films. J Biomater Sci Polym Ed. 2001;12(11):1191–205.
120. Dhawan S, Singla AK, Sinha VR. Evaluation of mucoadhesive properties of chitosan microspheres prepared by different methods. AAPS PharmSciTech. 2004;5(4):e67.
121. Peppas NA, Buri PA. Surface, interfacial and molecular aspects of polymer bioadhesion on soft tissues. J Control Release. 1985;2:257–75.
122. Patel MM, Smart JD, Nevell TG, Ewen RJ, Eaton PJ, Tsibouklis J. Mucin/poly(acrylic acid) interactions: a spectroscopic investigation of mucoadhesion. Biomacromolecules. 2003;4(5):1184–90.

123. Nikonenko NA, Bushnak IA, Keddie JL. Spectroscopic ellipsometry of mucin layers on an amphiphilic diblock copolymer surface. Appl Spectrosc. 2009;63(8):889–98.
124. Hu L, Sun Y, Wu Y. Advances in chitosan-based drug delivery vehicles. Nanoscale 2013;5(8):3103–11.
125. Sarmento B, das Neves J, editors. Chitosan-based systems for biopharmaceuticals: delivery, targeting and polymer therapeutics. 1st ed. Wiley; 2012.
126. Andrade F, Antunes F, Nascimento AV, da Silva SB, das Neves J, Ferreira D, Sarmento B. Chitosan formulations as carriers for therapeutic proteins. Curr Drug Discov Tech. 2011;8(3):157–72.
127. Bernkop-Schnurch A. Thiomers: a new generation of mucoadhesive polymers. Adv Drug Deliv Rev. 2005;57(11):1569–82.
128. Laffleur F, Bernkop-Schnurch A. Thiomers: promising platform for macromolecular drug delivery. Future Med Chem. 2012;4(17):2205–16.
129. Guggi D, Marschutz MK, Bernkop-Schnurch A. Matrix tablets based on thiolated poly(acrylic acid): pH-dependent variation in disintegration and mucoadhesion. Int J Pharm. 2004;274 (1–2):97–105.
130. Schmitz T, Grabovac V, Palmberger TF, Hoffer MH, Bernkop-Schnurch A. Synthesis and characterization of a chitosan-N-acetyl cysteine conjugate. Int J Pharm. 2008;347(1–2): 79–85.
131. Nema T, Jain A, Shilpi S, Gulbake A, Hurkat P, Jain SK. Insulin delivery through nasal route using thiolated microspheres. Drug Deliv. 2013;20(5):210–5.
132. Wang J, Tabata Y, Bi D, Morimoto K. Evaluation of gastric mucoadhesive properties of aminated gelatin microspheres. J Control Release. 2001;73(2–3):223–31.
133. Thanou M, Nihot MT, Jansen M, Verhoef JC, Junginger HE. Mono-N-carboxymethyl chitosan (MCC), a polyampholytic chitosan derivative, enhances the intestinal absorption of low molecular weight heparin across intestinal epithelia in vitro and in vivo. J Pharm Sci. 2001;90(1):38–46.
134. Tripathi P, Beaussart A, Alsteens D, Dupres V, Claes I, von Ossowski I, de Vos WM, Palva A, Lebeer S, Vanderleyden J, Dufrene YF. Adhesion and nanomechanics of pili from the probiotic *Lactobacillus rhamnosus* GG. ACS Nano. 2013;7(4):3685–97.
135. Vandamme TF, Brobeck L. Poly(amidoamine) dendrimers as ophthalmic vehicles for ocular delivery of pilocarpine nitrate and tropicamide. J Control Release. 2005;102(1):23–38.
136. Ivanov AE, Nilsson L, Galaev IY, Mattiasson B. Boronate-containing polymers form affinity complexes with mucin and enable tight and reversible occlusion of mucosal lumen by poly(vinyl alcohol) gel. Int J Pharm. 2008;358(1–2):36–43.
137. Perioli L, Ambrogi V, Giovagnoli S, Blasi P, Mancini A, Ricci M, Rossi C. Influence of compression force on the behavior of mucoadhesive buccal tablets. AAPS PharmSciTech. 2008;9(1):274–81.
138. Shemer A, Amichai B, Trau H, Nathansohn N, Mizrahi B, Domb AJ. Efficacy of a mucoadhesive patch compared with an oral solution for treatment of aphthous stomatitis. Drugs R D. 2008;9(1):29–35.
139. Donnelly RF, McCarron PA, Tunney MM, David Woolfson A. Potential of photodynamic therapy in treatment of fungal infections of the mouth. Design and characterisation of a mucoadhesive patch containing toluidine blue O. J Photochem Photobiol B. 2007;86(1): 59–69.
140. Modi P, Mihic M, Lewin A. The evolving role of oral insulin in the treatment of diabetes using a novel RapidMist System. Diabetes Metab Res Rev. 2002;18(1):S38–42.
141. Chaturvedi M, Kumar M, Pathak K. A review on mucoadhesive polymer used in nasal drug delivery system. J Adv Pharm Technol Res. 2011;2(4):215–22.
142. McInnes FJ, O'Mahony B, Lindsay B, Band J, Wilson CG, Hodges LA, Stevens HN. Nasal residence of insulin containing lyophilised nasal insert formulations, using gamma scintigraphy. Eur J Pharm Sci. 2007;31(1):25–31.
143. Coucke D, Schotsaert M, Libert C, Pringels E, Vervaet C, Foreman P, Saelens X, Remon JP. Spray-dried powders of starch and crosslinked poly(acrylic acid) as carriers for nasal delivery of inactivated influenza vaccine. Vaccine. 2009;27(8):1279–86.

144. Sensoy D, Cevher E, Sarici A, Yilmaz M, Ozdamar A, Bergisadi N. Bioadhesive sulfacetamide sodium microspheres: evaluation of their effectiveness in the treatment of bacterial keratitis caused by *Staphylococcus aureus* and *Pseudomonas aeruginosa* in a rabbit model. Eur J Pharm Biopharm. 2009;72(3):487–95.
145. de la Fuente M, Seijo B, Alonso MJ. Bioadhesive hyaluronan-chitosan nanoparticles can transport genes across the ocular mucosa and transfect ocular tissue. Gene Ther. 2008;15(9):668–76.
146. Varum FJ, McConnell EL, Sousa JJ, Veiga F, Basit AW. Mucoadhesion and the gastrointestinal tract. Crit Rev Ther Drug Carrier Syst. 2008;25(3):207–58.
147. Ahmed IS, Ayres JW. Bioavailability of riboflavin from a gastric retention formulation. Int J Pharm. 2007;330(1–2):146–54.
148. Rodes L, Coussa-Charley M, Marinescu D, Paul A, Fakhoury M, Abbasi S, Khan A, Tomaro-Duchesneau C, Prakash S. Design of a novel gut bacterial adhesion model for probiotic applications. Artif Cells Nanomed Biotechnol. 2013;41(2):116–24.
149. Chen S, Cao Y, Ferguson LR, Shu Q, Garg S. Evaluation of mucoadhesive coatings of chitosan and thiolated chitosan for the colonic delivery of microencapsulated probiotic bacteria. J Microencapsul. 2013;30(2):103–15.
150. El-Leithy ES, Shaker DS, Ghorab MK, Abdel-Rashid RS. Evaluation of mucoadhesive hydrogels loaded with diclofenac sodium-chitosan microspheres for rectal administration. AAPS PharmSciTech. 2010;11(4):1695–702.
151. Bassi P, Kaur G. Innovations in bioadhesive vaginal drug delivery system. Expert Opin Ther Pat. 2012;22(9):1019–32.
152. Pereira RR, Ribeiro Godoy JS, Stivalet Svidzinski TI, Bruschi ML. Preparation and characterization of mucoadhesive thermoresponsive systems containing propolis for the treatment of vulvovaginal candidiasis. J Pharm Sci. 2013;102(4):1222–34.
153. Khanvilkar K, Donovan MD, Flanagan DR. Drug transfer through mucus. Adv Drug Deliv Rev. 2001;48(2–3):173–93.
154. Norris DA, Sinko PJ. Effect of size, surface charge, and hydrophobicity on the translocation of polystyrene microspheres through gastrointestinal mucin. J Appl Polym Sci. 1997;63: 1481–92.
155. Norris DA, Puri N, Labib ME, Sinko PJ. Determining the absolute surface hydrophobicity of microparticulates using thin layer wicking. J Control Release. 1999;59(2):173–85.
156. Shen H, Hu Y, Saltzman WM. DNA diffusion in mucus: effect of size, topology of DNAs, and transfection reagents. Biophys J. 2006;91(2):639–44.
157. Ribbeck K. Do viruses use vectors to penetrate mucus barriers? Biosci Hypotheses. 2009;2(6):329–62.
158. Yuan H, Chen CY, Chai GH, Du YZ, Hu FQ. Improved transport and absorption through gastrointestinal tract by PEGylated solid lipid nanoparticles. Mol Pharmaceutics. 2013;10(5):1865–73.
159. Tang BC, Dawson M, Lai SK, Wang YY, Suk JS, Yang M, Zeitlin P, Boyle MP, Fu J, Hanes J. Biodegradable polymer nanoparticles that rapidly penetrate the human mucus barrier. Proc Natl Acad Sci U S A. 2009;106(46):19268–73.
160. Ferrari S, Kitson C, Farley R, Steel R, Marriott C, Parkins DA, Scarpa M, Wainwright B, Evans MJ, Colledge WH, Geddes DM, Alton EW. Mucus altering agents as adjuncts for nonviral gene transfer to airway epithelium. Gene Ther. 2001;8(18):1380–6.
161. Mert O, Lai SK, Ensign L, Yang M, Wang YY, Wood J, Hanes J. A poly(ethylene glycol)-based surfactant for formulation of drug-loaded mucus penetrating particles. J Control Release. 2012;157(3):455–60.
162. Ensign LM, Tang BC, Wang YY, Tse TA, Hoen T, Cone R, Hanes J. Mucus-penetrating nanoparticles for vaginal drug delivery protect against herpes simplex virus. Sci Transl Med. 2012;4(138):138ra179.
163. Suk JS, Boylan NJ, Trehan K, Tang BC, Schneider CS, Lin JM, Boyle MP, Zeitlin PL, Lai SK, Cooper MJ, Hanes J. N-acetylcysteine enhances cystic fibrosis sputum penetration and airway gene transfer by highly compacted DNA nanoparticles. Mol Ther. 2011;19(11):1981–9.

164. Boylan NJ, Suk JS, Lai SK, Jelinek R, Boyle MP, Cooper MJ, Hanes J. Highly compacted DNA nanoparticles with low MW PEG coatings: in vitro, ex vivo and in vivo evaluation. J Control Release. 2012;157(1):72–9.
165. Collnot EM, Ali H, Lehr CM. Nano- and microparticulate drug carriers for targeting of the inflamed intestinal mucosa. J Control Release. 2012;161(2):235–46.
166. Siegmund B. Targeted therapies in inflammatory bowel disease. Dig Dis. 2009;27(4):465–9.
167. Baumgart DC, Carding SR. Inflammatory bowel disease: cause and immunobiology. Lancet. 2007;369(9573):1627–40.
168. Lakatos PL. Recent trends in the epidemiology of inflammatory bowel diseases: up or down? World J Gastroenterol. 2006;12(38):6102–8.
169. Khor B, Gardet A, Xavier RJ. Genetics and pathogenesis of inflammatory bowel disease. Nature. 2011;474(7351):307–17.
170. Zhou SY, Fleisher D, Pao LH, Li C, Winward B, Zimmermann EM. Intestinal metabolism and transport of 5-aminosalicylate. Drug Metab Dispos. 1999;27(4):479–85.
171. Nakase H, Okazaki K, Tabata Y, Uose S, Ohana M, Uchida K, Matsushima Y, Kawanami C, Oshima C, Ikada Y, Chiba T. Development of an oral drug delivery system targeting immune-regulating cells in experimental inflammatory bowel disease: a new therapeutic strategy. J Pharmacol Exp Ther. 2000;292(1):15–21.
172. Mladenovska K, Raicki RS, Janevik EI, Ristoski T, Pavlova MJ, Kavrakovski Z, Dodov MG, Goracinova K. Colon-specific delivery of 5-aminosalicylic acid from chitosan-Ca-alginate microparticles. Int J Pharm. 2007;342(1–2):124–36.
173. Lamprecht A, Yamamoto H, Takeuchi H, Kawashima Y. Design of pH-sensitive microspheres for the colonic delivery of the immunosuppressive drug tacrolimus. Eur J Pharm Biopharm. 2004;58(1):37–43.
174. Lamprecht A, Ubrich N, Yamamoto H, Schafer U, Takeuchi H, Lehr CM, Maincent P, Kawashima. Design of rolipram-loaded nanoparticles: comparison of two preparation methods. J Control Release. 2001;71(3):297–306.
175. Lamprecht A, Yamamoto H, Takeuchi H, Kawashima Y. A pH-sensitive microsphere system for the colon delivery of tacrolimus containing nanoparticles. J Control Release. 2005;104(2): 337–46.
176. Makhlof A, Tozuka Y, Takeuchi H. pH-sensitive nanospheres for colon-specific drug delivery in experimentally induced colitis rat model. Eur J Pharm Biopharm. 2009;72(1):1–8.
177. Pertuit D, Moulari B, Betz T, Nadaradjane A, Neumann D, Ismaili L, Refouvelet B, Pellequer Y, Lamprecht A. 5-amino salicylic acid bound nanoparticles for the therapy of inflammatory bowel disease. J Control Release. 2007;123(3):211–8.
178. Jubeh TT, Nadler-Milbauer M, Barenholz Y, Rubinstein A. Local treatment of experimental colitis in the rat by negatively charged liposomes of catalase, TMN and SOD. J Drug Target. 2006;14(3):155–63.
179. Tirosh B, Khatib N, Barenholz Y, Nissan A, Rubinstein A. Transferrin as a luminal target for negatively charged liposomes in the inflamed colonic mucosa. Mol Pharm. 2009;6(4): 1083–91.
180. Serpe L, Canaparo R, Daperno M, Sostegni R, Martinasso G, Muntoni E, Ippolito L, Vivenza N, Pera A, Eandi M, Gasco MR, Zara GP. Solid lipid nanoparticles as anti-inflammatory drug delivery system in a human inflammatory bowel disease whole-blood model. Eur J Pharm Sci. 2010;39(5):428–36.
181. Bhol KC, Schechter PJ. Effects of nanocrystalline silver (NPI 32101) in a rat model of ulcerative colitis. Dig Dis Sci. 2007;52(10):2732–42.
182. Ensign LM, Cone R, Hanes J. Oral drug delivery with polymeric nanoparticles: the gastrointestinal mucus barriers. Adv Drug Deliv Rev. 2012;64(6):557–70.
183. Owen DA, Reid PE. Histochemical alterations of mucin in normal colon, inflammatory bowel disease and colonic adenocarcinoma. Histochem J. 1995;27(11):882–9.
184. Corfield AP, Carroll D, Myerscough N, Probert CS. Mucins in the gastrointestinal tract in health and disease. Front Biosci. 2001;6:D1321–57.

185. Shirazi T, Longman RJ, Corfield AP, Probert CS. Mucins and inflammatory bowel disease. Postgrad Med J. 2000;76(898):473–8.
186. Stremmel W, Braun A, Hanemann A, Ehehalt R, Autschbach F, Karner M. Delayed release phosphatidylcholine in chronic-active ulcerative colitis: a randomized, double-blinded, dose finding study. J Clin Gastroenterol. 2010;44(5):e101–7.
187. des Rieux A, Fievez V, Garinot M, Schneider Y-J, Préat V. Nanoparticles as potential oral delivery systems of proteins and vaccines: a mechanistic approach. J Control Release. 2006;116:1–27.
188. Wang X, Shah AA, Campbell RB, Wan KT. Glycoprotein mucin molecular brush on cancer cell surface acting as mechanical barrier against drug delivery. Appl Phys Lett. 2010;97:26370.

Chapter 4
Epithelial Permeation and Absorption Mechanisms of Biopharmaceuticals

Hanne Mørck Nielsen

4.1 Introduction

The potential to use biopharmaceuticals to treat a number of life threatening and often chronic diseases becomes increasingly evident and the importance of biopharmaceuticals with great therapeutic potential is continuously increasing due to the last decade's advances in biotechnology [1]. Thus, modifying the properties of the therapeutic aiming at noninjectable administration is possible to a greater extent. Also, the chemical technologies for production of peptides of various sizes, nucleotide-based drugs as well as for modifications of proteins produced by expression are advancing. Despite these advances and the obvious wish for non-injectable formulations, the vast majority of biopharmaceuticals is administered by injection, since the absorption into and across mucosal tissue constitutes a major challenge to be overcome.

Mucosal administration for delivery of sufficient amounts of the biopharmaceutical to the target site requires unique molecular properties making the therapeutic stable enough for a sufficient amount of time as well as providing the capabilities of the drug to penetrate into and permeate through the physiological barriers to reach the target receptor. However, easy and fast delivery is limited by the inherent physicochemical properties of biopharmaceuticals in relation to the physiology of the mucosal tissue, whereto the drug is administered.

Although minute amounts of the therapeutic peptide, protein, or nucleic acid or of the antigen is often sufficient to elicit a biological response, the inherent instability of these types of drugs as well as their large molecular size, high degree of ionization, and poor lipophilic properties constitute a delivery challenge that needs to be addressed. To some extent, chemical modifications may improve the properties of the parent compound with regards to stability and mucosal membrane permeability needed for improved non-injectable delivery or for compatibility with the excipients

H. M. Nielsen (✉)
Department of Pharmacy, University of Copenhagen,
Universitetsparken 2, 2100, Copenhagen, Denmark
e-mail: hanne.morck@sund.ku.dk

J. das Neves, B. Sarmento (eds.), *Mucosal Delivery of Biopharmaceuticals*, DOI 10.1007/978-1-4614-9524-6_4,

included in the delivery system. For the design of a successful drug delivery system, these cannot be separated. A variety of promising drug delivery technologies continuously emerge and are reported in literature; however, a basic understanding of how the mucosal permeation and delivery barriers are mechanistically affected by use of these technologies is crucial both to elucidate and control further pharmaceutical formulation optimization and to minimize the risk of side effects. This chapter in general terms addresses the mechanisms of transport into and through the mucosal barrier with an emphasis on the epithelial transport, providing examples of formulation approaches affecting these. Also, the chapter briefly deals with critical considerations when applying state-of-the-art methodologies for assessment of membrane permeability.

4.1.1 Main Challenges to Successful Non-Injectable Delivery of Biopharmaceuticals

Irrespective of whether the target site for the biopharmaceutical to be delivered is in the mucosal tissue as a surface bound receptor, in the cytoplasm of the epithelial cells, or to be reached via the systemic circulation after absorption through the mucosa, obtaining sufficient bioavailability of biopharmaceuticals at the target site is challenging. Even for insulin, a relatively small ($\approx$5.8 kDa) protein with a major target site in the liver, obtaining a biological response is difficult without pharmaceutical formulation strategies and high dosing.

From administration to delivery to target site, a number of barriers must be overcome (Fig. 4.1). These are often addressed separately in research, but must be combined for obtaining complete and in vivo representative results from research:

1. Upon administration of the dosage form, the first challenge is to ensure sufficient release of, e.g., protein from the administered dosage form at the intended site, which may be in the lumen in a specific region of the gastrointestinal tract, or upon interaction with the mucosa, if the biopharmaceutical is intended to be released prior to cellular uptake and/or epithelial transport.
2. Next, the drug may be compromised due to incompatibility with components in the local environment.
3. When released, the drug specifically prone to degradation by a variety of luminal or brush border enzymes [2].
4. Before reaching the epithelial surface, the drug or drug delivery system needs to diffuse through the aqueous and viscous boundary mucus layer, a physiological barrier to both macromolecular as well as particle diffusion [3].
5. The uniform layer of filamentous glycoproteins, the glycocalyx, which exists as an integral and dynamic part of the epithelial plasma membrane, comprises sialic acid residues that result in a negative charge to the epithelial layer at physiological pH and may limit epithelial uptake and transport. [4].
6. Once the drug has permeated the epithelium, a sufficient plasma half-life and delivery may be hampered by further degradation and/or binding to plasma proteins.

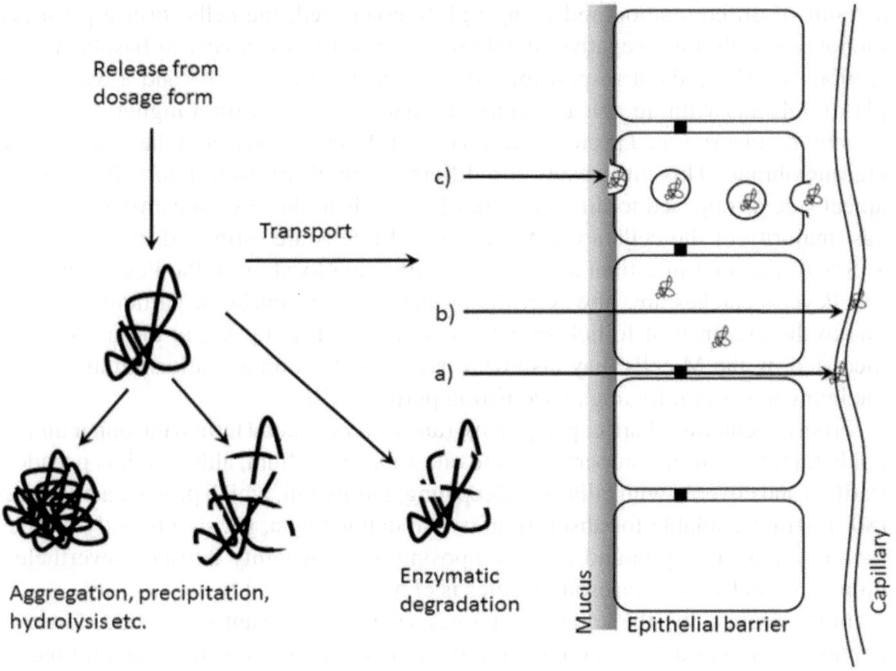

Fig. 4.1 Challenges to successful non-injectable transmucosal delivery of biopharmaceuticals include obtaining sufficient drug release from the dosage form, inactivation by, e.g., aggregation and hydrolysis, degradation by lumen and brush border enzymes, insufficient mucus penetration, and epithelial permeation by **a** passive paracellular or **b** transcellular mechanisms or **c** by transcytosis

However, most of these issues are addressed elsewhere in the book and will not be dealt with further in this chapter, which mainly addresses the transport into and across the epithelium.

4.1.2 *The Epithelial Barrier(s)*

Despite the different morphology of epithelial tissue and regional differences with regard to degree of stratification, differentiation, tightness, level of enzymatic activity, etc., some generalizations can be made with regard to the transport pathways and mechanisms important to address for sufficient (trans)epithelial delivery of a biopharmaceutical drug. For a more detailed description of a specific mucosal barrier, the reader is referred to the respective chapters in this book.

The intestinal monolayer of absorptive enterocytes, microfold (M) cells and mucus-producing goblet, and enteroendocrine cells has been estimated to cover in total an area of about 250 m^2 [5] in an adult and has an epithelial thickness of 20–30 μm comprising tightly bound columnar cells of approximately 10 μm in width. As a result of differentiation and being tightly connected, the cells form a polarized monolayer with a net negative membrane potential in the apical-to-basolateral direction, the H^+gradient responsible for an apical pH of 6.0–6.5 and a basolateral pH of 7.4, and with an apical membrane that is thicker, with a higher content of membrane-incorporated proteins, and with a different surface area than the basolateral membrane. This may result in a different rate of diffusion into cells from the apical side as opposed to diffusion out of the cells to the systemic circulation. The vast majority of the cells are enterocytes as M cells are estimated to cover about 1–5 % of the total intestinal surface area with main location in the Peyer's patches [6]. Peyer's patches are, however, the primary site for uptake of particulate matter due to the presence of follicle-assisted epithelium (FAE) [6], and the transcytosis mediated by the M cells may also be a target site for immunization/stimulation of the immune system, for, e.g., vaccination purposes [7].

From a membrane-barrier perspective (and in very general terms) the upper airway epithelium has similar properties as the intestinal epithelium, although it is pseudostratified and covered with cilia. The deep lung, the alveoli, which provide a huge area (80–140 m^2) available for absorption, is also defined by a, yet significantly thinner, monolayer of tightly bound cells composing a permeability barrier. Nevertheless pulmonary delivery of macromolecules is clinically relevant to consider [8].

In the oral cavity, the keratinized areas seem less relevant to consider for, e.g., peptide or protein delivery, compared to the non-keratinized sublingual and buccal tissue, which is similar to the vaginal epithelium in being stratified and with less absorptive pathways present resulting in that passive para- or transcellular diffusion are the main mechanisms by which drugs are transported across this mucosa. A general lower level of proteinases than in the intestinal tract mucosa may prolong the time that the drug is available for absorption, but the stratification of 10–20 cell layers in sublingual mucosa, and 40–50 cell layers in buccal tissue [9], and around 25–30 cell layers in vaginal mucosa [10] along with the fact that the cells may have less tight junctions (TJs) but a higher level of lipid intercellular matrix results in a relatively slow absorption rate across these mucosae.

4.2 Epithelial Transport Mechanisms

Transport into and across the mucosal epithelium may, once the diffusion through the mucus layer and the aqueous boundary layer on the surface of the epithelium has been overcome, occur by passive diffusion or active processes or a combination of the two [11]. Transport via the paracellular and transcellular route may thus be complementary. Passive transport mechanisms possess the characteristics of (1) being non-saturable, (2) not being subject to inhibition as no carriers are involved, (3) being less cell-type specific than carrier-medicated transport, and (4) being less dependent on molecular structure than active and receptor-mediated transport. On the other hand, active transport mechanisms are (1) saturable, (2) subject to inhibition, (3) cell-type specific as expression of carriers are a prerequisite for transport of substrates, and (4) dependent on the molecular structure of the transported compound.

4.2.1 Passive Transport Mechanisms—Diffusion

Drugs diffusing passively through the intercellular or intracellular routes with a rate driven mainly by the concentration gradient across the mucosa can be described by a simplified version of Fick's first law of diffusion describing the movement of molecules through a given cross-sectional (barrier) area during a given period of time, i.e., the flux J [amount/area × time]. The velocity of this diffusion is related to the diffusion coefficient, D [length × time] for diffusion of a molecule down a concentration gradient (ΔC) [amount/volume] in one plane over time (Eq. 4.1):

$$J = -D \times \Delta C \tag{4.1}$$

assuming that ΔC is constant and linear, i.e., time independent, this can be further simplified into Eq. 4.2, describing the correlation between steady state flux (J_{ss}) and the permeability coefficient (P) [length/time]:

$$P = \frac{J_{ss}}{\Delta C} \tag{4.2}$$

When determining the permeability across an epithelial barrier in vitro, the assumption is most often made that the diffusion throughout the biological barrier is constant, and Eq. 4.3 describes the permeability, calculated from the steady state rate of permeation (dQ/dt) [amount/time] across a given known epithelial area at sink condition, i.e., assumed constant ΔC. This equation is often applied for calculation of permeability coefficients from in vitro studies using cell culture models or excised mucosa as use of the permeability coefficient is advantageous in that it allows for comparison between different experimental setups since it adjusts for the applied concentration and barrier area.

$$P = \frac{dQ}{dt} \times \frac{1}{A} \times \frac{1}{\Delta C} \tag{4.3}$$

Importantly, measuring and calculating a permeability coefficient from in vitro experiments, using, e.g., mucosa or epithelium cultured on permeable supports, the assumption that the barrier is *one* barrier, is obviously erroneous, and it should be kept in mind that the overall P value can also be described by Eq. 4.4, illustrating that the permeability through the unstirred water layers (P_{uwl}) on the apical and basolateral sides resembling the lumen and the serosal side of the epithelium as well as through the surface adjacent glycocalyx and mucus layer (P_{mucus}) and the supporting filter (P_{filter}) are also important. The $P_{epithelium}$ may as well be divided into the paracellular permeability and the transcellular permeability. The contribution of each of the barriers depends on the properties of both the drug compound in relation to the properties of the barrier and should be considered for each experimental setup. For biopharmaceuticals, the most prominent barrier is very likely the epithelium, but for diffusion of drug delivery systems carrying the drug, the mucus layer and filter support may also constitute a barrier.

$$\frac{1}{P} = \frac{1}{P_{uwl}} + \frac{1}{P_{mucus}} + \frac{1}{P_{epithelium}} + \frac{1}{P_{filter}} \tag{4.4}$$

Overall, considerations to ensure that the assumptions for applying these simplified equations should be made, such as the presence of purely passive diffusion, no change in mucosa physiological properties (e.g., integrity), a perfect sink and thus constant ΔC, and no gradients of other compounds across the mucosa (such as electrical or hydrodynamic gradients) that may affect the diffusion of the compound of interest. In an in vitro experimental setup, these parameters are relatively easy to control and monitor, whereas in the in vivo situation, an approximation of the mucosal permeability is difficult resulting in use of and comparison to other readouts, e.g., plasma concentration and physiological response is necessary.

4.2.2 Active Transport Mechanisms

Transport kinetics for drug compounds that do not permeate an epithelium (only) by passive diffusion via the paracellular and transcellular routes in nature involve transcellular transport with binding and uptake via a carrier or receptor, trafficking through the epithelial cell and excretion at the basolateral side of the epithelium. Transport will thus depend on the capability of a substrate to occupy a binding site as well as the capability to accesses the opposite site of the barrier. As the kinetics are saturable, the constants of binding, cellular uptake, and saturation are used to describe the transport process, and data are often modeled using the Michaelis–Menten equation (Eq. 4.5):

$$J = \frac{J_{max} \times C}{K_m + C} \tag{4.5}$$

where the flux J is dependent on the maximum rate achieved at maximum, i.e., saturated substrate concentration, and K_m represents the concentration at the half V_{max} and may be expressed as the binding constant. The net flux is often a combination of active and passive flux, with the passive transport dominating at high concentrations C where the transport pathways are likely saturated (Eq. 4.6):

$$J = \frac{J_{max} \times C}{K_m + C} + P \times C \tag{4.6}$$

4.3 Epithelial Transport Pathways

4.3.1 Paracellular Transport

Passive intercellular diffusion is controlled by the limited paracellular area available for passage, and even though large variations among epithelia exist in terms of available area and tightness, this pathway is believed to be mainly relevant for smaller hydrophilic compounds. For example, the intercellular area makes up around 0.01 %

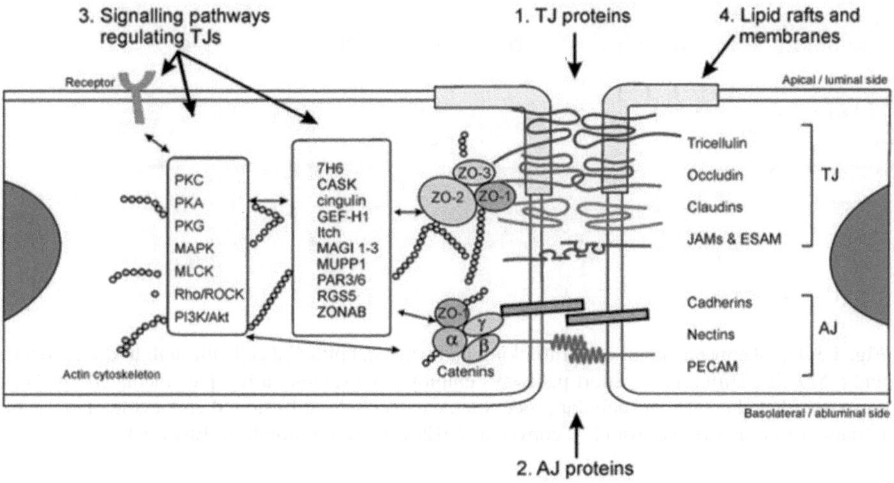

Fig. 4.2 Overview of the regulation of tight junction and adherent junctions interconnecting epithelial cells. (Reprinted from [23], copyright 2008, with permission from Elsevier)

of the intestinal epithelium, with a "pore" size of below 10 Å throughout the intestine of rats [12] and probably as low as 4 Å in human intestine [13]. In addition, paracellular diffusion of molecules with a radius exceeding 15 Å (corresponding to ≈3.5 kDa) is thus not considered relevant for therapeutic drug delivery [14], and even peptide or polypeptide drugs with a lower molecular weight are likely to be poorly transported via the paracellular route. Not only the molecular size, but also the net ionization state of the drugs [15] as well as competing diffusion being sensitive to local pH changes [16] play a role for paracellular permeability. Thus, it was shown that the overall negative charge of insulin (pI 5.3) at physiological pH limited the paracellular diffusion as opposed to diffusion of the (smaller) TRH (pI 6.3) under the same conditions [15]. This observation may augment targeting regional sites in the intestine as this show pH differences from approx. pH 5.5–8 or to locally modulate pH at the absorption site to increase the absorption. Another reason that an effect of pH on epithelial transport may be expected is the solubility (and thus likeliness of precipitation) of peptides and proteins at different pH values. Thus, changing the pI of the specific biopharmaceutical could be an approach. On the contrary, however, the conformational flexibility of peptide and protein drugs may potentiate transport of larger molecules than would be expected from their molecular weight [17] keeping in mind that the pharmacologically active conformation must be preserved or retained once the barrier has been transversed.

The junction proteins interconnecting the epithelial cells (Fig. 4.2) are characterized as either TJ, gap junction, or adherent junction (AJ) proteins or desmosomes with the TJ localized toward the apical side of the epithelium and the slightly looser gap junctions and AJ or desmosomes more toward the basolateral membrane, some of them localized in raft-like domains in the membrane [18]. The TJs consist of secreted extracellular proteins that tether the interface between the cells, the membrane-integrated proteins, and the cytoplasmic scaffolds and proteins associated

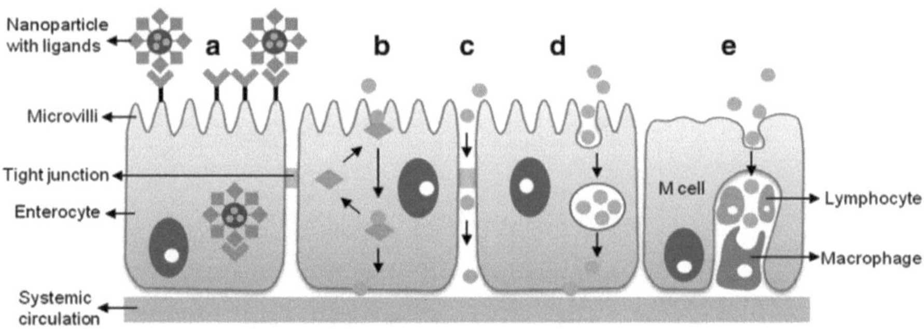

Fig. 4.3 Uptake mechanisms and trafficking pathways in epithelial cells through **a–d** enterocytes and **e** M-cells. Different transport pathways employ, for example, **a** receptor-mediated uptake, **b** carrier-mediated uptake and cellular processing, **c** paracellular diffusion, **d** endocytotic uptake and **e** transcytosis. (Reprinted from [7], copyright 2012, with permission from Elsevier)

with these [18]. Over the past decades, many TJ components have been identified, and recent progress has provided new insights into the proteins and interactions that dynamically regulate structure and function of the junctions [19, 20]. Some of the junction protein families that have been identified are the tricellulin, occludin, claudins, and cadherins, and these may directly or indirectly constitute a target for selective and reversible regulation of the paracellular pathway [21–24]. Overall, the junctional complexes are highly dynamic and undergo rapid remodeling during normal epithelial morphogenesis and under pathologic conditions. This remodeling of the protein complexes is responsible for maintaining functional junctions and is believed to be mediated by endocytotic internalization of junctional proteins [25].

4.3.2 Transcellular Transport

The drug or drug delivery system taken up by the absorptive cells by passive diffusion or carrier- or receptor-mediated transport will be exposed not only to the extracellular lumen and brush border enzyme activity, but also to the intracellular (more harsh) environment of the cytosol and lysosomes. Thus, for successful transepithelial delivery or delivery to a specific intracellular target, the biopharmaceutical must be physically shielded against the majority of these enzymes, and/or at least be directed toward a trafficking pathway circumventing, e.g., the lysosomes. This may be achieved by nanotechnology ensuring encapsulation of the drug and sufficient endosomal escape. As stated later, some chemical permeation enhancers interfere with the plasma membrane components resulting in increased transepithelial drug transport.

The uptake mechanisms and trafficking pathways in the epithelium, constituting a heterogonous population of cells with different properties, are numerous depending on the properties of the drug or drug delivery system applied (Fig. 4.3). In the following sections, the different mechanisms responsible for uptake into the epithelial cells will be addressed. This includes passive translocation across the plasma membrane as a result of diffusion, carrier-mediated uptake, endocytosis by receptor-mediated and nonreceptor-mediated uptake.

4.3.2.1 Passive Diffusion

Passive distribution of a biopharmaceutical across the plasma membrane into an epithelial cell requires that the properties of the dissolved active compound (or prodrug hereof) favor uptake and subsequent trafficking through the cell to the intracellular target or through the basolateral membrane. It is speculated to be a multistep process initiated by the partitioning of the solute into the membrane followed by either lateral diffusion either in or along the plasma membrane or via diffusion in the cell cytoplasm before reaching the basal side of the epithelium [26]. The exact pathway may thus be related to the properties of the drug, including molecular weight, hydrodynamic size, net charge, and thus hydrogen-bonding potential, as well as degree of lipophilicity.

The latter is illustrated by the fact that the cyclic structure of the undecapeptide cyclosporin A, a powerful immunosuppressant, does not only stabilize the peptide, but provides a higher partitioning into plasma membranes, as indicated by its logP value around 3. A direct correlation between partitioning and permeability does, however, not seem to be as predictive as the effect of the hydrogen-bonding potential as reported for short peptides [27]. Chemical modifications to stabilize therapeutic peptides and small proteins, such as the vasopressin analogue desmopressin and the insulin analogue HIM2, and at the same time to induce a higher degree of lipophilicity have been shown to increase the epithelial permeability of the peptide as compared to the native compound (reviewed in [27]).

4.3.2.2 Carrier-Mediated Epithelial Uptake and Transport

Absorptive transporters of nutrients in epithelial cells are mainly localized in the apical membrane of the epithelium (mainly the intestine and lung) comprising di-/tri-peptide, amino acid, bile salt, vitamin, mineral, hexose, and nucleoside transporters [28]. While it is recognized that they play a role for the epithelial absorption of low molecular drugs with molecular similarity to, e.g., di-and tri peptides [29], or β-lactams [30], their role in the cellular uptake and transepithelial delivery of biopharmaceuticals is anticipated to be limited.

4.3.2.3 Receptor-Mediated Epithelial Uptake and Transport

At the surface of epithelial cells, a number of receptors are expressed. Some of these are potentially relevant for (trans)cellular delivery of biopharmaceuticals and nanoencapsulated drugs. An example is the targeting of the transferrin receptors

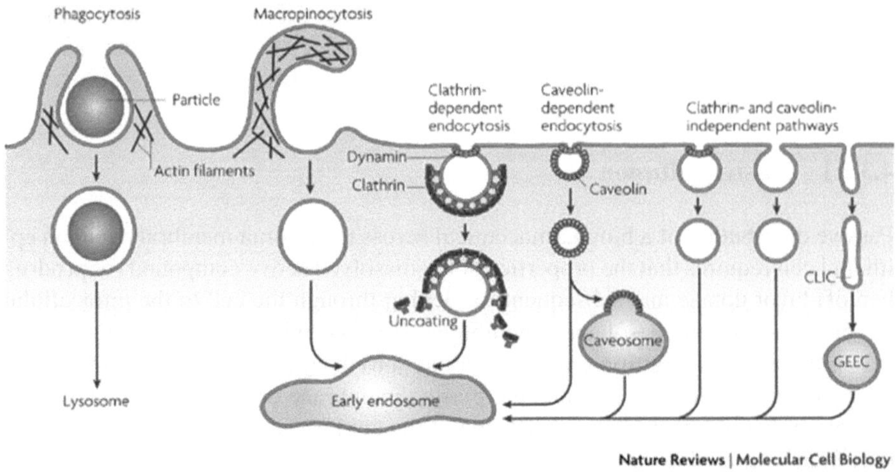

Fig. 4.4 Endocytotic uptake mechanisms and initial sorting in epithelial cells via either particle-dependent phagocytosis, or pinocytosis, which covers the clathrin- and calvolin-independent pathways (employed for complexes/particles up to ≈90 nm), the caveolin-mediated endocytosis (up to ≈60 nm), clathrin-dependent endocytosis (up to ≈120 nm), and macropinocytosis (< 1 μm). (Reprinted from [93] by permission from Macmillan Publishers Ltd, copyright 2007)

for increased alveolar [31] and oral [32] delivery of peptide and protein drugs. An example of another approach to target the transcellular pathway for delivery of biopharmaceuticals comprises the use of the receptor for, e.g., vitamin B12 present in the ileal enterocytes and thus covalent conjugation of vitamin B12 to peptides and proteins [33].

4.3.3 Active Transcellular Transport

4.3.3.1 Endocytosis and Exocytosis

The term endocytosis and exocytosis covers the transport of macromolecules by generally energy-dependent uptake into and excretion out of cells, respectively. The endocytotic pathways differ with the size of the endocytotic vesicle, the nature of the cargo, and the mechanisms of vesicle formation (Fig. 4.4) [34–35]. The pathways of endocytosis all share the initial step of interacting with the plasma membrane followed by a local inward curvature of the lipid bilayer membrane. To generate this membrane curvature, proteins or lipids enforce transversal asymmetry of the plasma membrane, which is thus believed to be a prerequisite for endocytosis to occur [36]. The process may be induced by binding of a ligand to a specific cell-surface receptor, or receptor clustering followed by internalization through vesicles into endosomal acidic compartments. The vesicle formation has been discussed to be dependent on the actin structural resistance to the plasma membrane bending [37]. The subsequent vesicle sorting into the cellular trafficking pathway is strongly dependent on the mechanism of endosomal uptake as well as the involved cargo and potential receptor [7].

Uptake can occur via both clathrin-coated pits and by clathrin-independent endocytosis. Clathrin is a ubiquitous cytosolic protein that can be recruited to membranes, and has the capability of forming different types of membrane lattices, e.g., the basket-like lattice that favors membrane deformation when the vesicles bud off into the intracellular environment [35]. Some toxins, such as diphtheria ($\approx$60 kDa) and anthrax toxin ($\approx$83 kDa) [38], exclusively use the clathrin-dependent pathway. This is achieved by binding to specific transmembrane domain protein receptors; —heparin-binding EGF-like growth factor precursor and anthrax toxin receptor, respectively. Further, the role of actin has been found important in clathrin-mediated endocytosis [39].

Clathrin-independent endocytosis can be subjected to regulation by a number of signalling pathways, and different cellular molecules involved can be, but not necessarily, dependent on lipid rafts, which are domains rich in cholesterol and sphingolipids [35] and likewise with regard to the involvement of, e.g., dynamin [40]. Different types of SNARE proteins (Soluble NSF (N-ethylmaleimide-sensitive factor) Attachment protein Receptors) are involved in the uptake and excretion independently of whether the uptake is lipid raft dependent or not [35]. So far, clathrin-independent apical endocytosis of the toxin ricin ($\approx$60–65 kDa) and the fluid-phase marker horseradish peroxidase (HRP) ($\approx$45 kDa) is the only endocytic process that seems to occur without guanosine triphosphate (GTP) hydrolysis. [41].

Caveolae have been identified in a large variety of tissue and cells; however, they are mainly abundant in endothelial cells, adipocytes, and type I pneumocytes in the alveoli. Thus, caveolae-mediated cellular uptake may be of less relevance with regard to transport across other types of mucosae. Caveolae domains high in cholesterol and sphingomyelin are reported to greatly increase the surface area of the cells supporting the hypothesis that caveolae are involved in macromolecular cellular transport. Involved in caveolin-mediated uptake, reported also to result in transcytosis, is the caveolin-1 scaffolding domain, which has since been shown to serve a dual role, acting both as an anchor holding various proteins within caveolae as well as a regulatory element capable of either inhibiting or enhancing a given protein's signaling activity [42].

Macropinocytosis is a bulk internalization process involving the formation of large F-actin coated vacuoles and macropinosomes, which are often larger than 1 μm. Macropinocytosis is a result of actin-driven formation of membrane protrusions, which collapses with the plasma membrane generating large endocytotic vesicles, and although this process is accompanied by seemingly chaotic membrane ruffling, it is a highly controlled mechanism [34]. Material internalized via this mechanism is believed to sort into the same pathway as the endosomes, and there is no consensus regarding the overall dynamics and fate of drug cargoes internalized by this mechanism [43].

4.3.3.2 Transcytosis

Transcytosis is a pathway by which both membrane-bound ligands and compounds in solution can be transported across epithelial cell layers [40]. The process is initiated by association of compounds, such as proteins, to the plasma membrane followed by endocytosis, intracellular trafficking, and excretion. As described above, the term endocytosis encompasses several diverse mechanisms by which cells internalize macromolecules and particles into transport vesicles budding off from the plasma membrane [34]. This vesicular transport of macromolecules from one side of a cell to the other thus constitutes a strategy to selectively move a variety of cargo material between two environments without altering the unique compositions of those environments [44].

Importantly, both apical as well as basolateral membrane protein recycling pathways are present in polarized cells, enabling delivery of a compound from the apical to the basolateral side and vice versa [35]. As the process is initiated by binding/association of a molecule to either the apical or the basolateral membrane protein transporters followed by intracellular trafficking via subsets of endosomal compartments depending on specific sorting signals [35, 40], the molecular mechanisms behind the uptake must be kept in mind in order to target or inhibit uptake by these processes.

4.3.3.3 Specialized-Cell Uptake and Transport: M Cells

Especially in FAE, the M cells are highly represented and may present a potential target for the delivery of biopharmaceuticals by particulate matter due to their high transcytosis capacity [7].

4.4 Enhancing Transepithelial Transport of Biopharmaceuticals

As more and more therapeutics with high molecular weight, high hydrogen-binding potential, and poor partitioning into the lipid plasma membranes emerge, an increasing interest in finding new safe excipients to enhance their absorption via paracellular and transcellular absorption is continuously supported. For many of these drugs, the pharmacological features justify mucosal delivery [45]; however, in order to be successful, the need for a comprehensive understanding of the structure and function of the mucosal barrier properties and how these are affected by the pharmaceutical formulations is recognized. And to obtain this, well-characterized, robust, and in vivo representative models are a requirement as well as the detailed characterization of the drug delivery system investigated.

Increasing the mucosal delivery by approaches such as stabilizing the drug by chemical modifications, co-addition of enzyme inhibitor or an encapsulating drug delivery system as well as increasing the concentration and/or retention at the site of absorption by use of formulation design using, e.g., mucoadhesive agents [46] will not be addressed directly in this chapter, but should not be neglected as a means of permeability enhancement. Recognizing that most of the investigated formulation approaches works by several different mechanisms, as illustrated in a recent review on lipid-based formulation approaches for oral delivery of peptides and proteins [47] and that transport across epithelia often occurs by multiple and parallel processes, the examples that are mentioned below are selected as they act by affecting the epithelial membrane permeability.

4.4.1 Paracellular Permeability Enhancement

In order to explain the mechanisms for novel as well as traditional chemical enhancers affecting the paracellular space of the epithelium, a comprehensive understanding of the function and structure of the barrier properties' junctions is recognized [21]. In order to maintain the functionality of the junctions, an either direct or indirect interference with the junctions resulting in loosening of the junctional protein complexes must be specific and transient. Insight into the mechanisms and rate of remodeling of the junctions will help elucidate the potential risk regarding this approach [25]. Approaches to increase epithelial permeability of biopharmaceuticals via the paracellular route thus includes some interaction with the junction proteins. This may relate to effects on the actin cytoskeleton [13].

Affecting the tightness of the junction proteins may be achieved by targeting different areas of the junctions, such as the signaling pathways as recently reviewed by Deli [23]. For this, there is an increasing interest in the use of peptide-based excipients (reviewed in [23, 48]) of which the most prominent examples are derived from bacterial enterotoxins, and they have shown great potential for enhancing the delivery of macromolecules as they all act in a dose-dependent, reversible, and nontoxic way [23] though differing in their mode of action. These peptides include toxin A and B derived from *Clostridium difficile* [49–50], the C-terminal part of *Clostridium perfringens* enterotoxin (C-CPE) [51] and AT1002 derived from the zonula occludens toxin (zot) secreted by *Vibrio cholerae* [52]. An example is the AT1002 hexamer amino acid fragment targeting the zonula occludens (ZO) kinases, which transiently opens intercellular TJs after binding to the zonulin receptor, resulting in increased epithelial transport of various molecular weight markers or low-bioavailability agents [23]. Also, interfering with the myosin light chain kinase (MLCK) is likely a part of the mechanism for permeation enhancers like bile acids and lysolipids [53–54]. Table 4.1 provides a few examples of some of the main classes of chemical enhancers and their suggested mechanism of action resulting in increased paracellular transport. Traditional chemical enhancers that have been employed for permeability enhancement of biopharmaceuticals include divalent ion chelators such as EDTA, EGTA as well as citric acid, fatty acids, bile salts, and polymers like chitosan (Table 4.1) [55].

Table 4.1 Mechanisms of enhancement induced by excipients applied to enhance transepithelial delivery of biopharmaceuticals. A few selected examples are presented

Type of excipient	Excipient(s)	Proposed mechanism	References
Peptide, polypeptide	Zonula occludens toxin, AT1002, C-CPE	Binding to tight junction proteins	[21, 23]
	Penetratin		[71]
Small compound	EDTA, EGTA	Depletion of divalent ion results inhibits tight junction redistribution and results in loss of epithelial integrity	[21, 23, 88]
	Citric acid	Enzyme inhibitor	[89]
Polymer	Chitosan		[21, 23, 66, 90]
	EGTA/DTPA-linked γ-polyglutamic acid	Depletion of Ca^2+ ions inhibits E-cadherin and other tight junction protein distribution and opens the tight junction	[91]
Lipid	Caprate	Tricellulin removal from tight junctions results in opening of tight junctions	[92]

EDTA Ethylenediaminetetraacetic acid, *EGTA* Ethylene glycol bis(2-aminoethyl ether)tetraacetic acid, *DPTA* Diethylenetriamine pentaacetic acid, *TEER* Transepithelial electrical resistance

Also, interfering with the plasma membrane, i.e., the lipid rafts associated with the membrane spanning junction proteins may result in increased paracellular permeability. Permeability enhancers, like β-cyclodextrin, are believed to bind and extract components like cholesterol from the membrane and by this change membrane fluidity and inhibit caveolae-mediated endocytosis [56]. This eventually influences the TJ structure and function likely correlated to a disturbance of the actin cytoskeleton.

4.4.2 Transcellular Permeability Enhancement

Irrespective of the specific mechanism involved in the translocation into the cell, the major steps are: (1) the interference with and partitioning into the plasma membrane, (2) the translocation into the cell interior, (3) trafficking through the cell, and (4) release at the basolateral side of the epithelium. Whether the capacity of the transcellular pathway is high enough to result in a biological response depends on the potency of the drug and potential enhancement ratio, among others.

4.4.2.1 Altering the Properties of the Drug Molecule

Approaches to increase the fraction of transcellular passage of biopharmaceuticals may thus rely on chemical conjugation of the drug increasing the lipophilic properties of the drug and making it less prone to degradation in the cytosol. As an example, modification of insulin with a single hexyl chain has been reported to result in not only higher resistance to degradation, but also facilitates improved transepithelial transport of the HIM2 insulin analogue [57]. The epithelial transport enhancement is likely the result of the altered physiochemical properties of the analogue as compared to native insulin, and improved interaction with the plasma membrane, which might lead to increased transcellular transport by passive diffusion or endocytotic processes. The effect may, however, also be a result of the increased stability and thus the existence of a higher gradient across the epithelial barrier.

4.4.2.2 Use of Lipid-Based Excipients

Coadministration of the biopharmaceutical with lipid surfactants, or incorporation hereof in a drug delivery system, also likely results in a direct effect on the plasma membrane bilayer integrity and fluidity, and thus potentially improved transepithelial delivery. Excipients with surface active properties, like the fatty acid salt, caprate [58], lysolipids [59], bile salts [60], and oils have been explored as enhancers [46, 61]. However, as previously indicated, a direct effect on the plasma membrane is likely to result in a secondary effect on the junctions interconnecting the epithelial cells. Another approach to mediate improved partitioning of a protein into a lipophilic phase is pursued by the use of precomplexing biopharmaceuticals with lipids [62–63]. Also, the mechanism of action of bile salts have been proposed to be due to their interaction with lipid-rich and protein-filled areas in the plasma membrane through formation of mixed micelles or a result of interaction with the polar head groups of the transmembrane proteins [61], both leading to increased membrane fluidity. Bile salts have also been suggested to stabilize proteins against enzyme-mediated degradation [60].

As mentioned above, excipients with surfactant-like properties interfere with the cell plasma membrane, and thus may merit permeability enhancement by transcellular diffusion of the drug. At a level where enhancement of a biopharmaceutical is significant, care should be taken that enhancement is not closely associated with cell damage. At lower concentrations, the net increase in transepithelial delivery is likely a result of increased paracellular flux due to affecting the intercellular junctions via interfering with the actins cytoskeleton [13].

4.4.2.3 Use of Polymer-Based Excipients

One type of polymers that have been employed to a large extent for transmucosal delivery of biopharmaceuticals are chitosan and their derivatives, most often coadministered or in particulate systems. The permeation enhancer effect of these cationic polymers has been reported to depend on whether the polymers are free in solution or in a particle structure [64–65]. In solution, the polymer may interact with TJ proteins [66], whereas when complexed to form a particle, the transcellular transport seems to be involved as the permeability pathway and transcytosis as the enhancement mechanism [64, 65, 67]. Similarly, endocytosis of a model protein was observed in vitro when formulated in particles using chondroitin as the polymer excipient, and the surface charge of the nanoparticles were reported to influence the cell uptake in the Caco-2 cell monolayer model [68]. Other excipients of the polymer-type include peptides and polypeptides, e.g., cell-penetrating peptides [69]. These have been reported to increase the mucosal absorption of peptide and protein drugs across, e.g., nasal and oral mucosa likely partly through a cellular internalization pathway [70–71]. Cell-penetrating peptides may internalize into epithelial cells, but are not likely to permeate the epithelium to a high extent [72], thus the mechanism of action is presumably partly by providing improved interaction between the therapeutic biopharmaceutical with the epithelial membrane.

4.4.3 Permeability Enhancement by Use of Particle Structures

As indicated above, particulate drug delivery systems have also been shown to increase the transmucosal delivery of biopharmaceuticals. The physical encapsulation of a drug in a nanoparticle first and foremost shields against degradation of the biopharmaceutical, whereas employment of targeting ligands may ensure optimal contact between the nanoparticle carrier and the biological surface leading to cellular uptake and processing. Nanotechnology has provided a "glimpse of hope" for mucosal delivery of biopharmaceuticals [7] and the design of nanoparticulate systems with well-defined characteristics (size, shape, surface coating, or grafting) may be the way forward for intracellular as well as transepithelial delivery of biopharmaceuticals.

Polymeric self-assembling micelles and complexes are typically of a size 20–50 nm and are taken up into cells via endocytosis, whereas larger nanoparticles or targeted nanoparticles binding to a specific receptor are endocytosed via mainly clathrin-dependent endocytosis as the nanoparticles are often larger than 100–200 nm. For larger structures or drug delivery systems, macropinocytosis may be a pathway for cellular uptake and processing. As demonstrated by the few examples given in the following, the efficacy of the particulate drug delivery systems may be a result of size, shape, grafting, and presence of targeting ligands, which may improve (1) stability of the biopharmaceutical, (2) mucus diffusion, (3) cellular uptake, and (4) translocation that altogether result in enhanced delivery.

4.4.3.1 Micelles and Complexes

As examples of successful micellar drug delivery systems, it was recently reported that by use of polyion complex micelles of a size around 70 nm composed of polyethylene glycol (PEG)-alginic acid polymers, the oral delivery of salmon calcitonin was increased. Elucidation of the mechanism in the in vitro Caco-2 cell culture model suggested that the employed mechanism was by transcytosis across the epithelium, as the paracellular integrity was not affected [73]. Also, exploiting the receptor-mediated uptake of micelles of approximately 35 nm was recently shown to be successful, as PEG-polycaprolactone (PCL) diblock polymers resulted in increased uptake and transport into and across intestinal epithelium (Caco-2 and rat jejunum) when the 7-mer targeting peptide was conjugated on the polymer. The uptake was concluded to follow clathrin-dependent uptake as well as unspecific endocytosis [74].

4.4.3.2 Particle Size and Shape

For particles of larger size than micellar structures, the size is obviously quite important, as was reported by He et al. [75], who investigated the absorption of polymeric nanoparticles of sizes between 300 nm to 1,000 nm in both in cell culture models, in ex vivo tissue models and in in vivo models. The improved delivery of the model payload formulated in the smaller particles was concluded to be due to both improved diffusion in the mucus as well as across the epithelium [75]. Recently, the transmucosal delivery of insulin was shown to depend on the size of the lipid-bile salt liposome vesicles indicating the importance of the delivery via the transepithelial pathway [76].

In addition to the size, the shape of the particles have also drawn attention, and it has been reported that rod-shaped antibody-coated particles were taken into cells specifically and nonspecifically to a higher degree as compared to spherical and disc-shaped nanoparticles presumably due to differences in binding to the surface of the cells [77].

4.4.3.3 Surface Coating

Grafting the particle surface with receptor-specific or unspecific ligands via direct conjugation or coating by electrostatic forces is a strategy that has been pursued by a variety of researchers in order to increase diffusion to the absorption site and/or adhesion to the cellular surface leading to endocytosis as recently reviewed [7]. Ligands such as folic acid, albumin, and cholesterol have been shown to facilitate uptake through caveolin-mediated endocytosis, whereas ligands for glycol receptors promote clathrin-mediated endocytosis [78]. Further, ligands targeting specific cells may be employed for mucosal drug delivery as, for example, the use of goblet cell-specific ligands on nanoparticles, which have been shown to improve the oral absorption of insulin in vitro, ex vivo, and in vivo. The mechanism by which the 9-mer peptide increased the delivery could be related to both improved mucus diffusion as well as increased transepithelial transport [79]. By use of a claudin-4 targeting peptide from *C. perfringens* toxin, the uptake of poly(lactic-co-glycolic acid) (PLGA)-based particles into M cell in the nose and stomach of mice was significantly enhanced [80]. M cell-homing peptides immobilized onto chitosan nanoparticles introduced superiority in transcytosis capacity depending on the peptide sequence as demonstrated in vitro in cell culture models and excised tissue [81]. Likewise, WGA-coated nanoparticles were shown to possess increased binding to cellular surfaces and enterocytes [78]. Targeting M cells by, e.g., lectin is a strategy often employed for oral vaccination purposes in the upper airways or via the intestine [82].

Importantly, it should be recognized that although functionalized nanoparticles are designed to show improved receptor binding or unspecific cell affinity, the targeting capabilities may be altered upon in vivo administration due to shielding by coronas formed by biomolecules adsorbed to the surface of the particle [83–84].

4.5 Critical Remarks on the Methodologies and Analysis of Epithelial Permeation and Absorption Mechanisms

4.5.1 Methodologies

In order to determine permeability rate and mechanisms of epithelial binding, uptake and transport, in vitro models in the form of excised tissue or cell culture models representative of the targeted epithelium are often employed. Numerous advantages of using in vitro cell culture models include the possibility for detailed monitoring of experimental conditions that can be altered during the experiment for specific purposes, while using a relatively simple but viable epithelium. However, it is important to keep the origin and the properties of the investigated epithelium in mind, such as the integrity, the degree of differentiation and thus the degree of, e.g., receptor expression. For example, the type and activity of extracellular and intracellular proteases, which are both specific with regards to species and investigated absorption site will be very dependent on the culturing conditions, the age of the cells etc. The growth conditions of a cell layer can also clearly influence the epithelial integrity (often measured as the transepithelial electrical resistance (TEER) value) as well as the uptake and transcytosis of a ligand taken up, e.g., by clathrin-independent endocytosis and the transfer of solute across epithelial cells. Knowing the extent to which an epithelial cell surface is polarized at a molecular level is crucial to provide a meaningful interpretation of the results of transcytosis studies. For example, expressions of the caveolin-1 protein and polarized formation of caveolae have been reported to be present only at the basolateral side of Caco-2 and MDCK (Madin-Darby canine kidney) cells [85].

A decrease in TEER and increased influx of small marker molecules are often indicative of an effect on the paracellular pathway, along with microscopic investigation of disassembly of junction proteins. For elucidating the transcellular transport mechanism for drugs or drug delivery systems via active transport mechanism, an interesting recent approach is the knockdown of the uptake using small interfering RNA (siRNA) targeted toward the endocytosis receptor [86]. More standardized,

Table 4.2 Inhibitors used for studying transcellular uptake and transport pathways. (From [44, 56])

Inhibitor	Function
Chlorpromazine	Inhibitor of clathrin-mediated endocytosis
Hypertonic sucrose	Inhibits formation of clathrin vesicles as it interferes with the clathrin lattices in the plasma membrane. Blocks occludin internalization in epithelial cells, which is important for caveolae-mediated transport of tight junction proteins and subsequent transient TEER decrease
Filipin	Inhibitor of caveolae-mediated endocytosis due to binding of cholesterol and caveolin disorganization
Methyl-β-cyclodextrin	Depletion of cholesterol from the plasma membrane alters the structure of cholestrol-rich domains in the membrane resulting in caveolae-mediated endocytosis
EIPA	Inhibitior of macrocytosis due to the blockage of the sodium-proton exchange

EIPA 5-(N-ethyl-N-isopropyl)-amiloride

however, is the use of marker molecules as well as inhibitors. However, a careful selection of inhibitors is crucial, recognizing the potential undesired effect on the properties of the epithelial cells and that complete specificity is not provided by using selected inhibiting agents or marker molecules. Examples of used inhibitors are given in Table 4.2.

When conducting experiments and interpreting data from permeability studies in vitro, it is essential to consider the mass balance, i.e., can the dosed amount of biopharmaceutical drug be recovered and can any loss be accounted for? It is necessary to account for all of the applied biopharmaceutical over the time course of the experiment, which involve that (1) the amount taken up from one side of the monolayer, (2) the amount appearing on the opposite side as well as (3) the intracellular/intraepithelial accumulation of the cargo, (4) any possible degradation/metabolism, and (5) the integrity of the biopharmaceutical, which has permeated the epithelium must be determined. This stands as an analytical challenge.

4.5.2 *Analytical Challenges*

For easiness of analysis, biopharmaceuticals are often radio labeled or fluorophore labeled. For practical reasons, and depending on the specific experimental setup, different labels and different analytical protocols are employed. Nevertheless, attaching a relatively bulky and hydrophobic fluorescent moiety to a hydrophilic cargo inevitably changes the properties of the molecules, and even radiolabeling can alter the biological properties of a drug. Thus, care should be taken to investigate the effect of the labeling on the properties of the biopharmaceutical in order to avoid erroneous or, at the least, questionable results. The label may influence the incorporation into a drug delivery system, the interactions with the coadministered excipients, the interactions with the mucosal barrier as well as the stability of the biopharmaceutical.

Further, if detection is done only by detecting the label, false positive permeability values for, e.g., peptide or protein permeability may be the result, as potential degradation of the drug is not accounted for. The analytical protocols employed must take into consideration that both the biochemical and biophysical properties including high-order structure as well as the biological activity of the molecule needs to be determined. For an ultimate evaluation of the delivery system, the immunochemical properties as well as purity, type and effect of impurities and contaminants must be determined [87].

4.6 Conclusions and Future Perspectives

Generally, there is a wish for more convenient and patient-friendly formulations to deliver drugs, including peptide, protein, and nucleotide-based drugs. With a pharmacological rationale for delivering a given biopharmaceutical via mucosal administration, the molecular properties of the drug in relation to the specific barrier properties will be decisive for the obtained biological response.

Numerous different approaches employing different formulation technologies and the use of several coexcipients are explored for delivery, and also considered in relation to novel device technologies. In the design of novel drug delivery systems, the trend points toward formulations that result in synergistic effects for (1) preventing degradation of the biopharmaceutical, (2) improving the diffusion through the mucus to the epithelial absorption site, and—not the least—(3) improving the permeability across the epithelial barrier. However, there is still much to be learned about the fundamental processes responsible for epithelial and thus also mucosal transport mechanisms for biopharmaceuticals. Thus, for achieving an improved understanding of the mechanisms leading to refined drug delivery systems for delivery of biopharmaceuticals, well-characterized in vitro and in vivo models representative of the human target mucosa are needed. For elucidating the transport mechanisms employed, modifications to the compound of interest and/or of the experimental setup are often needed, and it is important to emphasize that the conclusions of the overall delivery propensity and underlying permeability mechanisms are closely related to this as well as the methods used for detection and analysis of the results.

References

1. Mullard A. 2012 FDA drug approvals. Nat Rev Drug Discov. 2013;12(2):87–90.
2. Wang W. Oral protein drug delivery. J Drug Target. 1996;4(4):195–232.
3. Cone RA. Barrier properties of mucus. Adv Drug Deliv Rev. 2009;61(2):75–85.
4. Allen A, Hutton DA, Pearson JP, Sellers LA. Mucus glycoprotein structure, gel formation and gastrointestinal mucus function. Ciba Found Symp. 1984;109:137–56.
5. Kent M. Advanced Biology. Oxford: Oxford University Press; 2000.
6. Miller H, Zhang J, Kuolee R, Patel GB, Chen W. Intestinal M cells: the fallible sentinels? World J Gastroenterol. 2007;13(10):1477–86.

7. Yun Y, Cho YW, Park K. Nanoparticles for oral delivery: targeted nanoparticles with peptidic ligands for oral protein delivery. Adv Drug Deliv Rev. 2012;65(6):822–32.
8. Scheuch G, Kohlhaeufl MJ, Brand P, Siekmeier R. Clinical perspectives on pulmonary systemic and macromolecular delivery. Adv Drug Deliv Rev. 2006;58(9–10):996–1008.
9. Harris D, Robinson JR. Drug delivery via the mucous membranes of the oral cavity. J Pharm Sci. 1992;81(1):1–10.
10. Patton DL, Thwin SS, Meier A, Hooton TM, Stapleton AE, Eschenbach DA. Epithelial cell layer thickness and immune cell populations in the normal human vagina at different stages of the menstrual cycle. Am J Obstet Gynecol. 2000;183(4):967–73.
11. Sugano K, Kansy M, Artursson P, Avdeef A, Bendels S, Di L, Ecker GF, Faller B, Fischer H, Gerebtzoff G, Lennernaes H, Senner F. Coexistence of passive and carrier-mediated processes in drug transport. Nat Rev Drug Discov. 2010;9(8):597–614.
12. Pappenheimer JR. Physiological regulation of transepithelial impedance in the intestinal mucosa of rats and hamsters. J Membr Biol. 1987;100(2):137–48.
13. Rodgers LS, Fanning AS. Regulation of epithelial permeability by the actin cytoskeleton. Cytoskeleton (Hoboken). 2011;68(12):653–60.
14. Rubas W, Cromwell ME, Shahrokh Z, Villagran J, Nguyen TN, Wellton M, Nguyen TH, Mrsny RJ. Flux measurements across Caco-2 monolayers may predict transport in human large intestinal tissue. J Pharm Sci. 1996;85(2):165–9.
15. Rojanasakul Y, Wang LY, Bhat M, Glover DD, Malanga CJ, Ma JK. The transport barrier of epithelia: a comparative study on membrane permeability and charge selectivity in the rabbit. Pharm Res. 1992;9(8):1029–34.
16. Powell DW. Barrier function of epithelia. Am J Physiol. 1981;241(4):G275–88.
17. Paulett iGM, Gangwar S, Knipp GT, Nerurkar MM, Okumu FW, Tamura K, Siahaan TJ, Borchardt RT. Structural requirements for intestinal absorption of peptide drugs. J Control Release. 1996;41(1–2):3–17.
18. Tang VW. Proteomic and bioinformatic analysis of epithelial tight junction reveals an unexpected cluster of synaptic molecules. Biol Direct. 2006;1:37.
19. Shen L, Weber CR, Raleigh DR, Yu D, Turner JR. Tight junction pore and leak pathways: a dynamic duo. Annu Rev Physiol. 2011;73:283–309.
20. Weber CR. Dynamic properties of the tight junction barrier. Ann N Y Acad Sci. 2012;1257: 77–84.
21. Lemmer HJ, Hamman JH. Paracellular drug absorption enhancement through tight junction modulation. Expert Opin Drug Deliv. 2013;10(1):103–14.
22. Matsuhisa K, Kondoh M, Takahashi A, Yagi K. Tight junction modulator and drug delivery. Expert Opin Drug Deliv. 2009;6(5):509–15.
23. Deli MA. Potential use of tight junction modulators to reversibly open membranous barriers and improve drug delivery. Biochim Biophys Acta. 2009;1788(4):892–910.
24. Maher S, Brayden DJ, Feighery L, McClean S. Cracking the junction: update on the progress of gastrointestinal absorption enhancement in the delivery of poorly absorbed drugs. Crit Rev Ther Drug Carrier Syst. 2008;25(2):117–68.
25. Ivanov AI, Nusrat A, Parkos CA. Endocytosis of the apical junctional complex: mechanisms and possible roles in regulation of epithelial barriers. Bioessays. 2005;27(4):356–65.
26. Burton PS, Conradi RA, Hilgers AR. Mechanisms of peptide and protein absorption: (2) Transcellular mechanism of peptide and protein absorption: passive aspects. Adv Drug Deliv Rev. 1991;7(3):365–85.
27. Mahato RI, Narang AS, Thoma L, Miller DD. Emerging trends in oral delivery of peptide and protein drugs. Crit Rev Ther Drug Carrier Syst. 2003;20(2–3):153–214.
28. Nielsen CU, Steffansen B, Brodin B. Absorptive transporters. In: Steffansen B, Brodin B, Nielsen CU, editors. Molecular Biopharmaceutics. London: Pharmaceutical; 2010. pp. 193–212.
29. Nielsen CU, Brodin B, Jørgensen FS, Frokjaer S, Steffansen B. Human peptide transporters: therapeutic applications. Expert Opin Ther Pat. 2002;12(9):1329–50.

30. Dantzig AH. Oral absorption of β-lactam by intestinal peptide transport proteins. Adv Drug Deliv Rev. 1997;23(1–3):63–76.
31. Widera A, Kim KJ, Crandall ED, Shen WC. Transcytosis of GCSF-transferrin across rat alveolar epithelial cell monolayers. Pharm Res. 2003;20(8):1231–8.
32. Shah D, Shen WC. Transcellular delivery of an insulin-transferrin conjugate in enterocyte-like Caco-2 cells. J Pharm Sci. 1996;85(12):1306–11.
33. Petrus AK, Fairchild TJ, Doyle RP. Traveling the vitamin B12 pathway: oral delivery of protein and peptide drugs. Angew Chem Int Ed Engl. 2009;48(6):1022–8.
34. Conner SD, Schmid SL. Regulated portals of entry into the cell. Nature. 2003;422(6927):37–44.
35. Weisz OA, Rodriguez-Boulan E. Apical trafficking in epithelial cells: signals, clusters and motors. J Cell Sci. 2009;122(Pt 23):4253–66.
36. Ben-Dov N, Korenstein R. Enhancement of cell membrane invaginations, vesiculation and uptake of macromolecules by protonation of the cell surface. PLoS One. 2012;7(4):e35204.
37. Ben-Dov N, Korenstein R. Actin-cytoskeleton rearrangement modulates proton-induced uptake. Exp Cell Res. 2013;319(7):946–54.
38. Abrami L, Liu S, Cosson P, Leppla SH, van der Goot FG. Anthrax toxin triggers endocytosis of its receptor via a lipid raft-mediated clathrin-dependent process. J Cell Biol. 2003;160(3): 321–8.
39. Mooren OL, Galletta BJ, Cooper JA. Roles for actin assembly in endocytosis. Annu Rev Biochem. 2012;81:661–86.
40. Sandvig K, van Deurs B. Delivery into cells: lessons learned from plant and bacterial toxins. Gene Ther. 2005;12(11):865–72.
41. Garred O, Rodal SK, van Deurs B, Sandvig K. Reconstitution of clathrin-independent endocytosis at the apical domain of permeabilized MDCK II cells: requirement for a Rho-family GTPase. Traffic. 2001;2(1):26–36.
42. Cohen AW, Hnasko R, Schubert W, Lisanti MP. Role of caveolae and caveolins in health and disease. Physiol Rev. 2004;84(4):1341–79.
43. Jones AT. Macropinocytosis: searching for an endocytic identity and role in the uptake of cell penetrating peptides. J Cell Mol Med. 2007;11(4):670–84.
44. Tuma P, Hubbard AL. Transcytosis: crossing cellular barriers. Physiol Rev. 2003; 83(3):871–932.
45. Arbit E, Kidron M. Oral insulin: the rationale for this approach and current developments. J Diabetes Sci Technol. 2009;3(3):562–7.
46. Renukuntla J, Vadlapudi AD, Patel A, Boddu SH, Mitra AK. Approaches for enhancing oral bioavailability of peptides and proteins. Int J Pharm. 2013;447(1–2):75–93.
47. Li P, Nielsen HM, Mullertz A. Oral delivery of peptides and proteins using lipid-based drug delivery systems. Expert Opin Drug Deliv. 2012;9(10):1289–304.
48. van der Walle CF, Schmidt E. Modulation of the intestinal tight junctions using bacterial enterotoxins. In: van der Walle CF editor. Peptide and protein delivery. Oxford: Elsevier; 2011. pp. 195–219.
49. Hecht G, Pothoulakis C, LaMont JT, Madara JL. *Clostridium difficile* toxin A perturbs cytoskeletal structure and tight junction permeability of cultured human intestinal epithelial monolayers. J Clin Invest. 1988;82(5):1516–24.
50. Just I, Fritz G, Aktories K, Giry M, Popoff MR, Boquet P, Hegenbarth S, von Eichel-Streiber C. *Clostridium difficile* toxin B acts on the GTP-binding protein Rho. J Biol Chem. 1994;269(14):10706–712.
51. Takahashi A, Kondoh M, Masuyama A, Fujii M, Mizuguchi H, Horiguchi Y, Watanabe Y. Role of C-terminal regions of the C-terminal fragment of *Clostridium perfringens* enterotoxin in its interaction with claudin-4. J Control Release. 2005;108(1):56–62.
52. Song KH, Fasano A, Eddington ND. Effect of the six-mer synthetic peptide (AT1002) fragment of zonula occludens toxin on the intestinal absorption of cyclosporin A. Int J Pharm. 2008;351(1–2):8–14.
53. González-Mariscal L, Tapia R, Chamorro D. Crosstalk of tight junction components with signaling pathways. Biochim Biophys Acta. 2008;1778(3):729–56.

54. Hirase T, Kawashima S, Wong EY, Ueyama T, Rikitake Y, Tsukita S, Yokoyama M, Staddon JM. Regulation of tight junction permeability and occludin phosphorylation by Rhoa-p160ROCK-dependent and -independent mechanisms. J Biol Chem. 2001;276(13):10423–31.
55. Aungst BJ. Absorption enhancers: applications and advances. AAPS J. 2012;14(1):10–8.
56. Shen L, Turner JR. Actin depolymerization disrupts tight junctions via caveolae-mediated endocytosis. Mol Biol Cell. 2005;16(9):3919–36.
57. Park K, Kwon IC, Park K. Oral protein delivery: current status and future prospect. React Funct Polym. 2011;71(3):280–87.
58. Maher S, Leonard TW, Jacobsen J, Brayden DJ. Safety and efficacy of sodium caprate in promoting oral drug absorption: from in vitro to the clinic. Adv Drug Deliv Rev. 2009;61(15):1427–49.
59. Hovgaard L, Brøndsted H, Nielsen HM. Drug delivery studies in Caco-2 monolayers. II: Absorption enhancer effects of lysophosphatidylcholines. Int J Pharm. 1995;114(2):141–9.
60. Radwan MA, Aboul-Enein HY. The effect of absorption enhancers on the initial degradation kinetics of insulin by alpha-chymotrypsin. Int J Pharm. 2001;217(1–2):111–20.
61. Sayani AP, Chien YW. Systemic delivery of peptides and proteins across absorptive mucosae. Crit Rev Ther Drug Carrier Syst. 1996;13(1–2):85–184.
62. Yoo HS, Choi HK, Park TG. Protein-fatty acid complex for enhanced loading and stability within biodegradable nanoparticles. J Pharm Sci. 2001;90(2):194–201.
63. Cui F, Shi K, Zhang L, Tao A, Kawashima Y. Biodegradable nanoparticles loaded with insulin-phospholipid complex for oral delivery: preparation, in vitro characterization and in vivo evaluation. J Control Release. 2006;114(2):242–50.
64. Sadeghi AM, Dorkoosh FA, Avadi MR, Weinhold M, Bayat A, Delie F, Gurny R, Larijani B, Rafiee-Tehrani M, Junginger HE. Permeation enhancer effect of chitosan and chitosan derivatives: comparison of formulations as soluble polymers and nanoparticulate systems on insulin absorption in Caco-2 cells. Eur J Pharm Biopharm. 2008;70(1):270–78.
65. Woitiski CB, Sarmento B, Carvalho RA, Neufeld RJ, Veiga F. Facilitated nanoscale delivery of insulin across intestinal membrane models. Int J Pharm. 2011;412(1–2):123–31.
66. Chen MC, Mi FL, Liao ZX, Hsiao CW, Sonaje K, Chung MF, Hsu LW, Sung HW. Recent advances in chitosan-based nanoparticles for oral delivery of macromolecules. Adv Drug Deliv Rev. 2012;65(6):865–79
67. Prego C, Garcia M, Torres D, Alonso MJ. Transmucosal macromolecular drug delivery. J Control Release. 2005;101(1–3):151–62.
68. Hu CS, Chiang CH, Hong PD, Yeh MK. Influence of charge on FITC-BSA-loaded chondroitin sulfate-chitosan nanoparticles upon cell uptake in human Caco-2 cell monolayers. Int J Nanomedicine. 2012;7:4861–72.
69. Foged C, Nielsen HM. Cell-penetrating peptides for drug delivery across membrane barriers. Expert Opin Drug Deliv. 2008;5(1):105–17.
70. Khafagy el-S, Morishita M, Kamei N, Eda Y, Ikeno Y, Takayama K. Efficiency of cell-penetrating peptides on the nasal and intestinal absorption of therapeutic peptides and proteins. Int J Pharm. 2009;381(1):49–55.
71. Khafagy el-S, Morishita M. Oral biodrug delivery using cell-penetrating peptide. Adv Drug Deliv Rev. 2012;64(6):531–9.
72. Trehin R, Krauss U, Beck-Sickinger AG, Merkle HP, Nielsen HM. Cellular uptake but low permeation of human calcitonin-derived cell penetrating peptides and Tat(47–57) through well-differentiated epithelial models. Pharm Res. 2004;21(7):1248–56.
73. Li N, Li XR, Zhou YX, Li WJ, Zhao Y, Ma SJ, Li JW, Gao YJ, Liu Y, Wang XL, Yin DD. The use of polyion complex micelles to enhance the oral delivery of salmon calcitonin and transport mechanism across the intestinal epithelial barrier. Biomaterials. 2012;33(34):8881–92.
74. Du W, Fan Y, Zheng N, He B, Yuan L, Zhang H, Wang X, Wang J, Zhang X, Zhang Q. Transferrin receptor specific nanocarriers conjugated with functional 7peptide for oral drug delivery. Biomaterials. 2013;34(3):794–806.
75. He C, Yin L, Tang C, Yin C. Size-dependent absorption mechanism of polymeric nanoparticles for oral delivery of protein drugs. Biomaterials. 2012;33(33):8569–78.

76. Niu M, Lu Y, Hovgaard L, Guan P, Tan Y, Lian R, Qi J, Wu W. Hypoglycemic activity and oral bioavailability of insulin-loaded liposomes containing bile salts in rats: the effect of cholate type, particle size and administered dose. Eur J Pharm Biopharm. 2012;81(2):265–72.
77. Barua S, Yoo JW, Kolhar P, Wakankar A, Gokarn YR, Mitragotri S. Particle shape enhances specificity of antibody-displaying nanoparticles. Proc Natl Acad Sci U S A. 2013;110(9): 3270–5.
78. Hussain N. Ligand-mediated tissue specific drug delivery. Adv Drug Deliv Rev. 2000;43(2–3):95–100.
79. Jin Y, Song Y, Zhu X, Zhou D, Chen C, Zhang Z, Huang Y. Goblet cell-targeting nanoparticles for oral insulin delivery and the influence of mucus on insulin transport. Biomaterials. 2012;33(5):1573–82.
80. Rajapaksa TE, Stover-Hamer M, Fernandez X, Eckelhoefer HA, Lo DD. Claudin 4-targeted protein incorporated into PLGA nanoparticles can mediate M cell targeted delivery. J Control Release. 2010;142(2):196–205.
81. Yoo MK, Kang SK, Choi JH, Park IK, Na HS, Lee HC, Kim EB, Lee NK, Nah JW, Choi YJ, Cho CS. Targeted delivery of chitosan nanoparticles to Peyer's patch using M cell-homing peptide selected by phage display technique. Biomaterials. 2010;31(30):7738–47.
82. Jepson MA, Clark MA, Hirst BH. M cell targeting by lectins: a strategy for mucosal vaccination and drug delivery. Adv Drug Deliv Rev. 2004;56(4):511–25.
83. Monopoli MP, Aberg C, Salvati A, Dawson KA. Biomolecular coronas provide the biological identity of nanosized materials. Nat Nanotechnol. 2012;7(12):779–86.
84. Salvati A, Pitek AS, Monopoli MP, Prapainop K, Bombelli FB, Hristov DR, Kelly PM, Aberg C, Mahon E, Dawson KA. Transferrin-functionalized nanoparticles lose their targeting capabilities when a biomolecule corona adsorbs on the surface. Nat Nanotechnol. 2013;8(2):137–43.
85. Vogel U, Sandvig K, van Deurs B. Expression of caveolin-1 and polarized formation of invaginated caveolae in Caco-2 and MDCK II cells. J Cell Sci. 1998;111(Pt 6):825–32.
86. Al Soraj M, He L, Peynshaert K, Cousaert J, Vercauteren D, Braeckmans K, De Smedt SC, Jones AT. siRNA and pharmacological inhibition of endocytic pathways to characterize the differential role of macropinocytosis and the actin cytoskeleton on cellular uptake of dextran and cationic cell penetrating peptides octaarginine (R8) and HIV-Tat. J Control Release. 2012;161(1):132–41.
87. Berkowitz SA, Engen JR, Mazzeo JR, Jones GB. Analytical tools for characterizing biopharmaceuticals and the implications for biosimilars. Nat Rev Drug Discov. 2012;11(7):527–40.
88. Citi S. Protein kinase inhibitors prevent junction dissociation induced by low extracellular calcium in MDCK epithelial cells. J Cell Biol. 1992;117(1):169–78.
89. Fernandez D, Boix E, Pallares I, Aviles FX, Vendrell J. Structural and functional analysis of the complex between citrate and the zinc peptidase carboxypeptidase A. Enzyme Res. 2011;2011:128676.
90. Sung HW, Sonaje K, Liao ZX, Hsu LW, Chuang EY. pH-responsive nanoparticles shelled with chitosan for oral delivery of insulin: from mechanism to therapeutic applications. Acc Chem Res. 2012;45(4):619–29.
91. Su FY, Lin KJ, Sonaje K, Wey SP, Yen TC, Ho YC, Panda N, Chuang EY, Maiti B, Sung HW. Protease inhibition and absorption enhancement by functional nanoparticles for effective oral insulin delivery. Biomaterials. 2012;33(9):2801–11.
92. Krug SM, Amasheh M, Dittmann I, Christoffel I, Fromm M, Amasheh S. Sodium caprate as an enhancer of macromolecule permeation across tricellular tight junctions of intestinal cells. Biomaterials. 2013;34(1):275–82.
93. Mayor S, Pagano RE. Pathways of clathrin-independent endocytosis. Nat Rev Mol Cell Biol. 2007;8(8):603–12.

Part II
Delivery Strategies for Specific Mucosal Sites

Chapter 5
Oral Delivery of Biopharmaceuticals

Catarina Oliveira Silva, Bruno Sarmento and Catarina Pinto Reis

5.1 Introduction

For the past decade, oral drug delivery improved considerably in terms of drug innovations, new line extensions with better pharmacokinetic profiles, and thus, greater efficacy and patient compliance. In terms of R&D, the application of recent technologies and techniques, such as nanosizing or supercritical solutions, promoted this evolution as well as the recovery and readaptation of old fashion methods and well-known excipients, such as solid dispersions and conventional coating materials.

Approximately 85 % of the 50 most-sold pharmaceutical products in North America and Europe are given per os, as they represent safer, more comfortable, and cheaper ways to guarantee patient compliance and therapeutic efficiency, in comparison with other routes [1]. However, oral delivery brings as many advantages as challenges, since it shows several barriers, like drastic pH variations, food effect, mucus, mucins, enzymes, tight junctions (TJ), electrolytes, and water, which limit the absorption of peptides and proteins and other large and poor-water soluble biomolecules [1–2]. This chapter reviews the most recent strategies for improving the oral delivery of biopharmaceuticals based on two rationales: (1) increasing stability and (2) using safer and biocompatible systems.

Food and Drug Administration (FDA) and European Medicines Agency (EMA) hold a big role in regulatory affairs for evaluation of safety, effectiveness, and quality of the new products when entering the market. Nowadays, the majority of the

C. P. Reis (✉) · C. O. Silva
CBIOS—Laboratory of Nanoscience and Biomedical Nanotechnology,
Universidade Lusófona, Campo Grande 376, 1749-024 Lisbon, Portugal
e-mail: catarinapintoreis@gmail.com

B. Sarmento
NEWTherapies Group, INEB – Instituto de Engenharia Biomédica, Porto, Portugal

IINFACTS – Instituto de Investigação e Formação Avançada em Ciências e
Tecnologias da Saúde, Instituto Superior de Ciências da Saúde-Norte,
CESPU, Gandra, Portugal

J. das Neves, B. Sarmento (eds.), *Mucosal Delivery of Biopharmaceuticals*, DOI 10.1007/978-1-4614-9524-6_5,

drug delivery systems (DDS)- based oral formulations existent in the market are for biopharmaceuticals from class II (high permeability and low solubility) or class IV (low permeability and low solubility) [3]. The low number of approved technologies may indicate that characterization methods and standards need to be readjusted to a smaller scale or to the new polymers or materials developed in the laboratories. The good news is that recent studies, made at lab scale, focus on deeper characterization of the proposed systems, mainly physical-chemical properties (differential scanning calorimetry (DSC), Fourier transform infrared (FTIR), scanning/transmission electron microscopy (SEM/TEM), and mass spectrometry), ADME (Absorption, Distribution, Metabolism, Elimination), and bioavailability studies. In vivo stability studies are of extreme importance to evaluate, as soon as possible, if the new formulations are able to improve the behavior of conventional ones [4]. In addition, preference for natural origin and low-toxicity substances has created a great progress, especially in the area of drug development [2]. This approach, recently embraced by many researchers, has reduced the limitations in terms of approvals and doubts about toxicology profiles and in vivo behavior, since most excipients, solvents, and polymeric matrix materials used in nanocarriers' production were already used in conventional medicines [5].

Advances in this field may help new molecules with great pharmacological potential to go further in pipelines and improve the efficiency of older drugs, with a recognized action in therapies, which cannot be naturally absorbed by oral route [6]. One great example is insulin, since administration by oral route brings obvious advantages (less invasive therapy, high patient compliance, and less side effects) and also because by entering the gastrointestinal (GI) tract it reaches the liver as the primary site of action, when only 20 % of the injected insulin can target this organ [2, 7]. The main options for improving oral absorption of class II and class IV biopharmaceuticals will be further explored in this chapter.

5.2 Improvement of the Conventional Methods

Conventional methods considered to improve solubility and permeability of drugs are used by the pharmaceutical industries for the past years, including pH modification (e.g., tartaric acid), salt formulation, and micronization [8]. Traditionally, salts were used due to accumulated experience, but the best salt selection needs rigorous screening processes, and sometimes they are ineffective and influenced by microenvironmental pH variations [9]. In addition, salts are not suitable for every substance [10], and the presence of other ions (e.g., chloride) in the gastric fluids may compromise the solubility [8]. Until 2000, there were a low number of micelles and cyclodextrins in the market, and micronization via colloid mills or jet mill was used to increase the drug dissolution rate, by size reduction and surface area improvement [11, 12]. Still, size reduction does not change the solubility. Traditional techniques for the micronization method included crushing, grinding, bashing, cutting, and milling, which damage and alter the properties of the formulations [13], and cause the particle aggregation, increasing the need of surfactants [8].

In the last years, conventional techniques have been improved with the advent of "nanosizing" and the consequent increasing number of studies. One example is the wet media milling, optimized to reach the goals of nanosizing technology [6]. With this technological improvement, nanosuspensions may become a viable alternative to micronization [14]. However, not all substances benefit from a significant size reduction, as increased surface area for dissolution increases the drug exposure compared with the micronization, even when surfactants are used [15].

Other techniques that decrease the particle size and increase the solubility and the dissolution rate of drugs include the use of supercritical solutions or fluids (SCF) such as: rapid expansion of supercritical solutions (RESS), supercritical antisolvent precipitation (SAS), particle generation from gas-saturated solutions (PGSS) and new atomization processes. Characterization and effects on the products by these techniques were reviewed in the literature [13]. RESS can be used in thermosensitive drugs, and it is an environmentally green technique, since the solutions do not pollute or leave residues. However, it cannot be used in protein and peptides due to their hydrophilic nature and insolubility in SCF [13]. The second technique, PGSS, uses SCF for dissolving the solid in a solvent followed by its precipitation into solid materials. SAS is a good alternative in size reduction for inhalable powders and other substances that cannot be micronized normally and is now being scaled-up for major production effects [13]. Finally, SCF can be applied for preparation of micro- and nanoparticles, liposomes, cyclodextrins, and solid dispersions with better size control [13].

Solid dispersions—for improving the solubility of drugs—show many stability problems due to the formation of metastable forms and cocrystals, which are variable and can transform into the thermodynamically stable form (low energy form) during manufacturing, storage, and/or administration [8]. Thus, there are other ways to increase the solubility of drugs, besides reducing its size, as we will see ahead in the chapter. They were considered until now as nonconventional methods (e.g., self-emulsifying drug delivery systems (SEDDS), nanocrystals, nanosuspensions, and cyclodextrins), but are gaining greater impact and position in the biopharmaceutical market [8, 9, 16].

5.3 Nanocrystals, Self-Emulsifying Systems, and Other Non-Conventional Methods

Crystal forms affect several physicochemical characteristics of the drug such as: morphology, solubility, stability, and thermal and mechanical properties [9]. Polymorphisms can be observed in amorphous forms due to differences in dissolution or crystallization behaviors [9]. One of the biggest challenges—and a successful project in this area—was the development of Nanocrystal® technology, as line extensions for drugs with poor absorption or great in vivo variability. Briefly, nanocrystals are formed by the particle size reduction, which leads to an increased surface area and higher dissolution rate [8]. The first attempt was achieved via pearl milling by

Elan (1990), but took many hours or even days to be formulated [11]. Currently, production occurs mainly by bottom–up (from molecular level by precipitation) or top–down (milling and homogenization) methods [3]. Combination of both methods—formulation by wet milling and further incorporation of dried nanosuspension into solid dosage forms, or improvement of bottom–up processes—can easily control the crystal morphology and shape, enhancing the nanocrystal stability and in vivo performance [3, 6]. In 1999 appeared the first Nanocrystal® technology in the market developed by Wyeth in the form of Rapamune® (Sirolimus) [16]. This technology increased the pharmacokinetic parameters, efficacy, and safety of the drug by oral administration. Nowadays, there are five oral formulations in the market as line extensions that use the Nanocrystal® technology: Rapamune® (Sirolimus) from Wyeth, Emend® (Aprepitant) from Merck, Tricor® (Fenofibrate) from Abbott, Megace®ES (Megestrol acetate) from PAR Pharmaceutical, and Triglide® (Fenofibrate) from SkyePharma [15]. The main advantage is the enhancement of oral bioavailability by reduction of the food effect. In some cases, this effect is eliminated almost completely (Emend® and Tricor®), while in other cases improvement at this level is still needed. For example, Rapamune® could only decrease slightly the food effect and increase the bioavailability by 27 %, relative to the early formulation, but showed additional stability at room temperature and greater palatability [15, 16].

Aside from these technologies, lipid-based formulations have gained great attention for oral delivery of biopharmaceuticals with poor-water solubility. SEDDS or self-microemulsifying drug delivery systems (S(M)EDDS), as Gibaud and Attivi (2012) have defined them, are general isotropic solutions of oils and surfactants that form natural oil-in-water (O/W) emulsions [17]. These emulsions are then divided into two categories, according to the particle size: (1) S(M)EDDS and (2) self-nanoemulsifying drug delivery systems (SNEDDS) [17]. Solid dispersions and S(M)EDDS-based products exist since the 1970s–1980s [8]. Depending on the type of oil and surfactant, surfactant concentration, oil/surfactant ratio, and temperature of self-emulsification, these systems can be developed to increase the permeation and the drug dissolution rate [10]. Just as nanocrystals, S(M)EDDS can avoid the food effect since they do not depend, like other lipid formulations (e.g., conventional micro- or nanoemulsions), on the GI conditions for promoting their emulsification before the absorption [3]. Cosurfactants and/or cosolvents can be added to the formulation to facilitate the formation of emulsions before absorption [18, 19]. As reported before, the reduction of the particle size to micro or nano range may enhance the dissolution rate but, unless the drug is in its amorphous state or the formulation contains solubilizers, the solubility does not increase [20]. And, as amorphous forms show higher risk of instability and crystallization [14], the use of surfactants and cosurfactants is preferred. Simultaneously, as any other emulsion-like formulation, it needs pseudoternary (water/amphiphilic/oil) or quaternary (water/surfactant/cosurfactant/oil) diagrams for size-stability control. The main excipients used until now for these systems are described in Fig. 5.1.

The combination of two high hydrophilic-lipophilic balance (HLB) nonionic surfactants proved to significantly increase the microemulsion due to synergistic effect. There is also a possible relationship between the chain length of the surfactant and of

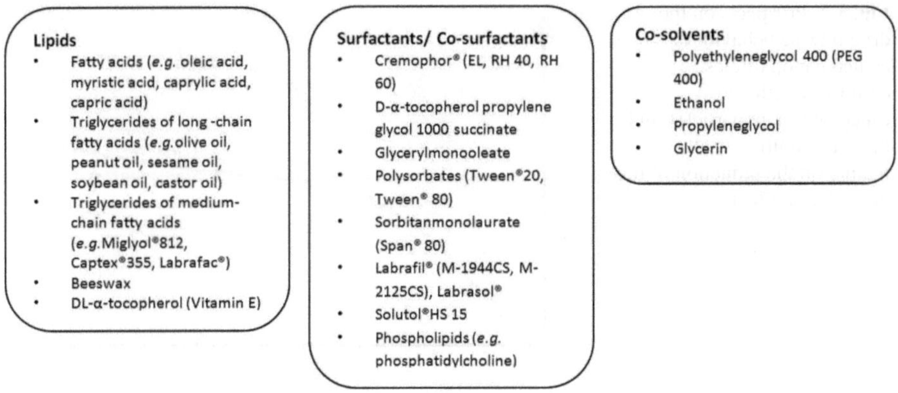

Fig. 5.1 Examples of the main components of S(M)EDDS formulations [4, 17, 21, 83]

the oil used, since mixtures of long chain fatty acid esters and medium chain triglyceride result in optimized emulsions [19]. Selecting from nondigestible lipids (e.g., mineral oils or sucrose polyesters) to digestive lipids (e.g., fatty acids, phospholipids, and cholesterol esters) is also important for avoiding the retention of lipophilic drugs and limitation of the absorption [21]. For example, medium-chain fatty acids (e.g., Mygliol® 812) are digested faster than long-chain fatty acids (e.g., soybean oil), but were only capable of delivering a small percentage of drug, while soybean oil-based systems released approximately 21–36 % of the drug fenofibrate under fasted and fed conditions in aqueous phase, increasing its absorption [22]. Some surfactants are also responsible for inhibition of enzymes (e.g., pancreatic lipase) [2, 22] or P-glycoprotein (P-gp) [23, 24], which may improve the system stability and consequent drug absorption in the GI tract. Currently, there are four S(M)EDDS formulations in the market: Neoral®, Neoral Sandimmune® and Sandimmun Neoral® (Cyclosporin A) from Novartis, Gengraf® (Cyclosporine—bioequivalent hard gelatin capsules to Neoral®) from Abbott Laboratories, Norvir® (Ritonavir) from Abbott Laboratories, and, since 2012, Fortovase® (Saquinavir) from Roche [17, 20]. Under development, are now S(M)EDDS of statins (e.g., simvastatin and lovastatin), anticancer drugs (e.g., paclitaxel and mitotane), antibiotics (e.g., beta-lactam AB and quaternary ammonium salts), and antihypertension drugs (e.g., nifedipine, nicarpine, and atenolol) [17]. Nevertheless, in vivo studies for S(M)EDDS are fundamental for their entrance in the market, due to the in vivo variability associated with these formulations [3]. Finally, S(M)EDDS products can be incorporated into liquid-filled capsules (e.g., gelatin-made) or converted to solid forms by spray drying, spray congealing, adsorption onto solid carriers, melt granulation, and hot-melt extrusion techniques [3, 9]. The future of these formulations may pass through the combination of multiple methods and materials: recently developed "super-SNEDDS" as supersaturated isotropic mixtures of oils, surfactants, co-solvents, and also co-polymers, like polyvinyl acetate (PVA) or hydroxypropyl methylcellulose (HPMC) increase drug absorption by inhibiting its early precipitation in the GI tract [25]. The addition of polymeric

Fig. 5.2 Prospects on the drug release behavior of coated nanoparticles, compared with bare (uncoated) nanoparticles and the drug solution itself, mainly on the reduction of the initial burst effect

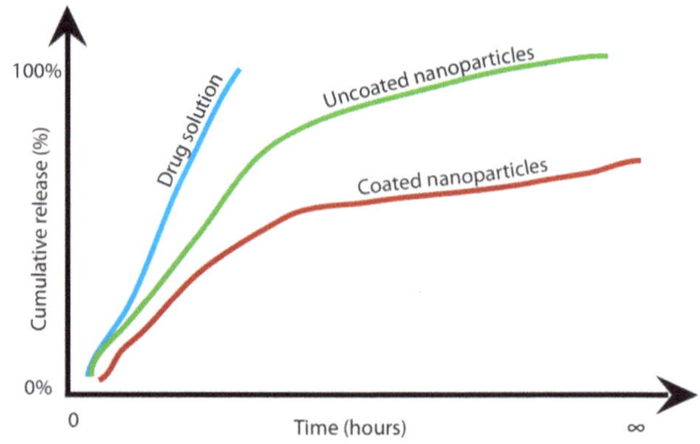

coatings or copolymers as well as conjugating different lipids may also eliminate the initial burst release, as represented in Fig. 5.2, that commonly occur in these systems, promoting a more sustained and prolonged release than the conventional tablets [26].

Other nonconventional methods include polymeric micelles, nanoemulsions, and cyclodextrin complexes. These systems need huge improvement and should be used in combination with others due to some limitations associated with drug exposure, fast drug release, and limited drug loading. Commercially available nanoemulsion-based formulations include Estrasorb® (Estradiol) from Novavax/Graceway, Flexogan® (camphor, menthol, and methyl salicylate) from AlphaRX, and Restasis® (Cyclosporine) from Allergan [27]. Cyclodextrins show low drug efficiency and only are suitable for drugs capable of complexation. Because the drug is partially encapsulated inside the cyclodextrin hydrophobic cavity, a new physical and chemical environment is created, and with it a new thermodynamic stability is achieved. There are no covalent bonds in the cyclodextrins formation, and thus the dissociation is similar to diffusion [27]. This advantage allows the drug solubilization to be a linear function, reducing aggregation. As for micelles, the hydrophobic core determines its capacity to solubilize poor water-soluble drugs [27]. Polymeric micelles interact greatly with cell membranes because of the steric hindrance from the shell polymer [27]. However, there is little information about the behavior of these formulations in the stomach (e.g., interaction with bile salts, enzymes, and food effect) because they are usually administered after fasting or directly in the duodenum [27]. As the other emulsion-based systems, after reduction of particle size, micelles can be incorporated in conventional dosage forms (e.g., tablets, capsules, or even pellets) by application of drying techniques such as lyophilization, spray drying, and ultracentrifugation [27]. These formulations may show stability problems that even cryoprotector cannot manage [28]. Moreover, in vivo stability is very important because an unstable formulation can lead to unexpected side effects and adverse reactions. Next, we will describe other particulate-based carriers used for oral drug delivery.

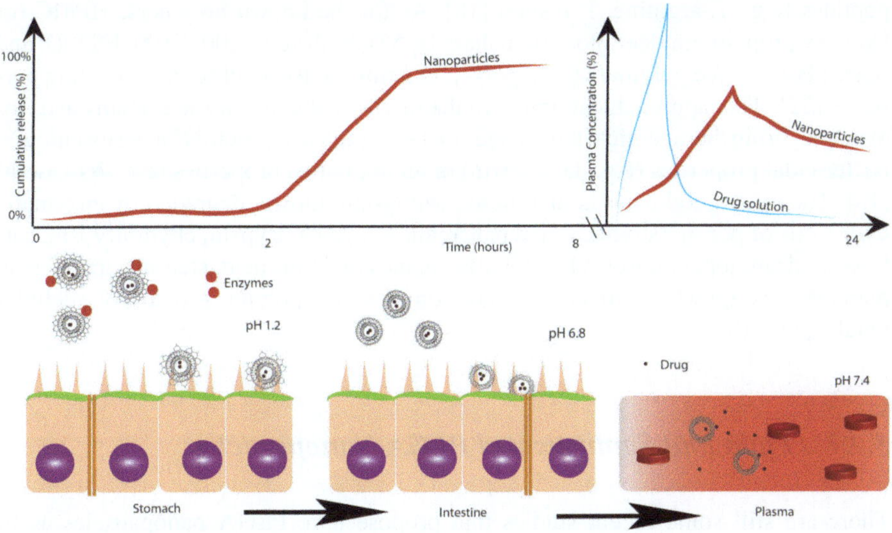

Fig. 5.3 Representation of drug release mechanism, bioavailability, and cellular interaction of sophisticated coated nanoparticles

5.4 Nanoparticulate-Based Carriers

Beside nanoparticles, other oral delivery systems are too large to pass through the epithelium. Ideal size for polymeric colloidal particles is between 300 and 600 nm to increase the cellular uptake and systemic circulation, but can induce more side effects [29, 30]. Smaller particles (less than 300 nm) may allow a lower release rate of the drug, which would prevent the degradation of the protein drug by reducing the exposure and degradation by intestinal enzymes [29] and can pass the cellular pathways and enter microfold (M) cells of the Peyer's patches [2]. On the contrary, particles with larger size can be used for regional or local delivery to intestinal tissues rather than systemic delivery [29].

Poly(lactic-co-glycolic acid) (PLGA) is the most accepted polymer by the pharmaceutical industry, due to its safe and low toxic profile, but laboratory studies have constantly demonstrated a low relative bioavailability [2]. Coating of PLGA particles is one of the solutions to obtain a sustained drug release, higher in vivo stability, and bioavailability, as demonstrated in Fig. 5.3. It is now accepted that nanoparticles for oral delivery should present a positive charge in order to interact with the negatively charged GI tract (− 50 mV) [31], namely with the hydrophilic glycoproteins from the mucus [2]. Still, cationic polymers are more toxic and may cause disruption of the mucosa. Thus, recent research groups are concerned with the use of natural origin polymers or others that have been used in pharmaceutical industry for many years and whose toxicology profile is greatly characterized. Natural polymers include: gelatin, collagen, cellulose, milk proteins (e.g., casein, whey proteins), albumin, elastin, silk fibroin, polysaccharides (e.g., chitosan, dextran, and alginate), and cell-penetrating peptides (e.g., L-arginine, L-lysine) [18]. As for the known polymers, HPMC (or hydroxy propylmethylcellulose phthalate, Hp55), Eudragit L100, S100, FS30D, and other pH-sensitive polymers (e.g., poly-γ-glutamic acid) are chosen for coating purposes [32]. This approach can improve the in vivo behavior of the systems and, for example, bring therapeutical advantages to the formulation, such as bacteriostatic and bactericidal properties (e.g., lactoferrin) or antioxidative properties (e.g., αs-casein) [18]. The other good news is that these coatings are mostly degraded by enzymatic hydrolysis of pepsin (stomach) and pancreatic enzymes (trypsin, chymotrypsin, carboxy, and aminopeptidases) [18]. Finally, ligands like folic acid, steroids, transferrin, mannose, and growth factors can also be conjugated to polymers to improve cellular uptake [18, 30].

5.4.1 Plain and Sophisticated PLGA Nanoparticles

There are still some recent studies that propose bare PLGA nanoparticles as final systems for delivering biopharmaceutical drugs (e.g., curcumin) with improved bioavailability compared with conventional solid formulation [33]. Although this improvement is verified, when comparing with coated PLGA nanoparticles, the advantages are greater. Lately, an increased attention was given to curcumin—a natural origin polyphenol with anticancer properties and low intrinsic toxicity—especially for incorporation into nanoparticulate systems, since its poor water-solubility influences its oral absorption.

Polymeric particles of PLGA and Poly(lactic-co-glycolic acid)-polyethylene glycol (PLGA-PEG) to deliver curcumin were evaluated in terms of bioavailability studies [34]. PEG influenced the release of curcumin, causing its desorption from the particle surface, with a faster release until 24 h and a sustained release over 9 days with a total 56.9 % drug released; PLGA nanoparticles did not show a biphasic behavior and the curcumin release was slower. There was a strong increase of the bioavailability and six-fold increase of half-life elimination time of curcumin when encapsulated inside PLGA-PEG nanoparticles [34]. Another study with curcumin-loaded PLGA nanoparticles demonstrated that these systems were able to successfully counteract the effects caused by diethylnitrosamine (DEN), a chemical hepatocarcinogen [35]. Nanoparticles also improved the drug therapeutic activity over the tumor, with less toxicity for other tissues. Further, daidzein, a water-insoluble isoflavone isolated from plants, with cardiovascular properties, but insoluble in water and in the majority of the organic solvents, demonstrated promising results in terms of sustained release and bioavailability when incorporated in phospholipid and cyclodextrin complexes encapsulated into PLGA nanoparticles [36].

Still, the negative surface charge of PLGA nanoparticles tends to limit the bioavailability [31]. Coating with chitosan is an approach that can enhance the in vivo cellular uptake and maintain a safe toxicology profile [37]. One study compared two systems (PLGA nanoparticles and chitosan-PLGA nanoparticles) loaded with insulin in terms of release, mucoadhesion, and in vivo activity [31]. As well as higher encapsulation efficiency, sustained drug release, and mucoadhesion with retention in the lumen of the GI tract for 7 h, chitosan-PLGA nanoparticles also showed better effect on blood glucose levels, enhancing the intestinal absorption of insulin, with good cell tolerability [31]. Chitosan-coated-PLGA nanoparticles, stabilized with lecithin, were also chosen for encapsulation of α-tocopherol [38]. Stability studies revealed that chitosan-PLGA nanoparticles were more stable than PLGA nanoparticles at pH 1.5, but were aggregated at pH 6.5, which may affect the mucoadhesion at the site of action. Nevertheless, higher and concentrated release of α-tocopherol was obtained in the intestinal part where the drug is preferentially absorbed [38].

With the nanocarriers evolution, oral route also became a possible pathway to reach other organs in the body. PLGA nanoparticles have been studied as site-specific targeting vehicles through decoration with specific ligands. Advantages include more comfortable therapies for the patient and, consequently, higher efficiency. Hydrophilic platelet-derived growth factor and hydrophobic simvastatin were encapsulated in a biodegradable microsphere made of poly-DL-lactide (PLA) core matrix and PLGA shell matrix for stimulation of osteogenesis [39]. The growth factor and simvastatin were encapsulated in the shell and core, respectively, and by the sequential release of the drugs for 14 days, there was a greater reduction of inflammation and apoptosis. Another study developed Tween® 80-coated PLGA nanoparticles that were able to orally deliver estradiol to the brain for Alzheimer's disease treatment over 72 h, but the drug was maintained mainly in the small intestine and liver [40]. Finally, insulin-loaded PLGA nanoparticles decorated with folate and PEG protected the drug from the GI tract enzymatic attack, reduced the initial burst release, and maintained the blood glucose levels for at least 24 h with basal levels returning only after 36 h [41]. The authors proposed also a double mechanism of absorption of the nanoparticles, involving absorption through M cells as well as folic acid receptor-mediated endocytosis in the intestinal wall [42]. In fact, folic-acid PLGA nanoparticles, besides increasing the system stability, are reported to enhance specific uptake of paclitaxel by cancer cells [43].

5.4.2 *Chitosan-Coated Nanoparticles*

Some polymer-based nanoparticles can form mucoadhesive interactions via glycoprotein chains of the mucus and prolong the carrier transit time [44]. Chitosan has been used for at least two decades as a suitable material for polymeric carriers, alone or with additives that can improve their performance [45, 46]. Nonmodified chitosan-coated carriers can enhance cellular uptake and drug absorption compared with the free drug but may show similarities to uncoated carriers like liposomes [47]. Still, this polysaccharide by itself is insoluble in the neutral or basic pH environment caused by deprotonation of its amino groups, and leading to the loss of its mucoadhesive characteristics and TJ opening activity, in the jejunum and ileum [48]. Thus, chitosan decoration (e.g., peptides, PEG, or quaternary amine groups) or chemically modified chitosan (e.g., trimethyl-chitosan and thiolation) alter the polymer mucoadhesive properties by turning its cationic charge as pH-independent [44, 49]. Chitosan can also be attached to protease inhibitors (e.g., aprotinin) to improve the bioavailability of biopharmaceutical drugs [50].

However, chitosan coating may significantly increase the size of the particles and may, eventually, establish too strong interactions with the mucosa causing its injury. For example, chitosan-coated ceramic nanocores for oral delivery of the enzyme serratiopeptidase more than doubled the size of the nanocores, due to its depositions on the surface [51], and in a recent study with gemcitabine-loaded chitosan nanoparticles, drug permeability across the rat intestinal sac actually increased three to five folds with this nanosystem, but also demonstrated the injury of the TJ between the intestinal cells [52]. By selecting the appropriate materials and methods, the nanoparticles size can be reduced, as achieved with other gemcitabine-loaded chitosan nanoparticles with a core of Pluronic® F-127, prepared by ionic gelation, which reached a size less than 200 nm without affecting the cytotoxicity and the mucoadhesion properties of the systems [53].

Modified chitosan-methylated N-(4-*N, N*-dimethylaminobenzyl) chitosan was applied as a coating for liposomes. Liposomes are naturally dissolved by bile salts and cannot protect the encapsulated drug from the digestive action of the GI tract. In this way, chitosan works as a drug carrier and a protector, improving the use of low-toxicity systems, like liposomes, for oral delivery of proteins and peptides [54]. N-trimethyl chitosan chloride-coated liposomes showed greater elimination half-time, absorption, and bioavailability due to long residence time, enzymatic protection, and facilitated transport through TJ [55]. Likely, N-carboxymethyl chitosan-coated solid lipid nanoparticles (SLNs) showed increased drug bioavailability, since this chemically modified chitosan is only soluble at pH > 5, protecting the lipid particles until they reach the site of action in the intestine [56]. Ultimately, chitosan-coated liposomes can also avoid the food effect. Chitosan-coated liposomes stayed for more time in the stomach but, as demonstrated by their higher recovery (i.e., > 90 %), the majority of the particles passed to the intestine even in the fed state (probably associated with the mucoadhesive function of chitosan) [57]. Another study showed that poly(anhydride) nanoparticles coated with chitosan of low molecular weight (e.g., 20 KDa) allow greater bioadhesion to the gut mucosa independently from thiolation [58].

Chitosan is actually one of the most selected polymer presently because of the better stability, low toxicity, simple and mild preparation methods, versatility of administration, and biocompatibility (as it is degraded by chitinases in the intestine after oral administration) [49]. Quaternary ammonium palmitoyl glycol chitosan (GCPQ) particles were developed for oral delivery of hydrophilic biomacromolecules such as leucine^5-enkephalin and quantified, for the first time, in terms of their fate after oral administration [59]. Results showed that the majority of the nanoparticles administered per os (85–90 %) remained in the GI tract after 1 h, while 1–6 % entered the blood circulation, reaching the liver, and even other unlikely organs, such as the brain and the skin. According to the authors, GCPQ particles seem to be absorbed via enterocytes and transported to the liver via gut lymphatic vessels or capillaries in the villi and from the liver back to GI tract [59]. Finally, the addition of thiol groups to chitosan can improve the carrier stability and a prolonged release over time, as this chemical modification is based on higher-crosslinking and lipophilicity that improve the cohesiveness of the nanosystem [60]. Thiomers also can interact with mucins (cysteine-rich subdomains of mucus glycoproteins) by disulfide bonds and act as temporary permeation enhancers, by opening the TJ, but may cause some defects of the lipid bilayer of the nanocarriers that increased the release [61].

Like chitosan, other polysaccharides can be used as coating materials as mentioned before. Cellulose acetate is a suitable polymer but it is poorly studied in terms rate-controlling polymer and ADME studies. Still, application of this polymer in solid dispersions increased curcumin's solubility, protected it from GI degradation, promoted its sustained release, and enhanced its bioavailability, compared with the plain drug [62].

5.4.3 pH-Sensitive Polymers

For many years, pH-sensitive polymers have been used for enteric coating in tablets, capsules, and granules. Lately, they were introduced in particulate carrier formulations to surpass the barrier for oral delivery of insulin [32], and other proteins or peptides, protecting them from being released in greater amount in the stomach. Mainly associated with PLGA nanoparticles, these polymers can enhance the in vivo performance of plain nanoparticles and take advantage of PLGA's matrix stability [32]. HPMCP-55 (also known as Hp55) is a pH-sensitive coating polymer admitted into the US National Formulary (US/NF), European Pharmacopeia (EP), and Japanese Pharmacopeia [63]. PLGA-Hp55 nanoparticles demonstrated to release insulin according to the GI pH values—at pH 1.2 less than 15 %, and at pH 7.4 about 90 %—and to rapidly reduce the glucose levels, with an insulin peak appearing after 3 h post administration [32]. Hp55 coating also reduced, in almost 30 %, the insulin initial burst release from the bare PLGA nanoparticles in the simulated gastric fluid and maintained the release of more than 60 % after 1 h in a simulated intestinal fluid, with an in vivo hypoglycemic effect prolonged for 24 h [63].

Other enteric materials like Eudragit L100 and Eudragit S100 are now commonly used as copolymers. The only drawback in this approach is that methacrylate or methacrylic acid polymers are not biodegradable. Eudragit L100 dissolves in the pH range of 6.0–6.5, which ensures drug release in the distal small intestine (ileum), while Eudragit S100 dissolves at pH > 7.0 and releases the drug into the colon [64]. Eudragit S100 particles recently demonstrated that they take longer (more than 6–8 h) to digest than HPMCP- and Eudragit L100-coated particles, exposed to the same conditions [64]. Liposomes coated with Eudragit S100 were capable of slowing down the drug release compared with the bare liposomes. However, it could not protect the degradation by bile salts (e.g., sodium taurocholate) as also observed with the liposomes [65, 66]. Another study has increased the drug protection and stability by formulating thiolated Eudragit L100 nanoparticles via addition of sulfhydryl groups [67]. This polymeric structure could be readapted by pH variations, permitting a local insulin release at the intestine and showing additional mucoadhesion, higher residence time, and insulin absorption, by the presence of the sulfhydryl groups. Moreover, the morphology of intestinal mucosa of rats showed no disruption of the epithelium after oral administration [67].

The application of these coating materials in the development of sophisticated nanocarriers can also broaden the therapeutics achievements by oral route. For example, the coating of PLGA nanoparticles with Eudragit FS30D demonstrated that these systems could resist the digestive action of GI tract and reach the large intestine causing an equal protective immunity to intracolorectal administration [68]. Finally, chitosan–poly(γ-glutamic acid) nanoparticles combined two pH-sensitive polymers for insulin encapsulation, and also the addition of a protease inhibitor and TJ opening agent, diethylenetriaminepentaacetic acid (DTPA), to increase the insulin absorption in the entire intestine, with a slow but prolonged hypoglycemic effect [69].

5.4.4 Other Coating Materials: Proteins and Peptides

Besides the mentioned earlier approaches, other materials—mainly proteins and peptides— appear to enhance the performance of oral DDS. These materials can be of natural origin or can be designed and synthesized at the laboratory scale for in vitro and in vivo preliminary testing. D-α-tocopheryl polyethylene glycol succinate (Vitamin E TPGS) has been intensively applied for developing several DDS. A synthesized biodegradable polymer made of PLA-vitamin E TPGS was developed for oral chemotherapy with a chemical detoxifier, methadone maintenance treatment (MMT) (Closite Na^+). In vivo pharmacokinetics experiments made in rats achieved 26.4 times longer half-life by oral delivery than intravenous administration of docetaxel (Taxotere®) [70]. In general, TPGS can be used as a surfactant and/or a component in liposomes for sustained and controlled drug delivery, and it is described to inhibit P-gp and enhance cellular uptake [71]. Although TPGS cannot achieve a targeting effect, it can be used as the linking agent as originally suggested by Zhang et al., mainly as two presentations: (1) TPGS-COOH as the blend matrix with the polymer and folate as the targeting agent or (2) TPGS-folate or another agent and the blend as matrix to decorate the nanoparticles [71].

Lectins are also proteins that can recognize and bind reversibly to carbohydrates of conjugates [72]. They are stable at low pH values and can be coupled to PLGA nanoparticles especially for vaccination [72, 73]. Lectin can work as a potent adjuvant for mucosal immunization by interacting greatly with mucins [73] and also establish hydrophobic interactions with the PLGA, which may delay the antigen release from the nanoparticles. Besides pectins, albumin [74, 75] also serve as coating proteins, for example, in alginate nanospheres loaded with insulin, protecting and preserving the protein from pepsin degradation.

As for other protein-like conjugations, we can find polymer-amino acid conjugations, such as the thiomer poly(acrylic acid)-cysteine nanoparticles to deliver insulin by oral route and that were able to, for the first time, release the insulin from the nanoparticulate system to be absorbed already within the stomach [76]. Actually, acrylic-based compounds are now known to be safe materials, which show mucoadhesive properties, facilitating the adherence to the GI mucosa and the sustained release of drugs [77]. Another research group used novel self-assembled nanoparticles made with analogs of the amino acid phenylalanine, namely dipeptide methionine-dehydrophenylalanine (MΔF), leucine-dehydrophenylalanine (LΔF) and isoleucine-dehydrophenylalanine (IΔF) for curcumin encapsulation [78]. Dipeptides nanostructures showed better results for drug encapsulation, efficiency and retention, and improved curcumin apoptotic properties for cancer cells, especially in terms of response time [78].

Finally, we cannot forget that cell-penetrating peptides continue to be extensively experimented as coating materials to improve the cellular uptake of particulate carriers. During the last two decades, polyaminoacids and polypeptides have been studied to construct DDS with well-defined structure, biocompatibility, and low toxicity [79–81]. Polyarginine is the most widely used polymer but other arginine- and histidine-rich molecules, and, most recently, penetratrin [82] are also used for cell penetration enhancement. Due to the large information associated with these materials, a complete section will be dedicated to them in this book.

5.4.5 Lipid-Based Systems

Liposomes are one of the oldest drug carriers studied and greatly upgraded since their discovery. These systems are essentially composed of several layers of lipids and surfactants and, due to the natural proximity with the skin lipids, they are studied mostly for topical and transdermal applications. However, liposomes, lipospheres, and lipid nanoparticles—including nanostructured lipid carriers (NLC) and SLNs—are interesting systems for oral delivery, since they can promote a sustained release and also interact greatly with the cell membranes, as the natural constituting lipids may enhance the cellular uptake, without increasing the toxicity [1]. In fact, the oral bioavailability of poor water-soluble drugs can be increased when taken with a lipid-rich meal, which made the researchers understand that lipid-based systems can be a good alternative as drug carriers [83].

Still, lipidic carriers suffer from a poor stability in vivo, mainly, degradation by bile salts and enzymes (e.g., pancreatin) [84], and during storage, as solid lipids are prone to polymorphic transitions due to its crystalline structure. As the lipid converts itself to its low-energy form (more stable crystalline form) the space for drug can vary, which may influence the encapsulation efficiency and the release over time [85]. Liquid lipid-made structure nanoparticles are more imperfect but may show higher stability and sustained drug release. These nanoparticles also show a concentration-dependent, time-dependent, and energy-dependent internalization by endocytosis into the cells—by clathrin-mediated and caveolae or lipid raft-mediated—and also avoided the passage of paclitaxel by P-gp and improved the intestinal epithelial permeability [86]. SLNs showed as well a time-dependent cellular uptake for the encapsulated idarubicin and were able of overcoming transporter-mediated efflux proteins like P-gp, breast cancer resistance protein (BCRP), and multidrug resistance protein 1 (MRPI) [87]. Finally, incorporation of Pluronic® F-127 as a liposomal modifier can increase the liposomes stability, when efficiently incorporated in the core, but also the mucus-penetrating properties and cellular uptake by caveolae-mediated uptake and clathrin-mediated endocytosis [88]. Last of all, another way to improve liposome permeability is to combine physical techniques like ultrasound, increasing the drug release rate, amount, and efficiency, especially in cancer treatment [89].

The selection of the right lipids and surfactants to be used is essential as well as long-term stability studies. Sometimes, surfactants may destabilize these lipidic systems. Liposomes stabilized only with glycerylcaldityltetraether showed constant membrane integrity in acidic conditions and protected the drug from bile salt degradation [90]; in contrast, when two surfactants were added, D-α-tocopheryl TPGS and cholylsarcosine, the liposomes destabilized. As for SLNs tested to deliver frankincense and myrrh oil with antimicrobial, anti-inflammatory, and antitumor activities, Compritol® 888 ATO was chosen as the lipid, while soybean lecithin and Tween® 80 were used as surfactants. Good compatibility with the drug was verified, but Compritol® 888 ATO changed its crystal shape to a less-ordered one, after the formation of the nanoparticles. Still, SLNs reduced the essential oil evaporation over 6 days of storage, increasing the drug stability [91]. Other lipids like cholesterol can improve drug encapsulation in the lipid-based systems and can protect the drug from digestive enzymes (e.g., trypsin and chymotrypsin) [18]. SLNs are also good for encapsulating poor water-soluble drugs and accept the addition of copolymers or surface ligands that may promote an increasing stability and a targeted delivery. Proteins can function as a platform for specific compounds with stabilization, therapeutical, or targeting functions (e.g., folic acid) or as a pH- or temperature-sensitive coating, controlling the drug release according to the environment [92].

Amphotericin B was encapsulated in polymeric lipid hybrid nanoparticles, with lecithin as the lipid and gelatin as the coating polymer [93]. These particles increased the bioavailability and reduced the initial burst release of the drug, but demonstrated low stability for little pH variations, which may be overcome by the addition of an enteric coating. In this line of action, a research group synthesized a biodegradable, pH-sensitive polymer with amino acids (L-lysine and L-leucine) and poly(ester amide) (PEA) to form microspheres and transport insulin by oral route [94]. These amino acid-based PEA systems would be degraded by enzymes (elastase and α-chymotrypsin) in the intestine and would protect the peptide drug from the harsh conditions of the stomach. The microspheres maintained almost the entire drug encapsulated in simulated gastric conditions (pH = 1.2), wrinkled at simulated intestinal conditions (pH = 6.8), rather than swelling like methacrylates polymers. At pH = 7.4, an initial burst release was observed and the particles showed a dose-dependent in vivo hypoglycemic effect, with a reduced fasting glucose levels down to almost 50 % in the first 5 h, which continued until 8 h [94].

Finally, in terms of preparation methods, it was found that in the case of SLNs, ultrasonography and high pressure homogenization (HPH) were both suitable methods and capable of forming stable lipid structures as well [95]. Still, HPH greatly improved the stability and the loading capacity and was more suitable for lab-scaling SLNs production. Nevertheless, it is important to develop adequate methods for producing stable lipid carriers at industrial scale. SLNs also showed promising results when incorporated in hydrogels for prolonged contact with the mucosa, sustained drug release and facilitated application [96], and also for a 2-year storage stability testing [97]. Still, another study concluded that for the same type and amount of surfactants and stabilizing agents NLCs were more stable than SLNs [85]. There are also some diverging opinions in terms of classification of some lipidic carriers (e.g., liposomes) as new chemical identities or as line extensions [14]. More clarifying studies are needed in order to move these systems forward in terms of validation and regulatory approval.

5.4.6 Porous Nanoparticles

Mesoporous silica carriers are good for entrapping drugs in the pores and deliver them in a targeted manner [3], but as this strategy is relatively new and there are no reports of scaling-up, some problems may occur in maintaining the nanoparticles intact, after exposure to pressure-based methods. Microporous scaffolds—silica particles included—can accommodate cells and release drugs at a controlled way, for both in vitro cell culture and in vivo implantation [98]. In order to take advantage of this material and improve its characteristics, ligands or coatings can also be added as carriers. For example, silica-lipid nanoparticles combine the solubilizing effect of lipids—through a lecithin-based emulsion system—and the stabilizing effects of silica for class II biopharmaceuticals, like celecoxib [99]. The particles showed a porous matrix structure and were stable at room temperature for at least 6 months. Porous silicon is apparently biocompatible and able to increase the solubility of certain drugs, and it can be adapted into the top–down production methods [100]. However, porous silicon nanoparticles are degraded at higher pH values and show reduced cell uptake [100], which may be difficult for their application as DDS to the intestinal mucosa. Still, when comparing different porous and solid particles, one study has demonstrated that in vivo behavior of porous silica nanoparticles was greater than the solid silica nanoparticles, in terms of the biopharmaceutical silymarin incorporation, sustained release for 72 h, absorption extent, and bioavailability [101]. Porous silica shows promising results for oral delivery with prolonged action—especially for drugs with narrow therapeutic window—and enhanced drug absorption by increased contact with mucosa [102]. The addition of surface-active proteins like hydrophobins increased cell viability, stomach retention, and mucosa bioadhesion for 2 h [102]. After entering the small intestine, the nanoparticles would lose their mucoadhesive properties and consequently be removed from the GI tract. Ultimately, porous silica nanoparticles can also be used as oral vaccine adjuvants, as they show high loading capacity for antigens and higher surface area due to their porous structure [103]. As silica is nonsoluble, it can form a depot and release the antigens in a controlled-rate manner, protect the drugs from the gastric medium, and improve oral immunization and mucosal immunization with significant levels of immunoglobulin G (IgG) and immunoglobulin A (IgA) antibodies [103]. This confirms the theoretical rational that vaccines by oral route can induce both mucous membrane and immune system response concomitantly [18].

5.5 Peptide-Mimetic Substances and Other Complexes and Ligands

In some cases, not even carriers or other DDS work for encapsulating specific proteins or peptides. Sometimes, it is actually difficult for hydrophobic polymers to encapsulate hydrophilic drugs and the readjustment of production techniques takes a huge amount of time, resources, and costs. The definition of protocols and guidelines for stable carrier production is one solution [5], but it may not be applied for all testing substances. Many alternatives to insulin have been investigated due to the difficulty in finding a suitable stable carrier for this drug for oral delivery. Incretin hormones, glucagon-like peptide 1, glucose-dependent insulinotropic polypeptide (GIP), peptide YY, and liraglutide— an incretin mimetic and long-acting glucagon-like peptide analogue with FDA approval for diabetes treatment—are some of the chosen ones [7]. Oral anticoagulants, like dabigatranetexilate, rivaroxaban (factor Xa inhibitors), apixaban (selective direct inhibitor of factor Xa), and edoxaban (oral direct factor Xa inhibitor) are also studied since this therapy by oral administration would become more comfortable for the patient and reduce the interactions with other drugs or food [104]. Still, there are some limitations in terms of the new drug absorption, clearance, high costs, and, in the case of the anticoagulants, lack of antidotes in case of major bleeding.

One main aspect in the stability of peptides or proteins is mainly the loss of tertiary and secondary protein structure. In fact, proteins show naturally good emulsification properties, especially, the ones composed of α-helical and/or random coil structures, but they may loose the structure and lose its structure and the property, after interfacial adsorption [5]. Thus, derivatives and new synthetic peptides have been developed to overcome these issues. Peptoids are a novel class of peptidomimetics that have different side chains attached to the backbone amide nitrogen instead of the α-carbon of common peptides. They are used as antimicrobial agents, molecular transporters for intracellular drug delivery, or ligands for tumor receptor binding [105]. In a recent study, three cationic amphipathic peptoids were evaluated in terms of their in vivo biodistribution profile. The tripeptoid appeared to be metabolically stable, with generally high in vivo accumulation and slower elimination than peptides, but with less absorption [105]. Deuterohemin-peptide conjugate (DhHP-6) is a microperoxidase with in vivo reactive oxygen species (ROS) scavenger properties, but also with a low resistance to enzymatic degradation. Several derivatives from DhHP-6 with specific multisite N-methyl modifications were synthesized, and it was observed that the N-methyl groups hindered the contact of enzymes and prevented the formation of H-bonds essential to link the peptide to the enzyme, increasing the peptide stability; however, this derivative showed low permeability both in vitro and ex vivo [106]. Another study compared a tripeptoid with a tetrapeptide in terms of oral absorption [107]. The peptoid appears to have advantages in terms of metabolic activity but also showed low oral absorption and rapid biliary excretion [107].

New purified subunits such as protein and peptide antigens are now being explored to substitute attenuated or inactivated microorganisms for oral vaccination. Nonpeptidic analogs, such as LDV peptidomimetic, triplicated the internalization by macrophages, compared with bare polymeric nanoparticles and also increased the immune response [108]. Other approaches like attaching fatty acids to carriers can increase the stabilization and the blood circulation of drugs, including peptides, and proteins, and then be naturally removed by enzymes (e.g., peptidases or esterases). This lipidization technique can improve the stability and pharmacokinetic profile of protein and peptide oral delivery [109]. One example is the formation of an insulin-sodium oleate complex in order to enhance the hydrophobicity of insulin and increase the encapsulation efficiency into PLGA nanoparticles [110]. The insulin complex was greatly incorporated into the nanoparticles (i.e., $> 90\,\%$), reduced the fasting plasma glucose for 24 h, and also improved the oral glucose tolerance, which was overall lower than the control group. Lipid-raft drug conjugates are also studied to maximize the amount of drug transport, with higher affinity toward a transporter and cellular accumulation [111].

5.6 Conclusions and Future Perspectives

This chapter presented the main advances in terms of methods for improving oral absorption of biopharmaceutical drugs. Although conventional methods are still widely used, drug carriers are gaining an outstanding weight in this field, as a result of developing nanoscale formulations and adapting the production methods to them. The number of pharmacologically active molecules with poor biopharmaceutical properties has increased over the past 15–20 years, and more than 50 % of all new drug candidates that enter the R&D pipeline fail because they are not optimized [3]. After oral administration, the degree of absorption is very variable and formulations must improve the drug bioavailability. There are, as seen in this chapter, hundreds of distinct ideas for increasing oral absorption of drugs but few of them are really getting through the FDA and EMA approvals for commercial aims. In fact, the few examples that we have in the market are acceptable for line extensions, which does not resolve the problem of the new drugs stopping at the middle of the R&D process, due to lack of stability or in vivo performance [14]. The pharmaceutical industry has a difficult task in deciding the best formulation from all of the new technologies and products suitable for oral delivery of drugs. Another problem is the still remaining gap between the development of new products, for example, highly sophisticated coated nanocarriers, and the methods and techniques used to evaluate their behavior. In vivo models, mainly rodents, show specific differences to humans (e.g., less barrier effect, higher density of Peyer's patches, and less acidic stomachs) that can compromise the results interpretation, is another example [30]. Cytotoxicity and cellular viability tests are also sometimes not appropriate and physical and chemical analysis of the materials do not go as deep as they should go.

We are now in the right direction for improving the safety, effectiveness, and acceptance of new ways for oral drug delivery. However, it is our job to try to minimize the gap between what is done in the laboratory and what is expected by the industry. The oral route continues to be one of the most studied administration route due to its enormous advantages, but there is still a lot to do in terms of stability and in vivo performance of biopharmaceuticals.

References

1. Elgart A, Cherniakov I, Aldouby Y, Domb AJ, Hoffman A. Lipospheres and pro-nano lipospheres for delivery of poorly water soluble compounds. Chem Phys Lipids. 2012;165(4): 438–53.
2. Reis CP, Damgé C. Nanotechnology as a promising strategy for alternative routes of insulin delivery. In: Düzgünes N, editor. Methods in enzymology. Vol. 508. Academic Press; 2012. pp. 271–94.
3. Singh A, Worku ZA, Van den Mooter G. Oral formulation strategies to improve solubility of poorly water-soluble drugs. Expert Opin Drug Deliv. 2011;8(10):1361–78.
4. Li P, Zhao L. Developing early formulations: practice and perspective. Int J Pharm. 2007;341(1–2):1–19.
5. van der Walle CF, Sharma G, Ravi Kumar M. Current approaches to stabilising and analysing proteins during microencapsulation in PLGA. Expert Opin Drug Deliv. 2009;6(2):177–86.
6. Merisko-Liversidge E, Liversidge GG. Nanosizing for oral and parenteral drug delivery: a perspective on formulating poorly-water soluble compounds using wet media milling technology. Adv Drug Deliv Rev. 2011;63(6):427–40.
7. Rekha MR, Sharma CP. Oral delivery of therapeutic protein/peptide for diabetes—future perspectives. Int J Pharm. 2013;440(1):48–62.
8. Kawabata Y, Wada K, Nakatani M, Yamada S, Onoue S. Formulation design for poorly water-soluble drugs based on biopharmaceutics classification system: basic approaches and practical applications. Int J Pharm. 2011;420(1):1–10.
9. Kawakami K. Modification of physicochemical characteristics of active pharmaceutical ingredients and application of supersaturatable dosage forms for improving bioavailability of poorly absorbed drugs. Adv Drug Deliv Rev. 2012;64(6):480–95.
10. Rahman MA, Hussain A, Hussain MS, Mirza MA, Iqbal Z. Role of excipients in successful development of self-emulsifying/microemulsifying drug delivery system (SEDDS/SMEDDS). Drug Dev Ind Pharm. 2012;39(1):1–19.
11. Müller RH, Jacobs C, Kayser O. Nanosuspensions as particulate drug formulations in therapy: rationale for development and what we can expect for the future. Adv Drug Deliv Rev. 2001;47(1):3–19.
12. Bansal K, Pant P, Rao P, Padhee K, Sathapathy A, Kochhar P. Micronization and dissolution enhacement of norethindrone. IJRPC. 2011;1:315–9.
13. Joshi JT. A review on micronization techniques. J Pharma Scie Technol. 2011;3(7):651–81.
14. van Hoogevest P, Liu X, Fahr A. Drug delivery strategies for poorly water-soluble drugs: the industrial perspective. Expert Opin Drug Deliv. 2011;8(11):1481–500.

15. Kesisoglou F, Mitra A. Crystalline nanosuspensions as potential toxicology and clinical oral formulations for BCS II/IV compounds. AAPS J. 2012;14(4):677–87.
16. Bosselmann S, Williams RO. Has nanotechnology led to improved therapeutic outcomes? Drug Dev Ind Pharm. 2012;38(2):158–70.
17. Gibaud S, Attivi D. Microemulsions for oral administration and their therapeutic applications. Expert Opin Drug Deliv. 2012;9(8):937–51.
18. Moutinho CG, Matos CM, Teixeira JA, Balcão VM. Nanocarrier possibilities for functional targeting of bioactive peptides and proteins: state-of-the-art. J Drug Target. 2011;20(2): 114–41.
19. Sprunk A, Strachan CJ, Graf A. Rational formulation development and in vitro assessment of SMEDDS for oral delivery of poorly water soluble drugs. Eur J Pharm Sci. 2012;46(5): 508–15.
20. Nishino Y, Kubota A, Kanazawa T, Takashima Y, Ozeki T, Okada H. Improved intestinal absorption of a poorly water-soluble oral drug using mannitol microparticles containing a nanosolid drug dispersion. J Pharm Sci. 2012;101(11):4191–200.
21. Nanjwade BK, Patel DJ, Udhani RA, Manvi FV. Functions of lipids for enhancement of oral bioavailability of poorly water-soluble drugs. Sci Pharm. 2011;79(4):705–27.
22. Mohsin K. Design of lipid-based formulations for oral administration of poorly water-soluble drug fenofibrate: effects of digestion. AAPS PharmSciTech. 2012;13(2):637–46.
23. Lv L, Tong C, Lv Q, Tang X, Li L, Fang Q, Yu J, Han M, Gao J. Enhanced absorption of hydroxysafflor yellow A using a self-double-emulsifying drug delivery system: in vitro and in vivo studies. Int J Nanomedicine. 2012;7:4099–107.
24. Zhang J, Peng Q, Shi S, Zhang Q, Sun X, Gong T, Zhang Z. Preparation, characterization, and in vivo evaluation of a self-nanoemulsifying drug delivery system (SNEDDS) loaded with morin-phospholipid complex. Int J Nanomedicine. 2011;6:3405–15.
25. Thomas N, Holm R, Müllertz A, Rades T. In vitro and in vivo performance of novel supersaturated self-nanoemulsifying drug delivery systems (super-SNEDDS). J Control Release. 2012;160(1):25–32.
26. Villar AMS, Naveros BC, Campmany ACC, Trenchs MA, Rocabert CB, Bellowa LH. Design and optimization of self-nanoemulsifying drug delivery systems (SNEDDS) for enhanced dissolution of gemfibrozil. Int J Pharm. 2012;431(1–2):161–75.
27. Lu Y, Park K. Polymeric micelles and alternative nanonized delivery vehicles for poorly soluble drugs. Int J Pharm. 2013;453(1):198–214.
28. Mou D, Chen H, Wan J, Xu H, Yang X. Potent dried drug nanosuspensions for oral bioavailability enhancement of poorly soluble drugs with pH-dependent solubility. Int J Pharm. 2011;413(1–2):237–44.
29. He C, Yin L, Tang C, Yin C. Size-dependent absorption mechanism of polymeric nanoparticles for oral delivery of protein drugs. Biomaterials. 2012;33(33):8569–78.
30. Ensign LM, Cone R, Hanes J. Oral drug delivery with polymeric nanoparticles: the gastrointestinal mucus barriers. Adv Drug Deliv Rev. 2012;64(6):557–70.
31. Zhang X, Sun M, Zheng A, Cao D, Bi Y, Sun J. Preparation and characterization of insulin-loaded bioadhesive PLGA nanoparticles for oral administration. Eur J Pharm Sci. 2012;45(5):632–8.
32. Wu Z, Ling L, Zhou L, Guo X, Jiang W, Qian Y, Luo K, Zhang L. Novel preparation of PLGA/HP55 nanoparticles for oral insulin delivery. Nanoscale Res Lett. 2012;7(1):299.
33. Xie X, Tao Q, Zou Y, Zhang F, Guo M, Wang Y, Wang H, Zhou Q, Yu S. PLGA Nanoparticles improve the oral bioavailability of curcumin in rats: characterizations and mechanisms. J Agric Food Chem. 2011;59(17):9280–9.
34. Khalil NM, do Nascimento TCF, Casa DM, Dalmolin LF, Mattos ACD, Hoss I, Romano MA, Mainardes RM. Pharmacokinetics of curcumin-loaded PLGA and PLGA-PEG blend nanoparticles after oral administration in rats. Colloids Surf B Biointerfaces. 2013;101:353–60.
35. Ghosh D, Choudhury ST, Ghosh S, Mandal AK, Sarkar S, Ghosh A, Saha KD, Das N. Nanocapsulated curcumin: oral chemopreventive formulation against diethylnitrosamine induced hepatocellular carcinoma in rat. Chem Biol Interact. 2012;195(3):206–14.

36. Ma Y, Zhao X, Li J, Shen Q. The comparison of different daidzein-PLGA nanoparticles in increasing its oral bioavailability. Int J Nanomedicine. 2012;7:559–70.
37. Semete B, Booysen LIJ, Kalombo L, Venter JD, Katata L, Ramalapa B, Verschoor JA, Swai H. In vivo uptake and acute immune response to orally administered chitosan and PEG coated PLGA nanoparticles. Toxicol Appl Pharmacol. 2010;249(2):158–65.
38. Murugeshu A, Astete C, Leonardi C, Morgan T, Sabliov CM. Chitosan/PLGA particles for controlled release of alpha-tocopherol in the GI tract via oral administration. Nanomedicine. 2011;6(9):1513–28.
39. Chang P-C, Lim LP, Chong LY, Dovban ASM, Chien L-Y, Chung M-C, Lei C, Kao M-J, Chen C-H, Chiang H-C, Kuo Y-P, Wang C-H. PDGF-simvastatin delivery stimulates osteogenesis in heat-induced osteonecrosis. J Dental Res. 2012;91(6):618–24.
40. Mittal G, Carswell H, Brett R, Currie S, Kumar MNVR. Development and evaluation of polymer nanoparticles for oral delivery of estradiol to rat brain in a model of Alzheimer's pathology. J Control Release. 2011;150(2):220–8.
41. Jain S, Rathi VV, Jain AK, Das M, Godugu C. Folate-decorated PLGA nanoparticles as a rationally designed vehicle for the oral delivery of insulin. Nanomedicine. 2012;7(9):1311–37.
42. Hattori Y, Maitani Y. Enhanced in vitro DNA transfection efficiency by novel folate-linked nanoparticles in human prostate cancer and oral cancer. J Control Release. 2004;97(1):173–83.
43. Roger E, Kalscheuer S, Kirtane A, Guru BR, Grill AE, Whittum-Hudson J, Panyam J. Folic acid functionalized nanoparticles for enhanced oral drug delivery. Mol Pharm. 2012;9(7):2103–10.
44. Yamanaka Y, Leong K. Engineering strategies to enhance nanoparticle-mediated oral delivery. J Biomater Sci Polym Ed. 2008;19(12):1549–70.
45. Sarmento B, Ribeiro A, Veiga F, Ferreira D, Neufeld R. Oral bioavailability of insulin contained in polysaccharide nanoparticles. Biomacromolecules. 2007;8(10):3054–60.
46. Sarmento B, Ribeiro A, Veiga F, Sampaio P, Neufeld R, Ferreira D. Alginate/chitosan nanoparticles are effective for oral insulin delivery. Pharm Res. 2007;24(12):2198–206.
47. Han H-K, Shin H-J, Ha DH. Improved oral bioavailability of alendronate via the mucoadhesive liposomal delivery system. Eur J Pharm Sci. 2012;46(5):500–7.
48. Sung H-W, Sonaje K, Liao Z-X, Hsu L-W, Chuang E-Y. pH-responsive nanoparticles shelled with chitosan for oral delivery of insulin: from mechanism to therapeutic applications. Acc Chem Res. 2012;45(4):619–29.
49. Garcia-Fuentes M, Alonso MJ. Chitosan-based drug nanocarriers: where do we stand? J Control Release. 2012;161(2):496–504.
50. Werle M, Makhlof A, Takeuchi H. Carbopol-lectin conjugate coated liposomes for oral peptide delivery. Chem Pharm Bull. 2010;58(3):432–4.
51. Rawat M, Singh D, Saraf S, Saraf S. Development and in vitro evaluation of alginate gel-encapsulated, chitosan-coated ceramic nanocores for oral delivery of enzyme. Drug Dev Ind Pharm. 2008;34(2):181–8.
52. Derakhshandeh K, Fathi S. Role of chitosan nanoparticles in the oral absorption of Gemcitabine. Int J Pharm. 2012;437(1–2):172–7.
53. Hosseinzadeh H, Atyabi F, Dinarvand R, Ostad S. Chitosan–Pluronic nanoparticles as oral delivery of anticancer gemcitabine: preparation and in vitro study. Int J Nanomedicine. 2012;7:1851–63.
54. Kowapradit J, Apirakaramwong A, Ngawhirunpat T, Rojanarata T, Sajomsang W, Opanasopit P. Methylated N-(4-N, N-dimethylaminobenzyl) chitosan coated liposomes for oral protein drug delivery. Eur J Pharm Sci. 2012;47(2):359–66.
55. Chen H, Wu J, Sun M, Guo C, Yu A, Cao F, Zhao L, Tan Q, Zhai G. N-trimethyl chitosan chloride-coated liposomes for the oral delivery of curcumin. J Liposome Res. 2012;22(2):100–9.
56. Venishetty VK, Chede R, Komuravelli R, Adepu L, Sistla R, Diwan PV. Design and evaluation of polymer coated carvedilol loaded solid lipid nanoparticles to improve the oral bioavailability: a novel strategy to avoid intraduodenal administration. Colloids Surf B Biointerfaces. 2012;95:1–9.

57. Sugihara H, Yamamoto H, Kawashima Y, Takeuchi H. Effects of food intake on the mucoadhesive and gastroretentive properties of submicron-sized chitosan-coated liposomes. Chem Pharm Bull (Tokyo). 2012;60(10):1320–3.
58. Llabot JM, Salman H, Millotti G, Bernkop-Schnürch A, Allemandi D, Manuel Irache J. Bioadhesive properties of poly(anhydride) nanoparticles coated with different molecular weights chitosan. J Microencapsul. 2011;28(5):455–63.
59. Lalatsa A, Garrett NL, Ferrarelli T, Moger J, Schätzlein AG, Uchegbu IF. Delivery of peptides to the blood and brain after oral uptake of quaternary ammonium palmitoyl glycol chitosan nanoparticles. Mol Pharm. 2012;9(6):1764–74.
60. Dünnhaupt S, Barthelmes J, Iqbal J, Perera G, Thurner CC, Friedl H, Bernkop-Schnürch A. In vivo evaluation of an oral drug delivery system for peptides based on S-protected thiolated chitosan. J Control Release. 2012;160(3):477–85.
61. Gradauer K, Vonach C, Leitinger G, Kolb D, Fröhlich E, Roblegg E, Bernkop-Schnürch A, Prassl R. Chemical coupling of thiolated chitosan to preformed liposomes improves mucoadhesive properties. Int J Nanomedicine. 2012;7:2523–34.
62. Wan S, Sun Y, Qi X, Tan F. Improved bioavailability of poorly water-soluble drug curcumin in cellulose acetate solid dispersion. AAPS PharmSciTech. 2012;13(1):159–66.
63. Cui F-D, Tao A-J, Cun D-M, Zhang L-Q, Shi K. Preparation of insulin loaded PLGA-Hp55 nanoparticles for oral delivery. J Pharm Sci. 2007;96(2):421–7.
64. Sharma M, Sharma V, Panda AK, Majumdar DK. Development of enteric submicron particle formulation of papain for oral delivery. Int J Nanomedicine. 2011;6:2097–111.
65. Barea MJ, Jenkins MJ, Gaber MH, Bridson RH. Evaluation of liposomes coated with a pH responsive polymer. Int J Pharm. 2010;402(1–2):89–94.
66. Barea M, Jenkins M, Lee Y, Johnson P, Bridson R. Encapsulation of liposomes within pH responsive microspheres for oral colonic drug delivery. Int J Biomater. 2012;2012:458712.
67. Zhang Y, Wu X, Meng L, Zhang Y, Ai R, Qi N, He H, Xu H, Tang X. Thiolated Eudragit nanoparticles for oral insulin delivery: preparation, characterization and in vivo evaluation. Int J Pharm. 2012;436(1–2):341–50.
68. Zhu Q, Talton J, Zhang G, Cunningham T, Wang Z, Waters R, Kirk J, Eppler B, Klinman D, Sui Y, Gagnon S, Belyakov I, Mumper R, Berzofsky J. Large intestine-targeted, nanoparticle-releasing oral vaccine to control genitorectal viral infection. Nature Med. 2012;18:1291–6.
69. Su F-Y, Lin K-J, Sonaje K, Wey S-P, Yen T-C, Ho Y-C, Panda N, Chuang E-Y, Maiti B, Sung H-W. Protease inhibition and absorption enhancement by functional nanoparticles for effective oral insulin delivery. Biomaterials. 2012;33(9):2801–11.
70. Feng S-S, Mei L, Anitha P, Gan CW, Zhou W. Poly(lactide)- Vitamin E derivative/montmorillonite nanoparticle formulations for the oral delivery of Docetaxel. Biomaterials. 2009;30(19):3297–306.
71. Zhang Z, Tan S, Feng S-S. Vitamin E TPGS as a molecular biomaterial for drug delivery. Biomaterials. 2012;33(19):4889–906.
72. Gupta M, Vyas SP. Development, characterization and in vivo assessment of effective lipidic nanoparticles for dermal delivery of fluconazole against cutaneous candidiasis. Chem Phys Lipids. 2012;165(4):454–61.
73. Mishra B, Patel BB, Tiwari S. Colloidal nanocarriers: a review on formulation technology, types and applications toward targeted drug delivery. Nanomedicine 2010;6(1):9–24.
74. Reis CP. Encapsulação de fármacos peptídicos pelo método de emulsificação/gelificação interna, PhD Thesis, Coimbra, Portugal; 2008.
75. Reis CP, Ribeiro AJ, Veiga F, Neufeld RJ, Damgé C. Polyelectrolyte biomaterial interactions provide nanoparticulate carrier for oral insulin delivery. Drug Deliv. 2008;15(2):127–39.
76. Deutel B, Greindl M, Thaurer M, Bernkop-Schnürch A. Novel insulin thiomer nanoparticles: in vivo evaluation of an oral drug delivery system. Biomacromolecules. 2007;9(1):278–85.
77. Cheddadi M, López-Cabarcos E, Slowing K, Barcia E, Fernández-Carballido A. Cytotoxicity and biocompatibility evaluation of a poly(magnesium acrylate) hydrogel synthesized for drug delivery. Int J Pharm. 2011;413(1–2):126–33.

78. Alam S, Panda J, Chauhan V. Novel dipeptide nanoparticles for effective curcumin delivery. Int J Nanomedicine. 2012;7:4207–21.
79. Chen Y, Yuan L, Zhou L, Zhang Z, Cao W, Wu Q. Effect of cell-penetrating peptide-coated nanostructured lipid carriers on the oral absorption of tripterine. Int J Nanomedicine. 2012;7:4581–91.
80. Brasseur R, Divita G. Happy birthday cell penetrating peptides: already 20 years. Biochim Biophys Acta (BBA). 2010;1798(12):2177–81.
81. González-Aramundiz J, Lozano MV, Sousa-Herves A, Fernandez-Megia E, Csaba N. Polypeptides and polyaminoacids in drug delivery. Expert Opini Drug Deliv. 2012;9(2):183–201.
82. Khafagy E-S, Morishita M. Oral biodrug delivery using cell-penetrating peptide. Adv Drug Delivery Rev. 2012;64(6):531–9.
83. Patel J, Patel A, Raval M, Sheth N. Formulation and development of a self-nanoemulsifying drug delivery system of irbesartan. J Adv Pharm Technol Res. 2011;2(1):9–16.
84. Hu S, Niu M, Hu F, Lu Y, Qi J, Yin Z, Wu W. Integrity and stability of oral liposomes containing bile salts studied in simulated and ex vivo gastrointestinal media. Int J Pharm. 2013;441(1-2):693–70.
85. Das S, Ng WK, Tan RBH. Are nanostructured lipid carriers (NLCs) better than solid lipid nanoparticles (SLNs): development, characterizations and comparative evaluations of clotrimazole-loaded SLNs and NLCs? Eur J Pharm Sci. 2012;47(1):139–51.
86. Zeng N, Gao X, Hu Q, Song Q, Xia H, Liu Z, Gu G, Jiang M, Pang Z, Chen H, Chen J, Fang L. Lipid-based liquid crystalline nanoparticles as oral drug delivery vehicles for poorly water-soluble drugs: cellular interaction and in vivo absorption. Int J Nanomedicine. 2012;7:3703–18.
87. Holpuch A, Hummel G, Tong M, Seghi G, Pei P, Ma P, Mumper R, Mallery S. Nanoparticles for local drug delivery to the oral mucosa: proof of principle studies. Pharm Res. 2012;27(7):1224–36.
88. Li X, Chen D, Le C, Zhu C, Gan Y, Hovgaard L, Yang M. Novel mucus-penetrating liposomes as a potential oral drug delivery system: preparation, in vitro characterization, and enhanced cellular uptake. Int J Nanomedicine. 2011;6:3151–62.
89. Gosangari SL, Watkin KL. Enhanced release of anticancer agents from nanoliposomes in response to diagnostic ultrasound energy levels. Pharm Dev Technol. 2012;17(3):383–8.
90. Parmentier J, Becker MMM, Heintz U, Fricker G. Stability of liposomes containing bio-enhancers and tetraether lipids in simulated gastro-intestinal fluids. Int J Pharm. 2011;405(1–2):210–7.
91. Shi F, Zhao J, Liu Y, Wang Z, Zhang Y, Feng N. Preparation and characterization of solid lipid nanoparticles loaded with frankincense and myrrh oil. Int J Nanomedicine. 2012;7:2033–43.
92. MaHam A, Tang Z, Wu H, Wang J, Lin Y. Protein-based nanomedicine platforms for drug delivery. Small. 2009;5(15):1706–21.
93. Jain S, Valvi PU, Swarnakar NK, Thanki K. Gelatin coated hybrid lipid nanoparticles for oral delivery of Amphotericin B. Mol Pharm. 2012;9(9):2542–53.
94. He P, Tang Z, Lin L, Deng M, Pang X, Zhuang X, Chen X. Novel biodegradable and pH-sensitive poly(ester amide) microspheres for oral insulin delivery. Macromol Biosci. 2012;12(4):547–56.
95. Silva AC, González-Mira E, García ML, Egea MA, Fonseca J, Silva R, Santos D, Souto EB, Ferreira D. Preparation, characterization and biocompatibility studies on risperidone-loaded solid lipid nanoparticles (SLN): high pressure homogenization versus ultrasound. Colloids Surf B Biointerfaces. 2011;86(1):158–65.
96. Silva AC, Amaral MH, E. González-Mira, Santos D, Ferreira D. Solid lipid nanoparticles (SLN)—based hydrogels as potential carriers for oral transmucosal delivery of Risperidone: preparation and characterization studies. Colloids Surf B Biointerfaces. 2012;93:241–8.
97. Silva AC, Kumar A, Wild W, Ferreira D, Santos D, Forbes B. Long-term stability, biocompatibility and oral delivery potential of risperidone-loaded solid lipid nanoparticles. Int J Pharm. 2012;436(1–2):798–805.

98. Wenk E, Merkle HP, Meinel L. Silk fibroin as a vehicle for drug delivery applications. J Control Release. 2011;150(2):128–41.
99. Tan A, Simovic S, Davey AK, Rades T, Prestidge CA. Silica-lipid hybrid (SLH) microcapsules: a novel oral delivery system for poorly soluble drugs. J Control Release. 2009;134(1):62–70.
100. Bimbo LM, Mäkilä E, Laaksonen T, Lehto V-P, Salonen J, Hirvonen J, Santos HA. Drug permeation across intestinal epithelial cells using porous silicon nanoparticles. Biomaterials. 2011;32(10):2625–33.
101. Cao X, Fu M, Wang L, Liu H, Deng W, Qu R, Su W, Wei Y, Xu X, Yu J. Oral bioavailability of silymarin formulated as a novel 3-day delivery system based on porous silica nanoparticles. Acta Biomaterialia. 2012;8(6):2104–12.
102. Sarparanta MP, Bimbo LM, Mäkilä EM, Salonen JJ, Laaksonen PH, Helariutta AMK, Linder MB, Hirvonen JT, Laaksonen TJ, Santos HA, Airaksinen AJ. The mucoadhesive and gastroretentive properties of hydrophobin-coated porous silicon nanoparticle oral drug delivery systems. Biomaterials. 2012;33(11):3353–62.
103. Wang T, Jiang H, Zhao Q, Wang S, Zou M, Cheng G. Enhanced mucosal and systemic immune responses obtained by porous silica nanoparticles used as an oral vaccine adjuvant: effect of silica architecture on immunological properties. Int J Pharm. 2012;436(1–2):351–8.
104. Hoffman R, Brenner B. The promise of novel direct oral anticoagulants. Best Pract Res Clin Haematol. 2012;25(3):351–60.
105. Seo J, Ren G, Liu H, Miao Z, Park M, Wang Y, Miller TM, Barron AE, Cheng Z. In vivo biodistribution and small animal PET of 64Cu-Labeled antimicrobial peptoids. Bioconjug Chem. 2012;23(5):1069–79.
106. Dong Q-G, Zhang Y, Wang M-S, Feng J, Zhang H-H, Wu Y-G, Gu T-J, Yu X-H, Jiang C-L, Chen Y, Li W, Kong W. Improvement of enzymatic stability and intestinal permeability of deuterohemin-peptide conjugates by specific multi-site N-methylation. Amino Acids. 2012;1–11.
107. Wang Y, Lin H, Tullman R, Jewell CF, Weetall ML, Tse FLS. Absorption and disposition of a tripeptoid and a tetrapeptide in the rat. Biopharm Drug Dispos. 1999;20(2):69–75.
108. Fievez V, Plapied L, des Rieux A, Pourcelle V, Freichels H, Wascotte V, Vanderhaeghen M-L, Jerôme C, Vanderplasschen A, Marchand-Brynaert J, Schneider Y-J, Préat V. Targeting nanoparticles to M cells with non-peptidic ligands for oral vaccination. Eur J Pharm Biopharm. 2009;73(1):16–24.
109. Hackett MJ, Zaro JL, Shen W-C, Guley PC, Cho MJ. Fatty acids as therapeutic auxiliaries for oral and parenteral formulations. Adv Drug Deliv Rev. 2012.
110. Sun S, Liang N, Piao H, Yamamoto H, Kawashima Y, Cui F. Insulin-S.O (sodium oleate) complex-loaded PLGA nanoparticles: formulation, characterization and in vivo evaluation. J Microencapsul. 2010;27(6):471–8.
111. Vadlapudi AD, Vadlapatla RK, Kwatra D, Earla R, Samanta SK, Pal D, Mitra AK. Targeted lipid based drug conjugates: a novel strategy for drug delivery. Int J Pharm. 2012;434(1–2):315–24.

Chapter 6
Buccal Delivery of Biopharmaceuticals: Vaccines and Allergens

Sevda Şenel, Merve Cansız and Michael J. Rathbone

6.1 Introduction

The term "biopharmaceutical" covers recombinant therapeutic proteins, monoclonal antibody-based products used for in vivo medical purposes, nucleic acid-based medicinal products, and engineered cell or tissue-based products [1, 2], which are produced by biotechnological methods involving bioprocessing. In the USA, to date the Food and Drug Administration (FDA) has not yet established the regulatory definition of a biopharmaceutical. However, this term is commonly used to refer to all therapeutic, prophylactic, and in vivo diagnostic products manufactured using live organisms or derived functional components [3]. In 2003, the FDA transferred the responsibility for regulating most therapeutic biologics, with certain exceptions (e.g., cell and gene therapy products and therapeutic vaccines) from the Office of Therapeutics Research and Review (OTRR) and the Center for Biologics Evaluation and Research (CBER), to the Office of New Drugs (OND), the Office of Pharmaceutical Science (OPS), and the Center for Drug Evaluation and Research (CDER) [4]. Among these agents were proteins intended for therapeutic use that are extracted from animals or microorganisms, including recombinant versions of these products, except clotting factors which included most of the recombinant proteins like monoclonal antibodies, cytokines, growth factors, and nonvaccine, nonallergenic immunomodulators [5]. These are now considered as more traditional biopharmaceutical drugs. The biological products, including cellular products, allergenic extracts, antitoxins, venoms, blood, blood components, plasma derived products, vaccines are reviewed in CBER [5]. In Europe, the European Medicines Agency (EMA) uses the term "Biological medicinal product" which is defined as a medicinal product whose active substance is made by or derived from a living organism including recombinant

S. Şenel (✉) · M. Cansız
Department of Pharmaceutical Technology, Faculty of Pharmacy,
Hacettepe University, 06100 Ankara, Turkey
e-mail: ssenel@hacettepe.edu.tr

M. J. Rathbone
School of Pharmacy, International Medical University, Kuala Lumpur, Malaysia

J. das Neves, B. Sarmento (eds.), *Mucosal Delivery of Biopharmaceuticals,* DOI 10.1007/978-1-4614-9524-6_6,

proteins, monoclonal antibodies, medicinal products derived from human blood and human plasma, immunological medicinal products, and advanced therapy medicinal products [6].

The complex protein structures of most biopharmaceuticals limit their delivery via the oral route due to their acidic and enzymatic degradation in the gastrointestinal tract. Furthermore, the high molecular weight of these drugs often results in poor absorption through the gastrointestinal membranes into the systemic circulation when administered orally. The most common route of administration for these therapeutic proteins is parenteral. Most of these proteins have short serum half-lives and need to be administered frequently or in high doses to be effective. Hence, administration of therapeutic proteins still remains a challenge, and studies are still continuing to develop improved delivery systems for these biopharmaceuticals which are capable of maintaining therapeutic levels without undesired effects. During the last two decades, a large number of new delivery technologies have been designed to deliver a biopharmaceutical, specifically to the target site at an effective concentration and at the right time, safely [7, 8]. Yet, there is still a need for systems which improve the stability and efficacy of the biopharmaceutical in the target tissue as well as enhance patient compliance by reducing the dosing frequency. Among the strategies for an improved therapy with biopharmaceuticals is their delivery via different routes of administration such as oral and nasal mucosae. In this chapter, we will focus on oral cavity mucosa as an alternative delivery route for biopharmaceuticals.

6.2 Oral Mucosa as a Site for Delivery

The usefulness of the oral cavity as a site for the delivery of drugs locally or as a platform for the delivery of drugs into the systemic circulation has been associated with an ongoing effort over many decades, which, in recent years have seen the successful development of a variety of oral mucosal drug delivery systems [9–11].

The general advantages of using the oral mucosa as a platform upon which to locate delivery technologies for systemic delivery of drugs include that it avoids first pass metabolism and the harsh environment of the gastrointestinal tract since drugs are absorbed directly into the systemic circulation. A further advantage is that drug delivery systems can be easily administered and precisely located onto the mucosa associated with the different regions of the oral cavity. Following administration, the delivery technologies are accessible and can, therefore, be removed to terminate delivery. A final advantage of the oral cavity as a route for drug delivery is that it is well accepted by patients. Its disadvantages result from the functional purposes of the different regions of the oral cavity linings (e.g., taste receptors, mobile tissues to facilitate eating and speaking) and physical/anatomical structures (e.g., the tongue, teeth, and salivary glands).

The oral cavity comprises different regions. These include the floor of the mouth (sublingual region), palatal mucosa, the inside of the cheeks (buccal region), and the gingival (gum region) [12]. Each region is associated with a different function.

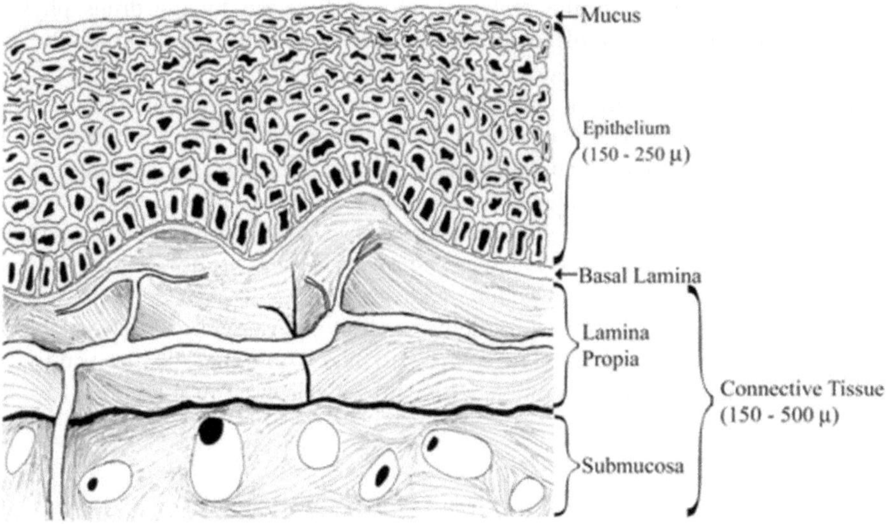

Fig. 6.1 Structure of the buccal mucosa [9]

Each region of the oral cavity offers different advantages and disadvantages as a site for drug delivery. When a rapid onset from the oral cavity is required, the sublingual mucosa is preferred over the buccal mucosa because it possesses a thinner membrane that is supplied by arteries that lie close to the surface and which are associated with high blood flow. However, if prolonged administration is required, this region is associated with several drawbacks. The surface of the sublingual region is constantly bathed by saliva. The constant arrival and removal of saliva to/from the surface of the sublingual membrane can promote unwanted high erosion rates of the delivery system followed by unwanted removal of the drug (or the entire delivery system) from the intended site of absorption. In contrast, the buccal mucosa offers several advantages as a platform for the delivery of drugs over prolonged periods. This is because it has a smooth surface that remains relatively immobile. In addition, the buccal mucosa is relatively permeable (compared to the transcutaneous route), robust (it has a tendency to withstand irritation or damage), and, in comparison with other mucosal tissues, is more tolerant to potential allergens. However, the buccal route is associated with several drawbacks including its relatively small area for absorption and its relatively thick mucosa which represents a barrier to drug absorption.

Despite the fact that development of an oral mucosal drug delivery system encounters many physiological, anatomical, and pharmaceutical challenges, the oral mucosa remains a viable portal for systemic drug delivery [9–11].

The oral mucosa consists of a laminate of several layers: a mucus layer that covers the epithelium; (in some regions) a keratinized layer on the surface of the epithelium; a basement membrane (basal lamina); a connective tissue (lamina propria); and a loose submucosa (Fig. 6.1) [13–15].

Drug absorption across the oral mucosa is influenced by the drugs physicochemical properties of molecular size, molecular weight, partition coefficient (lipophilicity), extent of ionization, and chemical nature.

Drug transport through the buccal mucosa involves two principal routes: transcellular (intracellular) and paracellular (intercellular) pathways. The transcellular route involves crossing of the cellular membranes which possess a polar and a lipid domain, whereas the paracellular route essentially involves passive diffusion through the extracellular lipid domain. Ionic drugs usually diffuse through the intercellular space via the paracellular route, whereas hydrophobic drugs are able to pass through cellular membranes via the transcellular pathway [16].

When the oral mucosa is used for delivery of biopharmaceuticals, especially for vaccines and allergens, the immune cells distributed within the oral mucosal tissue becomes an important consideration. This will be discussed in more detail in the following sections.

6.3 Delivery of Biopharmaceuticals

Among the studies performed on the delivery of biopharmaceuticals via the oral mucosa, almost all of the data reported on biotechnological products relate to recombinant vaccines and allergens. Therefore, in the remainder of this chapter, we will focus on the delivery of vaccines and allergens via the oral mucosa.

Most vaccines, both therapeutic and prophylactic, are given by parenteral injection, which stimulates the immune system to produce antibodies in the serum but generally fails to generate a mucosal antibody response. Mucosal immunization, which is referred to as a noninvasive route, is an attractive alternative to parenteral immunization, and by using an appropriate delivery system, it is possible to stimulate both the mucosal and systemic immune responses [17, 18]. In recent years, the potential of the oral mucosa, especially buccal and sublingual mucosa, has gained increasing interest for the administration of vaccines or allergens [19]. The sublingual delivery of prophylactic vaccines has been reviewed previously by the same authors [11] to which the reader is referred. In this chapter, we provide an updated list of the recent studies concerned with the sublingual delivery of vaccines, especially recombinant molecules (Table 6.1). The findings of these studies demonstrate that the sublingual route is an effective and safe route for vaccine delivery allowing protective immune responses. However, one must be aware that all these studies were performed in animal models.

Studies have been performed in humans, but these were only concerned with allergen-specific immunotherapies. In the following section, after a brief introduction to allergens and sublingual delivery of allergens, both the extracts and the recombinants as therapeutic vaccines will be reviewed.

Table 6.1 Recent studies on sublingually delivered vaccine antigens

Antigen	Adjuvant/delivery system	Animal model	Route	Response	References
Formalin-inactivated influenza A/PR/8 virus (H1N1)	mCTA-LTB	Female BALB/c mice, polymeric Ig receptor knockout (pIgR–/–) mice, MyD88–/– mice	s.l.	Induced systemic and mucosal antibody responses protection against a lethal i.n. challenge with influenza virus; induced systemic expansion of IFN-γ-secreting T cells and virus-specific cytotoxic T lymphocyte responses	[48]
Live influenza A/PR/8 virus (H1N1)				Provided heterosubtypic protection against respiratory challenge with H3N2 virus; A/PR/8 virus, whether live or inactivated, did not migrate to or replicate in the CNS after s.l. administration	
M2 gene derived from an H5N1 avian influenza virus SeV/ΔF/H5N1M2)	Recombinant F gene-deleted Sendai virus vector	5-week-old pigs	i.m or i.n.	Induced antibody response to M2e, which is cross-reactive to different avian, swine, and human influenza viruses	[49]
		Female C57BL/6 mice	s.l., i.m., sc, i.o	No cross-protection to a heterologous influenza virus in a mouse model	
Porphyromonas gingivalis (40k-OMP)	DNA plasmid encoding FL (pFL)	Female BALB/c mice	s.l.	Induced mucosal and serum immune responses to prevent oral infection by *P. gingivalis*	[50]
Human papillomavirus 16 L1(HPV16L1) protein	TLR agonists: poly(I:C), imiquimod monophosphoryl lipid A (MPL) Nucleotide-binding oligomerization-domain (NOD) agonists: murabutide peptidoglycans (PGN) muramyl di-peptide with a C18 fatty acid chain (L18-MDP); ɣ-polyglutamic acid; vitamin D3; cholera toxin B subunit	Female BALB/c mice	s.l., i.n., i.vag., t.d., i.m.	Induced effective mucosal and serum responses after i.n. and s.l. immunization	[51]

Table 6.1 (continued)

Antigen	Adjuvant/delivery system	Animal model	Route	Response	References
Chlamydia trachomatis mouse pneumonitis (CT-MoPn)	Cholera toxin	Female BALB/c mice	s.l., col.	Better protection against a respiratory challenge with *C. trachomatis* obtained with combined immunization compared systemic or mucosal immunizations alone	[52]
	CpG and Montanide		i.m., s.c.		
Chlamydia muridarum recombinant major outer membrane protein (rMOMP)	CpG and Montanide	Female BALB/c (H-2^d) mice	s.l. + i.n., i.vag. + col, i.m. + s.c.	Significant protection provided against a vaginal challenge with *Chlamydia* with combination of mucosal priming and systemic boosting immunization; immunizations by the s.l./ i.n. (2 ×) + i.m./s.c. (2 ×) schedule provided the best overall protection	[53]
α1-giardin	–	Female BALB/c mice	Oral + s.l., s.l.	Increased protection achieved by cholera toxin co-administration or sublingual antigen administration	[54]
Rat monoclonal IgG1 or IgG2a	–	Female BALB/c mice	s.l	Reduced allergic inflammation by sublingual immunoglobulin (SLIG)	[55]
Truncated version of a *Helicobacter pylori*-specific protein ($HpaA_{trunc}$) + rUreB	Cholera toxin	C57BL/6 mice	s.l., i.g.	Induced strong immune responses with the antigen combination to both antigens and a strong synergistic effect on protection, associated with synergistically increased expression of IL-17 in the stomach; sublingual immunization superior to intragastric immunization	[56]

Table 6.1 (continued)

Antigen	Adjuvant/delivery system	Animal model	Route	Response	References
Tetanus toxoid	*Escherichia coli* heat-labile enterotoxin (LT)	Female BALB/c mice	s.l.	Induced mucosal and systemic immune responses	[57]
	Mutant of LT lacking ADP ribosyltransferase activity (LTK63)			Higher responses obtained by LT-treated mice compared to LTK63 immunized mice	
Bacillus subtilis cells expressing tetanus toxin C fragment (TTFC)	Mutant heat labile toxin (mLT) of *E. coli*	Mice	s.l., i.n.	Similar local and systemic responses obtained with s.l. and i.n. immunizations; superior antibody response to TTFC obtained in absence of mLT	[58]
B. subtilis cells (either spore-based or vegetative cell-based) expressing tetanus toxin C fragment	Mutant heat labile toxin (mLT) of *E. coli*	Weaned piglets	s.l., i.n., oral	Induced local and systemic immune responses, similar to that obtained in mice	[59]
HIV-1 gp41 reverse transcriptase polypeptide coupled to the cholera toxin B subunit (CTB)	Cholera toxin B subunit	Female BALB/c mice	s.l., i.vag., t.d.	Induced B and T cell immune responses comprising secretory antibodies and cytotoxic CD8 T cells in the mouse genital tract	[60]
HIV-1 gp120cn54 protein	Nutritive immune-enhancing delivery system (NIDS) composed of vitamin A, a polyphenol-flavonoid (catechin hydrate) and mustard oil	Female BALB/c mice	s.l. + i.n., i.m.	Enhanced serum and vaginal antibody responses as well as cytokine responses following combined mucosal and systemic vaccination	[61]

Table 6.1 (continued)

Antigen	Adjuvant/delivery system	Animal model	Route	Response	References
Adenovirus serotype 5-based HIV-Gag (Ad5-HIV-Gag)	–	Male C57BL/6 mice	s.l., oral gavage	Induced antigen-specific cytotoxic T-lymphocyte responses in both systemic and mucosal compartments	[62]
Adenovirus serotype 5 vector expressing HIV-1 envelope glycoprotein (HIV-1_{MN} gp120)	–	Female BALB/c mice	s.l.	Elevated levels of HIV-1 envelope glycoprotein-specific serum IgA, and vaginal IgA and IgG	[63]
Herpes Simplex Virus Type-1 amplicon vector (HSV-1) expressing HIV-1 envelope glycoprotein (HIV-1_{MN} gp120)				No specific antibody responses in plasma and in vaginal washes with HSV-1 amplicon vector encoding HIV-1 Env	

CNS central nervous system, *col.* colonic, *HIV* human immunodeficiency virus, *i.g.* intragastric, *i.m.* intramuscular, *i.n.* intranasal, *i.o.* intraocular, *i.vag.* intravaginal, *mCTA/LTB* subunit of mutant cholera toxin E112K combined with the pentameric B subunit of heat-labile enterotoxin from enterotoxigenic *E. coli*, *OMP* outer membrane protein, *s.c.* subcutaneous, *s.l.* sublingual, *t.d.* transdermal, *TLR* toll-like receptor

6.4 Allergens

Allergy is a disease that is a consequence of Type I hypersensitivity reactions which are vigorous responses of the immune system triggered by the interaction of allergens with specific immunoglobulin E (IgE) antibodies leading to the release of inflammatory mediators including histamine, cytokines, and lipid mediators. Test allergens are used for clinical allergy diagnosis. Specific immune therapy with allergen products containing the same antigens is an immuno-modulatory treatment option which is intended to generate persistent relief from allergy symptoms [20].

In general, the diagnosis and immunotherapy currently applied to allergic diseases involve allergens obtained from allergen extracts, allergoids, and conjugates or allergens manufactured using recombinant DNA technology. In recent years, more and more allergens have been generated using recombinant DNA technology [21]. The ability to produce rationally designed hypoallergenic forms of allergens is leading to the development of novel and safe forms of allergy vaccines with improved efficacy. The initial clinical tests on recombinant allergen-based vaccine preparations have provided positive results. Due to the high number of allergens in an allergen extract or in an allergen extract mixture and the cross reactivity of the individual components, it is impossible to determine all relevant parameters for the allergens within a given extract or a defined allergen extract mixture. Recently, the concept of homologous groups, which are allergen extracts prepared from different species, different genera, or different families, have been introduced in the EMA guideline [20].

According to the FDA, allergenic extracts are used for the diagnosis and treatment of allergic diseases such as allergic rhinitis ("hay fever"), allergic sinusitis, allergic conjunctivitis, bee venom allergy, and food allergy. Allergenic extracts are injectable products that are manufactured from natural substances (such as molds, pollens, insect venoms, animal hair, and foods) known to elicit allergic reactions in susceptible individuals. Food extracts are only used to diagnose food allergies, but other allergenic extracts may be used for both diagnosis and treatment of allergic diseases [22].

6.5 Sublingual Therapy (SLIT)

In 2011, the scientific community celebrated 100 years of specific immune therapy, which involves the administration of allergen extracts to patients with the aim to cure allergic symptoms [23]. Various efforts have been made, mainly for safety reasons, to obtain alternative administration to the classic form of subcutaneous immunotherapy (SCIT). In the last two decades, sublingual administration has been recognized as a route of administration of allergens that is safer than subcutaneous injection, and there is increasing evidence that the therapeutic effects of sublingual immunotherapy (SLIT) are comparable with those of traditional SCIT. A recent meta-analysis of SLIT has shown that this approach is safe, has positive clinical effects, and provides prolonged therapeutic effects after discontinuation of treatment [24]. The sublingual administration of allergens is already being routinely used for conventional immunotherapy in Europe and has also demonstrated efficacy in respiratory allergic diseases, whereas there are currently no SLIT products approved in the USA [25]. The allergen products for SLIT are commercially available in two main pharmaceutical forms: a solution that is delivered by drop-counters, predosed actuators (mini-pumps) or disposable single-dose vials, and tablets with appropriate composition and formulation that dissolves in the mouth within 1–2 min following contact with saliva. In Europe, SLIT is prescribed in general for one or a few allergens, and mixtures are used less, though there is no immunological contraindication to give multiple allergens [26].

Sublingual vaccines based on biological extracts have been successfully used for the last two decades. These vaccines have shown safety and clinical efficacy in clinical trials as well as an increase in patient compliance. However, the risk of therapy-induced side effects limits their broad application. Recent work indicated that the epitope complexity of natural allergen extracts can be recreated using recombinant allergens, and hypoallergenic derivatives of these can be engineered to increase treatment safety. Most of the important allergens have been cloned, characterized, and produced as recombinant proteins, under good manufacturing practice conditions. These modified molecules are expected to improve the current practice of specific immunotherapy and form a basis for prophylactic vaccination [27].

6.6 Oral Mucosa and Allergens

Studies have shown that allergens administered via the sublingual route are not directly absorbed by the oral mucosa but are retained at the mucosal level, where the allergen molecules are captured by the antigen-presenting cells (APCs) and, following their migration in the draining lymph nodes, presented to T cells [28–30]. The cells directly related with immunologic properties of the oral mucosa are dendritic cells, (which are the APCs), T lymphocytes, and B lymphocytes. Dendritic cells, which are spread especially in superficial tissues, are part of the first-line defense of the body. They are critical in detecting all kinds of pathogens entering the organism. Oral dendritic cells (e.g., Langerhans cells) have the ability to take up and process antigens via several mechanisms such as receptor-mediated endocytosis or macropinocytosis [31]. Upon activation, they acquire the capacity to migrate into the tissue lymphoid organs in order to present the antigens to T cells and to induce an antigen-specific immune response leading to clinical tolerance [32]. Long-term changes that occur with immunotherapy include a decrease in mast cell sensitivity and a decrease in IgE production by mucosal B cells. The effector cells of allergic inflammation, i.e., eosinophils, mast cells, and basophiles, are normally absent or few in the oral mucosa (mostly located in submucosal areas) of allergic subjects, and in comparison with subcutaneous tissue, are less likely to give rise to anaphylactic reactions which account for the excellent tolerability of SLIT [29, 32].

Different regions of the oral cavity (e.g., the buccal mucosa) have been reported to represent an alternative to sublingual application for potent allergen uptake. Hence, in immune therapy, it is important to take into consideration how the immune cells are distributed within the entire oral mucosal tissue [33] which could be the basis for the development of new application forms of SLIT to assure allergen uptake within a defined and limited oral region to increase the efficacy and safety of SLIT.

6.7 Formulation Development and Clinical Studies on Allergens

As pointed out earlier, currently, the commercially available allergen (extract or recombinant) products are either in solution or tablet form. As with all other medicinal products, it is important to assure the quality, safety, and efficacy of the allergen products through international norms, standards, guidelines, and nomenclature. In an allergen product, the active substance can be an unmodified allergen extract, an allergoid, a conjugate as well as a purified natural or recombinant protein. Other excipients can be adsorbed or added to the allergen. Allergen extracts mainly consist of proteins and glycoproteins and contain various major and minor allergens as well as nonallergenic components. Because of the intrinsic variability of the natural source material, concentrations of individual allergens in such extracts may vary and standardization is therefore very important [20]. Active substances obtained by recombinant DNA technology consist of predefined allergenic polypeptides. The quantity and structure of these polypeptides can be determined, and these products should be standardized like other biological products consisting of purified proteins [20]. Biological potency is the basis of allergen standardization. Hence, more studies are required on the correlation between the concentration of individual allergen molecules and its translation into the potency of allergen products [34].

In order to improve the efficacy and safety of SLIT, besides chemical modification of allergens, adjuvants can be incorporated into the formulations or the contact of the allergen with the oral mucosa can be increased.

The magnitude of the immune response is critical for its success. In this respect, especially with recombinant allergens, low immune responses are obtained, hence, there is a need for an immunomodulator to enhance the immunogenic effect of the allergen. Among the adjuvants used, aluminum hydroxide is the only officially approved adjuvant with the longest clinical experience. In recent years, adjuvants which would enhance T-cell responses have been investigated especially for the recombinant products [35]. Immunomodulators targeting TLR-2, TLR-4 receptors with lipopolysaccharides (monophosphoryl lipid A) or TLR-9 receptors with unmethylated CpG dinucleotides, as well as 1,25-dihydroxyvitamin D3, selected probiotic strains have been investigated for SLIT [31, 36].

The other approach to enhance the immune response is to increase the allergen uptake by the APCs present in the oral mucosa by means of incorporating the antigen into the particulate systems such as micro/nanoparticles, virus-like particles, or vesicular systems (liposomes). Furthermore, the contact of these systems on the mucosa can be improved by utilization of mucoadhesive polymers, which would consequently result in an increased uptake [37–39]. Among the mucoadhesive polymers investigated for vaccine or allergen delivery, the most promising polymer has been shown to be chitosan which is a polymer obtained from the shells of the crustaceans and provides the additional benefit of exerting an adjuvant effect as well [40]. Recently, a European project (FP-7-SME) [41] on development of a new innovative mucoadhesive chitosan-based adjuvant, Viscogel, has been completed. Safety of the Viscogel has been shown in preclinical studies. Its adjuvant effect was also shown in vivo. Further investigations are needed to optimize the chitosan-based formulation factors which would affect the immune responses for sublingual delivery of antigens or allergens, but these initial findings are promising.

Almost all of the well-designed and double-blinded, placebo-controlled studies evaluated SLIT treatment with single-allergen extracts. Most meta-analyses published to date have evaluated immunotherapy with single allergen or extracts containing several cross-reactive allergens. While the first generation of sublingual vaccines currently used is based on natural biological extracts, new vaccines which rely upon selected recombinant allergens have also been studied in various patient groups. Recent clinical studies, performed using allergens (extract or recombinant), are summarized in Table 6.2. The clinical trials with recombinant allergen preparations that have either been completed or are ongoing have been extensively reviewed by Cromwell et al. [21], and the reader is referred to this publication for a detailed account.

Besides the crude allergen extracts, Bet v 1 from birch pollen (*Betulaverrucosa*), Phl p 5 from timothy grass, and Der p 1 from the house dust mite are among the most commonly investigated recombinant allergens [21, 42]. There is lack of evidence on multiallergen immunotherapy in polysensitized patients (mixture of noncross-reactive allergens) [43]. The findings of the first clinical study on immunotherapy using a cocktail of 5 recombinant grass pollen allergens has shown that the clinical efficacy was high with a good tolerance, together with the induction of strong allergen-specific immunoglobulin G (IgG) antibody responses [44].

Recently, systematic reviews of SLIT were published by Radulovic et al. [45]. The treatment effect within children was found to be similar to that seen in adults, especially when considering symptom scores. SLIT represented an attractive alternative to injection immunotherapy in this patient group. The efficacy and safety of SLIT has been shown in various meta-analyses, mainly for allergic rhinitis, less for asthma and some other allergic conditions [32, 46]. However, there are still many hurdles to overcome, like optimal dosage, duration of treatment, and long-term efficacy before SLIT can be suggested as the first-line treatment in allergy treatment [47].

6.8 Conclusions and Future Perspectives

The oral mucosa as a delivery route for biopharmaceuticals has been investigated for vaccines and has some applications in humans only for allergens. There are no examples reported for any other biopharmaceuticals. The preferred region for

Table 6.2 Examples of clinical studies on SLIT

Source	Allergen	Extract/ recombinant	Dosage form	Route	Patient group	Administration period	Response	References
Single								
Dermatophagoides farinae (house dust mite)	Der f 1	Extract	Solution	s.l.	Adults	12–18 months	Improved bronchial threshold to allergen challenge with high-dose	[64]
Birch pollen	Bet v 1	Extract	Solution	s.l.	Adults	12 months	Birch pollen may have no clinical effect on apple-induced oral allergy	[65]
Birch pollen	Bet v 1	Extract	Tablet	s.l.	Adults	2 weeks	Optimal tolerance range determined to be 12.5 μg–50 μg	[66]
Phleum pratense (timothy-grass pollen)	Phl p	extract	Solution	s.l.	Adults	1 week	Tolerability was acceptable, notably with the higher first dose of 50 SRU per day and the faster up-dosing over 10 days; no safety concerns were identified for different dosing protocols	[67]
Combination								
Orchard[a], meadow[b] rye[c], sweet vernal[d], and timothy[e] (5-grasspollens)		Extract	Solution	s.l.	Children	4 months a year, for 3 consecutive years	3 years of coseasonal immunisation improved seasonal allergic rhinitis symptoms and reduced the development of seasonal asthma in children with hay fever	[68]
		Extract	Solution, tablet	s.l.	Adults	Drops for 14 days; tablets for the rest of the study (total of 24 months)	Reduced symptoms; beneficial effects on nasal symptoms during the peak pollen season with higher doses	[69]

Table 6.2 (continued)

Source	Allergen	Extract/ recombinant	Dosage form	Route	Patient group	Administration period	Response	References
		Extract	Tablet	s.l.	Children	Started 4 months before pollen season, continued during pollen season	Optimal dose given in adults was effective and well tolerated in children and adolescents from the first pollen season	[70]
		Extract	Solution	s.l.	Children	3 months	Aqueous high dose is effective and well tolerated and has a significant effect on allergen-specific antibodies	[71]
		Extract	Tablet	s.l.	Adults	4 months	The efficacy and onset of action assessed for the first time in an allergen challenge chamber (ACC) and significant clinical efficacy obtained	[72]
		Extract	Tablet	s.l.	Elderly adults	36 months	Significant and well tolerated clinical improvement in the active group compared with the placebo group, particularly during the heating season	[73]
		Extract	Tablet	s.l	Adults	Started 6 months before pollen season, continued during pollen season	Clinically significant efficacy	[74]

Table 6.2 (continued)

Source	Allergen	Extract/ recombinant	Dosage form	Route	Patient group	Administration period	Response	References
Birch pollens, timothy grass	Bet v1, Phl p1	Extract	Solution	s.l.	Adults	–	Low reactivity of sublingual mucosa to allergens	[75]
	rBet v1, rPhl p1	Recombinant					Induced oral symptoms following sublingual challenges with recombinant allergens in patients with oral allergy syndrome (OAS) showing highsensitivity and specificity	
Dermatophagoides pteronyssinus and *D. farinae* (dust mites)	Der p 1, Der f 1	Extract	Solution	s.c., s.c + s.l.	Children	18 months	Rapid onset and potency in SCIT and safety and avoidance of injections in SLIT	[76]
Timothy grass, dust mite[f]	Phl p 1, Der f 1, Der f 2		Solution	s.l.	Children and adults	12 months	Clinically significant efficacy	[77]

[a] *Dactylis glomerata*
[b] *Poa pratensis*
[c] *Lolium perenne*
[d] *Anthoxanthum odoratum*
[e] *P. pratense*
[f] *D. farinae*

administration is the sublingual mucosa in which the immune system is prone to induce active tolerance mechanisms against allergens and antigens from the environment. The efficacy and safety of SLIT has been shown by numerous meta-analyses in both adults and children. This analysis suggested that SLIT was better than SCIT. Nevertheless, optimization of SLIT dosing and escalation protocols, relationship between immunologic changes, quality of life, and SLIT outcomes remain to be defined. From the formulation point of view, there is a need to develop improved allergen formulations other than the currently available drops or fast-dissolving tablets. Such improved formulations might include mucoadhesive gels and/or particulate (either alone or incorporated with an immunomodulator). Such approaches would aim to prolong the contact time and enhance the allergen uptake by immune cells. The outcome of an improved formulation would also provide high patient compliance.

References

1. Walsh G. Biopharmaceuticals: recent approvals and likely directions. Trends Biotechnol. 2005;23(11):553–8.
2. Walsh G. Post-translational modifications of protein biopharmaceuticals. Drug Discov Today. 2010;15(17–18):773–80.
3. Rader RA. What is a biopharmaceutical? Part 1: (Bio)technology-based definitions. BioExecutive International; 2005. pp. 60–5.
4. No authors. Federal register/ Vol. 68, No. 123/ Thursday, June 26, 2003/ Notices. http://www.fda.gov/ohrms/dockets/98fr/03–16108.pdf. Accessed 19 March 2013.
5. No authors. Federal Register/ Vol. 70, No. 56/ Thursday, March 24, 2005/ rules and regulations. http://www.gpo.gov/fdsys/pkg/FR-2005-03-24/pdf/05-5780.pdf. Accessed 19 March 2013.
6. No authors. Directive 2001/83/EC of the European parliament and of the Council of 6 November 2001 on the community code relating to medicinal products for human use. http://www.emea.europa.eu/docs/en_GB/document_library/Regulatory_and_procedural_guideline/2009/10/WC500004481.pdf. Accessed 19 March 2013.
7. Orive G, Gascon AR, Hernandez RM, Dominguez-Gil A, Pedraz JL. Techniques: new approaches to the delivery of biopharmaceuticals. Trends Pharmacol Sci. 2004;25(7):382–7.
8. Singh R, Singh S, Lillard JW. Past, present, and future technologies for oral delivery of therapeutic proteins. J Pharm Sci. 2008;97(7):2497–523.
9. Pather SI, Rathbone MJ, Şenel S. Current status and the future of buccal drug delivery systems. Expert Opin Drug Del. 2008;5(5):531–42.
10. Pather SI, Rathbone MJ, Şenel S. Oral transmucosal drug delivery. In: Rathbone M, et al., editors. Modified release drug delivery technology. 2nd ed. Vol. 1, Informa Healthcare; 2008. pp. 54–73.
11. Şenel S, Rathbone MJ, Cansız M, Pather I. Recent developments in buccal and sublingual delivery systems. Expert Opin Drug Del. 2012;9(6):615–28.
12. Şenel S, Kremer M, Nagy K, Squier C. Delivery of bioactive peptides and proteins across oral (buccal) mucosa. Curr Pharm Biotechnol. 2001;2(2):175–86.
13. Squier CA, Wertz P. Structure and function of the oral mucosa and implications for drug delivery. In: Rathbone MJ, editor. Oral mucosal delivery. New York: Marcel Dekker; 1996. pp. 1–26.
14. Harris D, Robinson JR. Drug delivery via the mucous-membranes of the oral cavity. J Pharm Sci. 1992;81(1):1–10.
15. Wertz PW, Squier CA. Cellular and molecular-basis of barrier function in oral epithelium. Crit Rev Ther Drug. 1991;8(3):237–69.

16. Scholz OA, Wolff A, Schumacher A, Giannola LI, Campisi G, Ciach T, Velten T. Drug delivery from the oral cavity. focus on a novel mechatronic delivery device. Drug Discov Today. 2008;13(5–6):247–53.
17. Arca HC, Günbeyaz M, Şenel S. Chitosan-based systems for the delivery of vaccine antigens. Expert Rev Vaccines. 2009;8(7):937–53.
18. Şenel S. Chitosan-based particulate systems for non-invasive vaccine delivery. Adv Polym Sci. 2011;243:111–138.
19. Kweon MN. Sublingual mucosa: a new vaccination route for systemic and mucosal immunity. Cytokine. 2011;54 (1):1–5.
20. No authors. EMEA/CHMP/BWP/304831/2007guideline on allergen products: production and quality issues. http://www.emea.europa.eu/docs/en_GB/document_library/Scientific_guideline/2009/09/WC500003333.pdf. Accessed 19 March 2013.
21. Cromwell O, Hafner D, Nandy A. Recombinant allergens for specific immunotherapy. J Allergy Clin Immun. 2011;127(4):865–72.
22. No authors. FDA—allergenics. http://www.fda.gov/BiologicsBloodVaccines/Allergenics/default.htm. Accessed 19 March 2013.
23. Linnemann DESL. One hundred years of immunotherapy: review of the first landmark studies. Allergy Asthma Proc. 2012;33(2):122–8.
24. Fujimura T, Okamoto Y, Taniguchi M. Therapeutic effects and biomarkers in sublingual immunotherapy: a review. J Allergy (Cairo). 2012;2012:381737.
25. Sikora JM, Tankersley MS. Perception and practice of sublingual immunotherapy among practicing allergists in the United States: a follow-up survey. Ann of Allergy Asthma Immunol. 2013;110(3):194–197.e194.
26. Frati F, La Grutta S, Bernardini R, Zampogna S, Scurati S, Puccinelli P, Riario-Sforza GG, Incorvaia C. Sublingual immunotherapy: administration, dosages, use. Int J Immunopathol Pharmacol. 2009;22(4 Suppl):13–6.
27. Valenta R. The future of antigen-specific immunotherapy of allergy. Nat Rev Immunol. 2002;2(6):446–53.
28. Moingeon P. Sublingual immunotherapy: from biological extracts to recombinant allergens. Allergy. 2006;61:15–9.
29. Frati F, Moingeon P, Marcucci F, Puccinelli P, Sensi L, Di Cara G, Incorvaia C. Mucosal immunization application to allergic disease. Sublingual immunotherapy. Allergy Asthma Proc. 2007;28(1):35–9.
30. Novak N, Bieber T, Allam JP. Immunological mechanisms of sublingual allergen-specific immunotherapy. Allergy. 2011;66(6):733–9.
31. Pfaar O, Cazan D, Klimek L, Larenas-Linnemann D, Calderon MA. Adjuvants for immunotherapy. Curr Opin Allergy Clin Immunol. 2012;12(6):648–57.
32. Calderon MA, Simons FER, Malling HJ, Lockey RF, Moingeon P, Demoly P. Sublingual allergen immunotherapy: mode of action and its relationship with the safety profile. Allergy. 2012;67(3):302–11.
33. Allam JP, Stojanovski G, Friedrichs N, Peng W, Bieber T, Wenzel J, Novak N. Distribution of Langerhans cells and mast cells within the human oral mucosa: new application sites of allergens in sublingual immunotherapy? Allergy. 2008;63(6):720–7.
34. Becker WM, Vogel L, Vieths S. Standardization of af lergen extracts for immunotherapy: where do we stand? Curr Opin Allergy Clin Immunol. 2006;6(6):470–5.
35. Moingeon P. Adjuvants for allergy vaccines. Hum Vacc Immunother. 2012;8(10):1492–8.
36. Kopp MV. Role of immunmodulators in allergen-specific immunotherapy. Allergy. 2011;66(6):792–7.
37. Razafindratsita A, Saint-Lu N, Mascarell L, Berjont N, Bardon T, Betbeder D, Van Overtvelt L, Moingeon P. Improvement of sublingual immunotherapy efficacy with a mucoadhesive allergen formulation. J Allergy Clin Immunol. 2007;120(2):278–85.
38. Passalacqua G, Lombardi C, Troise C, Canonica GW. Sublingual immunotherapy: certainties, unmet needs and future directions. Eur Ann Allergy Clin Immunol. 2009;41(6):163–170.

39. Moingeon P, Zimmer A, Van Overtvelt L, Tourdot S, Mascarell L. Oral mucosal immunity and allergy. Rev Fr Allergol. 2010;50(3):274–6.
40. Şenel S. Chitosan-based particulate systemsfor non-invąsive vaccine delivery. Adv Polym Sci. 2011;243:111–38.
41. No authors. Viscogel—A chitosan based adjuvant for prophylactic and therapeutic vaccination. http://cordis.europa.eu/projects/rcn/96803_en.html. Accessed 22 March 2013.
42. Bhalla PL, Singh MB. Biotechnology-based allergy diagnosis and vaccination. Trends Biotechnol. 2008;26(3):153–61.
43. Bahçeciler NN, Galip N, Çobanoğlu N. Multiallergen-specific immunotherapy in polysensitized patients: where are we? Immunotherapy. 2013;5(2):183–90.
44. Jutel M, Jaeger L, Suck R, Meyer H, Fiebig H, Cromwell O. Allergen-specific immunotherapy with recombinant grass pollen allergens. J Allergy Clin Immunol. 2005;116(3):608–13.
45. Radulovic S, Wilson D, Calderon M, Durham S. Systematic reviews of sublingual immunotherapy (SLIT). Allergy. 2011;66(6):740–52.
46. Calderon MA, Casale TB, Togias A, Bousquet J, Durham SR, Demoly P. Allergen-specific immunotherapy for respiratory allergies. From meta-analysis to registration and beyond. J Allergy Clin Immun. 2011;127(1):30–8.
47. Calderon MA, del Rio PR, Demoly P. Sublingual allergen immunotherapy in children. An evidence-based overview. Rev Fr Allergol. 2012;52(1):20–5.
48. Song JH, Nguyen HH, Cuburu N, Horimoto T, Ko SY, Park SH, Czerkinsky C, Kweon MN. Sublingual vaccination with influenza virus protects mice against lethal viral infection. Proc Nat Acad Sci U S A. 2008;105(5):1644–9.
49. Hikono H, Miyazaki A, Mase M, Inoue M, Hasegawa M, Saito T. Induction of a cross-reactive antibody response to influenza virus M2 antigen in pigs by using a Sendai virus vector. Vet Immunol Immunopathol. 2012;146(1)92–6.
50. Zhang T, Hashizume T, Kurita-Ochiai T, Yamamoto M. Sublingual vaccination with outer membrane protein of *Porphyromonas gingivalis* and Flt3 ligand elicits protective immunity in the oral cavity. Biochem Biophys Res Commun. 2009;390(3):937–41.
51. Cho HJ, Kim JY, Lee Y, Kim JM, Kim YB, Chun T, Oh YK. Enhanced humoral and cellular immune responses after sublingual immunization against human papillomavirus 16 L1 protein with adjuvants. Vaccine. 2010;28(14):2598–606.
52. Ralli-Jain P, Tifrea D, Cheng CM, Pal S, de la Maza LM. Enhancement of the protective efficacy of a *Chlamydia trachomatis* recombinant vaccine by combining systemic and mucosal routes for immunization. Vaccine. 2010;28(48):7659–66.
53. Carmichael JR, Pal S, Tifrea D, de la Maza LM. Induction of protection against vaginal shedding and infertility by a recombinant *Chlamydia* vaccine. Vaccine. 2011;29(32):5276–83.
54. Jenikova G, Hruz P, Andersson MK, Tejman-Yarden N, Ferreira PC, Andersen YS, Davids BJ, Gillin FD, Svard SG, Curtiss R, Eckmann L 3rd. Alpha1-giardin based live heterologous vaccine protects against *Giardia lamblia* infection in a murine model. Vaccine. 2011;29(51):9529–37.
55. Batard T, Zimmer A, Nony E, Bouley J, Airouche S, Luce S, Turfkruyer M, Tourdot S, Mascarell L, Moingeon P. Anti-inflammatory activity of sublingual immunoglobulin (SLIG) in a murine model of allergen-driven airway inflammation. Vaccine. 2012;30(38):5666–74.
56. Flach CF, Svensson N, Blomquist M, Ekman A, Raghavan S, Holmgren J. A truncated form of HpaA is a promising antigen for use in a vaccine against *Helicobacter pylori*. Vaccine. 2011;29(6):1235–41.
57. Negri DRM, Riccomi A, Pinto D, Vendetti S, Rossi A, Cicconi R, Ruggiero P, Del Giudice G, De Magistris MT. Persistence of mucosal and systemic immune responses following sublingual immunization, Vaccine. 2010;28(25):4175–80.
58. Amuguni JH, Lee S, Kerstein KO, Brown DW, Belitsky BR, Herrmann JE, Keusch GT, Sonenshein AL, Tzipori S. Sublingually administered *Bacillus subtilis* cells expressing tetanus toxin C fragment induce protective systemic and mucosal antibodies against tetanus toxin in mice. Vaccine. 2011;29(29–30):4778–84.
59. Amuguni H, Lee S, Kerstein K, Brown D, Belitsky B, Herrmann J, Keusch G, Sonenshein A, Tzipori S. Sublingual immunization with an engineered *Bacillus subtilis* strain expressing

tetanus toxin fragment C induces systemic and mucosal immune responses in piglets. Microbes Infect. 2012;14(5):447–56.

60. Hervouet C, Luci C, Cuburu N, Cremel M, Bekri S, Vimeux L, Maranon C, Czerkinsky C, Hosmalin A, Anjuere F. Sublingual immunization with an HIV subunit vaccine induces antibodies and cytotoxic T cells in the mouse female genital tract. Vaccine. 2010;28(34):5582–90.
61. Yu M, Vajdy M. A novel retinoic acid, catechin hydrate and mustard oil-based emulsion for enhanced. cytokine and antibody responses against multiple strains of HIV-1 following mucosal and systemic vaccinations. Vaccine. 2011;29(13):2429–36.
62. Appledorn DM, Aldhamen YA, Godbehere S, Seregin SS, Amalfitano A. Sublingual administration of an adenovirus serotype 5 (Ad5)-based vaccine confirms toll-like receptor agonist activity in the oral cavity and elicits improved mucosal and systemic cell-mediated responses against HIV antigens despite preexisting Ad5 immunity. Clin Vaccine Immunol. 2011;18(1):150–60.
63. Domm W, Brooks L, Chung HL, Feng CY, Bowers WJ, Watson G, McGrath JL, Dewhurst S. Robust antigen-specific humoral immune responses to sublingually delivered adenoviral vectors encoding HIV-1 Env: association with mucoadhesion and efficient penetration of the sublingual barrier. Vaccine. 2011;29(40):7080–9.
64. Bush RK, Swenson C, Fahlberg B, Evans MD, Esch R, Morris M, Busse WW. House dust mite sublingual immunotherapy: results of a US trial. J Allergy Clin Immunol. 2011;127(4):974–81.e971–7.
65. Kinaciyan T, Jahn-Schmid B, Radakovics A, Zwolfer B, Schreiber C, Francis JN, Ebner C, Bohle B. Successful sublingual immunotherapy with birch pollen has limited effects on concomitant food allergy to apple and the immune response to the Bet v 1 homolog Mal d 1. J Allergy Clin Immun. 2007;119(4):937–43.
66. Winther L, Poulsen LK, Robin B, Melac M, Malling H. Safety and tolerability of recombinant Bet v 1 (rBet v 1) tablets in sublingual immunotherapy (SLIT). J Allergy Clin Immun. 2009;123(2):S215.
67. Nittner-Marszalska M, Fayoux E, Chartier A, Strodl-Andersen J, Kuna P. Sublingual solution for immunotherapy: comparison of three different up-dosing schedules. Rev Fr Allergol. 2013;53(2):65–72.
68. Novembre E, Galli E, Landi F, Caffarelli C, Pifferi M, De Marco E, Burastero SE, Calori G, Benetti L, Bonazza P, Puccinelli P, Parmiani S, Bernardini R, Vierucci A. Coseasonal sublingual immunotherapy reduces the development of asthma in children with allergic rhinoconjunctivitis. J Allergy Clin Immunol. 2004;114(4):851–7.
69. Smith H, White P, Annila I, Poole J, Andre C, Frew A. Randomized controlled trial of high-dose sublingual immunotherapy to treat seasonal allergic rhinitis. J Allergy Clin Immunol. 2004;114(4):831–7.
70. Wahn U, Tabar A, Kuna P, Halken S, Montagut A, de Beaumont O, Le Gall M, Group SS. Efficacy and safety of 5-grass-pollen sublingual immunotherapy tablets in pediatric allergic rhinoconjunctivitis. J Allergy Clin Immunol. 2009;123(1):160–66.e163.
71. Wahn U, Klimek L, Ploszczuk A, Adelt T, Sandner B, Trebas-Pietras E, Eberle P, Bufe A, Group SS. High-dose sublingual immunotherapy with single-dose aqueous grass pollen extract in children is effective and safe: a double-blind, placebo-controlled study. J Allergy Clin Immunol. 2012;130(4):886–93.e885.
72. Horak F, Zieglmayer P, Zieglmayer R, Lemell P, Devillier P, Montagut A, Melac M, Galvain S, Jean-Alphonse S, Van Overtvelt L, Moingeon P, Le Gall M. Early onset of action of a 5-grass-pollen 300-IR sublingual immunotherapy tablet evaluated in an allergen challenge chamber. J Allergy Clin Immun. 2009;124(3):471–7.
73. Bozek A, Ignasiak B, Filipowska B, Jarzab J. House dust mite sublingual immunotherapy. a double-blind, placebo-controlled study in elderly patients with allergic rhinitis. Clin Exp Allergy. 2013;43(2):242–8.
74. Cox LS, Casale TB, Nayak AS, Bernstein DI, Creticos PS, Ambroisine L, Melac M, Zeldin RK. Clinical efficacy of 300IR 5-grass pollen sublingual tablet in a US study: the importance of allergen-specific serum IgE. J Allergy Clin Immun. 2012;130(6):1327.

75. Marcucci F, Sensi L, Di Cara G, Gidaro G, Incorvaia C, Frati F. Sublingual reactivity to rbEt v1 and rPhl p1 in patients with oral allergy syndrome. Int J Immunopathol Pharmacol. 2006;19(1):141–8.
76. KeleÅŸ S, Karakoç-Aydıner E, Özen A, İzgi AG, Tevetoğlu A, Akkoç T, Bahçeciler NN, Barlan I. A novel approach in allergen-specific immunotherapy: combination of sublingual and subcutaneous routes. J Allergy Clin Immunol. 2011;128(4):808–15, e807.
77. Swamy RS, Reshamwala N, Hunter T, Vissamsetti S, Santos CB, Baroody FM, Hwang PH, Hoyte EG, Garcia MA, Nadeau KC. Epigenetic modifications and improved regulatory T-cell function in subjects undergoing dual sublingual immunotherapy. J Allergy Clin Immun. 2012;130(1):215.

Chapter 7
Pulmonary Delivery of Biopharmaceuticals

Fernanda Andrade, Catarina Moura and Bruno Sarmento

7.1 Introduction

Since 1982, when recombinant human insulin (Humulin®, Eli Lilly) was approved and driven by the progresses seen in molecular biology, pharmaceutical industries have been increasingly investing in the research of biotechnology-based drugs. At the moment, around 900 biopharmaceuticals targeting more than 100 diseases are under development by US research companies (Fig. 7.1) [1]. This is explained by the overall improvement in quality of life and reduced burden of complex and challenging diseases achieved by these medicines in an extent sometimes not reached by conventional drugs. Consequently, many biopharmaceutical products have been granted market authorization over the years [2] and gain an increased share in global pharmaceutical market year-by-year [3].

Despite the high therapeutic efficiency demonstrated by many biopharmaceuticals, their administration in the native or active state remains a huge challenge for

F. Andrade (✉)
Laboratory of Pharmaceutical Technology, Faculty of Pharmacy,
University of Porto, Rua de Jorge Viterbo Ferreira,
228, 4050-313 Porto, Portugal
e-mail: fersilandrade@gmail.com

Nanoprobes & Nanoswitches Group, Institute for Bioengineering
of Catalonia (IBEC), Carrer de Baldiri Reixac, 10, 08028 Barcelona, Spain

C. Moura
Faculty of Engineering, University of Porto, Rua Doutor Roberto Frias, 4200-465 Porto, Portugal

B. Sarmento
NEW Therapies Group, INEB – Instituto de Engenharia Biomédica,
Rua do Campo Alegre, 823, 4150-180 Porto, Portugal

IINFACTS – Instituto de Investigação e Formação Avançada em Ciências e
Tecnologias da Saúde, Instituto Superior de Ciências da Saúde-Norte, CESPU,
Rua Central de Gandra, 1317, 4585-116 Gandra, Portugal
e-mail: bruno.sarmento@ineb.up.pt

J. das Neves, B. Sarmento (eds.), *Mucosal Delivery of Biopharmaceuticals,* DOI 10.1007/978-1-4614-9524-6_7,

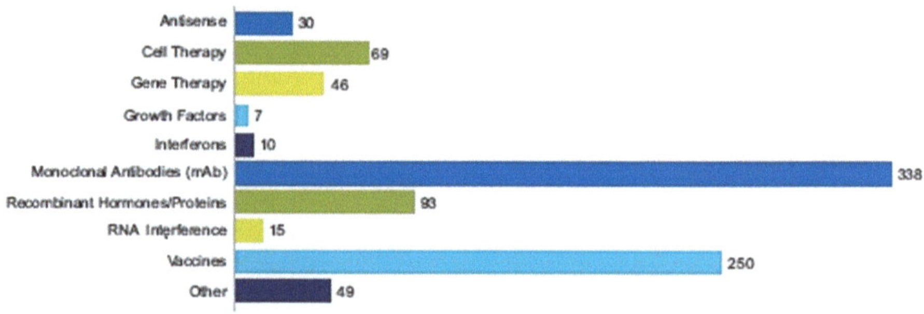

Fig. 7.1 Current biologic medicines under development by American companies. (With permission granted from ref. [1])

the pharmaceutical researchers. Characteristics such as hydrophilicity, high molecular weight (MW), reduced permeability, immunogenicity, and labile structure make difficult the formulation of medicines based on biopharmaceuticals and limit their administration almost to the parenteral route. In addition to the low acceptance and reduced compliance by patients to the treatment, the need for sterile formulations, cold chain transport and storage, administration by specialized personnel, and associated costs boosted industrial and academic researchers to seek for needle-free and user-friendly formulations for non-invasive administration [4, 5]. Among the different alternative routes, inhalation appears as a promising one for delivery of a variety of drugs including macromolecules [6].

In the present chapter, the characteristics of lungs and airways that make them suitable targets for delivery of biopharmaceuticals is discussed and the progresses seen in the last years regarding their administration by inhalation are summarized.

7.2 Inhalation as a Route for Delivery of Drugs

A demand for alternative routes for administration of biopharmaceuticals has been the flag of many research groups in the last decades. Although the oral route is preferred by patients due to its ease and convenience, the harsh gastric environment and the intestinal epithelia arise as strong barriers to efficient delivery of macromolecules [7]. Nasal, vaginal, transdermal, buccal, or ocular routes have also been investigated, but pulmonary administration attracts more attention.

Inhalation of compounds is documented since ancient cultures and essentially used to treat local diseases such as asthma [8]. However, in the last decades, pulmonary administration has been exploited for both local and systemic delivery of drugs and biopharmaceuticals [6] to treat diseases like tuberculosis [9], cystic fibrosis [10], diabetes [11], fungal infections [12], and as a vaccination platform [13, 14], owing to lung's physiological characteristics, easy administration, and social acceptance of inhalation.

Besides the advantages of inhalation, its complexity makes difficult the development of generic dosage forms for aerosol delivery, especially for biopharmaceuticals, as seen for other administration routes [15].

7.2.1 *Anatomo-Physiological Characteristics of Lungs and Airways*

Lungs present characteristics such as a surface area up to $140\,m^2$ and a thin alveolar epithelium of < 0.2 μm that in addition to the extensive blood supply (flow 5 L/min), the low enzymatic activity and efflux systems, and the avoidance of the hepatic first-pass metabolism, allow the higher absorption and bioavailability of biopharmaceuticals compared to other non-invasive routes [16, 17].

However, being the place of gas exchange and constant contact with the exterior, respiratory system developed defense mechanisms working as barrier for foreign particles that could impair the efficient delivery of drugs. The complex geometry and humidity of the airways hamper the passage of the larger particles to the deep lung, and the movement of the bronchial cilia transports the particles trapped in the mucus layer to the gastrointestinal tract. Particles or compounds capable of evading the mentioned barriers and reaching the deep lung have to face the alveolar macrophages, alveolar lining fluid, and the epithelium to attain the bloodstream [18].

7.2.2 *Pulmonary Absorption of Biopharmaceuticals*

Among the different mechanisms that influence the passage of compounds from the respiratory tract to the bloodstream and their bioavailability, the alveolar and airway epithelium is the major barrier to absorption of drugs. However, degradation by proteolysis is also relevant for peptides and proteins with small MW ($<$ 3,000 Da) [19].

The absorption of biopharmaceuticals through the respiratory tract is a complex and enigmatic process that involves various mechanisms that are not yet well characterized. It seems that the absorption is dependent on the hydrophilicity and the size of the biopharmaceuticals [20–22]. Different studies suggest that the rate of absorption is inversely proportional to the MW of the macromolecules. This influences not only the percentage of drug absorbed but also the time necessary for the absorption to occur. For example, the half-life time ($t_{1/2}$) of the alveolar absorption of macromolecules increases with their MW (insulin with MW: 5,250 Da and $t_{1/2}$: 225 min; dextran with MW: 20,000 Da and $t_{1/2}$: 688 min; dextran with MW: 75,000 Da and $t_{1/2}$: 1,670 min) [23].

There are three major mechanisms proposed for pulmonary absorption of biopharmaceuticals: paracellular diffusion, vesicular endocytosis or pinocytosis, and receptor-dependent transcytosis. Biopharmaceuticals with small MW are apparently absorbed by the paracellular route, diffusing through the tight junctions, while molecules with higher MW seem to suffer endocytosis [24]. Peptides can be absorbed by receptor-mediated transcytosis using the high-affinity peptide transporter 2 [25], while immunoglobulins are absorbed by a conjugation of pinocytosis with receptor-mediated transcytosis. After endocytosis, immunoglobulins bind with the fragment crystallizable (Fc) receptors that prevent the fusion with lysosomes and are released in the basolateral side of epithelial cells [26]. This immunoglobulin transport pathway is used to deliver biopharmaceuticals by conjugation of the therapeutic macromolecule with Fc portions of immunoglobulins [26, 27]. Other strategy proposed to enhance the pulmonary absorption of biopharmaceuticals is their coupling with specific peptide sequences that not only alter the biologic activity but also promote their translocation through the epithelium probably by receptor-mediated transport [28].

7.2.3 Formulation Requirements for Pulmonary Delivery of Biopharmaceuticals

Different aspects that can be related to the formulation, inhalation device, and patient present the capacity to influence the aerosolization and deposition of drugs and, consequently, their therapeutic efficacy. The airways' geometry, respiratory capacity (tidal volume, inspiratory flow rate, and breathing frequency), inhaler handling, smoking, and pathologies affecting the lungs will be responsible for therapeutic inter-individual variations. On the other hand, the choice of the inhaler will be dependent on the type of formulation—nebulizer for liquids, pressurized metered-dose inhaler (pMDI) for liquids and powders, and dry powder inhaler (DPI) for powders—and could greatly impact the aerosolization of drugs [29–31]. Among the high range of inhalers available, the choice of the appropriate device for a given formulation is a time-consuming and challenging step during the development of inhalable medicines [32, 33]. The ideal inhaler should generate an aerosol with a fine particle fraction (FPF) and reproducible drug dosage; guaranty protection and stability of the product during the shelf-life; be small, discrete, and user friendly to be accepted by patients. Regarding gene therapy, an important aspect related with the inhalation of genetic material is the ability of the vectors and the nanocomplexes to be delivered as aerosols. The sheer forces of nebulization usually tend to degrade the nanocomplexes, reducing their transfection efficiency. So, when designing a protocol for aerosol gene therapy, the evaluation of the nebulizer as well as the resistance of the vector to the nebulization conditions should also be taken into account [34, 35].

Formulation plays an important role in the performance of the inhaled drugs, in terms of stability, deposition, and absorption. It should maintain the drug in the active state and deliver it to a specific site of action to be absorbed or released for systemic or local action, respectively. In addition, the formulation must be stable upon storage. Since biopharmaceuticals are labile drugs, suitable to lose their activity through physical and chemical instability, their stability maintenance is a challenge and a series of considerations should be taken into account during their production and

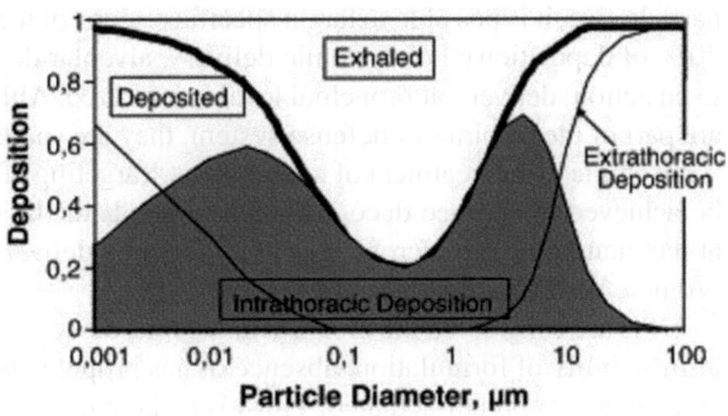

Fig. 7.2 Deposition pattern of particles in the respiratory tract after inhalation dependent on the mean aerodynamic diameter. (Reprinted with permission of Elsevier Limited from ref. [18])

storage. Temperature, pH, agitation, ionic strength, and presence of surfactants need to be controlled in order to avoid aggregation, degradation, or conformation lost [36, 37]. In the case of genetic material, additional care regarding the presence of DNAse or RNAse needs to be considered. The stability of biopharmaceuticals and their therapeutic performance can be improved through the incorporation of some excipients to the formulation. They include antioxidants, metal chelators, and enzyme inhibitors to reduce the activity of proteolytic enzymes, sugars, and salts to increase thermal stability, the non-ionic surfactants and polymers to reduce aggregation proteins, or cyclodextrins to promote the absorption [5, 36]. Despite the wide range of categories of excipients that could be used in the development of inhaled formulations, only few compounds that are biocompatible and easily metabolized or cleared are authorized for pulmonary delivery due to the low buffer capacity of lungs.

Among the different formulation characteristics, aerodynamic diameter plays a key role in the deposition pattern and therapeutic efficiency of the aerosolized particles. Aerodynamic diameter is the diameter of a unit density sphere that has the same terminal settling velocity in still air as the particle in consideration [38] and is defined by the following equation:

$$d_{\mathrm{ae}} = d_{\mathrm{geo}}\sqrt{\frac{\rho_{\mathrm{p}}}{\rho_{\mathrm{o}}\chi}} \tag{7.1}$$

where d_{geo} is the geometric diameter of an equivalent volume sphere of unit density, ρ_{p} and ρ_{o} are particle and unit densities, respectively, and χ is the dynamic shape factor.

After inhalation, depending on mass mean aerodynamic diameter (MMAD), particles will move through the airways and deposit in different parts of the respiratory track or be exhaled (Fig. 7.2). For deposition at the lower regions of lungs, particles in the range of 1–100 nm and 0.5–5 µm are required. Particles larger than 5 µm will impact the throat and be swallowed, while the middle-sized particles will be essentially exhaled [17, 39]. Different forces, namely inertial impaction, sedimentation, and diffusion, will govern the particles' fate and are related to the aerodynamic and hydrophilic properties of particles and shape of airways [18, 40]. By manipulating the particle size, it is possible to target specific regions of the respiratory tract (more than 50 % of deposition). For systemic delivery, alveolar deposition is needed, while for local action, delivery at bronchial level is preferred. Although alveolar macrophages are part of the respiratory defense system, they are sometimes the therapeutic target, for example in the treatment of tuberculosis. Targeting of alveolar macrophages could be achieved by surface decoration with ligands of the lectin-like receptors present at the membrane of macrophages [41, 42] or by delivering particles with a size that promote their phagocytosis [43, 44].

DPIs are considered the most advantageous devices for inhalation regarding long-term stability of formulation, absence of gas propellants, and patient's convenience, since they are breath-actuated. However, the engineering of solid particles with a narrow particle-size distribution and good flowability, suitable for aerosolization and lung deposition of biopharmaceuticals in the active state is challenging. Techniques like microcrystallization, micronization by jet- or ball-milling, lyophilization, spray drying, spray freeze-drying, or supercritical fluid technology can be used to produce particles [17, 45]. All the methods present advantages and disadvantages, and should be chosen according to the effect on the stability of the biopharmaceutical, the characteristics of particles required for a specific formulation, scale-up, cost-effectiveness, and safety issues [45].

As stated before, the capacity to produce an aerosol with a narrow particle-size distribution will influence the deposition pattern of the drugs. This is important to produce powders with good dispersibility. Solid particles are subject to cohesive and adhesive interactions with the surrounding environment that need to be broken during the aerosolization. Different forces are involved in particle's interactions and include electrostatic and van der Waals forces, capillary forces from the presence of residual water at the surface of particles, and mechanical interlocking due to surface roughness [46]. Distinct aerosolization properties could be obtained playing with these forces by specific particle engineering. For example, an efficient drying of the particles needs to be provided by the production method to reduce moisture and capillary forces, but extra drying should be avoided due to the formation of charges at the surface of particles that promote electrostatic interactions. One of the main factors affecting the particle's interactions is their surface area. The larger the surface area, the greater will be the interactions between particles, and the lower will be the flowability. Surface area is dependent on size, shape, and morphology of particles [46, 47]. Particles in the proper size range for inhalation possess high surface areas and are generally mixed with larger coarse carrier particles of excipients to improve their flow properties. The coarse carrier not only improves the dispersibility of particles but also provides bulk, which improves the handling and metering of the drug. Inhalation grade lactose is the most commonly used carrier in the development of DPIs [48]. There are marketed dozens of inhalation grade lactose with different characteristics (Fig. 7.3) (Flowlac®, Granulac®, Respitose®, Lactohale®, Inhalac®, among others) that should be carefully selected during the development of the formulation [49, 50]. The characteristics of the carriers and the adhesive forces between carrier and drug's particles influence the performance of the formulation and need to be assessed and optimized. Blending of drugs with carriers is a critical point during the development of a DPI and also an object of optimization [51–55].

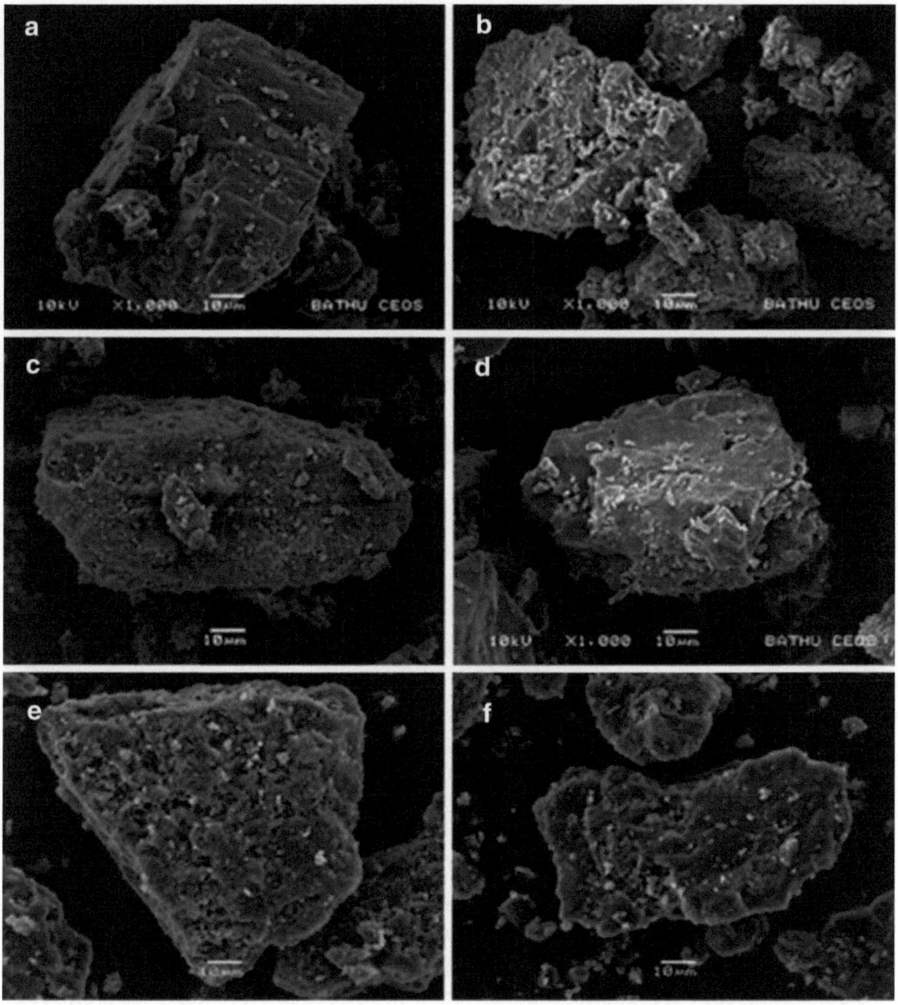

Fig. 7.3 SEM images of inhalation grades of, **a** lactose monohydrate (Lactohale LH200), **b** anhydrous lactose (Anhydrous 120MS), **c** lactose monohydrate (Respitose ML001), **d** lactose monohydrate (Monohydrate 120M), **e** anhydrous lactose (SuperTab. 21AN), and **f** anhydrous lactose (Lactopress Anhydrous). (Reprinted with permission of Elsevier Limited from ref. [48])

Although excipients usually comprise the greater part of the DPI's formulation, in few cases, it is not so. Pulmicort® Turbuhaler® (AstraZeneca) is an excipient-free budesonide marketed formulation used in clinics. Some engineered drug particles alone fulfill the requirements for inhalation, which is possible by the development of large porous or hollow particles (Fig. 7.4) [56–58]. Due to its small density, particles with high geometric size and, consequently, reduced cohesive forces, present appropriate aerodynamic diameters. For example, salbutamol particles prepared by

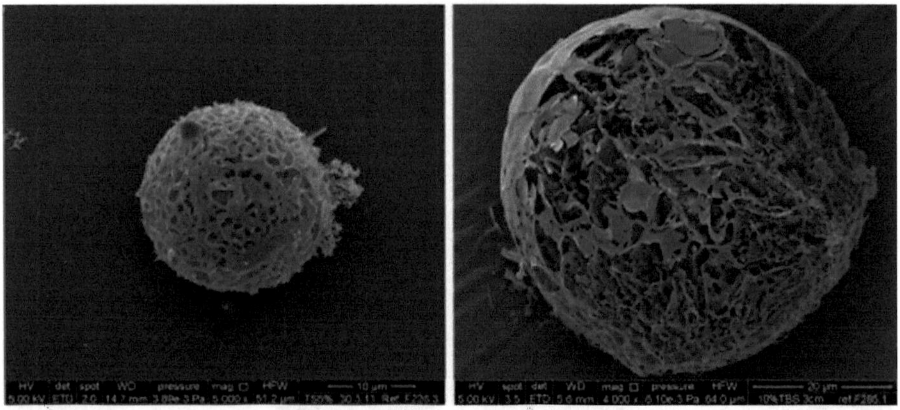

Fig. 7.4 Terbutaline sulphate large porous particles prepared by thermal ink-jet spray freeze-drying. (Reprinted with permission of Elsevier Limited from ref. [58])

thermal ink-jet spray freeze-drying with mean geometric diameter of 35 μm and mean aerodynamic diameter lower than 8.7 μm, present a percentage of FPF comparable to a salbutamol commercial formulation [57].

One drawback of therapeutic peptides and proteins is their inherent immunogenicity. Some studies show the development of antibodies against a particular protein after its administration [59, 60]. PEGylation is a strategy largely used to reduce the immunogenicity and increase the stability and half-life of proteins [61]. However, contrary to the general assumption, the immunogenicity's reduction by PEGylation is not linear and depends on the protein and administration route [62]. At the moment, some PEGylated molecules are in the market and used daily in the clinical practice (PEGasys®, Adagen®, Oncaspar®, or PEGIntron®).

Formulations based on biopharmaceuticals loaded into nano- or microparticles have been widely proposed in the last years as strategies to overcome the limitations of conventional formulations. The encapsulation of biopharmaceuticals potentially protects them from degradation, reduces their immunogenicity, improves their retention in lungs and permeation trough alveolar epithelium, or reduces their uptake by macrophages [16, 63]. Since a wide range of the developed nanoparticles falls within the particle's size range liable to suffer exhalation, the agglomeration of nanoparticles into micron-sized particles with proper aerodynamic characteristics that disaggregate after deposition have been exploited [64, 65]. Also, the development of mucoadhesive [66] and mucus-penetrating particulates has been proposed [67, 68] to increase the residence time of drugs into the lungs and improve their absorption and/or therapeutic efficacy. The last approach could have great impact in the pulmonary delivery of drugs to treat diseases with high mucus production like cystic fibrosis [69, 70]. One possible advantage of using agglomerates of nanoparticles instead of microparticles relies on the capacity of nanoparticles to easily evade mucociliary clearance and phagocytosis by alveolar macrophages. Some studies show that smaller particles are internalized at a lower extent than particles higher in size [71, 72].

7.3 State-of-the-Art Inhalation for Biopharmaceuticals

Despite the promise and advantages presented by pulmonary delivery, dornase alfa (Pulmozyme® from Genentech), a highly purified solution of recombinant human deoxyribonuclease I (rhDNase) used in the treatment of cystic fibrosis, is the only currently marketed biopharmaceutical formulation for inhalation. This could be attributed to the formulation and delivery challenges of inhalable drugs referred previously. However, the rational design of particles with appropriate characteristics for inhalation enabled the development of formulations that entered in clinical evaluation.

In this section, will be provided and discussed examples of formulations for pulmonary delivery of different classes of biopharmaceuticals that enroll distinct stages of pharmaceutical development.

7.3.1 Therapeutic Peptides and Proteins

Insulin has branded the history of the development of inhaled protein therapeutics with cycles of hope and disappointment. Attempts to develop an inhaled insulin formulation to substitute the current treatment of diabetes by subcutaneous injection have been made by dozens of researchers and companies all over the world during decades. The apogee of inhaled insulin was reached when US Food and Drug Administration (FDA) and European Medicines Agency (EMA) granted market authorization to Exubera® (Pfizer/Nektar) in 2006, but less than 2 years later, the product was withdrawn due to its failure in achieving the expected market success (detailed information in Chap. 21). This decision had as casualty the interruption of AERxiDMS® (Novo Nordisk/Aradigm) and AIR® (Eli Lilly/Alkermes)'s development, both at phase III clinical trials, few months after the decision [16, 73]. Although Pfizer argued that the decision was due to a commercial failure related to limitations of the formulation and the inhaler device hampering the acceptance by the patients and clinicians; the emergence of cases of lung cancer in patients treated with Exubera® raised questions regarding the immunological effects and safety of inhaled proteins [74]. Nonetheless, some companies continued the development of their products with different and improved technologies, hoping to succeed where Exubera® failed.

Currently, Afrezza® (MannKind), a DPI based on Technosphere® technology [75] using a next-generation inhaler device (Dreamboat®), is waiting for FDA approval. Since MannKind changed the inhaler device from MedTone® to Dreamboat® in the middle of the clinical studies, FDA is concerned about the equivalence of the two devices, which has been delaying the product approval. Technosphere® technology is based on fumaryldiketopiperazine large porous particles with a MMAD of 2–2.5 μm, suitable for delivery at deep lung, where small proteins can be absorbed onto the surface [75] (detailed information in Chap. 22). Pharmacodynamic and pharmacokinetic analysis of Afrezza® has shown a rapid absorption ($t_{max} = 12$–14 min), short onset of action (20–30 min), and action duration time (2–3 h) that mimic the physiological insulin requirements to cover prandial glucose absorption in type 2 diabetic patients [76]. In a pilot study, an optimal dose of inhaled insulin was able to control the postprandial glycemic levels of type 2 diabetes patients regardless of the meal carbohydrate content [77]. In addition, the absorption and pharmacokinetics of insulin after inhalation was not significantly altered in patients with mild-to-moderate chronic obstructive pulmonary disease (COPD) ($C_{max} = 34.7$ μU/mL and AUC = 2,037 μU/mL min) compared to healthy patients ($C_{max} = 39.5$ μU/mL and AUC = 2,279 μU/mL min) [78]. Concerning safety, in a 2-year study, Afrezza® showed to be well tolerated, promoting slight changes in lung function comparable to the usual treatment and mild, transient cough after inhalation (Fig. 7.5) [79]. However, the results of the study should be analyzed carefully, since a high percentage of treatment discontinuation owing to adverse events was higher in the Afrezza®-treated group.

Other inhalable insulin products are under clinical stage such as Aerodose (Phase II), Abbott Labs' inhaled insulin (Phase II), QDose (Phase I), Alveair® (Phase I), BioAir® (Phase I), or ProMaxx® (Phase I) [16, 73]. Insulin is commonly used as a drug model in the development of formulations for delivery of proteins, which explains the high number of studies at preclinical stage being published [64,80–89].

Besides Afrezza®, MannKind is also developing a glucagon-like peptide 1 (GLP-1) to treat type 2 diabetes by inhalation based on the same Technosphere® system. The product (MKC253) is currently at Phase I stage and has, so far, proved to be able to stimulate the insulin secretion and, consequently, reduce the postprandial glycemic levels without the gastrointestinal adverse effects usually seen in the subcutaneous or oral administration of GLP-1 and its analogs used in clinical practice [90].

Aerovance is a biopharmaceutical company that is also exploiting the potential of biopharmaceuticals' inhalation to treat local diseases. Presently, two DPI are at phase II clinical studies for the treatment of asthma (pitrakinra) and cystic fibrosis and COPD (bikunin). Pitrakinra is a recombinant human Interleukin-4 (IL-4) variant that efficiently inhibits both IL-4 and Interleukin-13 (IL-13) activity, reducing the inflammation in asthma and eczema [91, 92]. At first, in the studies both liquid and powder formulations were tested, but the last news available is related to a DPI (Aerovant®) formulation to treat exacerbations in patients with eosinophilic asthma. Bikunin is a truncated human SPINT2 serine protease inhibitor that presents the capacity to reduce the airway epithelial sodium ion channels activity, thereby reducing sodium hyper absorption in cystic fibrosis patients and COPD [93]. It is currently under development as two different products, with the name Aerolytic® and Pulmolytic®, for the treatment of cystic fibrosis and COPD, respectively.

Inhalable biopharmaceuticals to treat viral infections have also been proposed. DAS181 is a recombinant sialidase fusion protein that inactivates viral receptors on the cells of the human respiratory tract, thus preventing and treating infection by various influenza virus subtypes, including H5N1 and parainfluenza [94–96]. In a phase II clinical trial, DAS181 was able to reduce the lung viral load in patients infected with influenza B, H3N2, and H1N1 without significant side effects [94]. DAS181 was formulated using TOSAP® technology into dry powder microspheres for pulmonary delivery (Fludase®).

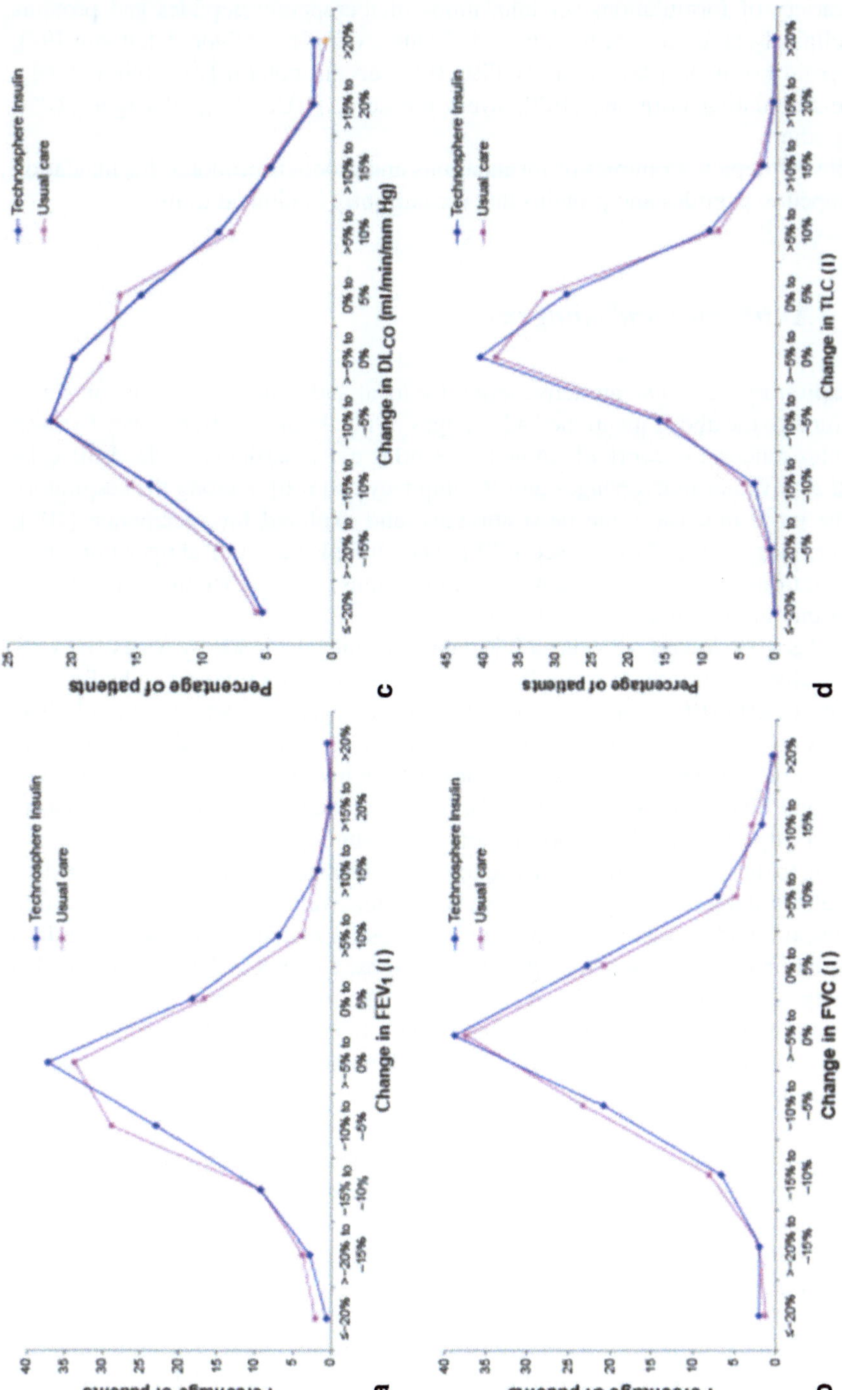

Fig. 7.5 Percentage of patients with change in **a** forced expiratory volume in 1 s (FEV1), **b** forced vital capacity (FVC), **c** lung diffusion capacity for carbon monoxide (DL_{CO}), and **d** total lung capacity (TLC) from baseline at the last measurement. (Reprinted with permission of John Wiley and Sons from ref. [79])

A variety of formulations for inhalations of therapeutic peptides and proteins at preclinical stage have been explored. Some examples include calcitonin [97], parathyroid hormone [98], detirelix [99, 100], erythropoietin [27], INF-α [101], follicle-stimulating hormone [102], cyclosporine A [103, 104], glucagon [105], among others.

Table 7.1 depicts examples of formulations and products proposed for inhalation of therapeutic peptides and proteins that are currently in clinical trials.

7.3.2 *Antibodies and Antigens*

The respiratory tract is an attractive target for local and systemic vaccination since, in addition to the above mentioned advantages of inhalation, it offers a way to reach a complex mucosal network of antigen-presenting cells, particularly, dendritic cells as well as alveolar macrophages and B lymphocytes [106]. Among the respiratory tract the nasal mucosa is the most attractive and explored for vaccination [107]; however, lungs could also be used [108]. The administration of antigens provides immunization and protection against viral infections [109], tuberculosis [110], and other bacterial infections.

AERAS-402/Crucell Ad35 is an injectable vaccine developed by Aeras and Crucell N.V. that is currently in phase II trials to boost the immunity primed by Bacillus Calmette-Guérin (BCG), the only available vaccine against tuberculosis [111]. The system is comprised by a recombinant adenovirus, serotype 35 (Ad35), expressing a fusion protein created from the sequences of the mycobacterial antigens Ag85A, Ag85B, and TB10.4. Driven by the good clinical results, the companies are exploring the possibility of an inhalable formulation of the same system [112].

Owing to their capacity to enhance antigen internalization and presentation to dendritic cells, polymeric nanoparticles have been extensively proposed as a vaccination platform to a wide range of administration routes. In a study conducted on guinea pigs, pulmonary delivery of dry powders composed of CRM-197 antigen-loaded poly lactic-co-glycolic-acid (PLGA) nanoaggregates provide higher mucosal immunization (higher immunoglobulin A (IgA) titers) than parenteral administration of the antigen and sufficient systemic protection (immunoglobulin G (IgG) antibodies) against diphtheria [113]. In addition, the inhaled nanoparticles increase the survival time of the animals after challenge with the toxic, compared to the intramuscular administration of the antigens (Fig. 7.6). The same group tested a similar system for the delivery of antigens to immunize against hepatitis B [114] and tuberculosis [115] with encouraging results.

Pulmonary administration of antibodies against cytokines, namely IL, involved in inflammation process has been tested to treat local diseases such as asthma or bronchitis. The administration of anti-IL-13 fragment antigen binding (Fab)′ region to mice by nebulization leads to a reduction in the pulmonary levels of proinflammatory markers, eosinophiles, and cell infiltration in lung tissue. By this, inhalation of anti-IL-13 Fab′ fragment was able to reduce the inflammation, bronchial responsiveness, and airway remodeling typical in asthma patients [116].

Table 7.1 Formulations for pulmonary administration of therapeutic peptides and proteins under development at clinical stage

Peptide/protein	Therapeutic indication	Name/system	Inhalation mode	Development stage	References
Bikunin	Cystic fibrosis and COPD	Aerolytic and pulmolytic	Nebulizer and DPI	Phase II	[140]
Cyclosporine A	Lung transplant rejection	Liquid solution	Nebulizer	Phase III	[141]
DAS181	Influenza virus infection	Fludase/TOSAP	DPI	Phase II	[94]
GLP-1	Diabetes mellitus	MKC253/Technosphere	DPI	Phase I	[90]
INF-γ	Cystic fibrosis, lung infection	Liquid solution	Nebulizer	Phase II	[142, 143]
Insulin	Diabetes mellitus	Afrezza/Technosphere	DPI	Phase III	[76]
Insulin	Diabetes mellitus	Aerodose/liquid solution	Nebulizer	Phase II	[144, 145]
Insulin	Diabetes mellitus	Dry crystals	MDI	Phase II	[146]
Insulin	Diabetes mellitus	QDose	DPI	Phase I	[147]
Insulin	Diabetes mellitus	Alveair/liquid solution	Nebulizer	Phase I	[148]
Insulin	Diabetes mellitus	BioAir/pegylated calcium phosphate nanoparticles	DPI	Phase I	[149]
Insulin	Diabetes mellitus	ProMaxx/microspheres	DPI	Phase I	[150]
Interleukin-2	Metastatic or unresectable solid tumors	Liquid solution	Nebulizer	Phase I	[151]
Pitrakinra	Asthma	Aerovant	DPI	Phase II	[92]
Sargramostin	Metastatic cancer, sarcoma	Liquid solution	Nebulizer	Phase II	[152]
α_1-antitrypsin	Cystic fibrosis	Liquid solution	Nebulizer	Phase I	[153]

Table 7.2 provides some examples of systems proposed for pulmonary delivery of antibodies and antigens.

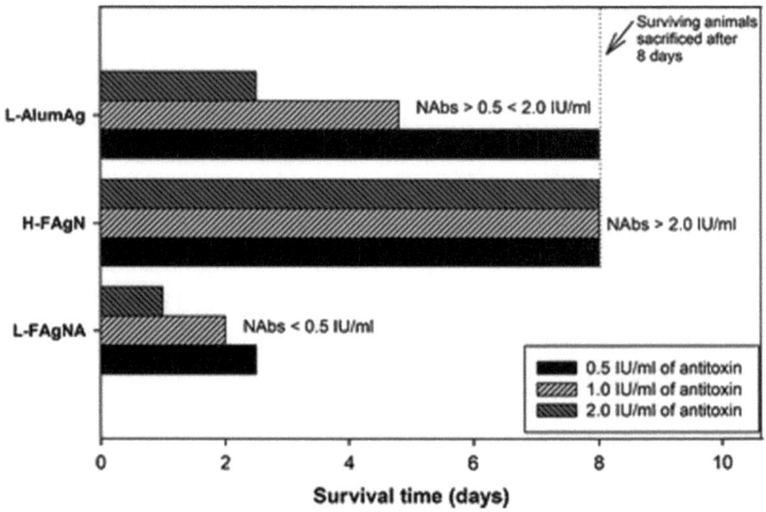

Fig. 7.6 Toxin-neutralizing test performed on the serum samples of vaccinated guinea pigs at three levels equivalent to 2.0, 1.0, and 0.5 international unit (IU) of diphtheria antitoxin/mL. Low-dose alum with adsorbed antigen (L-AlumAg) was administered by intramuscular injection, high-dose formalin-treated antigen in nanoparticles (H-FAgN) and low-dose formalin-treated antigen and nanoparticle admixture (L-FAgNA) were administered by inhalation. Nabs are neutralizing antibodies. (Reprinted with permission of Springer from ref. [113])

7.3.3 Genetic Material

The possibility to deliver an exogenous DNA encoding for an absent or defective gene as well as to silence specific genes via the RNA interference mechanism makes gene therapy a desired strategy for the treatment of various diseases, including the pulmonary disorders. The potential and usefulness of gene therapy becomes obvious when only one gene is involved, the case of cystic fibrosis and α_1-antitrypsin deficiency, because the problem is specifically solved by replacing the missing or defective gene. For multifactorial diseases such as COPD, asthma, interstitial lung diseases, or chronic infections, the gene replacement therapy would not be so effective. However, in these situations gene therapy still can be useful as a concomitant treatment to promote the temporary overexpression of protective genes or the suppression of injurious genes [117].

A wide range of vectors has been proposed to deliver the genetic material to the epithelium of lungs and airways. Viral vectors such as retrovirus, lentivirus, adenovirus, and parainfluenza virus are reported as efficient gene transfer vehicles; however, their associated immunogenicity, insertional mutagenesis, limited loading capacity, and labor-intensive or expensive procedures bring some constrains to their clinical use. Therefore, synthetic non-viral vectors such as cationic polymers and liposomes have emerged as attractive alternatives to gene delivery. Comparing with viral vectors, the non-viral vectors present greater affinity binding to the airway epithelium, but lower transfection efficiencies. Polyethylenimine (PEI) is the commonly used polymer in

Table 7.2 Formulations for pulmonary administration of antibodies and antigens under development

Antibody/antigen	Therapeutic indication	System	Inhalation mode	Development stage	References
Ad35 expressing 85A, 85B and TB10.4	Tuberculosis	Powder	–	In vitro	[112]
Anti-IL-13 Fab′ fragment	Asthma	Liquid solution	Nebulization	In vivo	[116]
CRM-197	Diphtheria	PLGA nanoparticles	Powder insufflation	In vivo	[113]
IgG1	Inflammation disorders	Powder	–	In vitro	[154, 155]
Influenza A Panama/2007/99	Influenza infection	Antigen microparticles	Powder insufflation	In vivo	[156]
Live-attenuated BCG	Tuberculosis	Powder	Powder insufflation	In vivo	[157]
Live-attenuated Newcastle virus	Newcastle disease	Powder	–	In vitro	[158]
rAg85B	Tuberculosis	PLGA microparticles	Powder insufflation	In vivo	[115]
rAg85B	Tuberculosis	Pluronic/Polypropylene sulfide nanoparticles	Intratracheal instillation	In vivo	[110]
rAg85B	Tuberculosis	Pluronic solution	Intratracheal instillation	In vivo	[159]
rHBsAg	Hepatitis B	PLGA/PEG nanoparticles	Powder insufflation	In vivo	[114]

gene delivery due to its good transfection efficiency; however, concerns regarding its safety have been impairing its use. The cytotoxicity usually related with the cationic non-viral vectors is due to polymers aggregation on cell surface as a result of their strong electrostatic charge [34, 117]. Regarding this aspect, various efforts have been made to find a secure gene delivery vector with the ideal properties to be administered via inhalation. Therefore, vectors based on different polymers, undergoing several modifications have been produced.

Cystic fibrosis is an autosomal recessively inherited disorder caused by mutation on the cystic fibrosis transmembrane conductance regulator (CFTR) gene. As a result of this mutation, the reduced thickness of the airway surface liquid, the dehydration of the secreted mucus, and the impaired mucociliary clearance facilitate the bacterial infections and the progressive debility of lung function. As the bronchiolar epithelium is the main therapeutic target, the aerosol delivery of the corrective gene to the airway epithelium has been experimented using several viral and non-viral vectors [118, 119]. Until now, dozens of clinical trials have been carried out in order to evaluate the usefulness of gene therapy in cystic fibrosis progression control. Nevertheless, all of the trials performed were clinically successful, and there is still much to be done in this direction [120–123]. One of the possible reasons for the failure is the presence of thickened mucus or mucus plugs that interferes with the correct transport of gene carriers to the airways' epithelium [35, 124]. In order to improve the therapeutic efficacy of the gene delivery approach, the concomitant administration of dornase alfa or N-acetylcysteine will reduce the mucus viscosity and improve the penetration and delivery of the vector [69].

A phase II randomized, double-blind and placebo-controlled clinical trial regarding the application of gene therapy to cystic fibrosis is still ongoing. This trial consists in assessing the safety and tolerability of repeated doses of aerosolized adeno-associated virus serotype type 2 containing the CFTR complementary DNA [125, 126]. The first results are positive and encouraging; however, the use of viral vectors is haunted by the potential development of oncogenicity and insertional mutagenesis in host cells. This fear leads many researchers to pursue for efficient non-viral vectors. The largest clinical trials regarding the pulmonary administration of non-viral gene therapy to treat cystic fibrosis are being conducted by the UK Cystic Fibrosis Gene Therapy Consortium and are at phase II. The system comprises cationic liposomes (GL67A) composed of dioleoylphosphatidylethanolamine:1,2-dimyristoyl-*sn*-glycero-3-phosphoethanolamine-*N*-[(poly ethylene glycol) 5000] (DOPE:DMPE-PEG_{5000}) combined with a CFTR plasmid (pGM169). Preclinical studies showed that the inhalation of the system is safe and able to induce the gene expression at clinical relevant levels [127, 128].

Cancer is the disorder in which gene therapy has its major percentage of application (64.3 %) [129]. Lung cancer is the leading cancer in men and presents an increase in the incidence in women. The delayed diagnosis and the increasingly new mutations and resistances to the current therapeutic strategies are what make this disease so worrying. Thus, great attention has been given to new therapeutic strategies, as for example, the development of gene-based immunotherapeutic vaccines for lung cancer treatment [34, 130].

Although the employment of aerosol administration applied to gene therapy still present some reserves due to security reasons, it has been reported its utility in the administration of tumor suppressor genes (e.g., p53, anti-VEGF, MMP-2,-9, IGF-IR) and immunotherapy (IL-6, 12, 1β, TNF-α) [34, 131–134]. As an example, chitosan genes (cytomegalovirus promoter encoding the murine interferon-β (pCMV-Muβ)) polyplexes to treat lung metastasis in mice were developed. The best results regarding the reduction in the number of pulmonary nodules and the survival time of animals were obtained with the intratracheal administration of powder in comparison with the intratracheal and intravenous administered solutions of the same components (Fig. 7.7) [135]. This could be due to higher concentrations of the gene owing to local delivery and retention of the powder particles in lung tissue. The thematic of inhaled gene therapy for lung cancer was extensively reviewed by Zarogoulidis et al. [34].

Another disorder in which inhaled gene therapy has gained a great importance as a new clinical option is the pulmonary arterial hypertension (PAH), a deadly disease caused by genetic and acquired abnormalities where apoptosis is inhibited and cell

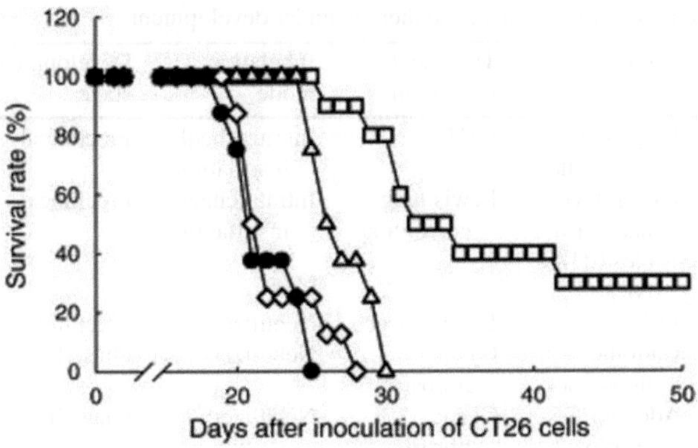

Fig. 7.7 The effect of pCMV-Muβ formulations on the survival rate of mice burdened with pulmonary metastasis. The pCMV-Muβ formulations, intravenous (◇), intratracheal solution (△), and intratracheal powder (□), were administered to mice the day after the inoculation with CT26 cells. Control mice were left untreated (•). Statistical differences ($p < 0.05$) in the mean survival time were observed for control vs. intratracheal solution, control vs. intratracheal powder, intravenous vs. intratracheal solution, intravenous vs. powder solution, and intratracheal solution vs. powder solution. (Reprinted with permission of Elsevier Limited from ref. [135])

proliferation of the vascular wall is increased. This disease is also characterized by the downregulation of the bone morphogenetic protein axis and voltage-gated potassium channels [136–138]. Survivin, an apoptosis inhibitor protein expressed in the pulmonary arteries, was inhibited through intratracheal administration of an adenovirus carrying a phosphorylation-deficient survivin mutant to rats. The survivin inhibition resulted in satisfactory reversion of PAH-associated physiologic irregularities and survival percentage of the animals [139].

Examples of vectors administered directly to the airways epithelium already developed for the treatment of cystic fibrosis, lung cancer, and PAH are listed in Table 7.3.

7.4 Conclusions and Future Perspectives

Pulmonary administration of biopharmaceuticals is hard work and challenging but still possible and could become a good non-invasive alternative to parenteral administration to treat both local and systemic diseases. The pathway of inhalable biopharmaceuticals to the success has been paved in the last decades by many research groups and companies, driven specially by the development of aerosol formulations of therapeutic peptides and proteins that reach clinical trials. Although the major slice of the development and clinical trials of inhalable biopharmaceuticals relies on peptides and proteins, vaccination and gene therapy by pulmonary administration have also been increasingly explored. Inhalable biopharmaceuticals may be truly useful to protect and/or treat diseases such as diabetes mellitus, tuberculosis, cystic

Table 7.3 Formulations for inhaled gene therapy under development

Gene	Vector	Therapeutic indication	Inhalation mode	Development stage	References
Adrenomedullin	Polyplexnanomicelles	PAH	Intratracheal instillation	Preclinical	[137]
AT2R or TRAIL	HIV-1 TAT nanoparticles (dTAT NP)	Lewis lung carcinoma	Intratracheal instillation	Preclinical	[160]
BC-819	PEI	Lung cancer	Nebulizer	Preclinical	[161]
CFTR	Cationic liposomes	Cystic fibrosis	Nebulizer	Phase II	[127, 128]
CFTR	Adenovirusserotype 2 (rAAV2)	Cystic fibrosis	Nebulizer	Phase II	[125, 126, 162]
Extracelular superoxide dismutase	Adenovirus	PAH	Intratracheal instillation	Preclinical	[163]
Kinase-deficient Akt1	Glucosylated PEI	Lung cancer	Nebulizer	Preclinical	[164]
Smallhairpin osteopontin	Lentivirus	Pulmonary metastasis of breast cancer	Nebulizer	Preclinical	[165]
Surviving mutant	Adenovirus	PAH	Intratracheal instillation	Preclinical	[139]
VEGF	Adenovirus	PAH	Intratracheal instillation	Preclinical	[136]

fibrosis, and lung cancer. A variety of formulations is presently at clinical stage, and it is expected that in the next years new inhaled biopharmaceuticals will join Pulmozyme®.

Despite the high absorption of some biopharmaceuticals through the alveolar epithelium reported in many studies, the development of innovative formulations and devices that efficiently deliver the drugs in the active state to the proper area of respiratory tract is the key factor and the most difficult task in pulmonary administration. Joint efforts between pharmaceutical and medical devices' companies must be made to successfully achieve clinical biopharmaceutical-based inhalable products.

Nanomedicine is a growing area and a higher number of studies regarding nanoparticles for drug delivery, including pulmonary administration of biopharmaceuticals, have been published in the last years over conventional formulations owing to its reported advantages. However, nanomedicines present also disadvantages such as complex structure and production techniques, the need for new and improved delivery devices, and possible higher costs of production. In addition, the toxicological concerns surrounding the use of nanomaterials in the development of medicines and the discussion at the regulatory agencies concerning the establishment of new tools and guidelines to specifically assess the safety, efficacy, and quality of nanomedicines at preclinical stage explains why few nanotechnology-based products were granted market authorization and why a low number of inhalable formulations enrolled for clinical trials. For that, conventional formulations continue to be the appropriate approach for some products, continuing to have a place in the development of new medicines.

Although some of the developed formulations for inhalation of biopharmaceuticals show little success in vivo or are unable of proper scale-up production with the current technologies available, the search for a suitable alternative to injection still continues. The results so far are promising and it is expected that with the technological advances new formulations with clinical relevance will appear.

Acknowledgments Fernanda Andrade gratefully acknowledges Fundação para a Ciência e a Tecnologia (FCT), Portugal, for financial support (SFRH/BD/73062/2010).

References

1. Overview: medicines in development—biologics. 2013.
2. Rader RA. FDA biopharmaceutical product approvals and trends in 2012. BioProcess International. 2013;11(3):18–27.
3. World Preview 2018.
4. Grenha A. Systemic delivery of biopharmaceuticals: parenteral forever? J Pharm Bioallied Sci. 2012;4(2):95.
5. Antosova Z, Mackova M, Kral V, Macek T. Therapeutic application of peptides and proteins: parenteral forever? Trends Biotechnol. 2009;27(11):628–35.
6. Laube BL. The expanding role of aerosols in systemic drug delivery, gene therapy, and vaccination. Respir Care. 2005;50(9):1161–76.
7. Pinto Reis C, Silva C, Martinho N, Rosado C. Drug carriers for oral delivery of peptides and proteins: accomplishments and future perspectives. Ther Deliv. 2013;4(2):251–65.
8. Sanders M. Inhalation therapy: an historical review. Prim Care Respir J. 2007;16(2):71–81.
9. Muttil P, Wang C, Hickey AJ. Inhaled drug delivery for tuberculosis therapy. Pharm Res. 2009;26(11):2401–16.
10. Christopher F, Chase D, Stein K, Milne R. rhDNase therapy for the treatment of cystic fibrosis patients with mild to moderate lung disease. J Clin Pharm Ther. 1999;24(6):415–26.
11. Cefalu W. Inhaled insulin: a proof-of-concept study. Ann Intern Med. 2001;134(9 Pt 1):795.
12. Rijnders BJ, Cornelissen JJ, Slobbe L, Becker MJ, Doorduijn JK, Hop WC, Ruijgrok EJ, Löwenberg B, Vulto A, Lugtenburg PJ, de Marie S. Aerosolized liposomal amphotericin B for the prevention of invasive pulmonary aspergillosis during prolonged neutropenia: a randomized, placebo-controlled trial. Clin Infect Dis. 2008;46(9):1401–8.
13. Hokey DA, Misra A. Aerosol vaccines for tuberculosis: a fine line between protection and pathology. Tuberculosis (Edinb). 2011;91(1):82–5.
14. Lu D, Garcia-Contreras L, Muttil P, Padilla D, Xu D, Liu J, Braunstein M, McMurray DN, Hickey AJ. Pulmonary immunization using antigen 85-B polymeric microparticles to boost tuberculosis immunity. AAPS J. 2010;12(3);338–47.
15. Apiou-Sbirlea G, Newman S, Fleming J, Siekmeier R, Ehrmann S, Scheuch G, Hochhaus G, Hickey A. Bioequivalence of inhaled drugs: fundamentals, challenges and perspectives. Ther Deliv. 2013;4(3):343–67.
16. Andrade F, Videira M, Ferreira D, Sarmento B. Nanocarriers for pulmonary administration of peptides and therapeutic proteins. Nanomedicine (Lond). 2011;6(1):123–41.

17. Pilcer G, Amighi K. Formulation strategy and use of excipients in pulmonary drug delivery. Int J Pharm. 2010;392(1–2):1–19.
18. Scheuch G, Kohlhaeufl MJ, Brand P, Siekmeier R. Clinical perspectives on pulmonary systemic and macromolecular delivery. Adv Drug Deliv Rev. 2006;58(9–10):996–1008.
19. Patton JS, Fishburn CS, Weers JG. The lungs as a portal of entry for systemic drug delivery. Proc Am Thorac Soc. 2004;1(4):338–44.
20. Folkesson H, Weström B, Karlsson B. Permeability of the respiratory tract to different-sized macromolecules after intratracheal instillation in young and adult rats. Acta Physiol Scand. 1990;139(2):347–54.
21. Conhaim R, Watson K, Lai-Fook S, Harms B. Transport properties of alveolar epithelium measured by molecular hetastarch absorption in isolated rat lungs. J Appl Physiol. 2001;91(4):1730–40.
22. Holter J, Weiland J, Pacht E, Gadek J, Davis W. Protein permeability in the adult respiratory distress syndrome. Loss of size selectivity of the alveolar epithelium. J Clin Invest. 1986;78(6):1513–22.
23. Siekmeier R, Scheuch G. Systemic treatment by inhalation of macromolecules—principles, problems, and examples. J Physiol Pharmacol. 2008;59 Suppl 6:53–79.
24. Kim K, Malik A. Protein transport across the lung epithelial barrier. Am J Physiol Lung Cell Mol Physiol. 2003;284(2):L247–59.
25. Groneberg D, Fischer A, Chung K, Daniel H. Molecular mechanisms of pulmonary peptidomimetic drug and peptide transport. Am J Respir Cell Mol Biol. 2004;30(3):251–60.
26. Bitonti AJ, Dumont JA. Pulmonary administration of therapeutic proteins using an immunoglobulin transport pathway. Adv Drug Deliv Rev. 2006;58(9–10):1106–18.
27. Bitonti A, Dumont J, Low S, Peters R, Kropp K, Palombella V, Stattel J, Lu Y, Tan C, Song J, Garcia A, Simister N, Spiekermann G, Lencer W, Blumberg R. Pulmonary delivery of an erythropoietin Fc fusion protein in non-human primates through an immunoglobulin transport pathway. Proc Natl Acad Sci U S A. 2004;101(26):9763–8.
28. Morris CJ, Smith MW, Griffiths PC, McKeown NB, Gumbleton M. Enhanced pulmonary absorption of a macromolecule through coupling to a sequence-specific phage display-derived peptide. J Control Release. 2011;151(1):83–94.
29. DeHaan WH, Finlay WH. In vitro monodisperse aerosol deposition in a mouth and throat with six different inhalation devices. J Aerosol Med. 2001;14(3):361–7.
30. Donovan MJ, Kim SH, Raman V, Smyth HD. Dry powder inhaler device influence on carrier particle performance. J Pharm Sci. 2012;101(3):1097–1107.
31. Coates MS, Fletcher DF, Chan HK, Raper JA. Effect of design on the performance of a dry powder inhaler using computational fluid dynamlics. Part 1: Grid structure and mouthpiece length. J Pharm Sci. 2004;93(11):2863–76.
32. Hess DR. Aerosol delivery devices in the treatment of asthma. Respir Care. 2008;53(6):699–723; discussion 695–723.
33. Labiris NR, Dolovich MB. Pulmonary drug delivery. Part II: the role of inhalant delivery devices and drug formulations in therapeutic effectiveness of aerosolized medications. Br J Clin Pharmacol. 2003;56(6):600–12.
34. Zarogouldis P, Karamanos NK, Porpodis K, Domvri K, Huang H, Hohenforst-Schimdt W, Goldberg EP, Zarogoulidis K. Vectors for inhaled gene therapy in lung cancer. Application for nano oncology and safety of bio nanotechnology. Int J Mol Sci. 2012;13(9):10828–62.
35. Manunta MD, McAnulty RJ, Tagalakis AD, Bottoms SE, Campbell F, Hailes HC, Tabor AB, Laurent GJ, O'Callaghan C, Hart SL. Nebulisation of receptor-targeted nanocomplexes for gene delivery to the airway epithelium. PLoS One. 2011;6(10):e26768.
36. Jeong SH. Analytical methods and formulation factors to enhance protein stability in solution. Arch Pharm Res. 2012;35(11):1871–86.
37. Frokjaer S, Otzen DE. Protein drug stability: a formulation challenge. Nat Rev Drug Discov. 2005;4(4):298–306.
38. de Boer AH, Gjaltema D, Hagedoorn P, Frijlink HW. Characterization of inhalation aerosols: a critical evaluation of cascade impactor analysis and laser diffraction technique. Int J Pharm. 2002;249(1–2):219–31.

39. Oberdörster G, Oberdörster E, Oberdörster J. Nanotoxicology: an emerging discipline evolving from studies of ultrafine particles. Environ Health Perspect. 2005;113(7);823–39.
40. Heyder J. Deposition of inhaled particles in the human respiratory tract and consequences for regional targeting in respiratory drug delivery. Proc Am Thorac Soc. 2004;1(4):315–20.
41. Moretton MA, Chiappetta DA, Andrade F, das Neves J, Ferreira D, Sarmento B, Sosnik A. Hydrolyzed galactomannan-modified nanoparticles and flower-like polymeric micelles for the active targeting of rifampicin to macrophages. J Biomed Nanotechnol. 2013;9(6):1–12.
42. Chono S, Tanino T, Seki T, Morimoto K. Efficient drug targeting to rat alveolar macrophages by pulmonary administration of ciprofloxacin incorporated into mannosylated liposomes for treatment of respiratory intracellular parasitic infections. J Control Release. 2008;127(1): 50–8.
43. Champion JA, Walker A, Mitragotri S. Role of particle size in phagocytosis of polymeric microspheres. Pharm Res. 2008;25(8):1815–21.
44. Chono S, Tanino T, Seki T, Morimoto K. Uptake characteristics of liposomes by rat alveolar macrophages: influence of particle size and surface mannose modification. J Pharm Pharmacol. 2007;59(1):75–80.
45. Shoyele SA, Cawthorne S. Particle engineering techniques for inhaled biopharmaceuticals. Adv Drug Deliv Rev. 2006;58(9–10):1009–29.
46. Telko MJ, Hickey AJ. Dry powder inhaler formulation. Respir Care. 2005;50(9):1209–27.
47. Crowder T, Rosati J, Schroeter J, Hickey A, Martonen T. Fundamental effects of particle morphology on lung delivery: predictions of Stokes' law and the particular relevance to dry powder inhaler formulation and development. Pharm Res. 2002;19(3):239–45.
48. Pitchayajittipong C, Price R, Shur J, Kaerger JS, Edge S. Characterisation and functionality of inhalation anhydrous lactose. Int J Pharm. 2010;390(2):134–41.
49. Kaialy W, Ticehurst M, Nokhodchi A. Dry powder inhalers: mechanistic evaluation of lactose formulations containing salbutamol sulphate. Int J Pharm. 2012;423(2):184–94.
50. Kaialy W, Alhalaweh A, Velaga SP, Nokhodchi A. Influence of lactose carrier particle size on the aerosol performance of budesonide from a dry powder inhaler. Powder Technology. 2012;227:74–85.
51. Schiavone H, Palakodaty S, Clark A, York P, Tzannis ST. Evaluation of SCF-engineered particle-based lactose blends in passive dry powder inhalers. Int J Pharm. 2004;281(1–2): 55–66.
52. Le VN, Hoang Thi TH, Robins E, Flament MP. Dry powder inhalers: study of the parameters influencing adhesion and dispersion of fluticasone propionate. AAPS PharmSciTech. 2012;13(2):477–84.
53. Le VN, Hoang Thi TH, Robins E, Flament MP. In vitro evaluation of powders for inhalation: the effect of drug concentration on particle detachment. Int J Pharm. 2012;424(1–2):44–9.
54. Dickhoff BH, de Boer AH, Lambregts D, Frijlink HW. The effect of carrier surface and bulk properties on drug particle detachment from crystalline lactose carrier particles during inhalation, as function of carrier payload and mixing time. Eur J Pharm Biopharm. 2003;56(2):291–302.
55. Dickhoff BH, de Boer AH, Lambregts D, Frijlink HW. The effect of carrier surface treatment on drug particle detachment from crystalline carriers in adhesive mixtures for inhalation. Int J Pharm. 2006;327(1–2):17–25.
56. Garcia-Contreras L, Fiegel J, Telko MJ, Elbert K, Hawi A, Thomas M, VerBerkmoes J, Germishuizen WA, Fourie PB, Hickey AJ, Edwards D. Inhaled large porous particles of capreomycin for treatment of tuberculosis in a guinea pig model. Antimicrob Agents Chemother. 2007;51(8):2830–6.
57. Mueannoom W, Srisongphan A, Taylor KM, Hauschild S, Gaisford S. Thermal ink-jet spray freeze-drying for preparation of excipient-free salbutamol sulphate for inhalation. Eur J Pharm Biopharm. 2012;80(1):149–55.
58. Sharma G, Mueannoom W, Buanz AB, Taylor KM, Gaisford S. In vitro characterisation of terbutaline sulphate particles prepared by thermal ink-jet spray freeze drying. Int J Pharm. 2013;447(1–2):165–70.

59. Rosenstock J, Cefalu W, Hollander P, Belanger A, Eliaschewitz F, Gross J, Klioze S, St Aubin L, Foyt H, Ogawa M, Duggan W. Two-year pulmonary safety and efficacy of inhaled human insulin (Exubera) in adult patients with type 2 diabetes. Diabetes Care. 2008;31(9):1723–8.
60. Perini P, Facchinetti A, Bulian P, Massaro AR, Pascalis DD, Bertolotto A, Biasi G, Gallo P. Interferon-beta (INF-beta) antibodies in interferon-beta1a- and interferon-beta1b-treated multiple sclerosis patients. Prevalence, kinetics, cross-reactivity, and factors enhancing interferon-beta immunogenicity in vivo. Eur Cytokine Netw. 2001;12(1):56–61.
61. Jevsevar S, Kunstelj M, Porekar VG. PEGylation of therapeutic proteins. Biotechnol J. 2010;5(1):113–28.
62. Gefen T, Vaya J, Khatib S, Harkevich N, Artoul F, Heller ED, Pitcovski J, Aizenshtein E. The impact of PEGylation on protein immunogenicity. Int Immunopharmacol. 2013;15(2):254–9.
63. Cryan S. Carrier-based strategies for targeting protein and peptide drugs to the lungs. AAPS J. 2005;7(1):E20–41.
64. Grenha A, Remuñán-López C, Carvalho E, Seijo B. Microspheres containing lipid/chitosan nanoparticles complexes for pulmonary delivery of therapeutic proteins. Eur J Pharm Biopharm. 2008;69(1):83–93.
65. Al-Qadi S, Grenha A, Carrión-Recio D, Seijo B, Remuñán-López C. Microencapsulated chitosan nanoparticles for pulmonary protein delivery: in vivo evaluation of insulin-loaded formulations. J Control Release. 2012;157(3):383–90.
66. Alpar HO, Somavarapu S, Atuah KN, Bramwell VW. Biodegradable mucoadhesive particulates for nasal and pulmonary antigen and DNA delivery. Adv Drug Deliv Rev. 2005;57(3):411–30.
67. Lai S, Wang Y, Hanes J. Mucus-penetrating nanoparticles for drug and gene delivery to mucosal tissues. Adv Drug Deliv Rev. 2009;61(2):158–71.
68. Tang BC, Dawson M, Lai SK, Wang YY, Suk JS, Yang M, Zeitlin P, Boyle MP, Fu J, Hanes J. Biodegradable polymer nanoparticles that rapidly penetrate the human mucus barrier. Proc Natl Acad Sci U S A. 2009;106(46):19268–73.
69. Suk JS, Boylan NJ, Trehan K, Tang BC, Schneider CS, Lin JM, Boyle MP, Zeitlin PL, Lai SK, Cooper MJ, Hanes J. N-acetylcysteine enhances cystic fibrosis sputum penetration and airway gene transfer by highly compacted DNA nanoparticles. Mol Ther. 2001;19(11):1981–9.
70. Suk JS, Lai SK, Wang YY, Ensign LM, Zeitlin PL, Boyle MP, Hanes J. The penetration of fresh undiluted sputum expectorated by cystic fibrosis patients by non-adhesive polymer nanoparticles. Biomaterials. 2009;30(13):2591–7.
71. Yu SS, Lau CM, Thomas SN, Jerome WG, Maron DJ, Dickerson JH, Hubbell JA, Giorgio TD. Size- and charge-dependent non-specific uptake of PEGylated nanoparticles by macrophages. Int J Nanomedicine. 2012;7:799–813.
72. Yue H, Wei W, Yue Z, Lv P, Wang L, Ma G, Su Z. Particle size affects the cellular response in macrophages. Eur J Pharm Sci. 2010;41(5):650–7.
73. Soares S, Costa A, Sarmento B. Novel non-invasive methods of insulin delivery. Expert Opin Drug Deliv. 2012;9(12):1539–58.
74. Opar A. Another blow for inhaled protein therapeutics. Nat Rev Drug Discov. 2008;7:189–90.
75. Angelo R, Rousseau K, Grant M, Leone-Bay A, Richardson P. Technosphere insulin: defining the role of Technosphere particles at the cellular level. J Diabetes Sci Technol. 2009;3(3): 545–54.
76. Pfützner A, Mann A, Steiner S. Technosphere/Insulin—a new approach for effective delivery of human insulin via the pulmonary route. Diabetes Technol Ther. 2002;4(5):589–94.
77. Zisser H, Jovanovic L, Markova K, Petrucci R, Boss A, Richardson P, Mann A. Technosphere insulin effectively controls postprandial glycemia in patients with type 2 diabetes mellitus. Diabetes Technol Ther. 2012;14(11):997–1001.
78. Potocka E, Amin N, Cassidy J, Schwartz SL, Gray M, Richardson PC, Baughman RA. Insulin pharmacokinetics following dosing with Technosphere insulin in subjects with chronic obstructive pulmonary disease. Curr Med Res Opin. 2010;26(10):2347–53.
79. Raskin P, Heller S, Honka M, Chang PC, Boss AH, Richardson PC, Amin N. Pulmonary function over 2 years in diabetic patients treated with prandial inhaled Technosphere Insulin or usual antidiabetes treatment: a randomized trial. Diabetes Obes Metab. 2012;14(2):163–73.

80. Liu J, Gong T, Fu H, Wang C, Wang X, Chen Q, Zhang Q, He Q, Zhang Z. Solid lipid nanoparticles for pulmonary delivery of insulin. Int J Pharm. 2008;356(1–2):333–44.
81. Nyambura B, Kellaway I, Taylor K. Insulin nanoparticles: stability and aerosolization from pressurized metered dose inhalers. Int J Pharm. 2009;375(1–2):114–22.
82. Huang X, Du Y, Yuan H, Hu F. Preparation and pharmacodynamics of low-molecular-weight chitosan nanoparticles containing insulin. Carbohydrate Polymers. 2009;76:368–73.
83. Yamamoto H, Hoshina W, Kurashima H, Takeuchi H, Kawashima Y, Yokoyama T, Tsujimoto H. Engineering of poly(DL-lactic-co-glycolic acid) nanocomposite particles for dry powder inhalation dosage forms of insulin with the spray-fluidized bed granulating system. Advanced Powder Technology. 2007;18(2):215–28.
84. Kawashima Y, Yamamoto H, Takeuchi H, Fujioka S, Hino T. Pulmonary delivery of insulin with nebulized DL-lactide/glycolide copolymer (PLGA) nanospheres to prolong hypoglycemic effect. J Control Release. 1999;62(1–2):279–87.
85. Grenha A, Seijo B, Remuñán-López C. Microencapsulated chitosan nanoparticles for lung protein delivery. Eur J Pharm Sci. 2005;25(4–5):427–37.
86. Zhang Q, Shen Z, Nagai T. Prolonged hypoglycemic effect of insulin-loaded polybutylcyanoacrylate nanoparticles after pulmonary administration to normal rats. Int J Pharm. 2001;218(1–2):75–80.
87. Huang Y, Wang C. Pulmonary delivery of insulin by liposomal carriers. J Control Release. 2006;113(1):9–14.
88. Bi R, Shao W, Wang Q, Zhang N. Spray-freeze-dried dry powder inhalation of insulin-loaded liposomes for enhanced pulmonary delivery. J Drug Target. 2008;16(9):639–48.
89. Chono S, Fukuchi R, Seki T, Morimoto K. Aerosolized liposomes with dipalmitoyl phosphatidylcholine enhance pulmonary insulin delivery. J Control Release. 2009;137(2):104–9.
90. Marino MT, Costello D, Baughman R, Boss A, Cassidy J, Damico C, van Marle S, van Vliet A, Richardson PC. Pharmacokinetics and pharmacodynamics of inhaled GLP-1 (MKC253): proof-of-concept studies in healthy normal volunteers and in patients with type 2 diabetes. Clin Pharmacol Ther. 2010;88(2):243–50.
91. Wenzel S, Wilbraham D, Fuller R, Getz EB, Longphre M. Effect of an interleukin-4 variant on late phase asthmatic response to allergen challenge in asthmatic patients: results of two phase 2a studies. Lancet. 2007;370(9596):1422–31.
92. Getz EB, Fisher DM, Fuller R. Human pharmacokinetics/pharmacodynamics of an interleukin-4 and interleukin-13 dual antagonist in asthma. J Clin Pharmacol. 2009;49: 1025–36.
93. Bridges RJ, Newton BB, Pilewski JM, Devor DC, Poll CT, Hall RL. Na^+ transport in normal and CF human bronchial epithelial cells is inhibited by BAY 39–9437. Am J Physiol Lung Cell Mol Physiol. 2001;281(1):L16–23.
94. Moss RB, Hansen C, Sanders RL, Hawley S, Li T, Steigbigel RT. A phase II study of DAS181, a novel host directed antiviral for the treatment of influenza infection. J Infect Dis. 2012;206(12):1844–51.
95. Chan RW, Chan MC, Wong AC, Karamanska R, Dell A, Haslam SM, Sihoe AD, Chui WH, Triana-Baltzer G, Li Q, Peiris JS, Fang F, Nicholls J.M. DAS181 inhibits H5N1 influenza virus infection of human lung tissues. Antimicrob Agents Chemother. 2009;53(9):3935–41.
96. Drozd DR, Limaye AP, Moss RB, Sanders RL, Hansen C, Edelman JD, Raghu G, Boeckh M, Rakita R.M. DAS181 treatment of severe parainfluenza type 3 pneumonia in a lung transplant recipient. Transpl Infect Dis. 2013;15(1):E28–32.
97. Kobayashi S, Kondo S, Juni K. Pulmonary delivery of salmon calcitonin dry powders containing absorption enhancers in rats. Pharm Res. 1996;13(1):80–3.
98. Shoyele SA, Sivadas N, Cryan S.A. The effects of excipients and particle engineering on the biophysical stability and aerosol performance of parathyroid hormone (1–34) prepared as a dry powder for inhalation. AAPS PharmSciTech. 2011;12(1):304–11.
99. Schreier H, McNicol K, Bennett D, Teitelbaum Z, Derendorf H. Pharmacokinetics of detirelix following intratracheal instillation and aerosol inhalation in the unanesthetized awake sheep. Pharm Res. 1994;11(7):1056–9.

100. Bennett D, Tyson E, Nerenberg C, Mah S, de Groot J, Teitelbaum Z. Pulmonary delivery of detirelix by intratracheal instillation and aerosol inhalation in the briefly anesthetized dog. Pharm Res. 1994;11(7):1048–55.
101. van Zandwijk N, Jassem E, Dubbelmann R, Braat M, Rumke P. Aerosol application of interferon-alpha in the treatment of bronchioloalveolar carcinoma. Eur J Cancer. 1990;26(6):738–40.
102. Low S, Nunes S, Bitonti A, Dumont J. Oral and pulmonary delivery of FSH-Fc fusion proteins via neonatal Fc receptor-mediated transcytosis. Hum Reprod. 2005;20(7):1805–13.
103. Iacono A, Johnson B, Grgurich W, Youssef J, Corcoran T, Seiler D, Dauber J, Smaldone G, Zeevi A, Yousem S, Fung J, Burckart G, McCurry K, Griffith B. A randomized trial of inhaled cyclosporine in lung-transplant recipients. N Engl J Med. 2006;354(2):141–50.
104. Keenan R, Iacono A, Dauber J, Zeevi A, Yousem S, Ohori N, Burckart G, Kawai A, Smaldone G, Griffith B. Treatment of refractory acute allograft rejection with aerosolized cyclosporine in lung transplant recipients. J Thorac Cardiovasc Surg. 1997;113(2):335–40; discussion 331–40.
105. Onoue S, Yamamoto K, Kawabata Y, Hirose M, Mizumoto T, Yamada S. Novel dry powder inhaler formulation of glucagon with addition of citric acid for enhanced pulmonary delivery. Int J Pharm. 2009;382(1–2):144–50.
106. Kunda NK, Somavarapu S, Gordon SB, Hutcheon GA, Saleem IY. Nanocarriers targeting dendritic cells for pulmonary vaccine delivery. Pharm Res. 2013;30(2):325–41.
107. Jabbal-Gill I. Nasal vaccine innovation. J Drug Target. 2010;18(10):771–86.
108. Blank F, Stumbles P, von Garnier C. Opportunities and challenges of the pulmonary route for vaccination. Expert Opin Drug Deliv. 2011;8(5):547–63.
109. Nembrini C, Stano A, Dane KY, Ballester M, van der Vlies AJ, Marsland BJ, Swartz MA, Hubbell JA. Nanoparticle conjugation of antigen enhances cytotoxic T-cell responses in pulmonary vaccination. Proc Natl Acad Sci U S A. 2011;108(44):E989–97.
110. Ballester M, Nembrini C, Dhar N, de Titta A, de Piano C, Pasquier M, Simeoni E, van der Vlies A. J, McKinney JD, Hubbell JA, Swartz MA. Nanoparticle conjugation and pulmonary delivery enhance the protective efficacy of Ag85B and CpG against tuberculosis. Vaccine. 2011;29(40):6959–66.
111. Abel B, Tameris M, Mansoor N, Gelderbloem S, Hughes J, Abrahams D, Makhethe L, Erasmus M, de Kock M, van der Merwe L, Hawkridge A, Veldsman A, Hatherill M, Schirru G, Pau MG, Hendriks J, Weverling GJ, Goudsmit J, Sizemore D, McClain JB, Goetz M, Gearhart J, Mahomed H, Hussey GD, Sadoff JC, Hanekom WA. The novel tuberculosis vaccine, AERAS-402, induces robust and polyfunctional CD4+ and CD8+ T cells in adults. Am J Respir Crit Care Med. 2010;181(12):1407–17.
112. Jin TH, Tsao E, Goudsmit J, Dheenadhayalan V, Sadoff J. Stabilizing formulations for inhalable powders of an adenovirus 35-vectored tuberculosis (TB) vaccine (AERAS-402). Vaccine. 2010;28(27):4369–75.
113. Muttil P, Pulliam B, Garcia-Contreras L, Fallon JK, Wang C, Hickey AJ, Edwards DA. Pulmonary immunization of guinea pigs with diphtheria CRM-197 antigen as nanoparticle aggregate dry powders enhance local and systemic immune responses. AAPS J. 2010;12(4):699–707.
114. Muttil P, Prego C, Garcia-Contreras L, Pulliam B, Fallon JK, Wang C, Hickey AJ, Edwards D. Immunization of guinea pigs with novel hepatitis B antigen as nanoparticle aggregate powders administered by the pulmonary route. AAPS J. 2010;12(3):330–7.
115. Lu D, Garcia-Contreras L, Muttil P, Padilla D, Xu D, Liu J, Braunstein M, McMurray DN, Hickey AJ. Pulmonary immunization using antigen 85-B polymeric microparticles to boost tuberculosis immunity. AAPS J. 2010;12(3):338–47.
116. Hacha J, Tomlinson K, Maertens L, Paulissen G, Rocks N, Foidart JM, Noel A, Palframan R, Gueders M, Cataldo DD. Nebulized anti-IL-13 monoclonal antibody Fab′ fragment reduces allergen-induced asthma. Am J Respir Cell Mol Biol. 2012;47(5):709–17.
117. Kolb M, Martin G, Medina M, Ask K, Gauldie J. Gene therapy for pulmonary diseases. Chest. 2006;130(3):879–84.

118. Conese M, Ascenzioni F, Boyd AC, Coutelle C, De Fino I, De Smedt S, Rejman J, Rosenecker J, Schindelhauer D, Scholte BJ. Gene and cell therapy for cystic fibrosis: from bench to bedside. J Cyst Fibros. 2011;10 Suppl 2:S114–28.
119. Ibrahim BM, Park S, Han B, Yeo Y. A strategy to deliver genes to cystic fibrosis lungs: a battle with environment. J Control Release. 2011;155(2):289–95.
120. Wagner JA, Nepomuceno IB, Messner AH, Moran ML, Batson EP, Dimiceli S, Brown BW, Desch JK, Norbash AM, Conrad CK, Guggino WB, Flotte TR, Wine JJ, Carter BJ, Reynolds TC, Moss RB, Gardner P. A phase II, double-blind, randomized, placebo-controlled clinical trial of tgAAVCF using maxillary sinus delivery in patients with cystic fibrosis with antrostomies. Hum Gene Ther. 2002;13(11):1349–59.
121. Bellon G, Michel-Calemard L, Thouvenot D, Jagneaux V, Poitevin F, Malcus C, Accart N, Layani MP, Aymard M, Bernon H, Bienvenu J, Courtney M, Döring G, Gilly B, Gilly R, Lamy D, Levrey H, Morel Y, Paulin C, Perraud F, Rodillon L, Sené C, So S, Touraine-Moulin F, Pavirani A. Aerosol administration of a recombinant adenovirus expressing CFTR to cystic fibrosis patients: a phase I clinical trial. Hum Gene Ther. 1997;8(1):15–25.
122. Crystal RG, McElvaney NG, Rosenfeld MA, Chu CS, Mastrangeli A, Hay JG, Brody SL, Jaffe HA, Eissa NT, Danel C. Administration of an adenovirus containing the human CFTR cDNA to the respiratory tract of individuals with cystic fibrosis. Nat Genet. 1994;8(1):42–51.
123. Harvey BG, Leopold PL, Hackett NR, Grasso TM, Williams PM, Tucker AL, Kaner RJ, Ferris B, Gonda I, Sweeney TD, Ramalingam R, Kovesdi I, Shak S, Crystal RG. Airway epithelial CFTR mRNA expression in cystic fibrosis patients after repetitive administration of a recombinant adenovirus. J Clin Invest. 1999;104(9):1245–55.
124. Tagalakis AD, McAnulty RJ, Devaney J, Bottoms SE, Wong JB, Elbs M, Writer MJ, Hailes HC, Tabor AB, O'Callaghan C, Jaffe A, Hart SL. A receptor-targeted nanocomplex vector system optimized for respiratory gene transfer. Mol Ther. 2008;16(5):907–15.
125. Moss RB, Rodman D, Spencer LT, Aitken ML, Zeitlin PL, Waltz D, Milla C, Brody AS, Clancy JP, Ramsey B, Hamblett N, Heald AE. Repeated adeno-associated virus serotype 2 aerosol-mediated cystic fibrosis transmembrane regulator gene transfer to the lungs of patients with cystic fibrosis: a multicenter, double-blind, placebo-controlled trial. Chest. 2004;125(2):509–21.
126. Moss RB, Milla C, Colombo J, Accurso F, Zeitlin PL, Clancy JP, Spencer LT, Pilewski J, Waltz DA, Dorkin HL, Ferkol T, Pian M, Ramsey B, Carter BJ, Martin DB, Heald AE. Repeated aerosolized AAV-CFTR for treatment of cystic fibrosis: a randomized placebo-controlled phase 2B trial. Hum Gene Ther. 2007;18(8):726–32.
127. McLachlan G, Davidson H, Holder E, Davies LA, Pringle IA, Sumner-Jones SG, Baker A, Tennant P, Gordon C, Vrettou C, Blundell R, Hyndman L, Stevenson B, Wilson A, Doherty A, Shaw DJ, Coles RL, Painter H, Cheng SH, Scheule RK, Davies JC, Innes JA, Hyde SC, Griesenbach U, Alton EW, Boyd AC, Porteous DJ, Gill DR, Collie DD. Pre-clinical evaluation of three non-viral gene transfer agents for cystic fibrosis after aerosol delivery to the ovine lung. Gene Ther. 2011;18(10):996–1005.
128. Alton E, Boyd A, Cheng S, Davies J, Davies L, Dayan A, Gill D, Griesenbach U, Higgins T, Hyde S, Innes A, McLachlan G, Porteous D, Pringle I, Scheule R, Sumner-Jones S. Repeat administration of GL67A/pGM169 is feasible, safe, and produces endogenous levels of CFTR expression after 12 doses. Thorax. 2012;67:A105.
129. J.W.a.S. Lda. The Journal of Gene Medicine. 2012. www.wiley.co.uk/genmed/clinical.
130. Siegel R, Naishadham D, Jemal A. Cancer statistics, 2012. CA Cancer J Clin. 2012;62(1): 10–29.
131. Frederiksen KS, Abrahamsen N, Cristiano RJ, Damstrup L, Poulsen HS. Gene delivery by an epidermal growth factor/DNA polyplex to small cell lung cancer cell lines expressing low levels of epidermal growth factor receptor. Cancer Gene Ther. 2000;7(2):262–8.
132. Kim HW, Park IK, Cho CS, Lee KH, Beck GR, Colburn NH, Cho MH. Aerosol delivery of glucosylated polyethylenimine/phosphatase and tensin homologue deleted on chromosome 10 complex suppresses Akt downstream pathways in the lung of K-ras null mice. Cancer Res. 2004;64(21):7971–6.

133. Zou Y, Tornos C, Qiu X, Lia M, Perez-Soler R. p53 aerosol formulation with low toxicity and high efficiency for early lung cancer treatment. Clin Cancer Res. 2007;13(16):4900–8.
134. Gautam A, Densmore CL, Melton S, Golunski E, Waldrep JC. Aerosol delivery of PEI-p53 complexes inhibits B16-F10 lung metastases through regulation of angiogenesis. Cancer Gene Ther. 2002;9(1):28–36.
135. Okamoto H, Shiraki K, Yasuda R, Danjo K, Watanabe Y. Chitosan-interferon-β gene complex powder for inhalation treatment of lung metastasis in mice. J Control Release. 2011;150(2):187–95.
136. Farkas L, Farkas D, Ask K, Möller A, Gauldie J, Margetts P, Inman M, Kolb M. VEGF ameliorates pulmonary hypertension through inhibition of endothelial apoptosis in experimental lung fibrosis in rats. J Clin Invest. 2009;119(5):1298–1311.
137. Harada-Shiba M, Takamisawa I, Miyata K, Ishii T, Nishiyama N, Itaka K, Kangawa K, Yoshihara F, Asada Y, Hatakeyama K, Nagaya N, Kataoka K. Intratracheal gene transfer of adrenomedullin using polyplex nanomicelles attenuates monocrotaline-induced pulmonary hypertension in rats. Mol Ther. 2009;17(7):1180–6.
138. Reynolds PN. Gene therapy for pulmonary hypertension: prospects and challenges. Expert Opin Biol Ther. 2011;11(2):133–43.
139. McMurtry MS, Archer SL, Altieri DC, Bonnet S, Haromy A, Harry G, Puttagunta L, Michelakis ED. Gene therapy targeting survivin selectively induces pulmonary vascular apoptosis and reverses pulmonary arterial hypertension. J Clin Invest. 2005;115(6):1479–91.
140. Aerovance Inc. A phase I/II study to investigate the efficacy and safety of AER 002 in cystic fibrosis given at 3 mg, 10 mg, and 30 mg doses in single then multiple ascending doses and to determine efficacy of the highest tolerable dose in a 4-week proof of concept study, 2005-000313-35. 2005.
141. APTPharmaceuticals. CIS001 extension study of cyclosporine inhalation solution (CIS002). NCT00938236.
142. Moss RB, Mayer-Hamblett N, Wagener J, Daines C, Hale K, Ahrens R, Gibson RL, Anderson P, Retsch-Bogart G, Nasr SZ, Noth I, Waltz D, Zeitlin P, Ramsey B, Starko K. Randomized, double-blind, placebo-controlled, dose-escalating study of aerosolized interferon gamma-1b in patients with mild to moderate cystic fibrosis lung disease. Pediatr Pulmonol. 2005;39(3):209–18.
143. Hallstrand TS, Ochs HD, Zhu Q, Liles WC. Inhaled IFN-gamma for persistent nontuberculous mycobacterial pulmonary disease due to functional IFN-gamma deficiency. Eur Respir J. 2004;24(3):367–70.
144. Kim D, Mudaliar S, Chinnapongse S, Chu N, Boies SM, Davis T, Perera AD, Fishman RS, Shapiro DA, Henry R. Dose-response relationships of inhaled insulin delivered via the Aerodose insulin inhaler and subcutaneously injected insulin in patients with type 2 diabetes. Diabetes Care. 2003;26(10):2842–7.
145. DancePharmaceuticals. A phase 1/2 trial investigating the pharmacokinetics, pharmacodynamics and safety of inhaled insulin in subjects with type 1 diabetes, 2012-002071-34. 2012.
146. Rave KM, Nosek L, de la Peña A, Seger M, Ernest CS, Heinemann L, Batycky RP, Muchmore DB. Dose response of inhaled dry-powder insulin and dose equivalence to subcutaneous insulin lispro. Diabetes Care. 2005;28(10):2400–5.
147. QDose. Investigating the pharmacokinetics and pharmacodynamics of recombinant human insulin administered by dry powder inhaler. NCT00426920. 2007.
148. de Galan B, Simsek S, Tack C, Heine R. Efficacy and safety of inhaled insulin in the treatment of diabetes mellitus. Neth J Med. 2006;64(9):319–25.
149. Garcia-Contreras L, Morçöl T, Bell SJ, Hickey AJ. Evaluation of novel particles as pulmonary delivery systems for insulin in rats. AAPS PharmSci. 2003;5(2):E9.
150. Heise T, Brugger A, Cook C, Eckers U, Hutchcraft A, Nosek L, Rave K, Troeger J, Valaitis P, White S, Heinemann L. PROMAXX inhaled insulin: safe and efficacious administration with a commercially available dry powder inhaler. Diabetes Obes Metab. 2009;11(5):455–9.

151. Eichelberg C, Andreas A, Heuer R, Huland H, Heinzer H, Huland E. Long-term tumor control with inhalational interleukin 2 therapy in cardiac high risk patients with metastatic renal carcinoma. J Urology. 2008;179(4):167.
152. Tazawa R, Nakata K, Inoue Y, Nukiwa T. Granulocyte-macrophage colony-stimulating factor inhalation therapy for patients with idiopathic pulmonary alveolar proteinosis: a pilot study; and long-term treatment with aerosolized granulocyte-macrophage colony-stimulating factor: a case report. Respirology. 2006;11 Suppl:S61–4.
153. CSL Behring. Safety and tolerability study of liquid alpha1 proteinase inhibitor (API) in subjects with cystic fibrosis. NCT01347190. 2012
154. Schüle S, Schulz-Fademrecht T, Garidel P, Bechtold-Peters K, Frieb W. Stabilization of IgG1 in spray-dried powders for inhalation. Eur J Pharm Biopharm. 2008;69(3):793–807.
155. Srinivasan AR, Shoyele SA. Self-associated submicron IgG1 particles for pulmonary delivery: effects of non-ionic surfactants on size, shape, stability, and aerosol performance. AAPS PharmSciTech. 2013;14(1):200–10.
156. Saluja V, Amorij JP, Kapteyn JC, de Boer AH, Frijlink HW, Hinrichs WL. A comparison between spray drying and spray freeze drying to produce an influenza subunit vaccine powder for inhalation. J Control Release. 2010;144(2):127–33.
157. Garcia-Contreras L, Wong YL, Muttil P, Padilla D, Sadoff J, Derousse J, Germishuizen WA, Goonesekera S, Elbert K, Bloom BR, Miller R, Fourie PB, Hickey A, Edwards D. Immunization by a bacterial aerosol, Proc Natl Acad Sci U S A. 2008;105(12):4656–60.
158. Huyge K, Van Reeth K, De Beer T, Landman WJ, van Eck JH, Remon JP, Vervaet C. Suitability of differently formulated dry powder Newcastle disease vaccines for mass vaccination of poultry. Eur J Pharm Biopharm. 2012;80(3):649–56.
159. Todoroff J, Ucakar B, Inglese M, Vandermarliere S, Fillee C, Renauld JC, Huygen K, Vanbever R. Targeting the deep lungs, Poloxamer 407 and a CpG oligonucleotide optimize immune responses to Mycobacterium tuberculosis antigen 85A following pulmonary delivery. Eur J Pharm Biopharm. 2013;84(1):40–8.
160. Kawabata A, Baoum A, Ohta N, Jacquez S, Seo GM, Berkland C, Tamura M. Intratracheal administration of a nanoparticle-based therapy with the angiotensin II type 2 receptor gene attenuates lung cancer growth. Cancer Res. 2012;72(8):2057–67.
161. Hasenpusch G, Pfeifer C, Aneja MK, Wagner K, Reinhardt D, Gilon M, Ohana P, Hochberg A, Rudolph C. Aerosolized BC-819 inhibits primary but not secondary lung cancer growth. PLoS One. 2011;6(6):e20760.
162. Flotte TR, Zeitlin PL, Reynolds TC, Heald AE, Pedersen P, Beck S, Conrad CK, Brass-Ernst L, Humphries M, Sullivan K, Wetzel R, Taylor G, Carter BJ, Guggino WB. Phase I trial of intranasal and endobronchial administration of a recombinant adeno-associated virus serotype 2 (rAAV2)-CFTR vector in adult cystic fibrosis patients: a two-part clinical study. Hum Gene Ther. 2003;14(11):1079–88.
163. Kamezaki F, Tasaki H, Yamashita K, Tsutsui M, Koide S, Nakata S, Tanimoto A, Okazaki M, Sasaguri Y, Adachi T, Otsuji Y. Gene transfer of extracellular superoxide dismutase ameliorates pulmonary hypertension in rats. Am J Respir Crit Care Med. 2008;177(2):219–26.
164. Minai-Tehrani A, Park YC, Hwang SK, Kwon JT, Chang SH, Park SJ, Yu KN, Kim JE, Shin JY, Kim JH, Kang B, Hong SH, Cho MH. Aerosol delivery of kinase-deficient Akt1 attenuates Clara cell injury induced by naphthalene in the lungs of dual luciferase mice. J Vet Sci. 2011;12(4):309–17.
165. Yu KN, Minai-Tehrani A, Chang SH, Hwang SK, Hong SH, Kim JE, Shin JY, Park SJ, Kim JH, Kwon JT, Jiang HL, Kang B, Kim D, Chae CH, Lee KH, Yoon TJ, Beck GR, Cho MH. Aerosol delivery of small hairpin osteopontin blocks pulmonary metastasis of breast cancer in mice. PLoS One. 2010;5(12):e15623.

Chapter 8
Nasal Delivery of Biopharmaceuticals

Eiji Yuba and Kenji Kono

8.1 Introduction

Nasal mucosa has a high degree of vascularization and high permeability, which enable systemic administration of biopharmaceuticals via this route. Considering the characteristics of nasal mucosa as a therapeutic target and a portal for drug delivery, various intranasal drug formulations have been developed commercially [1]. Intranasal delivery presents myriad benefits such as ease of administration, noninvasive needle-free administration, rapid onset of action, and the avoidance of gastrointestinal and hepatic first-pass effects. The limitations of intranasal administration are mainly associated with the transport of biopharmaceuticals across the nasal mucosa [2]. In this chapter, physicochemical, pharmaceutical, and physiopathological parameters related to intranasal drug delivery are discussed first. Subsequently, concrete examples of intranasal drug delivery formulations and technologies based on nanocarriers are described.

8.2 Biological Aspect of Intranasal Delivery

8.2.1 Physiological and Functional Characteristics of the Nasal Cavity

The major functions of the nasal cavity are breathing and olfaction. The nasal cavity also provides an important protective activity: filtering, heating, and humidifying of the inhaled air before reaching the lowest airways. The human nasal cavity has a

K. Kono (✉) · E. Yuba
Department of Applied Chemistry, Graduate School of Engineering,
Osaka Prefecture University, 1-1 Gakuen-cho, Naka-ku, Sakai, 599-8531 Osaka, Japan
e-mail: kono@chem.osakafu-u.ac.jp

E. Yuba
e-mail: yuba@chem.osakafu-u.ac.jp

J. das Neves, B. Sarmento (eds.), *Mucosal Delivery of Biopharmaceuticals,* DOI 10.1007/978-1-4614-9524-6_8,

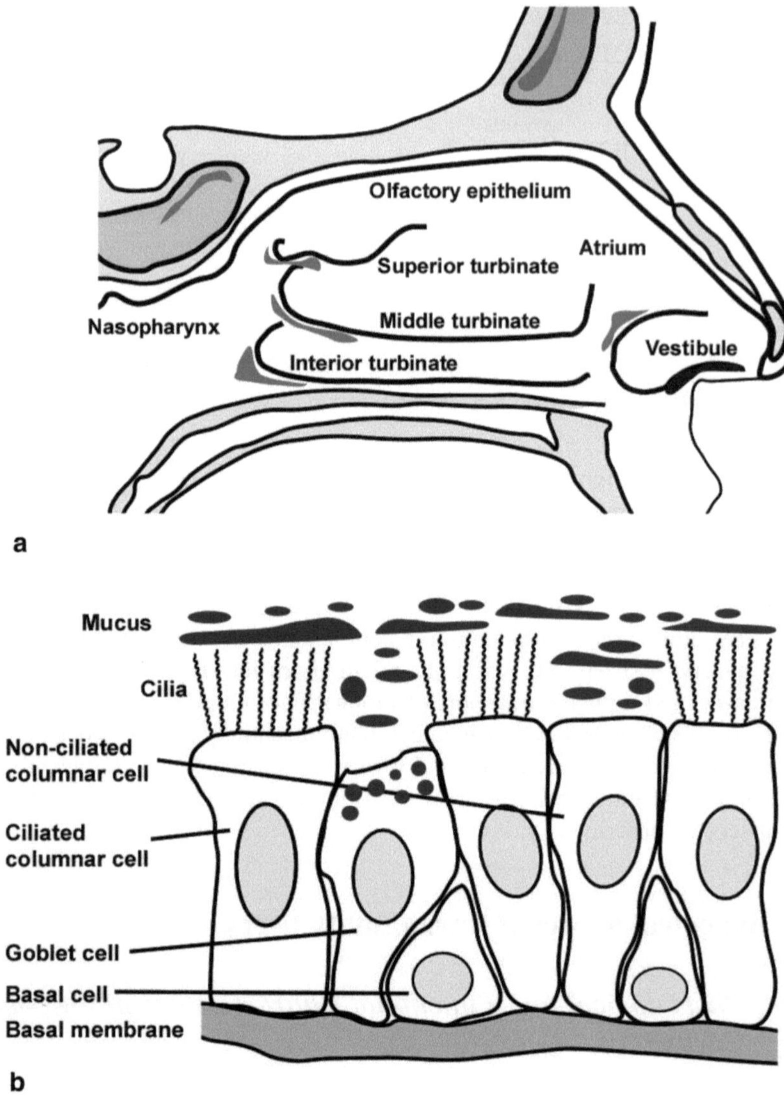

Fig. 8.1 **a** Nasal cavity and different areas: vestibule, atrium, inferior/middle/superior turbinates, olfactive region, and nasopharynx. **b** Nasal epithelium and its constitutive cells and mucus

total volume of 15–20 ml and a total surface area of approximately 150 cm^2 [3]. The nasal cavity consists of four areas: the nasal vestibule and atrium, respiratory region, and olfactory region, which are distinguished by their anatomic and histological characteristics [3] (Fig. 8.1a).

The vestibule and the atrium, which form the anterior part of the nasal cavity, are covered by stratified squamous and transitional nonciliated epithelial cells, respectively [4]. Nasal hairs, called vibrissae, filter the inhaled particles. This region shows low permeability because of its small relative surface area and low vasculature. It is of little interest for drug delivery applications.

The respiratory region, which occupies the largest part of the nose, has shelf-like structures called turbinates, which provide it with a large surface area [3]. Large surface area and rich vascularization lend the respiratory region high permeability, which is of great relevance for nasal drug delivery [5]. Most of the nasal cavity is covered by pseudostratified airway epithelium, comprising columnar cells, basal, and goblet cells (Fig. 8.1b). Each epithelial cell is interconnected by tight junctions on the apical side and by interdigitations of the cell membrane on the lateral side [6]. The apical region of the columnar epithelial cells is covered densely by microvilli, which significantly expanded the surface area of these cells toward the cavity. In addition, in the posterior part of the nasal cavity, the columnar cells possess actively biting hair-like structures called cilia, which are larger and less densely distributed surface expansions than microvilli are [6]. The active movement of these cilia is important for removing potentially harmful substances from the upper respiratory tract. Another key factor in this removal mechanism is the nasal mucus, which covers the epithelial cells described above and provides them with a protective physical barrier. The nasal mucus layer, which is only 5 μm thick, is organized into two different layers: an external, viscous and dense layer, and an internal, fluid and serous layer. The nasal mucus layer consists of 95 % of water, 2.5–3 % of mucin, and 2 % of electrolytes, proteins, lipids, enzymes, antibodies, sloughed epithelial cells, and bacterial products [7–9]. The presence of mucin in the nasal mucus layer is crucial because it might trap large molecular-weight drugs such as peptides and proteins [10]. Nasal mucus is partly produced by goblet cells but is mainly secreted by the serous and seromucous glands that are located in the connective tissue below the respiratory epithelium [11]. The active ciliar movement drives the overlying mucus layer continuously toward the nasopharynx. Consequently, inhaled particulates trapped in the mucus are cleared efficiently from the nasal passage.

The olfactory epithelium is a small region of specialized pseudostratified ciliated cells located on the upper part of the nasal cavity. This epithelium with interspersed neuronal terminations is involved directly in smell perception. The primary olfactory neurons are in contact with the environment in the nasal cavity. They communicate through their axons with the olfactory bulb in the brain [12]. Despite the small surface area (2–400 mm^2) of the olfactory region, two important considerations of this region are related to drug delivery. First, the basic smell-detecting capacity should not be compromised when administering any substance to the nasal mucosa. Second, the olfactory epithelium represents a unique pathway for direct nose-to-brain delivery via the olfactory bulb [13, 14]. Although some evidence suggests direct brain delivery by olfactory epithelial deposition, the mechanism of this pathway must be further elucidated.

8.2.2 Immunological Aspects

The mucosal surface of the nasal cavity is separated from the external environment by the epithelial barrier, which protects it using nonspecific defense mechanisms (mucosal secretion, mechanical cleaning, and others). In contrast, immunological functions of the mucosa-associated lymphoid tissue (MALT) provide specific protective mechanism [15]. Although these lymphoid tissues share numerous structural and functional characteristics throughout the body, each is adapted to its specific anatomical location. Nasal-associated lymphoid tissue (NALT) is an organized lymphoid aggregated and infiltrated to the overlying epithelium in the nasal cavity [16]. In humans, this tissue exists as a so-called diffuse NALT consisting of a collection of isolated subepithelial lymphoid follicles [16, 17]. In addition, highly organized lymphoid tissues exist in the human nasopharynx and oropharynx, incorporating the lingual, palatine, and nasopharyngeal tonsils (adenoids). This assembly of lymphoid tissues, denominated as Waldeyer's ring, plays an important role in primary respiratory immune defense [18]. Indeed, most particles entrapped in the mucus layer are carried to this region by the mucociliar clearance mechanism [3]. The NALT has various immunocompetent cells, including subepithelial B-lymphocytes, CD4 + and CD8 + T-lymphocytes, phagocytic antigen-presenting cells (APCs) such as macrophages and diverse subsets of dendritic cells (DCs) [19]. In addition, the overlying epithelium of mucosal follicles forms a specialized cell layer (i.e., follicle-associated epithelium) that has a loose structure which enables the contact between antigens and immune cells. More importantly, the follicle-associated epithelium also incorporates microfold cells (M cells) characterized by a basolateral cytoplasmic invagination that forms an intraepithelial pocket containing lymphocytes and some phagocytic cells [15]. The M cells possess a high capacity to transport various materials by transcellular vesicular transport to these underlying intraepithelial cells. Alternatively, in regions where organized follicles and M cells are absent, DCs can traffic close to the epithelial layer and establish contact with antigens through interaction with epithelial cells [20]. In contrast to soluble antigens, particulates can be taken up preferentially by M cells following nasal administration [21]. After their contact with pathogens, different subsets of intraepithelial or subepithelial APCs can stimulate local adaptive immune responses by presenting the antigen to neighboring lymphocytes via the major histocompatibility complex (MHC) molecules. Alternatively, DCs can migrate and carry the antigen to proximal draining lymph nodes and generate systemic immune responses [22] (Fig. 8.2). Immune responses in the mucosal tissue are dependent on characteristics of the antigen and also on the type of the APCs involved. Ideally, adaptive immune responses are expected to comprise both cellular and humoral immune responses against the pathogens. Cellular immune defense is mostly affected by cytotoxic T lymphocytes (CTL) and antibody-dependent cell-mediated cytotoxicity (ADCC) through natural killer cells (NK cells). Because these mechanisms can destroy specific cells directly, this type of immune response is crucial for the clearance of viruses and intracellular parasites [20, 23]. Humoral immune defense at the mucosal surface is mediated principally by the production of immunoglobulin A (IgA) following activation of B cells. IgA is found in mucosal secretions as dimeric or multimeric form contrast to other antibody isotypes, secretory IgA is resistant to enzymatic degradation, which makes it especially and uniquely suitable for mucosal defense [24]. The main role of the secretory antibody system is to inhibit invasion and colonization of pathogens in cooperation with the innate immune system [20]. In addition, mucosal immunization can result in the

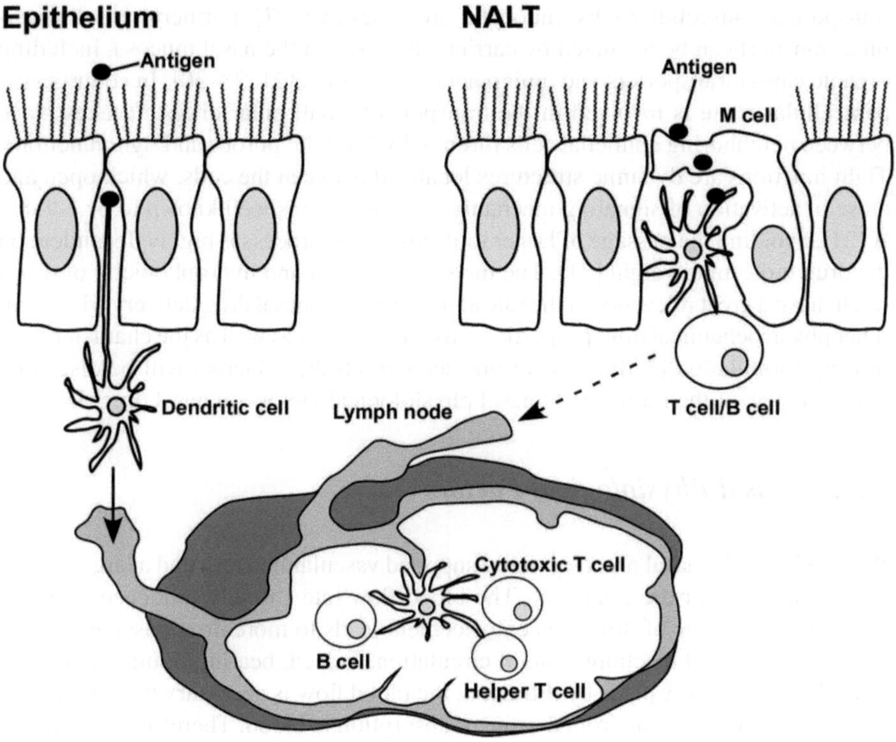

Fig. 8.2 Cellular mechanism of immune responses through nasal epithelium

production of serum IgA and serum IgG antibodies, which is related to the migration capacity of different subsets of immune cells that allows their contact with systemic inductive sites [25, 26]. Consequently, cells in the NALT can be involved in the close regulation of both cellular and humoral immune responses locally and also at distant sites. Because this is not always the case for other mucosal sites (e.g., oral or vaginal mucosal tissues), this feature makes the nasal route particularly attractive for drug administration.

8.3 Nasal Delivery of Biopharmaceuticals

8.3.1 Factors Influencing Nasal Drug Delivery

When a drug is administered nasally to induce systemic effects, it must pass through the mucus layer and epithelial barrier before reaching the site of action. The passage across the epithelium might occur by transcellular or paracellular mechanisms. The former includes passive diffusion through the interior of the cell. It is especially involved in the transport of hydrophobic drugs [5]. It is considered that compounds with a molecular weight that is greater than 1 kDa, such as peptides and proteins, are transported transcellularly by endocytic processes [10, 27]. Furthermore, transcellular transport can be mediated by carriers that exist in the nasal mucosa, including organic cation transporters and amino acid transporters [21, 28–30]. In contrast, the paracellular route is involved in the transport of small polar drugs. It takes place between neighboring epithelial cells through hydrophilic porous and tight junctions. Tight junctions are dynamic structures localized between the cells, which open and close in activation of signaling mechanisms. Their size is well known to be 3.9–8.4 Å [21], avoiding the passage of larger molecules, this process is highly dependent on the drug molecular weight [31]. The molecular weight and hydrophobicity of drugs might have a great relevance in the rate and extent of its nasal drug delivery. However, other physicochemical drug properties must be regarded as well as the characteristics of drug formulation [5, 10, 32]. In this section, all these factors will be discussed after a review of the influence of nasal physiological factors on nasal drug delivery.

8.3.2 Nasal Physiological Factors

Blood Flow The nasal mucosa, richly supplied vascular network and a large surface area, is suitable for drug delivery. The blood flow rate strongly influences the systemic nasal delivery of drugs, its enhancement leads to more drug passing through the membrane, and reaching general circulation. Indeed, bearing in mind that most drug absorption takes place by diffusion, the blood flow is necessary to maintain the gradient of concentration from the site of absorption to blood. Therefore, vasodilatation and vasoconstriction are well known as possibly determining the blood flow and the rate and extent of drug absorption. Several studies have evaluated this influence. For example, phenylephrine, a vasoconstrictor agent, inhibited the absorption of acetylsalicylic acid in nasal cavity [32]. Nasal absorption of dopamine was slow and incomplete, probably because of its own vasoconstrictive effect [33]. Based on these observations, we concluded that vasoconstriction decreases nasal drug absorption by diminishing the blood flow.

Mucociliar Clearance Mucociliar clearance (MCC) is the self-clearing mechanism of the bronchi. A nasal mucus layer plays an important role in defense of the respiratory tract because it protects the lungs from foreign substances, pathogens, and particles carried by inhaled air. These agents adhere to the mucus layer. Subsequently, they are transported all together to the nasopharynx and eventually to the gastrointestinal tract. This elimination, which is designated as MCC, also influences the nasal drug delivery markedly. The MCC system has been described as a conveyer belt wherein cilia provide the driving force, whereas mucus is a sticky fluid that collects and disposes foreign particles [34]. The efficiency of MCC thereby depends on the length, density, and beat frequency of cilia and on the amount and viscoelastic properties of mucus. All factors that increase mucus production, decrease mucus viscosity, or increase the ciliary beat frequency might increase MCC. If MCC decreases, then the residence time of the drug in nasal mucosa increases and consequently enhances the drug permeation. The clearance of a drug from the nasal cavity is also influenced by the deposition site. A drug deposited in a posterior area of the nose is cleared more rapidly from the nasal cavity than a drug deposited anteriorly because MCC is slower in the anterior part of the nose than in the more ciliated posterior part [34, 35]. However, the site of drug deposition in the nose is highly dependent on the dosage form. Nasal sprays deposit drugs more anteriorly than nasal drops do, resulting in a slower clearance for drugs administered from spray formulations [36]. Polar drugs are most affected by MCC because they are highly soluble in mucus and their passage across the membrane is slow. Consequently, all factors that influence the efficacy and speed of MCC can modify the drug absorption profile. For instance, environmental factors have a relevant influence in MCC. In addition, several pathological conditions exist in which MCC does not function properly [34, 37]. Furthermore, some components of drug formulations, such as preservatives and nasal absorption enhancers, might alter the MCC system [34].

Enzymatic Degradation Nasally administered drugs circumvent gastrointestinal and hepatic first-pass effects. However, they might be metabolized to a great degree in the lumen of the nasal cavity or during passage across the nasal epithelial barrier because of the presence of a broad range of metabolic enzymes in nasal tissues. Carboxyl esterases, aldehyde dehydrogenases, epoxide hydrolases, and glutathione S-transferases have been found in nasal epithelial cells. They are responsible for the degradation of drugs in nasal mucosa [38–40]. Cytochrome P450 isoenzymes are also present here. They have been reported as metabolizers of drugs such as cocaine, nicotine, alcohols, progesterone, and decongestants [41, 42]. Similarly, proteolytic enzymes (aminopeptidases and proteases) were found. They are believed to be the major barrier against the delivery of peptide drugs such as calcitonin, insulin, and desmopressin [43, 44]. Consequently, xenobiotic metabolizing enzymes existent in the nasal mucosa might affect the pharmacokinetic and pharmacodynamic profile of nasally applied drugs. In this context, although the nasal first-pass metabolism is usually weaker than hepatic and intestinal ones, it cannot be ignored.

Transporters and Efflux Systems The study of transporter systems present in the nasal tissue and their effects on the absorption of drugs into systemic circulation and the central nervous system (CNS) is a promising research area. Multidrug resistance transporters have already been identified in human nasal respiratory and olfactory mucosa, which might be involved in the transport of various hydrophobic and amphiphilic drugs [10]. P-glycoprotein (P-gp) is an efflux transporter that exists in the apical area of ciliated epithelial cells and in the submucosal vessels of the human olfactory region [30]. P-gp plays an important role in actively preventing the influx of drugs from the nasal membrane [21, 29, 30, 45].

8.3.3 Physicochemical Properties of Drugs

The influence of physicochemical characteristics of drug molecules on the rate and extent of gastrointestinal absorption is well understood. Therefore, in silico models have been developed to prioritize numerous drug candidates at the early phases of drug discovery. In the same way, but with some differences, the physicochemical properties of drugs (molecular weight, hydrophobicity, pKa, stability, and solubility) can influence nasal absorption.

Molecular Weight, Hydrophobicity, and pKa Hydrophobic drugs such as propranolol, progesterone, and fentanyl are, in general, well absorbed to the nasal cavity, presenting pharmacokinetic profiles similar to those obtained after intravenous administration and a nasal bioavailability near 100 %. Indeed, they are absorbed quickly and efficiently across the nasal membrane through transcellular mechanisms. However, it is important to state that this is true for hydrophobic compounds having molecular weight lower than 1 kDa. The extension of nasal absorption of hydrophobic drugs larger than 1 kDa is significantly less [31]. However, the rate and degree of nasal absorption of polar drugs are low and highly dependent on the molecular weight. The permeation of polar drugs with molecular weight of less than 300 Da is not influenced considerably by their physicochemical properties [31, 46–48]. In contrast, the rate of permeation is highly sensitive to molecular size if it is higher than 300 Da. An inverse relation exists between the rate of permeation and molecular weight [46, 49]. For some small polar molecules, only 10 % bioavailability is suggested. The value decreases to 1 % for large molecules such as proteins [48]. The nasal membrane is predominantly hydrophobic. Therefore, drug absorption is expected to diminish with a decrease in hydrophobicity [49, 50]. Consequently, polar drugs are not easily transported across the nasal membrane, thereby enhancing MCC. However, if hydrophobicity is too high, then the drug does not dissolve easily in the aqueous environment of the nasal cavity. Consequently, with accelerated MCC, the contact time with the nasal membrane diminishes, resulting in reduced permeation through the wall [51]. In general, the passage across biomembranes is affected not only by hydrophobicity and hydrophilicity, but also by the amount of drug existing as uncharged species. This effect depends on the drug pKa and the pH of the absorption site (5.0–6.5 in human nasal mucosa) [27, 52]. The nonionized fraction of a drug is more permeable than the ionized. For the nasal mucosa, a range of studies evaluating the effect of hydrophobicity and pH on the absorption of small drugs was performed [50, 53–55]. All results demonstrated that nasal absorption of weak electrolytes depends on their ionization degree. The greatest absorption occurs for the nonionized species. In this state, they present a higher apparent partition coefficient. Therefore, they are more hydrophobic. However, drugs such as acetylsalicylic acid [54] showed some permeability across the membrane even in environments in which they are expected to exist as ionized species. Based on these observations, for polar drugs, the partition coefficient is the major factor influencing permeability through nasal mucosa.

Stability During the development of new drug formulations, biological, chemical, and physical drug stability studies must be a matter of major importance in all processes. As discussed before, the environment of the nasal cavity can metabolize drugs using defensive enzymatic mechanisms, which might reduce the bioavailability of nasally administered drugs [38–40]. To overcome this difficulty, various strategies might be followed, mainly through the use of prodrugs [5, 10, 32, 56] and enzymatic inhibitors [57–59], as discussed later. However, many drugs might be physicochemically unstable because of hydrolysis, oxidation, isomerization, photochemical decomposition or polymerization reactions [10].

Solubility Drug dissolution is a prerequisite for drug absorption because only the molecularly dispersed form of a drug at the absorption site can cross the biomembranes. Therefore, the drug must be dissolved in the watery fluids of the nasal cavity before nasal absorption. Consequently, the appropriated aqueous drug solubility is of utmost importance to allow sufficient contact with the nasal mucosa and posterior absorption [55]. However, the absorption profile is influenced not only by drug solubility but also by the nature of pharmaceutical preparations, which must guarantee the delivery of drug at therapeutically relevant doses. Because the nasal cavity is small, the allowable volume of drug solution is low for intranasal drug administration [10]. For that reason, drugs that are poorly soluble in water and/or require high doses might constitute a problem. This can be overcome by enhancing the drug aqueous solubility [5, 10, 33, 60, 61].

8.3.4 Effect of Drug Formulation

Viscosity As the formulation viscosity increases, the contact time between the drug and nasal mucosa is enhanced. Thereby, the drug absorption potential increases. At the same time, high viscosity of formulations interferes with normal ciliary beating and/or MCC. Therefore, it increases the drug permeability. This phenomenon has been observed during nasal delivery of insulin [62], acyclovir [63], and metoprolol [64]. However, sometimes, enhancing the formulation viscosity does not enhance the drug absorption. For example, a study was performed to evaluate the influence of formulation viscosity on the retention time of metoclopramide hydrochloride in the nasal cavity and on its absorption [55]. The study showed that although the residence time was enhanced as viscosity increased, the drug absorption diminished. This observation has been attributed to a decrease in the drug diffusion from the formulation. However, it has also been reported that the solution viscosity might provide a longer therapeutic period for nasal formulations [55].

pH The extent of nasal absorption depends on the pKa of the drug and pH at the absorption site; contributing to that also is the pH of formulation. At this point, it should be stated that the pH of formulation must be selected with attention to drug stability. If possible, the greatest quantity of nonionized drug species should be assured. However, the pH of formulation can induce nasal mucosa irritation. For that reason, it should be similar to that found in human nasal mucosa (5.0–6.5) [27, 52]. Furthermore, pH often prevents bacterial growth [5]. To evaluate the effect of a pH solution on the integrity of nasal mucosa, the effect of pH was evaluated as 2–12 [65]. A study was performed in rats for which the nasal pH was 7.39 [54]. The results demonstrated that when pH was 3–10, minimal quantities of proteins and enzymes were released from cells, demonstrating no cellular damage. In contrast, if pH values were below 3 or above 10, damage was observed intracellularly and at the membrane level.

Pharmaceutical Form Nasal drops are the simplest and the most convenient nasal pharmaceutical form, but the exact amount of drug delivered is not easily quantified and often results in overdose [60]. Moreover, rapid nasal drainage can occur when using this dosage form. Solution and suspension sprays are preferred over powder sprays because the last one easily promoted the development of nasal mucosa irritation. Recently, gel devices have been developed for a more accurate drug delivery. They reduce postnasal drip and anterior leakage, fixing the drug formulation in nasal mucosa. This enhances the drug residence time and diminishes MCC. Thereby, it potentially increases the nasal absorption. Over the last few years, specialized systems such as polymeric particulates, lipid-based carriers have also been developed to improve nasal drug delivery as the following sections.

8.4 Drug Formulations for Nasal Drug Delivery

8.4.1 Prodrugs

Prodrugs (i.e., compounds that undergo transformation in the body before they can exert their pharmacological action) are useful to improve the stability and permeability of active principles that have no initially desired absorption properties. Hydrophilic groups can be added to improve the aqueous solubility of extremely hydrophobic molecules. Conversely, the addition of hydrophobic groups increases the hydrophobicity of polar molecules and thereby increases their ability to cross biological membranes. For example, this method has been used advantageously to facilitate the intranasal absorption of peptides (desmopressin acetate) and corticosteroids (beclomethasone dipropionate) [40], and can also provide these molecules with a degree of protection against degradation of enzymes and efflux proteins (by virtue of a lower binding affinity for these systems), as observed with esterified forms of acyclovir [56].

8.4.2 Solubilization Agents

The addition of excipients such as cyclodextrins increases the solubility and stability of active principles. Cyclodextrins are cyclic oligosaccharides with a hydrophilic outer surface and a hydrophobic internal cavity that can harbor hydrophobic molecules. They not only increase the solubility of hydrophobic drugs but also facilitate direct permeation through biological barriers because the overall hydrophobicity of the drug-cyclodextrin complex is higher than that of the molecule alone. This combination has been applied advantageously to intranasal administration of molecules such as midazolam [66] and granisetron [67].

8.4.3 Enzyme Inhibitors

For peptide and protein drugs, peptidase and protease inhibitors (such as bacitracin, boroleucine, amastatin, puromycin, and camostat) are useful to limit enzymatic degradation by epithelial cells. They have been found to increase the intranasal absorption of luteinizing hormone-releasing hormone, human growth hormone, encephalin, vasopressin, and desmopressin effectively [59, 68]. By transposing observations performed at the intestinal level [69, 70] to the nasal mucosa, it is reasonable to infer that the intranasal absorption of drugs that are substrates for the P450 cytochromes or efflux systems (such as P-gp) can be increased by simultaneously administering inhibitors of the latter systems. However, enzyme inhibitors are not (in view of their mechanism of action) at all involved in improving the penetration of active principles into epithelial cells; only a modest improvement in bioavailability is likely, except when inhibitors are used in association with absorption promoters.

8.4.4 Absorption Promoters

In theory, an ideal absorption promoter is expected to engender a rapid, transient, reversible, and reproducible increase in the nasal mucosa's absorption capacities. However, it should not itself cross the mucosa or exerts any systemic effect. The increase in absorption should be specific for the administered drug, so that the absorption of other, potentially toxic compounds from the pharmaceutical preparation or the environment is not enhanced. Finally, the absorption promoter should be free of local toxicity, allergenicity, and irritative activity and must be fully compatible with the other components in the pharmaceutical formulation. Although it is not fully understood how absorption promoters work, it is likely that they increase the epithelial cells' permeability by modifying the structure of the phospholipid bilayer membrane. This increase would engender an increase in membrane fluidity, the opening of the tight junctions, and a consequent increase in paracellular transport, with no change in mucociliary transport [71, 72]. The main classes of absorption promoters are cationic polymers (chitosan and cationic gelatin) and cyclodextrins. Although surfactants and bile salts have also been tested, their use has been abandoned in view of the local toxicity generated by chronic application of these compounds. Chitosan (CS, a polysaccharide obtained by deacetylation of the chitin from crustacean shells) is the most frequently used absorption promoter in pharmaceutical specialties. It is characterized by mucoadhesiveness (i.e., because of ionic interactions with the negatively charged sialic-acid groups in mucin) and the ability to open tight junctions guarding the paracellular pathway, thereby, facilitating the cellular permeability of biopharmaceuticals [3, 73]. Formulations containing CS in solution or as microspheres persist longer in the nasal mucosa in humans, which increases the drug-mucosa contact time [74]. Moreover, a study of cultured CaCo-2 intestinal epithelial cells confirmed that CS was able to open intercellular tight junctions transiently, thereby enabling hydrophilic molecules to cross the epithelium via the paracellular route [75]. Therefore, this polymer improves the intranasal bioavailability of insulin and morphine [76, 77]. Conversely, in a pharmacokinetic study, CS increased C_{max} and decreased T_{max} but also decreased the intranasal bioavailability of midazolam solubilized with a cyclodextrin [66]. Their utility as an absorption promoter might be mainly attributable to their ability to mask molecules by trapping them within their hydrophobic cavity and their ease of diffusion within the membrane [78, 79]. Finally, to improve the absorption of hydrophilic molecules or those with a higher molecular weight, tight junction modulators of various types have been described. Consequently, some lipid compounds (glycosylated sphingosines, oxidized lipids, and others) and peptides are able to open the tight junctions and facilitate the passage of drugs such as insulin. However, the cell toxicity varies [80]. Other compounds used to promote the transepithelial transport of drugs include N-acetylcysteine (which works by reducing the viscosity of the mucus and thus enabling drugs to gain better access to the epithelial cell surface) and nitroxide (NO) donors (which increase the paracellular passage of drugs via a mechanism that has yet to be fully described) [81, 82].

8.4.5 Particulate Systems

Different types of particulate systems have been developed in order to deliver various biopharmaceuticals by the intranasal route. Table 8.1 provides a selection of examples. In the following, we detail on polymeric and lipid-based carriers.

Polymeric Carriers The use of biodegradable polymeric nanocarriers as drug delivery system (DDS) for the mucosal delivery of drugs and antigens has received considerable attention [83, 84]. Using various polymeric materials and formulation processes facilitate the modulation of physicochemical properties (e.g., surface charge and mucoadhesiveness), drug loading, drug release profile, and biological behavior of nanoparticles (NPs) [85]. The most investigated polymers are chitosan (CS) and its derivatives (e.g., N-trimethyl chitosan (TMC), mono-N-carboxymethyl chitosan (MCC) [86], among others) as well as polyesters such as poly(lactic acid) (PLA) and poly(lactic-co-glycolic acid) (PLGA), and their copolymers. Other polymeric materials including poly(ε-caprolactone) (PCL), poly(acrylic acid) (PAA), poly(methacrylic acid) (PMAA), poly(alkyl cyanoacrylate) (PACA) [87], starch [88], dextran, alginate, to name a few, have also been used as carriers for the mucosal delivery of protein-based drugs and vaccines. To improve its mucoadhesive and/or permeation enhancing properties, various CS derivatives (e.g., TMC, MCC, CS-cysteine conjugates, CS-thioglycolic acid conjugates and CS-4-thio-butyl-amidine conjugates) have been synthesized [89]. Furthermore, a novel class of hybrid nanoparticles containing other polysaccharides (e.g., hyaluronic acid) or oligosaccharides (e.g., cyclodextrins and cyclodextrin derivatives [90, 91]) has been developed. These hybrid systems have shown improved physical properties and better pharmacological performance than conventional CS nanoparticles have [92]. CS-based nanocarriers can be prepared by ionic gelation with negatively charged polyanions (e.g., tripolyphosphate (TPP)) in aqueous media, which offers mild conditions for the entrapment of labile peptide and protein therapeutics [93]. The characteristics and potential of CS-based nanostructures (e.g., CS NPs, CS-coated oil nanodroplets, CS-coated lipid NPs) as vehicles for mucosal delivery of peptides and proteins were compared [94]. CS-based nanocarrier formulations have been

Table 8.1 Particulate systems for intranasal delivery of biopharmaceuticals

Formulation	Size (μm)	Biopharmaceutical	References
Chitosan	0.275	Insulin	[93]
	0.04–0.60	siRNA	[95]
	0.210	Hepatitis B surface antigen	[96]
Chitosan/alginate	0.643	Hepatitis B surface antigen	[97]
	1.324	Bovine serum albumin	[98]
Alginate	≤ 4	Tetanus toxoid	[99]
Poly(lactic acid)	1.15–1.27	$Caf1^{i}$ Lcr^{V}	[109]
	0.205–0.396	*S. equi* antigens	[110]
	0.167	Hepatitis B surface antigen	[111]
PLA-PEG	0.219–0.239	Hepatitis B surface antigen	[111]
PLA-PEG-PLA	0.176–0.216	Hepatitis B surface antigen	[111]
PEG-PLA-PEG	0.093–0.124	Hepatitis B surface antigen	[111]
PLA-PEG	0.2–10	Tetanus toxoid	[113]
PLGA/pluronic F68	0.163–0.184	β-galactosidase encoding gene	[114]
PLGA/tetronic 904	0.161–0.187	β-galactosidase encoding gene	[114]
PLGA	0.249	Diphtheria toxoid	[116]
PLGA/poly (ε-caprolactone)	0.267	Diphtheria toxoid	[116]
Poly(ε-caprolactone)	0.267	Diphtheria toxoid	[116]
PMMA-Eudragit L100/55	0.22	HIV-1 Tat	[117]
Poly(propylene sulfide)	0.050	Ovalbumin	[118]
γ-PGA	0.25–0.30	Ovalbumin	[119]
Polyethyleneimine	0.5–1.0	Influenza hemagglutinin herpes simplex virus type-2 glycoprotein D	[120]
Cationic cholesterylpullulan	0.040	Nontoxic subunit fragment of *C. botulinum* type-A neurotoxin	[121]
Liposome	2.3	Tetanus toxoid	[125]
	0.05–0.10	Monovalent subunit antigen from influenza	[126]
Liposome/SeV fusion protein	0.40	Ovalbumin	[128]
Liposome/carboxylated poly(glycidol)	0.1	Ovalbumin	[129]
Liposome/chitosan	0.774	Plasmid pRc/CMV-HBs(S)	[130]

developed for various drugs and antigens. Effective in vivo RNA interference was achieved in bronchiole epithelial cells of transgenic endogenous enhanced green fluorescent protein (EGFP) mice after nasal administration of CS/siRNA formulations [95]. Insulin-loaded NPs of a novel CS derivative, lauryl succinyl chitosan (LSC), comprising both hydrophilic (succinyl) and hydrophobic (lauryl) moieties was developed and exhibited improved release characteristics, mucoadhesion, and insulin permeability compared to native CS particles [86]. Hybrid nanoparticles of CS and cyclodextrins were found to exhibit increased encapsulation efficiency of insulin and heparin [91], permeation-enhancing properties and ability to transport insulin across the nasal barrier, engendering a considerable decrease in the plasma glucose levels [90]. In fact, CS and TMC, MCC NPs have been used extensively as nanocarriers for nasal delivery of antigens and were found to exhibit both systemic and mucosal immune responses. Immune responses were enhanced in the case of CS derivatives, probably because of better mucoadhesive properties that might result in the prolongation of the nasal residence time and promotion of the NPs uptake by M cells [96]. Alginate-coated CS and TMC NPs have also been developed as potential nanocarriers for nasal vaccination aiming to control the burst release of antigens from the NPs [97, 98]. Strong systemic IgG and mucosal IgA immune responses were induced in rabbits with intranasal administration of alginate microspheres containing tetanus toxoid and CpG-ODN [99].

PLA and PLGA are synthetic polymers that have been examined for use in the encapsulation of drugs and antigens. The well-documented biocompatibility and safety of these materials, together with their biodegradability and controlled release capacity (i.e., can sustain slow drug release rates up to several days, weeks or months [100]), has already led to their FDA approval for several clinical applications in humans, and also to a number of marketed products [3]. To improve the active targeting properties of a nanocarrier to specific cells, several surface functionalization approaches have been developed [101–103]. The most commonly used method for the preparation of PLA/PLGA NPs is the double emulsion-solvent evaporation method using dichloromethane or ethyl acetate as the polymer solvent. The aqueous protein solution is emulsified into the polymer solution by sonication followed by homogenization of the W/O emulsion in poly(vinyl alcohol) (PVA) or sodium cholate solution. Other methods such as nanoprecipitation and simple emulsification (O/W) have also been studied for the encapsulation of drugs and antigens [73]. Many protein- or nucleic acid- based therapeutics (e.g., insulin, thymopentin, helodermin, salmon calcitonin, decoy oligonucleotide, pDNA, anti-VEGF intraceptor (Fit23k) plasmid, and others) and vaccines (e.g., hepatitis B surface antigen (HBsAg), *Streptococcus equi* antigens, tetanus toxoid, diphtheria toxoid (DT), ovalbumin (OVA), to name a few) have been encapsulated successfully in PLA and PLGA NPs. PLGA NPs coencapsulated stabilizers and insulin maintained the protein integrity [104]. The presence of the stabilizers decreased the encapsulation efficiency and the sustained release of insulin resulting in prolonged reduction of blood glucose levels in streptozotocin-induced diabetic rats. In another approach, an insulin-lauryl sulfate complex was encapsulated in PLGA NPs by a spontaneous emulsion solvent diffusion method [105]. Reverse micelle-solvent evaporation method was developed to encapsulate an insulin-phospholipid complex in PLGA NPs aiming to decrease protein hydrophilicity [106]. Amine-modified comb-like PLLA (i.e., poly(vinyl-3-(diethylamino)-propylcarbamate-co-(vinyl acetate)-co-(vinyl alcohol))-*g*-poly-(*L*-lactic acid)) was developed allowing the formation of polymer-insulin nanocomplexes by spontaneous self-assembly after mixing of polymer and protein solutions [107]. The nanocomplexes with the higher lactide grafting showed the best protection against enzymatic degradation and the highest internalization and transport through Caco-2 monolayers. Deslorelinor transferrin-conjugated PLGA NPs were developed to enhance their permeability across the nasal mucosa [108]. The plasmid-loaded surface modified NPs were found to enhance the intranasal gene delivery at remote target cancer cells. With respect to vaccination, protection against bubonic and pneumonic plague following a single intranasal administration of rCaf1-loaded and $rLcr^V$-loaded PLA NPs to BALB/c mice [109] and glycol-CS-coated PLA NPs are promising as a delivery system for intranasal vaccination against strangles and do not require the coadministration of other adjuvants to achieve a balanced mucosal Th1/Th2 immune response fundamental for animal protection against *S. equi* infection [110]. Block copolymers of PLA and PEG have also been synthesized for the development of NPs encapsulating HBsAg for mucosal vaccination against hepatitis B [111, 112]. The block-copolymer-based NPs offered antigen stability during unfavorable conditions, prolonged release pattern, and enhanced mucosal uptake. Furthermore, immunological studies demonstrated the induction of systemic, mucosal, and moderate cellular immune responses, which are important to facilitate eradication of HBV-like viral infections [111]. Pegylated PLA NPs were also found to enhance the transport of tetanus toxoid across the nasal mucosa of conscious rats [113]. A new type of NPs consisting of blends of PLGA and polyethylene oxide (PEO) derivatives encapsulating pDNA was developed [114]. These NPs were found to transport pDNA across the nasal mucosa and to transfect the adequate cells resulting in significant systemic IgG antibody responses against the encoded protein. Moreover, the results of the immunization studies showed that the DNA-loaded NPs elicited a fast and strong response that was significantly more pronounced than that corresponding to the naked pDNA for up to 6 weeks. Furthermore, a single nasal immunization of BALB/c mice with PLGA NPs containing OVA and monophosphoryl lipid A (MPLA) induced a stronger IgG immune response than that induced by OVA solution or OVA-loaded PLGA NPs [115]. Moreover, significantly higher IgA titers were generated by the administration of the adjuvant-containing PLGA NPs compared to IgA stimulated by control formulations, thereby proving their ability to induce a mucosal immunity. Although PCL is a biodegradable hydrophobic polymer with similar degradation products such as PLGA (e.g., lactic and glycolic acid), PCL degrades more slowly than PLGA and therefore, does not generate an unfavorable low pH microenvironment for the entrapped biomolecules. Actually, PCL has not been explored extensively for the nanoencapsulation of protein-based drugs and vaccines, but its lack of toxicity makes it an interesting matrix for controlled release applications [116]. An intranasal vaccine delivery system consisting of DT-loaded PCL and PLGA/PCL NPs was developed [116]. Following intranasal administration, the NPs were shown to induce serum IgG antibody responses higher than PLGA and free DT. PCL NPs modified by different adjuvants as potential carriers were developed for *S. equi* antigens, with comparison of their ability to induce both systemic and local protective immunity after mucosal administration in a mouse model [116]. Results showed that the modified PCL NPs are useful to vaccinate animals against strangles because humoral, cellular, and mucosal immune responses were noticeably induced. Eudragit-coated PMMA NPs were developed as a delivery system for protein vaccine candidates. The NPs were shown to deliver and release HIV-1 Tat intracellularly and efficiently, protect it from oxidation, and preserve its biological activity, thereby increasing its shelf life, which is particularly noteworthy for vaccine applications [117]. Polypropylene sulfide (PPS) NPs conjugated with OVA and the TLR5 ligand flagellin were developed as a platform for nasal vaccination [118]. The NPs induced cytotoxic T lymphocytic responses in lung and spleen tissues and humoral response in mucosal airways as well as in the distant vaginal and rectal mucosal compartments. OVA-entrapping NPs comprising amphiphilic poly(γ-glutamic acid) (OVA/γ-PGA NPs) were used for intranasal vaccination to induce OVA-specific immune responses [119]. Mice vaccinated intranasally with OVA/γ-PGA NPs resisted challenge by E.G7-OVA tumor cells and lung metastasis of B16-OVA cells was suppressed significantly by three intranasal doses of OVA/γ-PGA NPs. Intranasal vaccination with OVA/γ-PGA NPs induced CTLs and interferon-γ-secreting cells, specific for OVA, efficiently in the spleen and lymph nodes. Polyethyleneimine (PEI) is a family of polycations used mainly for nucleic acid transfection reagents. Recently, nanoscale complexes comprising PEI with influenza hemagglutinin or herpes simplex virus type-2 (HSV-2) glycoprotein D showed potent mucosal adjuvant activity for viral subunit glycoprotein antigens [120]. A single intranasal administration of this PEI NPs elicited robust antibody-mediated protection from an otherwise lethal infection. A cationic type of cholesteryl-group-bearing pullulan (cCHP) was used for intranasal vaccine delivery system [121]. cCHP formed a nanometer-sized hydrogel ("nanogel") with a nontoxic subunit fragment of *Clostridium botulinum* type-A neurotoxin BoHc/A. Intranasally administered nanogels adhered continuously to the nasal epithelium and was effectively taken up by mucosal DCs. Vigorous botulinum-neurotoxin-A neutralizing serum IgG and secretory IgA antibody responses were induced without coadministration of mucosal adjuvant.

Lipid-Based Carriers Liposomes are vesicles that consist either of many, few, or just one phospholipid bilayer (i.e., large unilamellar vesicles (LUV), or small unilamellar vesicles (SUV)) [122]. Participation of nonionic surfactants instead of phospholipids in the bilayer formation results in niosomes. Research on liposome technology has progressed from conventional vesicles ("first-generation liposomes") to "second-generation liposomes," in which long-circulating liposomes are obtained by modification of the lipid composition and functionalization of the vesicle surface by various molecules, such as glycolipids, sialic acid, and PEG ("stealth" liposomes). "Stealth" liposomes can finally become targeted via conjugation of targeting ligands (e.g., monoclonal antibodies, proteins, folic acids, and others) with properly modified surface PEG molecules [123]. Liposomes are regarded as an interesting carrier for the administration of biomolecules through mucosal surfaces because they are versatile and tend to be innocuous (produced with natural and biodegradable compounds). Moreover, they provide protection to the encapsulated material. Their organized structure (an aqueous core encapsulated within one or more phospholipid bilayers) enables the association of drugs to both the aqueous phase and lipid hydrophobic phase. Drug release can usually be controlled, depending on the bilayer number and lipid compositions [124]. Related to the application of liposomes in mucosal vaccination, the efficiency of liposomes containing tetanus toxoid and CpG-ODN was evaluated [125]. Results of this study showed that, following nasal administration to rabbits, the liposomes induced high mucosal IgA and low systemic IgG responses. Additionally, cationic liposomes loaded with a monovalent subunit antigen derived from influenza A/New Caledonia/20/99-like (H1N1) and A/Panama/2007/99-like (H3N2) strains, when administered intranasally, are highly efficacious in inducing strong humoral and cellular systemic and local immune responses in mice, leading to protective immunity against the influenza virus [126]. Furthermore, HBsAg-loaded liposomes functionalized with *Ulex europaeus* agglutinin (UEA-1) were developed to increase transmucosal uptake by M cells [127]. The lectinized liposomes exhibited enhanced binding to M cells as compared with the nonlectinized ones. Moreover, lectinized liposomes induced higher secretary IgA level in mucosal secretions and cytokines level in the spleen homogenates. Fusogenic liposomes are used for activation of cellular immune responses because of their ability to deliver contents into cytosol of target cells (i.e., APCs: DCs, macrophages). Liposome-containing envelope glycoproteins of Sendai virus were developed as intranasal antigen-delivery vehicles [128]. This liposome efficiently delivered antigen to M cells, epithelial cells, and macrophages in NALT and induced antigen-specific CTL responses and Th1 and Th2 cell responses. These results reflect the intrinsic ability of Sendai virus, which naturally infects via mucosal epithelia. pH-Sensitive synthetic polymers were also used as fusogenic liposomes for activation of immune response via a mucosal surface. Carboxylated poly(glycidol)-modified liposomes were administered intranasally, inducing antigen-specific cellular immune responses [129] (Fig. 8.3). These results showed that surface-modified liposomes are a potential module for the development of effective mucosal vaccines. The potential of glycol-CS-coated liposomes as nasal vaccine delivery vehicles was studied [130]. Following intranasal administration, glycol-CS-coated liposomes elicited humoral, mucosal and cellular immune responses that were significant compared to naked DNA, thereby justifying the potential advantage of mucosal vaccination in the production of local antibodies at the sites where the pathogens enter the body. Solid lipid nanoparticles (SLNs) have attracted increasing attention as efficient and nontoxic alternative hydrophobic colloidal drug carriers. SLNs are made from solid lipids (e.g., triglycerides, fatty acids, among others) and are producible to incorporate either hydrophobic or hydrophilic drugs. Their colloidal dimensions and the controlled-release behavior enable drug protection and administration by various routes, thereby emphasizing their versatility. SLNs are prepared using various techniques such as high-pressure homogenization, microemulsion formation, precipitation, and as lipid nanopellets. Aiming to improve the encapsulation efficiency of hydrophilic protein drugs into SLNs, a reverse micelle-double emulsion method can be used for the synthesis of SLNs containing reverse micelles loaded with protein [122, 131]. Intranasal administration of HBsAg-loaded lipid microparticles to rats induced considerable mucosal immune responses as well as systemic immune responses [132].

8.5 Conclusions and Future Perspectives

To date, many drug formulations and nanoscaled carriers for intranasal delivery have been investigated. With the proper formulation and carrier design, the permeability and localization of drugs can be controlled. Intranasal drug delivery is promising for systemic delivery of orally inefficient drugs as well as an attractive alternative for noninvasive delivery of potent peptide and protein drugs. The needs for safer and

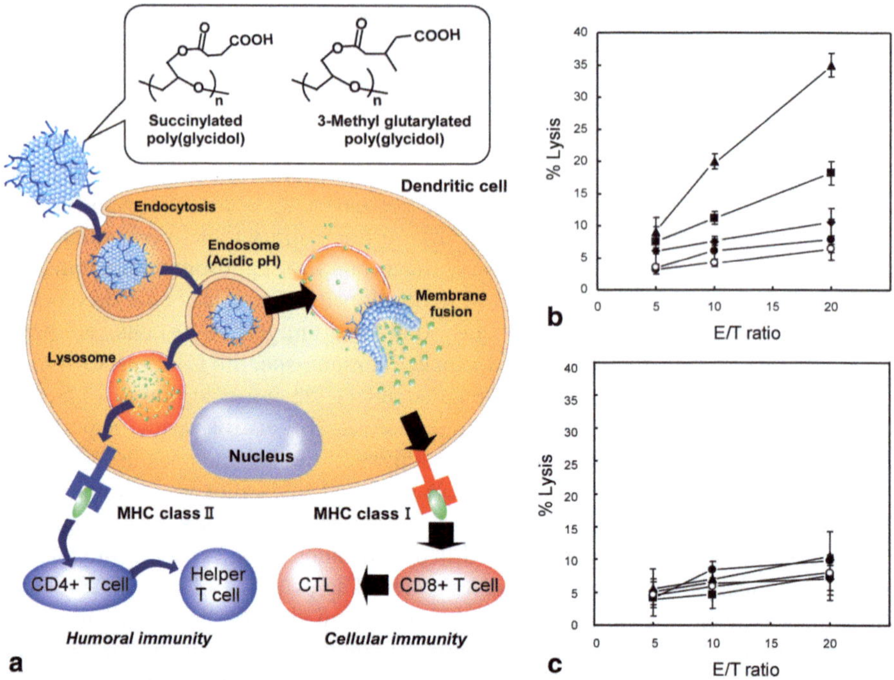

Fig. 8.3 **a** Design of pH-sensitive fusogenic polymer-modified liposomes for induction of cellular immunity. Structures of two types of pH-sensitive poly(glycidol) derivatives with a different side structures: succinylated poly(glycidol) (SucPG), 3-Methyl glutarylated poly(glycidol) (MGluPG) are shown. Expected mechanisms for induction of cellular immunity mediated by pH-sensitive polymer-modified liposomes are the following. The pH-sensitive polymer-modified liposomes are taken up by dendritic cells via an endocytic pathway and trapped in endosomes, which have a weak acidic environment. Then, the liposomes fuse with and/or destabilize endosomes and release antigenic molecules into cytosol efficiently, which results in the antigen presentation via MHC class I molecules and induction of the antigen-specific CTL. **b, c** OVA-specific CTL responses in spleen at Day 21 after nasal immunization with OVA solution (closed circles), polymer-unmodified liposomes (closed diamonds), SucPG liposomes (closed squares), and MGluPG liposomes (closed triangles). CTL responses were measured using a LDH assay at indicated E/T ratios. E.G7-OVA cells (**b**), and EL4 cells (**c**) were used as target cells. T cell responses from mice without treatment (open circles) were also shown as a negative control. (Adapted from [129], copyright 2009, with permission from Elsevier)

more effective nasal drug formulations and nanocarriers are crucial requirements to provide a promising future in the area of nasal drug delivery.

References

1. Pires A, Fortuna A, Alves G, Falcao A. Intranasal drug delivery: how, why and what for? J Pharm Sci. 2009;12:288–311.
2. Stanislas GD, Amparo B, Emmanuel N, Christophe F, Sabine BL, Louis-Jean C, Morgan LG, Marc F, Philippe D. Intranasal drug delivery: an efficient and non-invasive route for systemic administration focus on opioids. Pharmacol Ther. 2012;134:366–79.

3. Dahl R, Mygind N. Anatomy, physiology and function of the nasal cavities in health and disease. Adv Drug Del Rev. 1998;29:3–12.
4. Ugwoke MI, Agu RU, Verbeke N, Kinget R. Nasal mucoadhesive drug delivery: background, applications, trends and future perspectives. Adv Drug Del Rev. 2005;57:1640–65.
5. Arora P, Sharma S, Garg S. Permeability issues in nasal drug delivery. Drug Disc Today. 2002;7:967–75.
6. Busuttil A, More IA, McSeveney D. A reappraisal of the ultrastructure of the human respiratory nasal mucosa. J Anat. 1977;124:445–58.
7. Dondeti P, Zia H, Needham TE. Bioadhesive and formulation parameters affecting nasal absorption. Int J Pharm. 1996;127:115–33.
8. Verdugo P. Goblet cells secretion and mucogenesis. Ann Rev Physiol. 1990;52:157–76.
9. Lethem MI. The role of tracheobronchial mucus in drug administration to the airways. Adv Drug Deliv. 1993;11:19–27.
10. Costantino HR, Illum L, Brandt G, Johnson PH, Quay SC. Intranasal delivery: physicochemical and therapeutic aspects. Int J Pharm. 2007;337:1–24.
11. Lansley AB. Mucociliary clearance and drug delivery via the respiratory tract. Adv Drug Del Rev. 1993;11:299–327.
12. Mathison S, Nagilla R, Kompella UB. Nasal route for direct delivery of solutes to the central nervous system: fact or fiction? J Drug Target. 1998;5:415–41.
13. Dhanda DS, Frey WH, Leopold D, Kompella U. Nose to brain delivery: approaches for drug deposition in human olfactory epithelium. Drug Deliv Technol. 2005;5:1–9.
14. Illum L. Is nose-to-brain transport of drugs in man a reality? J Pharm Pharmacol. 2004; 56:3–17.
15. Neutra MR. M cells in antigen sampling in mucosal tissues. Curr Top Microbiol Immunol. 1999;236:17–32.
16. Debertin AS, Tschernig T, Tonjes H, Kleemann WJ, Troger HD, Pabst R. Nasal-associated lymphoid tissue (NALT): frequency and localization in young children. Clin Exp Immunol. 2003;134:503–7.
17. Debertin AS, Tschernig T, Schurmann A, Bajanowski T, Brinkmann B, Pabst R. Coincidence of different structures of mucosa-associated lymphoid tissue (MALT) in the respiratory tract of children: no indications for enhanced mucosal immunostimulation in sudden infant death syndrome (SIDS). Clin Exp Immunol. 2006;146:54–9.
18. Goeringer GC, Vidic B. The embryogenesis and anatomy of Waldeyer's ring. Otolaryngol Clin North Am. 1987;20:207–17.
19. Bienenstock J, McDermott MR. Bronchus- and nasal-associated lymphoid tissues. Immunol Rev. 2005;206:22–31.
20. Neutra MR, Kozlowski PA. Mucosal vaccines: the promise and the challenge. Nat Rev Immunol. 2006;6:148–58.
21. Illum L. Nanoparticulate systems for nasal delivery of drugs: a real improvement over simple systems? J Pharm Sci. 2007;96:473–83.
22. Iwasaki A. Mucosal dendritic cells. Ann Rev Immunol. 2007;25:381–418.
23. Storni T, Kundig TM, Senti G, Johansen P. Immunity in response to particulate antigen-delivery systems. Adv Drug Del Rev. 2005;57:333–55.
24. Lamm ME. Interaction of antigens and antibodies at mucosal surfaces. Ann Rev Microbiol. 1997;51:311–40.
25. MacPherson GG, Liu LM. Dendritic cells and Langerhans cells in the uptake of mucosal antigens. Curr Top Microbiol Immunol. 1999;236:33–53.
26. Kunkel EJ, Butcher EC. Plasma-cell homing. Nat Rev Immunol. 2003;3:822–9.
27. Dae-Duk K. Drug absorption studies: in situ, in vitro and in silico models. Chapter 9. Springer: USA; 2007.
28. Graff CL, Pollack GM. P-glycoprotein attenuates brain uptake of substrates after nasal instillation. Pharm Res. 2003;20:1225–30.
29. Graff CL, Pollack GM. Functional evidence for P-glycoprotein at the nose-brain barrier. Pharm Res. 2005;22:86–93.

30. Westin U, Piras E, Jansson B, Bergström U, Dahlin M, Brittebo E, Björk E. Transfer of morphine along the olfactory pathway to the central nervous system after nasal administration to rodents. Eur J Pharm Sci. 2005;24:565–73.
31. McMartin C. Analysis of structural requirements for the absorption of drugs and macromolecules from the nasal cavity. J Pharm Sci. 1987;76:535–40.
32. Huang CH, Kimura R, Nassar RB, Hussain A. Mechanism of nasal absorption of drugs. I: physicochemical parameters influencing the rate of in situ nasal absorption of drugs in rats. J Pharm Sci. 1985;74:608–11.
33. Kao HD, Traboulsi A, Itoh S, Dittert L, Hussain A. Enhancement of the systemic and CNS specific delivery of L-dopa by the nasal administration of its water soluble prodrugs. Pharm Res. 2000;17:978–84.
34. Merkus FW, Verhoef JC, Schipper NG, Marttin E. Nasal mucociliary clearance as a factor in nasal drug delivery. Adv Drug Deliv Rev. 1998;29:13–38.
35. Schipper N, Verhoef J, Merkus FW. The nasal mucociliary clearance: relevance to nasal drug delivery. Pharm Res. 1991;8:807–14.
36. Illum L. Nasal drug delivery: possibilities, problems and solutions. J Control Release. 2003;87:187–98.
37. Houtmeyers E, Gosselink R, Gayan-Ramirez G, Decramer M. Regulation of mucociliary clearance in health and disease. Eur Respir J. 1999;13:1177–88.
38. Bogdanffy MS. Biotransformation enzymes in the rodent nasal mucosa: the value of a histochemical approach. Environ Health Perspect. 1990;85:177–86.
39. Dahl AR, Lewis JL. Respiratory tract uptake of inhalants and metabolism of xenobiotics. Ann Rev Pharmacol Toxicol. 1993;33:383–407.
40. Mitra AK, Krishnamoorthy R. Prodrugs for nasal drug delivery. Adv Drug Deliv Rev. 1998;29:135–46.
41. Dimova S, Brewster ME, Noppe M, Jorissen M, Augustijns P. The use of human nasal in vitro cell systems during drug discovery and development. Toxicol In Vitro. 2005;19:107–22.
42. Sarkar MA. Drug metabolism in the nasal mucosa. Pharm Res. 1992;9:1–9.
43. Lee VH, Yamamoto A. Penetration and enzymatic barriers of peptide and protein absorption. Adv Drug Deliv Rev. 1990;4:171–207.
44. Harris AS. Intranasal administration of peptides: nasal deposition, biological response and absorption of desmopressin. J Pharm Sci. 1986;75:1085–8.
45. Kandimalla KK, Donovan MD. Transport of hydroxyzine and triprolidine across bovine olfactory mucosa: role of passive diffusion in the direct nose-to-brain uptake of small molecules. Int J Pharm. 2005;302:133–44.
46. Donovan M, Flynn G, Amidon G. Absorption of polyethylene glycols 600 through 2000: the molecular weight dependence of gastrointestinal and nasal absorption. Pharm Res. 1990;7:863–8.
47. Fisher A, Illum L, Davis S, Schacht E. Diiodo-L-tyrosine labelled dextrans as molecular size markers of nasal absorption in the rat. J Pharm Pharmacol. 1992;44:550–4.
48. Katdare A, Chaubal MV. Excipient development for pharmaceutical biotechnology and drug delivery systems. Taylor & Francis: USA; 2006.
49. Corbo DC. Characterization of the barrier properties of mucosal membranes. J Pharm Sci. 1990;79:202–6.
50. Corbo DC. Drug absorption through mucosal membranes: effect of mucosal route and penetrant hydrophilicity. Pharm Res. 1989;6:848–52.
51. Lipworth BJ, Jackson CM. Safety of inhaled and intranasal corticosteroids: lessons for the new millennium. Drug Saf. 2000;23:11–33.
52. Washington N, Steele RJ, Jackson SJ, Bush D, Mason J, Gill DA, Pitt K, Rawlins DA. Determination of baseline human nasal pH and the effect of intranasally administered buffers. Int J Pharm. 2000;198:139–46.
53. Shao Z, Park GB, Krishnamoorthy R, Mitra AK. The physicochemical properties, plasma enzymatic hydrolysis, and nasal absorption of acyclovir and its 2, ester prodrugs. Pharm Res. 1994;11:237–42.

54. Hirai S, Yashiki T, Matsuzawa T, Mima H. Absorption of drugs from the nasal mucosa of rats. Int J Pharm. 1981;7:317–25.
55. Zaki NM, Awad GA, Mortada ND, Abd ElHady SS. Rapid-onset intranasal delivery of metoclopramide hydrochloride. Part I. Influence of formulation variables on drug absorption in anesthetized rats. Int J Pharm. 2006;327:89–96.
56. Yang C, Gao H, Mitra AK. Chemical stability, enzymatic hydrolysis, and nasal uptake of amino acid ester prodrugs of acyclovir. J Pharm Sci. 2001;90:617–24.
57. Machida M. Effects of surfactants and protease inhibitors on nasal absorption of recombinant human granulocyte colonystimulating factor (rhG-CSF) in rats. Biol Pharm Bull. 1994;17:1375–8.
58. Morimoto K, Miyazaki M, Kakemi M. Effects of proteolytic enzyme inhibitors on nasal absorption of salmon calcitonin in rats. Int J Pharm. 1995;113:1–8.
59. Bernkop-Schnurch A. Use of inhibitory agents to overcome the enzymatic barrier to perorally administered therapeutic peptides and proteins. J Control Release. 1998;52:1–16.
60. Romeo VD, Meireles J, Sileno AP, Pimplaskar HK, Behl CR. Effects of physicochemical properties and other factors on systemic nasal delivery. Adv Drug Deliv Rev. 1998;29:89–116.
61. Hussain AA, Al-Bayatti AA, Dakkuri A, Okochi K, Hussain MA. Testosterone 17β-N, N-dimethylglycinate hydrochloride: a prodrug with a potential for nasal delivery of testosterone. J Pharm Sci. 2002;91:785–9.
62. Heidari A, Sadrai H, Varshosaz J. Nasal delivery of insulin using bioadhesive chitosan gels. Drug Deliv. 2006;13:31–8.
63. Alsarra IA, Hamed AY, Mahrous GM, El Maghraby GM, Al-Robayan AA, Alanazi FK. Mucoadhesive polymeric hydrogels for nasal delivery of acyclovir. Drug Dev Ind Pharm. 2009;35:352–62.
64. Kilian N, Müller DG. The effect of a viscosity and an absorption enhancer on the intra nasal absorption of metoprolol in rats. Int J Pharm. 1998;163:211–7.
65. Pujara CP, Shao Z, Duncan MR, Mitra AK. Effects of formulation variables on nasal epithelial cell integrity: biochemical evaluations. Int J Pharm. 1995;114:197–203.
66. Haschke M, Suter K, Hofmann S, Witschi R, Frohlich J, Imanidis G, et al. Pharmacokinetics and pharmacodynamics of nasally delivered midazolam. Br J Clin Pharmacol. 2010;69: 607–16.
67. Cho HJ, Balakrishnan P, Shim WS, Chung SJ, Shim CK, Kim DD. Characterization and in vitro evaluation of freeze-dried microparticles composed of granisetron-cyclodextrin complex and carboxymethylcellulose for intranasal delivery. Int J Pharm. 2010;400:59–65.
68. Morimoto K, Yamaguchi H, Iwakura Y, Miyazaki M, Nakatani E, Iwamoto T, et al. Effects of proteolytic enzyme inhibitors on the nasal absorption of vasopressin and an analogue. Pharm Res. 1991;8:1175–9.
69. Kato M. Intestinal first-pass metabolism of CYP3A4 substrates. Drug Metab Pharmacokinet. 2008;23:87–94.
70. Tachibana T, Kato M, Takano J, Sugiyama Y. Predicting drug-drug interactions involving the inhibition of intestinal CYP3A4 and P-glycoprotein. Curr Drug Metab. 2010;11:762–77.
71. Davis SS, Illum L. Absorption enhancers for nasal drug delivery. Clin Pharmacokinet. 2003;42:1107–28.
72. Karasulu E, Yavasoglu A, Evrensanal Z, Uyanikgil Y, Karasulu HY. Permeation studies and histological examination of sheep nasal mucosa following administration of different nasal formulations with or without absorption enhancers. Drug Deliv. 2008;15:219–25.
73. Sharma S, Mukkur TKS, Benson HAE, Chen Y. Pharmaceutical aspects of intranasal delivery of vaccines using particulate systems. J Pharm Sci. 2009;98:812–43.
74. Soane RJ, Frier M, Perkins AC, Jones NS, Davis SS, Illum L. Evaluation of the clearance characteristics of bioadhesive systems in humans. Int J Pharm. 1999;178:55–65.
75. Amidi M, Mastrobattista E, Jiskoot W, Hennink WE. Chitosan-based delivery systems for protein therapeutics and antigens. Adv Drug Deliv Rev. 2010;62:59–82.
76. Illum L, Farraj NF, Davis SS. Chitosan as a novel nasal delivery system for peptide drugs. Pharm Res. 1994;11:1186–9.

77. Illum L, Watts P, Fisher AN, Hinchcliffe M, Norbury H, Jabbal-Gill I, et al. Intranasal delivery of morphine. J Pharmacol Exp Ther. 2002;301:391–400.
78. Irie T, Wakamatsu K, Arima H, Aritomi H, Uekama K. Enhancing effects of cyclodextrins on nasal absorption of insulin in rats. Int J Pharm. 1992;84:129–39.
79. Shao Z, Krishnamoorthy R, Mitra AK. Cyclodextrins as nasal absorption promoters of insulin: mechanistic evaluations. Pharm Res. 1992;9:1157–63.
80. Duan X, Mao S. New strategies to improve the intranasal absorption of insulin. Drug Disc Today. 2010;15:416–27.
81. Baker G, Chetwin K, Hayward K, Bakirtzi K, Willman M. The effect of nitric oxide on the permeability of nasal epithelial cells from healthy and asthmatic donors. Med Sci Monit. 2003;9:276–82.
82. Matsuyama T, Morita T, Horikiri Y, Yamahara H, Yoshino H. Enhancement of nasal absorption of large molecular weight compounds by combination of mucolytic agent and nonionic surfactant. J Control Release. 2006;110:347–52.
83. Csaba N, Garcia-Fuentes M, Alonso MJ. Nanoparticles for nasal vaccination. Adv Drug Deliv Rev. 2009;61:140–57.
84. Kammona O, Kiparissides C. Recent advances in nanocarrier-based mucosal delivery of biomolecules. J Control Release. 2012;161:781–94.
85. Chen MC, Sonaje K, Chen KJ, Sung HW. A review of the prospects for polymeric nanoparticle platforms in oral insulin delivery. Biomaterials. 2011;32:9826–38.
86. Rekha MR, Sharma CP. Synthesis and evaluation of lauryl succinyl chitosan particles towards oral insulin delivery and absorption. J Control Release. 2009;135:144–51.
87. Mesiha MS, Sidhom MB, Fasipe B. Oral and subcutaneous absorption of insulin poly(isobutylcyanoacrylate) nanoparticles. Int J Pharm. 2005;288:289–93.
88. Jain AK, Khar RK, Ahmed FJ, Diwan PV. Effective insulin delivery using starch nanoparticles as a potential trans-nasal mucoadhesive carrier. Eur J Pharm Biopharm. 2008;69:426–35.
89. Bernkop-Schnürch A, Hornof M, Guggi D. Thiolated chitosans. Eur J Pharm Biopharm. 2004;57:9–17.
90. Teijeiro-Osorio D, Remuán-López C, Alonso MJ. New generation of hybrid poly/oligosaccharide nanoparticles as carriers for the nasal delivery of macromolecules. Biomacromolecules. 2009;10:243–9.
91. Krauland AH, Alonso MJ. Chitosan/cyclodextrin nanoparticles as macromolecular drug delivery system. Int J Pharm. 2007;340:134–42.
92. Goycoolea FM, Lollo G, Remuán-López C, Quaglia F, Alonso MJ. Chitosan/alginate blended nanoparticles as carriers for the transmucosal delivery of macromolecules. Biomacromolecules. 2009;10:1736–43.
93. Makhlof A, Tozuka Y, Takeuchi H. Design and evaluation of novel pH-sensitive chitosan nanoparticles for oral insulin delivery. Eur J Pharm Sci. 2011;42:445–51.
94. Prego C, Garcia M, Torres D, Alonso MJ. Transmucosal macromolecular drug delivery. J Control Release. 2005;101:151–62.
95. Howard KA, Rahbek UL, Liu X, Damgaard CK, Glud SZ, Andersen MØ, Hovgaard MB, Schmitz A, Nyengaard JR, Basenbacher F, Kjems J. RNA interference in vitro and in vivo using a chitosan/siRNA nanoparticle system. Mol Ther. 2006;14:476–84.
96. Mangal S, Pawar D, Garg NK, Jain AK, Vyas SP, Raman Rao DSV, Jaganathan KS. Pharmaceutical and immunological evaluation of mucoadhesive nanoparticles based delivery system(s) administered intranasally. Vaccine. 2011;29:4953–62.
97. Borges O, Cordeiro-da-Silva A, Tavares J, Santarem N, de Sousa A, Borchard G, Junginger HE. Immune response by nasal delivery of hepatitis B surface antigen and codelivery of a CpG ODN in alginate coated chitosan nanoparticles. Eur J Pharm Biopharm. 2008;69:405–16.
98. Li XY, Kong XY, Shi SA, Zheng XL, Guo G, Wei YQ, Qian ZY. Preparation of alginate coated chitosan microparticles for vaccine delivery. BMC Biotechnol. 2008;8:89.
99. Tafaghodi M, SA Sajadi Tabassi, Jaafari MR. Induction of systemic and mucosal immune responses by intranasal administration of alginate microspheres encapsulated with tetanus toxoid and CpG-ODN. Int J Pharm. 2006;319:37–43.

100. Mundargi RC, Babu VR, Rangaswamy V, Patel P, Aminabhavi TM. Nano/micro technologies for delivering macromolecular therapeutics using poly(D, L-lactide-co-glycolide) and its derivatives. J Control Release. 2008;125:193–209.
101. Yin YS, Chen DW, Qiao MX, Lu Z, Hu HY. Preparation and evaluation of lectin-conjugated PLGA nanoparticles for oral delivery of thymopentin. J Control Release. 2006;116:337–45.
102. Yin YS, Chen DW, Qiao MX, Wei XY, Hu HY. Lectin-conjugated PLGA nanoparticles loaded with thymopentin: ex vivo bioadhesion and in vivo biodistribution. J Control Release. 2007;123:27–38.
103. Mohamed F, van der Walle CF. Engineering biodegradable polyester particles with specific drug targeting and drug release properties. J Pharm Sci. 2008;97:71–87.
104. Kumar PS, Saini TR, Chandrasekar D, Yellepeddi VK, Ramakrishna S, Diwan PV. Novel approach for delivery of insulin loaded poly(lactide-co-glycolide) nanoparticles using a combination of stabilizers. Drug Deliv. 2007;14:517–23.
105. Shi K, Cui F, Yamamoto H, Kawashima Y. Optimized formulation of high-payload PLGA nanoparticles containing insulin-lauryl sulfate complex. Drug Dev Ind Pharm. 2009;35: 177–84.
106. Cui F, Qian F, Yin C. Preparation and characterization of mucoadhesive polymer-coated nanoparticles. Int J Pharm. 2006;316:154–61.
107. Simon M, Behrens I, Dailey LA, Wittmar M, Kissel T. Nanosized insulin-complexes based on biodegradable amine-modified graft polyesters poly[inyl-3-(diethylamino) propylcarbamate-co-(vinyl acetate)-co-(vinyl alcohol)]-graft-poly-(L-lactic acid): protection against enzymatic degradation, interaction with Caco-2 cell monolayers, peptide transport and cytotoxicity. Eur J Pharm Biopharm. 2007;66:165–72.
108. Sundaram S, Roy SK, Ambati BK, Kompella UB. Surface-functionalized nanoparticles for targeted gene delivery across nasal respiratory epithelium. FASEB J. 2009;23:3752–65.
109. Elvin SJ, Eyles JE, Howard KA, Ravichandran E, Somavarappu S, Alpar HO, Williamson ED. Protection against bubonic and pneumonic plague with a single dose microencapsulated sub-unit vaccine. Vaccine. 2006;24:4433–9.
110. Florindo HF, Pandit S, Gonçalves LMD, Alpar HO, Almeida AJ. New approach on the development of a mucosal vaccine against strangles: systemic and mucosal immune responses in a mouse model. Vaccine. 2009;27:1230–41.
111. Jain AK, Goyal AK, Gupta PN, Khatri K, Mishra N, Mehta A, Mangal S, Vyas SP. Synthesis, characterization and evaluation of novel triblock copolymer based nanoparticles for vaccine delivery against hepatitis B. J Control Release. 2009;136:161–9.
112. Jain AK, Goval AK, Mishra N, Vaidya B, Mangal S, Vyas SP. PEG-PLA-PEG block copolymeric nanoparticles for oral immunization against hepatitis B. Int J Pharm. 2010;387:253–62.
113. Vila A, Sanchez A, Evora C, Soriano I, McCallion O, Alonso MJ. PLA-PEG particles as nasal protein carriers: the influence of the particle size. Int J Pharm. 2005;292:43–52.
114. Csaba N, Sanchez A, Alonso MJ. PLGA: poloxamer and PLGA: poloxamine blend nanostructures as carriers for nasal gene delivery. J Control Release. 2006;113:164–72.
115. Sarti F, Perera G, Hintzen F, Kotti K, Karageorgiou V, Kammona O, Kiparissides C, Bernkop-Schnürch A. In vivo evidence of oral vaccination with PLGA nanoparticles containing the immunostimulant monophosphoryl lipid A. Biomaterials. 2011;32:4052–7.
116. Singh J, Pandit S, Bramwell VW, Alpar HO. Diphtheria toxoid loaded poly(-caprolactone) nanoparticles as mucosal vaccine delivery systems. Methods. 2006;38:96–105.
117. Voltan R, Castaldello A, Brocca-Cofano E, Altavilla G, Caputo A, Laus M, Sparnacci K, Ensoli B, Spaccasassi S, Ballestri M, Tondelli L. Preparation and characterization of innovative protein-coated poly(methylmethacrylate) core-shell nanoparticles for vaccine purposes. Pharm Res. 2007;24:1870–82.
118. Stano A, van der Vlies AJ, Martino MM, Swartz MA, Hubbella JA, Smeoni E. PPS nanoparticles as versatile delivery system to induce systemic and broad mucosal immunity after intranasal administration. Vaccine. 2011;29:804–12.
119. Matsuo K, Koizumi H, Akashi M, Nakagawa S, Fujita T, Yamamoto A, Okada N. Intranasal immunization with poly(γ-glutamic acid) nanoparticles entrapping antigenic proteins can induce potent tumor immunity. J Control Release. 2011;152:310–6.

120. Wegmann F, Gartlan KH, Harandi AM, Brinckmann SA, Coccia M, Hillson WR, Kok WL, Cole S, Ho LP, Lambe T, Puthia M, Svanborg C, Scherer EM, Krashias G, Williams A, Blattman JN, Greenberg PD, Flavell RA, Moghaddam AE, Sheppard NC, Sattentau QJ. Polyethyleneimine is a potent mucosal adjuvant for viral glycoprotein antigens. Nature Biotech. 2012;30:883–8.
121. Nochi T, Yuki Y, Takahashi H, Sawada S, Mejima M, Kohda T, Harada N, Kong IG, Sato A, Kataoka N, Tokuhara D, Kurokawa S, Takahashi Y, Tsukada H, Kozaki S, Akiyoshi K, Kiyono H. Nanogel antigenic protein-delivery system for adjuvant-free intranasal vaccines. Nat Mater. 2010;9:572–8.
122. Rawat M, Singh D, Saraf S, Saraf S. Lipid carriers: a versatile delivery vehicle for proteins and peptides. Yakugaku Zasshi. 2008;128:269–80.
123. Kiparissides C, Kammona O. Nanotechnology advances in controlled drug delivery systems. Phys Stat Solidi C. 2008;5:3828–33.
124. Grenha A, Remuán-López C, Carvalho ELS, Seijo B. Microspheres containing lipid/chitosan nanoparticles complexes for pulmonary delivery of therapeutic proteins. Eur J Pharm Biopharm. 2008;69:83–93.
125. Tafaghodi M, Jaafari MR, Sajadi Tabassi SA. Nasal immunization studies using liposomes loaded with tetanus toxoid and CpG-ODN. Eur J Pharm Biopharm. 2006;64:138–45.
126. Joseph A, Itskovitz-Cooper N, Samira S, Flasterstein O, Eliyahu H, Simberg D, Goldwaser I, Barenholz Y, Kedar E. A new intranasal influenza vaccine based on a novel polycationic lipid-ceramide carbamoyl-spermine (CCS) I. Immunogenicity and efficacy studies in mice. Vaccine. 2006;24:3990–4006.
127. Gupta PN, Vyas SP. Investigation of lectinized liposomes as M-cell targeted carrier-adjuvant for mucosal immunization. Colloids Surf B Biointerfaces. 2011;82:118–25.
128. Kunisawa J, Nakanishi T, Takahashi I, Okudaira A, Tsutsumi Y, Katayama K, Nakagawa S, Kiyono H, Mayumi T. Sendai virus fusion protein mediates simultaneous induction of MHC class I/II-dependent mucosal and systemic immune responses via the nasopharyngeal-associated lymphoreticular tissue immune system. J Immunol. 2001;167:1406–12.
129. Yuba E, Kojima C, Harada A, Tana, Watarai S, Kono K. pH-Sensitive fusogenic polymer-modified liposomes as a carrier of antigenic proteins for activation of cellular immunity. Biomaterials. 2010;31:943–51.
130. Khatri K, Goyal AK, Gupta PN, Mishra N, Mehta A, Vyas SP. Surface modified liposomes for nasal delivery of DNA vaccine. Vaccine. 2008;26:2225–33.
131. Liu J, Gong T, Fu H, Wang C, Wang X, Chen Q, Zhang Q, He Q, Zhang Z. Solid lipid nanoparticles for pulmonary delivery of insulin. Int J Pharm. 2008;356:333–44.
132. Saraf S, Mishra D, Asthana A, Jain R, Singh S, Jain NK. Lipid microparticles for mucosal immunization against hepatitis B. Vaccine. 2006;24:45–56.

Chapter 9
Ocular Delivery of Biopharmaceuticals

Holly Lorentz and Heather Sheardown

9.1 Introduction

Currently there is a worldwide need for novel, non-invasive ophthalmic biopharmaceutical treatments as there are many sight-threatening ocular diseases and conditions that are poorly understood, incurable, have ineffective treatment option, or require repeated invasive or surgical treatment. Furthermore, the majority of the current treatment options utilize conventional pharmaceuticals that, in many cases, only treat the symptoms of the disease and not the root of the problem. Therefore, the development of modern ocular biopharmaceutical treatment options could not only help treat some of these prevalent ocular conditions but they may also have the ability to prevent, repair, or cure the diseases themselves. Since topical ocular delivery treatments are considered to be the safest, least invasive, and most self-administrable, their development is highly sought [1–3]. However, there are many barriers blocking successful and effective topical delivery of biopharmaceuticals to the eye. These challenges include ocular anatomical barriers due to the extraordinary and complex structure of the eye, biopharmaceutical barriers related to the biopharmaceutical's properties, and patient barriers related to comfort, compliance, and self-administration of the treatment. Consequently, all of these aspects must be taken into effect when developing a new ocular biopharmaceutical delivery system. One way to overcome some of the eye's natural anatomical barriers is to take advantage of the ocular surface mucosal layer and use its structure to aid in biopharmaceutical adherence and penetration by incorporating mucoadhesive substances into the delivery system [4–7]. These mucoadhesive substances can be integrated into a range of different delivery systems and used in conjunction with a variety of biopharmaceuticals to make an effective device for ocular biopharmaceutical delivery.

This chapter will examine the main topics involved in ophthalmic biopharmaceutical mucosal delivery including: the biology of the eye, ocular surface, tear film and the mucin layer; ocular diseases, routes and delivery systems for ocular drug and biopharmaceutical delivery; challenges and barriers with ocular biopharmaceutical delivery;

H. Lorentz · H. Sheardown (✉)
Chemical Engineering, McMaster University,
1280 Main Street West, Hamilton, ON L8S 4L8, Canada
e-mail: sheardow@mcmaster.ca

J. das Neves, B. Sarmento (eds.), *Mucosal Delivery of Biopharmaceuticals*, DOI 10.1007/978-1-4614-9524-6_9,

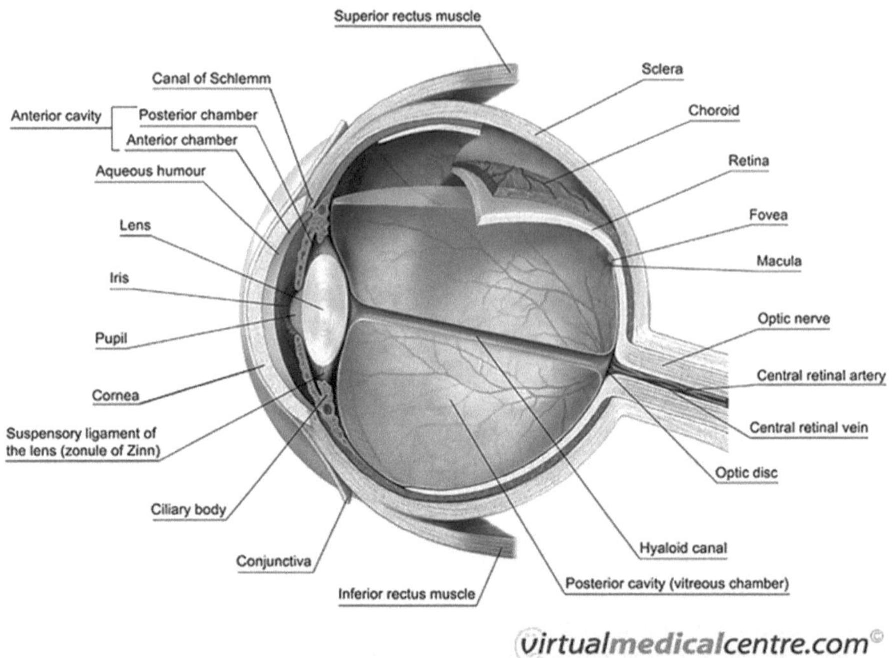

Fig. 9.1 Ocular anatomy. (Reprinted with kind permission of Virtual Medical Centre, Osborne Park, Australia, available from URL http://www.virtualmedicalcentre.com/anatomy/the-eye-and-vision/28)

strategies, technology and current research in ocular mucosal biopharmaceutical delivery; and future perspectives of ocular mucosal biopharmaceutical delivery.

9.2 Biology of the Eye, Ocular Surface, Tear Film and Mucin Layer: Functions, Components, and Structures

9.2.1 Anatomy of the Eye

The eyeball (Fig. 9.1) is enveloped by a three layer covering which wraps the internal structures. The innermost layer is the retina, middle is the uveal coat, and the outermost layer is the sclera [8–10]. The sclera is composed of tough fibrous tissue which covers the posterior section of the eyeball and continues into the anterior eye to form the clear transparent cornea [9].

Overall the eye is divided into two segments: the anterior segment and posterior segment.

Anterior Segment The anterior segment of the eye includes structures such as the lens, lachrymal system, iris, aqueous humor, ciliary body, pupil, conjunctiva, and the cornea. The cornea is a five-layered avascular tissue which protects the eye and is considered to be the most innervated tissue in the body [8, 11]. The five layers of the cornea are the epithelium, Bowman's layer, stroma, Descemet's membrane, and the endothelium [8–10, 12]. The epithelium is the anterior-most layer of the cornea consisting of 5–6 layers of cells, joined together by the presence of tight junctions and is able to regenerate following an injury [9, 10, 13]. The Bowman's layer is a cellular layer of the stroma that is not able to regenerate [8, 11]. The stroma is the thickest layer of the cornea, is highly hydrated, and is made up of parallelly arranged collagen fibers which provide the cornea with its transparent properties [8, 14]. The Descemet's membrane is an elastic membrane which covers the endothelium. The single cell layer of endothelial cells helps to maintain corneal clarity and regulate corneal hydration [8, 14].

The conjunctiva is a thin transparent membrane which covers the inside surface of the eyelids and extends onto the anterior surface of the eye to cover the sclera, meeting the corneal epithelium at the limbus. The conjunctiva is vascular and is composed of three layers: epithelium, substantia propria, and the submucosa and is divided into two main regions: the palpebral and bulbar conjunctiva. The surface epithelial cells of the conjunctiva are connected by tight junctions and have scattered goblet cells which produce mucus to lubricate the surface of the eye [9–10]. The substantia propria is a connective tissue containing blood vessels, lymphatics, and nerves and the submucosa attaches the conjunctiva to the underlying sclera [14].

The iris is the colored part of the eye which is located between the cornea and the lens and controls the size of the pupil. The iris has two main layers: the connective tissue rich stroma and the pigmented epithelium. The crystalline lens, located posterior to the iris, is attached to the ciliary body which contains the ciliary muscle that enables it to change its shape to allow for light to be focused on the retina. The lens separates the aqueous and vitreous humor and is composed of three main parts: lens fibers, lens epithelium, and the lens capsule. The aqueous humor is a clear jelly-like fluid that fills the anterior segment of the eye, controls intraocular pressure, removes waste and provides nutrients to the surrounding tissues [8].

The lachrymal system is a drainage system between both the ocular and nasal systems (Fig. 9.2). This system is responsible for three main functions: the secretion, distribution, and collection of tears [9, 15]. The lacrimal gland secretes tears due to the basic need to maintain the tear film, from reflex tearing due to a stimulation such as irritation or temperature, or due to emotional tearing [16]. In healthy individuals, basal tear production ranges from 0.5–2.2 μL/min, but this can increase to 300 μL/min for reflex tearing [9, 17].

Posterior Segment The posterior segment of the eye includes the retina, choroid, sclera, macula, fovea, optic nerve, and the vitreous humor [8]. The vitreous humor is a dense gelatinous substance that fills in the space between the posterior side of the lens and the retina. It was produced by the retina as an embryo and does not replenish [8].

The retina is a thin membrane which is composed of two layers: the outer pigmented epithelium and the inner neuro-epithelium. Overall, the retina is responsible for detecting light focused on the retina and converting it to nerve impulses which are sent through the optic nerve and into the brain. The retina is dense with photosensitive cells called cones and rods [8].

The choroid is located posterior to the retina and the uvea and is responsible for delivering oxygen and nourishment to the retina. It is composed of four layers, is

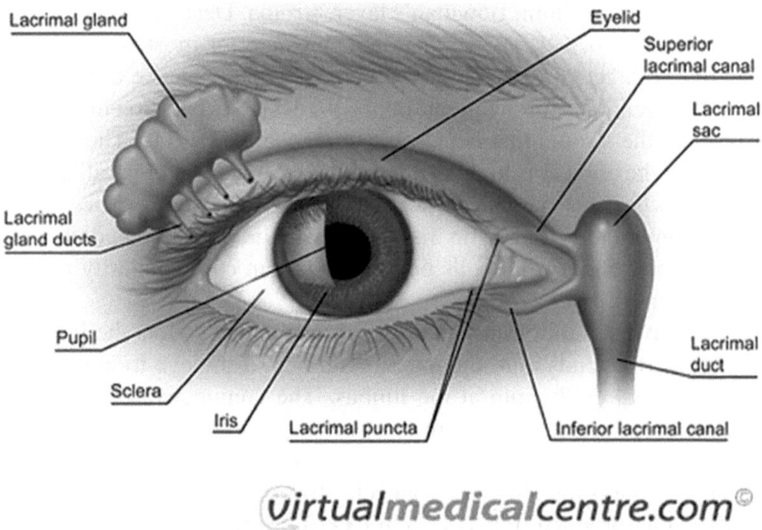

Fig. 9.2 Lacrimal drainage system. (Reprinted with kind permission of Virtual Medical Centre, Osborne Park, Australia, available from URL http://www.virtualmedicalcentre.com/anatomy/the-eye-and-vision/28)

heavily vascularized, pigmented, and contains connective tissue. The sclera is an extension of the cornea which begins at the limbus and continues posteriorly. The sclera's function is mainly protective and covers the bulk of the posterior part of the eyeball in a thick dense fibrous tissue and mucopolysaccharides [18–19]. Not only does this protect the internal sensitive structures but it also provides a site for attachment for the ocular muscles and maintains the shape of the eyeball [8].

9.2.2 Structure of the Tear Film

The tear film is a complex multilayered film that covers the anterior surface of the conjunctiva and cornea. It is thought to provide several unique roles and therefore its composition needs to be tightly regulated. The tear film is broadly described as having five main functions: it traps and washes potentially harmful foreign substances away through blinking; it evens out the tiny blemishes in ocular surface to provide smooth refractive surface; it provides moisture and lubrication for the conjunctival surfaces; it contains necessary gases and nutrients which maintain the health of the cornea; and finally, it contains various immunological and antibacterial agents to protect against ocular invasion and infection [20].

In the last 25 years the understanding of the arrangement of the tear film has undergone some revision [21–22], and much research has been completed analyzing the thickness of the tear film and its layers [23–29], the dynamics and organization of the layers [30–31], as well as the specific components of each layer [32–36]. The most current model of the tear film has moved away from the traditional three

Table 9.1 Tear film layers and functions. (Adapted from [222])

Layer	Origin	Components	Function of the layer
Lipid layer	Meibomian glands	Sterols, fatty acids, glycerides, esters, polar lipids	To prevent evaporation and to provide a barrier [223] To provide a smooth optical surface for the refraction light [224–225] To act as a lubricant to aid the eyelid movement [225] To form a barrier against tear film contamination [226] To provide a surfactant layer between the non-polar lipid layer and the aqueous layer [227] To prevent tear overflow [225]
Aqueous layer	Lacrimal glands	Proteins, lactoferrin, salts, glucose, urea, water	To create a favorable environment for the corneal epithelial cells, carry oxygen and nutrients to and from the cornea, and allow cell movement over the ocular surface [228] To wash away toxic substances and debris during blinking [229] To aid in antimicrobial activity through the tear film proteins (lipocalin, lactoferrin, lysozyme, and IgA) [228–230] Growth factors present in this tear film phase play a significant role in corneal physiology [229]
Mucin layer	Conjunctival goblet cells, Glands of Moll and Krasse	Glycoprotein	To act as a pathogen barrier using the ocular surface glycocalyx [231] Mucin is a lubricant, which allows the eyelid and conjunctiva to move smoothly over each another during blinking and ocular movements [232] Mucus threads protect the conjunctiva and cornea from injury by coating foreign bodies with a slippery mucus [231] Mucus aids in glycocalyx formation and wetting the ocular surface [233] Mucus helps overcome the hydrophobicity of the corneal surface [232]

layer model [37–39] and is currently described to include an outermost non-polar lipid layer, an inner polar lipid layer that contains intercalated proteins, an aqueous phase containing various proteins and gel-forming mucins, and finally a glycocalyx layer bordering on the corneal epithelium [21–22]. Current research on the tear film approximates its thickness to be 3 μm, with decreasing thicknesses present in those individuals with dry eye [23–29].

Just as the tear film as a whole has physiological and structural functions, each of the three broad layers of the tear film (lipid, aqueous, and mucin layers), as described above, have their own unique and critical functions as well (Table 9.1).

Ocular Mucin Conjunctival goblet cells, conjunctival epithelium, and the corneal epithelium are responsible for the secretion of mucin onto the ocular surface. Since the concept of mucosal drug and biopharmaceutical delivery involves an intimate relationship between the mucin and the delivery system, a more in-depth knowledge of mucin is required.

In a healthy individual, it is thought that more than one million goblet cells spread throughout the conjunctival epithelium produce mucins (MUC). Mucins are large glycoproteins which are mainly composed of a protein core, carbohydrates and are well glycosylated [40–41]. There are two main types of mucin: secreted mucins and membrane-associated mucins. Secreted mucins can be further divided into soluble and gel-forming mucins [42–44].

Ocular membrane-associated mucins, such as MUC1 and MUC16, are structured to have short cytoplasmic tails, a heavily glycosylated extracellular domain which can reach the glycocalyx, and a hydrophobic domain which spans the membrane and anchors the mucin. These membrane-associated mucins help create a hydrophilic barrier and may have their own signaling abilities [42–45]. Ocular secreted gel-forming mucins, such as MUC 2, 5AC, and 5B, are the largest glycoproteins, contribute to the viscoelastic properties of mucus, and help trap particles and bacteria [42, 43, 46]. Ocular secreted soluble mucins, such as MUC 7, are the smallest mucins found in the tear film.

The ocular mucus layer is composed of mucin, immunoglobulins, proteins, lipids, urea, salts, glucose, leukocytes, cellular debris, water, and enzymes [10, 46, 47]. Approximately 2.5 μl of mucus is produced every day, and it is thought that the mucus layer is replenished at least once daily [9, 48].

9.3 Ocular Disease, Routes and Delivery Systems for Ocular Drug and Biopharmaceutical Delivery

9.3.1 Ocular Diseases and Conditions Requiring Biopharmaceutical Intervention

The eye is a complex organ and therefore there is a plethora of ocular diseases and conditions that can impact the health and function of the eye and that subsequently may require a range of drugs or biopharmaceuticals administered directly to the eye. The range of ocular conditions that could be treated by ocular biopharmaceutical delivery includes: glaucoma, infections, dry eye, allergies, corneal neovascularization, corneal erosion, inflammation, and macular disorders. Some of these conditions are described below, but there are many other unmentioned ocular conditions that could benefit from biopharmaceutical research and treatments.

It is thought that 1–4 % of patients over 45 years old or over 60 million people worldwide suffer from glaucoma [49–51], a condition where there is decreased aqueous flow which causes increased intraocular pressure and progressive vision loss due to damage to the optic nerve [11]. It is the second leading cause of permanent blindness [52–53]. Glaucoma usually requires lifelong administration of ocular drug-containing eye drops to keep intraocular pressures low and, if this becomes insufficient, surgical intervention may be necessary. However, research on glaucoma gene therapy has been bringing hope of new treatment options. In glaucoma gene therapy, retinal ganglion cells could be targeted to prevent their apoptosis or the trabecular meshwork could be targeted to lower intraocular pressure [54].

Dry eye disease is a complex and multifaceted disease which is broadly defined to cause discomfort, inflammation, ocular surface changes, and alterations in tear film composition and structure [55]. It is thought to impact 7–33 % of people depending on their geographical location; with the USA and Australia having the lowest prevalence and Taiwan and Japan the highest [51, 56–59]. There are two main types of dry eye disease: aqueous-deficient and evaporative dry eye, each with a large list of mechanisms that can trigger the disease [55]. As dry eye is not thought to be curable, most ocular drug treatments are designed to manage the symptoms, usually using artificial tears, anti-inflammatories, and antibiotics [11, 60]. Since dry eye disease can be progressive, with a higher prevalence later in life, and is also linked to autoimmune diseases, these treatments can be required for a substantial period of time [59]. Current research is being conducted to see if viral vector gene transfer and other gene therapies can aid in treatment and diagnosis of lacrimal gland dysfunctions [61–62].

Ocular allergies are present in approximately 20–25 % of the general population and mainly affect the lids and the ocular surface [11, 51]. Sufferers usually have a predisposed allergic response that can be triggered by exposure to insect bites, medication, cosmetics, metals, pollen, dander, fungus, dust, trees, and grasses [11]. Symptoms can include swelling, itching, tearing, corneal erosion, redness, conjunctivitis and watering, and can be treated using a prescription of antihistamines, mast cell stabilizers, steroids, and/or artificial tears [11]. For those with seasonal and perennial allergies, long-term treatment may be necessary; however, new techniques with immunostimulatory oligonucleotides may be able to provide a new means of treatment [63].

Ocular infections can be bacterial, viral, protozoan and fungal in origin, and primarily impact the eyelids, cornea, and conjunctiva. Each of these types of ocular infections has a unique set of risk factors, pathogens, symptoms, and treatments; however, all can cause significant ocular tissue damage throughout the eye and possible blindness if not treated properly or in time. Treatment using anti-infectives, corticosteroids, mydriatics and local anesthetics are common and may be delivered systemically or topically depending on the condition [11]. Gene therapy and delivery of growth factors are also being studied as a treatment possibility [62]. The ocular anatomy is designed to protect against infections such as these and therefore compromised or injured ocular surfaces have a much higher risk of infection than healthy ones [11].

Viral infections, especially the human herpes simplex virus (HSV) are of great concern for patients and practitioners as it is the leading source of viral infection and infectious blindness [64–65]. Its symptoms can range from minor to life-threatening and it can cause other serious conditions including herpetic stromal keratitis (HSK) [64–66]. It is known that treatment with antiviral medications is only able to control the virus during its replication process and therefore is ineffective if the virus is not in this stage of development. Therefore, several different strategies are being explored for alternate treatment and prophylaxis of HSV and HSK including vaccines, peptide treatment, antisense oligonucleotide treatment, and naked DNA encoding cytokines [64, 66, 67].

Corneal neovascularization (CNV) is a condition where new blood vessels grow from the limbus into the cornea which can jeopardize sight [68–70]. CNV can be induced due to a number of different corneal events or conditions including contact lens wear, infection, inflammation, ocular injury, or allergic eye disease [68–70]. It is estimated that 4 % of the American population have CNV, with over 1 million developing it every year [71]. Management of CNV can be quite challenging for practitioners as current drug-based corticosteroid treatment and surgical-based treatments can be not only unsuccessful but may also cause an overabundance of negative ocular conditions [70, 72]. Currently, various gene therapies and treatments delivering antivascular endothelium growth factor (VEGF) are being researched [70, 73, 74].

Macular degeneration is prevalent in 11–28 % of patients over 65 years old [51]. It occurs when a section of the retina, called the macula, which is responsible for crisp central vision, is damaged due to the abnormal growth of deposits or the abnormal growth of choroidal blood vessels. This can cause slight or severe loss of central vision, retinal detachment and blindness [11]. For some types of macular degeneration no treatment is possible, but for others, intravitreal injections of antiangiogenics or surgery may be of some benefit, but these methods have their own set of risks and complications [11]. One possible treatment option for macular degeneration, and other retinal conditions like diabetic retinopathy and retinitis pigmentosa, is to deliver DNA, genes, oligonucleotides, and proteins to the cornea or directly to the retina [75–76].

9.3.2 Routes for Ocular Biopharmaceutical Delivery

There are several routes to deliver drugs and biopharmaceuticals to the targeted part of the eye. The main routes of ocular pharmaceutical administration include topical, systemic/oral, periocular and intravitreal. Each of these routes of delivery has their own list of advantages, challenges, and ocular targets.

Topical applications are most commonly used in the form of eye drops, gels or ointments and are usually used to target the anterior segment, including the various layers of the cornea, conjunctiva, sclera, iris or ciliary body. There are several benefits to this type of ocular delivery including the fact that it is non-invasive, can be administered by the patients themselves, and it can easily target the anterior segment. Despite these benefits, topical applications are very inefficient with less than 5 % of the drug or biopharmaceutical penetrating through the physiological barriers present in the anterior eye. Topical delivery has been used to treat a range of ocular diseases and conditions such as ocular infections, allergies, glaucoma, dry eye, and is still the most popular method for ocular drug delivery but not as popular in biopharmaceutical delivery [1–3].

Systemic delivery, such as oral or intravenous, is very patient compliant and non-invasive, however, is challenged with significant ocular barriers which limits its bioavailability to less than 2 % and can require high dosage concentrations which can cause toxicity [1]. Although this method is used for some ocular drug [77–80] and biopharmaceutical [63, 81] administration, it is not the most popular, efficient or safe route.

Periocular delivery covers a whole range of delivery routes including injections and implants at the peribulbar, retrobulbar, subtenon, and subconjunctival locations. These routes are more invasive and less patient compliant, but are more efficient at delivering drugs and biopharmaceuticals to their target tissue especially if the posterior segment of the eye is desired. These injection sites avoid some of the main barriers to delivery but could cause hemorrhages, among other complications [1–2].

Intravitreal injections or implants are the most invasive type of ocular drug and biopharmaceutical delivery and therefore carry the greatest risk for the patient including hemorrhage, retinal detachment, and cataract development. For these reasons patient compliance is low, however, intravitreal injections deliver the pharmaceutical directly to the retina and vitreous and therefore higher concentrations can be maintained. This is virtually impossible with other ocular delivery routes [58–59].

9.3.3 Drug and Biopharmaceutical Ocular Delivery Systems

There have been many different delivery systems developed for ocular biopharmaceutical and drug delivery including: liquid, particulate, liposomal and niosomal, emulsions, gels, and ocular insert delivery systems [82]. All of these systems have a distinct set of advantages and disadvantages in their use and many can be used for a variety of delivery routes.

Liquid topical applications are a conventional method of ocular drug delivery; however, their low bioavailability has encouraged researchers to incorporate viscosity enhancing agents to increase the residence time on the ocular surface and in the conjunctival cul-de-sac [83–86]. Viscosity enhancing agents alone have had limited success, but once paired with mucoadhesive agents, the liquid system has been shown to increase drug penetration and effectiveness [10, 84, 86–89].

Micelle, liposomal, and niosomal formulations are vesicular drug delivery systems where the drug or biopharmaceutical is trapped within the lipid vesicle and can therefore be carried across cell membranes. Lipid vesicles have been found to be useful for immunology, genetic engineering, and drug delivery [90]. Lipid vesicles are known to be biocompatible, stable, biodegradable, can be synthesized in a variety of sizes and layers, and are usually delivered in a liquid formulation [82]. Despite these advantages, these vesicular systems are known to be expensive, difficult to produce, and have limited drug and biopharmaceutical entrapment [64, 90]. Micelles are vesicles that are composed of a single layer of lipid molecules. Niosomes are non-ionic surfactant and cholesterol vesicles, whereas liposomes are phospholipid and cholesterol-based bilayer vesicles. Both liposomes and niosomes can be utilized to deliver lipophilic and hydrophilic substances, however, niosomes are more stable as they do not undergo phospholipid oxidation or hydrolysis and are more economical [91–96]. Lipoplexes are liposomes or micelles designed to entrap DNA and polyplexes are polymer vesicles that entrap DNA [97].

Emulsions are mixtures of liquids that do not mix together, and thus form dispersions. Conventional emulsions are usually oil droplets in water that are over a micrometer in diameter. Nanoemulsions are usually formed using high energy devices such as high pressure homogenizers to form transparent oil droplets that are smaller than 300 nm in diameter and are kinetically stable but not in a state of equilibrium [98–99]. Microemulsions are micronized droplet dispersions formed spontaneously by combining specific ratios of aqueous, oil, surfactant and cosurfactant phases together to form a thermodynamically equilibrated system [98, 100]. One of the main advantages of an emulsion ocular drug or biopharmaceutical delivery system is that it can provide a slower release rate; however, it can cause blurred vision and possible oil entrapment for the patient [90].

Particulate systems such as microparticles and nanoparticles are desirable due to their potential to deliver drugs and biopharmaceuticals to the anterior eye, posterior eye, and target specific tissues. The particles are synthesized from degradable or non-degradable polymers that entrap, encapsulate or are bonded to a drug and require a distinct manufacturing technique such as homogenization, emulsion technology or supercritical fluid technology [2, 90]. The differences between microparticles and nanoparticles are based on their size; where microparticles are over 1 μm and nanoparticles have a diameter less than 1 μm. Size is particularly important as microparticles too large can cause irritation and foreign body sensation [101]. These small particulate systems have higher patient acceptance and are able to release a drug for a longer period of time, however, they are costly to develop and manufacture and can be easily washed away from the ocular surface without the incorporation of a mucoadhesive polymer [90]. Micro and nanoparticles also have the flexibility to be incorporated into other delivery forms like gels, suspensions, and tablets [102].

In situ gelling systems and gels are specifically designed to increase the residence time of the incorporated drug or biopharmaceutical on the ocular surface and decrease drainage from tearing [103]. In situ gelling systems have the benefit of being a liquid dosage form during instillation but undergo a phase transition into a gel triggered by temperature, pH, UV light or due to the presence of ions [104–105]. It is important that this gel is able to endure the action of blinking and not cause irritation and blurring of vision. Drawbacks to phase transition systems are that the eyelids can become sticky and matted [90].

Ocular or ophthalmic inserts are a group of biodegradable and non-biodegradable delivery systems which include contact lenses, tablets placed in the conjunctival cul-de-sac, collagen shields, punctal plugs, sclera plugs, and intravitreal implants. There are a range of advantages and disadvantages of these systems depending on their ocular location, route of administration required, risk of adverse events, propensity for visual blurring, discomfort and irritation. The one benefit to these types of inserts is that often they can allow for controlled release over a longer period of time which can reduce the need for repeated applications [2, 90].

9.4 Challenges and Barriers with Ocular Biopharmaceutical Delivery

9.4.1 Anatomical Challenges

The anatomy of the eye is complex and is designed to specifically keep out bacteria and harmful substances, not allowing penetration through to the intraocular tissues. Therefore, the greatest barrier to ocular biopharmaceutical or drug delivery is the anatomy of the eye itself. An understanding of the barriers, how they can limit delivery and how they could be exploited is therefore critical to the development of the next generation of drug delivery systems.

Topical formulations applied to the surface of the eye have numerous anatomical barriers that reduce the biopharmaceutical's bioavailability to $< 5\,\%$, such as the tear film, the cornea, and the conjunctiva. The tear fluid volume that resides on the ocular surface is 7–9 μL, whereas most eye drops dispensed have a volume that is approximately 5–6 times this [2]. These large increases in fluid at the ocular surface cause drainage through the nasolacrimal duct, overflow out of the eye and onto the face, reflex blinking, and possibly increased tear secretion if the eye drops contain an irritant [2, 106, 107]. These processes dilute the biopharmaceutical within minutes [108]. The tear film structure itself also acts as a barrier against biopharmaceutical absorption. The mucin forms a hydrophilic layer covering the corneal epithelium and helps remove debris, pathogens and therefore ocular biopharmaceutical formulations from the surface of the eye [109].

If the biopharmaceutical is able to remain on the ocular surface throughout the diluting and draining experienced by the tear film, it then faces its next barrier; the cornea. The corneal surface available for absorption is only about 1 cm^2, which is a fairly small surface when compared to the 17 cm^2 available at the conjunctiva [110]. The numerous layers of the cornea each act as another barrier to biopharmaceutical permeation and delivery. The corneal epithelium is the first line of defense against pathogens that exist in the tear film and therefore represents a significant defense mechanism for foreign substance permeation. Not only does the epithelium have five layers of cells but the cells are joined by tight junctions and gap junctions, which keep the layer very well sealed [2]. Therefore, the biopharmaceuticals must penetrate beyond the epithelium by either penetrating between or through the epithelial cells [2]. The stroma, which is mainly aqueous, can provide resistance for biopharmaceutical penetration depending on the lipophilicity of the biopharmaceutical. The Bowman's and Descemet's membranes are not thought to provide much resistance to biopharmaceutical permeation [2]. Since the endothelium is only a single layer thick with intercellular junctions, it is considered to be significantly more permeable than the epithelium and thus leakier [111–114].

The conjunctiva and sclera provide another route for ocular biopharmaceutical absorption but also another distinct set of barriers. Although the conjunctiva has a larger surface area, the cornea is still thought to be the primary absorption site, except when certain types of biopharmaceuticals are being transported. Since the structure of the conjunctiva is markedly different than the cornea, it has different properties for ocular biopharmaceutical delivery. The epithelium provides a barrier for biopharmaceutical diffusion due to the presence of tight junctions at the surface [115]. Substances crossing this barrier can be absorbed by the circulatory or lymphatic systems present in the substantia propria and substantially reduce their bioavailability to the targeted ocular tissues or they can permeate through the conjunctiva into the sclera and further into the posterior segment of the eye [116–118]. The sclera's avascular collagenous nature is thought to make it less permeable when compared to the conjunctiva but more permeable than the well-sealed cornea [14, 119].

Systemically delivered biopharmaceuticals targeting the anterior or posterior segment of the eye have a new set of barriers to cross; the blood-aqueous barrier and the blood-retinal barrier. The blood-aqueous barrier protects the anterior segment of the eye and has two well-defined layers: the non-pigmented ciliary epithelium and the iris/ciliary body blood vessels endothelium [14]. Both of these layers of cells are well sealed by tight junctions and control the passage of biopharmaceuticals into the aqueous humor and into the posterior segment of the eye [14, 120, 121]. Despite these safeguards, this blood-aqueous barrier is not foolproof due to fenestrated capillaries of the ciliary [122–123].

The blood-retinal barrier is composed of two cell types in the posterior segment of the eye. It is the retinal pigment epithelium and the retinal capillary endothelium that provide this barrier due to their tight junctions and non-fenestrated capillaries [14, 122, 124, 125]. This barrier, if healthy, can be very efficient at restricting penetration to the posterior segment and thus systemic administration could require large doses to be effective, however, this can cause undesired side effects in other parts of the body [14, 126].

9.4.2 Biopharmaceutical Driven Challenges

Another aspect that needs to be taken into account is the characteristics of the biopharmaceutical that is to be delivered to the eye. Hydrophilicity or lipophilicity of the pharmaceutical and its delivery system as well as its size and charge will have an impact regarding its permeability, penetration, and success especially for topically administered biopharmaceuticals.

Hydrophilic and lipophilic biopharmaceuticals and their delivery systems have their own unique set of penetration challenges. Lipophilic biopharmaceutical systems are able to more easily penetrate through the corneal epithelium and endothelium due to their high concentration of cellular lipoidal membranes [2, 127]. In contrast, hydrophilic compounds are more easily able to permeate through the aqueous stroma and the sclera [2, 8, 127]. When comparisons between tissues are made, it is thought that hydrophilic biopharmaceutical systems are more permeable in the conjunctiva than the cornea or sclera [128–129]. The biopharmaceutical systems which have the greatest chance for penetration through the cornea and conjunctiva are amphiphilic biopharmaceutical systems, as they contain both a lipophilic and hydrophilic component [115].

The molecular mass of the biopharmaceutical and its delivery system can also greatly impact its delivery to the targeted tissue [130]. Smaller molecules are able to better penetrate the corneal epithelium, stroma, and the sclera, whereas larger macromolecules are able to penetrate the endothelium and the conjunctiva [8, 12, 131, 132]. Biopharmaceutical system molecule size in the vitreous also has an effect, as large substances have longer residence time but a slower distribution rate [133].

A pH that is too different than the natural tear film, which is approximately 7.4, can cause increased tearing and irritation on instillation. This can dilute the formulation significantly and greatly minimize the bioavailability of the biopharmaceutical [127, 134]. The components in tears also must be taken into effect when formulating a new biopharmaceutical delivery system, as tear film components like buffers and proteins can bind to the biopharmaceutical or change its ionization state, altering its effective concentration and bioavailability [127]. Ionized biopharmaceutical systems have greater difficulty crossing the corneal epithelium and endothelium due to tight junctions [127], however, negatively charged biopharmaceutical systems are more permeable in the sclera as they will not bind to the glycoproteins like positively charged biopharmaceutical systems will [8, 19].

DNA, RNA, and oligonucleotide delivery are known to suffer from poor biological stability due to degradation by nucleases in vivo, thus delivery of these biopharmaceuticals requires the nucleotide sequence to be protected from the environment until it can be delivered [135]. Proteins and peptides are similarly impacted by enzymatic degradation in the conjunctiva and therefore should also be protected [136]. Another issue with the delivery of biopharmaceuticals is that they can be immunogenic and therefore also require a carrier [137].

9.4.3 Patient-Driven Challenges

The anatomical and biopharmaceutical related barriers facing ocular delivery can provide a substantial challenge to scientists developing new ocular biopharmaceutical delivery systems; however, there are other challenges which must be taken into account. The patients requiring these drugs and biopharmaceuticals provide a new set of challenges for scientists as not only do they expect the ocular delivery systems to be effective, they desire the system to cause no local or systemic adverse events, to require few applications, be easy to handle and dispense, cause little to no visual interference, no ocular discomfort or foreign body sensation, no blockage of puncti or canaliculi, be as non-invasive as possible and be inexpensive [2, 138]. Studies have shown that the more instillations or injections required and the more invasive the procedure, the greater the degree of patient non-compliance [2].

Specifically, compliance to eye drops has been extensively studied in glaucoma patients as they require lifelong medication on a daily basis. At its face, the instillation of eye drops seems simple, however, effective instillation and treatment require correct intervals between instillations, the correct number of instillations per day, the correct placement of the drop in the eye, the ability to eject the drop from the bottle, and the ability to control eyelid movements to ensure drop enters the eye [139–140]. All of these things can be difficult for those who are older, have poor control over their hands, and/or have poor memory [139]. It is thought that less than half of glaucoma patients are able to maintain their instillation schedule to keep their intraocular pressures lower [138, 141, 142]. Many patients also do not fully understand the risk of their non-compliance to glaucoma medications until their vision is significantly compromised [143].

Although not all of these barriers can be overcome by the ocular biopharmaceutical delivery system alone, they must be taken into effect if they are to be successful in effectively treating the patient.

9.5 Strategies, Technology and Current Research in Ocular Mucosal Biopharmaceutical Delivery

9.5.1 Topical Delivery

As discussed, there are many routes that can be utilized for ocular biopharmaceutical delivery; however, several of the routes are extremely invasive, especially those targeting the posterior segment of the eye and its corresponding diseases. These invasive systems often include surgically implanted devices or intravitreal injections, which can have detrimental side effects, decreased patient compliance, and limited success [58–59]. For these reasons, there is special interest in ocular biopharmaceutical delivery aimed at targeting the anterior segment of the eye or at transporting the pharmaceutical from the anterior to the posterior segment without the need for invasive methods.

There has been much research aimed at developing new topical methods to deliver biopharmaceuticals for the treatment of a variety of diseases. These topical methods, although they come into contact with the ocular mucosa, are not specifically designed to interact with the mucosal layer to enhance delivery. Within topical biopharmaceutical delivery research, the main themes of research have included the treatment of AMD, glaucoma, HSV, and delivering insulin, peptides, hormones, and vaccines.

Gene Therapy Gene therapy describes a group of therapies targeted at delivering nucleotides, modifying gene translation, modifying gene or protein expression to help treat ophthalmic conditions [144]. These treatments can include the delivery siRNA, antisense oligonucleotides, ribozymes, and aptamers [144]. Gene therapy allows for enhanced targeted and prolonged treatment over conventional drug therapies. Although gene therapy is often thought of as a posterior segment treatment; anterior eye and ocular surface diseases can also benefit from this technology [144–145]. Often corneal gene therapy involves transduction of corneal cells, stromal keratocytes, scleral cell with plasmid DNA via topical application, gene-gun, electroporation or iontophoresis, or vectors (liposome, nanoparticle, dendrimer, viral) [145]. Gene therapy comes with its own set of challenges as genes delivered by viral vectors can elicit unwanted immunogenic responses and non-viral vectors suffer from low transfection rates [62, 146].

In 2001, Noisakran and Carr utilized plasmid DNA which encodes for antiviral type I interferon (IFN), specifically IFN-α1 and compared it with recombinant IFN-αA to treat HSV infection in a mouse model [147]. Topical application of the plasmid DNA with the IFN-α1 transgene in a lipoidal solution was found to give protection against the HSV infection, however, the transgene was found to travel beyond the eye. The recombinant IFN-αA did not protect against HSV infection when administered after infection, but both methods were found to be successful if administered prior to HSV infection [147].

In 2007, Toropainen et al. experimented with liposomal and polymeric non-viral vectors in order to transfect human corneal epithelial cells and thus promote secretion and delivery of proteins to ocular tissues [62]. Topically administered plasmid liposome vectors were also tested in rabbit eyes. The authors found that transfection efficiency was impacted both by the type of carrier used and the stage of cell differentiation. Overall, increased protein secretion was found to persist for several days following ocular surface transfection and therefore could act therapeutically to treat ocular surface diseases [62].

More recently, polyethylenimine conjugated gold nanoparticles (PEI2-GNPs) have been examined as tools for gene transfer using cornea in vitro and rabbit in vivo studies [148]. The PEI2-GNPs were found to penetrate quickly into the cornea, remain within corneal tissue in vivo for over a month and once the GNPs were emptied, they clear naturally from the ocular tissue. Cellular toxicity studies found low toxicity, and ocular assessment of treated animals found no evidence of irritation and only minor immune responses. This study supports evidence that PEI2-GNPs could be a useful vector in corneal gene therapy [148].

In another unique study, investigators studied the use of adeno-associated virus (AAV) vectors to transfer genes to the lacrimal gland as means of treating incurable lacrimal gland dysfunctions such as Sjögren's syndrome and maintain the tear film's health and stability [61]. Using a mouse model, AAV gene transfer was examined and characterized to find that DNA can be delivered safely to the lacrimal gland using AAV vectors, however, topical applications were not as efficient as injections [61].

Oligonucleotide and Aptamer Therapy Oligonucleotides (ON) are short single strands of DNA or RNA and antisense oligonucleotides (AS-ONs) are the complimentary strands of those sequences. AS-ONs are designed to bind to mRNA which ceases their translation and thus protein production [149]. Aptamers are short chains of nucleic acids that can bind to proteins, organisms, and nucleic acids and inhibit gene expression [149–152]. The delivery of nucleic acids can be seen in Fig. 9.3.

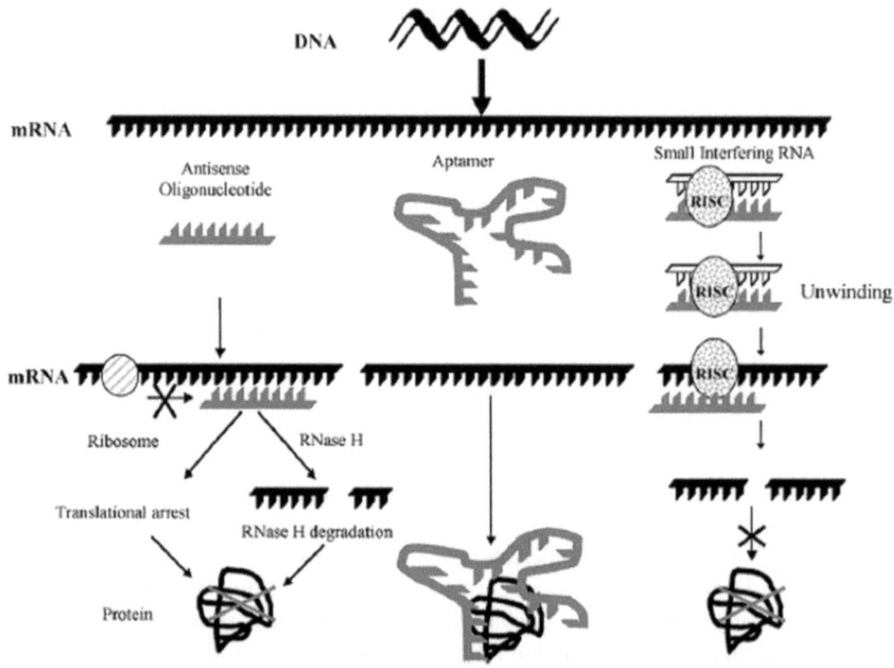

Fig. 9.3 Action of antisense oligonucleotides, aptamers and siRNA. (Reprinted from reference [149], copyright 2006, with permission from Elsevier)

Oligonucleotide therapy has become more common in the literature over the last 15 years and has been tested to treat a variety of ocular conditions [63, 66, 153]. In 2000, Magone et al. utilized topical and systemic administration of immunostimulatory oligonucleotides in an allergic conjunctivitis mouse model [63]. Both administration methods were found to be an effective means of reducing cellular infiltration and hypersensitivity and were found to outperform commercially available corticosteroids [63]. Topical treatment with antisense oligonucleotides (AS-ON) targeting the tumor necrosis factor (TNF-α) was investigated by Wasmuth et al. in 2003 as a method to treat herpetic stromal keratitis from HSV [66]. A variety of in vitro and in vivo mouse studies were completed to find that treatment with the AS-ON-TNF-α decreased the incidence and seriousness of the disease, the number of inflammatory cells, and cytokine expression, while remaining on the corneal surface for more than 10 days [66]. In 2005, immunostimulatory oligodeoxynucleotides along with HSV-1 glycoprotein D were mixed and delivered topically as a vaccine to activate the ocular mucosal immune system [153]. Repeated applications were found to elicit virus-neutralizing immunoglobulins in tears and serum, production of peptides and T cells, and local IFN-γ and IL-2 responses. Overall, this mixed vaccine which is activated by the ocular mucosal immune system has implications to treat diseases such as HSV [153].

In addition to those applications, AS-ONs have also been explored as a treatment in graft rejection and neovascularization [76, 154, 155]. Based on previous successful in vitro and in vivo studies [156–157], Kain et al. have begun Phase I clinical trials to assess GS-101 AS-ON eye-drop treatment for its ability to inhibit expression of the insulin receptor substrate-1 and prevent angiogenesis and neovascularization [154]. Both short-term and long-term experiments were completed on healthy participants and no adverse events or ocular irritancy from GS-101 treatment was found [154]. Aganirsen is an AS-ON that is known to reduce corneal neovascularization by inhibiting insulin receptor substrate (IRS)-1 [76]. In a 2012 study, topical AS-ON therapy with aganirsen on choroidal neovascularization was tested in monkeys and a rat model was utilized to examine its use on oxygen induced retinopathy [76]. The researchers found that topical administration of aganirsen did result in retinal delivery, IRS-1 expression in the retina was found, neovascular lesions in monkeys did diminish, and the incidence of corneal neovascularization decreased significantly in the animal models. All of these results suggest a safe and effective treatment [76]. Also, in 2012, a phosphorodiamidate morpholino AS-ON AVI-5126 was examined in a cornea transplant rat model to determine if it could improve the chances of graft survival following storage [155]. The authors found that storage of the corneas in the AVI-5126 did not degrade the corneas after 1 month, and the AVI-5126 significantly increased graft survival. Posttransplantation application of the AVI-5126 solution was able to enter the cornea and not cause adverse reactions [155].

siRNA Therapy Small interfering RNAs (siRNA) are double strands of nucleic acids that target mRNA and are used to treat angiogenic diseases [149, 158]. Several researchers have examined the use of siRNAs for topical delivery to treat various ocular conditions including hypertension in open angle glaucoma [159–160]. Topically applied siRNAs was recently demonstrated to have potential for ocular hypertension treatment due to glaucoma in vivo in a rabbit and in Phase I clinical trial data [159–160]. In their in vivo rabbit model, ocular hypertension was stimulated using water loading, with the siRNAs targeting β-2-adrengic receptor and carbonic anhydrase IV being topically applied as eye drops in saline. Overall, this siRNA treatment prevented hypertension by keeping intraocular pressure (IOP) low, in most cases, lower than commercially available drugs [159]. In the Phase I trial, the siRNA targeting the β-2-adrengic receptor was topically applied in saline eye drops to healthy participants. Both a single application regime and a multiple application regime were tested and found to cause no ocular surface or iris tolerance or irritancy complications [160].

An interesting study published by Johnson et al. examined the use of cell-penetrating peptides for ocular drug delivery (POD) of a variety of molecules including siRNA, plasmid DNA, and quantum dots [161]. These PODs have protein transduction domains and the $GGG(ARKKAAKA)_4POD$ was found to be able to cross the plasma membrane and enter cells within a few minutes and have inherent antimicrobial properties. Delivery of plasmid DNA and siRNA via POD was found to achieve over 50 % expression and topical administration resulted in corneal epithelial, scleral, and choroidal penetration. Ocular injections were also found to be successful. This technique may be beneficial to treat a variety of ocular conditions [161].

Protein Peptide Therapy The ocular topical delivery of proteins and peptides has gained a lot of interest recently. The delivery of peptide drugs such as insulin, p-nitrophenyl P-cellopentaoside (PNP), luteinizing hormone-releasing hormone (LHRH), calcitonin and thyrotropin-releasing hormone (TRH) has been successfully shown via topical administration to the ocular surface [162–164]. In an early study exploring peptide delivery to the ocular surface, Sasaki et al. examined the in vitro penetration of PNP, TRH, and LHRH in dissected rabbit corneal and conjunctival tissues [162]. It was found that conjunctival membranes had the highest permeability to the peptide drugs when compared to the cornea, TRH had the highest corneal penetration, removing the corneal epithelial layer by scraping increased penetration of the peptide drugs, and as the molecular weight of the peptide drugs increased the permeability decreased but to varying amounts depending on the tissue [162]. In a more recent publication, Kompella et al. explored the possibility of protein delivery using a nanoparticle functionalized with transferrin, a protein, and deslorelin, an LHRH agonist using an ex vivo bovine eye model [165]. Through histology and immunostaining, it was confirmed that the cornea is a significant barrier to nanoparticle delivery, however, the use of deslorelin and transferrin to target surface cells significantly increased permeability through both the corneal and conjunctival layers. This method may provide enhanced targeted topical delivery for biopharmaceuticals [165]. In 2003, Ahsan et al. explored the use of a complex skin moisturizer called sucrose cocoate to improve ocular and nasal peptide delivery of insulin and calcitonin in rats [164]. Increases in insulin and decreases in blood glucose were found following nasal and ocular delivery containing sucrose cocoate, however, nasal delivery had greater insulin bioavailability. When calcitonin was delivered to the ocular surface with the aid of sucrose cocoate, no significant changes in calcitonin and calcium levels were found, however, nasal administration did correlate to significant increases in calcitonin and decreases in calcium levels [164].

In a couple of early studies, Yamamoto et al. and Pillion et al. explored topical delivery of insulin to the ocular surface as an alternative delivery route to the widely used injection delivery system [163, 166]. Yamamoto examined the use of absorption enhancers in topical systemic ocular delivery of insulin to rabbits [163]. Bile salts such as sodium glycocholate (GC), sodium taurocholate (TC), and sodium deosycholate (DC) were tested along with polyoxyethylene-9-lauryl ether (POELE), a surfactant as absorption enhancers for insulin to find that all added promoters significantly increased the bioavailability of insulin. Overall, the use of POELE resulted in the greatest increase in insulin bioavailability, the concentration of GC used did impact its ability to enhance insulin penetration, the nasal mucosa aided in significant systemic absorption, and the conjunctival uptake was dependent on the enhancer [163]. Pillion et al. looked at insulin delivery from topical applied eye drops, using *Quillaja* saponins and derivatives, which are amphipathic quillaic acids extracted from the *Quillaja saponaria* tree, to aid in penetration [166–167]. Rats received both eye and nasal insulin drops with and without a variety of *Quillaja* saponins and their derivatives. It was found that the saponins had a range of abilities to enhance insulin delivery which roughly correlated with their surfactant strength, degree of hemolysis, and maybe even their functional groups. One interesting outcome was that one of the derivatives, DS-1, did not cause an immune response in the animals when the parent saponin, QS-21, did. For this reason, DS-1 could be a beneficial substance for ocular and nasal peptide delivery [166].

Bevacizumab and ranibizumab are two particular monoclonal antibodies that are of great interest for the ocular community due to their ability to neutralize VEGF, inhibit angiogenesis, and treat corneal neovascularization. Various short and long-term studies examining the use of bevacizumab topically found that it is tolerated well by patients with few to no adverse events [73, 74, 168–170], it was able to decrease and inhibit corneal neovascularization [73, 74, 168, 170], and in some cases the treatment effects were seen after bevacizumab use had ended [73]. Bevacizumab has also been found as a useful treatment for persistent pterygium [171] and for ocular surface neovascularization, corneal opacification, and conjunctival injections caused by Stevens–Johnson syndrome [172–173]. Despite the success of topical applications of bevacizumab, some studies have found better intraocular tissue penetration when intravitreal and subconjunctival injections are utilized [174] and possibly give better results with ranibizumab [170].

Throughout the last 20 years, a range of other proteins and peptides have been delivered topically to the eye to treat a range of ocular conditions [64, 175–178]. In 1995, liposomes containing antirat immunoglobulin G (IgG) monoclonal antibodies that target CD4 + were studied as a topical treatment for corneal graft rejection [175]. Treatment with these specialized liposomes were found to significantly decrease the rejection rate and increase graft survival time, most likely due to increased delivery and bioavailability [175]. In 1996, Rafferty et al. examined the use of encapsulated cytokine interleukins (IL-5 and IL-6) and dinitrophenylated bovine serum albumin (DNP-BSA) in a biodegradable microparticle as a means to deliver a controlled release vaccine [176]. The poly(DL-lactide-*co*-glycolide) (PLGA) microparticles were extensively characterized. Both topical and intraperitoneal administration were found to increase serum IgG and tear IgA for a month and a half after the secondary administration and for 5 months after the tertiary administration, indicating that this system as a potential for long-term topical vaccination [176]. In 2006, Cortesi et al. examined the use of cationic phosphatidylcholine-based liposomes used to encapsulate HSV immunogenic peptides (secretory HSV-1 gB1s, DTK1, and DTK2) to assess their protective ability in a rabbit model [64]. This study found that all peptides could be encapsulated into liposomes with a DTK encapsulation efficiency of approximately 30 %, the diameter of all liposomes with encapsulated peptides were close to 320 nm, and the peptides released in a similar fashion as when it is free in solution. Inhibition of the HSV-1 infection was found with neutralization studies, and during animal studies protection against lethal infection and reactivation were found; however, full protection against the disease was not found [64].

More recently, in 2010, Hu et al. examined the use of delivering peptides topically to mucosal membranes of transgenic rabbits to elicit immune responses instead of using a gene-gun delivery technique [177]. Various adjuvants were utilized with the peptide vaccine both nasally and ocularly to see if they could increase immunity to cottontail rabbit papillomavirus. Mucosal delivery with an adjuvant provided partial protection in transgenic rabbits, however, when mucosal delivery was followed by a lower dose DNA vaccine delivered by gene-gun complete protection was obtained [177]. In 2013, Jose et al. investigated the use of a hydrophilic antiviral peptide derived from HIV protein called TAT-Cd_0 to obstruct HSV type 1 infection in mice [178]. Three different concentrations of the peptide were utilized with four different liquid vehicles to find that the symptoms and severity of HSV keratitis was reduced. Overall, the aqueous-based vehicles, higher concentrations of TAT-Cd_0, and administration of the peptide treatment shortly following infection were the most efficacious at significantly lowering vascularization, symptoms, and replication of the virus. Interestingly, the methylcellulose vehicle which had an increased viscosity was not the most efficient [178].

As seen from the topical biopharmaceutical research discussed, the bulk of the biopharmaceutical treatments were targeting anterior eye and ocular surface diseases, as many biopharmaceuticals cannot penetrate to the posterior segment by topical applications alone. In some of these studies, alternate locations of biopharmaceutical delivery were assessed and in some cases found to be more successful. Therefore, it is clear that topical delivery of biopharmaceuticals is not always ideal, but many of these studies have found encouraging results indicating the possibility of future biopharmaceutical treatments to help treat debilitating diseases such as HSV, glaucoma, and Sjögren's syndrome using a topical non-invasive method.

Conventional liquid drops were utilized in a large number of these studies; however, more complex biopharmaceutical delivery systems such as nanoparticles, liposomes, or gels may be able to improve the bioavailability, penetration, ocular comfort, and therefore success of the biopharmaceutical. In addition, many of the publications described have specified that they have delivered biopharmaceuticals, such as proteins and nucleic acids, mucosally, but in fact they were only delivered topically. Although it is true that topical applications will come into contact with the ocular mucosa, there was no specific interaction between the biopharmaceutical or its delivery system and the mucin directly; therefore, mucosal delivery was not achieved. For many of these formulations, non-mucosal topical applications may therefore suffer from reduced residence time, penetration, action, bioavailability, and effectiveness. Incorporation of a mucoadhesive polymer into the biopharmaceutical delivery system could improve the effectiveness of these treatments.

9.5.2 Topical Delivery with Mucoadhesives

One of the main aims currently in topical ocular drug and biopharmaceutical delivery is to increase the residence time of the drug at the ocular surface, to increase drug uptake, diffusion and transport. This can be done by changing the characteristics of the delivery system, for example: by making a liquid system more viscous, but this creates its own set of problems for the patient, including blurred vision, which may jeopardize its success [82]. However, if the delivery system could specifically bind to the ocular surface, this could increase biopharmaceutical residence and release time, decrease the concentration and volume required, and also decrease the frequency of treatment administration. This is the aim of ocular mucosal drug and biopharmaceutical delivery, by which the specific structural components of the mucin layer of the tear film are exploited to become an integral part in ocular delivery, which creates a more effective, efficient, and user friendly ocular delivery system.

In order to efficiently use the ocular mucin layer in ocular biopharmaceutical delivery, there must be an incorporated substance or polymer in the delivery system which will preferentially interact with the ocular mucin. Some of the main mucoadhesive materials utilized are chitosan (CS), alginate, hyaluronic acid (HA), poly(acrylic acid), cellulose, poloxamer, pectin, xanthan gum, gellan gum, carbomer, tamarind seed polysaccharide, and boronic acid [6, 10, 13, 47, 82, 87, 179–181]. One of the main categories of mucoadhesive polymers is polysaccharides. Polysaccharides are composed of individual carbohydrate molecules that are connected together by glycosidic bonds to form long and possibly branching chains [182]. Mucoadhesive polysaccharides include chitosan, cellulose and its derivatives, alginate, hyaluronic acid, xanthan gum, gellan gum, and other gums [82]. Another group of mucoadhesive polymers are the acrylates, which include poly(acrylic acid), Carbopol®, Eudragit®, and polycarbophil [82]. These substances can be high-molecular-weight polymers of acrylic acid with various side chains and functional groups, and their mucoadhesive properties come from their ability to form hydrogen bonds [82, 183]. Thiomers, polymers with a thiol functional group, can form covalent bonds with mucin and thus are a strong group of mucoadhesives [82]. For this reason, other mucoadhesive polymers, such as chitosan, cellulose, and polyacrylic acids have been modified to incorporate a thiol group, thus forming an even stronger mucoadhesive molecule [82]. In addition to these mucoadhesive substances, polyesters, polyethylene oxides, poloxamers, and boronic acids have also been studied [82].

The main factors that impact the success of a mucoadhesive agent are its molecular weight, charge, spatial conformation, functional groups, hydration, pH, and concentration [184–185]. Generally, the better mucoadhesive polymers are those that have hydrophilic functional groups, a larger molecular weight with long polymer chains, anionic or cationic, and are at a pH lower than the pKa [184–187]. These polymers are able to interact with the mucus through both chemical bonds and entanglement, thus forming networks. Of course, beyond the polymer's ability to bond with the mucus, the mucoadhesive polymer must also be non-toxic to the ocular surface, non-irritating as not to induce tearing, should not degrade during storage, and be cost effective [6, 188].

Many of the traditional topical ocular drug and biopharmaceutical delivery systems have been adapted to incorporate a mucoadhesive component including liquid, particulate, liposomal and niosomal, emulsions, gels, and ocular insert delivery systems [82]. Only a handful of researchers have taken advantage of mucoadhesive polymers to interact and increase residence time of the biopharmaceuticals on the ocular surface. Similar to ocular drug delivery, topical delivery of biopharmaceuticals has been expanded to incorporate many different delivery systems, especially liposomes, gels, microparticles, and nanoparticles. These unique systems allow for greater biopharmaceutical uptake by the ocular tissues.

Lipoidal Systems In 2001 and 2007, a poloxamer-based micelle delivery system was investigated for ocular gene delivery [189–190]. The 2001 study examined the use of non-ionic PEO-PPO-PEO micelles in gene therapy to deliver the *LacZ* gene for β-galactosidase using plasmid DNA in both rabbits and mice. Micelle characterization and gene expression were completed to find 160 nm micelles with a-4.4 mV zeta potential that have the highest gene expression after 2–3 days of topical administration. Expression was detected in a variety of rabbit and mouse intraocular tissues and increased presence of plasmid DNA was found with the use of ethylenediaminetetraacetic acid (EDTA) and cytochalasin B permeation enhancers that can open tight junctions [190].

The 2007 study extended this work to examine the use of *LacZ* gene delivery with cornea-specific promoters (keratin 12 and keratocan) in both mouse and rabbit eyes [189]. The three plasmid micelles were characterized by critical micelle concentration, dynamic light scattering, atomic force microscopy, and electrophoresis and transgene expression was examined using real-time polymerase chain reaction (PCR) and staining. Polymeric micelles which contained the various plasmids were round in shape, were 140–190 nm in size, had a zeta potential of approximately −9 to −12 mV, were more stable after freeze-thaw cycles with less conformational changes, and experienced less degradation from nucleases than the plasmid alone. Following a 2-day eye drop treatment with polymer micelles incorporating keratin 12 and keratocan, cornea and stromal gene expression was found in both animal models. The use of EDTA to aid in opening tight junctions was successful; however, the use of arginine-glycine-aspartic acid peptide decreased expression [189].

In 1998, Bochot et al. developed a system to deliver oligonucleotides trapped within liposomes that were incorporated within a poloxamer 407 thermo-sensitive gel [191]. This system was designed to protect the oligonucleotides from degradation, decrease their toxicity, and increase their efficacy [192–194]. For this study, a model oligonucleotide (pdT16) was radiolabeled with^{33}P, pdT16 cholesterol-phospholipid (CH:PC) or cholesterol-phospholipid-1, 2-distearoyl-sn-glycero-3 phosphatidylethanolamine-*N*-(Poly(ethyleneglycol)-2000) (CH:PC:PEG-DSPE) liposomes were prepared, liposomal characterization, and in vitro release studies were completed. The vesicles were found to be 400 nm in diameter, entrapment efficacy of pdT16 within the liposomes was 15 %, PEG-DSPE incorporation into liposomes reduced liposome aggregation and leakage of the oligonucleotide, poloxamer dissolution and pdT16 release were both significantly slowed by increased poloxamer concentrations, and the incorporation of PEG-DSPE into the liposomes reduced the pdT16 release especially in a 2 % poloxamer solution [191]. In a follow-up study, Bochot et al. looked at the ocular distribution of the same model oligonucleotide pdT16 in several different systems including a poloxamer solution, gel, liposome, and the liposomal system embedded into the poloxamer gel after instillation in rabbit eyes [195]. Overall, radioactive tracking of the pdT126 oligonucleotide into ocular tissues found the highest concentrations in the conjunctiva and cornea, but also high concentrations in the sclera. In this study, the poloxamer solution correlated to the highest corneal and conjunctival concentration of pdT16, but the gel system had the highest concentration of pdT16 in the sclera and iris. This study highlighted non-corneal absorption and the possible downfalls of liposomal systems for oligonucleotide delivery in ocular tissues [195].

The delivery of protein bioactives or protein/peptide drugs across mucosal surfaces can be difficult due to hydrolytic breakdown and permeability, and thus most of these drugs are delivered using invasive methods [196]. Hence, some recent research has focused on developing a non-invasive nasal and ocular delivery system for peptides, such as insulin, using multivesicular liposomes coated in chitosan or carbopol [196]. These multivesicular liposomes have a larger aqueous space and are larger in diameter than conventional liposomes, which allow for sustained drug delivery, mucosal penetration, and protection against drug degradation. In this study, the degree of mucoadhesion using rat intestines, in vitro insulin release, enzyme degradation, and pharmacological evaluation of insulin delivery both by ocular and nasal liposome administration were evaluated. The conventional liposomes released insulin for a 24-h period, whereas the incorporation of a mucoadhesive covering was found to sustain insulin delivery for 1 week or more in vitro. Both chitosan and carbopol multivesicular liposomes were found to decrease blood glucose levels for significantly longer than non-coated and conventional liposomes; however, chitosan was found to be more effective than carbopol and the nasal delivery route was slightly better than the ocular route. Overall, the ability to use multivesicular liposomes for the non-invasive ocular delivery of insulin into the blood stream is promising [196].

In a 2008 study, polyethyleneimine (PEI) polyplexes covered in HA were explored as a means to deliver DNA in non-viral gene therapy via CD44 receptors into corneal epithelial cells [197]. The HA coating of DNA/PEI polyplexes, was designed to protect the polyplexes from non-specific binding and to increase ocular residence time via mucoadhesive interactions between the mucosal layer and the HA. Throughout this study, it was found that purification steps significantly decreased the size and zeta potential of the polyplexes, HA coating of polyplexes did not jeopardize the successful condensation of DNA by PEI or the stability of the complexes. Overall, the lowest-molecular-weight HA (< 10 kDa) complexes were the most stable, were associated with higher transfection efficacy, and were found to increase the corneal epithelial cell CD44 receptor uptake [197].

In the last few years, the use of AS-ON has been examined using a poloxamer-based nanoemulsion system to deliver anti-VEGF as a treatment for ocular neovascularization associated with AMD [198–199]. In these studies, various parameters were assessed including nanoemulsion characterization, cellular toxicity and proliferation, in vitro transfection, AS-ON release, AS-ON stability, pharmacokinetic and ocular tissue distribution, and in vivo rat assessments of the ability of the specific AS-ONs to reduce neovascularization. The authors of these papers were able to synthesize a triglyceride containing cationic DOTAP nanoemulsion that was deemed to be non-toxic for two different cell lines, and when this formulation incorporated fluorescently tagged AS-ON specific for VEGFR-2-(17 MER), it was able to enter retinal cells and their nuclei [198]. AS-ON rabbit experiments found that the chosen nanoemulsion was able to shield the AS-ON from degradation for several days and provided enhanced penetration into the retinal cells. Topical penetration of the AS-ON into the anterior and posterior segments of the eye was poor, indicating that the AS-ON would not reach the retina in any meaningful concentrations with the chosen nanoemulsions [198]. When the ability of the AS-ON nanoemulsion to reduce neovascularization was tested in rat and mouse models, both topical and subconjunctival injections were found to significantly reduce corneal neovascularization; however, the subconjunctival injections were found to be slightly better [199].

Gels In 1996, a polyacrylic acid (PAA) based gel was examined for its ability to deliver catalytic RNA or ribozymes to the mouse eye [200]. PAA was used due to its ability to gel at a neutral pH and collapse back into a liquid due to cation presence. Ribozymes were radiolabeled and incorporated into PAA liquid and gel formulations with various concentrations based on formulation pH. Mice were given topical instillations and then were killed, eyes were removed, ribozyme was quantified using a PhosphoImager and autoradiography was completed with tissue sections and staining. Ribozyme retention was found to be significantly greater when incorporated into the PAA and greater accumulation was found when instilled as a liquid and not a gel. Ribozymes were found to penetrate into the outer corneal epithelium after 10 min, after 30 min they were found in the deeper epithelial layers and concentration peaked at that time, and after 3 h ribozyme levels were still detectable. PAA did not degrade the ribozymes or impact their catalytic activity after release from the polymer. It is hypothesized that the PAA not only increases the residence time but also facilitates the ocular retention and penetration [200].

In 2002, Kim et al. described a mucoadhesive system to deliver the polypeptide human epithelial growth factor (rhEGF) to the ocular surface [201]. The successful delivery of rhEGF can be complicated due to its instability and chemical degradation in pharmaceutical formulations which can limit its ability to stimulate corneal epithelial cell proliferation and differentiation. In order to combat this, they stabilized the rhEGF with a hydroxy-β-cyclodextrin (HP-β-CD) which is known to be biocompatible [202–204] and then dispersed it into an in situ-gelling poloxamer which will allow for mucoadhesion on the ocular surface and therefore prolong retention time. Material characterization, stability, gelation temperature, bioadhesive force, in vitro release, in vivo ocular bioavailability in rabbits was analyzed. It was found that higher molar ratios of HP-β-CD increased the stability of the rhEGF, decreased the bioadhesive force, decreased the viscosity, decreased the release of rhEGF, and increased the concentration precorneal retention time following instillation in rabbits [201].

Micro- and Nanoparticles Particulate systems have been a popular choice for ocular delivery of drugs and biopharmaceuticals and much recent research has focused on creating and optimizing new particulate systems. In the past few years, however, only a couple of papers have been published discussing the prospect of mucoadhesive nanoparticles for ocular biopharmaceutical delivery.

Four such papers, published between 2008 and 2011, discuss the prospect of plasmid DNA loaded HA-CS nanoparticles as a possible treatment for chronic ocular surface diseases [205–208]. In the first paper, characterization of the nanoparticles, efficiency of transfecting, and mucosal interactions using confocal microscopy were all examined [205]. Nanoparticles were found to be between 100–215 nm with zeta potentials between − 30 mV (for higher HA ratios) and + 40 mV (CS alone). Typical HA-CS oligomer nanoparticles had a zeta potential of + 20–25 mV. Nanoparticle transfection was analyzed using an in vitro human corneal epithelial cell model and in vivo rabbit instillations and found that low-molecular-weight CS nanoparticles had the highest levels of expression and were able to permeate and assimilate in corneal and conjunctival cells. Not only were they able to reach critical transfection levels but levels were maintained for 1 week. In terms of mucosal interaction, the nanoparticles were found to interact on the ocular surface and integrate into the cells [205].

In the second paper, by de la Fuente et al. [206], the HA-CS nanoparticles loaded with pDNA encoding green fluorescent protein (pEGFP) or pDNA encoding β-galactosidase (pβ-gal) plasmids were examined with both human corneal and conjunctival cell lines and their interaction with the CD44 receptor. Once again the particle size ranged from 100–235 nm and zeta potentials ranged from − 30 to + 30 mV depending on the ratio of HA to CS and the presence of the oligomer. Increased amounts of HA were thought to cause the negative zeta potentials and increased particle size. After immunostaining was completed, both conjunctival and corneal cell lines were found to have CD44 expression. When both cell lines were exposed to increased concentrations of nanoparticles it was found that cell viability suffered, however, these concentrations were higher than the amount needed for transfection. When HA concentration was examined, decreased HA correlated with increased cell death was found. Efficiency of transfection was found to be influenced not only by the polymer ratios of the nanoparticles but also the pDNA itself. When the internalization of the nanoparticles was imaged using confocal microscopy it was found that the nanoparticles entered ocular surface cells endocytically through CD44 receptors. Overall, the HA-CS nanoparticles were found to be a novel method of gene therapy for the ocular surface [206].

In the 2010 study, the HA-CS-based nanoparticles were tested for their in vivo ocular irritancy and uptake on rabbits. Fluoresceinamine-labeled nanoparticles were administered topically to rabbits' eyes, the rabbits were then killed and the tissues were excised. The results showed that the nanoparticle fluorescent signal was found intercellularly in both corneal and conjunctival tissue; however, conjunctival uptake was higher and goblet cells were also found to be stained. For in vivo irritancy testing, the rabbits were given instillations every 30 min for 6 h, and follow-up ocular grading found only a grade 1 ocular discharge but no other signs of discomfort, irritation, edema, or redness. The ocular tissues were also examined post HA-CS nanoparticle instillation for morphological changes; however, no significant changes were found and no alterations in tear film production or drainage were found [208].

In the 2011 study, the HA-CS plasmid nanoparticles were examined specifically to determine the mechanism of internalization within the corneal and conjunctival cell lines and if the plasmids were able to reach the cell nucleus [207]. This information aids in the understanding and optimization of this particulate's pathway,

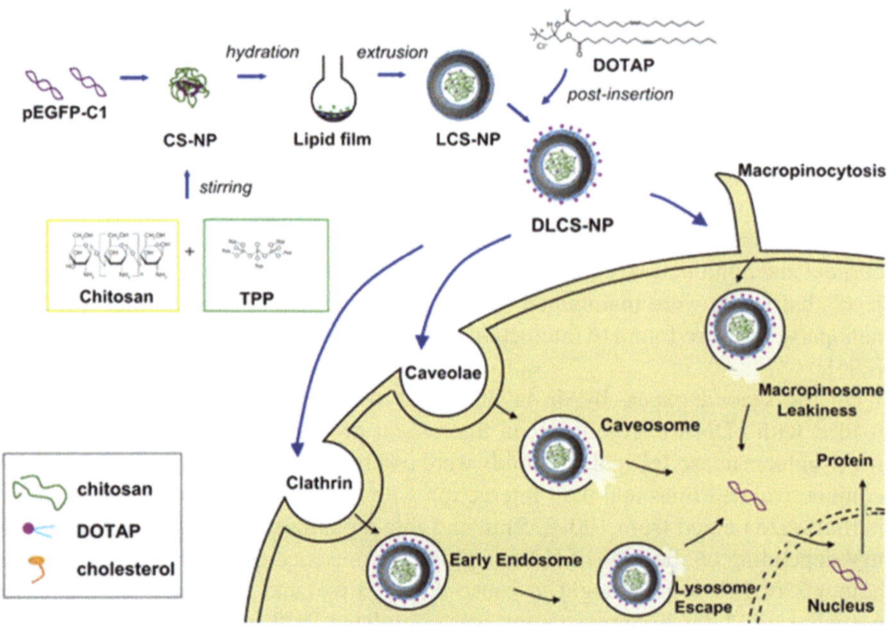

Fig. 9.4 DLCS-NP synthesis, cellular pathways, and entrance into the nucleus. (Reprinted from reference [209], copyright 2012, with permission from Elsevier)

bioavailability, and stability. Once again the nanoparticles were found within both conjunctival and corneal cells and the plasmid was able to leave the nanoparticle structure and make its way into the nucleus itself. Intracellular presence of the nanoparticles did decrease with time and uptake was significantly decreased by lowering temperatures, by blocking HA receptors by numerous methods, and sodium azide presence. Overall, this study found that the HA-CS plasmid nanoparticles were not cytotoxic and that nanoparticle uptake was facilitated by HA receptors [207]. This series of manuscripts outlines a safe, reliable method to deliver genetic material to corneal and conjunctival cells for the treatment of ocular surface disease.

In 2012, a unique nanoparticle system was introduced which has a chitosan plasmid core nanoparticle (CS-NP) that was then encased by a tri-lipid shell (LCS-NP) and then a cationic phospholipid (1,2-dioleoyl-3-trimethylammonium-propane) was inserted to create a dual cationic core-shell liponanoparticle (DLCS-NP) [209]. DLCS-NPs were characterized, examined for their cytotoxicity, cellular uptake, intercellular fate, and in vitro and in vivo transfection. Average plasmid loaded DLCS-NPs were approximately 250 nm in diameter, had zeta potential of + 44 mV, nanoparticles were spherical in shape with the characteristic liposome shell visible by electron microscopy, and no significant cytotoxicity was noted. Cellular uptake for the DLCS-NPs was found to be several folds higher than the CS-NPs or the LCS-NPs and was found to involve several different intercellular pathways (Fig. 9.4) and endolysosome escape. In vitro and in vivo rabbit transfection experiments found significantly greater expression for the DLCS-NPs than the more simply structured NPs. Overall, this type of cationic core-shell liponanoparticle was found to provide increased cellular uptake, and transfection [209]. Although the mucoadhesive polymer chitosan was incorporated into this nanoparticle system, its core location may not necessarily interact with the mucosal layer on the ocular surface, however, if the lipoidal shell opened prior to ocular surface penetration then the chitosan may aid in delivering the plasmid DNA into the cell.

As can be seen from the above research, that only a handful of studies to date have incorporated mucoadhesive polymers into the delivery system of biopharmaceuticals to the eye. In addition to this, it is clear that when a mucoadhesive is incorporated, discussing or analyzing the interactions at the mucosal membrane are not always thought necessary or even a primary goal. It would ultimately be interesting to analyze these mucoadhesive interactions and their specific involvement in the delivery of biopharmaceuticals to the ocular surface as the preliminary evidence suggests that this method has significant promise for increasing cellular uptake and residence time. Despite this, mucoadhesive polymer containing delivery systems have been found to be a valuable, efficient, and a successful technique to topical biopharmaceutical delivery. Topical ocular biopharmaceutical delivery research is currently in its infancy, and as it grows and develops, it could become a powerful and valuable treatment option for many difficult and painful ocular conditions.

9.6 Conclusions and Future Perspectives

This chapter has discussed in detail, ocular anatomy and its role in ocular biopharmaceutical delivery, barriers to biopharmaceutical delivery, routes available for ocular delivery, types of delivery systems that can be manufactured, ocular diseases that are of interest, benefits and types of mucoadhesives that can be incorporated, differences between topical and mucosal biopharmaceutical delivery, and the current state of research in ocular mucosal biopharmaceutical delivery. Overall, the development of effective and safe ocular mucosal biopharmaceutical delivery devices have numerous challenges to overcome and extensive testing to go through before any products are available to patients and doctors; however, it is worth the effort and cost if we are able to help treat, prevent, or cure a disease. Researchers have only scratched the surface of true ophthalmic mucoadhesive biopharmaceutical delivery and there are still many unexplored avenues to pursue in this worldwide battle against ocular disease and blindness.

It is true that ophthalmic mucoadhesive biopharmaceutical research has only just begun and there are infinite areas for future research and development. The sheer number of mucoadhesive polymers, biopharmaceuticals, and ocular delivery systems provide an overwhelming combination of potential mucoadhesive biopharmaceutical delivery devices. To date, much of the research has concentrated on poloxamers, chitosan and HA as the main mucoadhesive polymers [189–194, 196–199, 205–208], however, there are many other mucoadhesive substances that could be successfully incorporated especially: boronic acids, alginates, thiomers, dendrimers, glycoproteins, celluloses, or even mixtures of two or more [6, 7, 82]. Many of these additional mucoadhesive polymers can be successfully incorporated into various ocular drug delivery systems. For example, phenylboronic acid functionalized poly(D, L-lactide)-*b*-dextran nanoparticles [210] and thiolated nanostructured lipid carriers [211] have been found to be promising technologies to improve ocular surface residence time, sustained release characteristics, and penetration of drugs into ocular tissues [210–211].

There are also many other types of delivery systems which could be utilized, combined or improved upon. In situ gelling systems are convenient and easy to use and can be designed to gel according to ocular temperatures, pH or ion presence [103–105]. Contact lenses are being examined for their ability to store and release pharmaceuticals onto the surface of the eye [212–213] and there are a wide range of inserts that could be examined including biodegradable and non-biodegradable punctual plugs [214–215], collagen shields [216–217], and conjunctival cul-de-sac inserts [218–219]. More complexly structured nanoparticles are very promising in that they can combine a lipoidal component, mucoadhesive component, polymer component and be biodegradable [220–221]. In the future, the most powerful technology could actually be the combination of various delivery systems, for example, nanoparticles could easily be delivered within a secondary system such as a contact lens, in situ gelling system, emulsion, or insert depending on the target tissue or target condition.

Just as we can expand to utilize more mucoadhesive polymers and delivery systems, there are many different types of biopharmaceuticals that are yet to be examined for their mucosal ophthalmic applications. As our understanding of DNA and ocular disease grows and we continue to advance our techniques in biotechnology, there is no doubt that many new biologics will be incorporated into modern medicine. However, as we develop this technology there is a continuing need to develop newer and better predictive in vitro assays and animal models to assess the efficacy and safety of these technologies prior to clinical trials.

One of the greatest challenges in the future will be to develop topical mucosal biopharmaceutical treatments that are able to specifically target tissues in the posterior section of the eye to help treat and cure retinal and optic nerve disorders. Development of this type of treatment could eliminate the need for intravitreal injections and other invasive procedures; however, due to the complex ocular anatomy and its defense mechanisms, effective posterior delivery from topical treatment will require significant development and optimization.

References

1. Gaudana R, Ananthula HK, Parenky A, Mitra AK. Ocular drug delivery. AAPS J. 2010;12(3):348–60.
2. Ghate D, Edelhauser HF. Ocular drug delivery. Expert Opin Drug Deliv. 2006;3(2):275–87.

3. Davis JL, Gilger BC, Robinson MR. Novel approaches to ocular drug delivery. Curr Opin Mol Ther. 2004;6(2):195–205.
4. Sandri G, Rossi S, Ferrari F, Bonferoni MC, Zerrouk N, Caramella C. Mucoadhesive and penetration enhancement properties of three grades of hyaluronic acid using porcine buccal and vaginal tissue, Caco-2 cell lines, and rat jejunum. J Pharm Pharmacol. 2004;56(9): 1083–90.
5. Sandri G, Rossi S, Ferrari F, Bonferoni MC, Muzzarelli C, Caramella C. Assessment of chitosan derivatives as buccal and vaginal penetration enhancers. Eur J Pharm Sci. 2004;21 (2–3):351–9.
6. Asane GS, Nirmal SA, Rasal KB, Naik AA, Mahadik MS, Rao YM. Polymers for mucoadhesive drug delivery system: a current status. Drug Dev Ind Pharm. 2008;34(11):1246–66.
7. Khutoryanskiy VV. Advances in mucoadhesion and mucoadhesive polymers. Macromol Biosci. 2011;11(6):748–64.
8. Stjernschantz J, Astin M. Anatomy and physiology of the eye, physiological aspects of ocular drug therapy. In: Edman P, editor. Biopharmaceutics in ocular drug delivery. Boca Raton: CRC; 1993. pp. 1–25.
9. Robinson JC. Ocular anatomy and physiology releavent to ocular drug delivery. In: Mitra AK, editor. Ophthalmic drug delivery systems. New York: Marcel Dekker; 1993. pp. 29–57.
10. Greaves JL, Wilson CG. Treatment of diseases of the eye with mucoadhesive delivery systems. Adv Drug Deliv Rev. 1993;11:349–83.
11. Kanski JJ. Clinical ophthalmology: a systematic approach. 6th edn. New York: Butterworth-Heinemann/Elsevier, Edinburgh; 2007.
12. Sunkara G, Kompella UB. Membrane transport processes in the eye. In: Mitra AK, editor. Ophthalmic drug delivery systems. New York: Marcel Dekker; 2003. pp. 13–58.
13. Kaur IP, Smitha R. Penetration enhancers and ocular bioadhesives: two new avenues for ophthalmic drug delivery. Drug Dev Ind Pharm. 2002;28(4):353–69.
14. Barar J, Javadzadeh AR, Omidi Y. Ocular novel drug delivery: impacts of membranes and barriers. Expert Opin Drug Deliv. 2008;5(5):567–81.
15. Ayub M, Thale AB, Hedderich J, Tillmann BN, Paulsen FP. The cavernous body of the human efferent tear ducts contributes to regulation of tear outflow. Invest Ophthalmol Vis Sci. 2003;44(11):4900–7.
16. Murube J, Murube L, Murube A. Origin and types of emotional tearing. Eur J Ophthalmol. 1999;9:77–84.
17. Mishima S, Gasset A, Klyce SD Jr. Baum JL. Determination of tear volume and tear flow. Invest Ophthalmol. 1966;5(3):264–276.
18. Hamalainen KM, Kananen K, Auriola S, Kontturi K, Urtti A. Characterization of paracellular and aqueous penetration routes in cornea, conjunctiva, and sclera. Invest Ophthalmol Vis Sci. 1997;38(3):627–34.
19. Kim SH, Lutz RJ, Wang NS, Robinson MR. Transport barriers in transscleral drug delivery for retinal diseases. Ophthalmic Res. 2007;39(5):244–54.
20. Craig J. Structure and function of the preocular tear film. In: Korb D, Craig J, Doughty M, Guillon J, Smith G, Tomlinson A, editors. The tear film: structure, function and clinical examination. UK: Butterworth-Heinemann; 2002.
21. Green-Church KB, Butovich I, Willcox M, Borchman D, Paulsen F, Barabino S, Glasgow BJ. The international workshop on meibomian gland dysfunction: report of the subcommittee on tear film lipids and lipid-protein interactions in health and disease. Invest Ophthalmol Vis Sci. 2011;52(4):1979–93.
22. Butovich IA, Millar TJ, Ham BM. Understanding and analyzing meibomian lipids—a review. Curr Eye Res. 2008;33(5):405–20.
23. Azartash K, Kwan J, Paugh JR, Nguyen AL, Jester JV, Gratton E. Pre-corneal tear film thickness in humans measured with a novel technique. Mol Vis. 2011;17:756–67.
24. Wang J, Aquavella J, Palakuru J, Chung S, Feng C. Relationships between central tear film thickness and tear menisci of the upper and lower eyelids. Invest Ophthalmol Vis Sci. 2006;47(10):4349–55.

25. Wang J, Fonn D, Simpson TL, Jones L. Precorneal and pre- and postlens tear film thickness measured indirectly with optical coherence tomography. Invest Ophthalmol Vis Sci. 2003;44(6):2524–8.
26. King-Smith PE, Fink BA, Fogt N. Three interferometric methods for measuring the thickness of layers of the tear film. Optom Vis Sci. 1999;76(1):19–32.
27. King-Smith PE, Fink BA, Fogt N, Nichols KK, Hill RM, Wilson GS. The thickness of the human precorneal tear film: evidence from reflection spectra. Invest Ophthalmol Vis Sci. 2000;41(11):3348–59.
28. King-Smith PE, Fink BA, Hill RM, Koelling KW, Tiffany JM. The thickness of the tear film. Curr Eye Res. 2004;29(4–5):357–68.
29. King-Smith PE, Fink BA, Nichols JJ, Nichols KK, Hill RM. Interferometric imaging of the full thickness of the precorneal tear film. J Opt Soc Am A Opt Image Sci Vis. 2006;23(9):2097–104.
30. Khanal S, Millar TJ. Nanoscale phase dynamics of the normal tear film. Nanomedicine. 2010;6(6):707–13.
31. Kulovesi P, Telenius J, Koivuniemi A, Brezesinski G, Rantamaki A, Viitala T, Puukilainen E, Ritala M, Wiedmer SK, Vattulainen I, Holopainen JM. Molecular organization of the tear fluid lipid layer. Biophys J. 2010;99(8):2559–67.
32. Mudgil P, Torres M, Millar TJ. Adsorption of lysozyme to phospholipid and meibomian lipid monolayer films. Colloids Surf B Biointerfaces. 2006;48(2):128–37.
33. Tsai PS, Evans JE, Green KM, Sullivan RM, Schaumberg DA, Richards SM, Dana MR, Sullivan DA. Proteomic analysis of human meibomian gland secretions. Br J Ophthalmol. 2006;90(3):372–7.
34. Millar TJ, Tragoulias ST, Anderton PJ, Ball MS, Miano F, Dennis GR, Mudgil P. The surface activity of purified ocular mucin at the air-liquid interface and interactions with meibomian lipids. Cornea. 2006;25(1):91–100.
35. Jauhiainen M, Setala NL, Ehnholm C, Metso J, Tervo TM, Eriksson O, Holopainen JM. Phospholipid transfer protein is present in human tear fluid. Biochemistry. 2005;44(22): 8111–6.
36. Saaren-Seppala H, Jauhiainen M, Tervo TM, Redl B, Kinnunen PK, Holopainen JM. Interaction of purified tear lipocalin with lipid membranes. Invest Ophthalmol Vis Sci. 2005;46(10):3649–56.
37. Holly FJ, Lemp MA. Tear physiology and dry eyes. Surv Ophthalmol. 1997;22:69–87.
38. Wolff E. The muco-cutaneous junction of the lid margin and the distribution of the tear fluid. Trans Ophthalmol Soc UK. 1946;66:291–308.
39. Wolff E. The anatomy of the eye and orbit. 4th ed. London: H.K. Lewis; 1954.
40. Hicks SJ, Carrington SD, Kaswan RL, Adam S, Bara J, Corfield AP. Demonstration of discrete secreted and membrane-bound ocular mucins in the dog. Exp Eye Res. 1997;64(4):597–607.
41. Gipson IK, Inatomi T. Mucin genes expressed by the ocular surface epithelium. Prog Retin Eye Res. 1997;16:81–98.
42. Argueso P, Gipson IK. Epithelial mucins of the ocular surface: structure, biosynthesis and function. Exp Eye Res. 2001;73(3):281–9.
43. Danjo Y, Hazlett LD, Gipson IK. C57BL/6 mice lacking Muc1 show no ocular surface phenotype. Invest Ophthalmol Vis Sci. 2000;41(13):4080–4.
44. Gipson IK. The ocular surface: the challenge to enable and protect vision: the Friedenwald lecture. Invest Ophthalmol Vis Sci. 2007;48(10):4390, 4391–8.
45. Singh PK, Hollingsworth MA. Cell surface-associated mucins in signal transduction. Trends Cell Biol. 2006;16(9):467–76.
46. Nichols BA, Chiappino ML, Dawson CR. Demonstration of the mucous layer of the tear film by electron microscopy. Invest Ophthalmol Vis Sci. 1985;26:464–473.
47. Krisnamoorthy R, Mitra AK. Mucoadhesive polymers in ocular drug delivery. In: Mitra AK, editor. Ophthalmic drug delivery systems. New York: Marcel Dekker; 1993. pp. 199–221.
48. Robinson JR, Mlynek GM. Bioadhesive and phase-change polymers for ocular drug delivery. Adv Drug Deliv Rev. 1995;16:45–50.

49. Quigley HA. Number of people with glaucoma worldwide. Br J Ophthalmol. 1996;80(5): 389–93.
50. Quigley HA, Broman AT. The number of people with glaucoma worldwide in 2010 and 2020. Br J Ophthalmol. 2006;90(3):262–7.
51. Clark AF, Yorio T. Ophthalmic drug discovery. Nat Rev Drug Discov. 2003;2(6):448–59.
52. Blomdahl S, Calissendorff BM, Tengroth B, Wallin O. Blindness in glaucoma patients. Acta Ophthalmol Scand. 1997;75(5):589–91.
53. Munier A, Gunning T, Kenny D, O'Keefe M. Causes of blindness in the adult population of the Republic of Ireland. Br J Ophthalmol. 1998;82(6):630–3.
54. Borras T. Gene therapy strategies in glaucoma and application for steroid-induced hypertension, Saudi. J Ophthalmol. 2011;25:353–62.
55. No authors. The definition and classification of dry eye disease: report of the definition and classification subcommittee of the international dry eye workshop. Ocul Surf. 2007;5(2): 75–92.
56. Lin PY, Tsai SY, Cheng CY, Liu JH, Chou P, Hsu WM. Prevalence of dry eye among an elderly Chinese population in Taiwan: the Shihpai eye study. Ophthalmology. 2003;110(6): 1096–101.
57. Shimmura S, Shimazaki J, Tsubota K. Results of a population-based questionnaire on the symptoms and lifestyles associated with dry eye. Cornea. 1999;18(4):408–11.
58. McCarty CA, Bansal AK, Livingston PM, Stanislavsky YL, Taylor HR. The epidemiology of dry eye in Melbourne, Australia. Ophthalmology. 1998;105(6):1114–9.
59. No authors. The epidemiology of dry eye disease: report of the Epidemiology Subcommittee of the International Dry Eye WorkShop. Ocul Surf. 2007;5(2):93–107.
60. No authors. Management and therapy of dry eye disease: report of the management and therapy subcommittee of the international dry eye workshop. Ocul Surf. 2007;5(2):163–78.
61. Rocha EM, Di Pasquale G, Riveros PP, Quinn K, Handelman B, Chiorini JA. Transduction, tropism, and biodistribution of AAV vectors in the lacrimal gland. Invest Ophthalmol Vis Sci. 2011;52(13):9567–72.
62. Toropainen E, Hornof M, Kaarniranta K, Johansson P, Urtti A. Corneal epithelium as a platform for secretion of transgene products after transfection with liposomal gene eyedrops. J Gene Med. 2007;9:208–16.
63. Magone MT, Chan CC, Beck L, Whitcup SM, Raz E. Systemic or mucosal administration of immunostimulatory DNA inhibits early and late phases of murine allergic conjunctivitis. Eur J Immunol. 2000;30(7):1841–50.
64. Cortesi R, Argnani R, Esposito E, Dalpiaz A, Scatturin A, Bortolotti F, Lufinob M, Guerrini R, Cavicchioni G, Incorvaia C, Menegatti E, Manservigi R. Cationic liposomes as potential carriers for ocular administration of peptides with anti-herpetic activity. Int J Pharm. 2006;317:90–100.
65. Bernstein DI, Stanberry LR. Herpes simplex virus vaccines. Vaccine. 1999;17(13–14): 1681–9.
66. Wasmuth S, Bauer D, Yang Y, Steuhl KP, Heiligenhaus A. Topical treatment with antisense oligonucleotides targeting tumor necrosis factor-alpha in herpetic stromal keratitis. Invest Ophthalmol Vis Sci. 2003;44(12):5228–34.
67. Daheshia M, Kuklin N, Kanangat S, Manickan E, Rouse BT. Suppression of ongoing ocular inflammatory disease by topical administration of plasmid DNA encoding IL-10. J Immunol. 1997;159(4):1945–52.
68. Ellenberg D, Azar DT, Hallak JA, Tobaigy F, Han KY, Jain S, Zhou Z, Chang JH. Novel aspects of corneal angiogenic and lymphangiogenic privilege. Prog Retin Eye Res. 2010;29(3): 208–48.
69. Qazi Y, Maddula S, Ambati BK. Mediators of ocular angiogenesis. J Genet. 2009;88(4): 495–515.
70. Mohan RR, Tovey JC, Sharma A, Schultz GS, Cowden JW, Tandon A. Targeted decorin gene therapy delivered with adeno-associated virus effectively retards corneal neovascularization in vivo. PLoS One. 2011;6(10):e26432.

71. Lee P, Wang CC, Adamis AP. Ocular neovascularization: an epidemiologic review. Surv Ophthalmol. 1998;43(3):245–69.
72. Aydin E, Kivilcim M, Peyman GA, Esfahani MR, Kazi AA, Sanders DR. Inhibition of experimental angiogenesis of cornea by various doses of doxycycline and combination of triamcinolone acetonide with low-molecular-weight heparin and doxycycline. Cornea. 2008;27(4):446–53.
73. Cheng SF, Dastjerdi MH, Ferrari G, Okanobo A, Bower KS, Ryan DS, Amparo F, Stevenson W, Hamrah P, Nallasamy N, Dana R. Short-term topical bevacizumab in the treatment of stable corneal neovascularization. Am J Ophthalmol. 2012;154(6):940–8 e941.
74. Dastjerdi MH, Al-Arfaj KM, Nallasamy N, Hamrah P, Jurkunas UV, Pineda R, 2nd, Pavan-Langston D, Dana R. Topical bevacizumab in the treatment of corneal neovascularization: results of a prospective, open-label, noncomparative study. Arch Ophthalmol. 2009;127(4):381–9.
75. Binder C, Read SP, Cashman SM, Kumar-Singh R. Nuclear targeted delivery of macromolecules to retina and cornea. J Gene Med. 2011;13(3):158–70.
76. Cloutier F, Lawrence M, Goody R, Lamoureux S, Al-Mahmood S, Colin S, Ferry A, Conduzorgues JP, Hadri A, Cursiefen C, Udaondo P, Viaud E, Thorin E, Chemtob S. Antiangiogenic activity of aganirsen in nonhuman primate and rodent models of retinal neovascular disease after topical administration. Invest Ophthalmol Vis Sci. 2012;53(3):1195–203.
77. Kampougeris G, Antoniadou A, Kavouklis E, Chryssouli Z, Giamarellou H. Penetration of moxifloxacin into the human aqueous humour after oral administration. Br J Ophthalmol. 2005;89(5):628–31.
78. Coppens M, Versichelen L, Mortier E. Treatment of postoperative pain after ophthalmic surgery. Bull Soc Belge Ophtalmol. 2002;(285):27–32.
79. Rajpal, Srinivas A, Azad RV, Sharma YR, Kumar A, Satpathy G, Velpandian T. Evaluation of vitreous levels of gatifloxacin after systemic administration in inflamed and non-inflamed eyes. Acta Ophthalmol. 2009;87(6):648–52.
80. Chong DY, Johnson MW, Huynh TH, Hall EF, Comer GM, Fish DN. Vitreous penetration of orally administered famciclovir. Am J Ophthalmol. 2009;148(1):38–42 e31.
81. Niccoli L, Nannini C, Benucci M, Chindamo D, Cassara E, Salvarani C, Cimino L, Gini G, Lenzetti I, Cantini F. Long-term efficacy of infliximab in refractory posterior uveitis of Behcet's disease: a 24-month follow-up study. Rheumatology (Oxford). 2007;46(7):1161–4.
82. Ludwig A. The use of mucoadhesive polymers in ocular drug delivery. Adv Drug Deliv Rev. 2005;57(11):1595–639.
83. Lee VHL, Robinson JR. Topical ocular drug delivery: recent developments and future challenges. J Ocul Pharmacol. 1986;2:67–108.
84. Saettone MF, Giannaccini B, Teneggi A, Savigni P, Tellini N. Vehicle effects on ophthalmic bioavailability: the influence of different polymers on the activity of pilocarpine in rabbit and man. J Pharm Pharmacol. 1982;34(7):464–6.
85. Trueblood JH, Rossomondo RM, Wilson LA, Carlton WH. Corneal contact times of ophthalmic vehicles. Evaluation by microscintigraphy. Arch Ophthalmol. 1975;93(2):127–30.
86. Saettone MF, Giannaccini B, Ravecca S, La Marca F, Tota G. Polymer effects on ocular bioavailability—the influence of different liquid vehicles on the mydriatic response of tropicamide in humans and rabbits. Int J Pharm. 1984;25:187–202.
87. Saettone MF, Burgalassi S, Chetoni P. Ocular bioadhesive drug delivery systems. In: Mathiowitz E, Chickering DE, Lehr CM, editors. Bioadhesive drug delivery systems. Fundametnals, novel approaches and development. New York: Marcel Dekker; 1999. pp. 601–40.
88. Saettone MF, Giannaccini B, Guiducci A, Marca F, Tota G. Polymer effects on ocular bioavailability: II. The influence of benzalkonium chloride on the mydriatic response of tropicamide in different polymeric vehicles. Int J Pharm. 1985;25:73–83.
89. Hui HW, Robinson JR. Ocular drug delivery of progesterone using a bioadhesive polymer. Int J Pharm. 1985;26:203–13.

90. Dave V, Sharma S, Yadav S, Paliwal S. Advancement and tribulations in ocular drug delivery. Int J Drug Deliv. 2012;4:1–8.
91. Pham TT, Jaafar-Maalej C, Charcosset C, Fessi H. Liposome and niosome preparation using a membrane contactor for scale-up. Colloids Surf B Biointerfaces. 2012;94:15–21.
92. Kaur IP, Aggarwal D, Singh H, Kakkar S. Improved ocular absorption kinetics of timolol maleate loaded into a bioadhesive niosomal delivery system. Graefes Arch Clin Exp Ophthalmol. 2010;248(10):1467–72.
93. Carafa M, Santucci E, Alhaique F, Coviello T, Murtas E, Riccieri FM, Lucania G, Torrisi MR. Preparation and properties of new unilamellar non-ionic/ionic surfactant vesicles. Int J Pharm. 1998;160:51–9.
94. Kaur IP, Garg A, Singla AK, Aggarwal D. Vesicular systems in ocular drug delivery: an overview. Int J Pharm. 2004;269(1):1–14.
95. Saettone MF, Perini G, Carafa M, Santucci E, Alhaique F. Non-ionic surfactant vesicles as ophthalmic carriers for cyclopentolate—a preliminary evaluation. STP Pharma Sci. 1996;6(1):94–8.
96. Uchegbu IF, Vyas SP. Non-ionic surfactant based vesicles (niosomes) in drug delivery. Int J Pharm. 1998;172(1–2):33–70.
97. Tros de Ilarduya C, Sun Y, Duzgunes N. Gene delivery by lipoplexes and polyplexes. Eur J Pharm Sci. 2010;40(3):159–70.
98. Anton N, Vandamme TF. Nano-emulsions and micro-emulsions: clarifications of the critical differences. Pharm Res. 2011;28(5):978–85.
99. Gallarate M, Chirio D, Bussano R, Peira E, Battaglia L, Baratta F, Trotta M. Development of O/W nanoemulsions for ophthalmic administration of timolol. Int J Pharm. 2013;440(2): 126–34.
100. Kesavan K, Kant S, Singh PN, Pandit JK. Mucoadhesive chitosan-coated cationic microemulsion of dexamethasone for ocular delivery: in vitro and in vivo evaluation. Curr Eye Res. 2013;38(3):342–52.
101. Chiang CH, Tung SM, Lu DW, Yeh MK. In vitro and in vivo evaluation of an ocular delivery system of 5-fluorouracil microspheres. J Ocul Pharmacol Ther. 2001;17(6):545–53.
102. Choy YB, Park JH, McCarey BE, Edelhauser HF, Prausnitz MR. Mucoadhesive microdiscs engineered for ophthalmic drug delivery: effect of particle geometry and formulation on preocular residence time. Invest Ophthalmol Vis Sci. 2008;49(11):4808–15.
103. Lin HR, Sung KC. Carbopol/pluronic phase change solutions for ophthalmic drug delivery. J Control Release. 2000;69(3):379–88.
104. Rathore KS. In-situ gelling ophthalmic drug delivery system: an overview. Int J Pharm Pharm Sci. 2010;2:30–4.
105. Srividya B, Cardoza RM, Amin PD. Sustained ophthalmic delivery of ofloxacin from a pH triggered in situ gelling system. J Control Release. 2001;73(2–3):205–11.
106. Ananthula HK, Vaishya RD, Barot M, Mitra AK. Duane's ophthalmology. In: Tasman W, Jaeger EA, editors. Bioavailability. Philadelphia: Lippincott Williams & Wilkins; 2009.
107. Urtti A, Salminen L. Minimizing systemic absorption of topically administered ophthalmic drugs. Surv Ophthalmol. 1993;37(6):435–56.
108. Sasaki H, Yamamura K, Mukai T, Nishida K, Nakamura J, Nakashima M, Ichikawa M. Enhancement of ocular drug penetration. Crit Rev Ther Drug Carrier Syst. 1999;16(1):85–146.
109. Gipson IK, Argueso P. Role of mucins in the function of the corneal and conjunctival epithelia. Int Rev Cytol. 2003;231:1–49.
110. Watsky MA, Jablonski MM, Edelhauser HF. Comparison of conjunctival and corneal surface areas in rabbit and human. Curr Eye Res. 1988;7(5):483–6.
111. Greaves JL, Wilson CG. Treatment of diseases of the eye with mucoadhesive delivery systems. Adv Drug Deliv Rev. 1993;11:349–83.
112. Huang HS, Schoenwald RD, Lach JL. Corneal penetration behavior of beta-blocking agents II: assessment of barrier contributions. J Pharm Sci. 1983;72:1272–9.
113. Huang HS, Schoenwald RD, Lach JL. Corneal permeation behavior of beta-blocking III: in vitro—in vivo correlations. J Pharm Sci. 1983;72:1279–81.

114. Sunkara GKU. Membrane transport processes in the eye. In: Mitra AK, editor. Ophthalmic drug delivery systems. New York: Marcel Dekker; 2003. pp. 13–58.
115. Saha P, Kim KJ, Lee VH. A primary culture model of rabbit conjunctival epithelial cells exhibiting tight barrier properties. Curr Eye Res. 1996;15(12):1163–9.
116. Lee TW, Robinson JR. Drug delivery to the posterior segment of the eye III: the effect of parallel elimination pathway on the vitreous drug level after subconjunctival injection. J Ocul Pharmacol Ther. 2004;20(1):55–64.
117. Robinson MR, Lee SS, Kim H, Kim S, Lutz RJ, Galban C, Bungay PM, Yuan P, Wang NS, Kim J, Csaky KG. A rabbit model for assessing the ocular barriers to the transscleral delivery of triamcinolone acetonide. Exp Eye Res. 2006;82(3):479–87.
118. Kothuri MK, Pinnamaneni S, Das NG, Das SK. Microparticles and nanoparticles in ocular drug delivery. In: Mitra AK, editor. Ophthalmic drug delivery systems. 2nd edn. New York: Marcel Dekker; 2003. pp. 437–66.
119. Ambati J, Canakis CS, Miller JW, Gragoudas ES, Edwards A, Weissgold DJ, Kim I, Delori FC, Adamis AP. Diffusion of high molecular weight compounds through sclera. Invest Ophthalmol Vis Sci. 2000;41(5):1181–5.
120. Bill A. The blood-aqueous barrier. Trans Ophthalmol Soc UK. 1986;105(Pt 2):149–155.
121. Freddo TF. Shifting the paradigm of the blood-aqueous barrier. Exp Eye Res. 2001;73(5): 581–92.
122. Hornof M, Toropainen E, Urtti A. Cell culture models of the ocular barriers. Eur J Pharm Biopharm. 2005;60(2):207–25.
123. Schlingemann RO, Hofman P, Klooster J, Blaauwgeers HG, Van der Gaag R, Vrensen GF. Ciliary muscle capillaries have blood-tissue barrier characteristics. Exp Eye Res. 1998;66(6):747–54.
124. Maurice DM, Mishima S. Ocular pharmacokinetics. In: Sears ML, editor. Handbook of experimental pharmacology, vol. 69. Berlin: Springer; 1984. pp. 16–119.
125. Gardner TW, Antonetti DA, Barber AJ, Lieth E, Tarbell JA. The molecular structure and function of the inner blood-retinal barrier. Penn State Retina Research Group. Doc Ophthalmol. 1999;97(3–4):229–37.
126. Selvin BL. Systemic effects of topical ophthalmic medications. South Med J. 1983;76(3): 349–58.
127. Achouri D, Alhanout K, Piccerelle P, Andrieu V. Recent advances in ocular drug delivery. Drug Dev Ind Pharm. 2013;39(11):1599–617.
128. Ahmed I, Gokhale RD, Shah MV, Patton TF. Physicochemical determinants of drug diffusion across the conjunctiva, sclera, and cornea. J Pharm Sci. 1987;76(8):583–6.
129. Sasaki H, Igarashi Y, Nagano T, Nishida K, Nakamura J. Different effects of absorption promoters on corneal and conjunctival pernetration of ophthalmic beta-blockers. Pharm Res. 1995;12(8):1146–50.
130. El Sanharawi M, Kowalczuk L, Touchard E, Omri S, Kozak Y de, Behar-Cohen F. Protein delivery for retinal diseases: from basic considerations to clinical applications. Prog Retin Eye Res. 2010;29(6):443–65.
131. Geroski DH, Edelhauser HF. Transscleral drug delivery for posterior segment disease. Adv Drug Deliv Rev. 2001;52(1):37–48.
132. Rabinovich-Guilatt L, Couvreur P, Lambert G, Dubernet C. Cationic vectors in ocular drug delivery. J Drug Target. 2004;12(9–10):623–33.
133. Mitra AK, Anand BS, Duvvuri S. Drug delivery to the eye. In: Fischbarg J, editor. The biology of the eye. New York: Academic; 2006. pp. 307–51.
134. Shell JW. Pharmacokinetics of topically applied ophthalmic drugs. Surv Ophthalmol. 1982;26(4):207–18.
135. Das SK, Miller KJ. Gene, oligonucleotide, and ribozyme therapy in the eye. In: Mitra AK, editor. Ophthalmic drug delivery systems. New York: Marcell Dekker; 1993. pp. 609–57.
136. Hayakawa E, Chien DS, Inagaki K, Yamamoto A, Wang W, Lee VH. Conjunctival penetration of insulin and peptide drugs in the albino rabbit. Pharm Res. 1992;9(6):769–75.

137. Porteus MH, Connelly JP, Pruett SM. A look to future directions in gene therapy research for monogenic diseases. PLoS Genet. 2006;2(9):e133.
138. Schwartz GF, Quigley HA. Adherence and persistence with glaucoma therapy. Surv Ophthalmol. 2008;53(Suppl 1):S57–68.
139. Lavik E, Kuehn MH, Kwon YH. Novel drug delivery systems for glaucoma. Eye (Lond). 2011;25(5);578–86.
140. Winfield AJ, Jessiman D, Williams A, Esakowitz L. A study of the causes of non-compliance by patients prescribed eyedrops. Br J Ophthalmol. 1990;74(8):477–80.
141. Sleath B, Robin AL, Covert D, Byrd JE, Tudor G, Svarstad B. Patient-reported behavior and problems in using glaucoma medications. Ophthalmology. 2006;113(3):431–6.
142. Rotchford AP, Murphy KM. Compliance with timolol treatment in glaucoma. Eye. 1998;12:234–6.
143. Ashburn FS Jr. Goldberg I, Kass MA. Compliance with ocular therapy. Surv Ophthalmol. 1980;24(4):237–248.
144. Del Amo EM, Urtti A. Current and future ophthalmic drug delivery systems. A shift to the posterior segment. Drug Discov Today. 2008;13(3–4):135–43.
145. Williams KA, Coster DJ. Gene therapy for diseases of the cornea—a review. Clin Experiment Ophthalmol. 2010;38(2):93–103.
146. Bertelmann E, Ritter T, Vogt K, Reszka R, Hartmann C, Pleyer U. Efficiency of cytokine gene transfer in corneal endothelial cells and organ-cultured corneas mediated by liposomal vehicles and recombinant adenovirus. Ophthalmic Res. 2003;35(2):117–24.
147. Noisakran S, Carr DJ. Topical application of the cornea post-infection with plasmid DNA encoding interferon-alpha1 but not recombinant interferon-alphaA reduces herpes simplex virus type 1-induced mortality in mice. J Neuroimmunol. 2001;121(1–2):49–58.
148. Sharma A, Tandon A, Tovey JC, Gupta R, Robertson JD, Fortune JA, Klibanov AM, Cowden JW, Rieger FG, Mohan RR. Polyethylenimine-conjugated gold nanoparticles: gene transfer potential and low toxicity in the cornea. Nanomedicine. 2011;7(4):505–13.
149. Fattal E, Bochot A. Ocular delivery of nucleic acids: antisense oligonucleotides, aptamers and siRNA. Adv Drug Deliv Rev. 2006;58(11):1203–23.
150. Nimjee SM, Rusconi CP, Sullenger. BA. An emerging class of therapeutics. Annu Rev Med. 2005;56:555–83.
151. Proske D, Blank M, Buhmann R, Resch A. Aptamers—basic research, drug development, and clinical applications. Appl Microbiol Biotechnol. 2005;69(4):367–74.
152. Pestourie C, Tavitian B, Duconge F. Aptamers against extracellular targets for in vivo applications. Biochimie. 2005;87(9–10):921–30.
153. Nesburn AB, Ramos TV, Zhu X, Asgarzadeh H, Nguyen V, BenMohamed L. Local and systemic B cell and Th1 responses induced following ocular mucosal delivery of multiple epitopes of herpes simplex virus type 1 glycoprotein D together with cytosine-phosphate-guanine adjuvant. Vaccine. 2005;23(7):873–83.
154. Kain H, Goldblum D, Geudelin B, Thorin E, Beglinger C. Tolerability and safety of GS-101 eye drops, an antisense oligonucleotide to insulin receptor substrate-1: a 'first in man' phase I investigation. Br J Clin Pharmacol. 2009;68(2):169–73.
155. Hosseini A, Lattanzio FA Jr. Samudre SS, DiSandro G, Sheppard JD Jr., Williams PB. Efficacy of a phosphorodiamidate morpholino oligomer antisense compound in the inhibition of corneal transplant rejection in a rat cornea transplant model. J Ocul Pharmacol Ther. 2012;28(2):194–201.
156. Al-Mahmood S, Colin S, Farhat N, Thorin E, Steverlynck C, Chemtob S. Potent in vivo antiangiogenic effects of GS-101 (5′-TATCCGGAGGGCTCGCCATGCTGCT-3′), an antisense oligonucleotide preventing the expression of insulin receptor substrate-1. J Pharmacol Exp Ther. 2009;329(2):496–504.
157. Andrieu-Soler C, Berdugo M, Doat M, Courtois Y, BenEzra D, Behar-Cohen F. Downregulation of IRS-1 expression causes inhibition of corneal angiogenesis. Invest Ophthalmol Vis Sci. 2005;46(11):4072–8.

158. Hadj-Slimane R, Lepelletier Y, Lopez N, Garbay C, Raynaud F. Short interfering RNA (siRNA), a novel therapeutic tool acting on angiogenesis. Biochimie. 2007;89(10):1234–44.
159. Jimenez A, Mediero A, Loma P, Pintor J, Peral A, Gónzalez V. Efficacy of topically administered siRNAs in glaucoma treatment: in vivo results in hypertensive model. Invest Ophthalmol Vis Sci. 2009;50:E-Absract 4054.
160. Ruz V, Moreno-Montaés J, Sadaba B, González V, Jiménez AI. Phase I study with a new siRNA: SYL040012. Tolerance and effect on intraocular pressure. Invest Ophthalmol Vis Sci. 2011;52:E-Abstract 223.
161. Johnson LN, Cashman SM, Kumar-Singh R. Cell-penetrating peptide for enhanced delivery of nucleic acids and drugs to ocular tissues including retina and cornea. Mol Ther. 2008;16(1):107–14.
162. Sasaki H, Ichikawa M, Yamamura K, Nishida K, Nakamura J. Ocular membrane permeability of hydrophilic drugs for ocular peptide delivery. J Pharm Pharmacol. 1997;49(2):135–9.
163. Yamamoto A, Luo AM, Dodda-Kashi S, Lee VH. The ocular route for systemic insulin delivery in the albino rabbit. J Pharmacol Exp Ther. 1989;249(1):249–55.
164. Ahsan F, Arnold JJ, Meezan E, Pillion DJ. Sucrose cocoate, a component of cosmetic preparations, enhances nasal and ocular peptide absorption. Int J Pharm. 2003;251(1–2):195–203.
165. Kompella UB, Sundaram S, Raghava S, Escobar ER. Luteinizing hormone-releasing hormone agonist and transferrin functionalizations enhance nanoparticle delivery in a novel bovine ex vivo eye model. Mol Vis. 2006;12:1185–98.
166. Pillion DJ, Amsden JA, Kensil CR, Recchia J. Structure-function relationship among *Quillaja* saponins serving as excipients for nasal and ocular delivery of insulin. J Pharm Sci. 1996;85(5):518–24.
167. Kensil CR, Soltysik S, Patel U, Marciani DJ. Structure-Function Relationship in Adjuvants from *Quillaja Saponaria* Molina. Vaccines. 1992;92:35–40.
168. Koenig Y, Bock F, Horn F, Kruse F, Straub K, Cursiefen C. Short- and long-term safety profile and efficacy of topical bevacizumab (Avastin) eye drops against corneal neovascularization. Graefes Arch Clin Exp Ophthalmol. 2009;247(10):1375–82.
169. Bock F, Onderka J, Rummelt C, Dietrich T, Bachmann B, Kruse FE, Schlotzer-Schrehardt U, Cursiefen C. Safety profile of topical VEGF neutralization at the cornea. Invest Ophthalmol Vis Sci. 2009;50(5):2095–102.
170. Stevenson W, Cheng SF, Dastjerdi MH, Ferrari G, Dana R. Corneal neovascularization and the utility of topical VEGF inhibition: ranibizumab (Lucentis) vs bevacizumab (Avastin). Ocul Surf. 2012;10(2):67–83.
171. Wu PC, Kuo HK, Tai MH, Shin SJ. Topical bevacizumab eyedrops for limbal-conjunctival neovascularization in impending recurrent pterygium. Cornea. 2009;28(1):103–4.
172. Uy HS, Yu EN, Sua AS. Histologic findings of bevacizumab-treated human conjunctiva in Stevens-Johnson syndrome. Cornea. 2011;30(11):1273–6.
173. Uy HS, Chan PS, Ang RE. Topical bevacizumab and ocular surface neovascularization in patients with Stevens-Johnson syndrome. Cornea. 2008;27(1):70–3.
174. Nomoto H, Shiraga F, Kuno N, Kimura E, Fujii S, Shinomiya K, Nugent AK, Hirooka K, Baba T. Pharmacokinetics of bevacizumab after topical, subconjunctival, and intravitreal administration in rabbits. Invest Ophthalmol Vis Sci. 2009;50(10):4807–13.
175. Pleyer U, Milani JK, Dukes A, Chou J, Lutz S, Ruckert D, Thiel HJ, Mondino BJ. Effect of topically applied anti-CD4 monoclonal antibodies on orthotopic corneal allografts in a rat model. Invest Ophthalmol Vis Sci. 1995;36(1):52–61.
176. Rafferty DE, Elfaki MG, Montgomery PC. Preparation and characterization of a biodegradable microparticle antigen/cytokine delivery system. Vaccine. 1996;14(6):532–8.
177. Hu J, Cladel N, Balogh K, Christensen N. Mucosally delivered peptides prime strong immunity in HLA-A2.1 transgenic rabbits. Vaccine. 2010;28(21):3706–13.
178. Jose GG, Larsen IV, Gauger J, Carballo E, Stern R, Brummel R, Brandt CR. A cationic peptide, TAT-Cd degrees, inhibits herpes simplex virus type 1 ocular infection in vivo. Invest Ophthalmol Vis Sci. 2013;54(2):1070–9.

179. Le Bourlais CA, Treupel-Acar L, Rhodes CT, Sado PA, Leverge R. New ophthalmic drug delivery systems. Drug Dev Ind Pharm. 1995;21:19–59.
180. Le Bourlais CA, Acar L, Zia H, Sado PA, Needham T, Leverge R. Ophthalmic drug delivery systems—recent advances. Prog Retin Eye Res. 1998;17:33–58.
181. Lee JW, Park JH, Robinson JR. Bioadhesive-based dosage forms: the next generation. J Pharm Sci. 2000;89(7):850–66.
182. Garrett R, Grisham CM. Biochemistry. Brooks Cole. 2012.
183. Leung SS, Robinson JR. The contribution of anionic polymer structural features to mucoadhesion. J Control Release. 1988;5:223–31.
184. Shaikh R, Raj Singh TR, Garland MJ, Woolfson AD, Donnelly RF. Mucoadhesive drug delivery systems. J Pharm Bioallied Sci. 2011;3(1):89–100.
185. Maniyar AH, Patil RM, Kale MT, Jain DK, Baviskar DT. A new polymeric controlled drug delivery. Int J Pharm Sci Rev Res. 2011;8(2):54–60.
186. Mortazavi SA, Smart JD. Factors influencing gel-strengthening at the mucoadhesive-mucus interface. J Pharm Pharmacol. 1994;46(2):86–90.
187. Riley RG, Smart JD, Tsibouklis J, Dettmar PW, Hampson F, Davis JA, Kelly G, Wilber WR. An investigation of mucus/polymer rheological synergism using synthesised and characterised poly(acrylic acid)s. Int J Pharm. 2001;217(1–2):87–100.
188. Patil SB, Murthy RSR, Mahajan HS, Wagh RD, Gattani SG. Means of improving drug delivery. Pharm Times. 2006;38(4):25–30.
189. Tong YC, Chang SF, Liu CY, Kao WW, Huang CH, Liaw J. Eye drop delivery of nano-polymeric micelle formulated genes with cornea-specific promoters. J Gene Med. 2007;9(11):956–66.
190. Liaw J, Chang SF, Hsiao FC. In vivo gene delivery into ocular tissues by eye drop of poly(ethylene oxide)-poly(propylene oxide)-poly(ethylene oxide) (PEO-PPO-PEO) polymeric micelles. Gene Ther. 2001;8:999–1004.
191. Bochot A, Fattal E, Gulik A, Couarraze G, Couvreur P. Liposomes dispersed within a thermosensitive gel: a new dosage form for ocular delivery of oligonucleotides. Pharm Res. 1998;15(9):1364–9.
192. Ropert C, Malvy C, Couvreur P. Inhibition of the Friend retrovirus by antisense oligonucleotides encapsulated in liposomes: mechanism of action. Pharm Res. 1993;10(10):1427–33.
193. Pleyer U, Lutz S, Jusko WJ, Nguyen KD, Narawane M, Ruckert D, Mondino BJ, Lee VH, Nguyen K. Ocular absorption of topically applied FK506 from liposomal and oil formulations in the rabbit eye. Invest Ophthalmol Vis Sci. 1993;34(9):2737–42.
194. Tremblay C, Barza M, Szoka F, Lahav M, Baum J. Reduced toxicity of liposome-associated amphotericin B injected intravitreally in rabbits. Invest Ophthalmol Vis Sci. 1985;26(5):711–8.
195. Bochot A, Mashhour B, Puisieux F, Couvreur P, Fattal E. Comparison of the ocular distribution of a model oligonucleotide after topical instillation in rabbits of conventional and new dosage forms. J Drug Target. 1998;6(4):309–13.
196. Jain AK, Chalasani KB, Khar RK, Ahmed FJ, Diwan PV. Muco-adhesive multivesicular liposomes as an effective carrier for transmucosal insulin delivery. J Drug Target. 2007;15(6):417–27.
197. Hornof M, de la Fuente M, Hallikainen M, Tammi RH, Urtti A. Low molecular weight hyaluronan shielding of DNA/PEI polyplexes facilitates CD44 receptor mediated uptake in human corneal epithelial cells. J Gene Med. 2008;10(1):70–80.
198. Hagigit T, Abdulrazik M, Orucov F, Valamanesh F, Hagedorn M, Lambert G, Behar-Cohen F, Benita S. Topical and intravitreous administration of cationic nanoemulsions to deliver antisense oligonucleotides directed towards VEGF KDR receptors to the eye. J Control Release. 2010;145(3):297–305.
199. Hagigit T, Abdulrazik M, Valamanesh F, Behar-Cohen F, Benita S. Ocular antisense oligonucleotide delivery by cationic nanoemulsion for improved treatment of ocular neovascularization: an in-vivo study in rats and mice. J Control Release. 2012;160(2):225–31.

200. Ayers D, Cuthbertson JM, Schroyer K, Sullivan SM. Polyacrylic acid mediated ocular delivery of ribozymes. J Control Release. 1996;8:167–75.
201. Kim EY, Gao ZG, Park JS, Li H, Han K. rhEGF/HP-beta-CD complex in poloxamer gel for ophthalmic delivery. Int J Pharm. 2002;233(1–2):159–67.
202. Loftssona T, Jarvinen T. Cyclodextrins in ophthalmic drug delivery. Adv Drug Deliv Rev. 1999;36(1):59–79.
203. Loftsson T, Stefansson E. Effect of cyclodextrins on topical drug delivery to the eye. Drug Dev Ind Pharm. 1997;23(5):473–81.
204. Rajewski RA, Stella VJ. Pharmaceutical applications of cyclodextrins. 2. In vivo drug delivery. J Pharm Sci. 1996;85(11):1142–69.
205. de la Fuente M, Seijo B, Alonso MJ. Bioadhesive hyaluronan-chitosan nanoparticles can transport genes across the ocular mucosa and transfect ocular tissue. Gene Ther. 2008;15(9):668–76.
206. de la Fuente M, Seijo B, Alonso MJ. Novel hyaluronic acid-chitosan nanoparticles for ocular gene therapy. Invest Ophthalmol Vis Sci. 2008;49(5):2016–24.
207. Contreras-Ruiz L, de la Fuente M, Parraga JE, Lopez-Garcia A, Fernandez I, Seijo B, Sanchez A, Calonge M, Diebold Y. Intracellular trafficking of hyaluronic acid-chitosan oligomer-based nanoparticles in cultured human ocular surface cells. Mol Vis. 2011;17:279–90.
208. Contreras-Ruiz L, de la Fuente M, Garcia-Vazquez C, Saez V, Seijo B, Alonso MJ, Calonge M, Diebold Y. Ocular tolerance to a topical formulation of hyaluronic acid and chitosan-based nanoparticles. Cornea. 2010;29(5):550–8.
209. Jiang M, Gan L, Zhu CL, Dong Y, Liu JP, Gan Y. Cationic core-shell liponanoparticles for ocular gene delivery. Biomaterials. 2012;33(30):7621–30.
210. Liu S, Jones L, Gu FX. Development of mucoadhesive drug delivery system using phenylboronic acid functionalized poly(D, L-lactide)-b-dextran nanoparticles. Macromol Biosci. 2012;12(12):1622–6.
211. Shen J, Deng Y, Jin X, Ping Q, Su Z, Li L. Thiolated nanostructured lipid carriers as a potential ocular drug delivery system for cyclosporine A: improving in vivo ocular distribution. Int J Pharm. 2010;402(1–2):248–53.
212. Bengani LC, Hsu KH, Gause S, Chauhan A. Contact lenses as a platform for ocular drug delivery. Expert Opin Drug Deliv. 2013.
213. Guzman-Aranguez A, Colligris B, Pintor J. Contact lenses: promising devices for ocular drug delivery. J Ocul Pharmacol Ther. 2013;29(2):189–99.
214. Gupta C, Chauhan A. Ophthalmic delivery of cyclosporine A by punctal plugs. J Control Release. 2011;150(1):70–6.
215. Chee SP. Moxifloxacin punctum plug for sustained drug delivery. J Ocul Pharmacol Ther. 2012;28(4):340–9.
216. Bucolo C, Mangiafico S, Spadaro A. Methylprednisolone delivery by Hyalobend corneal shields and its effects on rabbit ocular inflammation. J Ocul Pharmacol Ther. 1996;12(2):141–9.
217. Willoughby CE, Batterbury M, Kaye SB. Collagen corneal shields. Surv Ophthalmol. 2002;47(2):174–82.
218. Sasaki H, Nagano T, Sakanaka K, Kawakami S, Nishida K, Nakamura J, Ichikawa N, Iwashita J, Nakamura T, Nakashima M. One-side-coated insert as a unique ophthalmic drug delivery system. J Control Release. 2003;92(3):241–7.
219. Gupta C, Chauhan A. Drug transport in HEMA conjunctival inserts containing precipitated drug particles. J Colloid Interface Sci. 2010;347(1):31–42.
220. Zhang L, Chan JM, Gu FX, Rhee JW, Wang AZ, Radovic-Moreno AF, Alexis F, Langer R, Farokhzad OC. Self-assembled lipid–polymer hybrid nanoparticles: a robust drug delivery platform. ACS Nano. 2008;2(8):1696–702.
221. Shi J, Xiao Z, Votruba AR, Vilos C, Farokhzad OC. Differentially charged hollow core/shell lipid-polymer-lipid hybrid nanoparticles for small interfering RNA delivery. Angew Chem Int Ed Engl. 2011;50(31):7027–31.

222. Lorentz H. Modeling in vitro lipid deposition on silicone hydrogel and conventional hydrogel contact lens materials Waterloo. University of Waterloo, PhD Thesis. 2012.
223. Craig JP, Tomlinson A. Importance of the lipid layer in human tear film stability and evaporation. Optom Vis Sci. 1997;74(1):8–13.
224. Driver PJ, Lemp MA. Meibomian gland dysfunction. Surv Ophthalmol. 1996;40(5):343–367.
225. McCulley JP, Shine WE. The lipid layer: the outer surface of the ocular surface tear film. Biosci Rep. 2001;21(4):407–18.
226. Tiffany JM. The lipid secretion of the meibomian glands. Adv Lipid Res. 1987;22:1–62.
227. Nagyova B, Tiffany JM. Components responsible for the surface tension of human tears. Curr Eye Res. 1999;19(1):4–11.
228. Hodges RR, Dartt. DA. Physiology and biochemistry of the tear film. In: Kracmer JH, Mannis MJ, Holland EJ, editors. Cornea. Philadelphia: Lippincott Williams & Wilkins; 2005. pp. 577–602.
229. Flanagan JL, Willcox MD. Role of lactoferrin in the tear film. Biochimie. 2009;91(1):35–43.
230. Glasgow BJ, Marshall G, Gasymov OK, Abduragimov AR, Yusifov TN, Knobler CM. Tear lipocalins: potential lipid scavengers for the corneal surface. Invest Ophthalmol Vis Sci. 1999;40(13):3100–7.
231. Corfield AP, Carrington SD, Hicks SJ, Berry M, Ellingham R. Purification, metabolism and functions. Prog Retin Eye Res. 1997;16(4):627–56.
232. Tiffany JM. Composition and biophysical properties of the tear film: knowledge and uncertainty. Adv Exp Med Biol. 1994;350:231–8.
233. Dilly PN. Structure and function of the tear film. Adv Exp Med Biol. 1994;350:239–47.

Chapter 10
Vaginal Delivery of Biopharmaceuticals

José das Neves

10.1 Introduction

The administration of drugs in the vagina is an ancient practice and is thought to be as old as medicine itself. The vaginal route of drug delivery is mainly considered in current days as an interesting alternative to oral therapy, particularly for the management of local genital diseases and female reproductive conditions [1–4]. Despite presenting several limitations, such as physiological changes during the reproductive cycle or cultural issues related with genital manipulation and sexuality, the direct administration of drugs in the vagina may be of value in preventing deleterious events such as systemic side effects. Also, the vaginal mucosa has been shown able to absorb different active molecules and yield sufficient blood plasma levels in order to allow systemic activity [5–6]. Most of the drugs currently used for vaginal delivery present low molecular weight (MW) and are obtained by chemical synthesis. These can be readily absorbed depending on their chemical properties, namely solubility. Nevertheless, several efforts have been conducted in order to study the ability of the vaginal route as an alternative for the delivery of biomolecules such as peptides, proteins, and genetic material. In this case, the objective may still be a local or systemic effect. This chapter reviews the most important features of the vagina related with drug delivery and different strategies adopted to optimize therapy with biopharmaceuticals through this particular delivery route.

J. das Neves (✉)
IINFACTS – Instituto de Investigação e Formação Avançada em Ciências e Tecnologias da Saúde, Instituto Superior de Ciências da Saúde-Norte, CESPU, Rua Central de Gandra, 1317, 4585-116 Gandra, Portugal
e-mail: j.dasneves@gmail.com

Faculty of Pharmacy, University of Porto, Porto, Portugal

J. das Neves, B. Sarmento (eds.), *Mucosal Delivery of Biopharmaceuticals*, DOI 10.1007/978-1-4614-9524-6_10,

10.2 The Vaginal Route of Drug Delivery

The vagina is an S-shaped canal extending from the cervix to the introitus. Its functions include the reception of the penis during sexual intercourse and the passage of menstrual and other fluids from the upper genital tract to the exterior of the body. The vaginal canal also plays an important role as the passageway of the child during labor. The mucosa of the vagina is covered by a nonkeratinized stratified squamous epithelium which is dependent on hormonal status. In particular, its thickness varies alongside the menstrual cycle thus influencing drug permeability. The surface area available for drug absorption is variable and has been estimated around 50–600 cm^2 [7]. The mucosa is covered by a thin layer of acidic mucus which provides lubrication and an additional barrier to drug permeation [8]. Lactic acid resulting from the fermentation of host glycogen by commensal bacteria, in particular lactobacilli, is the main responsible for its acidic pH around 4–5 [9]. Depletion of normal microbiota results in increased pH (up to around neutrality) and facilitates infection [10]. Among other components, the vaginal fluid possesses relatively low enzymatic activity; however, the presence of different enzymes, particularly peptidases, may impact on the possible delivery of biopharmaceuticals by this route [11]. For further detailed information on the anatomy, histology, and physiology of the vagina as related to drug delivery, the reader is referred to previous reviews by the author and colleagues [12–13] and others [14–15].

The vaginal route has been traditionally used for managing local conditions such as infection or for contraception purposes (spermicides). However, the good absorption profile of several active compounds through the vaginal mucosa led to the development of products for systemic drug delivery such as rings containing hormonal contraceptives and tablets or inserts containing labor inducers (e.g., dinoprostone, misoprostol) [16–18]. Current focus of the field of vaginal drug delivery is mainly on microbicides, which have been defined as products intended to be used around the time of sexual intercourse in order to prevent the transmission of HIV and, potentially, other sexually transmitted pathogens (e.g., HSV-2) [19–20]. Despite all hopes and a recently successful hallmark clinical trial [21], vaginal microbicides still require further evidence in order to be considered as a real option in the fight against HIV [22].

As all drug delivery routes, the vagina presents advantages and limitations. A summary of the most commonly referred is presented in Table 10.1. Since systemic exposure is usually limited, either by poor drug absorption or the use of lower doses, systemic adverse effects can be abbreviated. The administration of vaginal products is usually simple and painless, thus allowing for self-administration and patient comfort. Avoidance of trauma or infection at the administration site is usually considered as an advantage over parenteral administration. The avoidance of the hepatic first-pass effect is also regarded as advantageous since different biopharmaceuticals may undergo extensive hepatic degradation, which limits their use by the oral route.

The main disadvantage of this route is undoubtedly its gender specificity. This limits its potential usefulness to roughly half of the human population, a fact that can

Table 10.1 General advantages and limitations of the vagina as a drug delivery route

Pros	Cons
Reduction of systemic side effects	Gender specific
Allows self-administration	Erratic drug absorption due to physiological changes
Noninvasive	Cultural issues (e.g., genital manipulation)
Painless	Possible interference with sexual intercourse
Avoids hepatic first-pass effect observed for oral route	Possible onset of local adverse effects (e.g., irritation)
Usually economical	Discomfort

be of great importance from a marketing point of view. The variability of the vaginal physiology throughout the menstrual cycle, as well as in the overall reproductive life cycle of women (i.e., premenarche, fertile years, menopause), is able to influence the permeability of drugs across the mucosa. These changes may affect the consistency and continuity of therapy. Cultural myths and taboos also limit the acceptance of vaginal products by women, particularly when managing conditions not directly involving genitalia or reproduction. The interference with sexual intercourse is usually related with lubrication conferred by vaginal products. Even if generally regarded as positive, since it facilitates penetration and alleviates painful coitus, "dry sex" may be preferred by both women and men in some cultural settings [23]. The presence in the vagina of relatively high amounts of liquid masses of varying viscosity resulting from vaginal products (e.g., gels or melting of suppositories) frequently results in leakage. This leads to discomfort and, alongside possible local adverse effects such as burning sensation or itching resulting from irritation, may contribute to poor compliance.

10.3 Limitations and Possibilities for Vaginal Delivery of Biopharmaceuticals

Even if the therapeutic objectives of vaginal delivery of active biomolecules may span those usually intended for other routes (e.g., diabetes and osteoporosis management with insulin and calcitonin, respectively), it is in the field of reproductive health and sexually transmitted disease management that biopharmaceuticals may encounter their most clear usefulness. In particular, the discovery of antibodies or proteins/peptides with activity against HIV (or potentially other pathogens) for microbicide development has seen great developments in recent years [24–29]. Another interesting strategy for the prevention of vaginal HIV transmission (and potentially other pathogens) is the development of mucosal vaccines [30–32]; moreover, immunotherapy with antifungal antibodies may also be a potential approach for the vaginal treatment and prevention of infection, namely candidiasis [33]. With the advent of RNA interference as a potentially useful tool in medicine, the possibility of delivering siRNA through the vaginal route has also been proposed, particularly for managing local conditions or prevention of sexually transmitted diseases [34]. Furthermore, antibodies targeting sperm have been proposed as potential contraceptive compounds for intravaginal use [35–36].

As stated above, the vaginal mucosa is able to be permeated by active molecules, particularly those of low MW (< 300 Da) and presenting some degree of hydrophobicity [37–38]. Different low MW compounds have been shown to permeate vaginal tissue at comparable or even higher rates than across buccal or intestinal mucosae [39–42]. This can also be the case of some higher MW molecules. For instance, vasopressin, a nonapeptide (MW $= 1.1$ kDa), was shown able to permeate vaginal mucosa similarly to the buccal mucosa [40]. However, high MW usually results in poor permeability across the vaginal mucosa. As an example, the vaginal administration of nafarelin in an acetate buffered liquid vehicle (pH ≈ 5) to women yielded negligible serum concentrations of this peptide (MW $= 1.3$ kDa) in contrast with its nasal administration [43]. The inability of biopharmaceuticals to permeate the vaginal mucosa may be advantageous in those cases where systemic absorption is undesirable. For instance, Cole et al. [44] showed that RC-101 peptide (MW $= 1.9$ kDa), a retrocyclin analog being developed as a vaginal microbicide, was not able to cross an in vitro vaginal epithelium model. This cyclic peptide is active against HIV-1 by inhibiting target-cell entry and exerts its activity at the level of the vaginal epithelium. In this specific case, permeation may lead to undesirable systemic levels.

Stability in cervicovaginal fluids is an important issue regarding biopharmaceuticals as these may be degraded in the vaginal canal before even permeating the mucosa. In particular, enzymatic degradation of bioactive peptides and proteins in the vaginal milieu (including the mucosa) may be a limitation. Even if less pronounced than in other mucosae, vaginal enzymatic activity has been shown considerable in both animal models and humans [11, 45–47]. Thus, enzymatic inhibition may be an interesting strategy for promoting the stability of biopharmaceuticals in the vaginal environment and thus potentially enhance vaginal permeation/absorption. Also, other factors such as typical acidic pH and the interaction with components present in cervicovaginal fluids may impact the stability of biopharmaceuticals and their influence should be assessed during development [48–49]. Interactions of biomolecules with the mucin mesh comprising mucus may be of importance, namely in determining their diffusion across cervicovaginal fluids. Even so, previous studies indicate that different macromolecules (e.g., peptides and proteins with MW ranging 1.2–970 kDa) may diffuse almost unhindered through human cervical mucus [50–51].

Variations to normal fluids present in the vagina should also be considered. For instance, Sassi et al. [52] showed that RC-101 peptide was not significantly degraded at pH values in the range of 3–7; however, deleterious effects on the molecule were observed upon exposure to increasing levels of hydrogen peroxide but which could be minimized by the presence of ethylenediaminetetraacetic acid (EDTA), thus suggesting oxidative degradation. Of particular notice, this study indicated that RC-101 was highly unstable in human vaginal fluid collected from women with bacterial vaginitis. Higher content of hydrolytic enzymes and electrostatic interactions of the peptide with the membrane of bacteria were suggested as being implicated in the degradation of RC-101 [52].

One additional issue related with the use of biopharmaceuticals for managing vaginal conditions is cost. This is particularly true in the field of microbicides. Different biomolecules were shown as interesting candidates for the development of vaginal anti-HIV microbicides [53–57] but production costs may not be reasonable to sustain a final product with suitable cost for long-term use, particularly in low income and developing countries. However, efforts undertaken to produce some of these active molecules (e.g., 2G12 antibody, bovine colostrum-derived antibodies, and griffithsin) at affordable prices have shown that this is indeed possible [58–60].

10.4 Strategies for Vaginal Delivery of Biopharmaceuticals

10.4.1 Conventional and Novel Vaginal Dosage Forms

The use of buffered solutions and hydrophilic gels is simple and common practice when considering in vivo evaluation of biopharmaceuticals intended for vaginal delivery [61–65]. For example, Dereuddre-Bosquet et al. [66] have recently shown that a gel containing a CD4 peptide mimetic (mini-CD4 M48U1 presented on a stable 27 amino-acid scaffold) was able to protect cynomolgus macaques (5 out of 6 animals) from intravaginal challenge with a simian-human immunodeficiency virus strain (SHIV162P3). In contrast, animals treated with the placebo formulation were all infected. The active gel comprised 0.03 % (w/w) of the mini-CD4 peptide and the following excipients: 1.5 % hydroxyethylcellulose (HEC), 2.5 % glycerol, 0.1 % sorbic acid and water. Moreover, the gel presented adequate values of pH (4.6) and osmolality (around 290 mOsm/kg) for vaginal administration. Studies using gels have also been undertaken in the field of vaccines. Curran et al. [67–69] developed different formulations to administer intravaginally the HIV-1 envelope glycoprotein CN54gp140 in order to elicit an immune response. In one case, these researchers developed a gel based on HEC, polycarbophil, and polyvinylpyrollidone presenting potentially suitable rheological, mucoadhesive, syringeability (i.e., the ability to be extruded from a syringe/applicator), and active molecule release properties. Temperature-dependent stability issues were identified, with CN54gp140 glycoprotein being degraded, for example, by around 79 % when gels were stored at 37 °C for 9 days; this may limit the usefulness of these formulations. Notwithstanding, gels were shown able to elicit specific immune responses to CN54gp140 glycoprotein after vaginal administration to rabbits, namely by increasing immunoglobulin A (IgA) and immunoglobulin G (IgG) levels in lavages, which was associated with the ability of developed formulations to increase retention and intimate contact of the peptide with the mucosa [67]. Another interesting approach for delivering biomolecules is the development of stimuli-sensitive gels. For instance, temperature-sensitive gels, which are liquid at room temperature, may be appropriate for easy vaginal administration and good vaginal distribution; still, upon gelling at around 37 °C, gels reduce leakage and enhance drug residence. Polymers such as poloxamers [triblock copolymers composed of poly(ethylene oxide)-poly(propylene oxide)-poly(ethylene oxide)] seem to be particularly interesting for formulating thermosensitive vaginal gels [70–71] and their use has been shown advantageous in the development of vaginal vaccines [72–73]

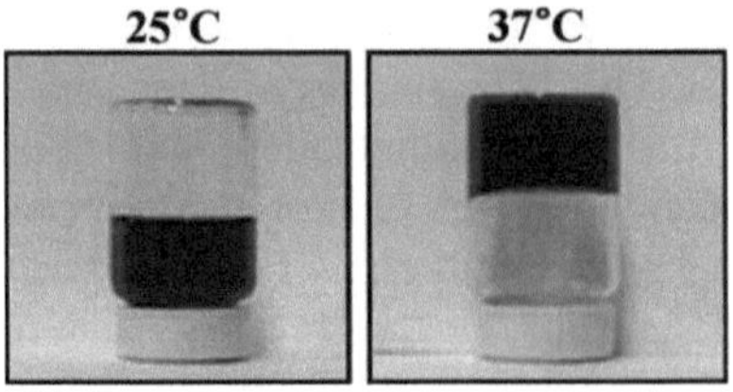

Fig. 10.1 Thermosensitive vehicle containing HPV16 L1 protein antigen (1 mg/mL) used for vaginal immunization. The formulation [poloxamer 188 (20 % w/v), poloxamer 407 (12 % w/v), and polycarbophil (20 % w/v; mucoadhesive polymer)] was placed in vials and incubated at different temperatures for 5 min. Then, vials were placed upside down in order to evidence its liquid (at 25 °C) or gel (at 37 °C) state. Bromophenol blue was incorporated (0.4 % w/v) in order to provide the vehicle with some opacity (*dark gray*). (Reprinted from [73], copyright 2005, with permission from Elsevier)

(Fig. 10.1) and peptide-based anti-HIV microbicides [74]. Another varying factor in the vaginal environment that can be used for producing "smart" dosage forms is pH. Gels responsive to pH variations (e.g., due to intravaginal ejaculation or bacterial vaginitis) have been developed for vaginal delivery of low MW drugs [75] and may also be an interesting approach to the delivery of macromolecules.

Stability problems may arise from using aqueous-based dosage forms such as hydrophilic gels, and anhydrous or solid systems may be preferential in the formulation of biopharmaceuticals. The group of Curran [68] proposed the preparation of freeze-dried solid dosage forms (tablets and rods) from developed gels [67] in order to abbreviate stability issues of CN54gp140 glycoprotein, alongside several variations, including the substitution of HEC by sodium carboxymethylcellulose (NaCMC) [68]. This last modification in combination with freeze-drying showed to be particularly promising regarding stability: loss of CN54gp140 glycoprotein was equal to or less than around 20 % after 150 days at 37 °C. Moreover, rods containing CN54gp140 glycoprotein were shown well tolerated upon vaginal administration to mice and boosted systemic specific antibody responses in subcutaneously primed animals. Another approach involving the use of solid systems was proposed by Gunaseelan et al. [76]: they developed subliming solid matrices for the vaginal delivery of the anti-HIV C5A peptide (MW = 2.5 kDa), which is chemically unstable, particularly in the presence of water. Matrices were made of hydrophobic cyclododecane (which maintained moisture away from the interstices of the system) and allowed the peptide to be released in a sustained fashion, ranging from days to months, due to surface erosion achieved through sublimation and not by conventional matrix hydrolysis or dissolution. Suppositories may also be a convenient solid dosage form for the vaginal delivery of biomolecules, particularly in veterinary practice. For example, Loehr and colleagues [77] used a suppository formulation, based on commercially available mixture of hydrogenated coco-glycerides (Witepsol® H-15), for the vaginal administration to cows of a plasmid DNA encoding for gD glycoprotein from bovine herpes virus-1 (BHV-1). The system proved to be effective in inducing protective systemic and distal mucosal (e.g., at the nasal mucosa) immune responses (IgG and IgA).

Polymeric films are dosage forms associated with several advantages for vaginal delivery including easiness of application, extended coverage of the mucosa, mucoadhesion, and possibility of rapid release of incorporated drugs [78]. Their use for the delivery of different active agents, including biomolecules, has been found suitable. For instance, vaginal films containing RC-101 were developed and shown to possess suitable physicochemical and technological properties for vaginal administration [79]. Apart from the active molecule, optimized films comprised polyvinyl alcohol and hydroxypropyl methylcellulose (HPMC) as matrix-forming polymers and EDTA as antioxidative. The stability of the active molecule was not affected for 90 days at 25 °C upon incorporation in the optimized film, and its safety for vaginal delivery and maintenance of activity against HIV-1 was observed in vitro and ex vivo in explants of female macaques (*Macaca nemestrina*) and women.

Rings are well established dosage forms for vaginal drug delivery, in particular of hormonal agents intended for contraception and replacement therapy in post-menopausal women [17, 80]. Interest on these delivery systems has also been renewed recently by researchers in the field of microbicides [81–82]. However, efforts have been generally limited to the development of rings intended for the delivery of hydrophobic, low MW drugs. Typically, molecules presenting high MW are not able to diffuse through the elastomeric polymer matrix of vaginal rings thus limiting their application to biopharmaceuticals. One simple strategy to circumvent this problem is the incorporation of hydrophilic substances in the matrix composition, which allow creation of additional channels upon contact with aqueous fluids through which biomolecules can diffuse. For instance, Radomsky et al. [83] developed anti-human chorionic gonadotropin (hCG) antibody-loaded rings by previously preparing freeze-dried particles of antibody/Ficoll (an hydrophilic polysaccharide) and incorporating these last in a miniaturized poly(ethylene-*co*-vinyl acetate) ring matrix. Obtained systems allowed for sustained in vitro release of bovine serum albumin (BSA; used as a model macromolecule) or antibody up to 1 month. Also, adequate antibody distribution for up to nine days in the vaginal lumen of mice was observed after administration of these systems. A similar system was also developed for the delivery of plasmid DNA to mice, but in this case the delivery system assumed the form of a disk ($\approx$ 1.0 mm in thickness and 2.5 mm in diameter) [84]. Alongside controlled release of the genetic material and maintenance of its activity, disks allowed enhanced transfection of vaginal tissues as compared to naked plasmid DNA. In a subsequent study by the same group, poly(ethylene-*co*-vinyl acetate) disks were also successfully used for developing protein-based intravaginal vaccines and inducing mucosal immune responses in mice [85]. Even if these last approaches seemed successful, the addition of hydrophilic substances frequently results in rings (or disks) presenting poor mechanical properties. Thus, a few studies on possible modifications to standard rings have been conducted in order to allow their use for the vaginal delivery of biomacromolecules. For example, Malcolm and collaborators [86–87] proposed the so-called insert vaginal ring (Fig. 10.2) comprising a standard silicone ring body bearing different cavities that can accommodate various drug loaded inserts (e.g., modified silicone rods, directly compressed tablets or freeze-dried gels). In all cases, proof-of-concept was established using BSA as a model protein; the release of BSA was

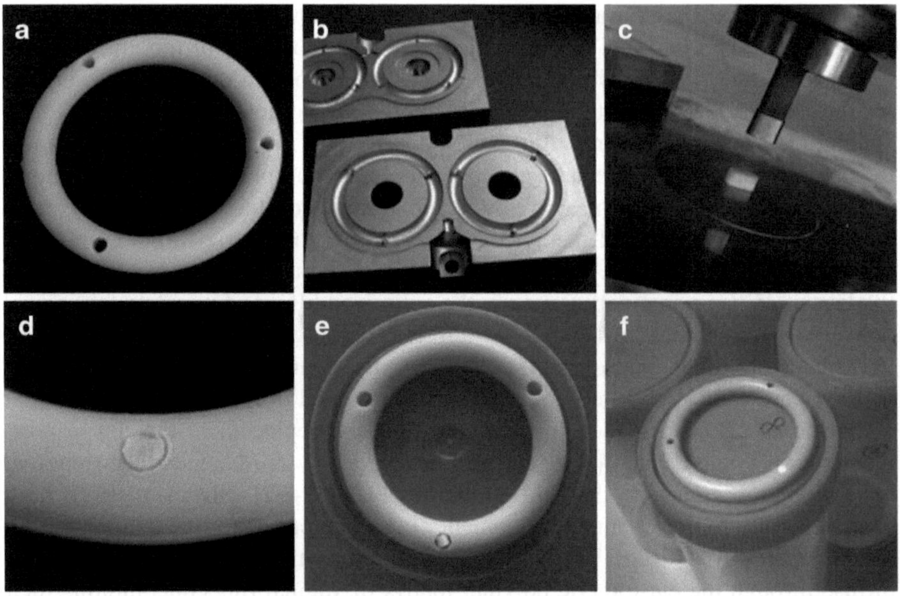

Fig. 10.2 The "insert vaginal ring": **a** silicone insert vaginal ring, **b** injection moulds for insert vaginal ring manufacture, **c** directly compressed insert (tablet) manufacture, **d** modified silicone insert, **e** directly compressed tablet insert, and **f** freeze-dried insert. (Reprinted from [86], copyright 2010, with permission from Elsevier)

able to be modulated from hours to more than one month by modifying the properties of inserts. Freeze-dried gels were further tested for the incorporation of an anti-HIV monoclonal antibody (2F5) and seemed particularly suitable for the delivery of such a labile compound due to the mild preparation processing conditions involved. The versatility of this type of rings may also allow the incorporation of multiple active molecules in the same system and facilitate dosage adjustments. Altogether, vaginal rings stand as promising delivery systems biopharmaceuticals and future developments are expected.

Even if well established vaginal dosage forms and novel ones provide convenient ways to administer active biomolecules, they might not be adequate in all cases. Different formulations strategies may be required in order to optimize performance because of intrinsic properties of biomolecules such as poor permeability or stability. Indeed, stability should be assessed and optimized early in the formulation stage. Compatibility with ingredients and properties of dosage forms/delivery systems are important issues that should not be neglected when formulating biopharmaceuticals as even small changes may induce important loss of activity on a short time period [88]. Also, safety issues must not be overlooked and the normal physiology of the vagina should be preserved, namely in terms of cervicovaginal fluid pH and osmolarity, natural microbiota and epithelial integrity.

10.4.2 Use of Permeability Enhancers and Enzymatic Inhibitors

The use of permeability/absorption enhancers and enzymatic inhibitors, in particular of aminopeptidases, may be a convenient strategy for optimizing the vaginal delivery of peptides and proteins [46, 89–90]. The use of simple organic acids in relatively high concentrations was shown helpful in a rat model in enhancing the vaginal absorption of peptides, namely leuprolide (MW = 1.2 kDa), by acidification/chelation mechanisms [91–93]. For instance, citric acid was able to induce a transient loosening of the vaginal epithelial barrier, thus increasing the amount of permeated leuprolide. Benzalkonium chloride, a cationic surfactant, was also shown to be effective in increasing the ex vivo permeation of cyclosporine (MW = 1.2 kDa) across human vaginal mucosa at a concentration of 0.01 % [94]. However, quaternary ammonium surfactants are well known for their toxicity even at such low concentrations and therefore their vaginal use should be limited [95]. In another study, Değim et al. [96] showed that chitosan-based gels may provide suitable vehicles for vaginal delivery of insulin (MW = 5.8 kDa), particularly when dimethyl-β-cyclodextrin was included in the formulation as a permeability enhancer. This last formulation was able to provide prolonged decrease of blood glucose levels in rabbits as mediated by the ability of the gel to release insulin in a sustained fashion and increase intravaginal drug residence, combined with the permeation enhancement effect of both dimethyl-β-cyclodextrin and chitosan. Hydrogen peroxide used in the low micromolar range has been recently proposed as an effective vaginal permeation enhancer of macromolecules [97]. In vitro experiments using a commercially available model of the vaginal epithelium (EpiVaginal™) were successful in showing that hydrogen peroxide enhanced the permeability of insulin by around one log. Mechanism behind this effect was determined to be the reversible disassembling of intercellular tight junctions. Overall, and even if the use of permeability enhancers seems interesting, their effects on the integrity of vaginal epithelium and overall histology may limit their usefulness, particularly when chronic administration is intended [98].

As for the use of enzymatic inhibitors, Nakada et al. [99] tested the influence of different peptidase inhibitors, namely bestatin, leupeptin, and pepstatin A, in the vaginal absorption of calcitonin (MW = 3.5 kDa) in rats. The absorption of this peptide was enhanced and correlated with its decreased degradation as shown in vitro. Other enzymatic inhibitors (sodium glycocholate, aprotinin and *p*-chloromercuriphenylsulfonic acid) were also shown to be useful in promoting the in vitro stability of insulin in vaginal homogenates of rabbits [45]. In another study, thiolated polymers such as thiolated carbopol 974P were shown helpful in reducing the in vitro activity of aminopeptidase against luteinizing hormone-releasing hormone (LH-RH) [100]. Enzymatic inhibition seemed to be associated to the degree of thiol-modification. In conjunction with the ability of thiolated carbopol 974P to be used in the development of suitable vaginal gels or solid dosage forms (e.g., tablets), this polymer may provide an interesting tool in the development of products for vaginal delivery of peptides/proteins. Despite these examples on the potential of enzymatic inhibition, the long-run use of such compounds may be deleterious to the mucosal environment and further safety studies are required.

10.4.3 Micro- and Nanocarriers

The use of microcarriers may be a suitable strategy for delivering biomolecules. In a study by Richardson et al. [101], mucoadhesive starch microspheres containing insulin were shown effective in reducing glucose plasma levels in sheep after vaginal administration and when compared to insulin in solution, particularly in the presence of lysophosphatidylcholine used as a permeability enhancer. The same system was also used as a carrier for a glycoprotein fragment from the influenza virus hemagglutinin (MW = 40 kDa) and was shown effective in inducing an Ig-based immune response in sheep after vaginal administration [102]. Levels of serum IgG and vaginal IgA were significantly higher when compared to the ones obtained from sheep treated intravaginally with the hemagglutinin fragment in solution or in solution containing lysophosphatidylcholine. These results emphasize the role of starch microspheres, in particular their mucoadhesive properties, in the generation of the immune response. In a subsequent study by the same group, hyaluronan esters-based microspheres were developed for the vaginal delivery of salmon calcitonin [103–104]. Stability of the peptide was enhanced while its vaginal administration in rats (100 IU/kg) allowed decreasing plasma calcium levels down to levels similar to the ones obtained for subcutaneous injection of 10 IU/kg. Moreover, microspheres were shown effective in preventing bone loss in ovariectomized rats after daily vaginal administration for 60 days (50 IU/kg/day), with results being comparable to those of animals treated with daily intramuscular calcitonin (10 IU/kg/day) [105]. In all cases, the mucoadhesive nature of microspheres was claimed by the authors as crucial for the observed results.

Among other drug carriers, liposomes have seen a great deal of development when considering the vaginal delivery of biopharmaceuticals. As an example, Kish-Catalone et al. [106–107] incorporated −2 RANTES, an analogue of chemokine (C-C motif) ligand 5 (CCL5 or RANTES—Regulated on Activation, Normal T cell Expressed and Secreted; MW = 8 kDa), into commercially available Novasome liposomes (typically 200–700 nm) and tested their ability to prevent SHIV vaginal transmission in macaques (*Macaca fascicularis*). Importantly, −2 RANTES retained antiviral activity after incorporation and allowed sustained release up to 2 h. Loaded liposomes administered intravaginally before viral challenge provided enhanced protection as compared to −2 RANTES administered in phosphate buffered saline. Differences were allegedly related with the ability of liposomes to enhance the vaginal retention and mucosal coverage by –2 RANTES. Also, liposomes showed no evidence of cervicovaginal toxicity as assessed in murine and rabbit models. In another study, Ning et al. [108] developed nonionic surfactant-based liposomes (niosomes) for the intravaginal delivery of insulin. In a rat model, these sorbitan monooleate-based nanocarriers (≈ 250 nm) showed relative bioavailability values of 8–10 % when compared to subcutaneous administration of insulin solution, presumably due to the permeability enhancing effect of niosomes. Liposomes may also provide interesting delivery systems for vaccine development. In one recent study, Gupta and colleagues [69] described different types of liposomes (≈ 120–160 nm) to deliver the HIV-1 envelope glycoprotein CN54gp140. Liposomes were further

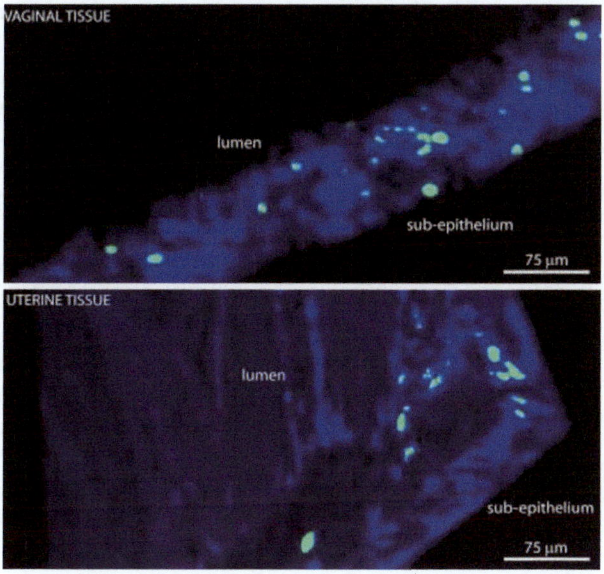

Fig. 10.3 Mucosal tissue penetration of PLGA nanoparticles loaded with a fluorescent dye, coumarin-6 (*green signal*), in mice as assessed by multiphoton microscopy: vaginal (*upper panel*) and uterine (*lower panel*) tissues. *Blue color* is from Hoescht dye (DNA labeling). (Adapted from [110] by permission from Macmillan Publishers Ltd, copyright 2009)

incorporated in HEC-based gels and freeze-dried in order to obtain rod-shaped devices that may be useful in developing suitable dosage forms to be administered in the vagina. Noticeably, rods allowed obtaining enhanced mucoadhesive properties of reconstituted HEC gels in a simulated vaginal fluid.

In another study, Ham et al. [109] studied the possibility of using polymeric nanoparticles (around 250 nm) to deliver another antiviral RANTES analogue, PSC-RANTES (MW = 7.9 kDa). Nanoparticles composed of poly(lactic-*co*-glycolic acid) (PLGA) were able to mediate the penetration of PSC-RANTES deep into human ectocervical epithelium in vitro, thus reaching areas where HIV-target cells (i.e., CCR5-bearing immune populations) are mainly located. In contrast, PSC-RANTES was mainly retained at the superficial layers of the epithelium when tested in solution. Further, encapsulation of PSC-RANTES did not alter its activity against HIV infection as tested in vitro. PLGA nanoparticles have also been tested for the vaginal delivery of small interfering RNA (siRNA) by Saltzman and collaborators [110–111]. In particular, natural polyamines such as spermidine or putrescine were used to precomplex siRNA and increase its subsequent association to PLGA nanoparticles. These nanocarriers (100–300 nm) were shown able to be taken up by different cell lines, contrasting with the inability of polyamine-siRNA complexes, and release their payload in vitro in a sustained fashion up to several weeks. Nanoparticles distributed throughout the genital tract of mice and penetrated deeply the mucosa after vaginal delivery (Fig. 10.3). Further, PLGA nanoparticles containing siRNA against *egfp* allowed knockdown of gene expression for up to 2 weeks after one single vaginal instillation in transgenic green fluorescence protein (GFP) mouse model [110]. In a subsequent study by the same group, siRNA directed against nectin-1, a transmembrane glycoprotein used by HSV-2 for cell infection, was preassociated to spermidine and loaded into PLGA-based nanoparticles (150–200 nm). When delivered intravaginally, nanoparticles were able to significantly increase the survival of mice challenged with the virus [111].

Apart from the use of viral vectors and some nanocarriers (as exemplified above), transfection agents are required for delivering genetic material to the cell interior. Typically, these encompass the use of cationic polymers or lipids/liposomes which are able to condense nucleic acids due to electrostatic interactions and form polyplexes or lipoplexes, respectively. Obtained complexes may also protect genetic material from degradation and allow its stabilization; on the other hand, enhanced toxicity may be a disadvantage namely when comparing with polymeric nanoparticles (e.g., PLGA-based ones) [34]. In one study, Eszterhas and coworkers [112] used a commercially available cationic polymeric transfection agent (INTERFERin®) for delivering siRNA targeting the expression of CD4 and CCR5. The nanosystem (around 50 nm) was shown able to silence receptor and coreceptor expression and partially prevent the infection of human cervical explants by HIV-1, thus showing potential to be used in the development of vaginal microbicides. As for the case of lipoplexes, Palliser et al. [113] used a commercially available lipid-based transfection agent (Oligofectamine™) to deliver siRNA targeting two HSV-2 genes (*UL27* and *UL29*). Experiments conducted using a mouse model of HSV-2 infection showed that lipolexes were able to be taken up by vaginal tissue and significantly protect animals from intravaginal viral infection (60–80 % survival vs. around 20 % in untreated mice at 14 days post challenge), without significant toxicity being observed, namely inflammation. Also noteworthy, the combination of both siRNAs (but not either alone) was able to provide protection even when administered 3–6 h after viral challenge. In another study, Wu et al. [114] tested the ability of poly(ethylene glycol) (PEG)-modified cationic liposomes to deliver intravaginally a model siRNA targeting lamin A/C protein expression. These last lipoplexes (200–350 nm) were incorporated in alginate scaffolds and delivered to mice: PEG-modified lipoplexes were shown more efficient in reducing the expression of lamin A/C when compared to non-PEGylated lipoplexes based on dioleoyl trimethylammonium propane liposomes. The rationale for using PEG-modified systems was based on the work of Hanes and collaborators [115–117], who showed that dense PEGylated nanosystems (polymeric nanoparticles up to 500 nm) were able to tackle the mucus fluid barrier present at the cervix and vagina. This strategy confers a muco-inert hydrophilic and nonionic surface to the system, thus avoiding mucoadhesive interactions with mucin fibers, and allows nearly unhindered diffusion through mucus. In this way, nanosystems are able to reach underlying epithelial cells and improve transfection. In order to complement the above information on siRNA delivery, the reader is referred to Chap. 15 of this book.

10.4.4 Other Strategies

Modification of biomolecules may be an interesting approach to their vaginal delivery. For example, Wheeler and coworkers [57] used CD4 aptamer-siRNA chimeras in order to deliver siRNA to HIV-target cells (i.e., cells bearing the CD4 receptor) present at the vaginal mucosa. Different siRNAs were tested, namely targeting HIV *gag* and *vif* or host *CCR5*. Further, chimeras were synthesized using 2'-fluoro-pyrimidines or cholesterol-modified siRNA, which provided enhanced stability to siRNA when in human vaginal fluid. Intravaginal instillation of chimeric RNAs prior to viral challenge was shown effective in inhibiting HIV infection in polarized human cervicovaginal explants and in a humanized HIV-susceptible mouse model (Bone-Marrow Liver Thymic mice—BLT mice). Also, no immune response was apparent in treated mice. Interestingly, chimeras were shown to possess a dual mechanism of action: adding up to gene silencing provided by siRNA, the CD4 aptamer alone was also able to inhibit HIV infection to some extent.

Another interesting approach to the vaginal delivery of biomolecules is the use of modified microbiota [118–122]. Commensal bacteria are genetically modified in order to produce active molecules. For instance, human commensal *Streptococcus gordonii* was successfully engineered in order to secrete or display a microbicidal single-chained antibody (H6) [123]. Modified *S. gordonii* was able to colonize the rat vagina after vaginal instillation. This treatment was as efficacious as fluconazole therapy in reducing *Candida albicans* fungal burden and resolving infection. Also, in a recent report, Lagenaur and coworkers [122] modified *Lactobacillus jensenii* in order to express cyanovirin-N, a 11 kDa protein that is able to inhibit HIV-1 cell entrance. After five consecutive daily administrations in a HEC gel, recombinant *L. jensenii* was able to persistently colonize the vagina of rhesus macaques (*Macaca mulatta*) for 3–6 weeks. Colonization protocol also involved the use of an antibiotic (azithromycin) in order to reduce endogenous lactobacilli and facilitate the proliferation of engineered microbiota. Inhibitory levels of cyanovirin-N were observed in cervicovaginal fluids after as little as 24 h post colonization. Further, infection of colonized animals after repeated SHIVSF162P intravaginal challenge was reduced by 63 % as compared to controls.

The use of cell-penetrating peptides, i.e., sequences of 30 or less amino acids that are able to promote direct or endocytosis-mediated cell penetration [124], and needle-free injectors has been proposed for the vaginal delivery of DNA vaccines [125–126]. This innovative strategy seems particularly interesting in triggering both systemic and local Ig-based immune responses in a relatively low invasive fashion. The reader is referred to Chap. 16 for further details.

10.5 Conclusions and Future Perspectives

The vagina may constitute an interesting route for delivering biopharmaceuticals, particularly when local conditions are to be considered. The field of anti-HIV microbicides has been particularly prolific in helping on the discovery and development of peptides/proteins and siRNA molecules with antiviral activity. Also, the fields of vaginal immunotherapy and vaccine development have seen recent developments. However, challenges related with the vaginal histology and physiology may impact on the activity of biopharmaceuticals and different formulation approaches are usually required. Strategies such as the use of conventional and novel vaginal dosage forms, namely films and rings, and permeability enhancers and enzymatic inhibitors have been proposed. Micro- and nanocarriers may also be an interesting approach as their usefulness as systems for vaginal drug delivery has been well established. Future work needed in the field involves the full development of products that can be used by women and the conduction of clinical trials that attest their value in therapy and prevention. In particular, safety issues should be addressed including their impact on reproduction.

Acknowledgments This work was supported by Fundação para a Ciência e a Tecnologia, Portugal (grant VIH/SAU/0021/2011).

References

1. Alexander NJ, Baker E, Kaptein M, Karck U, Miller L, Zampaglione E. Why consider vaginal drug administration? Fertil Steril. 2004;82(1):1–12.
2. Bartusevicius A, Barcaite E, Nadisauskiene R. Oral, vaginal and sublingual misoprostol for induction of labor. Int J Gynaecol Obstet. 2005;91(1):2–9.
3. Nurbhai M, Grimshaw J, Watson M, Bond C, Mollison J, Ludbrook A. Oral versus intra-vaginal imidazole and triazole anti-fungal treatment of uncomplicated vulvovaginal candidiasis (thrush). Cochrane Database Syst Rev. 2007;(4):CD002845.
4. Roumen FJ. The contraceptive vaginal ring compared with the combined oral contraceptive pill: a comprehensive review of randomized controlled trials. Contraception. 2007;75(6):420–9.
5. Benziger DP, Edelson J. Absorption from the vagina. Drug Metab Rev. 1983;14(2):137–68.
6. Hussain A, Ahsan F. The vagina as a route for systemic drug delivery. J Control Release. 2005;103(2):301–13.
7. Katz DF, Henderson MH, Owen DH, Plenys AM, Walmer DK. What is needed to advance vaginal formulation technology? In: Rencher WF, editor. Vaginal microbicide formulations workshop. Philadelphia: Lippincott-Raven Publishers; 1998. pp. 90–9.
8. Cone RA. Barrier properties of mucus. Adv Drug Deliv Rev. 2009;61(2):75–85.
9. Boskey ER, Cone RA, Whaley KJ, Moench TR. Origins of vaginal acidity: high D/L lactate ratio is consistent with bacteria being the primary source. Hum Reprod. 2001;16(9):1809–13.
10. Caillouette JC, Sharp CF, Jr., Zimmerman GJ, Roy S. Vaginal pH as a marker for bacterial pathogens and menopausal status. Am J Obstet Gynecol. 1997;176(6):1270–75.
11. Acartürk F, Parlatan ZI, Saracoglu OF. Comparison of vaginal aminopeptidase enzymatic activities in various animals and in humans. J Pharm Pharmacol. 2001;53(11):1499–504.
12. das Neves J, Amaral MH, Bahia MF. Vaginal drug delivery. In: Gad SC, editor. Pharmaceutical Manufacturing Handbook: Production and Processes. Hoboken: Wiley; 2008. pp. 809–78.
13. das Neves J, Palmeira-de-Oliveira R, Palmeira-de-Oliveira A, Rodrigues F, Sarmento B. Vaginal mucosa and drug delivery. In: Khutoryanskiy VV, editor. Mucoadhesive Materials and Drug Delivery Systems. Chichester: Wiley; 2014. (In press).
14. Vermani K, Garg S. The scope and potential of vaginal drug delivery. Pharm Sci Technol Today. 2000;3(10):359–64.
15. Woolfson AD, Malcolm RK, Gallagher R. Drug delivery by the intravaginal route. Crit Rev Ther Drug Carrier Syst. 2000;17(5):509–55.
16. Hofmeyr GJ, Gulmezoglu AM. Vaginal misoprostol for cervical ripening and induction of labour. Cochrane Database Syst Rev. 2003;(1):CD000941.
17. Johansson ED, Sitruk-Ware R. New delivery systems in contraception: vaginal rings. Am J Obstet Gynecol. 2004;190(4 Suppl 1): S54–9.

18. Vollebregt A, van't Hof DB, Exalto N. Prepidil compared to propess for cervical ripening. Eur J Obstet Gynecol Reprod Biol. 2002;104(2):116–9.
19. Ariën KK, Jespers V, Vanham G. HIV sexual transmission and microbicides. Rev Med Virol. 2011;21(2):110–33.
20. Shattock RJ, Rosenberg Z. Microbicides: topical prevention against HIV. Cold Spring Harb Perspect Med. 2012;2(2):a007385.
21. Abdool Karim Q, Abdool Karim SS, Frohlich JA, Grobler AC, Baxter C, Mansoor LE, Kharsany AB, Sibeko S, Mlisana KP, Omar Z, Gengiah TN, Maarschalk S, Arulappan N, Mlotshwa M, Morris L, Taylor D. Effectiveness and safety of tenofovir gel, an antiretroviral microbicide, for the prevention of HIV infection in women. Science. 2010;329(5996):1168–74.
22. Obiero J, Mwethera PG, Wiysonge CS. Topical microbicides for prevention of sexually transmitted infections. Cochrane Database Syst Rev. 2012;6:CD007961.
23. Braunstein S, van de Wijgert J. Preferences and practices related to vaginal lubrication: implications for microbicide acceptability and clinical testing. J Womens Health. 2005;14(5):424–33.
24. Emau P, Tian B, O'Keefe BR, Mori T, McMahon JB, Palmer KE, Jiang Y, Bekele G, Tsai CC. Griffithsin, a potent HIV entry inhibitor, is an excellent candidate for anti-HIV microbicide. J Med Primatol. 2007;36(4–5):244–53.
25. Welch BD, Francis JN, Redman JS, Paul S, Weinstock MT, Reeves JD, Lie YS, Whitby FG, Eckert DM, Hill CP, Root MJ, Kay MS. Design of a potent D-peptide HIV-1 entry inhibitor with a strong barrier to resistance. J Virol. 2010;84(21):11235–44.
26. Eade CR, Wood MP, Cole AM. Mechanisms and modifications of naturally occurring host defense peptides for anti-HIV microbicide development. Curr HIV Res. 2012;10(1):61–72.
27. Hooven TA, Randis TM, Hymes SR, Rampersaud R, Ratner AJ. Retrocyclin inhibits Gardnerella vaginalis biofilm formation and toxin activity. J Antimicrob Chemother. 2012;67(12):2870–2.
28. Veselinovic M, Neff CP, Mulder LR, Akkina R. Topical gel formulation of broadly neutralizing anti-HIV-1 monoclonal antibody VRC01 confers protection against HIV-1 vaginal challenge in a humanized mouse model. Virology. 2012;432(2):505–10.
29. Morellato-Castillo L, Acharya P, Combes O, Michiels J, Descours A, Ramos OH, Yang Y, Vanham G, Arien KK, Kwong PD, Martin L, Kessler P. Interfacial cavity filling to optimize CD4-mimetic miniprotein interactions with HIV-1 surface glycoprotein. J Med Chem. 2013;56(12):5033–47.
30. Bogers WM, Bergmeier LA, Ma J, Oostermeijer H, Wang Y, Kelly CG, P. Ten Haaft, Singh M, Heeney JL, Lehner T. A novel HIV-CCR5 receptor vaccine strategy in the control of mucosal SIV/HIV infection. AIDS. 2004;18(1):25–36.
31. Kanazawa T, Takashima Y, Okada H. Vaginal DNA vaccination against infectious diseases transmitted through the vagina. Front Biosci (Elite Ed.). 2012;4:2340–53.
32. Shin H, Iwasaki A. A vaccine strategy that protects against genital herpes by establishing local memory T cells. Nature. 2012;491(7424):463–7.
33. Magliani W, Conti S, Cassone A, De Bernardis F, Polonelli L. New immunotherapeutic strategies to control vaginal candidiasis. Trends Mol Med. 2002;8(3):121–6.
34. Yang S, Chen Y, Ahmadie R, Ho EA. Advancements in the field of intravaginal siRNA delivery. J Control Release. 2013;167(1):29–39.
35. Castle PE, Whaley KJ, Hoen TE, Moench TR, Cone RA. Contraceptive effect of sperm-agglutinating monoclonal antibodies in rabbits. Biol Reprod. 1997;56(1):153–9.
36. Norton EJ, Diekman AB, Westbrook VA, Flickinger CJ, Herr JC. RASA, a recombinant single-chain variable fragment (scFv) antibody directed against the human sperm surface: implications for novel contraceptives. Hum Reprod. 2001;16(9):1854–60.
37. Corbo DC, Liu JC, Chien YW. Characterization of the barrier properties of mucosal membranes. J Pharm Sci. 1990;79(3):202–6.
38. van der Bijl P, van Eyk AD. Human vaginal mucosa as a model of buccal mucosa for in vitro permeability studies: an overview. Curr Drug Deliv. 2004;1(2):129–35.

39. van der Bijl P, van Eyk AD, Thompson IO. Permeation of 17beta-estradiol through human vaginal and buccal mucosa. Oral Surg Oral Med Oral Pathol Oral Radiol Endod. 1998;85(4):393–8.
40. van der Bijl P, van Eyk AD, Thompson IO, Stander IA. Diffusion rates of vasopressin through human vaginal and buccal mucosa. Eur J Oral Sci. 1998;106(5):958–62.
41. van der Bijl P, Penkler L, van Eyk AD. Permeation of sumatriptan through human vaginal and buccal mucosa. Headache. 2000;40(2):137–41.
42. van der Bijl P, van Eyk AD. Comparative in vitro permeability of human vaginal, small intestinal and colonic mucosa. Int J Pharm. 2003;261(1–2):147–52.
43. Chan RL, Henzl MR, LePage ME, LaFargue J, Nerenberg CA, Anik S, Chaplin MD. Absorption and metabolism of nafarelin, a potent agonist of gonadotropin-releasing hormone. Clin Pharmacol Ther. 1988;44(3):275–82.
44. Cole AL, Herasimtschuk A, Gupta P, Waring AJ, Lehrer RI, Cole AM. The retrocyclin analogue RC-101 prevents human immunodeficiency virus type 1 infection of a model human cervicovaginal tissue construct. Immunology. 2007;121(1):140–5.
45. Yamamoto A, Hayakawa E, Lee VH. Insulin and proinsulin proteolysis in mucosal homogenates of the albino rabbit: implications in peptide delivery from nonoral routes. Life Sci. 1990;47(26):2465–74.
46. Chun IK, Chien YW. Transmucosal delivery of methionine enkephalin. I: solution stability and kinetics of degradation in various rabbit mucosa extracts. J Pharm Sci. 1993;82(4):373–8.
47. Acartürk F, Robinson JR. Vaginal permeability and enzymatic activity studies in normal and ovariectomized rabbits. Pharm Res. 1996;13(5):779–83.
48. Zeitlin L, Olmsted SS, Moench TR, Co MS, Martinell BJ, Paradkar VM, Russell DR, Queen C, Cone RA, Whaley KJ. A humanized monoclonal antibody produced in transgenic plants for immunoprotection of the vagina against genital herpes. Nat Biotechnol. 1998;16(13):1361–4.
49. Castle PE, Karp DA, Zeitlin L, Garcia-Moreno EB, Moench TR, Whaley KJ, Cone RA. Human monoclonal antibody stability and activity at vaginal pH. J Reprod Immunol. 2002;56(1–2):61–76.
50. Saltzman WM, Radomsky ML, Whaley KJ, Cone RA. Antibody diffusion in human cervical mucus. Biophys J. 1994;66(2 Pt 1):508–15.
51. Olmsted SS, Padgett JL, Yudin AI, Whaley KJ, Moench TR, Cone RA. Diffusion of macromolecules and virus-like particles in human cervical mucus. Biophys J. 2001;81(4):1930–7.
52. Sassi AB, Bunge KE, Hood BL, Conrads TP, Cole AM, Gupta P, Rohan LC. Preformulation and stability in biological fluids of the retrocyclin RC-101, a potential anti-HIV topical microbicide. AIDS Res Ther. 2011;8:27.
53. Veazey RS, Shattock RJ, Pope M, Kirijan JC, Jones J, Hu Q, Ketas T, Marx PA, Klasse PJ, Burton DR, Moore JP. Prevention of virus transmission to macaque monkeys by a vaginally applied monoclonal antibody to HIV-1 gp120. Nat Med. 2003;9(3):343–6.
54. Lederman MM, Veazey RS, Offord R, Mosier DE, Dufour J, Mefford M, Piatak M Jr., Lifson JD, Salkowitz JR, Rodriguez B, Blauvelt A, Hartley O. Prevention of vaginal SHIV transmission in rhesus macaques through inhibition of CCR5. Science. 2004;306(5695):485–7.
55. Van Herrewege Y, Morellato L, Descours A, Aerts L, Michiels J, Heyndrickx L, Martin L, Vanham G. CD4 mimetic miniproteins: potent anti-HIV compounds with promising activity as microbicides. J Antimicrob Chemother. 2008;61(4):818–26.
56. Zeitlin L, Pauly M, Whaley KJ. Second-generation HIV microbicides: continued development of griffithsin. Proc Natl Acad Sci U S A. 2009;106(15):6029–30.
57. Wheeler LA, Trifonova R, Vrbanac V, Basar E, McKernan S, Xu Z, Seung E, Deruaz M, Dudek T, Einarsson JI, Yang L, Allen TM, Luster AD, Tager AM, Dykxhoorn DM, Lieberman J. Inhibition of HIV transmission in human cervicovaginal explants and humanized mice using CD4 aptamer-siRNA chimeras. J Clin Invest. 2011;121(6):2401–12.
58. Ramessar K, Rademacher T, Sack M, Stadlmann J, Platis D, Stiegler G, Labrou N, Altmann F, Ma J, Stoger E, Capell T, Christou P. Cost-effective production of a vaginal protein microbicide to prevent HIV transmission. Proc Natl Acad Sci U S A. 2008;105(10):3727–32.

59. O'Keefe BR, Vojdani F, Buffa V, Shattock RJ, Montefiori DC, Bakke J, Mirsalis J, d'Andrea AL, Hume SD, Bratcher B, Saucedo CJ, McMahon JB, Pogue GP, Palmer KE. Scaleable manufacture of HIV-1 entry inhibitor griffithsin and validation of its safety and efficacy as a topical microbicide component. Proc Natl Acad Sci U S A. 2009;106(15):6099–104.
60. Kramski M, Center RJ, Wheatley AK, Jacobson JC, Alexander MR, Rawlin G, Purcell DF. Hyperimmune bovine colostrum as a low-cost, large-scale source of antibodies with broad neutralizing activity for HIV-1 envelope with potential use in microbicides. Antimicrob Agents Chemother. 2012;56(8):4310–9.
61. Morimoto K, Takeeda T, Nakamoto Y, Morisaka K. Effective vaginal absorption of insulin in diabetic rats and rabbits using polyacrylic acid aqueous gel bases. Int J Pharm. 1982;12(2–3):107–11.
62. Wang B, Dang K, Agadjanyan MG, Srikantan V, Li F, Ugen KE, Boyer J, Merva M, Williams WV, Weiner DB. Mucosal immunization with a DNA vaccine induces immune responses against HIV-1 at a mucosal site. Vaccine. 1997;15(8):821–5.
63. Veazey RS, Klasse PJ, Schader SM, Hu Q, Ketas TJ, Lu M, Marx PA, Dufour J, Colonno RJ, Shattock RJ, Springer MS, Moore JP. Protection of macaques from vaginal SHIV challenge by vaginally delivered inhibitors of virus-cell fusion. Nature. 2005;438(7064):99–102.
64. das Neves J, Bahia MF. Gels as vaginal drug delivery systems. Int J Pharm. 2006;318(1–2):1–14.
65. Li L, Ben Y, Yuan S, Jiang S, Xu J, Zhang X. Efficacy, stability, and biosafety of sifuvirtide gel as a microbicide candidate against HIV-1. PLoS One. 2012;7(5):e37381.
66. Dereuddre-Bosquet N, Morellato-Castillo L, Brouwers J, Augustijns P, Bouchemal K, Ponchel G, Ramos OHP, Herrera C, Stefanidou M, Shattock R, Heyndrickx L, Vanham G, Kessler P, Le Grand R, Martin L. MiniCD4 microbicide prevents HIV infection of human mucosal explants and vaginal transmission of SHIV162P3 in cynomolgus macaques. PLoS Pathog. 2012;8(12):e1003071.
67. Curran RM, Donnelly L, Morrow RJ, Fraser C, Andrews G, Cranage M, Malcolm RK, Shattock RJ, Woolfson AD. Vaginal delivery of the recombinant HIV-1 clade-C trimeric gp140 envelope protein CN54gp140 within novel rheologically structured vehicles elicits specific immune responses. Vaccine. 2009;27(48):6791–8.
68. Donnelly L, Curran RM, Tregoning JS, McKay PF, Cole T, Morrow RJ, Kett VL, Andrews GP, Woolfson AD, Malcolm RK, Shattock RJ. Intravaginal immunization using the recombinant HIV-1 clade-C trimeric envelope glycoprotein CN54gp140 formulated within lyophilized solid dosage forms. Vaccine. 2011;29(27):4512–20.
69. Gupta PN, Pattani A, Curran RM, Kett VL, Andrews GP, Morrow RJ, Woolfson AD, Malcolm RK. Development of liposome gel based formulations for intravaginal delivery of the recombinant HIV-1 envelope protein CN54gp140. Eur J Pharm Sci. 2012;46(5):315–22.
70. Bilensoy E, Rouf MA, Vural I, Sen M, Hincal AA. Mucoadhesive, thermosensitive, prolonged-release vaginal gel for clotrimazole:beta-cyclodextrin complex. AAPS PharmSciTech. 2006;7(2):E38.
71. Date AA, Shibata A, Goede M, Sanford B, La Bruzzo K, Belshan M, Destache CJ. Development and evaluation of a thermosensitive vaginal gel containing raltegravir + efavirenz loaded nanoparticles for HIV prophylaxis. Antiviral Res. 2012;96(3):430–6.
72. Oh YK, Park JS, Yoon H, Kim CK. Enhanced mucosal and systemic immune responses to a vaginal vaccine coadministered with RANTES-expressing plasmid DNA using in situ-gelling mucoadhesive delivery system. Vaccine. 2003;21(17–18):1980–8.
73. Han IK, Kim YB, Kang HS, Sul D, Jung WW, Cho HJ, Oh YK. Thermosensitive and mucoadhesive delivery systems of mucosal vaccines. Methods. 2006;38(2):106–11.
74. Bouchemal K, Frelichowska J, Martin L, Lievin-Le Moal V, Le Grand R, Dereuddre-Bosquet N, Djabourov M, Aka-Any-Grah A, Koffi A, Ponchel G. Note on the formulation of thermosensitive and mucoadhesive vaginal hydrogels containing the miniCD4 M48U1 as anti-HIV-1 microbicide. Int J Pharm. 2013;454(2):649–52.
75. Gupta KM, Barnes SR, Tangaro RA, Roberts MC, Owen DH, Katz DF, Kiser PF. Temperature and pH sensitive hydrogels: an approach towards smart semen-triggered vaginal microbicidal vehicles. J Pharm Sci. 2007;96(3):670–81.

76. Gunaseelan S, Gallay PA, Bobardt MD, Dezzutti CS, Esch T, Maskiewicz R. Sustained local delivery of structurally diverse HIV-1 microbicides released from sublimation enthalpy controlled matrices. Pharm Res. 2012;29(11):3156–68.
77. Loehr BI, Rankin R, Pontarollo R, King T, Willson P, Babiuk LA, van Drunen Littel-van den Hurk S. Suppository-mediated DNA immunization induces mucosal immunity against bovine herpesvirus-1 in cattle. Virology. 2001;289(2):327–33.
78. Machado RM, Palmeira-de-Oliveira A, Martinez-de-Oliveira J, Palmeira-de-Oliveira R. Vaginal films for drug delivery. J Pharm Sci. 2013;102(7):2069–81.
79. Sassi AB, Cost MR, Cole AL, Cole AM, Patton DL, Gupta P, Rohan LC. Formulation development of retrocyclin 1 analog RC-101 as an anti-HIV vaginal microbicide product. Antimicrob Agents Chemother. 2011;55(5):2282–9.
80. Ballagh SA. Vaginal rings for menopausal symptom relief. Drugs Aging. 2004;21(12):757–66.
81. Kiser PF, Johnson TJ, Clark JT. State of the art in intravaginal ring technology for topical prophylaxis of HIV infection. AIDS Rev. 2012;14(1):62–77.
82. Malcolm RK, Fetherston SM, McCoy CF, Boyd P, Major I. Vaginal rings for delivery of HIV microbicides. Int J Womens Health. 2012;4:595–605.
83. Radomsky ML, Whaley KJ, Cone RA, Saltzman WM. Controlled vaginal delivery of antibodies in the mouse. Biol Reprod. 1992;47(1):133–40.
84. Shen H, Goldberg E, Saltzman WM. Gene expression and mucosal immune responses after vaginal DNA immunization in mice using a controlled delivery matrix. J Control Release. 2003;86(2–3):339–48.
85. Kuo-Haller P, Cu Y, Blum J, Appleton JA, Saltzman WM. Vaccine delivery by polymeric vehicles in the mouse reproductive tract induces sustained local and systemic immunity. Mol Pharm. 2010;7(5):1585–95.
86. Morrow RJ, Woolfson AD, Donnelly L, Curran R, Andrews G, Katinger D, Malcolm RK. Sustained release of proteins from a modified vaginal ring device. Eur J Pharm Biopharm. 2011;77(1):3–10.
87. Pattani A, Lowry D, Curran RM, McGrath S, Kett VL, Andrews GP, Malcolm RK. Characterisation of protein stability in rod-insert vaginal rings. Int J Pharm. 2012;430(1–2):89–97.
88. Hubert P, Evrard B, Maillard C, Franzen-Detrooz E, Delattre L, Foidart JM, Noel A, Boniver J, Delvenne P. Delivery of granulocyte-macrophage colony-stimulating factor in bioadhesive hydrogel stimulates migration of dendritic cells in models of human papillomavirus-associated (pre)neoplastic epithelial lesions. Antimicrob Agents Chemother. 2004;48(11):4342–8.
89. Richardson JL, Illum L. (D) Routes of delivery: case studies: (8) the vaginal route of peptide and protein drug delivery. Adv Drug Deliv Rev. 1992;8(2–3):341–66.
90. Sayani AP, Chun IK, Chien YW. Transmucosal delivery of leucine enkephalin: stabilization in rabbit enzyme extracts and enhancement of permeation through mucosae. J Pharm Sci. 1993;82(11):1179–85.
91. Okada H, Yamazaki I, Ogawa Y, Hirai S, Yashiki T, Mima H. Vaginal absorption of a potent luteinizing hormone-releasing hormone analog (leuprolide) in rats I: absorption by various routes and absorption enhancement. J Pharm Sci. 1982;71(12):1367–71.
92. Okada H, Yamazaki I, Yashiki T, Mima H. Vaginal absorption of a potent luteinizing hormone-releasing hormone analogue (leuprolide) in rats II: mechanism of absorption enhancement with organic acids. J Pharm Sci. 1983;72(1):75–8.
93. Okada H, Yamazaki I, Yashiki T, Shimamoto T, Mima H. Vaginal absorption of a potent luteinizing hormone-releasing hormone analogue (leuprolide) in rats. IV: Evaluation of the vaginal absorption and gonadotropin responses by radioimmunoassay. J Pharm Sci. 1984;73(3):298–302.
94. van der Bijl P, van Eyk AD, Gareis AA, Thompson IO. Enhancement of transmucosal permeation of cyclosporine by benzalkonium chloride. Oral Dis. 2002;8(3):168–72.
95. Gali Y, Delezay O, Brouwers J, Addad N, Augustijns P, Bourlet T, Hamzeh-Cognasse H, Ariën KK, Pozzetto B, Vanham G. In vitro evaluation of viability, integrity and inflammation in genital epithelia upon exposure to pharmaceutical excipients and candidate microbicides. Antimicrob Agents Chemother. 2010;54(12):5105–14.

96. Değim Z, Değim T, Acartürk F, Erdoğan D, Özoğul C, Köksal M. Rectal and vaginal administration of insulin-chitosan formulations: an experimental study in rabbits. J Drug Target. 2005;13(10):563–72.
97. Fatakdawala H, Uhland SA. Hydrogen peroxide mediated transvaginal drug delivery. Int J Pharm. 2011;409(1–2):121–7.
98. Richardson JL, Illum L, Thomas NW. Vaginal absorption of insulin in the rat: effect of penetration enhancers on insulin uptake and mucosal histology. Pharm Res. 1992;9(7):878–83.
99. Nakada Y, Miyake M, Awata N. Some factors affecting the vaginal absorption of human calcitonin in rats. Int J Pharm. 1993;89(3):169–75.
100. Valenta C, Marschutz M, Egyed C, Bernkop-Schnürch A. Evaluation of the inhibition effect of thiolated poly(acrylates) on vaginal membrane bound aminopeptidase N and release of the model drug LH-RH. J Pharm Pharmacol. 2002;54(5):603–10.
101. Richardson JL, Farraj NF, Illum L. Enhanced vaginal absorption of insulin in sheep using lysophosphatidylcholine and a bioadhesive microsphere delivery system. Int J Pharm. 1992; 88(1–3):319–25.
102. O'Hagan DT, Rafferty D, Wharton S, Illum L. Intravaginal immunization in sheep using a bioadhesive microsphere antigen delivery system. Vaccine. 1993;11(6):660–4.
103. Richardson JL, Ramires PA, Miglietta MR, Rochira M, Bacelle L, Callegaro L, Benedetti L. Novel vaginal delivery systems for calcitonin: I. Evaluation of HYAFF/calcitonin microspheres in rats. Int J Pharm. 1995;115(1):9–15.
104. Rochira M, Miglietta MR, Richardson JL, Ferrari L, Beccaro M, Benedetti L. Novel vaginal delivery systems for calcitonin: II. Preparation and characterization of HYAFF® microspheres containing calcitonin. Int J Pharm. 1996;144(1):19–26.
105. Bonucci E, Ballanti P, Ramires PA, Richardson JL, Benedetti LM. Prevention of ovariectomy osteopenia in rats after vaginal administration of Hyaff 11 microspheres containing salmon calcitonin. Calcif Tissue Int. 1995;56(4):274–9.
106. Kish-Catalone TM, Lu W, Gallo RC, DeVico AL. Preclinical evaluation of synthetic -2 RANTES as a candidate vaginal microbicide to target CCR5. Antimicrob Agents Chemother. 2006;50(4):1497–509.
107. Kish-Catalone T, Pal R, Parrish J, Rose N, Hocker L, Hudacik L, Reitz M, Gallo R, Devico A. Evaluation of −2 RANTES vaginal microbicide formulations in a nonhuman primate simian/human immunodeficiency virus (SHIV) challenge model. AIDS Res Hum Retroviruses. 2007;23(1):33–42.
108. Ning M, Guo Y, Pan H, Yu H, Gu Z. Niosomes with sorbitan monoester as a carrier for vaginal delivery of insulin: studies in rats. Drug Deliv. 2005;12(6):399–407.
109. Ham AS, Cost MR, Sassi AB, Dezzutti CS, Rohan LC. Targeted delivery of PSC-RANTES for HIV-1 prevention using biodegradable nanoparticles. Pharm Res. 2009;26(3):502–11.
110. Woodrow KA, Cu Y, Booth CJ, Saucier-Sawyer JK, Wood MJ, Saltzman WM. Intravaginal gene silencing using biodegradable polymer nanoparticles densely loaded with small-interfering RNA. Nat Mater. 2009;8(6):526–33.
111. Steinbach JM, Weller CE, Booth CJ, Saltzman WM. Polymer nanoparticles encapsulating siRNA for treatment of HSV-2 genital infection. J Control Release. 2012;162(1):102–10.
112. Eszterhas SK, Ilonzo NO, Crozier JE, Celaj S, Howell AL. Nanoparticles containing siRNA to silence CD4 and CCR5 reduce expression of these receptors and inhibit HIV-1 infection in human female reproductive tract tissue explants. Infect Dis Rep. 2011;3(2):e11.
113. Palliser D, Chowdhury D, Wang QY, Lee SJ, Bronson RT, Knipe DM, Lieberman J. An siRNA-based microbicide protects mice from lethal herpes simplex virus 2 infection. Nature. 2006;439(7072):89–94.
114. Wu SY, Chang HI, Burgess M, McMillan NA. Vaginal delivery of siRNA using a novel PEGylated lipoplex-entrapped alginate scaffold system. J Control Release. 2011;155(3):418–26.
115. Lai SK, O'Hanlon DE, Harrold S, Man ST, Wang YY, Cone R, Hanes J. Rapid transport of large polymeric nanoparticles in fresh undiluted human mucus. Proc Natl Acad Sci U S A. 2007;104(5):1482–7.

116. Wang YY, Lai SK, Suk JS, Pace A, Cone R, Hanes J. Addressing the PEG mucoadhesivity paradox to engineer nanoparticles that "slip" through the human mucus barrier. Angew Chem Int Ed Engl. 2008;47(50):9726–9.
117. Ensign LM, Tang BC, Wang YY, Tse TA, Hoen T, Cone R, Hanes J. Mucus-penetrating nanoparticles for vaginal drug delivery protect against herpes simplex virus. Sci Transl Med. 2012;4(138):138ra179.
118. Medaglini D, Oggioni MR, Pozzi G. Vaginal immunization with recombinant gram-positive bacteria. Am J Reprod Immunol. 1998;39(3):199–208.
119. Rao S, Hu S, McHugh L, Lueders K, Henry K, Zhao Q, Fekete RA, Kar S, Adhya S, Hamer DH. Toward a live microbial microbicide for HIV: commensal bacteria secreting an HIV fusion inhibitor peptide. Proc Natl Acad Sci USA. 2005;102(34):11993–8.
120. Liu X, Lagenaur LA, Simpson DA, Essenmacher KP, Frazier-Parker CL, Liu Y, Tsai D, Rao SS, Hamer DH, Parks TP, Lee PP, Xu Q. Engineered vaginal lactobacillus strain for mucosal delivery of the human immunodeficiency virus inhibitor cyanovirin-N. Antimicrob Agents Chemother. 2006;50(10):3250–9.
121. Vangelista L, Secchi M, Liu X, Bachi A, Jia L, Xu Q, Lusso P. Engineering of *Lactobacillus jensenii* to secrete RANTES and a CCR5 antagonist analogue as live HIV-1 blockers. Antimicrob Agents Chemother. 2010;54(7):2994–3001.
122. Lagenaur LA, Sanders-Beer BE, Brichacek B, Pal R, Liu X, Liu Y, Yu R, Venzon D, Lee PP, Hamer DH. Prevention of vaginal SHIV transmission in macaques by a live recombinant *Lactobacillus*. Mucosal Immunol. 2011;4(6):648–57.
123. Beninati C, Oggioni MR, Boccanera M, Spinosa MR, Maggi T, Conti S, Magliani W, De Bernardis F, Teti G, Cassone A, Pozzi G, Polonelli L. Therapy of mucosal candidiasis by expression of an anti-idiotype in human commensal bacteria. Nat Biotechnol. 2000;18(10):1060–4.
124. Lindgren M, Langel U. Classes and prediction of cell-penetrating peptides. Methods Mol Biol. 2011;683:3–19.
125. Kanazawa T, Takashima Y, Shibata Y, Tsuchiya M, Tamura T, Okada H. Effective vaginal DNA delivery with high transfection efficiency is a good system for induction of higher local vaginal immune responses. J Pharm Pharmacol. 2009;61(11):1457–63.
126. Kanazawa T, Tamura T, Yamazaki M, Takashima Y, Okada H. Needle-free intravaginal DNA vaccination using a stearoyl oligopeptide carrier promotes local gene expression and immune responses. Int J Pharm. 2013;447(1–2):70–4.

Part III
Case Studies of Mucosal Delivery of Biopharmaceuticals

Chapter 11
Nanoparticles-in-Microsphere Oral Delivery Systems (NiMOS) for Nucleic Acid Therapy in the Gastrointestinal Tract

Shardool Jain and Mansoor Amiji

11.1 Introduction

Gene therapy is based on the concept of introducing genetic material in specific cells of the body with the intention to either upregulate or downregulate the expression of the target gene and subsequently regulate protein synthesis. The genetic material can be in the form of plasmid DNA, small interfering RNA (siRNA) and microRNA (miRNA) duplexes, or single stranded antisense oligonucleotides (ODN) [1, 2]. This form of therapy is considered as an alternative to traditional chemotherapy with fewer side effects and longer expression so that therapeutic levels can be sustained for a significantly longer period of time. However, in order to ensure safe delivery of the nucleic acid construct, the genetic material needs to be protected from potentially degrading enzymes and unfavorable pH conditions. These barriers become even more pronounced when one is trying to achieve delivery of nucleic acid via the oral route of administration [3]. Thus, a delivery strategy to overcome this problem can aid in protecting the payload.

From the perspective of gene delivery, viral and non-viral vectors have been explored; however, potential adverse side effects of viral vectors such as immunogenicity, carcinogenicity or large-scale production have hindered their progress into clinic [4]. In contrast, nonviral counterparts can be designed to overcome some of the viral vector issues but suffer from low transfection efficiency. The non-viral vectors mainly comprises lipid- or polymer-based nanoparticles/microparticles. Advantages of such systems over one another and their preparation methods have been reviewed elsewhere [5–7]. This review will highlight some of the prominent examples from literature to emphasis the emergence of a new type of delivery system where the payload can be encapsulated into nanoparticles that are in turn loaded into microparticles. The potential advantages of such systems over traditional delivery platforms

M. Amiji (✉) · S. Jain
Department of Pharmaceutical Sciences, School of Pharmacy,
Northeastern University, Boston, MA, USA
e-mail: m.amiji@neu.edu

J. das Neves, B. Sarmento (eds.), *Mucosal Delivery of Biopharmaceuticals*, DOI 10.1007/978-1-4614-9524-6_11,

are dual protection, encapsulation of multiple payloads, and sustained release of the nucleic acid in the cell for efficient gene transfection of silencing.

11.2 Oral Gene Therapy

Oral gene therapy has the ability to significantly impact the local and systemic diseases such as inflammatory bowel disease (IBD), gastric or duodenal ulcers, gastrointestinal infections, and septic-shock. In addition, DNA vaccination can also be envisioned as a gene therapy arm that can impart mucosal and systemic immunity against some of these pathological conditions. The unique features of gastrointestinal tract (GI) such as large surface area of the gut epithelium allow for particle uptake and nucleic acid expression, and access to luminal site of inflammation via both oral and rectal routes of administration can be utilized for the effective delivery of the payload [3]. The following sections will highlight some of the key pathological conditions along with key targets that can be utilized for gene therapy approach.

11.2.1 Treatment of Localized GI Diseases

One of the main areas of research with oral gene therapy has been visualized in the treatment of local gastrointestinal diseases such as IBD, *Helicobacter pylori* bacterial infection, and peptic ulcer disease, and periodontal diseases. IBD is a pathological condition that comprises two separate disorders: Crohn's disease and ulcerative colitis [8]. Although, these two conditions some common features, the main difference lies in between the events that lead up to the disease. The pathogenesis of the disease mainly involves hyper-activation of mucosal immune response against the normal luminal flora [9]. As a consequence, the resident macrophages, dendritic cells, and T-cells secrete proinflammatory cytokines that can lead to cascade of events resulting into IBD. Conventional chemotherapy for the treatment of the disease mostly includes treatment with anti-inflammatory drugs such as 5-aminosalicylic acid (ASA), azathioprine (prodrug), methotrexate, cyclosporine, and corticosteroids [10]. However, these drugs are nonspecific in their action and are required in high doses to maintain the therapeutic effects, which ultimately cause serious side effects such as diarrhea, abdominal pain, difficulty in breathing and swallowing. Similarly, biological therapies including antibody-based treatment have also been explored as an alternative and are usually utilized as a last resort option to treat refractory disease. In particular, anti-TNF-α (e.g., Remicade®) monoclonal antibody and other anti-inflammatory biological agents have been utilized with marginal success in the clinical setting. However, systemic side effects and patients tending to lose response or become intolerant are major causes of concern with the application of these biologics [11]. Thus, gene therapy approach can be considered as a viable option to restore the pro- vs. anti-inflammatory cytokine balance and in process improve the local intestinal architecture.

Nakase et al. (2003) reported developing gelatin microspheres containing murine Interleukin 10 (IL-10) plasmid DNA, which can be released in a sustained manner at local site of action while retaining its biological activity [12]. They administered these microspheres rectally to IL-10 knockout mice to investigate whether this treatment could ameliorate colitis. The colitis was induced in Balb/C mice via treatment with 5 % dextran sodium sulfate. Colonic inflammation was remarkably reduced in GM-IL-10-treated mice as compared to mice treated with IL-10 alone. Also, the expression of CD40 on Mac-1-positive cells was significantly decreased upon treatment with GM-IL-10 microspheres in comparison to IL-10 alone. The success of this therapy was also marked by a decrease in histological score, myeloperoxidase activity, and nitric oxide production compared with those treated with free agents. Additionally, the gene expressions of cytokines Tumor necrosis factor-α (TNF-α), Interleukin-1β (IL-1β), and Interferon-γ (IFN-γ) were downregulated in treated animals. Serum IL-10 levels and systemic macrophages were unchanged after treatment. This study suggests that local macrophages in the intestine play a critical role in the initiation of chronic colitis in the animal model of IBD. A drug delivery system using these microspheres containing immunomodulatory IL-10 (GM-IL-10) might be useful for treatment of patients with IBD.

Similarly, from the perspective of local delivery, another avenue of research has been focused on treatment of periodontal re-generation. The disease is marked by inflammatory reactions elicited by bacterial biofilms on the gingival tissues. The deposition of the biofilm ultimately results in the loss of alveolar bone, cementum, and periodontal ligament [13]. The surgical reconstructive procedures such as bone allografts, autografts, or cell occlusive barrier method to restore the lost tooth support have shown limited success in terms of healing response. As an alternative to these procedures, delivery of growth factors such as platelet-derived growth factor (PDGF) or bone morphogenetic protein (BMP) have been explored to stimulate bone growth and fill out the periodontal defects [14]. From the perspective of gene therapy, most of the work has been either done with viral vectors or biodegradable scaffolds that can act as slow releasing depot systems for plasmid DNA release and transfection. Although there is limited data on the development of nanoparticle/microparticle-based delivery systems for periodontal gene therapy, the proof-of-concept studies performed in cell culture systems provide encouraging evidence on the transfection ability of a nonviral delivery system for PDGF gene delivery [15, 16]. For example, Elangovan et al. (2012) utilized calcium phosphate-based nanocomplexes (NCaPP) for PDGF-B plasmid DNA delivery into fibroblasts [17]. The particles were reported to be 30–50 nm in size. Cytotoxicity studies revealed that these particles were relatively nontoxic to the cells as compared to positively charged polymer such as poly-(ethyleneimine) (PEI). The initial transfection studies conducted with reporter GFP plasmid DNA (20 μg dose) showed that these nanocomplexes were superior in their ability to deliver the payload to the fibroblasts as compared to commercially available transfection reagent, Lipofectin®. Afterwards, using RT-PCR and PDGF-B-specific enzyme-linked immunosorbent assay (ELISA), the group evaluated the transfection profile of these particles with plasmid DNA (20 μg dose) encoding for PDGF-B gene. It was noteworthy here that in order to eliminate the effect of exogenous PDGF-B factor in the normal cell culture media, these experiments were conducted in depleted 1 % serum conditions. It was reported that highest transgene expression of 67 pg/ml was observed at 48 h posttransfection with detectable level up to 96 h in the case of calcium-phosphate nanocomplexes. Lastly, results of the cell proliferation assay indicated that cells treated with NCaPP-PDGF-B nanoparticles continued to divide and proliferate even in the 1% serum conditions due to sufficient levels of transfected and secreted PDGF-B growth factor. In conclusion, the group demonstrated the efficacy of an inorganic nanoparticle system for gene delivery. The authors also highlighted that successful treatment using gene therapy in this case would require nanoparticles being further encapsulated in a biodegradable scaffold that can be placed right into the defective cavity and can act as a reservoir for the release of particles.

11.2.2 Systemic Protein Therapy upon Oral Transfection with Plasmid DNA

Bowman et al. (2008) attempted to deliver a therapeutic plasmid DNA encoding for FVIII (factor VIII) gene using chitosan nanoparticles for systemic absorption through oral route of administration in a mice model of hemophilia A [18]. The authors stated that the main advantage of this approach would be to achieve sustained systemic transgene expression through repeated administration by the oral route. In addition, chitosan polymer was chosen because of its mucoadhesive and permeating enhancer properties. The average particle size was reported to be 300 nm with a zeta-potential of 10 mV at pH = 5.7. The *in vitro* studies conducted in COS-7 cells indicated that functional FVIII protein and mRNA were successfully detected upon transfection with chitosan nanoparticles containing plasmid DNA. It was also mentioned that the transgene levels, obtained at 72 h post transfection, were significantly higher as compared to cells treated with naked plasmid DNA; however, the levels observed upon Lipofectamine® treatment were higher than that of chitosan nanoparticles. The authors also reported that treatment with Lipofectamine® also caused a higher cytotoxicity than chitosan/pDNA complex and reduced the total protein content to about 60 % as compared to 90–110 % observed with chitosan/pDNA treatment. The tissue biodistribution studies were performed with intent to detect the plasmid DNA copy number into different tissues upon nanoparticle treatment. For these experiments, chitosan nanoparticles with a high (600 μg-broken down into five feeding cycles of 120 μg each) and low (50 μg) dose were administered. Naked plasmid DNA at the above-mentioned doses was used as a control in these studies. The results of these studies revealed that apart from stomach, ileum, and Peyer's patches, plasmid DNA was also detected in liver, spleen, and other systemic tissues at 72 h post treatment and up to 2 weeks after last feeding. However, the authors did not observe any differences between the plasmid copy number for high and low dose of chitosan nanoparticles.

Moreover, the levels achieved in both these instances were similar to naked plasmid DNA treatment. Therapeutic studies were conducted in hemophilic mice using human FVIII BDD MLP plasmid DNA. The studies were performed with different doses of DNA (50, 250, and 600 μg). For these experiments, nanoparticles or naked plasmid DNA were mixed with strawberry Jell-O® brand gelatin and were fed to mice overnight. Tissue samples and plasma was collected for the transgene expression using PCR (polymerase chain reaction) and human FVIII activity. In addition, authors performed a phenotypic tail-clip test to show that effective transfection was able to cease the bleeding upon transecting the tail and led to thrombin generation. The naked plasmid DNA was largely ineffective across all the different doses with FVIII expression reaching 1 % at the highest dose. However, in comparison a more expected response was observed with the chitosan nanoparticle treatment group. The lower-to-intermediate doses showed an average FVIII expression of 1.5 %, whereas administration of nanoparticles containing 600 μg plasmid DNA resulted in about 3 % expression efficiency. In addition, modest levels of thrombin generation and antihuman FVIII antibodies were reported upon chitosan nanoparticle administration containing the highest plasmid DNA dose. However, the group reported that FVIII expression persisted for almost a month in mice with this nanoparticle treatment. It was reported that phenotypic bleeding correction was observed in 65 % of the mice treated with either medium or high doses of chitosan-DNA nanoparticles. Although modest protein levels and high variability in gene transfer was observed; nonetheless, it was encouraging to note that detectable levels of the secreted therapeutic protein were observed upon oral delivery. Based on these studies, the authors were optimistic that further improvements in the formulation design can enhance the protein expression and lead to successful gene therapy product.

11.2.3 Oral DNA Vaccination

Vaccination is a proven strategy in the prevention of the infectious diseases and cancer. They can be developed from various sources to generate an immune response and at the same time potentiate the harmful effects associated with an actual infection. Conventional vaccines include live attenuated or inactivated pathogens, antigenic peptides, proteins, and polysaccharides while novel approaches are based on generation of vaccines from genetic material [19]. Some of the vaccine strategies mentioned here have suffered from poor bioavailability, primarily due to the delivery issues. For example, subunit vaccines, such as antigenic proteins, peptides, and polysaccharides, are not ideal candidates for oral and rectal administration as such therapeutics are prone to enzymatic degradation in the GI. DNA vaccines also suffer from similar problems of degradation by harsh pH environment and enzymes [20]. The studies conducted with the naked plasmid DNA have shown that intravenous or intramuscular administration of the DNA elicited a weak immune response due to the restrictive movement, degradation by macrophages, and negligible uptake by myocytes [21, 22]. Besides all the inherent problems, the advantages associated with these modalities in comparison to the live attenuated or inactivated pathogen-based vaccines have prompted researchers to exploit the nanocarrier-based approach to enhance the efficacy of such vaccines. In addition, the nanocarriers may also act as adjuvants to further enhance the immune response by protecting the antigen, modulating cytokine release, activating CD8+ CTL responses or delivering the antigen to target tissue [23].

The mucosal delivery most commonly involves gastrointestinal, urogenital, and respiratory tracts. The delivery of vaccines via the mucosal route is preferred as it can not only generate systemic immune response but can also provide local immune protection [24]. Additionally, mucosal surfaces are considered to be the most common route for pathogen entry into the body and hence, targeting such sites can prevent the invasion by the foreign antigen. The delivery systems for mucosal vaccines have been primarily designed to target mucosal-associated lymphoid tissues (MALT) of the Peyer's patches in the gut and respiratory tract. The tissue is separated from the lumen by the follicle-associated epithelium (FAE), which is composed of enterocytes and specialized microfold (M) cells. These cells are capable of transcytosis of foreign matter from the apical to basal side of the membrane. M-cell basolateral membrane contains a pocket that is deeply invaginated with lymphocytes and antigen-presenting cells (APCs) such as macrophages [25]. Therefore, the strategic placement of macrophages at this site allows the sampling and processing of the foreign antigens which can ultimately lead to activation of T-cells and B-cells and hence, the generation of cellular and humoral immune responses. Therefore, targeting the M-cells for mucosal vaccine delivery can be considered as a passive and effective way to deliver the antigen to APCs.

11.3 Oral RNA Interference Therapy

Gene silencing has emerged as another promising tool to alleviate the expression of a "mal-function" protein by degrading the encoding corresponding messenger RNA. The principle behind the function of RNAi is reviewed elsewhere and interested readers are encouraged to refer to the work of Meister and Tuschl (2004) [26], Mello and Conte (2004) [27], and Filipowicz et al. (2005) [28]. Some of the major issues associated with RNAi-based therapeutics are duplex design and selection, route of administration, off-target effects, and delivery [29]. With the advent of science and computational tools, some of the issues related to design of such synthetic nucleic acid constructs have been met [30, 31]. In addition, in order to limit the off-target effects various chemical modifications have been proposed [32, 33]. However, the effective delivery of the siRNA to the targeted site remains a challenge, and these problems become even more pronounced with regards to the oral delivery, due to the harsh environment and presence of degrading enzymes. Literature review also seems to suggest that oral route is not a preferred choice for siRNA delivery as very limited reports pertaining to particle-mediated delivery are available. The following sections will again highlight some of the key targets that researchers have sought after to demonstrate the efficacy of siRNA-based nanosystems in pathological conditions from the perspective of oral delivery.

11.3.1 Treatment of Localized GI Diseases

Chronic inflammation can occur due to an imbalance between the anti and proinflammatory cytokines. In previous section, the advantage of using IL-10 plasmid gene delivery in the treatment of IBD was discussed. Similarly, gene-silencing approach targeting TNF-α has also been explored. Initial attempts to successfully deliver siRNA via oral route were made by Wilson et al. (2010) [34]. The group attempted to deliver TNF encoding siRNA incorporated into thioketal-based nanoparticles (TKNs) in mouse model of ulcerative colitis. It was mentioned that TKNs were formulated from a novel polymer, poly-(1-4-phenyleneacetone dimethylene thioketal) (PPADT) that would degrade selectively in the presence of reactive oxygen species (ROS). Therefore, abnormally high levels of ROS at the site of intestinal inflammation would trigger the release of siRNA upon TKN degradation. With a very simple test, utilizing gel permeation chromatography, the group showed that when PPADT is incubated with a superoxide solution, the molecular weight of the polymer (9,000 Daltons) decreased to 800 Da in 8 h. However, neither 0.5 N HCl nor 0.5 N NaOH treatments had an effect on the polymer. In order to formulate the particles, the siRNA (TNF-α specific or scrambled) (187.5 μM in nuclease-free water) was initially mixed with DOTAP (1,2-dioleoyl-3-trimethylammonium-propane), a cationic lipid (9.45 mM in 0.4 ml dichloromethane (DCM). The mixture was then added to methanol (MeOH) and the resulting single-phase suspension was vortexed for 60 s. Next, the DCM (0.4 ml) and nuclease-free water (0.4 ml) were added and solution was further centrifuged for 1,000 g for 10 min to separate out the two phases. Subsequently, the organic phase containing siRNA/DOTAP complex were removed and added to 50 mg of PPADT. The organic siRNA/DOTAP/PPADT mixture was then added to 10 ml of 5 % solution of polyvinyl alcohol (PVA) in pH 7.4 (PBS) and the biphasic mixture was homogenized at 17,500 rpms for 60 s. The resulting oil-in-water emulsion was then added to 60 ml of 1 % PVA solution and stirred in an open container to remove the residual DCM. The particles were then isolated via centrifugation and washed three times to remove excess PVA. The resulting formulation was reported to be ∼600 nm in size. The siRNA loading of targeted and scrambled sequence in the particles was reported to be 4.7 and 4.1 μg per mg of particles, respectively.

RAW-264.7 macrophages were used to demonstrate the efficacy of siRNA-loaded particles in a cell culture setting. Briefly, 23 μg siRNA/ml was added to 10^7 cells/well of 12 well cell culture plates and incubated for 4 h. Afterwards, the media was removed, cell were washed, and stimulated with lipopolysaccharide (LPS, 5 μg/ml). Next, after 24 h, 0.1 ml of media from each treatment well was removed and TNF-α levels were analyzed using an ELISA. For these experiments, siRNA/DOTAP complex, empty TKN particles, and PBS saline were used as other relevant controls. The results indicated that TNF-α-specific siRNA-TKN particle treatment significantly reduced the expression of the cytokine ($p < 0.05$) as compared to other controls including scrambled sequence and siRNA/DOTAP complexes.

For *in vivo* studies, the colitis was induced by dextran sodium sulfate (DSS) administration in female C57BL/6 mice. DSS was added to the drinking water of all the animals, and mice in all the groups received DSS-treated water for 7 days. Particles containing 0.23 mg siRNA/Kg were administered via oral gavage once daily for the first 5 days. The animals were euthanized on day 7 and colonic tissue was collected for histology, myeloperoxidase activity (MPO), and detecting of proinflammatory cytokines such as TNF-α, IL-6, -1, and IFN-γ.

TNF-α/TKN treatment led to significant decrease in the cytokine levels including scrambled siRNA/TKN treatment group ($p < 0.001$). However, the group also reported that TNF-α/TKN treatment also alleviated the expression of other proinflammatory cytokines. This trend was not observed with other treatment groups. Additionally, histology and MPO activity score seemed to indicate that targeted sequence nanoparticle treatment resulted in intact epithelium, well-defined crypt structures, relatively low neutrophil infiltration, and significantly reduced MPO activity. The other treatment controls were reported to be ineffective and led to uncontrolled inflammation and colonic damage. However, the specificity of the TNF-α siRNA/TKN treatment was questionable as it also led to downregulation of other cytokines. It is even more suspicious that control nanoparticle did not cause such an effect. The authors also did not provide any arguments to support the data nor did they address any off-target effects resulting from siRNA sequence. It was concluded that superior stability of nanoparticles in GI fluids and site-specific siRNA release were some of the key attributes behind the success of the thioketal-based formulation.

11.3.2 Systemic Gene Silencing upon Oral siRNA Administration

Aouadi et al. (2009) investigated the possibility of employing gene-silencing approach in an animal model of septic shock using a multicompartmental macrophage targeted delivery system [35]. Toward this end, the group engineered a natural, multicompartmental delivery system made up of β-1.3-α-glucan (GeRPs) containing siRNA. It was mentioned that glucan-based particles can specifically interact with dectin-1 (also known as CLEC7A) receptor on the surface of macrophages. The group hypothesized that such an approach can be used to target the underlying gut-lymphoid lymphatic tissue (GALT) macrophages upon oral administration of GeRPs. It is very well known that GALT macrophages can traffic away from gut and infiltrate reticuloendothelial system tissues such that over time some percentage of total macrophage population will contain ingested GeRPs containing therapeutic siRNA. Therefore, targeting GALT macrophages can be regarded as a valid approach to impart protection against systemic inflammatory condition such as septic shock.

With regards to the formulation design, hollow β-1, 3-α-glucan shells (2–4 μm) were first purified by treating baker's yeast with a series of alkaline, acid, and solvent extractions to remove cytoplasm and other cell wall-associated polysaccharides. The, empty particles were first diluted with sterile saline and then incubated with 1 nmole Endo-porter for 1 h at 20–25 °C. The pmoles of siRNA was added and incubated at same temperature for 2 h. Afterwards, polyethylenimine (PEI) (5 μg in saline) was added while vortexing and incubating for 20 min to trap siRNA. The PEI was then quenched by addition of 0.6 ml of complete DMEM-media. The group tested the delivery efficacy of the formulation with siRNA sequences specific for Map4k4 (a germinal center protein kinase that facilitates TNF signaling) in both peritoneal exudate cells and LPS-induced septic shock animal model. The group conducted a series of experiments separately to demonstrate that Map4k4 defines a new proinflammatory pathway that is capable of activating TNF-α and works independently of the traditional JNK1/2, p38, and NF-κB route. Later on, target validation was achieved by silencing Map4k4 expression and which in turn decreased the expression of TNF-α in LPS macrophages. 5′RACE (rapid amplification of cDNA ends) and nested PCR was used to analyze the Map4k4 and TNF levels at the transcriptional level. Thus, the group then decided to also include Map4k4 siRNA sequence in their experimental design and further test this hypothesis in an *in vivo* model. The siRNA dose used for these studies was reported to be 20 μg/kg. Readers interested in the siRNA sequences are encouraged to refer to the paper or contact the author(s) directly to procure the information.

For the *in vitro* experiments, the peritoneal exudate cell macrophages from C57BL6/J male mice were isolated after 1–5 days following intraperitoneal (IP) injection of thioglycollate broth. The particles containing the targeted sequence (Map4k4) were incubated with cells for 48 h at a 10:1 particle/cell ratio. The siRNA dose used for these studies was 40 pmoles. The cells were then stimulated with LPS for an additional 6 h. Interestingly, Map4k4 silencing decreased the TNF-α expression at the mRNA and protein level by 50 and 30 %, respectively, as compared to scrambled siRNA sequence. Furthermore, the unloaded GeRPs and phosphate buffered saline (PBS) treatment failed to have any silencing effect.

Next, the animal studies were conducted in 10-week-old C57BL6/J male mice. Particles (4×10^9) containing either Map4k4 specific or scrambled sequences were administered via oral gavage for days 1–8, consecutively. Subsequently, 25 mg of D-galactosamine (D-GalN) and 0.25 μg Escherichia coli LPS was injected via IP route. Animals were monitored for the survival assessment for a period of 24 h and then PECs were collected for analysis. Additionally, blood samples were also collected at 1.5 and 4 h post LPS/D-GalN injection for secreted TNF-α measurements. It was mentioned that these time points were chosen because circulation TNF-α levels tend to peak at 1.5 h post LPS/D-GalN administration and tend to normalize to basal levels after 4 h. Notably, an 80 % decrease in the TNF-α mRNA expression was reported in mice orally gavaged with Map4k4 siRNA containing GeRPs as compared to scrambled sequence. Interestingly, same amount of knockdown of IL-1β was also reported with this targeted sequence treatment. However, levels of IL-10 and chemokine receptor 2 (Ccr2) remained unchanged. The serum and peritoneal fluid TNF-α levels observed at 1.5 h post LPS/D-GalN challenge were also significantly reduced in comparison to scrambled siRNA treatment and these results were in agreement with the PCR data. In addition, the unloaded GeRPs, scrambled Map4k4 siRNA sequence containing particles, and PBS saline treatments were largely ineffective in suppressing TNF levels. Lastly, the survival data generated at 24 h post LPS lethality challenge indicated that 90 % of the mice in the scrambled siRNA sequence treatment group died between 4 and 8 h post challenge. In contrast, 50 % of the animals treated with Map4k4- specific siRNA containing particles survived for 8 h (11/22 mice) and 40 % (8–9/22 mice) survived the LPS challenge long term. The group also stated that regardless of the GeRPs treatment, serum levels of liver enzymes (aspartate aminotransferase and alanine aminotransferase) were within the normal range, and serum insulin and glucose levels were unaffected. Supplementary data with TNF-α-specific siRNA containing particles was also reported in this paper and was found to be equally effective. Also, GALT macrophage containing fluorescent GeRPs trafficking to other organs such as liver, spleen, and lungs was also demonstrated. Overall, the succinct design of the formulation and ability to target GALT macrophages was successfully demonstrated here. Additionally, the authors were able to show that GALT macrophages can act as Trojan horse carrier for their particles and traffic them to other organs and in turn provide protection against septic shock induced upon LPS administration.

11.4 Nanoparticles-in-Microsphere Oral System (NiMOS)

As previously discussed, several polymer-based formulations have been used for the delivery of drug or biologic therapeutic moiety to the intestine mucosa. Ideally, targeting this particular region requires particles below the size range of 10 μm [36, 37]. The concept of nanoparticle-in-microsphere formulation as a multicompartment delivery system is attractive as the microsphere coating on the exterior can protect the payload from enzymatic degradation and harsh environmental conditions, such as acidic pH of the stomach. To accomplish the delivery objectives, Bhavsar et al. [38] proposed to formulate a NiMOS platform, using poly (ε-caprolactone) (PCL) as the external microsphere encapsulation gelatin nanoparticles containing the plasmid DNA. The authors mentioned that the PCL was chosen because of its inherent hydrophobic nature to enhance the mucosal uptake, delayed degradation profile, and resistance to acidic pH. On the other hand, type B gelatin was chosen to synthesize the nanoparticles for condensing, hydrogel-type system to encapsulate nucleic acids. The rationale behind selection of gelatin was based on its hydrophilic properties that can increase the entrapment efficiency of the polymer and at the same time avoid high-energy sources required to formulate hydrophobic particles. In addition, the group mentioned that gelatin has been extensively used as an excipient in various oral and parenteral formulations.

The process of making type B gelatin nanoparticles encapsulating plasmid DNA was achieved by controlled precipitation technique described by Kaul and Amiji [39, 40]. However, authors employed a mathematical surface response 3^3-factorial design model, which constituted of 27 overall experiments to optimize the NiMOS particles. The variables selected for this formulation were amount of gelatin nanoparticles (X_3) (average particle size = 100 nm) in internal phase, PCL concentration

(X_1), and the homogenization speed (X_2). Following quadratic equation was generated upon performing the regression analysis on the particle size data collected upon changing the variables in a simultaneous order:

$$\begin{aligned} Y = & 11.789 + 4.804X_1 - 0.684X_2 - 4.646X_3 + 0.562X_1^2 + 2.482X_2^2 \\ & + 1.543X_3^2 - 0.870X_1X_2 + 0.955X_2X_3 - 2.70X_1X_3 - 0.431X_1X_2X_3 \end{aligned}$$

Surface plots were then constructed to visualize the impact of changing the variables on the size of the particles. In order to form these particles, a double-emulsion technique was employed. Briefly, fluorescein-5-isothiocyanate (FITC) labeled gelatin nanoparticles (dye labeled gelatin was used with the intent to perform confocal microscopy on the microspheres) of varying amounts (10, 30, and 60 mg) were suspended in 0.5 ml of distilled water and homogenized with PCL of varying concentrations (1, 5, and 10 % w/v) in dichloromethane (DCM) at 5,000, 7,000, and 9,000 rpm using a homogenizer to form a stable emulsion like system. Afterwards, the suspension was homogenized with 20 ml of 0.1 % (w/v) polyvinyl alcohol (PVA) in deionized distilled water for 5 min and then magnetically sterilized until the DCM is evaporated. The microspheres were then collected via centrifugation, washed and lyophilized for particle size and morphology analysis.

The results of these studies indicated that the particle size was strongly dependent on the variables selected. The quadratic equation generated by regression analysis to predict the particle size based on the three variables was validated by coulter counter particle size measurements and scanning electron micrograph imaging. As shown in Fig. 11.1, with X1 = 3 % w/v, X2 = 31 mg of fluorescein isothiocyanate conjugated gelatin nanoparticles, and X3 = 9,000 rpms, the model predicted the value to be 8 μm whereas the coulter size analysis revealed a particle size of 9.5 μm. Overall, it was reported that a higher concentration of the PCL (10 % w/v) caused an increase in particle size. The authors indicated that reason for such a response could be attributed to the increase in collision frequency of the formed particles. On the other hand, speed of homogenization and amount of gelatin nanoparticles constituting the internal phase had a negative effect on the size of the particles. In the former case, it was intuitive that the higher homogenization speed will lead to lower particle size; however, it was interesting to note that higher amount of gelatin particles decreased the overall size. It was mentioned that solid internal phase formed by gelatin nanoparticle might not cause the emulsion droplets to fuse with each other and hence, decrease the distortion capacity of the droplets themselves on colliding.

Next, upon successfully formulating the NiMOS formulation, the group looked to determine the encapsulation efficiency of these particles using a reporter plasmid DNA system encoding for enhanced green fluorescent protein (EGFP-N1) [41]. The gelatin nanoparticles containing the plasmid DNA was formed via ethanol precipitation method. Afterwards, the nanoparticles were homogenized with 0.5 % (w/v) PCL in dichloromethane to form a stable dispersion system. The encapsulation efficiency of the plasmid DNA in gelation nanoparticles and subsequently in NiMOS formulation was determined in two separate experiments. In the former case, a known

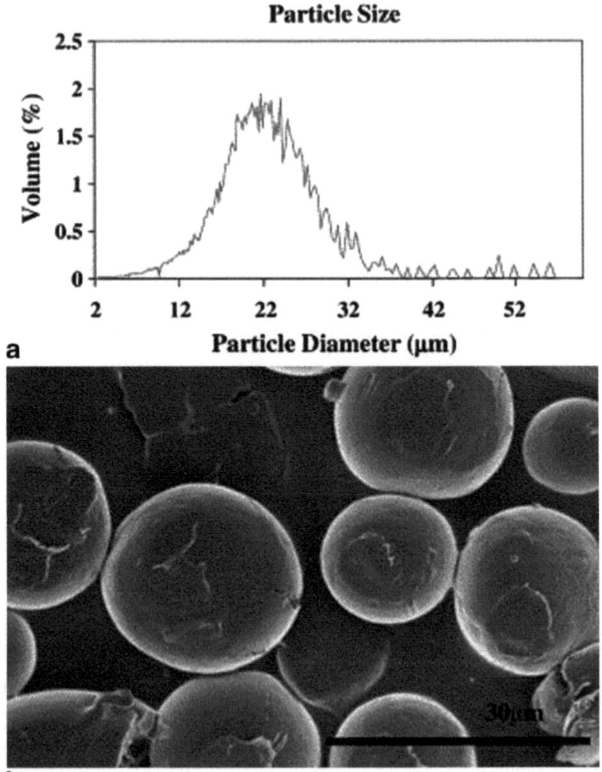

Fig. 11.1 Particle size analysis plot (**a**), and scanning electron micrograph (**b**) of the nanoparticle-in-microsphere oral delivery system with less than 10 μm particle diameter. (Reprinted from [38], copyright (2005), with permission from Elsevier)

amount of nanoparticles were dissolved in phosphate buffered saline (PBS, pH-7.4) containing 0.2 mg/ml protease and incubated at 37 °C for 1 h. The rationale behind protease treatment was to degrade the gelatin matrix that will result in the release of the plasmid DNA in the suspension. The amount of DNA in the resulting suspension was then quantified using PicoGreen® assay and reported as μg DNA/mg of nanoparticles (in terms of loading capacity) and percent DNA loaded vs. the initial amount of DNA added (in terms of loading efficiency). In the latter case, the loading capacity and efficiency for NiMOS formulation was determined by first dissolving the PCL matrix in DCM followed by separation of the gelatin nanoparticles upon addition of equal amount of distilled water. Plasmid DNA was then extracted from the gelation nanoparticles in the same fashion. Based on these experiments, it was reported that the average loading capacity and efficiency in the gelatin particles was 8.21 μg DNA/mg of nanoparticles and 93.2 %, respectively. On the other hand, the average plasmid DNA loading was significantly lower in the case of NiMOS particles and was reported to be 1.73 μg DNA/mg of nanoparticles and 46.2 %, respectively. The other key experiments carried out in this study were the evaluation of DNA release profile in the presence of protease and lipase enzymes and stability of encapsulated DNA in the gelatin and NiMOS particles using gel electrophoresis. The results of these experiments showed that NiMOS formulation not only had a more controlled release profile where the 100 % of the payload was recovered at ∼ 8 h of incubation, whereas most of the plasmid DNA was released from the gelatin nanoparticles in about 3 h. More importantly, the DNA stability studies showed that plasmid DNA is preserved in its native supercoiled form in gelatin and NiMOS formulations and was protected from degradation upon treatment with DNA degrading enzyme (DNAse I). However, treatment of gelatin nanoparticles with protease, followed by DNAse treatment resulted in degradation of the payload. This phenomenon was not observed in the case of NiMOS formulation. Lastly, by means of *in vivo* transfection studies with reporter plasmid DNA (EGFP-N1), the group showed that external PCL matrix of the NiMOS formulation was essential in protecting the plasmid DNA in stomach in the presence of degrading proteolytic enzymes. At the same time, high lipase activity in the small and large intestine resulted in degradation of PCL matrix and subsequent release of gelatin nanoparticles at the targeted site. Interestingly, only in the case of NiMOS formulation-treated animals, the GFP expression was detected in the small and large intestine up to 5 days post oral administration.

Similarly, in a different set of studies, the group used the same formulation to encapsulate TNF-α-specific siRNA (small interfering RNA) as a therapeutic intervention for treatment of IBD in mice model [42]. The particles containing siRNA were formed in very similar fashion with slight modifications. The particle size of the NiMOS microspheres was reported to be between 2.4 and 3 μm. The siRNA loading efficiency was determined in the same way as with plasmid DNA-loaded particles. The average loading efficiency of siRNA in gelatin nanoparticles and NiMOS microspheres was reported to be 90.2 and 55.2 %, respectively. The siRNA stability studies upon proteolytic and RNase treatment also highlighted the superior ability of the NiMOS microparticles to protect the nucleic acid cargo from enzymatic degradation.

11.4.1 Oral IL-10 Plasmid DNA Delivery with NiMOS

Bhavsar and Amiji (2008) employed nanoparticle-in-microsphere oral system (NiMOS) to evaluate the potential of murine interleukin-10 (IL-10) gene therapy for the treatment of IBD [43]. The nanoparticles containing the murine IL-10 plasmid DNA were made of type-B gelatin polymer and were encapsulated into the matrix of the poly (epsilon-caprolactone) microsphere system. NiMOS system was reported to be 2–5 μm in size, with the overall DNA loading efficiency of 46 %.

For the current study, an acute colitis model was established upon rectal administration of the hapten, trinitrobenzenesulfonic acid (TNBS), in female Balb/c mice. Acute colitis-bearing mice were randomly divided into two treatment groups: NiMOS formulation containing 100 μg of IL-10 plasmid DNA and gelatin nanoparticles containing 100 μg of IL-10 plasmid DNA. In addition, the third group comprised animals with no treatment. The formulations were administered by the oral gavage in fasted conscious mice. It was reported that 4 days post particle administration, the animals were euthanized and large intestines were excised for evaluation of expression of IL-10 transgene and therapeutic efficacy.

RT-PCR and IL-10 specific ELISA techniques were employed to determine the expression of the IL-10 protein. On the other hand, the therapeutic efficacy was determined by taking into consideration number of factors such as inflammatory cytokine and chemokine profiling, macroscopic evaluation was performed by determining loss in body weight, stool inconsistency, and rectal bleeding. In addition, colon length and weight were also incorporated as measurement criteria. The group also reported that histological analysis on the excised colon was also performed for mucosal architectural change, cellular infiltration, external muscle thickening, presence of crypt abscess, goblet cell depletion, signs of edema, surface epithelial cell hyperplasia, and signs of epithelial regeneration.

The results of the body-weight measurements and clinical score assigned on the parameters of loss in body weight, stool inconsistency, and rectal bleeding indicated that treatment with the NiMOS formulation led to the complete restoration of the body weights after 4 days of particle administration. It was also mentioned that prior to the administration of the formulation, the animals lost 10 % of their original body weight. In comparison, the loss in the body weight of the animals in the gelatin nanoparticle treatment group was as high as 25 % of the original weight after 4 days of particle administration. Based on the clinical score adopted by the group, animals in the NiMOS treatment group were assigned a clinical score of less than 1, indicating normal activity whereas the gelatin nanoparticle group was assigned a score of 3. It was also mentioned that the same trend was observed with the colon length and weight measurements where treatment with only NiMOS-DNA formulation led to increase in colon length and weight to normal baseline levels.

Lastly, tissue myeloperoxidase (MPO) activity was also determined as a measure of inflammation that is based on the infiltration of the neutrophils. As per the results obtained for the RT-PCR and IL-10 specific ELISA, the levels of the IL-10 observed in the case of the NiMOS treatment group ($\sim$ 180 pg/mg of tissue) were significantly higher ($p < 0.005$) than the levels observed in gelatin nanoparticles (25 pg/mg of tissue) and animals with colitis (no treatment) group. The group attributed the results to the ability of the NiMOS formulation to protect the payload from gastrointestinal barriers and exposure of the particles to the underlying cells of the GI tract in case of colitis. Next, the group compared the therapeutic effect of the IL-10 by measuring the levels of the cytokine and chemokines such as IL-1, 12, TNF-α, IFN-γ, MIP-1α, MCP-1, and RANTES. The results of this study indicated that levels of these cytokines, except RANTES, were significantly lower in the NiMOS treatment group in comparison to the controls including gelatin nanoparticles. The group also acknowledged that decrease in the levels of chemokines such as MIP-1α and MCP-1 may not be a consequence of direct effects of IL-10 cytokine expression. Also, no change in the levels of the RANTES among the different treatment groups was attributed to the acute model of colitis employed in this study.

Similarly, tissue histology along with the MPO activity as a measure of tissue cell infiltration was also evaluated post treatment (Fig. 11.2). These results also agreed with the previous observations where treatment with DNA containing NiMOS formulation led to the restoration of the normal colon architecture and MPO activity was significantly reduced, almost reaching normal levels. In comparison, it was

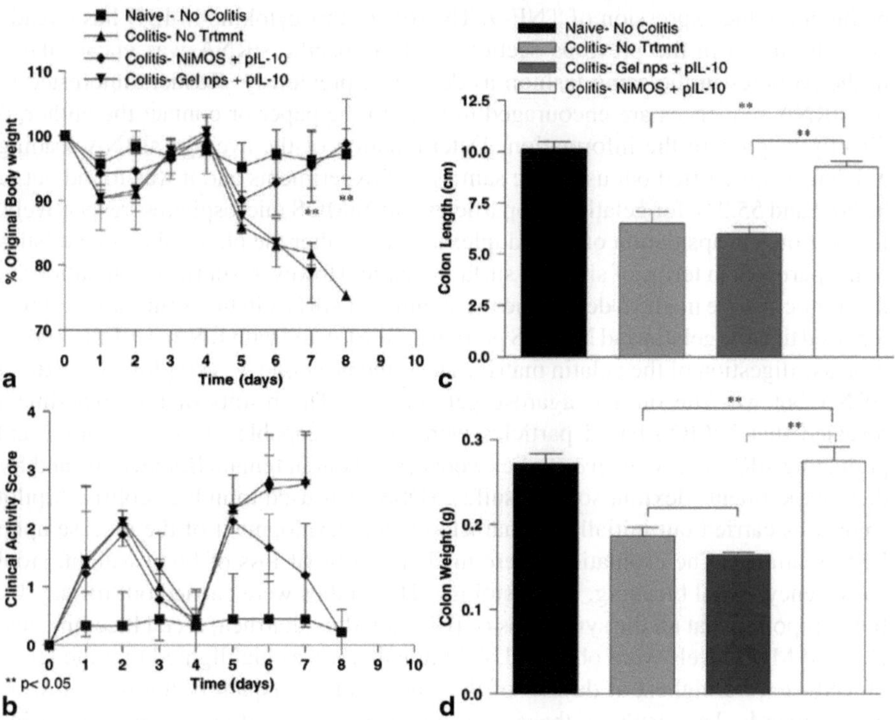

Fig. 11.2 Changes in body weight, clinical activity score, and the lengths and weights of colonic tissue upon oral administration of murine interleukin (IL)-10-expressing plasmid DNA in nanoparticles-in-microsphere oral system (NiMOS). The body weight change was used as a marker of therapeutic efficacy achieved with locally expressed IL-10 over the course of 8 days (**a**). The clinical activity scores in control and treatment animals as measured using an aggregate of body weight changes, rectal bleeding and stool consistency (**b**). Additionally, the colon length (**c**), and colon weights (**d**) were also measured. Each conscious animal received a 100 μg oral dose of pORF5-mIL-10 in gelatin nanoparticles or NiMOS. Mean ± S.D. ($n = 4$). (Reprinted from [43], copyright (2008), by permission from Macmillan Publishers Ltd)

reported that lack of ability of the gelatin nanoparticles to successfully deliver the transgene to the cells of the colon inflicted colitis resulted in no change in the ongoing inflammation process occurring in the colon which was marked by loss in protective epithelial layer and heavy infiltration of the immune cells (as indicated by the high MPO activity score).

11.4.2 TNF-α and Cyclin-D1 Gene Silencing with Oral siRNA Delivery

As an extension of this work, same formulation was used to deliver the TNF-α specific siRNA for the treatment of IBD [42]. The underlying idea behind this approach was to again reinstate the balance between pro- and anti-inflammatory cytokines by blocking the expression of TNF-α. The role of this cytokine in IBD has already been discussed in the previous section. TNF-α specific siRNA was encapsulated in the particles in the same fashion as described previously. Readers interested in the siRNA sequences are encouraged to refer to the paper or contact the author(s) directly to procure the information. Determination of the average siRNA loading efficiency was carried out using the same set of experiments and it was found out to be 90.2 and 55.2 % for gelatin nanoparticles and NiMOS microspheres, respectively. In addition, encapsulation of such duplexes did not alter the physical characteristics of the particles in terms of size and surface charge. However, subtle modifications to the protocol were made to determine the stability of siRNA in the formulation. It was reported that the gelatin and NiMOS particles were treated with RNAse A followed by protease digestion of the gelatin matrix. The same principle was employed to extract siRNA but was run on 4 % agarose gels instead. The results of this experiment revealed that NiMOS-based particles were equally capable of encapsulating and protecting siRNA. The *in vivo* studies were carried out in female Balb/c mice and for these experiments dextran sodium sulfate (DSS) was used to induce colitis. A pilot study was carried out initially to characterize the development of the disease upon DSS treatment. The evaluations were made in terms of loss of body weight, stool consistency, rectal bleeding, and histology. The studies were carried out till day 14. It was reported that all the symptoms of IBD including diarrhea, rectal bleeding, and elevated MPO levels were observed. Additionally, it was highlighted that the levels of TNF-α were highest at day 10 of the study and subsequently normalized from there onwards. The results of these experiments were critical in deciding the dosing regimen of the siRNA treatment. Therefore, based on these observations control and siRNA loaded particles, at a dose of 1.2 mg/kg, were administered orally at days 3, 5, and 7. The treatment efficacy in terms of TNF-α silencing and subsequent alleviation in inflammation was determined by collecting the samples on day 10 and 14. The evaluations were made both qualitatively and quantitatively in terms of tissue histology, MPO activity, multiplex cytokine/chemokine-specific ELISA, and real-time PCR.

As reported, TNF-α levels at the transcriptional and translational levels were two/three fold lower at days 10 and 14 in the TNF-α siRNA containing particles as compared to other treatment groups including DSS control, blank and scrambled-siRNA duplex sequence containing NiMOS microparticles. The multiplex ELISA panel results revealed that proinflammatory cytokines such as IL-1β showed a 3, 2, and 1.5-fold decrease upon TNF-α siRNA NiMOS treatment as compared to other groups at day 14 of the treatment. However, a completely different cytokine profile was observed on day 10 where the lowest levels of cytokines such as IL-2, 5, 6, 12p70, and IFN-γ were observed in the inactive NiMOS containing scrambled siRNA sequence. However, by day 14, the levels of these cytokines were lower in TNF-α siRNA NiMOS group (not statistically significant all the cases). The authors stated that these results could be attributed to the off-target effects of the scrambled TNF-α siRNA sequence, which might have contributed to the lower levels of these cytokines on day 10 and described these effects as an unspecified reaction.

Furthermore, the results of histology indicated that typical signs of IBD such as cellular infiltration, goblet depletion, and irregular mucosal structure were observed in the DSS control, blank, and scrambled NiMOS groups; whereas TNF-α specific siRNA containing particles showed a more robust response and intestinal morphology resembled that of naïve mice that were not administered with DSS. Moreover, as shown in Fig. 11.3, the other tests showed that treatment with TNF-α specific siRNA NiMOS minimized loss in body weight (8 % of original B.W.) and decrease in colon length (6 cm) whereas treatment with DSS control, blank, and scrambled NiMOS formulation caused a 19, 26, and 18 % loss of original body weight. The colon length of naïve mice was reported to be 8 cm. Lastly, MPO activity score of the TNF-α siRNA NiMOS group was comparable to normal rats whereas these values were significantly higher in the rest of the groups.

Later on, the same group delivered a dual siRNA sequence encoding for TNF-α and cyclin D1 (Ccnd1) [44]. Although the exact role of cyclin D1 is not known, it is a key regulator in cell cycle and is involved in the progression from G1-S phase. It has been reported that it is over expressed in many human cancers and inflammatory diseases and hence, serves as a potential target for gene silencing therapy in IBD. As part of the experimental design, acute colitis was induced upon DSS (3.5 % w/w) administration. Then, the animals were randomly distributed into different groups. The animals were fasted overnight and then the particles containing either TNF-α alone, Cyclin D1 (CyD1) alone, or a combination of two sequences were administered at day 3, 5, and, 7 via oral route. DSS (no treatment), blank particles, and scrambled siRNA sequences were used as controls. The animals were euthanized at day 10 and 12 and samples were collected for same set of analysis as in previous study.

Interestingly, combined siRNA treatment led to a much lower TNF-α mRNA expression as compared to TNF-α siRNA NiMOS treatment alone at day 10. However, scrambled siRNA NiMOS and CyD1 siRNA particles were relatively ineffective in reducing the TNF-α levels. These results also indicated that specificity of the treatment as CyD1 siRNA treatment did not alter the TNF-α levels and instead led to upregulation of the same. The same trend was observed in the opposite scenario where TNF-α siRNA alone was not able to alleviate the expression of cyclin D1 at day 10 or 12. Also, it was reported that at day 12, the level of TNF-α was comparable among all the three silencing groups. Similarly, experiments were performed to quantify the protein levels of TNF-α and cyclin D1 via ELISA and western blot analysis, respectively. These results also corroborated with PCR results. It was shown that expression of TNF-α and a multitude of other proinflammatory cytokines such as IL-1α, IL-1β, IL-5, and IL-17 and chemokines (MIP-1α, MCP-1, and GM-CSF) were decreased on day 12 as compared to DSS control. Interestingly, the CyD1-siRNA treatment was also very effective in downregulating the expression profile of these cytokines and chemokines, and in some instances was either comparable or better than the combined therapy. The authors mentioned that potential dilution effect in the combined treatment can be accounted for the difference between the two groups. However, these observations can also mean that cyclin D1 can be a more potent target in inflammation and silencing its expression may be more beneficial than

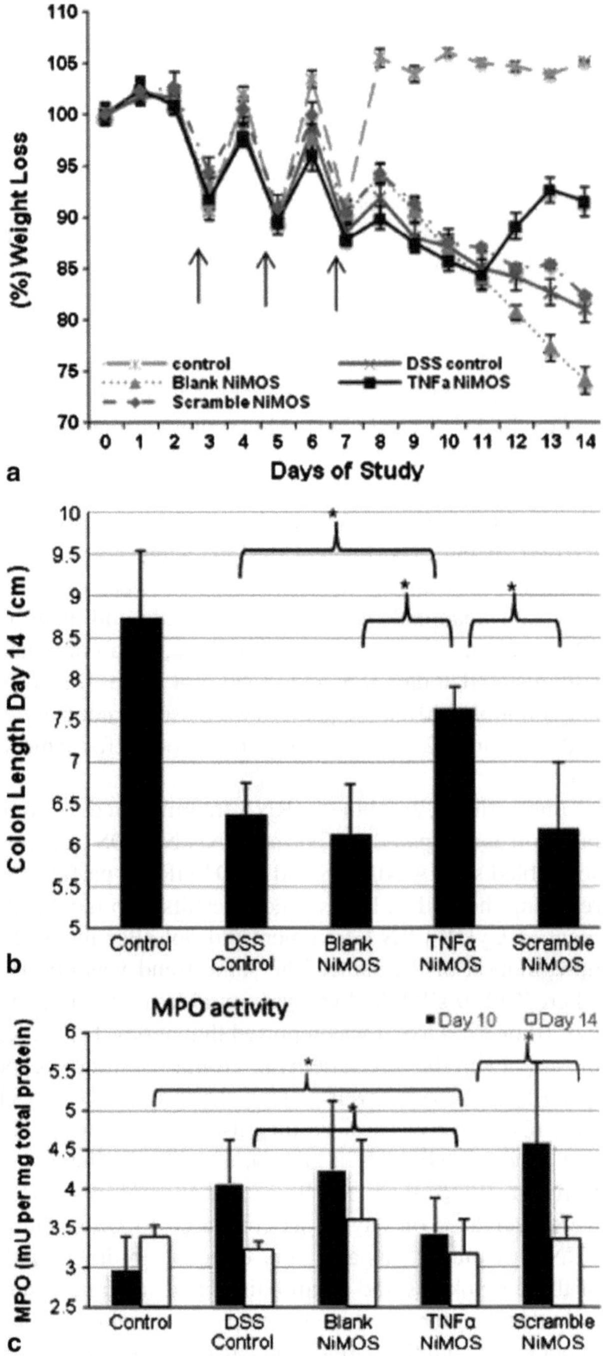

Fig. 11.3 **a** Percent change in body weight of Balb/c mice upon continuous exposure to DSS for development of acute colitis model. **b** Measurement of colon length on day 14. **c** Tissue myeloperoxidase activity in the large intestine normalized to the total protein content of each sample. $p <\leftrightarrow 0.05 =$ *, only significant differences are shown. (Reprinted from [42], copyright (2010), with permission from Elsevier)

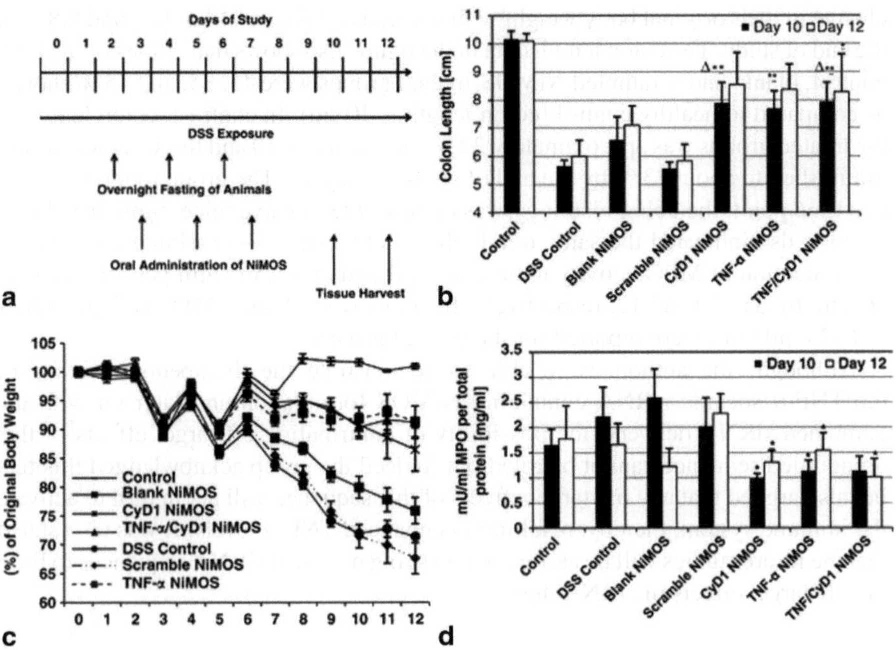

Fig. 11.4 Macroscopic assessment of anti-inflammatory therapeutic efficacy. **a** Timeline of the study. Animals were continuously exposed to 3.5 %(wt) dextran sulfate sodium (DSS) throughout the course of the study. Oral administration of short interfering RNA (siRNA)-containing and blank nanoparticles-in-microsphere oral system (NiMOS) was performed on days 3, 5, and 7 followed by tissue harvest on days 10 and 12, as indicated by the medium and long *arrows*, respectively. **b** Determination of colonic length in control and test groups at both end time points of the study. Silencing NiMOS test groups showed an increase in colon length compared with animals from control groups except the healthy control mice. **c** Percent change of original body weight of Balb/c mice upon continuous exposure to DSS for development of acute colitis model for the duration of the study (12 days). Weight loss was most severe in the DSS control group as well as animals receiving blank or scrambled siRNA sequence NiMOS. The test groups consisting of cyclin D1 (CyD1), tumor necrosis factor-α (TNF-α), and TNF-α/CyD1 siRNA-encapsulating NiMOS exhibited significantly less change in original body weight. **d** Myeloperoxidase (MPO) activity in the large intestine normalized to the total protein content of each sample. Administration of silencing NiMOS led to a reduction in MPO activity in all three test groups on both time points tested whereas elevated levels were measured in the control and DSS control group, as well as groups receiving blank and inactive siRNA sequence-containing NiMOS. Levels represent concentrations obtained from samples on days 10 and 12 of the study (3 and 5 days after administration). Values are expressed as mean ± s.d. ($n = 4$–5). $^{\Delta}P < 0.05$, vs. DSS control; $^{*}P < 0.05$, vs. Scramble; $^{**}P < 0.01$, vs. Scramble; Statistical comparison was performed on data sets of DSS control vs. TNF-α, Cyclin D1, and TNF-α/Cyclin D1 combination NiMOS, and between TNF-a, Cyclin D1, and TNF-α/Cyclin D1 combination vs. Scramble NiMOS group. Only significant differences are shown. (Reproduced from [44], copyright (2011), by permission from Macmillan Publishers Ltd)

that of TNF-α. In addition, as shown in Fig. 11.4, other measures of efficacy such as changes in body weight of colitis-induced mice, colon length, and MPO activity also showed that CyD1, TNF-α, and TNF-α/Cyclin D1 NiMOS showed significantly less change in their original body weight with respective values of 14.5, 8.7, and 8.8 % at the end of study. The colon length data in the figure also shows that colons of the DSS control, blank, and scrambled NiMOS-treated groups were 45, 35, and 45 % shorter as compared to healthy control (colon length ~ 10 cm). In contrast, colon length in the treated groups was approximately 22 % shorter at day 10 and these values further diminished to about 13 % by the end of study on day 12. The group reported that at this time point, the colon histology was comparable to naïve mice. Similarly, MPO activity also indicated the same trend where active siRNA-containing NiMOS led to a reduction in MPO activity to around 1 mU/ml and 1.5mU/ml per total protein content by day 10 and 12, respectively. In comparison, higher MPO activity values (1.6–2.6 mU/ml) were reported for the control groups.

Although, the authors were able to demonstrate the therapeutic efficacy of the TNF-α specific siRNA containing NiMOS formulation and later on with the combined siRNA delivery, the possibility of contributing off-target effects of this unmodified sequence cannot be ruled out. Indeed the group acknowledged this fact but also argued that any off-target effects of this sequence will act in turn to activate the immune system, thereby, resulting in enhanced TNF-α levels. It was also stated that the future studies will be carried out by suing modified siRNA sequences to limit the off-target effects of siRNA therapy.

11.5 Other Multicompartmental Delivery Systems

11.5.1 Water-in-Oil-in-Water (W/O/W) Multiple Emulsion Formulation

Apart from NiMOS, our group has been actively involved in developing and characterizing other multicompartmental delivery systems that have proven to be effective both *in vitro* and *in vivo*. An example of such a system is the multicompartmental system, water-in-oil-in-water (W/O/W) multiple emulsion system that was developed with squalane oil for peptide and DNA vaccine delivery in melanoma immunotherapy. The main motivation behind the design of this system was to develop a carrier that can, not only act as a suitable vehicle for immunogen delivery, but target antigen presenting cells (APC) to aid in enhancing the immune response. Previous results from our laboratory and others have also demonstrated the advantage of squalene oil as an adjuvant and the ability of such formulations to get efficiently internalized by APC at the site of injection [45–47]. In this study, the main objective was to investigate the effectiveness of squalene oil-based multiple emulsion containing gp100 antigen as a vaccine delivery system for melanoma immunotherapy [48]. The efficacy was evaluated both as a prophylactic and active immunotherapy in B16 murine melanoma model. Further details about gp100 protein and its use as an immunogen has been reviewed elsewhere [49, 50].

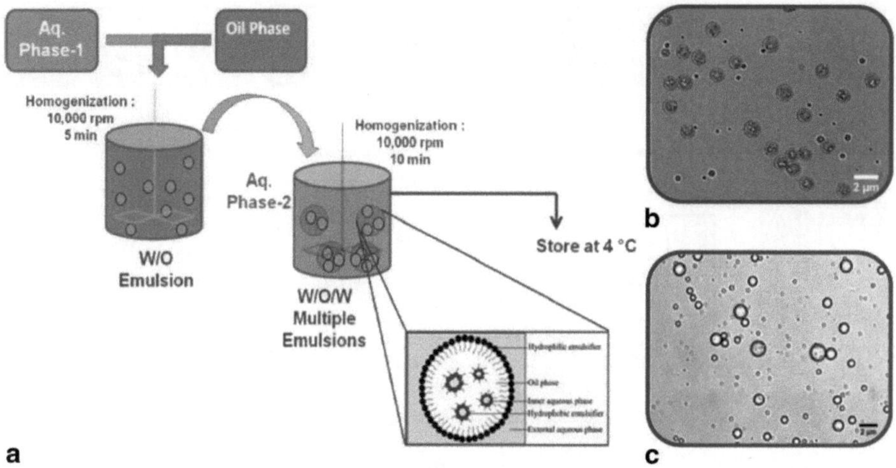

Fig. 11.5 Two-step emulsification method and the bright-field images of the water-in-oil-in-water (W/O/W) multiple emulsion system. (**a**) The W/O/W multiple emulsions-based vaccine formulation was prepared by two-step emulsification method. The first emulsification step involved homogenization of gp100 peptide containing internal aqueous-phase with squalane oil-phase containing Span 80™ to form W/O primary emulsion. The second emulsification step involved homogenization of W/O primary emulsion with outer aqueous-phase containing Pluronic® F127 to form the W/O/W multiple emulsions. (**b**) Staining with Evans blue and Sudan red 7B ascertained the phase configuration of the W/O/W multiple emulsions. An internal aqueous-phase (*blue*) encapsulating, dispersed oil-phase (*pink*) that was stabilized in an outer aqueous phase (*gray*). (**c**) The bright-field image of the W/O/W multiple emulsions system with encapsulated gp-100 peptide antigen. (Reproduced from [48], copyright (2012), with kind permission from Springer Science + Business Media, Inc)

Figure 11.5 represents the schematic diagram to formulate the emulsion. Briefly, gp100 peptide solution (3 mg/ml) was emulsified with squalene oil-Span 80™ mixture (9:1) using a homogenizer (Silverson Model: L4RT-A) at 10,000 rpms for 5 min. The resulting primary water-in-oil (W/O) emulsion was further emulsified with Pluronic® F127 solution (0.5 % w/v) using the homogenizer at the same speed for 10 min to produce the multiple water-oil-water (W/O/W) emulsion. The average particle diameter of the oil droplets was reported to be 1.6 μm with a polydispersity index of 0.4. In addition, the average surface charge of dispersed droplets was reported as − 37.9 mV. As indicated in the figure, the oil droplets were round with an internal aqueous phase, a discrete oil phase, and a stabilized outer aqueous phase. The authors used a water and oil specific dyes to characterize the multiple phases of this system.

Next, the *in vivo* experiments were performed in female C57BL/6 mice. A B16-F10 (pigmented melanoma) murine cell line was used to induce tumor in these mice. It was mentioned that approximately 100,000 cells, suspended in 0.1 ml PBS, were injected in the hind flank. For the prophylactic immunization, a total of five treatment groups were used including saline control, W/O incomplete Freund's adjuvant emulsion with and without gp100 peptide, W/O/W squalene oil multiple emulsion (SME) with and without gp100. A total of eight mice per group were utilized for this study.

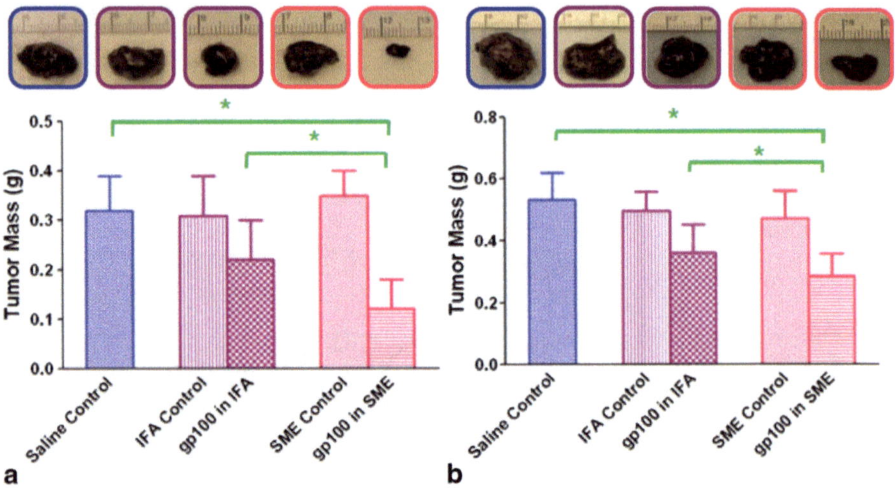

Fig. 11.6 Tumor mass and a representative picture of excised tumors. Animals were euthanized and tumor tissues were excised when mean tumor volume for saline-treated group of mice reached to 1,000 mm^3. Excised tumors were freed of skin remnants, rinsed in phosphate buffered saline and dried on blotting paper. The mass and a representative picture of excised tumors from five treatment groups: (1) saline control (*blue*), (2) W/O IFA emulsion control (*purple*), (3) gp100 in W/O IFA emulsion (*purple*), (4) W/O/W squalane oil multiple emulsions (SME) control (*pink*), and (5) gp100 in W/O/W SME (*pink*) in prophylactic (**a**), and active (**b**) treatment modes are shown. Results are presented as mean ± SD, $n = 8$ ($^*p < 0.05$). (Reproduced from [48], copyright (2012), with kind permission from Springer Science + Business Media, Inc)

All the mice under this protocol received three subcutaneous (s.c.) immunizations of 0.1 ml control or vaccine (50 μg dose/inj.) at a 2-week interval. Ten days after the last immunization, tumor challenge was initiated by s.c. implantation of B16 cells. In contrast, for the active immunization, cells were administered at day 0. Then, at days 1, 4, and 11, mice received either control or vaccine formulations containing the same dose as for the prophylactic therapy. All the animals were killed once tumor volume reached to 1,000 mm^3. Serum samples and tumor tissue were collected for Th1 cytokine analysis and immunohistochemical analysis, respectively. Also, mean tumor volume and tumor growth delay times were recorded as additional measures of efficacy.

Prophylactic treatment with W/O/W emulsion vs. IFA emulsion resulted in about three-fold and two-fold reduction in the mean tumor volume, respectively (Fig. 11.6). In the same vein, a 2.6-fold and 1.3-fold reduction in the mean excised tumor mass was observed for multiple emulsion and IFA emulsion, respectively as compared to saline control. Similarly, under the active immunization treatment regimen, it was reported that multiple emulsion treatment led to about two-fold and 1.8-fold reduction in mean tumor volume and excised tumor mass, respectively, as compared to saline control. Whereas, the IFA emulsion treatment resulted in 1.5-fold and 1.8-fold reduction of the same.

A multiplex ELISA looking at Th1 cytokines panel such as IL-2, 12, IFN-γ, and TNF-α was utilized to compare the CD8+ T-cell responses upon immunization. Overall, regardless of the treatment regimen, the cytokine levels of the immunized mice with the gp100 peptide were significantly higher as compared to control groups. However, the differences between the cytokine induction of W/O/W and IFA-based emulsion were minimal. The immunohistochemical staining of the tumor sections was performed to determine the presence of CD4+ and CD8+ T-cell infiltrates at the tumor site. The results of these studies confirmed the presence of these cells in tumor vicinity and their involvement in the cell-mediated antitumor immunity, which in turn could have protected the mice against tumor challenge. Lastly, body weight of mice was also recorded during the treatment as a measure of formulation tolerability. It was reported that all the animals in the multiple emulsion control and vaccine group tolerated the formulation as no appreciable decrease in body weight or overt systemic toxicity in mice was observed.

These studies provide evidence on the ability of the squalene emulsion-based vaccine containing the gp100 peptide to act as a potent vaccine. The study demonstrated that this system is capable of mounting an immune response for effective antitumor activity. The group reported that future studies with this formulation would focus on delivering variety of immunogens such as plasmid DNA or exosomes to determine if the T-cell response can be further enhanced.

11.5.2 Solid Nanoparticles-in-Emulsion (NiE) for Anti-inflammatory Gene Therapy

Another such system is the multicompartmental nanoparticles-in-emulsion formulation that was tested for macrophage-specific anti-inflammatory gene delivery [51]. The role of macrophages in inflammatory disease states have been described elsewhere [52, 53]. The three major goals of these cells are phagocytosis, antigen presentation, and modulation of immune response. With regards to particle-mediated delivery a lot of work has been done on different passive and active approaches to target these cells [54–56]. Moreover, it has been reported that particle size and shape can also enhance the delivery to these cells. For example, Schafer et al. (1992) looked at poly(methyl-methacrylate), poly(alkyl cyanoacrylate), and human serum albumin particle uptake by human macrophages [57]. The results indicated that nanoparticles made from the same material but of larger diameter were phagocytosed to a larger extent. For example, phagocytosis of the nanoparticles made from human serum albumin of 1.5 μm in diameter was higher in comparison to 200 nm particles made from the same material. Thus, microparticle-based delivery system can be effective in enhancing macrophage-specific uptake and the same principle was applied in this study. In order to formulate the nanoparticle-in-emulsion (NiE), inner core was made of gelatin nanoparticles that were formed as discussed in the previous section. However, the outer shell was made of oil phase. The advantages of using emulsion,

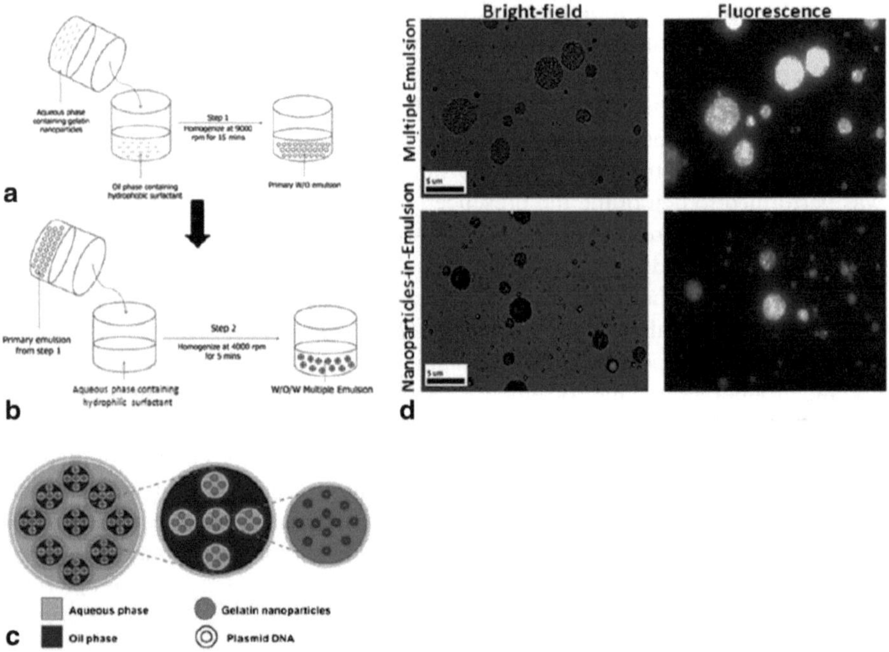

Fig. 11.7 Schematic illustration of the method for preparation of water-in-oil-in-water (W/O/W) multiple emulsion (**a**) and incorporation of plasmid DNA-containing type B gelatin nanoparticles to form nanoparticles-in-emulsion (NiE) multi-compartmental delivery system (**b**). The formulation has internal aqueous phase with suspended gelatin nanoparticles, the middle oil phase, and the external aqueous phase (**c**). Bright-field and fluorescence microscopy show that the oil droplets of NiE had a diameter of less than 5 μm and the rhodamine-labeled gelatin nanoparticles were completely encapsulated in the internal aqueous phase of formulation (**d**). (Reproduced from [51], copyright (2012), with kind permission from Springer Science + Business Media, Inc)

as an outer core, are multifold. Firstly, the liquid formulation results in ease of administration and can be given either orally or parenterally. Secondly, oil phase of the emulsion can be used to trap small molecular weight anti-inflammatory agent that are hydrophobic in nature. Thirdly, the oil phase can also impart protection to the encapsulated DNA in the nanoparticles from enzymatic degradation. Lastly, the oil droplet size and surface characteristics can be tailored for macrophage-specific phagocytosis. Figure 11.7 shows the schematic diagram of the NiE formulation. The W/O/W emulsion was prepared using a two-step emulsification process [47]. Briefly, gelatin nanoparticles containing plasmid DNA were suspended in the aqueous phase and then was mixed and homogenized with oil phase (safflower oil) at 9,000 rpm for 15 min using a Silverson L4RT® homogenizer to make the primary emulsion. Then, to the 4 ml of the primary W/O emulsion, another 4 ml of aqueous phase containing 20 μL of water-soluble surfactant Tween®80 was added. The preparation was again homogenized at 4,000 rpm for 4 min.

A multiple emulsion (ME) containing plasmid DNA in the inner aqueous phase was used as a control and was prepared using the same process. As shown by microscopy, the particle size of the formulation was reported to be less than 5 μm. The average plasmid DNA loading efficiency of the ME and NiE were reported to be 54.4 and 70.3 %, respectively. The DNA stability studies were performed by first destabilizing the emulsion with 1/10 volume of 5M NaCl followed by centrifugation at 20,000 rpm for 30 min which resulted in the separation of the oil and water phases. Subsequently, plasmid DNA was extracted from gelatin nanoparticles in the aqueous phase. The extracted plasmid DNA was then subjected to agarose gel electrophoresis and it was shown that encapsulated payload is preserved in the native supercoiled form and is not affected by the homogenization process. The *in vitro* studies were conducted in a relevant J774A.1 macrophage cell line. The initial studies were focused on particle uptake (with rhodamine-labeled particles) and transfection studies using murine IL-10 plasmid DNA (pORF-IL-10). These studies revealed that upon incubation of NiE particles with the cells, there was an appreciable amount of uptake within 60 min and the signal intensity from the rhodamine-labeled particles increased up to 120 min. Transfection studies were also conducted in a time-dependent (24 to 144 h) fashion and Lipofectin®, a commercially available transfection reagent was used as a positive control. It was reported that after 24 h, the average IL-10 levels in the case of NiE particles were 200 pg/ml as compared to 110 pg/ml for ME, and 75 pg/ml for gelatin nanoparticles, whereas the Lipofectin® treatment resulted in IL-10 levels of $\sim$85 pg/ml. Additionally, after 96 h, the level of IL-10 cytokine ($\sim$160 pg/ml) in the NiE group was still significantly higher as compared to other controls. The same trend was confirmed via RT-PCR to detect the presence of IL-10 transcript. Furthermore, therapeutic efficacy of the IL-10 transgene product was demonstrated in LPS-stimulated macrophages. The rationale behind this approach was to show that IL-10 production would alleviate the levels of proinflammatory markers such as TNF-α and IL-1β. With regards to the experimental design, the particles containing murine IL-10 plasmid DNA were incubated with macrophages and 6 h before the completion of transfection time, the cells were stimulated with LPS to induce the production of proinflammatory cytokines. Again, these studies were carried out in a time-dependent fashion from 24 to 96 h. The results were evaluated via ELISA and RT-PCR. As shown in Fig. 11.8, the NiE treatment led to significant reduction in the levels of TNF-α and IL-1β across all transfection time points. For example, it was stated that 48 h post transfection, the levels of TNF-α were 300 pg/ml in NiE as compared to elevated levels of 650, 1,000, 1,100, and 1,700 pg/ml in the ME, Lipofectin®, gelatin nanoparticles, and naked plasmid DNA, respectively. Similarly, IL-1β levels were influenced upon IL-10 transfection. The nanoparticle-in-emulsion treatment decreased IL-1β levels to $\approx$35 pg/ml at 24 h and remained at that level for up to 72 h. Overall, these studies provided evidence on the potential of the nanoparticle-in-emulsion based nonviral vector to act as a sustained gene delivery system for local or systemic therapy by enhancing the macrophage-specific uptake.

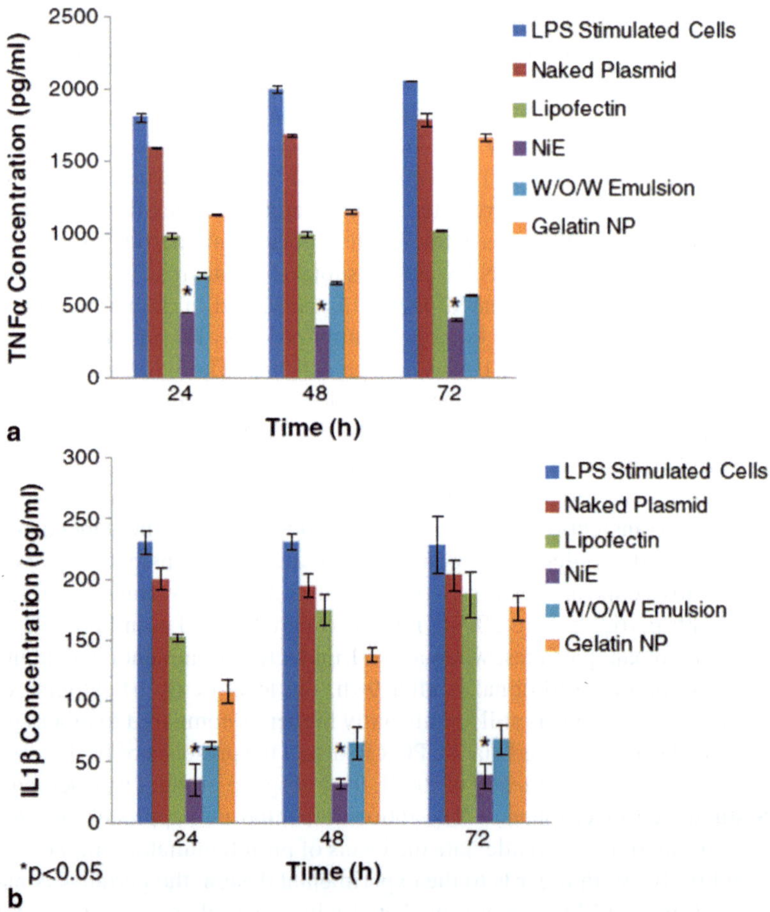

Fig. 11.8 Quantitative analysis to confirm downregulation of proinflammatory cytokines, tumor necrosis factor-alpha (TNF-α) (**a**) and inter-leukin 1-beta (IL-1β) (**b**), proteins by ELISA in lipopolysaccharide-treated J774A.1 adherent alveolar macrophage cells that had been transfected with murine interleukin-10 (mIL-10) plasmid DNA. The plasmid was administered in type B gelatin nanoparticles, water-in-oil-in-water (W/O/W) multiple emulsion, and nanoparticles-in-emulsion (NiE) multi-compartmental delivery system. Plasmid DNA complexed with commercially available cationic lipid transfection reagent, Lipofectin®, and administered as free (naked plasmid) were used as control. (Reproduced from [51], copyright 2012, with kind permission from Springer Science + Business Media, Inc)

11.6 Conclusions and Future Perspectives

Oral gene and RNAi therapy holds lot of promise for the treatment of both local and systemic diseases as well as for prevention of countless diseases through mucosal DNA vaccination. Oral administration is also a very attractive concept based on ease of self-administration and the associated patient compliance. For gene or RNAi therapy, delivery of nucleic acid constructs to the right tissues and cells and subsequent efficient intracellular availability is a major hurdle for translation of these experimental approaches into clinically viable therapeutic strategies. Barriers such as protection of the labile payload during transit, targeted availability at the site of interest, cellular uptake and processing, transgene expression, followed by protein production and posttranslational modification are all necessary requisites. Similarly, RNAi requires double stranded siRNA or miRNA duplexes to be delivered to the cytoplasm and subsequently interact with RISC for mRNA degradation.

In this review, we described our effort to develop oral gene and RNAi therapy using multicompartmental formulations such as NiMOS, W/O/W, and NiE. The main advantages of such systems are high entrapment efficiency, ability to encapsulate multiple payloads, stabilization of cargo in the GI fluid from variable pH and enzymatic conditions, and ability to target specific cells, such as GALT macrophages and enterocytes. Additionally, from the perspective of oral vaccine therapy, example of nanoparticle in microemulsion system made from squalene oil was desirable for its adjuvant properties, which resulted in further enhancing the humoral and cytotoxic immune response against the tumor.

Based on the concept of multi-compartmental particles, it will also be interesting to make and compare such formulations with pH sensitive or M-cell-specific lectin-based polymers that can facilitate cargo release at targeted site or improve transcytosis /phagocytosis of particles across the intestinal epithelium for systemic absorption, respectively. In addition, extension of these systems to other disease conditions such gastric cancer, HIV-1, periodontal defects, and bacterial infections such as *H. pylori* will also be desirable.

However, these systems are still at the stage of infancy and lot more data needs to be generated in order to better understand the mechanistic aspects of their interaction with the tissue of interest. For example, it will be critical to show the transport mechanism of these systems across the intestinal epithelium, clearance mechanism of such particles, or long-term toxicity, or complement activation or whether nanoparticle administration in the case of inflammation is able to convert macrophage phenotype from classical activated (M1) to alternate activated state (M2). Answering these questions in relevant preclinical models will ultimately help in realizing the transition of such platforms into clinic.

Acknowledgments The research discussed in this review was partially supported by a grant (R01-DK080477) from the National Institute of Diabetes, Digestive Diseases, and Kidney Diseases of the National Institutes of Health. We are also grateful to Dr. David Nguyen in Professor Robert Langer's lab at MIT (Cambridge, MA, USA) for the use of the Coulter particle size analysis instrument and to Mr. William Fowle who assisted with scanning electron microscopy at the Nano-Instrumentation Facility at Northeastern University (Boston, MA, USA).

References

1. Kawakami S, Higuchi Y, Hashida M. Nonviral approaches for targeted delivery of plasmid DNA and oligonucleotide. J Pharma Sci. 2008;97(2):726–45.
2. Goverdhana S, Puntel M, Xiong W, Zirger JM, Barcia C, Curtin JF, Soffer EB, Mondkar S, King GD, Hu J, Sciascia SA, Candolfi M, Greengold DS, Lowenstein PR, Castro MG. Regulatable gene expression systems for gene therapy applications: progress and future challenges. Mol Ther. 2005;12(2):189–211.

3. Bhavsar MD, Amiji MM. Polymeric nano- and microparticle technologies for oral gene delivery. Expert Opin Drug Deliv. 2007;4(3):197–213.
4. Kay MA, Glorioso JC, Naldini L. Viral vectors for gene therapy: the art of turning infectious agents into vehicles of therapeutics. Nat Med. 2001;7(1):33–40.
5. Mahapatro A, Singh DK. Biodegradable nanoparticles are excellent vehicle for site directed in-vivo delivery of drugs and vaccines. J Nanobiotechnol. 2011;9:55.
6. Dass CR, Choong PF. Chitosan-mediated orally delivered nucleic acids: a gutful of gene therapy. J Drug Target. 2008;16(4):257–61.
7. Petkar KC, Chavhan SS, Agatonovik-Kustrin S, Sawant KK. Nanostructured materials in drug and gene delivery: a review of the state of the art. Crit Rev Ther Drug. 2011;28(2):101–64.
8. Xavier RJ, Podolsky DK. Unravelling the pathogenesis of inflammatory bowel disease. Nature. 2007;448(7152):427–34.
9. Wu F, Dassopoulos T, Cope L, Maitra A, Brant SR, Harris ML, Bayless TM, Parmigiani G, Chakravarti S. Genome-wide gene expression differences in Crohn's disease and ulcerative colitis from endoscopic pinch biopsies: insights into distinctive pathogenesis. Inflamm Bowel Dis. 2007;13(7):807–21.
10. Perrier C, Rutgeerts P. Cytokine blockade in inflammatory bowel diseases. Immunotherapy. 2011;3(11):1341–52.
11. Talley NJ, Abreu MT, Achkar JP, Bernstein CN, Dubinsky MC, Hanauer SB, Kane SV, Sandborn WJ, Ullman TA, Moayyedi P. An evidence-based systematic review on medical therapies for inflammatory bowel disease. Am J Gastroenterol. 2011;106(Suppl 1):S2–25; quiz S26.
12. Nakase H, Okazaki K, Tabata Y, Chiba T. Biodegradable microspheres targeting mucosal immune-regulating cells: new approach for treatment of inflammatory bowel disease. J Gastroenterol. 2003;38(Suppl 15):59–62.
13. Kaigler D, Cirelli JA, Giannobile WV. Growth factor delivery for oral and periodontal tissue engineering. Expert Opin Drug Deliv. 2006;3(5):647–62.
14. Sood S, Gupta S, Mahendra A. Gene therapy with growth factors for periodontal tissue engineering—a review. Med Oral Patol Oral Cir Bucal. 2012;17(2):e301–10.
15. Zhang Y, Wang Y, Shi B, Cheng X. A platelet-derived growth factor releasing chitosan/coral composite scaffold for periodontal tissue engineering. Biomaterials. 2007;28(8):1515–22.
16. Peng L, Cheng X, Zhuo R, Lan J, Wang Y, Shi B, Li S. Novel gene-activated matrix with embedded chitosan/plasmid DNA nanoparticles encoding PDGF for periodontal tissue engineering. J Biomed Mater Res A. 2009;90(2):564–76.
17. Elangovan S, Jain S, Tsai PC, Margolis HC, Amiji M. Nano-sized calcium phosphate particles for periodontal gene therapy. J Periodontol. 2013;84(1):117–25.
18. Bowman K, Sarkar R, Raut S, Leong KW. Gene transfer to hemophilia A mice via oral delivery of FVIII-chitosan nanoparticles. J Control Release. 2008;132(3):252–9.
19. Azad N, Rojanasakul Y. Vaccine delivery—current trends and future. Curr Drug Deliv. 2006;3(2):137–46.
20. Chadwick S, Kriegel C, Amiji M. Delivery strategies to enhance mucosal vaccination. Expert Opin Biol Ther. 2009;9(4):427–40.
21. Srivastava IK, Singh M. DNA vaccines: focus on increasing potency and efficacy. Int J Pharm Med. 2005;19:15–28.
22. Singh M, O'Hagan DT. Recent advances in vaccine adjuvants. Pharm Res. 2002;19(6):715–28.
23. Singh M, Chakrapani A, O'Hagan D. Nanoparticles and microparticles as vaccine-delivery systems. Expert Rev Vaccines. 2007;6(5):797–808.
24. Shahiwala A, Vyas TK, Amiji MM. Nanocarriers for systemic and mucosal vaccine delivery. Recent Pat Drug Deliv Formul. 2007;1(1):1–9.
25. Clark MA, Jepson MA, Hirst BH. Exploiting M cells for drug and vaccine delivery. Adv Drug Deliv Rev. 2001;50(1–2):81–106.
26. Meister G, Tuschl T. Mechanisms of gene silencing by double-stranded RNA. Nature. 2004;431(7006):343–9.
27. Mello CC, Conte D Jr. Revealing the world of RNA interference. Nature. 2004;431(7006): 338–42.

28. Filipowicz W, Jaskiewicz L, Kolb FA, Pillai RS. Post-transcriptional gene silencing by siRNAs and miRNAs. Curr Opin Struct Biol. 2005;15(3):331–41.
29. Behlke MA. Progress towards in vivo use of siRNAs. Mol Ther. 2006;13(4):644–70.
30. Amarzguioui M, Prydz H. An algorithm for selection of functional siRNA sequences. Biochem Biophys Res Commun. 2004;316(4):1050–8.
31. Reynolds A, Leake D, Boese Q, Scaringe S, Marshall WS, Khvorova A. Rational siRNA design for RNA interference. Nat Biotechnol. 2004;22(3):326–30.
32. Peacock H, Kannan A, Beal PA, Burrows CJ. Chemical modification of siRNA bases to probe and enhance RNA interference. J Org Chem. 2011;76(18):7295–300.
33. Bramsen JB, Kjems J. Chemical modification of small interfering RNA. Methods Mol Biol. 2011;721:77–103.
34. Wilson DS, Dalmasso G, Wang L, Sitaraman SV, Merlin D, Murthy N. Orally delivered thioketal nanoparticles loaded with TNF-alpha-siRNA target inflammation and inhibit gene expression in the intestines. Nat Mater. 2010;9(11):923–8.
35. Aouadi M, Tesz GJ, Nicoloro SM, Wang M, Chouinard M, Soto E, Ostroff GR, Czech MP. Orally delivered siRNA targeting macrophage Map4k4 suppresses systemic inflammation. Nature. 2009;458(7242):1180–4.
36. Benoit MA, Baras B, Gillard J. Preparation and characterization of protein-loaded poly(epsilon-caprolactone) microparticles for oral vaccine delivery. Int J Pharm. 1999;184(1):73–84.
37. Florence AT. The oral absorption of micro- and nanoparticulates: neither exceptional nor unusual. Pharm Res. 1997;14(3):259–66.
38. Bhavsar MD, Tiwari SB, Amiji MM. Formulation optimization for the nanoparticles-in-microsphere hybrid oral delivery system using factorial design. J Control Release. 2006;110(2):422–30.
39. Kaul G, Amiji M. Long-circulating poly(ethylene glycol)-modified gelatin nanoparticles for intracellular delivery. Pharm Res. 2002;19(7):1061–7.
40. Kaul G, Amiji M. Tumor-targeted gene delivery using poly(ethylene glycol)-modified gelatin nanoparticles: *in vitro* and *in vivo* studies. Pharm Res. 2005;22(6): 951–61.
41. Bhavsar MD, Amiji MM. Gastrointestinal distribution and in vivo gene transfection studies with nanoparticles-in-microsphere oral system (NiMOS). J Control Release. 2007;119(3):339–48.
42. Kriegel C, Amiji M. Oral TNF-alpha gene silencing using a polymeric microsphere-based delivery system for the treatment of inflammatory bowel disease. J Control Release. 2011;150(1):77–86.
43. Bhavsar MD, Amiji MM. Oral IL-10 gene delivery in a microsphere-based formulation for local transfection and therapeutic efficacy in inflammatory bowel disease. Gene Ther. 2008;15(17):1200–9.
44. Krigel C, Amiji M. Dual TNF-α/Cyclin D1 gene silencing with an oral polymeric microparticle system as a novel strategy for the treatment of inflammatory bowel disease. Clin Transl Gastroenterol. 2011;2(e2):1–12.
45. Bozkir A, Hayta G. Preparation and evaluation of multiple emulsions water-in-oil-in-water (w/o/w) as delivery system for influenza virus antigens. J Drug Target. 2004;12(3):157–64.
46. Dupuis M, Murphy TJ, Higgins D, Ugozzoli M, van Nest G, Ott G, McDonald DM. Dendritic cells internalize vaccine adjuvant after intramuscular injection. Cell Immunol. 1998;186(1): 18–27.
47. Shahiwala A, Amiji MM. Enhanced mucosal and systemic immune response with squalane oil-containing multiple emulsions upon intranasal and oral administration in mice. J Drug Target. 2008;16(4):302–10.
48. Kalariya M, Ganta S, Amiji M. Multi-compartmental vaccine delivery system for enhanced immune response to gp100 peptide antigen in melanoma immunotherapy. Pharm Res. 2012;29(12):3393–403.
49. Kawakami Y, Dang N, Wang X, Tupesis J, Robbins PF, Wang RF, Wunderlich JR, Yannelli JR, Rosenberg SA. Recognition of shared melanoma antigens in association with major HLA-A alleles by tumor infiltrating T lymphocytes from 123 patients with melanoma. J Immunother. 2000;23(1):17–27.

50. Sikora AG, Jaffarzad N, Hailemichael Y, Gelbard A, Stonier SW, Schluns KS, Frasca L, Lou Y, Liu C, Andersson HA, Hwu P, Overwijk WW. IFN-alpha enhances peptide vaccine-induced CD8+ T cell numbers, effector function, and antitumor activity. J Immunol. 2009;182(12):7398–407.
51. Attarwala H, Amiji M. Multi-compartmental nanoparticles-in-emulsion formulation for macrophage-specific anti-inflammatory gene delivery. Pharm Res. 2012;29(6):1637–49.
52. Fujiwara N, Kobayashi K. Macrophages in inflammation. Curr Drug Targets-Inflamm Allergy. 2005;5:281–6.
53. Ross JA, Auger MJ. The biology of the macrophage. In: Burke B, Lewis CE, editors. The Macrophage, 2 edn. New York: Oxford University Press; 2002. pp. 16–23.
54. Gupta S, Dube A, Vyas SP. Antileishmanial efficacy of amphotericin B bearing emulsomes against experimental visceral leishmaniasis. J Drug Target. 2007;15(6):437–44.
55. Herre J, Gordon S, Brown GD. Dectin-1 and its role in the recognition of beta-glucans by macrophages. Mol Immunol. 2004;40(12):869–76.
56. Jain SK, Gupta Y, Jain A, Saxena AR, Khare P. Mannosylated gelatin nanoparticles bearing an anti-HIV drug didanosine for site-specific delivery. Nanomedicine. 2008;4(1):41–8.
57. Schafer V, von Briesen H, Andreesen R, Steffan AM, Royer C, Troster S, Kreuter J, Rubsamen-Waigmann H. Phagocytosis of nanoparticles by human immunodeficiency virus (HIV)-infected macrophages: a possibility for antiviral drug targeting. Pharm Res. 1992;9(4):541–6.

Chapter 12
Bacteria-Based Vectors for Oral Gene Therapy

Yong Bai, Rachael Burchfield, Sangwei Lu and Fenyong Liu

12.1 Introduction

Human cytomegalovirus (HCMV) belongs to the *Betaherpesvirinae* genus of the Herpesviridae family. It is the most structurally complex herpesvirus and has the largest genome among all sequenced human herpesviruses [1]. During natural infection, HCMV can replicate in many differentiated host cell types, such as epithelial cells, endothelial cells, smooth muscle cells, mesenchymal cells, hepatocytes, granulocytes, and monocyte-derived macrophages [2]. As shown in Fig. 12.1, the infectious virion initializes the infection by either endocytosis or fusion with the cell membrane, which releases the nucleocapsid and some tegument proteins into the cytoplasm. After viral entry, the nucleocapsid is uncoated and transported across the cytoplasm. When docking at the nuclear pores, the nucleaocapsid injects the viral DNA through nuclear pores into the nucleus, where replication and capsid assembly take place. Once virus progeny is mature and starts to egress and release, infected cells can continue to produce viral particles for several days [1].

A number of ideas for using catalytic nucleic acids to inactivate viral genes inactivation has been enthusiastically proposed and tested [3–5]. Among the different nucleotide-based gene interference technologies, Ribonuclease P (RNase P), especially its derivative M1GS catalytic RNA, has proved to be very effective and specific in blocking viral gene expression and replication in cultured cells [5–14]. M1RNA can be engineered to cleave tRNA-like substrates and other target RNAs, including specific mRNAs [15–18]. A sequence-specific ribozyme, M1GS, was constructed by attaching to M1RNA an additional small RNA (guide sequence, GS) which contains a sequence complementary to a target mRNA and a 3′ proximal CCA (Fig. 12.2).

S. Lu (✉) · Y. Bai · F. Liu
School of Public Health, University of California, Berkeley, CA 94720, USA
e-mail: sangwei@berkeley.edu

F. Liu · R. Burchfield · S. Lu
Program in Comparative Biochemistry, University of California, Berkeley, CA 94720, USA
e-mail: liu_fy@uclink4.berkeley.edu

J. das Neves, B. Sarmento (eds.), *Mucosal Delivery of Biopharmaceuticals*, DOI 10.1007/978-1-4614-9524-6_12,

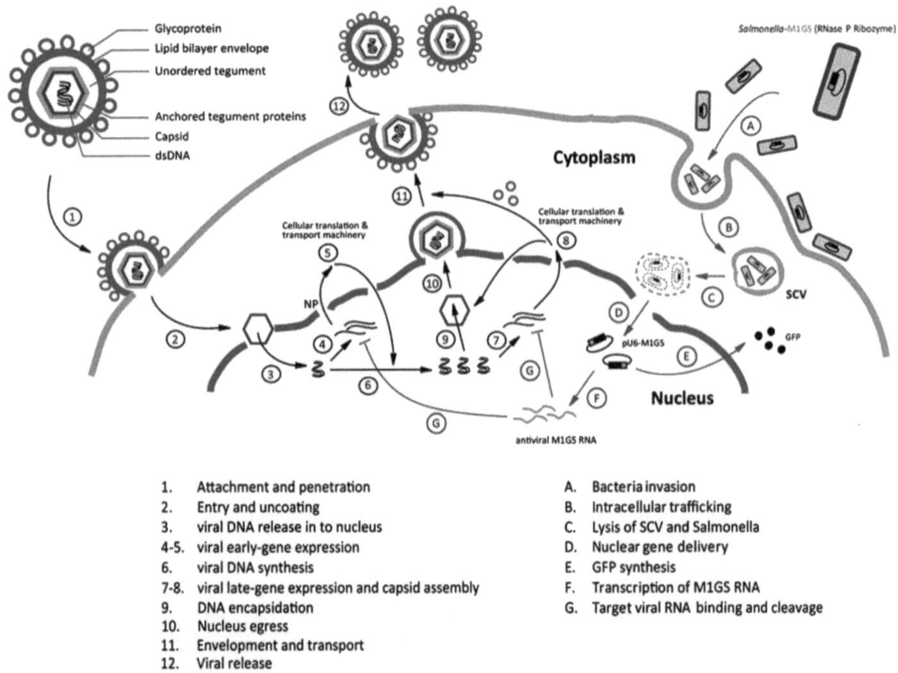

Fig. 12.1 Schematic diagram illustrating CMV viral replication pathway inside the host cells (*1–12*), and *Salmonella*-assisted gene therapy (*A–G*). *Salmonella* (*rod*) transformed with pU6-M1GS (oval inside *Salmonella*) infects a macrophage and resides in SCV (*Salmonella*-containing vacuole). After the bacteria undergo intracellular lysis, the plasmid contents are released and transferred into nucleus where they can be expressed by cellular machinery. M1GS RNAs will bind with and cleave the target viral mRNA to block the viral replication

This guide sequence complementarily binds to its target mRNA and directs covalently attached M1RNA to the cleavage site and catalyzes the hydrolytic reaction. Previous studies demonstrated that M1GS RNA are effective in cleaving both viral and cellular mRNAs and blocking their expression in cultured cells, including inhibition of gene expression of human influenza, and herpes viruses [19–23]. Compared with other nucleic acid-based interference approaches, M1GS ribozyme possesses several unique features such as high catalytic efficiency, high target specificity, no detectable cytotoxicity, and low target sequence requirement. Thus, M1GS catalytic RNA is a powerful gene-targeting tool which exhibits promising antiviral activity for future clinical application.

Before the clinical application of M1GS in human, animal study is required to test its antiviral effect in vivo, and furthermore, several technical barriers have to be overcome. The first and inevitable issue is how to achieve the delivery and expression of ribozyme M1GS RNA in live animals with a safe, efficient, and tissue-specific way. The hunt for a new delivery system led us to *Salmonella*, an invasive bacterium. During the past 20 years, *Salmonella* has been investigated as an efficient delivery

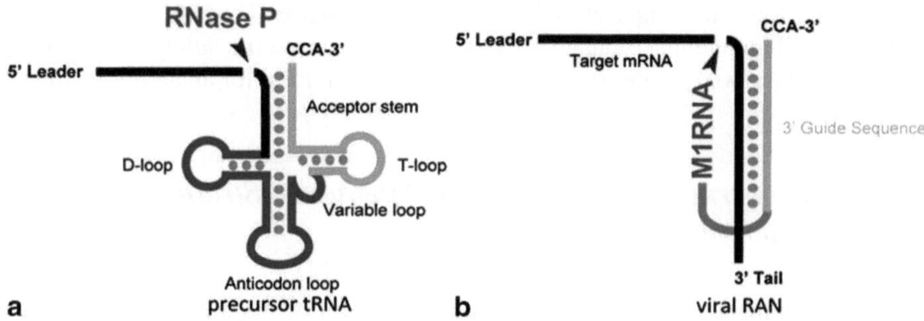

Fig. 12.2 **a** Schematic representation of natural substrates for RNase P. **b** A hybridized complex of a target RNA (e.g., viral mRNA) and M1GS RNA. The *arrow* shows the site of the cleavage by RNase P and M1 RNA

system for a variety of different molecules. *Salmonella* has several advantages as an M1GS delivery vector. First, attenuated *Salmonella* strains are low cost, easy to prepare, and use oral route administration [24–26]. Second, *Salmonella* has been widely used as a vaccine and as a delivery vector in cancer therapy and has proved to be safe for human use [24, 27–31]. Third, methods for genetic manipulations are readily available and it is feasible to generate new attenuated strains [32]. Fourth, plasmids carrying M1GS sequence are easy to construct and transform into *Salmonella* strains. Fifth, oral infection of animals (SCID mice) by *Salmonella* can be consistently performed without ill effect on animals. Sixth, and most importantly, *Salmonella* will bring the constructs carrying M1GS to the tissues where cytomegalovirus (CMV) resides. After oral infection, the bacteria will pass through the gastrointestinal tract, evade the attack from immune system, and reach the target organs for self-replication and further infection. Seventh, mutagenesis strategy provides high potential to generate novel attenuated mutants with low pathogenicity and high delivery efficiency leading to good therapeutical effect.

In this chapter, we mainly focused the case study on how to improve the application of M1GS antiviral methodology in tissue-cultured cells and live animals (mice) by using engineered attenuated *Salmonella* strains as the delivery vehicle. Experimental design, important findings, and future direction are summarized and discussed. For further information, including methods, materials, and detail experimental procedures with figures, please review the related articles published by our lab [33–36].

12.2 In Vitro Study on Using *Salmonella*-Based M1GS Delivery System

During natural infection, CMV can infect and replicate in monocyte-derived macrophages. During latent infection, CMV can be detected in macrophage progenitor cells in the bone marrow. For *Salmonella*, macrophages represent the major in vivo reservoir following bacterial systemic dissemination. Thus, we decided to use differentiated human macrophage cells (THP-1) to evaluate the *Salmonella*'s ability to mediate the ribozyme delivery into cultured cells and test for an antiviral effect.

12.2.1 Construction of *Salmonella* Attenuated Strain for M1GS Ribozyme Delivery

M1GS ribozyme was constructed to target the region of the mRNA encoding the HCMV capsid scaffolding protein (CSP). CSP completely overlaps with and is within the 3′ coding sequence of another viral capsid protein, assembling [1]. Both CSP and assembling are essential for HCMV capsid formation and replication [1]. Using dimethyl sulfate (DMS), an in vivo mapping approach [7] was employed to determine the accessibility of the region of the CSP mRNA in HCMV-infected cells. A highly accessible region was selected as the cleavage site for M1GS RNA. The functional M1GS ribozyme was constructed by covalently linking the 3′ terminus of M1GS ribozyme with a guide sequence that is complementary to the targeted mRNA sequence. Here, an in vitro selection procedure can be used to generate highly efficient M1RNA variant which showed the most activity in cleaving the target viral RNA [37]. In vitro cleavage assay showed that incubation of the substrate CSP mRNA sequence with functional M1GS ribozyme yielded efficient cleavage.

12.2.2 *Salmonella*-assisted Expression of M1GS Ribozymes in Cultured Macrophage

The DNA sequence coding for M1GS ribozyme were cloned into vector pU6, which contains a green fluorescent protein (GFP) expression cassette, and placed under the control of the small nuclear U6 RNA promoter. This promoter, which is transcribed by RNA polymerase III, has previously been shown to express M1GS RNA and other RNAs steadily [7, 38]. The DNAs of construct containing the M1GS sequence were transformed into auxotrophic *Salmonella* strain SL7207, which is attenuated in virulence and pathogenesis in vivo and has been shown to function efficiently as a gene delivery carrier for the expression of several transgenes in mammalian cells [30–31].

First, growth analyses of *Salmonella* strains were performed and indicated that the presence of the ribozyme sequence did not result in an impaired viability of the bacterial carrier. Northern analysis showed that neither the GFP nor the M1GS RNA transcript was detected in *Salmonella* carrying ribozyme constructs. Furthermore, no GFP signal was observed when *Salmonella* was examined under a fluorescence microscope. These results suggested that M1GS RNA, which was under the control of the U6 expression cassette, was not expressed in the *Salmonella* vector. Second, to determine whether the bacteria can efficiently deliver the M1GS sequences into human cells, differentiated macrophage THP-1 cells were infected with *Salmonella* SL7207 carrying pU6-M1GS. One day after infection, more than 80 % of cells were GFP positive, indicating efficient gene transfer mediated by *Salmonella*. To examine the ribozyme expression after *Salmonella*-mediated gene transfer, total RNAs were isolated from *Salmonella*-infected cells and the levels of M1GS RNAs were detected. Northern analysis showed that the M1GS RNAs appeared to be exclusively expressed in the nuclei as they were detected only in the nuclear but not the cytoplasmic RNA fractions. So M1GS RNA expressed by the U6 promoter are primarily localized in the nuclei where the ribozyme-mediated cleavage happened [14, 38].

12.2.3 Blocking MCMV Infection in Cultured Macrophage by Salmonella-Mediated M1GS Ribozymes

Differentiated THP-1 cells were first treated with *Salmonella* carrying the pU6-M1GS plasmids. The *Salmonella*-containing cells were then isolated by FACS analysis based on GFP expression, and infected with HCMV. Total RNAs were isolated from the infected cells and the expression levels of target CSP and assembling viral RNAs were determined by Northern analysis. A reduction of about 90 ± 8 and 90 ± 9 % in the expression levels of CSP and assembling mRNA was observed in cells that expressed functional M1GS, respectively. Western blotting showed that the CSP protein levels in cells that were treated with *Salmonella* carrying the functional M1GS sequence-containing plasmid were also reduced. A reduction of about 87 % in the protein level of CSP was observed in cells treated with *Salmonella* carrying the functional M1GS sequence. These results suggest that the significant reduction of CSP expression in cells treated with the functional M1GS-containing *Salmonella* is due to *Salmonella*-mediated gene delivery of the ribozyme.

To determine whether *Salmonella*-mediated gene delivery of ribozymes inhibits the growth of HCMV, the bacteria-infected cultures were harvested at 1-day intervals through 7 days post infection and determined the viral titers of these samples. A reduction of at least 5,000-fold in viral yield was observed in cells that were treated with *Salmonella* carrying the vector containing the functional M1GS sequence. In contrast, no significant reduction was found in control group.

To further determine the antiviral specificity of the *Salmonella*-mediated gene delivery of M1GS against the CSP mRNA, two sets of experiments were performed. First, the expression of other viral genes was tested. Northern analysis showed that the inhibition of CSP assembling expression did not affect the expression of other viral genes, including immediate-early (α), early (β), and late (γ) genes [1]. Western blot also showed no significant difference in the levels of these genes among *Salmonella*-treated cells, suggesting that the *Salmonella*-mediated delivery of functional M1GS specifically inhibits the expression of its target and does not affect overall viral gene expression. Second, viral genomic replication and capsid maturation in treated cells were tested. Total DNA was isolated from HCMV-infected cell lysates, and the level of intracellular viral DNA was determined by PCR detection. DNA samples isolated from HCMV-infected cell lysates were also treated with DNase I. The encapsidated viral DNAs are resistant to DNase I digestion, whereas those that are not packaged in the capsid will be susceptible to degradation. When the DNA samples from cell lysates that were not treated with DNase I were assayed, no significant difference in the level of total intracellular (both encapsidated and uncapsidated) viral DNA was observed. However, when the samples were first treated with DNase I and then assayed, the "encapsidated" DNA was hardly detected in cells that were treated with *Salmonella* carrying the M1GS expression plasmid. These observations suggest that *Salmonella*-mediated gene delivery of ribozymes against the CSP mRNA does not affect the replication of viral DNA but blocks DNA encapsidation and capsid formation.

12.3 In Vivo Study on Using *Salmonella*-Based M1GS Delivery System

Murium cytomegalovirus (MCMV) shares many similar features with its human counterpart, HCMV. As such, MCMV infecting mice provides a good animal model for studying CMV pathogenesis in vivo, and thus has been extensively used for developing and screening novel antiviral agents [1, 39–41]. CB17 SCID mice which lack functional T and B lymphocytes have been shown to be extremely susceptible to MCMV infection [1, 42, 43] and represent an excellent animal model to study CMV pathogenesis in immunocompromised hosts. Analysis of viral replication in these mice can be used for studying whether new therapeutic approaches block CMV opportunistic infection and prevent viral-associated diseases in immunocompromised hosts. To determine the effect of M1GS ribozymes on the replication and infection of MCMV in vivo, we applied hydrodynamic transfection [44–46] of plasmid LXSN-M1GS DNA to SCID mice, provided first evidence that M1GS RNA is effective in inhibiting viral gene expression and blocking viral infection and pathogenesis, leading to improved survival of animals [47]. This modified intravenous injection method is useful to demonstrate the feasibility of delivering novel antiviral compounds into animals and to test their activity in vivo [44–46]; however, it is not suitable for clinical applications. To achieve safe, stable, and efficient delivering purpose, *Salmonella*-based delivery system was developed and tested in tissue-cultured cell line and living animals.

12.3.1 Salmonella-Assisted Inhibition of MCMV Infection in Mouse Macrophage

M1GS ribozymes were constructed to target the mRNA coding for MCMV protein M80.5. The coding sequence of M80.5 is completely within the 3′ coding sequence of viral protease (PR). Thus, the ribozyme would be expected to target both M80.5

and PR, which are essential for MCMV capsid assembly and replication [1]. An in vivo mapping approach with dimethyl sulphate (DMS) [7] was used to determine the accessibility of the region of the M80.5 mRNA in MCMV-infected cells and have chosen a highly accessible region as the cleavage site for M1GS RNA. Functional ribozyme M1GS was constructed by linking the 3′ terminus of M1 RNA with a guide sequence that is complementary to the targeted M80.5 mRNA sequence. In vitro cleavage of a M80.5 mRNA substrate by M1GS was observed and this ribozyme is ready for further tests.

As described in Sect. 1.2 of Chap. 15, DNA sequences encoding M1GS RNA was cloned into vector pU6, and was transformed into a new *Salmonella* delivery strain, SL101, for gene delivery studies. SL101 was derived from auxotrophic strain SL7207 [48] and, in addition, contained a deletion of ssrA/B genes. SsrA/B regulates the expression of *Salmonella* Pathogenicity Island-2 (SPI-2) genes, which are important for bacteria intracellular survival in macrophages and virulence in vivo [49]. Deletion of ssrA/B is expected to further reduce the virulence of *Salmonella* and facilitate intracellular lysis of bacteria and release of the transgene construct, leading to efficient expression of the delivered gene in target cells. The presence of the ribozyme sequence did not affect the viability of the bacterial carrier and when cultured in vitro, neither the GFP nor M1GS transcript was detected in *Salmonella* carrying ribozyme constructs. When mouse J774 macrophages were infected with *Salmonella* carrying pU6-M1GS constructs, more than 80 % of cells were GFP and Northern analysis confirmed M1GS expression in these cells. These results demonstrated efficient gene transfer mediated by *Salmonella*. The level of M1GS RNAs in cells treated with SL101 carrying pU6-M1GS was about 3-fold higher than those with SL7207 carrying the same construct, suggesting that SL101 is a more effective delivery vector, possibly as a result of more efficient intracellular lysis of *Salmonella* and release of pU6-M1GS due to the deletion of ssrA/B, leading to a higher level of gene expression.

The *Salmonella*-containing cells were isolated by FACS analysis based on GFP expression and infected with MCMV. The expression levels of M80.5/PR mRNAs were determined by Northern analysis. A reduction of $81 + 6$ and $81 + 8\,\%$ in the level of the target M80.5 and PR mRNA was observed in cells treated with SL101 carrying pU6-M1GS while no significant reduction was observed in control groups. Furthermore, western analysis detected a reduction of about 85 % in the protein level of M80.5 in cells treated with SL101 carrying functional pU6-M1GS construct. It also showed that the inhibition of M80.5/PR expression did not affect the expression of other viral genes, including immediate-early (α), early (β), and late (γ) genes (1). No significant difference in the levels of these genes among *Salmonella*-treated cells, suggesting that the *Salmonella*-mediated delivery of M1GS specifically inhibits the expression of its target, and does not affect overall viral gene expression. To test if *Salmonella*-mediated gene delivery of anti-M80.5 ribozyme effectively inhibited MCMV growth, the infected cultures were harvested at 1-day intervals through 5 days post infection, and viral titers of these samples were determined. A reduction of at least 2,500-fold in viral yield was observed in cells treated with *Salmonella*-carrying pU6-M1GS, while no significant reduction was found in control groups.

12.3.2 Inhibition of MCMV Infection in Mice by Salmonella-Mediated Delivery of M1GS Ribozymes

To study *Salmonella*-assisted delivery of M1GS in vivo, we intragastrically inoculated SCID mice with SL101 carrying pU6-M1GS constructs. Gene delivery mediated by SL101 was efficient in vivo as substantial amounts of M1GS and GFP-positive cells were detected in the liver and spleen of the *Salmonella*-treated mice. Furthermore, SL101 exhibited much less virulence in vivo than the parental strain SL7207 and a wild-type strain ST14028s. All mice infected with SL101 (1×10^9 CFU/mouse) remained alive even after 70 days post inoculation. In contrast, mice inoculated with a much lower dose of ST14028s (1×10^3 CFU/mouse) and SL7207 (5×10^5 CFU/mouse) died within 7–15 days, respectively. Thus, SL101 appeared to be efficient in gene transfer and exhibited little virulence/pathogenicity in vivo.

To study the antiviral effect of *Salmonella*-assisted oral delivery of M1GS in vivo, SCID mice were intraperitoneally infected with MCMV, followed by oral inoculation of *Salmonella*-carrying ribozyme constructs 36 h later. To further allow sustained expression of M1GSs, we repeated oral inoculation of *Salmonella* every 5 days. Three sets of experiments were carried out to study the effect of *Salmonella*-mediated delivery of M1GSs on MCMV virulence and infection in vivo. First, the survival rate of the animals was determined. In MCMV-infected mice treated with SL101 expressing functional M1GS, life span improved significantly as no animals died within 50 days post infection. In contrast, control groups had no effect on animal survival compared with untreated animals, as all mice died within 25 days post infection with MCMV. Second, viral replication in various organs of the animals was studied during a 21-day infection period before the onset of mortality of the infected animals. Twenty-one days post infection, the viral titers in the spleen and liver of animals treated with functional M1GS containing SL101 were lower than those from animals receiving SL101 carrying control constructs by 400 and 600 fold, respectively. Third, viral gene expression in the tissues was also examined. Fourteen days post infection, substantial expression of viral M80.5/PR mRNAs as well as M80.5 protein was readily detectable in livers and spleens of mice receiving SL101 carrying control plasmids, while little expression of M80.5/PR was detected in mice treated with SL101 carrying functional M1GS constructs. Thus, *Salmonella*-assisted oral delivery of M1GS blocked MCMV infection in the treated mice.

12.4 Conclusions and Future Perspectives

The experimental results showed that attenuated *Salmonella* strain SL7207 can efficiently deliver M1GS sequences into cultured human cells, leading to the expression of M1GS catalytic RNA and effective inhibition of HCMV infection. For the purpose of delivering M1GS sequence into live animals, the newly constructed attenuated *Salmonella* strain, SL101, was generated from its parental SL7207 strain and contained the deletion of ssrA/B genes. Study showed that SL101 efficiently delivered antiviral M1GS into targeted organs, leading to a substantial expression of M1GS RNA without causing significant adverse effects in the animals. Compared with the control groups, mice received SL101 carrying functional M1GS sequence showed reduced viral gene expression, decreased viral titers, and greatly improved survival.

The final antiviral outcome using *Salmonella*-mediated ribozyme expression depends on the combination of the dose of bacteria, the route of delivery, the virulence of vector bacteria, and the genetic materials carried by the bacteria, in our case, M1GS ribozyme sequence. Although newly constructed SL101 was highly efficient for gene delivery in mice and exhibiting low virulence, there is still room for improvement before applying this technique to primate study. First and foremost, the virulence of the vector can be further reduced. Lower bacterial pathogenicity will make a higher inoculation dose feasible. As we know, different bacterial components and bacterial virulent factors can activate various immune responses, including innate immunity, adaptive innate immunity, and antigen-specific immunity. Some of these defenses are beneficial to the host while others are harmful. To reduce the potential cytotoxicity, further mutations can be introduced into bacterial vectors to inactivate specific bacterial components [24]. For example, many cytokines in the host are activated by and respond to *Salmonella* lipopolysaccharides (LPS), such as platelet-activating factor, chemokines, eicosanoids, TNFα, IL1β, IL6, IL12, and IFNγ [50]. During *Salmonella* infection, these cytokines are involved in the endotoxic septic shock syndrome characterized from bacteremia to multiorgan failure. Even though attenuated auxotrophic strain was used in the study, cytokine factors had to be seriously taken into consideration. To reduce the proinflammatory immune response caused by LPS, deletion mutants can be introduced into LPS biosynthesis pathway. For example, msbB, a lipid A biosynthesis (KDO) 2-(lauroyl)-lipid IVA acyltransferase, plays an important role in LPS biosynthesis pathway. MsbB mutant prevents the addition of a terminal myristyl group to the lipid-A domain and reduces the induction of proinflammatory cytokines and nitric oxide synthase [51]. This mutation suppressed *Salmonella* virulence in vivo and mice inoculated orally with this mutant have a better survival rate compared with those infected by the wild-type strain [52, 53]. Moreover, msbB mutant has been used in Phase I study of the intravenous administration for cancer therapy which makes this mutant more reliable and feasible as delivery vector for M1GS in vivo [54]. In future, attenuate strain MsbB/ssrAB mutants in the SL7207 background will provide more opportunities for us to screen better delivery vectors for antiviral gene therapy. One more concern here is that newly constructed attenuated strains of *Salmonella* may not be very efficient in disseminating via blood stream to get to the target organs. Related assays have to be considered and carried out during the screening process.

Acknowledgments Gratitude goes to Dr. Phong Trang, Dr. Hao Gong, and Dr. Xiaohong Jiang for suggestions and technical assistance. This research has been supported by NIH (AI041927, AI091356, and DE014842).

References

1. Mocarski ES, Shenk T, Pass RF. Cytomegaloviruses. In: Knipe DM, editor Fields virology. Philadelphia: Lippincott; 2007. p. 2701–72.
2. Landolfo S, Gariglio M, Gribaudo G, Lembo D. The human cytomegalovirus. Pharmacol Ther. 2003;98(3):269–97.
3. Pal BK, Scherer L, Zelby L, Bertrand E Rossi JJ. Monitoring retroviral RNA dimerization in vivo via hammerhead ribozyme cleavage. J Virol. 1998;72(10):8349–53.
4. Yu M, Leavitt MC, Maruyama M, Yamada O, Young D, Ho AD, Wong-Staal F. Intracellular immunization of human fetal cord blood stem/progenitor cells with a ribozyme against human immunodeficiency virus type 1. Proc Natl Acad Sci U S A. 1995;92(3):699–703.
5. Sarver N, Cantin EM, Chang PS, Zaia JA, Ladne JA, Stephens DA, Rossi JJ. Ribozymes as potential anti-HIV-1 therapeutic agents. Science. 1990;247(4947):1222–5.
6. Kim K, Trang P, Umamoto S, Hai R, Liu F. RNase P ribozyme inhibits cytomegalovirus replication by blocking the expression of viral capsid proteins. Nucleic Acids Res. 2004;32(11):3427–34.
7. Liu F, Altman S. Inhibition of viral gene expression by the catalytic RNA subunit of RNase P from Escherichia coli. Genes Dev. 1995;9(4):471–80.
8. Plehn-Dujowich D, Altman S. Effective inhibition of influenza virus production in cultured cells by external guide sequences and ribonuclease P. Proc Natl Acad Sci U S A. 1998;95(13):7327–32.
9. Su YZ, Li HJ, Li YQ, Chen HJ, Tang DS, Zhang X, Jiang H, Zhou TH. In vitro construction of effective M1GS ribozymes targeting HCMV UL54 RNA segments. Acta Biochim Biophys Sin (Shanghai). 2005;37(3):210–4.
10. Trang P, Kim K, Liu F. Developing RNase P ribozymes for gene-targeting and antiviral therapy. Cell Microbiol. 2004;6(6):499–508.
11. Kim K, Umamoto S, Trang P, Hai R, Liu F. Intracellular expression of engineered RNase P ribozymes effectively blocks gene expression and replication of human cytomegalovirus. RNA. 2004;10(3):438–47.
12. Trang P, Kim K, Zhu J, Liu F. Expression of an RNase P ribozyme against the mRNA encoding human cytomegalovirus protease inhibits viral capsid protein processing and growth. J Mol Biol. 2003;328(5):1123–35.
13. Trang P, Kilani A, Lee J, Hsu A, Liou K, Kim J, Nassi A, Kim K, Liu F. RNase P ribozymes for the studies and treatment of human cytomegalovirus infections. J Clin Virol. 2002;25(Suppl 2):S63–S74.
14. Trang P, Lee M, Nepomuceno E, Kim J, Zhu H, Liu F. Effective inhibition of human cytomegalovirus gene expression and replication by a ribozyme derived from the catalytic RNA subunit of RNase P from Escherichia coli. Proc Natl Acad Sci U S A. 2000;97(11):5812–7.
15. Altman S, Kirsebom LA. Ribonuclease P. In: Gesteland RF, Atkins JF, editors. The RNA world. 2nd ed. Cold Spring Harbor: Cold Spring Harbor Laboratory Press; 1999.
16. Frank DN, Pace NR. Ribonuclease P: unity and diversity in a tRNA processing ribozyme. Annu Rev Biochem. 1998;67:153–80.
17. Guerriertakada C, Li Y, Altman S. Artificial regulation of gene-expression in escherichia-coli by Rnase-P. Proc Natl Acad Sci U S A. 1995;92(24):11115–9.
18. Forster AC, Altman S. External guide sequences for an RNA enzyme. Science. 1990;249(4970):783–6.
19. Raj SML, Liu F. Engineering of RNase P ribozyme for gene-targeting applications. Gene. 2003;313:59–69.
20. Liu FY, Altman S. Inhibition of viral gene expression by the catalytic RNA subunit of RNase P from *Escherichia coli*. Genes Dev. 1995;9(4):471–80.
21. Plehn-Dujowich D, Altman S. Effective inhibition of influenza virus production in cultured cells by external guide sequences and ribonuclease P. Proc Natl Acad Sci U S A. 1998;95(13):7327–32.

22. Cobaleda C, Sanchez-Garcia I. In vivo inhibition by a site-specific catalytic RNA subunit of RNase P designed against the BCR-ABL oncogenic products: a novel approach for cancer treatment. Blood. 2000;95(3):731–7.
23. Kilani AF, Trang P, Jo S, Hsu A, Kim J, Nepomuceno E, Liou K, Liu FY. Ribozymes selected RNase P in vitro to cleave a viral mRNA effectively inhibit its expression in cell culture. J Biol Chem. 2000;275(14):10611–22.
24. Clairmont C, Lee KC, Pike J, Ittensohn M, Low KB, Pawelek J, Bermudes D, Brecher SM, Margitich D, Turnier J, Li Z, Luo X, King I, Zheng LM. Biodistribution and genetic stability of the novel antitumor agent VNP20009, a genetically modified strain of *Salmonella typhimurium*. J Infect Dis. 2000;181(6):1996–2002.
25. Levine MM, Herrington D, Murphy JR, Morris JG, Losonsky G, Tall B, Lindberg AA, Svenson S, Baqar S, Edwards MF, et al. Safety, infectivity, immunogenicity, and in vivo stability of two attenuated auxotrophic mutant strains of *Salmonella typhi*, 541Ty and 543Ty, as live oral vaccines in humans. J Clin Invest. 1987;79(3):888–902.
26. Hone DM, Attridge SR, Forrest B, Morona R, Daniels D, LaBrooy JT, Bartholomeusz RC, Shearman DJ, Hackett J. A galE via (Vi antigen-negative) mutant of *Salmonella typhi* Ty2 retains virulence in humans. Infect Immun. 1988;56(5):1326–33.
27. Vassaux G, Nitcheu J, Jezzard S, Lemoine NR. Bacterial gene therapy strategies. J Pathol. 2006;208(2):290–8.
28. Darji A, Guzman CA, Gerstel B, Wachholz P, Timmis KN, Wehland J, Chakraborty T, Weiss S. Oral somatic transgene vaccination using attenuated *S. typhimurium*. Cell. 1997;91(6):765–75.
29. Grillot-Courvalin C, Goussard S, Courvalin P. Bacteria as gene delivery vectors for mammalian cells. Curr Opin Biotechnol. 1999;10(5):477–81.
30. Paglia P, Terrazzini N, Schulze K, Guzman CA, Colombo MP. In vivo correction of genetic defects of monocyte/macrophages using attenuated *Salmonella* as oral vectors for targeted gene delivery. Gene Ther. 2000;7(20):1725–30.
31. Yang N, Zhu X, Chen L, Li S, Ren D. Oral administration of attenuated. *S. typhimurium* carrying shRNA-expressing vectors as a cancer therapeutic. Cancer Biol Ther. 2008;7(1):145–51.
32. Datsenko KA, Wanner BL. One-step inactivation of chromosomal genes in Escherichia coli K-12 using PCR products. Proc Natl Acad Sci U S A. 2000;97(12):6640–5.
33. Bai Y, Sunkara N, Liu F. Targeting mRNAs by engineered sequence-specific RNase P ribozymes. Methods Mol Biol. 2012; 848:357–68.
34. Bai Y, Gong H, Li H, Vu GP, Lu S, Liu F. Oral delivery of RNase P ribozymes by *Salmonella* inhibits viral infection in mice. Proc Natl Acad Sci U S A. 2011; 108(8):3222–7.
35. Bai Y, Rider PJ, Liu F. Catalytic M1GS RNA as an antiviral agent in animals. Methods Mol Biol. 2010; 629:339–53.
36. Bai Y, Li H, Vu GP, Gong H, Umamoto S, Zhou T, Lu S, Liu F. *Salmonella*-mediated delivery of RNase P-based ribozymes for inhibition of viral gene expression and replication in human cells. Proc Natl Acad Sci U S A. 2010; 107(16):7269–74.
37. Kilani AF, Trang P, Jo S, Hsu A, Kim J, Nepomuceno E, Liou K, Liu F. RNase P ribozymes selected in vitro to cleave a viral mRNA effectively inhibit its expression in cell culture. J Biol Chem. 2000;275(14):10611–22.
38. Bertrand E, Castanotto D, Zhou C, Carbonnelle C, Lee NS, Good P, Chatterjee S, Grange T, Pictet R, Kohn D, Engelke D, Rossi JJ. The expression cassette determines the functional activity of ribozymes in mammalian cells by controlling their intracellular localization. RNA. 1997;3(1):75–88.
39. Collins TM, Quirk MR, Jordan MC. Biphasic viremia and viral gene expression in leukocytes during acute cytomegalovirus infection of mice. J Virol. 1994;68(10):6305–11.
40. Hudson JB. The murine cytomegalovirus as a model for the study of viral pathogenesis and persistent infections. Arch Virol. 1979;62(1):1–29.
41. Katzenstein DA, Yu GS, Jordan MC. Lethal infection with murine cytomegalovirus after early viral replication in the spleen. J Infect Dis. 1983;148(3):406–11.
42. Abenes G, Chan K, Lee M, Haghjoo E, Zhu J, Zhou T, Zhan X, Liu F. Murine cytomegalovirus with a transposon insertional mutation at open reading frame m155 is deficient in growth and virulence in mice. J Virol. 2004;78(13):6891–9.

43. Pollock JL, Virgin HW. Latency, without persistence, of murine cytomegalovirus in the spleen and kidney. J Virol. 1995;69(3):1762–8.
44. Liu F, Song Y, Liu D. Hydrodynamics-based transfection in animals by systemic administration of plasmid DNA. Gene Ther. 1999;6(7):1258–66.
45. Song E, Lee SK, Wang J, Ince N, Ouyang N, Min J, Chen J, Shankar P, Lieberman J. RNA interference targeting Fas protects mice from fulminant hepatitis. Nat Med. 2003;9(3):347–51.
46. Zhang G, Budker V, Wolff JA. High levels of foreign gene expression in hepatocytes after tail vein injections of naked plasmid DNA. Hum Gene Ther. 1999;10(10):1735–7.
47. Bai Y, Trang P, Li H, Kim K, Zhou T, Liu F. Effective inhibition in animals of viral pathogenesis by a ribozyme derived from RNase P catalytic RNA. Proc Natl Acad Sci U S A. 2008;105(31):10919–24.
48. Hoiseth SK, Stocker BA. Aromatic-dependent *Salmonella typhimurium* are non-virulent and effective as live vaccines. Nature. 1981;291(5812):238–9.
49. Walthers D, Carroll RK, Navarre WW, Libby SJ, Fang FC, Kenney LJ. The response regulator SsrB activates expression of diverse *Salmonella* pathogenicity island 2 promoters and counters silencing by the nucleoid-associated protein H-NS. Mol Microbiol. 2007;65(2):477–93.
50. Van Amersfoort ES, Van Berkel TJ, Kuiper J. Receptors, mediators, and mechanisms involved in bacterial sepsis and septic shock. Clin Microbiol Rev. 2003;16(3):379–414.
51. Khan SA, Everest P, Servos S, Foxwell N, Zahringer U, Brade H, Rietschel ET, Dougan G, Charles IG, Maskell DJ. A lethal role for lipid A in *Salmonella* infections. Mol Microbiol. 1998;29(2):571–9.
52. Lee SR, Kim SH, Jeong KJ, Kim KS, Kim YH, Kim SJ, Kim E, Kim JW, Chang KT. Multi-immunogenic outer membrane vesicles derived from an MsbB-deficient *Salmonella* enterica serovar typhimurium mutant. J Microbiol Biotechnol. 2009;19(10):1271–9.
53. Low KB, Ittensohn M, Le T, Platt J, Sodi S, Amoss M, Ash O, Carmichael E, Chakraborty A, Fischer J, Lin SL, Luo X, Miller SI, Zheng L, King I, Pawelek JM, Bermudes D. Lipid A mutant *Salmonella* with suppressed. virulence and TNFalpha induction retain tumor-targeting in vivo. Nat Biotechnol. 1999;17(1):37–41.
54. Toso JF, Gill VJ, Hwu P, Marincola FM, Restifo NP, Schwartzentruber DJ, Sherry RM, Topalian SL, Yang JC, Stock F, Freezer LJ, Morton KE, Seipp C, Haworth L, Mavroukakis S, White D, MacDonald S, Mao J, Sznol M, Rosenberg SA. Phase I study of the intravenous administration of attenuated. *Salmonella typhimurium* to patients with metastatic melanoma. J Clin Oncol. 2002;20(1):142–52.

Chapter 13
Self-Assembled Polysaccharide Nanogels for Nasal Delivery of Biopharmaceuticals

Tomonori Nochi, Yoshikazu Yuki, Kazunari Akiyoshi and Hiroshi Kiyono

13.1 Introduction

Nanotechnology is an innovative bioengineering to create functional nanometer-sized biomaterials that are useful in the multiple fields of life science (e.g., medicine, pharmaceutics) [1–2]. Among the several types of nanometer-sized biomaterials, cholesteryl group-bearing pullulan (CHP) nanogels described herein possess unique structural characteristics. Specifically, the CHP nanogels, which consist of hydrophilic long-chain polymers (named pullulan) associated with hydrophobic cholesterol, are self-assembly formed hydrogels with a three-dimensional network structure ($\approx 30\,\text{nm}$) [3]. The most outstanding characteristic of CHP nanogels is its chaperon-like activity [4–6]. Bioactive proteins, such as cytokines, enzymes, and vaccine antigens, can be incorporated in the CHP nanogels by mostly hydrophobic interaction in the hydrated polymer network where the cholesterol moieties in the CHP nanogels associate mutually [4–6]. The numbers of protein that can be entrapped into one CHP nanogel mostly depend on the molecular weight of protein [1].

H. Kiyono (✉) · Y. Yuki
Division of Mucosal Immunology, Institute of Medical Science,
The University of Tokyo, 4-6-1 Shirokanedai, Minato-ku, Tokyo 108-8639, Japan
e-mail: kiyono@ims.u-tokyo.ac.jp

International Research and Development Center for Mucosal Vaccines,
Institute of Medical Science, The University of Tokyo, Tokyo, Japan

Core Research for Evolutional Science and Technology,
Japan Science and Technology Agency, Tokyo, Japan

T. Nochi
Laboratory of Functional Morphology, Graduate School of Agricultural Science,
Tohoku University, Sendai, Miyagi, Japan

K. Akiyoshi
Department of Polymer Chemistry, Graduate School of Engineering,
Kyoto University, Kyoto, Japan

ERATO Bio-nanotransporter Project, Japan Science and Technology Agency, Kyoto, Japan

J. das Neves, B. Sarmento (eds.), *Mucosal Delivery of Biopharmaceuticals,* DOI 10.1007/978-1-4614-9524-6_13,

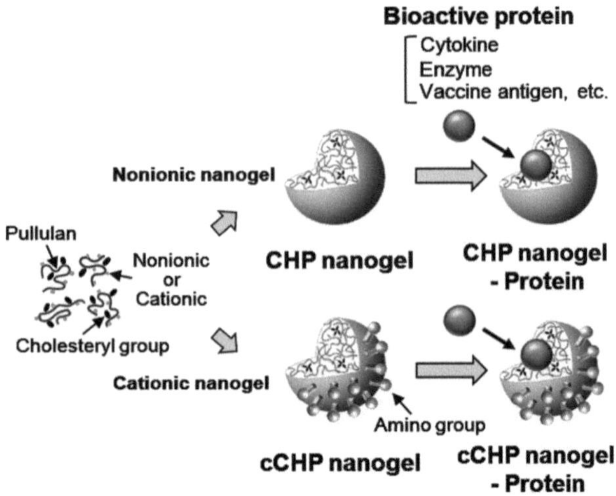

Fig. 13.1 Unique characteristics of CHP and cCHP nanogels. Cholesterol-bearing pullulan self-assembly form hydrogels with a three-dimensional network structure (CHP nanogels). Cationic type of CHP nanogels (cCHP nanogels) are generated by adding amine groups to CHP nanogels for enhancing the ability of binding to host cells

For example, in one CHP nanogel, one molecule of bovine serum albumin (molecular weight: 66,000 Da) is incorporated while five molecules of insulin (molecular weight: 5,735 Da) are entrapped [1]. In addition, the hydrophobicity of guest protein is also involved in the efficiency of incorporation in CHP nanogels (Fig. 13.1, [1]).

The molecular chaperon activity of CHP nanogels is critical when the CHP nanogels are used as a protein carrier. Once the guest protein incorporated in CHP nanogels is injected *in vivo*, the entrapped protein is released gradually from the hydrogels by replacing it to other excess amount of endogenous proteins [3]. Importantly, the protein released from CHP nanogels still possesses intact bioactivity [4–8]. Therefore, the use of CHP nanogels enables us to sustain the protein spontaneously in the targeted tissues, resulting that the long-term effect of protein delivery and release can be expected by using CHP nanogels. One of the clinical trials based on the chaperon activity of CHP nanogels is the tumor immunotherapy with cancer antigen (i.e., HER2 or NY-ESO-1) that is incorporated in CHP nanogels [9–10]. The cancer patients who received the cancer antigen with CHP nanogels subcutaneously induced not only tumor antigen-specific antibody response but also CD4 and CD8 types of cellular responses effectively [9–10]. To this end, CHP nanogels would be the potential artificial chaperon that can be used as a novel delivery vehicle for the protein-based immunotherapy.

Nasal mucosa is an important immune tissue, especially as a part of the mucosal immune system providing a first line of defense to protect the host from respiratory infectious diseases, since nasal immune system induces antigen-specific immune responses against respiratory pathogens (e.g., influenza virus, *Streptococcus pneumoniae*) [11–12]. One of the practical approaches based on the concept of this nasal immune system is to develop nasal vaccine because it induces antigen-specific immune responses in the respiratory tissues (e.g., nasal mucosa, trachea and lung) in addition to systemic compartments (e.g., spleen) [11–12]. Since 2003, nasal flu vaccine named FluMist®, which is a nasal spray type of live-attenuated influenza vaccine, has been developed and applied for the clinical use in the USA [13–14]. Nasal vaccination with FluMist® induces not only influenza virus-specific immunoglobulin A (IgA) antibody response but also $CD8^+$ T cell-mediated cellular immune responses to influenza virus-infected cells in the respiratory tissues, resulting that it enables us to inhibit the viral dissemination from respiratory mucosa to systemic tissues [13–14]. In 2011, European Union also approved the use of live-attenuated nasal influenza vaccine named Fluenz® [15]. Currently, the nasal influenza vaccine has been used in Canada, South Korea, Hong Kong, Macau, and Israel in addition to the USA and European countries [16]. However, even though FluMist® and Fluenz® are composed of live-attenuated influenza virus with low virulence, there is still a risk to cause an infection particularly when given in people whose immune system is weakened or disturbed [11, 14]. Therefore, it is only allowed to use FluMist® for all healthy, nonpregnant women whose age is between 2 and 49 years (2 and 59 years in Canada) and to use Fluenz® for only children aged 2–17 years in European countries, so people suffering from certain chronic diseases and weak immune system are not eligible to receive the vaccine [17]. Moreover, most countries have not yet allowed the usage of the live-attenuated nasal influenza vaccine because of its safety concerns. Therefore, further considerable approaches that are capable of safely and effectively inducing antigen-specific immune responses in the respiratory immune system without the use of live-attenuated virus needs to be developed [11, 14]. In that sense, nasal immunization with a subunit antigen, such as a purified or recombinant bacterial/viral molecule with antigenicity has been expected to be the safest strategy without causing any undesired and unfavorable biological reactions (e.g., sever inflammation and high fever) [11, 14]. However, dissimilar to live-attenuated vaccine, nasal vaccination with a subunit antigen itself generally induces insufficient level of antigen-specific immune responses because of the absence of its associated bacterial/viral immunomodulatory molecules that enhance the innate and acquired immune responses [11, 14]. Therefore, coadministration with an adjuvant such as a bacterial toxin [e.g., cholera toxin (CT) and heat-labile enterotoxin (LT)] [18–19] or innate immunity-enhancing CpGDNA [20] has been shown to be necessary for accelerating the vaccine-induced immune responses [21]. However, a previous human clinical trial performed in Switzerland for developing a nasal influenza vaccine with inactivated influenza virus together with a small amount of LT was withdrawn because the coadministered LT was suspected of causing Bell's palsy in a few vaccinated subjects [22]. Therefore, further scientific and technological innovations that will facilitate the development of safe but effective nasal vaccine strategies, such as an antigen-delivery system to nasal immune system, which may not require co-administration of adjuvant, are high priority in immunology, vaccinology and biopharmaceuticals.

13.2 Attempt for the Adaptation of CHP Nanogels for Nasal Vaccine Development

CHP nanogels are useful biomaterials for tumor immunotherapy with HER2 or NY-ESO-1 as described above [9–10], we initially investigated the efficacy of CHP nanogels as an antigen delivery vehicle for a subunit type of nasal vaccine. When a

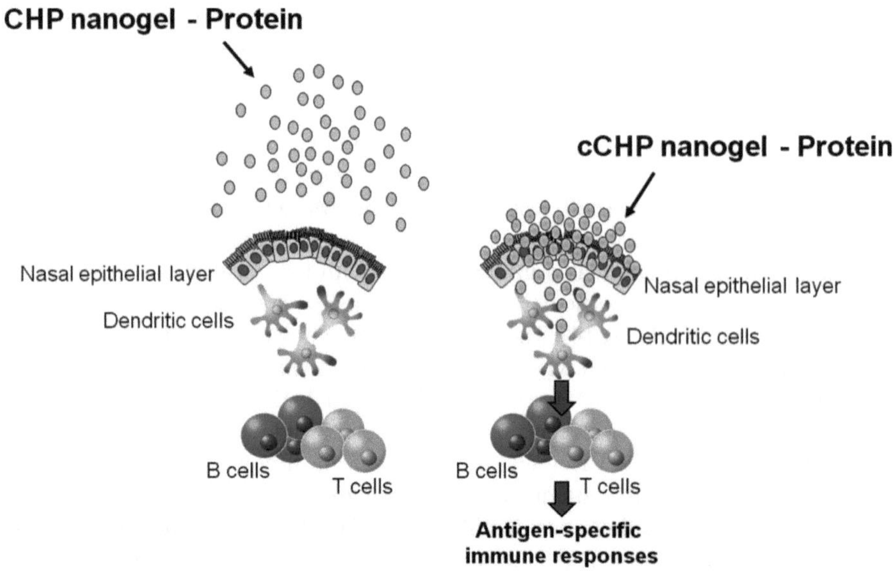

Fig. 13.2 cCHP nanogels are appropriate delivery vehicle for nasal vaccine development. cCHP (not CHP) nanogel-protein complex highly associates with nasal epithelium, resulting in effective uptaking of vaccine antigen by nasal dendritic cells for the initiation of antigen-specific immune responses

nontoxic receptor-binding portion (heavy-chain C terminus) of *C. botulinum* type-A neurotoxin subunit antigen Hc (BoHc/A) , which has been known as an effective vaccine antigen for the induction of antigen-specific neutralizing antibodies against *C. botulinum* neurotoxin [23], was immunized nasally in mice after incorporation in CHP nanogels, however, unexpectedly, no remarkable difference to nasal immunization with naked BoHc/A for inducing antigen-specific antibody responses was found [24]. Because the nasal epithelial layer covers nasal immune tissues tightly, we hypothesized that the most BoHc/A given nasally could not penetrate into the nasal epithelium, thus was unable to reach to the nasal immune system even though BoHc/A was incorporated in CHP nanogels (Fig. 13.2). Therefore, we added 15 amino groups per 100 glucose units in CHP nanogels to develop a cationic type of CHP nanogels (cCHP nanogels) for enhancing the ability of binding to the nasal epithelium whose surface is anionic because of excess amount of negatively charged membranous proteins [24–25]. Dynamic light scattering (DLS) and fluorescence resonance energy transfer (FRET) analyses showed that cCHP nanogels possess similar structural characteristics to CHP nanogels because cCHP nanogels maintain nanoscale size uniformity even after incorporating BoHc/A [24–25]. However, importantly, dissimilar to CHP nanogels, cCHP nanogels highly associate with the several types of cells (e.g., HeLa, CHO-K1, cos-7) and are capable of delivering the protein antigen efficiently into the cells *in vitro* without causing any cellular damages [25]. Taken together, these results suggest that cCHP nanogels could be a possible candidate antigen delivery vehicle for nasal vaccine development.

13.3 cCHP Nanogels-Based Antigen Delivery System for Nasal Vaccine

Our initial *in vivo* study using cCHP nanogels for nasal vaccine development was to investigate how long the vaccine antigen could be sustained at the nasal mucosa when the antigen incorporated in cCHP nanogels was nasally administered. The positron emission tomography (PET) and the direct radioisotope counting analyses with nasal tissue, harvested from mice nasally administered with radioisotope-conjugated BoHc/A in the presence or absence of cCHP nanogels, indicated that BoHc/A administered with cCHP nanogels (cCHP-BoHc/A) stayed at nasal tissue for more than 48 h while naked BoHc/A disappeared within just 6 h from nasal tissues [24]. In the nasal epithelium, BoHc/A was released from cCHP nanogels and immediately taken up by nasal dendritic cells, leading to the initiation of antigen-specific immune responses ([24], Fig. 13.2). Importantly, mice nasally immunized with cCHP-BoHc/A induced significantly high levels of BoHc/A-specific antibody responses with neutralizing activity in both nasal tissue and systemic compartments [24]. In contrast, the mice immunized with CHP-BoHc/A or naked BoHc/A induced extremely low levels of antigen-specific immune responses, thus were unable to protect from lethally challenged *C. botulinum* type-A neurotoxin [24].

When the development of nasal vaccine is planned, a critical issue is to overcome its safety concerns about the potential dissemination of vaccine antigens to the central nervous system (CNS), such as olfactory bulb and brain [14, 26]. Cholera toxin, which has been used extensively as a nasal adjuvant in animal studies, has been shown to not only accumulate in itself but also direct coadministered vaccine antigen into the CNS [27]. Our *in vivo* tracer study with radioisotope-conjugated BoHc/A incorporated in cCHP nanogels showed that no transition of BoHc/A into the olfactory bulbs or brain was observed over a 2-day period after nasal administration [24]. These results indicate that cCHP nanogels possess no risk of redirecting the vaccine antigen into the CNS when administered nasally and, therefore, can be used as a safe delivery vehicle for nasal vaccines.

13.4 Universal Usage of cCHP Nanogels for Nasal Vaccine Development

When a novel strategy for mucosal antigen delivery into the targeted tissues is developed by using biomaterials, it is critical to confirm its broad utility in addition to the safety as described above. Our subsequent analysis using alternative vaccine antigen Pneumococcal surface protein A (PspA), which is commonly expressed by all capsular serotypes of *S. pneumoniae* and has been extensively used for an injectable type of pneumococcal subunit vaccine antigen [28–29], indicated that nasal immunization with PspA incorporated in cCHP nanogels (cCHP-PspA) induced extremely high levels of antigen-specific neutralizing antibody responses in the respiratory tissue as well as systemic compartments [30]. We also confirmed the safety of cCHP nanogel-based pneumococcal nasal vaccine with PspA as demonstrated by no transition of PspA into the olfactory bulbs or brain over a 2-day period after nasal administration [30]. Moreover, our additional study to investigate the efficacy of cCHP nanogel-based nasal vaccine with other type of antigen, such as toxoid that is prepared with toxin inactivated by formalin treatment, indicated that nasal immunization with tetanus toxoid (TT) incorporated in cCHP nanogels (cCHP-TT) induced significantly high levels of antigen-specific antibody responses compared with nasal immunization with necked TT [24]. These findings indicate that the cCHP nanogels can be used universally for both subunit antigen- and toxoid-based nasal vaccine.

Another interesting finding we have noticed is that PspA-specific Th17 response is induced in nasal mucosa by nasal immunization with cCHP-PspA [30]. The detail mechanism of inducing Th17 cell development by nasal immunization with cCHP-PspA is still unknown. However, Th17 cells play an important role in preventing pneumococcal nasal colonization in mice immunized nasally with whole antigen derived from *Streptococcus pneumonia* [31–32] and also has a role in inducing autoimmunity [33], such that further continuous study should be carefully performed to investigate the positive or negative effects of cCHP nanogel-based nasal vaccine for the human clinical use.

13.5 Conclusions and Future Perspectives

cCHP nanogels are stable biomaterials that can be stored for a prolonged time without any degradation [1], but it has not been carefully examined yet about the stability of cCHP nanogels after incorporating the vaccine antigen. Our current study is to investigate how vaccine antigen incorporated in cCHP nanogels can be stored without loss of the bioactivity and antigenicity until just before the immunization. Although it has been demonstrated that vaccine antigen does not reach to the CNS after nasal administration with cCHP nanogels, the influence to olfactory sense after the use of cCHP nanogel-based antigen-delivery system has not been carefully addressed yet. Because olfactory epithelial cells are highly distributed in nasal epithelial layer where cCHP nanogels associate upon nasal administration, a further physiological study should be performed to deny the possibility of cCHP nanogel-induced side effects in the olfactory system. Also, an immunological study to demonstrate the efficacy of cCHP nanogel-based nasal vaccine in humans needs to be performed. Taken all together, it is important to continue immunological, physiological, and pharmacological investigations to demonstrate the stability, safety, and efficacy of cCHP nanogel-based nasal vaccine for human application.

Acknowledgments This work was supported by the Japan Society for the Promotion of Science grants for the Ministry of Education, Culture, Sports, Science, and Technology of Japan for

Scientific Research (to H.K.), for the Leading-Edge Research Infrastructure Program (to H.K.), and for the Young Researcher Overseas Visits Program for Vitalizing Brain Circulation (to H.K.).

References

1. Sasaki Y, Akiyoshi K. Nanogel engineering for new nanobiomaterials: from chaperoning engineering to biomedical applications. Chem Rec. 2010;10(6):366–76.
2. Wagner V, Dullaart A, Bock AK, Zweck A. The emerging nanomedicine landscape. Nat Biotechnol. 2006;24(10):1211–7.
3. Akiyoshi K, Kobayashi S, Shichibe S, Mix D, Baudys M, Kim SW, Sunamoto J. Self-assembled hydrogel nanoparticle of cholesterol-bearing pullulan as a carrier of protein drugs: complexation and stabilization of insulin. J Control Release. 1998;54(3):313–20.
4. Akiyoshi K, Sasaki Y, Sunamoto J. Molecular chaperone-like activity of hydrogel nanoparticles of hydrophobized pullulan: thermal stabilization with refolding of carbonic anhydrase B. Bioconjug Chem. 1999;10(3):321–4.
5. Nomura Y, Ikeda M, Yamaguchi N, Aoyama Y, Akiyoshi K. Protein refolding assisted by self-assembled nanogels as novel artificial molecular chaperone. FEBS Lett. 2003;553(3):271–6.
6. Nomura Y, Sasaki Y, Takagi M, Narita T, Aoyama Y, Akiyoshi K. Thermoresponsive controlled association of protein with a dynamic nanogel of hydrophobized polysaccharide and cyclodextrin: heat shock protein-like activity of artificial molecular chaperone. Biomacromolecules. 2005;6(1):447–52.
7. Ikeda K, Okada T, Sawada S, Akiyoshi K, Matsuzaki K. Inhibition of the formation of amyloid beta-protein fibrils using biocompatible nanogels as artificial chaperones. FEBS Lett. 2006;580(28–29):6587–95.
8. Boridy S, Takahashi H, Akiyoshi K, Maysinger D. The binding of pullulan modified cholesteryl nanogels to Abeta oligomers and their suppression of cytotoxicity. Biomaterials. 2009;30(29):5583–91.
9. Uenaka A, Wada H, Isobe M, Saika T, Tsuji K, Sato E, Sato S, Noguchi Y, Kawabata R, Yasuda T, Doki Y, Kumon H, Iwatsuki K, Shiku H, Monden M, Jungbluth AA, Ritter G, Murphy R, Hoffman E, Old LJ, Nakayama E. T cell immunomonitoring and tumor responses in patients immunized with a complex of cholesterol-bearing hydrophobized pullulan (CHP) and NY-ESO-1 protein. Cancer Immun. 2007;7:9.
10. Kageyama S, Kitano S, Hirayama M, Nagata Y, Imai H, Shiraishi T, Akiyoshi K, Scott AM, Murphy R, Hoffman EW, Old LJ, Katayama N, Shiku H. Humoral immune responses in patients vaccinated with 1–146 HER2 protein complexed with cholesteryl pullulan nanogel. Cancer Sci. 2008;99(3):601–7.
11. Kunisawa J, Nochi T, Kiyono H. Immunological commonalities and distinctions between airway and digestive immunity. Trends Immunol. 2008;29(11):505–13.
12. Kiyono H, Fukuyama S. NALT- versus Peyer's-patch-mediated mucosal immunity. Nat Rev Immunol. 2004;4(9):699–710.
13. Belshe R, Lee MS, Walker RE, Stoddard J, Mendelman PM. Safety, immunogenicity and efficacy of intranasal, live attenuated influenza vaccine. Expert Rev Vaccines. 2004;3(6):643–54.
14. Yuki Y, Kiyono H. Mucosal vaccines: novel advances in technology and delivery. Expert Rev Vaccines. 2009;8(8):1083–97.
15. Carter NJ, Curran MP. Live attenuated influenza vaccine (FluMist(R); Fluenz): a review of its use in the prevention of seasonal influenza in children and adults. Drugs. 2011;71(12):1591–622.
16. Ambrose CS, Yi T, Falloon J. An integrated, multistudy analysis of the safety of Ann Arbor strain live attenuated influenza vaccine in children aged 2–17 years. Influenza Other Respir Viruses. 2011;5(6):389–97.

17. Ambrose CS, Walker RE, Connor EM. Live attenuated influenza vaccine in children. Semin Pediatr Infect Dis. 2006;17(4):206–12.
18. Freytag LC, Clements JD. Bacterial toxins as mucosal adjuvants. Curr Top Microbiol Immunol. 1999;236:215–36.
19. Ryan EJ, McNeela E, Pizza M, Rappuoli R, L. O'Neill, Mills KH. Modulation of innate and acquired immune responses by *Escherichia coli* heat-labile toxin: distinct pro- and anti-inflammatory effects of the nontoxic AB complex and the enzyme activity. J Immunol. 2000;165(10):5750–9.
20. Fukuiwa T, Sekine S, Kobayashi R, Suzuki H, Kataoka K, Gilbert RS, Kurono Y, Boyaka PN, Krieg AM, McGhee JR, Fujihashi K. A combination of Flt3 ligand cDNA and CpG ODN as nasal adjuvant elicits NALT dendritic cells for prolonged mucosal immunity. Vaccine. 2008;26(37):4849–59.
21. Yuki Y, Kiyono H. New generation of mucosal adjuvants for the induction of protective immunity. Rev Med Virol. 2003;13(5):293–310.
22. Mutsch M, Zhou W, Rhodes P, Bopp M, Chen RT, Linder T, Spyr C, Steffen R. Use of the inactivated intranasal influenza vaccine and the risk of Bell's palsy in Switzerland. N Engl J Med. 2004;350(9):896–903.
23. Byrne MP, Smith TJ, Montgomery VA, Smith LA. Purification, potency, and efficacy of the botulinum neurotoxin type A binding domain from *Pichia pastoris* as a recombinant vaccine candidate. Infect Immun. 1998;66(10):4817–22.
24. Nochi T, Yuki Y, Takahashi H, Sawada S, Mejima M, Kohda T, Harada N, Kong IG, Sato A, Kataoka N, Tokuhara D, Kurokawa S, Takahashi Y, Tsukada H, Kozaki S, Akiyoshi K, Kiyono H. Nanogel antigenic protein-delivery system for adjuvant-free intranasal vaccines. Nat Mater. 2010;9(7):572–8.
25. Ayame H, Morimoto N, Akiyoshi K. Self-assembled cationic nanogels for intracellular protein delivery. Bioconjug Chem. 2008;19(4):882–90.
26. Yuki Y, Nochi T, Harada N, Katakai Y, Shibata H, Mejima M, Kohda T, Tokuhara D, Kurokawa S, Takahashi Y, Ono F, Kozaki S, Terao K, Tsukada H, Kiyono H. *In vivo* molecular imaging analysis of a nasal vaccine that induces protective immunity against botulism in nonhuman primates. J Immunol. 2010;185(9):5436–43.
27. van Ginkel FW, Jackson RJ, Yuki Y, McGhee JR. Cutting edge: the mucosal adjuvant cholera toxin redirects vaccine proteins into olfactory tissues. J Immunol. 2000;165(9):4778–82.
28. Nabors GS, Braun PA, Herrmann DJ, Heise ML, Pyle DJ, Gravenstein S, Schilling M, Ferguson LM, Hollingshead SK, Briles DE, Becker RS. Immunization of healthy adults with a single recombinant pneumococcal surface protein A (PspA) variant stimulates broadly cross-reactive antibodies to heterologous PspA molecules. Vaccine. 2000;18(17):1743–54.
29. Crain MJ, Waltman WD 2nd, Turner JS, Yother J, Talkington DF, McDaniel LS, Gray BM, Briles DE. Pneumococcal surface protein A (PspA) is serologically highly variable and is expressed by all clinically important capsular serotypes of *Streptococcus pneumoniae*. Infecti Immun. 1990;58(10):3293–9.
30. Kong IG, Sato A, Yuki Y, Nochi T, Takahashi H, Sawada S, Mejima M, Kurokawa S, Okada K, Sato S, Briles DE, Kunisawa J, Inoue Y, Yamamoto M, Akiyoshi K, Kiyono H. Nanogel-based PspA intranasal vaccine prevents invasive disease and nasal colonization by *Streptococcus pneumoniae*. Infecti Immun. 2013;81(5):1625–34.
31. Lu YJ, Gross J, Bogaert D, Finn A, Bagrade L, Zhang Q, Kolls JK, Srivastava A, Lundgren A, Forte S, Thompson CM, Harney KF, Anderson PW, Lipsitch M, Malley R. Interleukin-17A mediates acquired immunity to pneumococcal colonization. PLoS Pathog. 2008;4(9):e1000159.
32. Malley R, Trzcinski K, Srivastava A, Thompson CM, Anderson PW, Lipsitch M. CD4 + T cells mediate antibody-independent acquired immunity to pneumococcal colonization. Proc Natl Acad Sci U S A. 2005;102(13):4848–53.
33. Bettelli E, Oukka M, Kuchroo VK. T(H)-17 cells in the circle of immunity and autoimmunity. Nat Immunol. 2007;8(4):345–50.

Chapter 14
Pheroid^{TM} Vesicles and Microsponges for Nasal Delivery of Biopharmaceuticals

Lissinda H. du Plessis and Awie F. Kotzé

14.1 Introduction

The anatomy and physiology of the nasal passage offers numerous practical advantages for the introduction of therapeutic peptides into the systemic circulation. The highly vascular nasal mucosa makes rapid absorption of the administered drug possible and, furthermore, ensures that the drug avoids degradation in the gastrointestinal tract and first-pass metabolism in the liver associated with oral administration. Nasal administration and intravenous administration of therapeutic peptides often exhibit similar concentration-time profiles that suggest a rapid onset of pharmacological activity after nasal administration [1–3]. The absorption of foreign material in the nose is prevented by a physical barrier or the mucus and epithelium, a temporal barrier also known as mucociliary clearance and a chemical or enzymatic barrier. These barriers may influence drug permeation as nasally administered drugs have to pass through these barriers. A drug can permeate epithelial membranes either passively by the paracellular pathway or both passively and actively via the transcellular pathway. Some other transport mechanisms include carrier-mediated transport. The method of transport depends on the lipophilicity of the compound: if the lipophilicity is increased the absorption of the compound increases through the nasal mucosa via the transcellular passive diffusion pathway [4]. The patented Pheroid^{TM} technology, which is a unique colloidal drug delivery system, has numerous advantages in nasal delivery. A Pheroid^{TM} is a stable structure within a novel therapeutic system which can be manipulated in terms of morphology, structure, size, and function. Pheroid^{TM} consist mainly of plant and essential fatty acids and can entrap, transport, and deliver pharmacologically active compounds and other useful substances to the desired site of action [5]. In this chapter the application of Pheroid^{TM} technology in the delivery of peptides via the nasal route will be reviewed. The possibility of delivery of peptide drugs via other mucosal routes with Pheroid^{TM} technology will also be discussed.

L. H. du Plessis (✉) · A. F. Kotzé
Unit for Drug Research and Development, North-West University,
Potchefstroom 2520, South Africa
e-mail: Lissinda.DuPlessis@nwu.ac.za

J. das Neves, B. Sarmento (eds.), *Mucosal Delivery of Biopharmaceuticals*, DOI 10.1007/978-1-4614-9524-6_14,

14.2 Pheroid™ Technology

Pheroid™ is a lipid-based technology consisting of plant and essential fatty acids, primarily used to control the rate and period of drug delivery, and offers the ability to optimize the absorption of and profile of certain drugs to match their pharmacokinetics to a target indication. It is a colloidal system consisting of two separate phases: the lipid-based dispersed phase with submicron- and micron-sized stable structures and the continuous phase. The continuous phase consists of aqueous and nitrous oxide (N_2O) gas phase. The morphology, structure, size, and function of these dispersed structures can be manipulated. The particles have a diameter of between 200 nm and 2 μm and can entrap active pharmaceutical compounds that can possibly enhance the therapeutic effect [6–7].

Colloidal drug carriers have numerous advantages to deliver drugs. Colloidal systems can be grouped into three groups. The first group is lyophilic or reversible colloids, the second is lyophobic or irreversible colloids, and the last is association colloids [8]. Examples of colloidal dosage forms include liposomes, emulsions and microemulsions, polymeric microspheres and macromolecular microspheres [9]. Pheroid™ formulations are essentially a colloidal system incorporating different features of the different colloidal dosage forms [6]. Pheroid™ Technology as a delivery system have proven improved delivery of various types of drugs, reduced time to onset of action, decreased minimal effective drug concentrations, enhanced therapeutic efficacy, and decreased cytotoxicity. It has the ability to cross most barriers in the body such as keratinized tissue, cell membranes, intestinal lining, and the vascular system that provides it the capability to target specific treatment areas. Pheroid™ is safe and do not elicit immunological responses.

The following characteristics make Pheroid™ technology unique:

- Structure and classification of the different types of Pheroid™ formulations
- Toxicity profile of Pheroid™
- Mechanism of uptake
- Pharmacokinetic properties—drug entrapment, protection of drugs, absorption, and volume of distribution

14.3 Structure and Classification of the Different Types of Pheroid™ Formulations

Pheroid™ technology consists of different types of Pheroid™ formulations that depends on the composition and method of manufacturing. The three main formulations are:

- Pheroid™ vesicles
- Pheroid™ microsponges
- Pro-Pheroid™ in depots or reservoirs

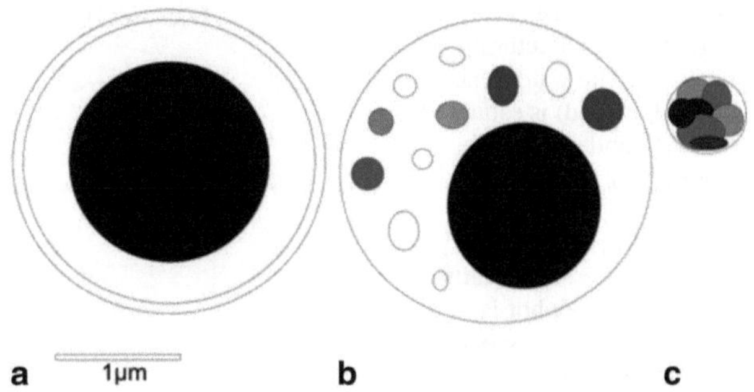

Fig. 14.1 A schematic representation of the different types of Pheroid™ formulation. **a** Pheroid™ vesicle with a bilayer and a aqueous core, **b** Pheroid™ microsponge with the central hydrophilic aqueous space a thick sponge-like membrane, and **c** a Pro-Pheroid™

Pheroid™ consists of essential fatty acids that are a natural ingredient of the body dispersed in a liquid and N_2O gas phase. It contains a lipid bilayer, but do not include phospholipids of cholesterol into the membrane structure. It contains an aqueous core (vesicles) or various small reservoirs (microsponges and Pro-Pheroid™) [6].

Every Pheroid™ formulation has a unique composition that can be manipulated for the intended use. The vesicle is a lipid-bilayer structure in the range of 0.5 μm to 1.5 μm [5]. The microsponge formulation contains various small depots and range in size between 1.5 μm and 5 μm. Both the vesicle and the microsponge structure give the Pheroid™ the ability to entrap hydrophilic compounds in the center and hydrophobic compounds in the membrane [6]. Fig. 14.1 shows the different types of Pheroid™ formulations.

The main components of the Pheroid™ lipid phase are ethyl esters of the essential fatty acids linoleic acid and linolenic acid, as well as the *cys*-form of oleic acid [5]. The essential fatty acids cannot be manufactured by the human body but are important for normal cellular functions. Their functions include homeostasis of energy, modulating the immune system, maintaining the integrity of cell membranes, and regulating some components of programmed cell death [10]. The functions of α-tocopherol include antioxidant, signal transduction and cell metabolism [11].

The aqueous phase is saturated with N_2O [7]. N_2O is a volatile anesthetic gas that is both lipid and water soluble [12] and has at least three known functions in the formulation: assisting with the self-assembly process of Pheroid™ vesicles and the miscibility of the fatty acids in the dispersed medium, as well as playing a role in the stability of Pheroid™ vesicles or microsponges that are formed [6].

The basic Pheroid™ manufacturing procedure is illustrated in Fig. 14.2. The procedure is similar to the manufacturing of simple emulsions [5]. The Pheroid™ can be manipulated to change its structural and functional features, through the following formulation changes:

- Changing the fatty acid composition or concentrations
- The addition of nonfatty acids or phospholipids such as cholesterol
- The addition of cryoprotectants
- The addition of charge-inducing agents
- Changing the hydration medium (ionic strength, pH)
- Changing the method of preparation
- Changing the character and the concentration of the active compound

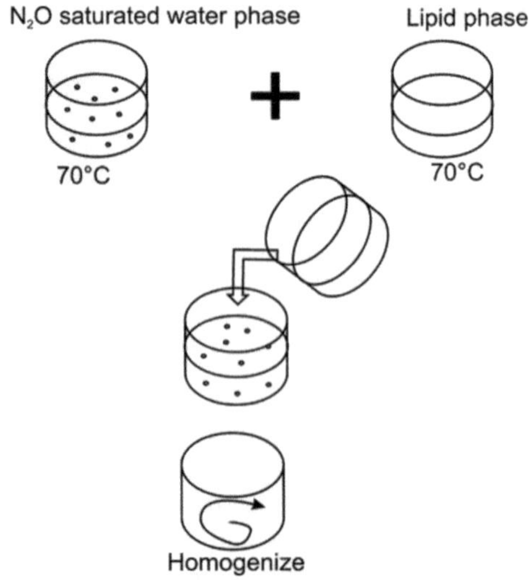

Fig. 14.2 The basic Pheroid™ manufacturing procedure. The water phase (can be buffered) is saturated with N_2O under pressure (1.6pKa) for 4 days. Thereafter it is heated to 70 °C. The oil phase consisting of essential fatty acids and Cremophor is heated to 70 °C in a separate container. It is cooled and dl-α tocopherol is added. The cooled oil phase is added to the gassed water phase and is homogenized

The three types of Pheroid™ formulations are summarized, according to their composition, manufacturing methods, and structural features in Table 14.1.

14.4 Toxicity

During an extensive in vivo toxicity test on Sprague Dawley rats, no signs of toxicity was observed with the administration of oral pro-Pheroid™ formulation at a concentration of 50 mg/kg. No genotoxicity was observed and all the formulations were well tolerated. Because Pheroid™ consists of fatty acids that are natural ingredients, no immune response is elicited [13]. Pheroid™ are not cytotoxic at a concentration of up to 8 % lipids. However, hemolysis in red blood cells is observed at 0.5 %.

Table 14.1 A summary of the different types of Pheroid™ formulation with their corresponding characteristics

Type of Pheroid™	Description
Vesicles	Basic Pheroid™ vesicle with a highly elastic double layer membrane with the lipids packed loosely; contains 96 % water, 2.8 % essential fatty acids, 1 % Cremophor, and 0.2 % dl-α tocopherol
Microsponges	Small, sponge-like vesicles can entrap hydrophobic molecules in their membrane, while small hydrophilic molecules can be entrapped within the aqueous central area; contains eicosapentaenoic acid and docosahexaenoic acid (0.25:0.25 % w/v) that is added to the oil phase
Pro-Pheroid™	Pro-Pheroid™ spheres, or reservoirs, ranging from 5 to 100 μm in diameter, serving as a depot are prepared without adding the water phase

14.5 Mechanism of Uptake

Various proposed mechanisms of uptake have been proposed for Pheroid™ vesicles. One of the mechanisms is protein-mediated transfer by intracellular fatty acid binding proteins through active transport. Another proposed mechanism of uptake of the Pheroid™ into the cell is the high affinity of the fatty acids for the cell membranes and is transported by transcellular passive diffusion. It is able to target the subcellular level to some extent depending on the formulation. There is a difference in the absorption between the Pheroid™ vesicles and microsponges. The addition of eicosapentaenoic acid and docosahexaenoic acid to the formulation changes the lipid composition of the bilayer [7].

14.6 Pharmacokinetic Properties of Pheroid™ Formulations

The kinetic properties of the delivery system include the drug absorption, distribution, and elimination including metabolism and excretion. Entrapment of the drug and the protection given by the delivery system has a direct influence on the bioavailability and absorption [5]. The following unique pharmacokinetic properties of Pheroid™ will be discussed:

- Drug entrapment in the delivery system
- Protection of the drug by the delivery system
- Absorption
- Volume of distribution

14.6.1 Drug Entrapment

Pheroid™ can entrap drugs that are water and lipid soluble with high efficacy (85–90 %). Pheroid™ has the ability to entrap one drug in the interior space and another in the membrane. This characteristic decreases drug interaction making combination therapy possible with a single preparation. The Pheroid™ can be manipulated in size, charge, lipid composition, and membrane packing to optimize formulation to ensure that the maximum effect is achieved. Other lipid-based delivery systems can either entrap water soluble or lipid soluble drugs [5].

14.6.2 Protection of the Drug

Many drugs have reduced therapeutic effects because of the partial degradation of the drug before it reaches the specific target [14]. Pheroid™ protects the entrapped drug against degradation before it reaches the target site. No leakage of the drug out of the delivery system is observed. Other lipid-based delivery systems can protect the drug after IV administration but leakage is present [5].

14.6.3 Absorption

Bioavailability is the measurement of the rate and extent of active drug that reaches the systemic circulation. Pheroid™ increases the absorption and bioavailability in oral, topical, and nasal preparations. An increase in the bioavailability leads to a reduction in the minimal inhibitory concentration with increased therapeutic efficacy. Other lipid-based delivery systems can also improve the bioavailability whereas some can decrease it [5].

14.6.4 Volume of Distribution

The volume of distribution of a drug represents the volume that must be taken into consideration when calculating the drug concentration needed. A drug with a large volume of distribution means that there is more drug concentrated in the extravascular than intravascular tissue. Drugs that have high protein binding have lower volume of distribution. Drugs with high volume of distribution can be entrapped in the Pheroid™ to reduce the volume of distribution. This leads to an increase of drug concentration at the target site. The narrow therapeutic index is enhanced with less toxicity [5].

14.7 Delivery of Biopharmaceuticals with Pheroid™

14.7.1 Vaccines as Preventative Therapy

Vaccination of smallpox led to the elimination of smallpox. Prophylactic vaccination is, therefore, an important strategy in improving human health. The new generation vaccines are typically derived from purified pathogen subunits in an attempt to reduce harmful side effects but are less efficacious at eliciting a human immune response. It is necessary for effective vaccine adjuvants to enhance the immunogenicity and immunostimulatory properties of these vaccines. The vaccine adjuvants can be grouped in two classes, the immunostimulatory or immunomodulatory adjuvants and vaccine delivery systems [15]. The vaccines formulated with Pheroid™ technology investigated were a rabies vaccine, a hepatitis B vaccine, and a diphtheria vaccine [5].

Rabies is a virus-based vaccine. Carnivores and certain bat species host the rabies virus. Infections of humans from rabid animal bites are fatal. Postexposure vaccination prevents the death of millions of people each year. The inactivated virus is used in the vaccine formulation [16]. Animal studies were conducted to prove the efficacy of the Pheroid^{TM} formulated with the rabies vaccine. The inactivated virus in combination with Pheroid^{TM} was compared with the commercially available rabies vaccine with alum as adjuvant. The formulations were administered intraperitoneal on day 1 and again on day 7. After 14 days the mice were challenged by intracerebral injection of the virus. The Pheroid^{TM} formulations showed a significant increase in antibody response (9-fold) compared to the commercial vaccine. Unvaccinated mice died within 6 days [5]. The nonrecombinant hepatitis B vaccines are a peptide-based vaccine. A surface molecule of the virus is used as antigen. An antibody response needs to be elicited to ensure the effectiveness of the vaccine [16]. In combination with Pheroid^{TM} there was a 7-fold increase in the antibody production [5]. After the success of these initial vaccine studies with the formulation injected intraperitoneally and subcutaneously it was decided to investigate intranasal vaccination.

The Pheroid^{TM} drug delivery system was investigated as vehicle for the nasal and oral delivery of the diphtheria toxoid as antigen. In this study, immune responses were compared by measuring neutralizing antibodies against the diphtheria toxoid. Alum-based parenterally administered diphtheria toxoid was used as reference in a mouse model. Experimental animals were randomly assigned to 6 treatment groups, each of which consisted of 10 SPF balb/c female mice. Pathogen-free experimental animals were randomly assigned to treatment groups and one treatment of equal dosage was administered per animal by nasal vaccination over 3 consecutive days in weeks one and three. Each group contained its own negative control group, i.e., the unloaded Pheroid^{TM} delivery system. In each case, two positive controls were used, a diphtheria toxoid in phosphate buffered saline (PBS) administered and a diphtheria toxoid adsorbed to alum, which was injected subcutaneously. Micrometer Pheroid^{TM} vesicles were compared to nanometer Pheroid^{TM} vesicles, both containing the diphtheria toxoid. Half of the mice (five) were killed and blood samples collected for IgG (Immunoglobulin G) determination in week 4. The other half was similarly treated in week 6. After the mice were bled, nasal washing was performed and nasal lavages were collected and analyzed. The IgG titers (systemic immune response) and IgA (Immunoglobulin A) titers (local immune response) were determined using an enzyme-linked immunosorbent assay (ELISA). In the study conducted to determine the systemic immune response, it was seen that the diphtheria toxoid associated to all the above mentioned formulations produced a systemic immune response. When these formulations were compared to the positive controls, it was seen that the mice vaccinated with the diphtheria toxoid formulated with Pheroid^{TM} vesicles produced a significantly higher immune response than the mice vaccinated with diphtheria toxoid in PBS. As expected, the negative controls (unloaded adjuvant) showed no antibody response, either locally or systemically. The Pheroid^{TM} -based nasal vaccines showed immunogenicity equal to that of the alum-based parenteral vaccines. The size of the particles used had an effect on the antibody response with the micrometer range Pheroid^{TM} vesicles being significantly more effective in inducing both a systemic and local immune response [5, 17]. The Pheroid^{TM} appears to have a dual role in vaccination; firstly as delivery system for disease specific antigens, and secondly as immunostimulatory adjuvant. It was proven that it also complies with international requirements in terms of safety.

14.7.2 Peptide Drug Delivery

Protein drugs are used in neurological, endocrinological, and hematological diseases and disorders. Peptide and other protein drugs are poorly absorbed after oral delivery and mucosal routes are extensively investigated to improve the bioavailability of these drugs [18]. The possible enhancement of the absorption of peptide drugs using Pheroid™ technology has been investigated [7, 19–20]. These drugs include:

- Calcitonin for the treatment of osteoporosis
- Insulin for the treatment of insulin-dependent diabetes mellitus
- Human growth hormone important for the maintenance of optimal cellular performance

Calcitonin Calcitonin is necessary for regulating the calcium concentration in plasma levels. The nasal administration of calcitonin in combination with Pheroid™ microsponges and vesicles were investigated in rats [7]. There were no apparent adverse systemic events in the rats following nasal calcitonin administration and the Pheroid™ formulations appeared to be well tolerated. The mean plasma concentration-time profiles of nasally administered calcitonin were similar for Pheroid™ vesicles and Pheroid™ microsponges, but Pheroid™ vesicles resulted in a higher C_{max}. These profiles were characterized by an early plasma calcitonin peak, indicative of rapid absorption and distribution of the drug, followed by a slower decline due to elimination. Pheroid™ vesicles and microsponges with 10 IU/kg sCT intranasal increased the area under the curve (AUC) significantly when compared to the control and caused absolute bioavailability of 123.8 and 107.9 %, respectively (Table 14.2). The differences in absorption between the Pheroid™ vesicles and Pheroid™ microsponges were hypothetically attributed to the differences in the chemical composition of the membranes. During the preparation of Pheroid™ microsponges the addition of eicosapentaenoic acid and docosahexaenoic to the formulation had an effect on the formation of the vesicles that changed the Pheroid™ size [7].

Insulin Insulin-dependent diabetes mellitus is a fatal disorder when left untreated. This chronic disorder is treated by injection of recombinant human insulin subcutaneously. The in vivo effect of the nasal administration of insulin was determined using Sprague Dawley rats. The insulin formulated in either Pheroid™ microsponges or Pheroid™ vesicles showed a decrease in glucose levels after nasal administration [19]. Insulin (8.0 IU/kg) in Pheroid™ vesicles showed a decline in blood glucose levels 44.4 % after 3 h. There was a clear correlation between blood glucose levels and blood plasma insulin levels. Pheroid™ microsponges with insulin resulted in a decrease in blood glucose of 45.1 % after 3 h. The absolute bioavailability significantly increased to 132.4 % and 116.1 % for the Pheroid™ vesicles and Pheroid™

Table 14.2 Comparison of peptide drugs formulated with Pheroid™ vesicles and Pheroid™ microsponges

		Calcitonin (10 IU/kg)[a]	Insulin (8 IU/kg)[b]	Growth hormone (3.6 IU/kg)[c]
AUC	Pheroid™ vesicles	24048 ± 1739.5[d]	24283 ± 2178.7[e]	4070 ± 1854.2[f]
	Pheroid™ microsponges	20962 ± 1191.6	21291 ± 2194.8	1235 ± 409.1
Absolute bioavailability (%)	Pheroid™ vesicles	123.8	132.4	128.5
	Pheroid™ microsponges	107.9	116.1	38.9
Size (μm)	Pheroid™ vesicles	1.1	1.7	3.0
	Pheroid™ microsponges	1.6	8.2	4.3

[a] Adapted from [7]
[b] Adapted from [19]
[c] Adapted from [20]. Units of area under the curve (AUC) are in
[d] pg.hr/ml
[e] μg.hr/ml
[f] μIU.hr/ml

microsponges, respectively (Table 14.2) [19]. As with calcitonin the difference in absorption could be attributed to the difference in membrane composition and size of the Pheroids™.

Human Growth Hormone Human growth hormone is used in the treatment of short stature in children. This polypeptide hormone in combination with Pheroid™ showed an increase in plasma concentration after nasal administration. In this in vivo study male Sprague Dawley rats were used. Pheroid™ technology shows great potential to enhance absorption [20]. The individual plasma concentration-time profiles after nasal administration of Pheroid™ vesicles and Pheroid™ microsponges showed that Pheroid™ vesicles increased the absorption of recombinant human growth hormone (rhGH) to a greater extent than Pheroid™ microsponges. Pheroid™ vesicles not only increased the bioavailability of rhGH (3.6 IU/kg) that are administered via the nasal route but a delay in absorption of rhGH was observed. This could possibly be attributed to increased residence time of the formulation in the nasal cavity [20]. The absolute bioavailability increased to 128.5 for Pheroid™ vesicles but only 38.9 for Pheroid™ microsponges (Table 14.2).

14.8 Conclusions and Future Perspectives

Drug delivery systems like the patented Pheroid™ technology are primarily used to control drug delivery with target specific delivery. It has a unique submicron emulsion-type formulation of fatty acids capable of encapsulating various drugs.

The ability to entrap both water and lipid soluble drugs makes nasal delivery of biopharmaceuticals possible. The structure, morphology, size, and function can be manipulated according to the intended use. It has reduced cytotoxicity with little toxicity at the prescribed dose. Pheroid™ vesicles have proven to be a successful adjuvant for vaccines and especially the nasal delivery of the diphtheria vaccine. Pheroid™ vesicles and Pheroid™ microsponges have the ability to enhance the nasal absorption of calcitonin, with a resulting decrease in the plasma calcium levels. Compared to subcutaneous and oral administration, nasal administration of insulin in Pheroid™ formulations seems to be the best route of administration for these formulations. Even at low concentrations blood glucose levels were lowered to such an extent, and in such a short time, that it complies with the requirements of such a formulation intended to lower blood glucose levels effectively. Pheroid™ vesicles have been found to be more effective absorption enhancers than the Pheroid™ microsponges. As the vesicles and microsponges differs only in (a) the composition of the particles with the presence of two additional fatty acids in the microsponges, (b) the size of the particle with the microsponge being twice the size of the vesicle, and (c) the steric structure of the particle, the results showed the importance of any or all, or a combination of these factors in intranasal administration. The high systemic absorption of rhGH and the increase in the therapeutic window opens up the possibility of intranasal administration of rhGH with the added advantage of increased intervals between dosages without any increase in the dose of the rhGH. The particle size of drug delivery systems plays an important role with regard to mucoadhesion and residence time in the nasal cavity. The differences in size between the Pheroid™ vesicles and Pheroid™ microsponges could, therefore, be responsible for the differences observed in the bioavailability of the peptide drugs. The differences in chemical composition of the membrane can attribute to the difference in the absorption enhancing properties of the Pheroid™ vesicles and Pheroid™ microsponges.

Future studies needs to elucidate the specific reasons behind the differences observed between the Pheroid™ vesicles and Pheroid™ microsponges. There is also an increasing need to evaluate Pheroid™ technology formulated with peptide drugs in other mucosal routes including rectal, intravaginal, and buccal.

References

1. Hussain AA. Intranasal drug delivery. Adv Drug Deliv Rev. 1998;29(1–2):39–49.
2. Illum L. Nanoparticulate systems for nasal delivery of drugs: a real improvement over simple systems? J Pharm Sci. 2007;96(3):473–83.
3. Pires A, Fortuna A, Alves G, Falcão A. Intranasal drug delivery: how, why and what for? J Pharm Pharm Sci. 2009;12(3):288–311.
4. Arora P, Sharma S, Garg S. Permeability issues in nasal drug delivery. Drug Discov Today. 2002;7(18):967–75.
5. Grobler AF. Pharmaceutical applications of Pheroid™ technology, Potchefstroom. South Africa: North-West University, Ph.D Thesis; 2009. p. 493.

6. Grobler A, Kotzé A, Du Plessis J. The design of a skin-friendly carrier for cosmetic compounds using Pheroid™ technology. In: Wiechers JW, editors. Science and applications of skin delivery systems. Wheaton: Allured Publishing Corporation; 2007.
7. du Plessis LH, Lubbe J, Strauss T, Kotzé AF. Enhancement of nasal and intestinal calcitonin delivery by the novel Pheroid fatty acid based delivery system, and by N-trimethyl chitosan chloride. Int J Pharm. 2010;385(1–2):181–6.
8. Martin A. Physical pharmacy: physical chemical principles in the pharmaceutical sciences. 4th ed. Philadelphia: Lippincott Williams & Wilkins; 1993.
9. Mishra B, Patel BB, Tiwari S. Colloidal nanocarriers: a review on formulation technology, types and applications toward targeted drug delivery. Nanomedicine. 2010;6(1):9–24.
10. Arterburn LM, Hall EB, Oken H. Distribution, interconversion, and dose response of n-3 fatty acids in humans. Am J Clin Nutr. 2006;83(6 Suppl):1467S–1476S.
11. Traber MG, Packer L. Vitamin E: beyond antioxidant function. Am J Clin Nutr. 1995;62 (6 Suppl):1501S–1509S.
12. Eger EI 2nd. The pharmacology of inhaled anesthetics. Semin Anesth Perioper Med Pain. 2005;24(2):89–100.
13. Elgar D. Evaluation of the preclinical effects of perorally administered pro-Pheroids™, Potchefstroom. South Africa: North-West University, Ph.D Thesis; 2008; p. 139.
14. Vogelson CT. Advances in drug delivery systems. Mod Drug Discov. 2001;4(4):49–52.
15. Sharma S, Mukkur TK, Benson HA, Chen Y. Pharmaceutical aspects of intranasal delivery of vaccines using particulate systems. J Pharm Sci. 2009;98(3):812–43.
16. Gonik B. Passive immunization: the forgotten arm of immunologically based strategies for disease containment. Am J Obstet Gynecol. 2011;205(5):444 e1–6.
17. Truter EM. Chitosan derived formulations and Emzaloid™ technology for mucosal vaccination against diphtera: nasal efficacy in mice, Potchefstroom. South Africa: North-West University, MSc. Thesis; 2005. p. 156.
18. Chen MC, Sonaje K, Chen KJ, Sung HW. A review of the prospects for polymeric nanoparticle platforms in oral insulin delivery. Biomaterials. 2011;32(36);9826–38.
19. Oberholzer ID. Peroral and nasal delivery of insulin with Pheroid™ technology, Potchefstroom. South Africa: North-West University, Ph.D Thesis; 2009. p. 228.
20. Steyn D, du Plessis L, Kotzé A. Nasal delivery of recombinant human growth hormone: in vivo evaluation with Pheroid technology and N-trimethyl chitosan chloride. J Pharm Pharm Sci. 2010;13(2):263–73.

Chapter 15
Delivery Strategies for Developing siRNA-Based Vaginal Microbicides

Joseph A. Katakowski and Deborah Palliser

15.1 Introduction

RNA interference (RNAi) is a mechanism of posttranscriptional gene silencing, initially described in plants and worms and more recently in mammals. The RNAi pathway uses small stretches of RNA, typically 21–23 nucleotides in length, to bind mRNA with full or partial homology resulting in reduced expression of the targeted gene. RNAi is a ubiquitous pathway and the small double-stranded RNA species are derived from endogenous and exogenous sources. Indeed RNAi was first described as a mechanism used by plants for protection against foreign genetic elements such as viruses or transposons (reviewed in [1]). RNAi triggers that have been identified include small interfering RNAs (siRNAs), microRNAs (miRNAs) and PIWI-interacting RNAs (piRNAs). piRNAs were first identified as being required for spermatogenesis. piRNAs are mainly expressed in the germline where they are involved in maintenance of genome integrity, although recent studies have uncovered roles for piRNAs in additional cell types such as neurons [2]. As piRNAs have an expression pattern that is mainly restricted to reproductive tissue they will not be discussed further.

15.1.1 A Brief Overview of RNAi

miRNAs are derived from long primary transcripts ($>$ 70 nt long; pri-miRNA). These are processed into a $\approx$70 nt transcript by a complex formed by the RNase III enzyme Drosha and its RNA-binding cofactor DGCR8. This transcript, termed the precursor miRNA (pre-miRNA) forms a hairpin structure containing bulges in areas of nucleotide mismatch. Pre-miRNA exit from the nucleus is mediated by Exportin V and

D. Palliser (✉) · J. A. Katakowski
Department of Microbiology and Immunology, Albert Einstein College of Medicine, 1300 Morris Park Avenue, Bronx, NY 10461, USA
e-mail: deborah.palliser@einstein.yu.edu

J. das Neves, B. Sarmento (eds.), *Mucosal Delivery of Biopharmaceuticals,* DOI 10.1007/978-1-4614-9524-6_15,

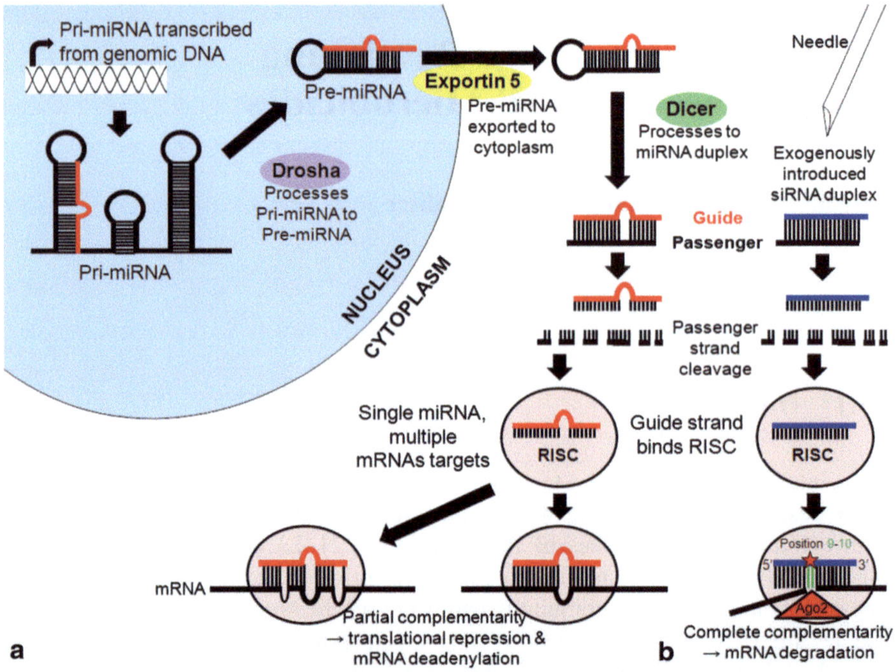

Fig. 15.1 Schematic representation of the miRNA and siRNA pathway. **a** Primary miRNA transcript (pri-miRNA) is cleaved by the RNase III enzyme Drosha and its cofactor DGCR8 to generate the precursor miRNA (pre-miRNA). The pre-miRNA is exported from the nucleus to the cytoplasm via Exportin V. The pre-miRNA is processed by Dicer and its cofactor TRBP to form the mature miRNA duplex. The miRNA is loaded into RISC followed by cleavage of the passenger strand. The guide strand binds target mRNA *via* imperfect homology resulting in translational repression and deadenylation of the target gene followed by mRNA degradation [1]. **b** Introduction of an siRNA duplex into the cytoplasm results in loading onto RISC, cleavage of the passenger strand, followed by guide strand binding to its target mRNA [4]. Ago2 cleaves the mRNA at the position corresponding to nucleotides 9 and 10 of the guide strand (depicted by star)

RanGTP and is succeeded by further cleavage of the pre-miRNA by the RNase III enzyme Dicer and its cofactor (TAR RNA binding protein) TRBP. The processed transcript consists of a ≈21 base pair duplex that has a 5′ phosphate, 3′ hydroxyl, and 2 nt overhang at the 3′ ends. The "diced" 21-mer duplex associates with the argonaute protein, Ago2—the resulting complex termed the RNA-induced silencing complex (RISC). At this point one of the strands of the duplex (the passenger strand) is cleaved by Ago2 and ejected from RISC. The strand remaining in RISC can now bind to its target mRNA resulting in translational repression or mRNA cleavage. The mechanism of suppression utilized is dependent on the degree of homology between the miRNA and its mRNA target—only miRNAs that are highly homologous to the target mRNA mediate target cleavage (summarized in Fig. 15.1 and [1]). Furthermore, as only partial homology is required for miRNA-mediated gene silencing a single miRNA can repress the translation of many genes. Conversely, expression of a single gene can be controlled by multiple miRNAs. Indeed it has been estimated that > 60 % of human genes are regulated by miRNAs [3].

siRNAs most often refers to chemically synthesized 21-mer duplexes that are introduced into cells to induce gene-specific silencing (reviewed in [1, 4]). The siRNAs utilize much of the same pathway as miRNAs, albeit they bypass a requirement for Drosha and mostly for Dicer [5]. Unlike miRNA duplexes, which display partial homology between the passenger and guide strands, resulting in a bulged 21-mer structure, siRNAs are typically synthesized as perfectly complementary duplexes. Furthermore, whereas it has been shown that either miRNA strand can be loaded into RISC, siRNA strand loading displays a strand bias [6–7]. The strand that is preferentially loaded into RISC has been shown to be more thermodynamically unstable at its 5′ end. For recognition of target mRNA 100 % homology of the corresponding siRNA guide strand is typically required. This prerequisite has been exploited in the design of siRNAs used for silencing single nucleotide polymorphisms (SNPs), while leaving the wild-type gene intact [8].

Therefore, the RNAi pathway is an endogenous pathway that is ubiquitously expressed and manipulation of this pathway can result in downregulation of one or more genes. These characteristics translate into a potential use of RNAi to target any gene in any cell type to modulate gene expression. This makes RNAi a powerful tool in studies using reverse genetics. Furthermore, the ability to target any gene in any cell has made RNAi an attractive therapeutic modality. Current clinical trials using siRNAs mainly focus on diseases that are confined to organs that are relatively accessible, such as the liver (e.g., NCT01617967). Topical surfaces are also a site that could be viewed as readily accessible and sites including the eye, lung, rectum, and vagina have been successfully targeted.

15.1.2 siRNAs as a Vaginal Microbicide

As outlined above due to the specificity of siRNAs and the accessibility of the vaginal tract, siRNAs are an attractive modality for use as a microbicide. Vaginal application of many drugs, including proteins and small molecules as well as siRNAs, has been shown to result in limited systemic uptake, thereby reducing potential toxic effects [9]. Although these attributes suggest that siRNAs could be useful for vaginal use, there are several limitations that need to be addressed.

One issue is the physiology of the vaginal tract itself. The vagina is designed to limit uptake of microbes present in the vaginal cavity and this is achieved in various ways (reviewed in [10]). A layer of mucus covers the vaginal epithelium and it confers several protective properties. The gel-like consistency acts as a physical barrier to any potential pathogens. Antimicrobial proteins such as immunoglobulins and β defensins as well as nucleases are also present and the pH of the vaginal cavity is acidic. The structure of the epithelium also changes during the menstrual cycle, which needs to be considered for optimal delivery of any drug to the vagina.

Overall, successful siRNA delivery must utilize a vehicle that will not be trapped in the vaginal mucus and is not degraded through the actions of antimicrobial proteins, nucleases, or the acidic environment. In addition to satisfying these criteria, siRNA must be delivered to the appropriate cell type, and uptake must result in siRNA access to the cytosol, where the RNAi machinery resides. Finally, vaginal application of the siRNA or the delivery vehicle cannot result in toxic effects. *In toto*, fulfillment of these requirements is not trivial. However, as discussed below, there are various strategies that have been used to successfully deliver siRNA to the vaginal mucosa.

15.2 Strategies for Vaginal siRNA Delivery

15.2.1 What are the Main Considerations for designing an siRNA-Based Microbicide?

For development of siRNAs as an active component in a microbicide, most studies have focused on three viral diseases: HIV-1, HSV-2, and HPV. We will summarize the progress that has been made with each of these.

HIV-1 Heterosexual sex is the predominant route for HIV-1 transmission [11]. HIV-1 primarily infects CD4 + T cells, macrophages, and dendritic cells and this can occur at the mucosal surface of the vagina, ectocervix, or endocervix [12]. Therefore, an effective microbicide must gain access to these regions of the genital tract, as well as the pertinent cell types that are located in the subepithelial tissue (Fig. 15.2). The intact vaginal mucosa creates an effective barrier from HIV-1 infection. The acidity of the vaginal lumen and the presence of cervical mucus also resist pathogen access, as do antiviral proteins such as secretory leucocyte protease inhibitor. Any disruption of the epithelial membrane, caused for example by trauma (e.g., microabrasions induced by sex) or inflammatory responses (e.g., genital ulcerative disease caused by viral infection such as HSV-2), increases the susceptibility to HIV-1 infection [13]. Therefore, a microbicide must be designed to be resistant to the acidic environment and antiviral proteins and be able to traverse the mucosal layer.

As HIV-1 infects T cells, macrophages, and dendritic cells, delivering siRNAs to these hard-to-transfect cells represents a challenge. Studies evaluating in vivo siRNA-mediated approaches for targeting HIV-1 have focused on utilizing reagents that are capable of directing siRNA delivery to these cell types. Antibodies specific for a cell-surface receptor expressed by one of these cell types or by cells infected with HIV-1 have been modified to carry siRNA to these cells. Binding of the antibody/siRNA complex to the appropriate cell can result in uptake and subsequent siRNA-mediated gene silencing. This approach has been used for targeting siRNAs to HIV-1-infected cells. An antibody fused to protamine, or any arginine-rich peptide, can bind siRNAs via charge-charge interactions. Song et al. used an HIV-specific gp120 antibody, linked to protamine to deliver siRNAs targeting a combination of oncogenes to a gp120-expressing melanoma. Not only was tumor shrinkage observed when the

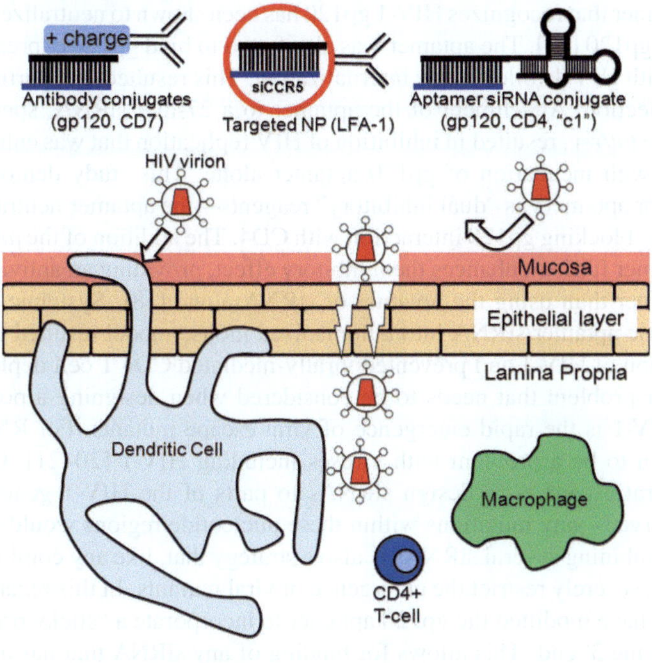

Fig. 15.2 Approaches for targeting HIV-1 at the vaginal mucosa. HIV-1 infects CD4 + T cells, macrophages, and dendritic cells present in the vaginal submucosa. These cells are present in particularly high numbers during inflammatory responses, brought about by pathogenic infection, reactivation of viruses such as HSV-2, and microabrasions caused during sex [79]. Intact epithelium represents a barrier to viral infection. RNAi has been used to target HIV-1 by constructs including antibody/siRNA conjugates, aptamer/siRNA conjugates, and LFA-1-coated NPs [14–15, 19, 24, 26, 77]. To date, only CD4 aptamers have been used intravaginally for targeting HIV-1

antibody/siRNA complex was injected intratumorally, but systemic injection of the antibody/siRNA was also found to restrict tumor growth [14]. A similar approach was used to target siRNAs to T cells via the CD7 receptor. The CD7 antibody was linked to nine arginines and incubated with siRNAs specific for HIV-1 viral genes and the HIV-1 coreceptor CCR5. Systemic injection of these complexes into humanized mice resulted in siRNA uptake and gene silencing only in CD7 + cells. When mice injected with antibody/siRNAs were challenged with HIV-1, viral replication was not detected and CD4 T cell numbers were maintained [15]. Injection of CD7 antibody/siRNA was also effective for controlling an ongoing HIV infection.

A similar approach uses structured nucleic acids, termed aptamers, for directed siRNA delivery. Aptamers are selected by incubating nucleic acid libraries with a protein or cell type of interest (reviewed in Yan and Levy [16]). After several rounds of selection receptor-specific aptamers often have binding affinities in the nanomolar to picomolar range. Aptamers are easily synthesized and are amenable to chemical modifications, thereby allowing facile addition of various cargoes, including siRNAs. Chemical modifications can also serve to limit aptamer degradation *via* nucleases as well as preventing induction of immune responses. Indeed the inability of aptamers to induce immune responses makes them attractive therapeutic modalities.

An aptamer that recognizes HIV-1 gp120 has been shown to neutralize the virus by binding to gp120 [17]. The aptamer was also found to bind gp120 expressed on cells infected with HIV-1 followed by internalization. This resulted in a partial reduction in viral infection. Attachment of the aptamer to a 27-mer siRNA, specific for the HIV-1 gene *tat/rev*, resulted in inhibition of HIV replication that was enhanced when compared with incubation of gp120 aptamer alone. This study demonstrated the potential for aptamers as "dual inhibitory" reagents—the aptamer neutralizes HIV-1 infection by blocking gp120 interaction with CD4. The addition of the *tat/rev* siRNA to the aptamer further enhances the inhibitory effect, providing an antiviral response that is greater than using the aptamer or siRNA alone [18]. Systemic injection of this chimeric aptamer/siRNA into a humanized mouse model resulted in inhibition of replication of HIV-1 and prevented virally-mediated CD4 T cell depletion [19].

A major problem that needs to be considered when designing a potential therapy for HIV-1 is the rapid emergence of viral escape mutants. For RNAi this has been shown to be a problem with viruses including HIV-1 [20–21]. One method to limit viral escape is to design siRNAs to parts of the HIV-1 genome that are well conserved—any mutations within these nucleotide regions would reduce viral fitness. Combining several siRNAs is also a strategy that, like any combination therapy, should severely restrict the emergence of viral mutants. In this regard Rossi and colleagues have modified the gp120 aptamer to incorporate a "sticky bridge" of GC residues at the 3′ end. This allows for binding of any siRNA that has a GC tract at the 3′ end of the passenger or guide strand. Three siRNAs targeting HIV-1 *tat/rev*, together with two HIV-1 host-dependency factors, CD4 and TNPO3, were bound to the gp120 aptamer. This aptamer/siRNA was internalized by HIV-1-infected cells resulting in dicer processing of the siRNA and inhibition of HIV-1 replication in vitro and in vivo [22–24].

How effective these approaches would be for vaginal siRNA delivery is not known. The ability of siRNA(s) attached to an antibody or aptamer to inhibit viral infection following systemic administration may not translate to successful inhibition following intravaginal application. Design of vaginally applied siRNAs must take into account the vaginal environment, which, as outlined above, is very different to the systemic route. Although few of these applications have been tried intravaginally some aptamer-based modalities are taken up across the vaginal mucosa. A set of aptamers that was selected on different cell lines was found to share a motif that promoted cellular internalization. These aptamers were internalized by a variety of murine and human cell lines, including primary cells. Intravaginal application of one of these aptamers, termed C1, resulted in uptake of the aptamer by cells of the epithelium and lamina propria. The ability of this aptamer to transduce a large range of cell types makes it an attractive candidate for delivery of cargoes, including siRNAs to the vaginal mucosa [25].

An aptamer targeting CD4 has been used to deliver siRNAs to the vaginal tract. A major advantage of using an aptamer specific for a host receptor is the ability to deliver siRNAs prior to viral infection. Such a strategy could inhibit viral infection and thereby prevent transmission of the virus. As observed with gp120 aptamer, the CD4 aptamer by itself partially inhibited HIV-1 infection in vitro and in vivo. This

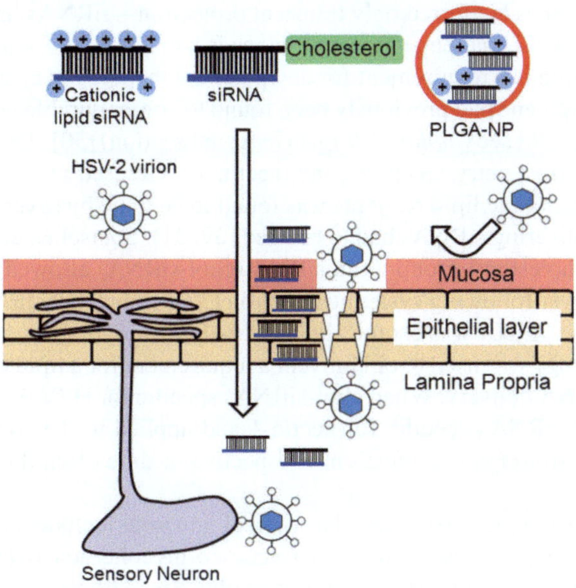

Fig. 15.3 Approaches for targeting HSV-2 at the vaginal mucosa. HSV-2 infects epithelial cells and neurons. Cationic lipid complexed siRNAs have been used to prevent HSV-2 infection [27–28]. However, protection is transient and can be variable, likely due to the inflammatory responses associated with the cationic delivery vehicle [29, 31, 34]. Vaginal application of chol-siRNAs and PLGA-NP/siRNAs has been reported to protect mice from HSV-2 infection [29, 34]. siRNA uptake has been observed in epithelial cells as well as deep into the lamina propria

was likely due to the CD4 aptamer binding CD4 resulting in blocking the binding to this receptor by HIV-1. However, treatment of cervicovaginal explants with the CD4 aptamer attached to siRNAs targeting the HIV-1 genes *gag* and *vif* and the cellular coreceptor *CCR5* resulted in inhibition of viral replication that was up to fourfold more effective than using CD4 aptamer alone. Application of the CD4 aptamer/siRNAs to the vaginal mucosa of humanized mice resulted in protection from viral infection and maintenance of CD4 T cell counts. CD4 aptamer treatment by itself conferred partial protection—plasma viremia and depletion of CD4 + T cells was delayed. However, after 4 weeks HIV-1 viral replication was detectable and CD4 + T cell numbers started to decline [26].

HSV-2 Due to a lack of useful murine models of vaginal HIV-1 infection, HSV-2 was the initial disease model used to show proof-of-concept for siRNA uptake and functionality across the vaginal mucosa (Fig. 15.3). Following application to the vaginal mucosa siRNAs complexed in a cationic lipid were taken up by epithelial and lamina propria cells [27–28]. Administration of siRNAs specific for HSV-2 viral genes protected mice from infection by HSV-2 [27]. Preliminary data showed no indication of toxicity, however, more comprehensive analyses showed an influx of CD45 + cells into the vaginal mucosa. Use of a cationic lipid was also shown to enhance viral infection, clearly a concern for a potential therapeutic agent [29]. Treatment with siRNAs targeting viral genes conferred only transient protection—siRNAs had to be applied within hours of viral challenge. As compliance is a major issue for microbicide use, durable protection is a requirement for any potential therapeutic agent. Targeting an endogenous host gene had previously been found to confer durable gene knockdown (for possible reasons see *Choice of Target Genes* subsection) [30]. Therefore, siRNAs specific for the host entry receptor gene, nectin-1 were used in subsequent studies [29]. As use of cationic lipid reagents was found to be toxic by several groups, other methods for delivering siRNAs had to be used [29, 31]. Soutschek et al. showed that addition of a cholesterol group to an siRNA (chol-siRNA) allowed delivery of the siRNA to the liver following systemic injection [32]. To prevent siRNA degradation by nucleases the siRNA was chemically modified (see *siRNA Modifications* subsection). This strategy was used to circumvent a requirement for a lipid delivery reagent for vaginal siRNA delivery. When chol-siRNAs specific for HSV-2 viral genes were combined with siRNAs specific for nectin-1 and applied to the vaginal tract mice were protected from HSV-2 infection, irrespective of time of viral challenge for up to 1 week [29].

Additional strategies to target the vaginal mucosa include encapsulation of siRNAs into nanoparticles (NPs). Poly(lactic-*co*-glycolic acid) (PLGA) NPs are FDA-approved, biodegradable, and biocompatible. In addition to a favorable safety profile, PLGA has been used for controlled release of DNA, making it an attractive candidate for siRNA delivery across an extended time frame [33]. Coencapsulation of siRNAs with cationic spermidine resulted in high loading efficiencies of up to 50 pmol/mg of PLGA. Uptake of siRNAs was observed throughout the vaginal epithelium and submucosa following vaginal application of PLGA. This uptake was accompanied by specific gene silencing in the vagina, cervix, and the uterine horns for at least 14 days. In contrast to siRNA delivery using cationic lipids the PLGA NPs did not result in any overt histological abnormalities or activation of inflammatory genes [31]. A follow-up study showed the feasibility of the PLGA system for vaginal siRNA delivery. siRNAs specific for the HSV-2 host entry receptor nectin-1, were encapsulated into PLGA NPs and applied to the vaginal mucosa. Mice that received the PLGA NPs were protected from HSV-2 infection for up to 28 days [34].

HPV Unlike HIV-1 and HSV-2 a prophylactic vaccine targeting HPV is available [35]. Although the vaccine is effective incidence rates for HPV are high and therapies are required. HPV is the major risk factor for cervical cancer, but also plays a role in cancers such as vulvar, anal, penile, and oropharyngeal carcinomas. The majority of cervical cancers are caused by HPV types 16 and 18 and expression of E6 and E7 oncogenes is required to achieve and maintain cellular transformation [36–37]. Therefore, most studies have explored the potential for silencing E6 or E7 *via* RNAi (Fig. 15.4 and reviewed in [38]).

Targeting E6 and E7 in vitro resulted in a reduction in cell growth, either due to cell senescence or induction of apoptosis—some controversy surrounds which mechanism is involved [39–40]. In vivo, use of the HPV16 E6/E7 and *ras*-transformed tumor TC-1 has been used extensively as a tumor flank model to demonstrate efficacy of various forms of RNAi in inhibiting and reducing tumor growth. In one

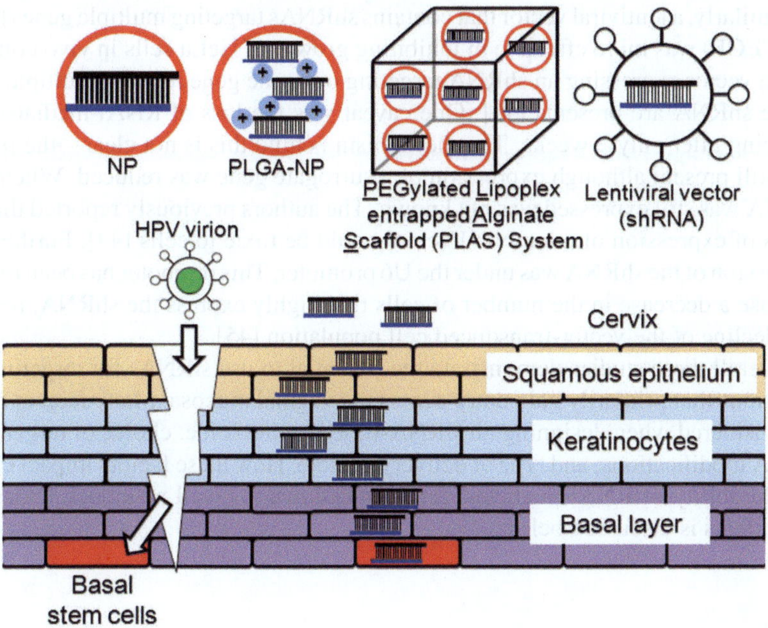

Fig. 15.4 Approaches for targeting HPV at the cervicovaginal epithelium. HPV infects epithelial stem cells located in the basal epithelial cell layer of the cervix. As observed with infection of other viral pathogens, HPV cannot penetrate the intact epithelium. Following infection of stem cells, virions replicate episomally. Expression of E6 and E7 delays cell-cycle arrest and differentiation resulting in further viral replication [80]. Due to a lack of suitable animal models most RNAi-based HPV studies have used flank cervical tumor models to evaluate HPV-specific siRNAs. Lentiviral systems and siRNAs encapsulated in NPs have shown efficacy [41, 43]. The PLAS and PLGA NPs have been shown to deliver siRNAs to the cervicovaginal mucosa making them potential siRNA delivery vehicles for targeting HPV [31, 70]

study siRNAs were designed to silence expression of E6 and E7. In addition, the siRNAs were not chemically modified which resulted in potent stimulation of TLR7/8. As these siRNAs were capable of both knocking down gene expression and simultaneously activating immune responses, they were termed "bifunctional" siRNAs. Transfection of TC-1 cells with liposomes containing the siRNAs prior to subcutaneous injection resulted in decreased growth of the tumor. Systemic injection of liposome-encapsulated siRNAs also inhibited TC-1 growth. When the siRNA was either chemically modified to negate TLR7/8 activation or the siRNA was immunostimulatory, but not E6/E7-specific, some inhibition of tumor growth was observed. However, the reduction in tumor size was not as great as that observed with the bifunctional siRNA [41]. Use of bifunctional siRNAs and similar strategies that combine immune activation with siRNA-mediated knock down have been shown to be effective in other disease models [42]. This combinatorial approach for targeting diseases such as tumors and viruses should serve to induce more potent responses, while reducing the possibility of emergence of viral/tumor escape mutants.

Similarly, a lentiviral vector that contains shRNAs targeting multiple genes (E6/E7 and VEGF) was more effective in inhibiting growth of HeLa cells in vivo compared with a vector expressing an shRNA targeting only one gene (even if multiple copies of the shRNA are present) [43]. One caveat was the loss of RNAi-mediated gene silencing after only 2 weeks. The mechanism behind this is not clear—the plasmid was still present although expression of a surrogate gene was reduced. Whether the shRNA was still expressed was not known. The authors previously reported that high levels of expression of a lentiviral vector could be toxic to cells [44]. Furthermore, expression of the shRNA was under the U6 promoter. This promoter has been reported to cause a decrease in the number of cells that highly express the shRNA, resulting in a decline of the vector-transduced cell population [45].

Overall these studies demonstrate the potential to use siRNAs for targeting viral infections that primarily gain entry across the vaginal mucosa. Many factors need to be considered when designing an siRNA-based microbicide: choice of target genes, siRNA modifications, and type of delivery vehicle. How these factors impact efficacy and durability of RNAi-mediated gene knockdown, as well as induce undesirable side effects is discussed below.

15.2.2 Choice of Target Genes

On the face of it, this may seem like a trivial point: targeting endogenous genes always carries the possible risk of toxicity, whereas an siRNA specific for a viral gene would eliminate this problem. However, studies suggest that targeting endogenous genes may achieve effective and durable protection [27–30]. Why this occurs is not known, although some data have shown that siRNAs targeting endogenous genes persist for longer times when compared to siRNAs specific for genes that are not expressed (e.g., viral genes in the absence of an active infection). For example, when macrophages were transfected with siRNAs targeting the HIV-1 coreceptor CCR5 they were protected from infection with HIV-1 for up to 15 days. Transfection with siRNA targeting p24, an HIV-1 viral gene, conferred maximal protection for only 5 days. Thereafter, the protective effect progressively declined over time: when cells were infected at 15 days posttransfection HIV-1 levels were equivalent to those observed with cells given control siRNAs [30]. The loss of protection coincided with reduction in detectable siRNA. One explanation for this observation is an inherent difference in the intracellular stability of the siRNA. However, this does not seem to be the case. Studies have shown that modifying siRNAs to reduce their susceptibility to nuclease-mediated degradation (and thereby increase their half-life) may not result in increased efficacy of gene silencing [46]. Instead, the sequence of the siRNA as well as the composition of the overhangs have been shown to significantly impact the duration of gene knockdown in vivo. In particular, a dTdT overhang has been reported to negatively impact gene silencing over time [47]. Why this occurs is not known. One possibility is the presence of thymidine may expose the siRNA to DNase activity. Alternatively, the sequence of the siRNA guide strand, including the composition of the overhang may affect its ability to bind RISC. In this regard we found that transfection of an siRNA that targeted an endogenous gene (nectin-1) provided longer-lasting protection from viral infection (HSV-2) when compared with an siRNA targeting a viral gene (UL29, an HSV-2 gene). Presence of the guide strand of each siRNA was determined over time and the amount of UL29 guide strand decreased more rapidly over time when compared with nectin-1 siRNA. Interestingly, the initial intracellular level of the nectin-1 siRNA was significantly higher than that seen for UL29. This difference persisted over time with higher levels of nectin-1 siRNA detected out to day 13 [29]. Overall, these studies could suggest that targeting an endogenous gene may result in more durable gene silencing. Alternatively, durability of gene silencing is determined not by the presence of an endogenous gene, but by the efficacy and stability of an siRNA for incorporation into RISC. The studies described only looked at a handful of siRNAs. Therefore, to determine which of the proposed factors significantly impacts siRNA efficacy and durability of gene silencing, a systematic analysis of siRNAs that vary by sequence and are subject to various chemical modifications (see *siRNA modifications* section) is required.

15.2.3 siRNA Modifications

siRNA-Mediated Immune Responses Due to their short length it was originally believed that siRNAs would not induce interferon-related responses. However, it soon became apparent that siRNAs were capable of activating immune responses *via* the Toll-like receptor (TLR) and retinoic acid inducible gene-I (RIG-I) pathways (Fig. 15.5). These receptors were termed pattern recognition receptors (PRRs) because they recognize particular molecular signatures expressed mainly by pathogens (reviewed in [48–49]). TLRs are expressed by various cell types and are well represented on immune cells. They are located either on the cell surface or within endosomes or lysosomes. RIG-I is part of a family of cytosolic RIG-I-like receptors. siRNAs can activate the immune response in a number of ways. The nucleotide sequence of the siRNA has been shown to be one determinant of immune activation. For example, single strand RNA viruses that contain G- and U-rich RNA sequences activate the TLR7/8 pathway [50–51]. TLR7/8 activation by ssRNA viruses results in secretion of large amounts of type I interferons [49]. siRNA duplexes that are GU-rich and those that contain the 5′-UGU-3′ motif have been identified as being highly immunostimulatory [52]. Although the same group showed that avoidance of GU-rich sequences could select for siRNA duplexes that elicited little immunostimulation, it was also apparent that the 5′-UGU-3′ motif was not the only sequence recognized by TLR7/8. In fact one study showed that many unmodified siRNA duplexes show some immunostimulatory activity [53]. Furthermore, Diebold et al. demonstrated that the presence of uridine and a ribose sugar backbone was necessary and sufficient for TLR7 activation [50].

siRNAs have also been reported to bind TLR3. TLR3 was identified as a receptor for long double-stranded RNA (dsRNA), a replicative intermediate that is produced

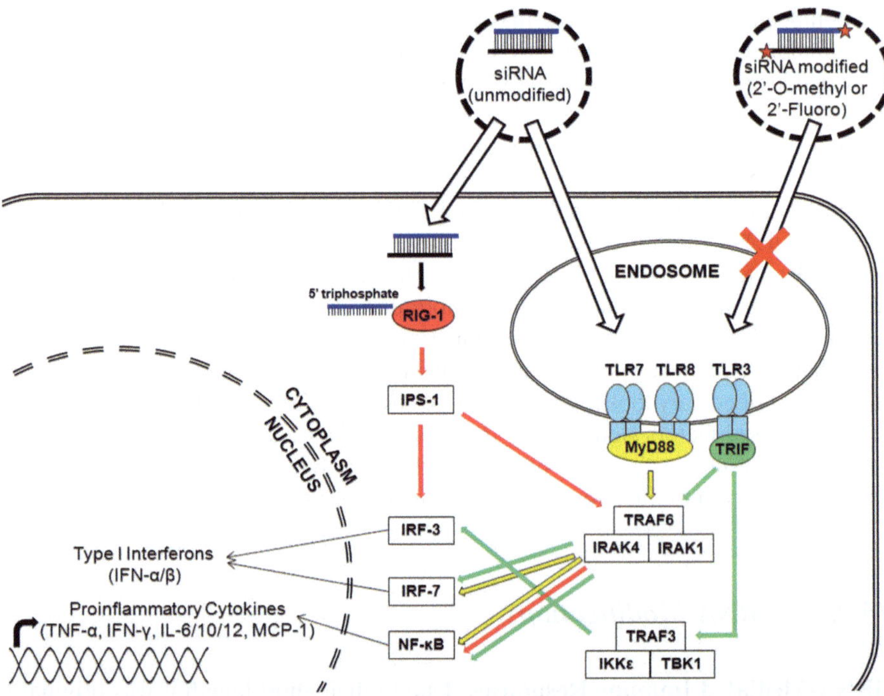

Fig. 15.5 Induction of inflammatory responses by siRNAs. Unmodified siRNAs can induce activation of immune responses through ligation of TLRs 3 and 7/8 resulting in production of type I interferons and proinflammatory cytokines [49]. RIG-I can be activated by structural motifs such as presence of uncapped 5′-triphosphates. Backbone modification of siRNAs by introduction of 2′-*O*-Me or 2′-F nucleotides is often sufficient to abrogate immune responses [63]

by most viruses at some stage of their replication cycle [54]. Ligation of TLR3 leads to signaling through the TRIF adaptor protein resulting in production of type I interferons. A positive feedback loop leads to activation of additional transcription factors including NF-κB, ATF, and c-Jun resulting in induction of inflammatory cytokines [49]. Kleinman et al. found that siRNAs injected into a mouse eye restricted angiogenesis, the formation of new blood vessels that occurs in age-related macular degeneration, irrespective of siRNA specificity. Their results suggested that the siRNAs were activating TLR3 resulting in production of inflammatory cytokines including IFNγ and IL12 and that this production of cytokines was leading to restriction of blood vessel formation [55]. This study clearly highlights the need to evaluate each siRNA for its immune-stimulatory capacity (also emphasized by Robbins et al. [53]) and demonstrates how a biological effect can be mistakenly attributed to an RNAi-mediated gene-specific event. However, it should be noted that this study also showed that a specific siRNA (targeting vascular endothelial growth factor; VEGF) attached to cholesterol—thereby enabling siRNA uptake by choroidal endothelial cells—was capable of causing RNAi-mediated VEGF knockdown in the absence of any detectable immune stimulation.

Additional pathways can recognize siRNAs, resulting in immune response activation. RIG-I is a cytoplasmic helicase protein that recognizes viral RNA [56]. It binds RNA that contains uncapped 5′-triphosphates [57–58]. This structure can be found in certain viral RNAs as well as blunt-ended and in vitro transcribed siRNA. RIG-I ligation results in activation of IRF3 and NF-κB leading to the production of IFNγ and inflammatory cytokines, respectively [59–60].

Chemical Modifications to Negate Immune Responses and Stabilize siRNAs Despite the ability of siRNAs to induce multiple immune pathways, chemical modifications of the ribose sugar backbone is usually sufficient to ameliorate these responses. In particular introduction of 2′-*O*-Methyl (2′-OMe)-modified nucleotides can inhibit both TLR and RIG-I responses. However, introduction of modified nucleotides can alter siRNA efficacy. Therefore, each siRNA must be carefully evaluated not only to determine any reduction in immune activation, but also to assess any change in RNAi-mediated gene knockdown. Studies have shown that modification of as few as two nucleotides in the siRNA sense strand is sufficient to minimize immune response activation [61]. Many other modifications have been tested, including 2′-fluoro (2′-F), a modification routinely used in aptamer synthesis. 2′-F modification has been reported to reduce immune stimulation [62]. However, sequence-specificity, number of bases modified, as well as the position of these modified bases dictate whether immune stimulation is reduced [63].

Incorporation of chemical modifications is also required for protecting siRNAs from nuclease-mediated degradation. The stability of unmodified siRNAs in serum or vaginal fluids is of the order of seconds [29, 32]. The modifications used to stabilize siRNAs are essentially identical to those used to reduce immune responses. As is the case with incorporation of modified bases to negate immunity, care must be taken to maintain siRNA potency when modifying nucleotides to achieve increased stability [64]. The main goal for stabilizing siRNAs is to protect them from nucleases present in bodily fluids; therefore, use of a delivery vehicle may obviate a requirement for using modified siRNAs. However, once the siRNA is delivered into the cell, intracellular nucleases could limit potency of an unmodified siRNA. Alternatively, nucleotide modifications could act by enhancing siRNA incorporation into RISC, thereby improving siRNA potency. Finally, as noted above, TLRs 3 and 7/8 and RIG-I are located in the endosome/lysosome and cytoplasm, respectively. As siRNAs will come into contact with these receptors modified siRNAs should abrogate induction of immune responses.

15.2.4 Delivery Agent

To date the major factor that limits the use of siRNA as a viable therapeutic agent is a lack of suitable delivery agents. As outlined above several strategies have been tried, but problems such as induction of inflammatory responses, transient protection, and the potential for toxicity when targeting endogenous genes all need to be addressed. The first study to demonstrate siRNA uptake across the vaginal mucosa made use of cationic lipids to complex siRNAs [27]. Although this work showed that siRNAs complexed with cationic lipids could confer protection from HSV-2 infection, the associated toxicity of the lipid carrier, as well as the transient nature of viral protection clearly limit the utility of this approach [29, 31, 34]. It should be noted that unmodified siRNAs were used in most of these studies. Wu et al. used phosphorothioate-modified siRNAs to protect siRNAs from nuclease degradation, but whether these modifications were sufficient to ameliorate possible immune responses is unknown [29]. Indeed in a comprehensive study Robbins et al. found that several siRNAs published in various studies showed some level of immune activity [53]. How this activation affects a biological outcome will depend on the level of immune induction elicited as well as the disease model studied.

For siRNA delivery to the vaginal (and rectal) mucosa a delivery vehicle must avoid activation of any inflammatory responses. Considering the history of microbicide development this is not an easy goal to achieve. One of the main problems is a lack of biomarkers that predict safety of candidate microbicides. The microbicides nonoxynol-9 (N-9) and cellulose sulfate (CS) designed to prevent HIV-1 infection illustrate this point. Both microbicides were originally deemed safe. However, not only did they fail to prevent HIV-1 infection in clinical trials, their use was associated with an increased susceptibility to viral infection [65–66]. Follow-up studies with N-9 suggested that it could occur due to induction of inflammatory cytokines, however, when CS was evaluated no cytokines were detected. Recent studies have started to evaluate other biomarkers that could be used in preclinical safety studies. Changes in epithelial cell barrier integrity following treatment with candidate microbicides have been reported as a sensitive method for predicting vehicle toxicity [67]. Susceptibility of mice to vaginal HSV-2 following application of potential microbicides has also been used as a biomarker. This study showed that excipients commonly used as "inert" vehicles for delivery of cargos to the vaginal mucosa can have toxic effects. These included K-Y warming jelly and 30 % glycerin, used in the 1 % tenofovir gel in the CAPRISA 004 trial [68].

These studies demonstrate a major problem for delivery of siRNA (or any cargo) safely and effectively to the vaginal mucosa. Even products that are FDA-approved and have been deemed safe have been demonstrated to affect vaginal integrity resulting in increased susceptibility to viral infection. One example is PLGA NPs, which are able to induce toxic responses [69]. Whether this will limit their utility for vaginal use will need to be tested. Other platforms that are being developed for vaginal delivery include the PEGylated Lipoplex-entrapped Alginate Scaffold (PLAS) system, which entraps a gene delivery vector within a biodegradable alginate scaffold. Particles are PEGylated to avoid mucosal entrapment and the alginate scaffold constitutes an extended release system for the particles. When PLAS particles, encapsulating siRNAs were applied to the vaginal mucosa gene-specific knockdown was observed. Therefore, PLAS particles may be useful as a durable release platform [70]. As mucous is efficient at trapping many types of particles, an NP platform that can penetrate the mucosal barrier could be useful for siRNA delivery. Recent work by several groups including Hanes and colleagues demonstrate that coating conventional NPs, previously shown to be inefficient at delivering cargos to the vaginal mucosa, with PEG (termed mucous-penetrating particles; MPPs) resulted in rapid uptake across the vaginal epithelium and uniform distribution throughout the vagina [71–72]. Vaginal application of MPPs formulated with acyclovir monophosphate protected mice from HSV-2 infection [72]. The MPPs did not cause inflammation in the vaginal tract—in contrast to uncoated NPs. Furthermore, the MPPs remained detectable after 24 h, making them promising candidates for delivery of durable siRNA

One of the few NP platforms that has been comprehensively tested for safe and effective systemic siRNA delivery is lipid nanoparticles (LNPs) containing ionizable cationic lipids [73–74]. The lipids are cationic at low pH, thereby allowing siRNA binding, but are charge-neutral at physiological pH. They have an excellent safety profile and have been shown to be effective for reducing liver tumors in patients [75]. Incorporation of a modified form of the ionizable lipid into LNPs results in gene knockdown at an ED_{50} as low as 0.005 mg/kg [76]. Although LNPs have not yet been evaluated for vaginal delivery, the low doses required for gene silencing following systemic injection combined with their favorable safety profile makes them an attractive candidate for vaginal delivery.

In certain circumstances targeted delivery may be required for siRNAs to gain access to appropriate cells present in the vaginal mucosa. For example, HIV-1 infects CD4 + T cells, macrophages, and dendritic cells. These cells, present in the vaginal submucosa may not be accessible to delivery agents that lack a targeting moiety. CD4-specific aptamers have been shown to protect mice from vaginal transmission of HIV-1. Several other targeted strategies (e.g., CD7 antibody/siRNA conjugate, LFA-1 antibody-coated NPs, gp120 aptamer/siRNA conjugates) have demonstrated an ability to inhibit viral infection following systemic delivery [15, 24, 77]. Whether these strategies could be useful for preventing vaginal transmission needs to be tested.

15.3 Conclusions and Future Perspectives

RNAi is considered an attractive therapeutic modality due its ability to specifically reduce target gene expression. As discussed, RNAi can be used to inhibit replication of HIV-1, HSV-2, and HPV. Furthermore, vaginal application of siRNAs targeting HIV-1 or HSV-2 viral genes or host-encoded viral entry receptor genes in mouse models of vaginal HIV-1 and HSV-2 infection, respectively, has demonstrated the ability of siRNAs to prevent viral disease in vivo.

Recent studies have imparted invaluable information regarding criteria including siRNA and delivery vehicle design as well as providing relevant novel safety platforms. These findings will need to be considered when developing strategies for vaginal delivery of siRNAs.

Although topical application of siRNA to the vaginal tract can result in siRNA uptake resulting in gene-specific silencing, studies have highlighted the difficulties in achieving effective and durable gene knockdown in the absence of toxicity. Encouragingly, since the first proof-of-concept study demonstrated uptake of topically applied siRNA across the vaginal mucosa resulting in effective and specific gene knockdown, several studies have developed strategies for optimizing delivery.

To achieve optimal siRNA-mediated silencing several factors need to be considered. From the perspective of siRNA design, it remains unclear whether durable gene silencing will only be achieved when an endogenous gene is targeted. Alternatively, siRNAs need to be designed that are resistant to nuclease degradation, and are also preferentially and stably loaded into RISC—irrespective of the presence of target mRNA. Use of chemical modifications has been demonstrated to stabilize siRNAs as well as negate immune stimulation. Currently it is not known whether further modification of siRNAs to enhance RISC loading will enhance gene knockdown in vivo.

As with any RNAi-mediated therapeutic modality the major issue is delivery. The pH of the vaginal tract and the presence of mucus significantly affect the ability of an siRNA to gain access to the cells of the vaginal mucosa (reviewed in [78]). Promising strategies are being developed to overcome these problems [71–72]. Use of nontargeted delivery vehicles may be sufficient for siRNA delivery to vaginal epithelial cells. Indeed, studies have demonstrated that lipoplexed siRNA can be detected deep into the lamina propria [27–28]. However, for certain cell types a targeted approach may be required. This may be the case for HIV-1, which has been successfully targeted using vaginally applied CD4-specific aptamers bound to HIV-specific siRNAs. This study also demonstrated a lack of uptake by CD4 + cells present in the vaginal mucosa following topical application of chol-siRNAs [26]. It should also be remembered that cellular access does not necessarily equate with gene-specific knockdown. As the RISC machinery resides in the cytoplasm, siRNA uptake must result in cytoplasmic access—as several delivery vehicles are taken up *via* the endocytic pathway, how cytoplasmic access is achieved becomes a major consideration.

With the history of failed clinical trials for microbicides, in particular N-9 and CS, the safety of any potential siRNA and its component delivery vehicle cannot be understated. Recent studies have developed additional biomarkers that may serve as predictors of safety [67–68]. Evaluation of additional biomarkers will be required to effectively prescreen potential microbicide candidates prior to initiating clinical trials.

Overall, studies looking at the basic biology of RNAi-mediated gene silencing have identified several factors that contribute to effective gene knockdown in the absence of toxicity. Therefore, it is reasonable to predict that effective, durable, and safe gene silencing in the vaginal mucosa is an achievable goal.

References

1. Carthew RW, Sontheimer EJ. Origins and mechanisms of miRNAs and siRNAs. Cell. 2009;136(4):642–55.
2. Peng JC, Lin H. Beyond transposons : the epigenetic and somatic functions of the Piwi-piRNA mechanism. Current Opin Cell Biol. 2013;25(2):190–4.
3. Friedman RC, Farh KK, Burge CB, Bartel DP. Most mammalian mRNAs are conserved targets of microRNAs. Genome Res. 2009;19(1):92–105.
4. Snead NM, Rossi JJ. Biogenesis and function of endogenous and exogenous siRNAs. Wiley Interdiscip Rev RNA. 2010;1(1):117–31.
5. Murchison EP, Partridge JF, Tam OH, Cheloufi S, Hannon GJ. Characterization of Dicer-deficient murine embryonic stem cells. Proc Natl Acad Sci U S A. 2005;102(34):12135–40.
6. Khvorova A, Reynolds A, Jayasena SD. Functional siRNAs and miRNAs exhibit strand bias. Cell. 2003;115(2):209–16.
7. Schwarz DS, Hutvagner G, Du T, Xu Z, Aronin N, Zamore PD. Asymmetry in the assembly of the RNAi enzyme complex. Cell. 2003;115(2):199–208.
8. Schwarz DS, Ding H, Kennington L, Moore JT, Schelter J, Burchard J, Linsley PS, Aronin N, Xu Z, Zamore PD. Designing siRNA that distinguish between genes that differ by a single nucleotide. PLoS Genet. 2006;2(9):e140.
9. Hussain A, Ahsan F. The vagina as a route for systemic drug delivery. J Control Release. 2005;103(2):301–13.
10. Wira CR, Fahey JV, Sentman CL, Pioli PA, Shen L. Innate and adaptive immunity in female genital tract: cellular responses and interactions. Immunol Rev. 2005;206:306–35.
11. Shattock RJ, Moore JP. Inhibiting sexual transmission of HIV-1 infection. Nat Rev Microbiol. 2003;1(1):25–34.
12. Shen R, Richter HE, Smith PD. Early HIV-1 target cells in human vaginal and ectocervical mucosa. Am J Reprod Immunol. 2011;65(3):261–7.
13. Cutler B, Justman J. Vaginal microbicides and the prevention of HIV transmission. Lancet Infect Dis. 2008;8(11):685–97.
14. Song E, Zhu P, Lee SK, Chowdhury D, Kussman S, Dykxhoorn DM, Feng Y, Palliser D, Weiner DB, Shankar P, Marasco WA, Lieberman J. Antibody mediated in vivo delivery of small interfering RNAs via cell-surface receptors. Nat Biotechnol. 2005;23(6):709–17.
15. Kumar P, Ban HS, Kim SS, Wu H, Pearson T, Greiner DL, Laouar A, Yao J, Haridas V, Habiro K, Yang YG, Jeong JH, Lee KY, Kim YH, Kim SW, Peipp M, Fey GH, Manjunath N, Shultz LD, Lee SK, Shankar P. T cell-specific siRNA delivery suppresses HIV-1 infection in humanized mice. Cell. 2008;134(4):577–86.
16. Yan AC, Levy M. Aptamers and aptamer targeted delivery. RNA Biol. 2009;6(3):316–20.
17. Khati M, Schuman M, Ibrahim J, Sattentau Q, Gordon S, James W. Neutralization of infectivity of diverse R5 clinical isolates of human immunodeficiency virus type 1 by gp120-binding 2'F-RNA aptamers. J Virol. 2003;77(23):12692–8.
18. Zhou J, Li H, Li S, Zaia J, Rossi JJ. Novel dual inhibitory function aptamer-siRNA delivery system for HIV-1 therapy. Mol Ther. 2008;16(8):1481–9.
19. Neff CP, Zhou J, Remling L, Kuruvilla J, Zhang J, Li H, Smith DD, Swiderski P, Rossi JJ, Akkina R. An aptamer-siRNA chimera suppresses HIV-1 viral loads and protects from helper CD4(+) T cell decline in humanized mice. Sci Transl Med. 2011;3(66):66ra66.
20. Das AT, Brummelkamp TR, Westerhout EM, Vink M, Madiredjo M, Bernards R, Berkhout B. Human immunodeficiency virus type 1 escapes from RNA interference-mediated inhibition. J Virol. 2004;78(5):2601–5.
21. Gitlin L, Karelsky S, Andino R. Short interfering RNA confers intracellular antiviral immunity in human cells. Nature. 2002;418(6896):430–4.
22. Zhou J, Swiderski P, Li H, Zhang J, Neff CP, Akkina R, Rossi JJ. Selection, characterization and application of new RNA HIV gp 120 aptamers for facile delivery of Dicer substrate siRNAs into HIV infected cells. Nucleic Acids Res. 2009;37(9):3094–109.

23. Zhou J, Rossi JJ. Cell-specific aptamer-mediated targeted drug delivery. Oligonucleotides. 2011;21(1):1–10.
24. Zhou J, Neff CP, Swiderski P, Li H, Smith DD, Aboellail T, Remling-Mulder L, Akkina R, Rossi JJ. Functional in vivo delivery of multiplexed anti-HIV-1 siRNAs via a chemically synthesized aptamer with a sticky bridge. Mol Ther. 2013;21(1):192–200.
25. Magalhaes ML, Byrom M, Yan A, Kelly L, Li N, Furtado R, Palliser D, Ellington AD, Levy M. A general RNA motif for cellular transfection. Mol Ther. 2012;20(3):616–24.
26. Wheeler LA, Trifonova R, Vrbanac V, Basar E, McKernan S, Xu Z, Seung E, Deruaz M, Dudek T, Einarsson JI, Yang L, Allen TM, Luster AD, Tager AM, Dykxhoorn DM, Lieberman J. Inhibition of HIV transmission in human cervicovaginal explants and humanized mice using CD4 aptamer-siRNA chimeras. J Clin Invest. 2011;121(6):2401–12.
27. Palliser D, Chowdhury D, Wang QY, Lee SJ, Bronson RT, Knipe DM, Lieberman J. An siRNA-based microbicide protects mice from lethal herpes simplex virus 2 infection. Nature. 2006;439(7072):89–94.
28. Zhang Y, Cristofaro P, Silbermann R, Pusch O, Boden D, Konkin T, Hovanesian V, Monfils PR, Resnick M, Moss SF, Ramratnam B. Engineering mucosal RNA interference in vivo. Mol Ther. 2006;14(3):336–42.
29. Wu Y, Navarro F, Lal A, Basar E, Pandey RK, Manoharan M, Feng Y, Lee SJ, Lieberman J, Palliser D. Durable protection from Herpes Simplex Virus-2 transmission following intravaginal application of siRNAs targeting both a viral and host gene. Cell Host Microbe. 2009;5(1):84–94.
30. Song E, Lee SK, Dykxhoorn DM, Novina C, Zhang D, Crawford K, Cerny J, Sharp PA, Lieberman J, Manjunath N, Shankar P. Sustained small interfering RNA-mediated human immunodeficiency virus type 1 inhibition in primary macrophages. J Virol. 2003;77(13): 7174–81.
31. Woodrow KA, Cu Y, Booth CJ, Saucier-Sawyer JK, Wood MJ, Saltzman WM. Intravaginal gene silencing using biodegradable polymer nanoparticles densely loaded with small-interfering RNA. Nat Mater. 2009;8(6):526–33.
32. Soutschek J, Akinc A, Bramlage B, Charisse K, Constien R, Donoghue M, Elbashir S, Geick A, Hadwiger P, Harborth J, John M, Kesavan V, Lavine G, Pandey RK, Racie T, Rajeev KG, Rohl I, Toudjarska I, Wang G, Wuschko S, Bumcrot D, Koteliansky V, Limmer S, Manoharan M, Vornlocher HP. Therapeutic silencing of an endogenous gene by systemic administration of modified siRNAs. Nature. 2004;432(7014):173–8.
33. Campolongo MJ, Luo D. Drug delivery: old polymer learns new tracts. Nat Mater. 2009;8(6):447–8.
34. Steinbach JM, Weller CE, Booth CJ, Saltzman WM. Polymer nanoparticles encapsulating siRNA for treatment of HSV-2 genital infection. J Control Release. 2012;162(1):102–10.
35. Bonanni P, Boccalini S, Bechini A. Efficacy, duration of immunity and cross protection after HPV vaccination: a review of the evidence. Vaccine. 2009;27(Suppl 1):A46–53.
36. Walboomers JM, Jacobs MV, Manos MM, Bosch FX, Kummer JA, Shah KV, Snijders PJ, Peto J, Meijer CJ, Munoz N. Human papillomavirus is a necessary cause of invasive cervical cancer worldwide. J Pathol. 1999;189(1):12–9.
37. von Knebel Doeberitz M, Rittmuller C, zur Hausen H, Durst M. Inhibition of tumorigenicity of cervical cancer cells in nude mice by HPV E6-E7 anti-sense RNA. Int J Cancer. 1992;51(5):831–4.
38. Singhania R, Khairuddin N, Clarke D, McMillan NA. RNA interference for the treatment of papillomavirus disease. Open Virol J. 2012;6:204–15.
39. Putral LN, Bywater MJ, Gu W, Saunders NA, Gabrielli BG, Leggatt GR, McMillan NA. RNA interference against human papillomavirus oncogenes in cervical cancer cells results in increased sensitivity to cisplatin. Mol Pharmacol. 2005;68(5):1311–9.
40. Yoshinouchi M, Yamada T, Kizaki M, Fen J, Koseki T, Ikeda Y, Nishihara T, Yamato K. In vitro and in vivo growth suppression of human papillomavirus 16-positive cervical cancer cells by E6 siRNA. Mol Ther. 2003;8(5):762–8.
41. Khairuddin N, Gantier MP, Blake SJ, Wu SY, Behlke MA, Williams BR, McMillan NA., siRNA-induced immunostimulation through TLR7 promotes antitumoral activity against HPV-driven tumors in vivo. Immunol Cell Biol. 2012;90(2):187–96.

42. Poeck H, Besch R, Maihoefer C, Renn M, Tormo D, Morskaya SS, Kirschnek S, Gaffal E, Landsberg J, Hellmuth J, Schmidt A, Anz D, Bscheider M, Schwerd T, Berking C, Bourquin C, Kalinke U, Kremmer E, Kato H, Akira S, Meyers R, Hacker G, Neuenhahn M, Busch D, Ruland J, Rothenfusser S, Prinz M, Hornung V, Endres S, Tuting T, Hartmann G, 5'-Triphosphate-siRNA: turning gene silencing and Rig-I activation against melanoma. Nat Med. 2008;14(11):1256–63.
43. Gu W, Payne E, Sun S, Burgess M, McMillan NA. Inhibition of cervical cancer cell growth in vitro and in vivo with dual shRNAs. Cancer Gene Ther. 2011;18(3):219–27.
44. Gu W, Putral L, Hengst K, Minto K, Saunders NA, Leggatt G, McMillan NA. Inhibition of cervical cancer cell growth in vitro and in vivo with lentiviral-vector delivered short hairpin RNA targeting human papillomavirus E6 and E7 oncogenes. Cancer Gene Ther. 2006;13(11):1023–32.
45. An DS, Qin FX, Auyeung VC, Mao SH, Kung SK,Baltimore D, Chen IS. Optimization and functional effects of stable short hairpin RNA expression in primary human lymphocytes via lentiviral vectors. Mol Ther. 2006;14(4):494–504.
46. Layzer JM, McCaffrey AP, Tanner AK, Huang H, Kay MA, Sullenger BA. In vivo activity of nuclease-resistant siRNAs. RNA. 2004;10(5):766–71.
47. Strapps WR, Pickering V, Muiru GT, Rice J, Orsborn S, Polisky BA, Sachs A, Bartz SR. The siRNA sequence and guide strand overhangs are determinants of in vivo duration of silencing. Nucleic Acids Res. 2010;38(14):4788–97.
48. Iwasaki A, Medzhitov R. Toll-like receptor control of the adaptive immune responses. Nat Immunol. 2004;5(10):987–95.
49. Kawai T, Akira S. The roles of TLRs, RLRs and NLRs in pathogen recognition. Int Immunol. 2009;21(4):317–37.
50. Diebold SS, Massacrier C, Akira S, Paturel C, Morel Y, Reis e Sousa C. Nucleic acid agonists for Toll-like receptor 7 are defined by the presence of uridine ribonucleotides. Eur J Immunol. 2006;36(12):3256–67.
51. Heil F, Hemmi H, Hochrein H, Ampenberger F, Kirschning C, Akira S, Lipford G, Wagner H, Bauer S. Species-specific recognition of single-stranded RNA via toll-like receptor 7 and 8. Science. 2004;303(5663):1526–9.
52. Judge AD, Sood V, Shaw JR, Fang D, McClintock K, MacLachlan I. Sequence-dependent stimulation of the mammalian innate immune response by synthetic siRNA. Nat Biotechnol. 2005;23(4):457–62.
53. Robbins M, Judge A, Ambegia E, Choi C, Yaworski E, Palmer L, McClintock K, MacLachlan I. Misinterpreting the therapeutic effects of small interfering RNA caused by immune stimulation. Hum Gene Ther. 2008;19(10):991–9.
54. Alexopoulou L, Holt AC, Medzhitov R, Flavell RA. Recognition of double-stranded. RNA and activation of NF-kappaB by Toll-like receptor 3. Nature. 2001;413(6857):732–8.
55. Kleinman ME, Yamada K, Takeda A, Chandrasekaran V, Nozaki M, Baffi JZ, Albuquerque RJ, Yamasaki S, Itaya M, Pan Y, Appukuttan B, Gibbs D, Yang Z, Kariko K, Ambati BK, Wilgus TA, DiPietro LA, Sakurai E, Zhang K, Smith JR, Taylor EW, Ambati J. Sequence- and target-independent angiogenesis suppression by siRNA via TLR3. Nature. 2008;452(7187):591–7.
56. Kato H, Takeuchi O, Sato S, Yoneyama M, Yamamoto M, Matsui K, Uematsu S, Jung A, Kawai T, Ishii KJ, Yamaguchi O, Otsu K, Tsujimura T, Koh CS, Reis e Sousa C, Matsuura Y, Fujita T, Akira S. Differential roles of MDA5 and RIG-I helicases in the recognition of RNA viruses. Nature. 2006;441(7089):101–5.
57. Hornung V, Ellegast J, Kim S, Brzozka K, Jung A, Kato H, Poeck H, Akira S, Conzelmann KK, Schlee M, Endres S, Hartmann G. 5'-Triphosphate RNA is the ligand for RIG-I. Science. 2006;314(5801):994–7.
58. Pichlmair A, Schulz O, Tan CP, Naslund TI, Liljestrom P, Weber F, Reis e Sousa C. RIG-I-mediated antiviral responses to single-stranded RNA bearing 5'-phosphates. Science. 2006;314(5801):997–1001.
59. Hiscott J, Lin R, Nakhaei P, Paz S. MasterCARD : a priceless link to innate immunity. Trends Mol Med. 2006;12(2):53–6.

60. Yoneyama M, Kikuchi M, Matsumoto K, Imaizumi T, Miyagishi M, Taira K, Foy E, Loo YM, Gale M Jr., Akira S, Yonehara S, Kato A, Fujita T. Shared and unique functions of the DExD/H-box helicases RIG-I, MDA5, and LGP2 in antiviral innate immunity. J Immunol. 2005;175(5):2851–8.
61. Judge AD, Bola G, Lee AC, MacLachlan I. Design of noninflammatory synthetic siRNA mediating potent gene silencing in vivo. Mol Ther. 2006;13(3):494–505.
62. Cekaite L, Furset G, Hovig E, Sioud M. Gene expression analysis in blood cells in response to unmodified and 2'-modified siRNAs reveals TLR-dependent and independent effects. J Mol Biol. 2007;365(1):90–108.
63. Robbins M, Judge A, MacLachlan I. siRNA and innate immunity. Oligonucleotides. 2009;19(2):89–102.
64. Choung S, Kim YJ, Kim S, Park HO, Choi YC. Chemical modification of siRNAs to improve serum stability without loss of efficacy. Biochem Biophys Res Commun. 2006;342(3):919–27.
65. Van Damme L, Govinden R, Mirembe FM, Guedou F, Solomon S, Becker ML, Pradeep BS, Krishnan AK, Alary M, Pande B, Ramjee G, Deese J, Crucitti T, Taylor D. Lack of effectiveness of cellulose sulfate gel for the prevention of vaginal HIV transmission. N Engl J Med. 2008;359(5):463–72.
66. Van Damme L, Ramjee G, Alary M, Vuylsteke B, Chandeying V, Rees H, Sirivongrangson P, Mukenge-Tshibaka L, Ettiegne-Traore V, Uaheowitchai C, Karim SS, Masse B, Perriens J, Laga M. Effectiveness of COL-1492, a nonoxynol-9 vaginal gel, on HIV-1 transmission in female sex workers: a randomised controlled trial. Lancet. 2002;360(9338):971–7.
67. Mesquita PM, Cheshenko N, Wilson SS, Mhatre M, Guzman E, Fakioglu E, Keller MJ, Herold BC. Disruption of tight junctions by cellulose sulfate facilitates HIV infection model of microbicide safety. J Infect Dis. 2009;200(4):599–608.
68. Moench TR, Mumper RJ, Hoen TE, Sun M, Cone RA, Microbicide excipients can greatly increase susceptibility to genital herpes transmission in the mouse. BMC Infect Dis. 2010;10:331.
69. Xiong S, George S, Yu H, Damoiseaux R, France B, Ng KW, Loo JS. Size influences the cytotoxicity of poly (lactic-co-glycolic acid) (PLGA) and titanium dioxide (TiO(2)) nanoparticles. Arch Toxicol. 2012;87(6):1075–86.
70. Wu SY, Chang HI, Burgess M, McMillan NA. Vaginal delivery of siRNA using a novel PEGylated lipoplex-entrapped alginate scaffold system. J Control Release. 2011;155(3):418–26.
71. Cu Y, Booth CJ, Saltzman WM. In vivo distribution of surface-modified PLGA nanoparticles following intravaginal delivery. J Control Release. 2011;156(2):258–64.
72. Ensign LM, Tang BC, Wang YY, Tse TA, Hoen T, Cone R., Hanes J. Mucus-penetrating nanoparticles for vaginal drug delivery protect against herpes simplex virus. Sci Transl Med. 2012;4(138):138ra179.
73. Morrissey DV, Lockridge JA, Shaw L, Blanchard K, Jensen K, Breen W, Hartsough K, Machemer L, Radka S, Jadhav V, Vaish N, Zinnen S, Vargeese C, Bowman K, Shaffer CS, Jeffs LB, Judge A, MacLachlan I, Polisky B. Potent and persistent in vivo anti-HBV activity of chemically modified siRNAs. Nat Biotechnol. 2005;23(8):1002–7.
74. Zimmermann TS, Lee AC, Akinc A, Bramlage B, Bumcrot D, Fedoruk MN, Harborth J, Heyes JA, Jeffs LB, John M, Judge AD, Lam K, McClintock K, Nechev LV, Palmer LR, Racie T, Rohl I, Seiffert S, Shanmugam S, Sood V, Soutschek J, Toudjarska I, Wheat AJ, Yaworski E, Zedalis W, Koteliansky V, Manoharan M, Vornlocher HP, MacLachlan I. RNAi-mediated gene silencing in non-human primates. Nature. 2006;441(7089):111–4.
75. Tabernero J, Shapiro GI, Lorusso PM, Cervantes A, Schwartz GK, Weiss GJ, Paz-Ares L, Cho DC, Infante JR, Alsina M, Gounder MM, Falzone R, Harrop J, Seila White AC, Toudjarska I, Bumcrot D, Meyers RE, Hinkle G, Svrzikapa N, Hutabarat RM, Clausen VA, Cehelsky J, Nochur SV, Gamba-Vitalo C, Vaishnaw AK, Sah DW, Gollob JA, Burris HA 3rd. First-in-man trial of an RNA interference therapeutic targeting VEGF and KSP in cancer patients with liver involvement. Cancer Discov. 2013;3(4):406–17.
76. Jayaraman M, Ansell SM, Mui BL, Tam YK, Chen J, Du X, Butler D, Eltepu L, Matsuda S, Narayanannair JK, Rajeev KG, Hafez IM, Akinc A, Maier MA, Tracy MA, Cullis PR, Madden

TD, Manoharan M, Hope MJ. Maximizing the potency of siRNA lipid nanoparticles for hepatic gene silencing in vivo. Angew Chem Int Ed Engl. 2012;51(34):8529–33.
77. Kim SS, Peer D, Kumar P, Subramanya S, Wu H, Asthana D, Habiro K, Yang YG, Manjunath N, Shimaoka M, Shankar P. RNAi-mediated CCR5 silencing by LFA-1-targeted nanoparticles prevents HIV infection in BLT mice. Mol Ther. 2010;18(2):370–6.
78. Yang S, Chen Y, Ahmadie R, Ho EA. Advancements in the field of intravaginal siRNA delivery. J Control Release. 2013;167(1):29–39.
79. Hladik F, Hope TJ. HIV infection of the genital mucosa in women. Curr HIV/AIDS Rep. 2009;6(1):20–8.
80. Frazer IH. Prevention of cervical cancer through papillomavirus vaccination. Nat Rev Immunol. 2004;4(1):46–54.

Chapter 16
Delivery Strategies for Developing Vaginal DNA Vaccine Combining Cell-Penetrating Peptide and Jet Injection

Takanori Kanazawa and Hiroaki Okada

16.1 Introduction

The application of DNA immunization as new generation vaccines has been well studied since its invention, and a variety of such vaccines have undergone clinical trials [1–3] or are used in veterinary practice [4, 5]. DNA vaccines were first reported in the early 1990s as a novel method for vaccination [6–9]. The DNA vaccines have several advantages, which include simplicity of manufacture, biological stability, and cost-effectiveness. The safety of DNA vaccines is their most important advantage, as no live virus or viral fragments are utilized in the preparation of this type of vaccine. In addition, different genes can be combined simultaneously, resulting in multivalent vaccines. Another important benefit is the induction of not only humoral immunity but also cellular immunity. The earliest Phase I clinical trial of a DNA vaccine was of an HIV-1 candidate tested in individuals infected with HIV-1, followed by studies in volunteers who were not infected with HIV-1 [10]. Other prophylactic and therapeutic DNA vaccine trials followed, including trials that tested DNA vaccines against cancer, influenza, malaria, hepatitis B, and HIV-1 [2, 11–14]. Currently, clinical trials with a DNA HIV-1 vaccine have already started using an adenovirus-based prime-boost vaccine and vaccinia virus Ankara. Presently, about 90 human clinical DNA vaccine trials are underway [3]. Furthermore, in the past 3 years, four DNA products have been licensed for animal use: one against West Nile virus in horses [15], one against infectious hematopoietic necrosis virus in schooled salmon [16], one for treatment of melanoma in dogs [17], and, the most recent licensure, growth hormone releasing hormone (GHRH) product for fetal loss in swine [18].

Several reports have established that mucosal transmission is the initial step toward systemic infection, such as HIV, and thus inhibition of viral mucosal transmission

T. Kanazawa (✉) · H. Okada
Laboratory of Pharmaceutics and Drug Delivery,
Department of Pharmaceutical Science,
School of Pharmacy, Tokyo University of Pharmacy and Life Sciences,
1432-1 Horinouchi, Hachioji, Tokyo, 192-0392, Japan
e-mail: kanazawa@toyaku.ac.jp

J. das Neves, B. Sarmento (eds.), *Mucosal Delivery of Biopharmaceuticals,* DOI 10.1007/978-1-4614-9524-6_16,

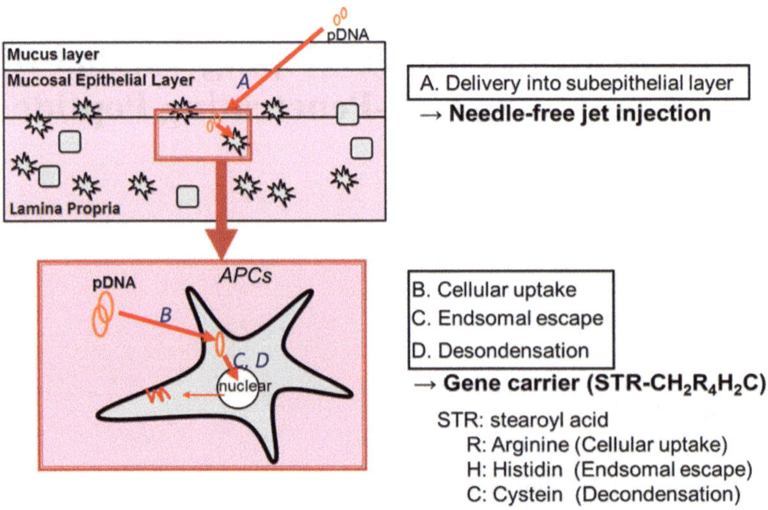

Fig. 16.1 Strategies to overcome the barriers of mucosal DNA delivery for mucosal DNA vaccination. (Adapted from [51], copyright 2013, with permission from Elsevier)

would appear to be the most efficient approach to prevent infection [19–23]. In general, the genital tract, which includes the uterus, cervix, and vagina, is the most common entry site for viral infections that are transmitted through heterosexual intercourse, including human immunodeficiency virus (HIV), papillomavirus (HPV) associated with cervical carcinoma, and herpes simplex virus. Thus, the development of safe, easy-to-use, effective, stable, and inexpensive vaccines against these viral infections is urgently needed. To prevent or respond to these infections, strong vaginal immunity is therefore required. In our previous study [24, 25], we confirmed that immunization using an efficient vaginal gene delivery system with pCMV-OVA promoted local immunoglobulin A (IgA) production in the vaginal mucosa of mice to a greater extent than intradermal or nasal immunization routes. We believe that a strong vaginal immune response can be obtained by inducing expression of antigens encoded by DNA vaccines in local vaginal tissue. To improve transfection efficiency in antigen-presenting cells (APCs) in the vaginal subepithelial layer, it is important that the vaccine is delivered across the various barriers, such as vaginal epithelial layer and cellular membrane (Fig. 16.1). This requires the development of a less invasive and more effective gene delivery device into both subepithelial layer and APCs such as dendritic cells.

In this chapter, we will introduce our studies on the effects of the menstrual cycle for vaginal vaccination and efficient vaginal DNA vaccination methods using needle-free injector and novel cell-penetrating peptide for development of DNA vaccines that penetrate the vaginal epithelial layer and cellular membrane.

16.2 Effects of the Menstrual Cycle for Vaginal Delivery

The vaginal mucosa is under constant exposure to infectious agents, and is consequently surveyed by a network of dendritic cells to induce mucosal immunity [26].

Unlike other mucosal tissues, the female reproductive tract undergoes dramatic hormone-dependent changes over the course of the menstrual cycle. One potential mechanism relates to the thickness and leakiness of the vaginal epithelial layer. With the increase in serum estrogen levels, the epithelial layer thickness increases during the estrous stage. Subsequently, during the metestrous stages, with the increase in serum progesterone levels and decrease in estrogen, the superficial layers of the vaginal epithelium are delaminated, and become maximally thin and leaky at the diestrous stage.

Vaginal absorption of relatively large and water-soluble compounds, such as peptides and proteins, has been systemically determined in rats to be very poor and significantly influenced by the menstrual cycle [27–30].

We examined the transfection efficiency into vaginal mucosa during the estrous cycle, which has the four stages of proestrus, estrus, metestrus, and diestrus, in mice [31]. The estrous stage was assessed by daily morning microscopic observation of vaginal smears taken as a swab and stained with Giemsa solution, after which pCMV-Luc was electroporated at the vaginal surface at around 09:00 in the morning, following the smear check. The vaginal membrane was electroporated at its surface immediately after vaccine administration of phosphate buffer saline (PBS) or one of the plasmid DNA (pDNAs), using an electroporator. A custom-designed needle electrode, consisting of two parallel needles (anode and cathode) 5 mm in length and 5 mm apart, each consisting of 3 platinum needles of 1 mm in diameter, was used to apply 15 pulses of electricity at 250 V/cm for 5 min. These electroporation parameters were established in our previous study as the optimal conditions for greatest gene transfection efficacy with minimal vaginal irritation [25, 31]. As shown in Fig. 16.2a, at metestrus and diestrus, luciferase gene expression was threefold higher than at proestrus and estrus, indicating that the transfection efficiency in vaginal mucosa was clearly affected by the estrous cycle. The mucosa of the vagina consists of epithelial cell layers that form a barrier to absorption of water-soluble and large molecules. Histological observation (Fig. 16.2b) indicated that the difference in transfection efficiency during the four menstrual stages might be explained by a change in the membrane structure. At metestrus and diestrus, these epithelial cell layers are very thin compared with those at the other stages, and at diestrus they are extremely porous.

Antigen presentation is known to be most reduced at the estrous stage of the estrous cycle, at which time estrogen levels are most elevated and ovulation takes place [32], whereas the number of APCs has been found to be maximal in the vagina, and the number of layers of epithelial cells lining the vagina of rodents decreases dramatically, at the diestrous stage, which would be expected to enhance uptake of luminal antigens. Indeed, uptake of proteins and the ability of vaginal immunization to induce

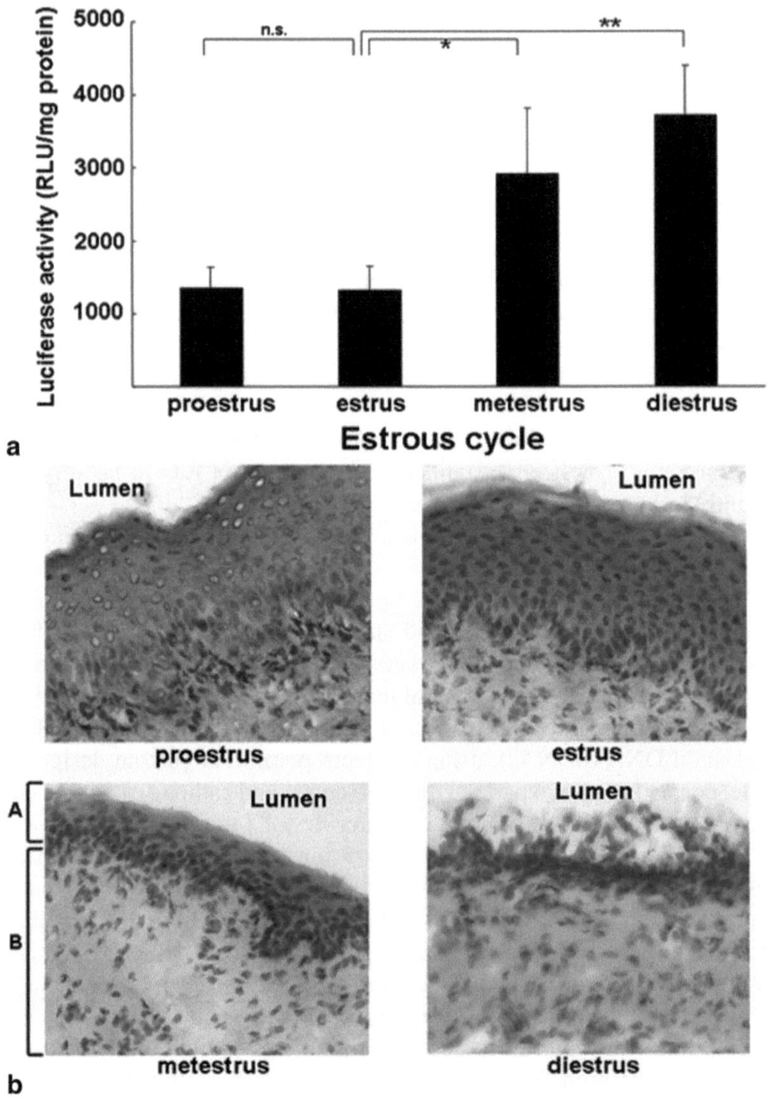

Fig. 16.2 Effects of the estrous cycle on transfection of pCMV-Luc into vaginal mucosa in diestrous mice. **a** Luciferase activity in the vaginal mucosal membrane was determined 24 h after vaginal administration of pCMV-Luc (20 μg) at various estrous stages via electroporation (250 V/cm, 5 min, 15 pulses) in mice pretreated with 5 % citric acid solution for 2 h. Each data point represents the mean ± S.E. ($n = 4$). $^{*}p < 0.05$, $^{**}p < 0.01$, n.s. $p > 0.05$, *t-test*). **b** Histological observation of a section of the vaginal mucosal membrane in mice during different stages of the estrous cycle. The stage of the estrous cycle of the mice was determined using a morning smear test. Vaginal tissue was collected and 10-μm frozen sections stained with hematoxylin and eosin. *A* epithelium, *B* stroma (subepithelium), *Lumen* vaginal lumen. (Adapted from [31], copyright 2008, with permission from Elsevier)

specific antigen responses in mice are optimal when preparations are administered during diestrus [33]. Thus, vaccine strategies for protection against sexually transmitted diseases must take into account that sex hormones affect immune responses. It has also been reported that the immune-associated cells in the vaginal submucosal membrane increase at diestrus [26].

These findings indicate that vaginal DNA vaccination at diestrus, the late luteal phase, and early follicle phase in humans, would be most suitable for practical therapy.

16.3 Needle-Free Injectors for Vaginal DNA Vaccination

A strong vaginal immune response can be obtained by inducing strong gene expression of antigen-coding DNA vaccines in local vaginal tissue. In order to improve transfection efficiency in the vaginal subepithelial layer, it is important to break through the vaginal epithelial layer. This requires the development of less invasive and more effective delivery methods into the subepithelial layer across the vaginal epithelial layer.

Needle-free jet injection has been extensively investigated as a method to immunize laboratory animals, such as mice [34, 35], rabbits [36, 37], pigs and dogs [38], and monkeys, through the transdermal route. In addition, jet injection has been tested subcutaneously in several human clinical trials [39] and is already produced commercially for daily injection of insulin and growth hormone. The vast majority of studies in animals have demonstrated an enhancement in resulting immune responses with jet injection over conventional needle-syringe injection [40].

We first confirmed that the luciferase activity in rat skin inoculated with the transdermal needle-free jet injector was strikingly greater than that by needle-syringe injection [25, 41]. This result because of that the pCMV-Luc solution via the needle-syringe typically forms a sphere of fluid at the injection spot of the tissue, whereas the pCMV-Luc solution through the needle-free jet injector disperses more widely into the dermal tissue, likely due to the high pressure of the fluid stream. This wide distribution by the needle-free injector possibly achieves markedly higher luciferase activity in rat skin. In addition, the luciferase activity following injection by needle-syringe injection with electroporation was higher than that by needle-syringe injection alone, whereas the luciferase activity following administration by the needle-free injector with electroporation did not differ from that by the needle-free injector alone [41]. Thus, needle-free injection provides a similarly wide and effective delivery of pDNA into local tissue cells to electroporation. These results indicate that the needle-free injector can deliver pDNA widely in dermal tissue and might deliver to a number of APCs, which induce immune responses. OVA-specific interferon-γ (IFN-γ) production as well as OVA-specific immunoglobulin G_{2a} (IgG_{2a}) production levels in mice immunized with the needle-free injector were also significantly greater than those by conventional needle-syringe injection. This was due to a wider distribution of pDNA solution in the dermal tissue injected through the needle-free jet injector, resulting in a higher contact incidence between the pCMV-OVA and APCs, such as APCs and lymphocytes found in dermal tissue.

Therefore, we next developed the mucosal needle-free injector device as a vaginal vaccination tool [41] because we expected that needle-free mucosal vaccination can induce strong mucosal immune responses. This injection device, which has an injector angle of 45°, was designed for use in the human vagina from the entrance to inject in a right-angled direction into the middle site of the tract. The use of this mucosal needle-free injector greatly promoted the gene expression in rabbit vagina rather than the use of the conventional needle-syringe injection [25, 41]. Moreover, intravaginal vaccination using this mucosal needle-free injector significantly promoted vaginal IgA secretion and IFN-γ mRNA expression in lymphocytes compared to conventional needle-syringe injection [41]. These results demonstrate that the needle-free injector can be used not only as an intradermal vaccination device but also as a mucosal vaccination device. This study has demonstrated for the first time that a needle-free injector can be used for effective local mucosal vaccination.

16.4 Combination of Needle-Free Injector and Cell-Penetrating Peptide for Vaginal DNA Vaccination

16.4.1 Cell-Penetrating Peptide for DNA Delivery into Dendritic Cells

Dendritic cells, which originate in the bone marrow, are professional antigen-capturing cells and APCs, and these processes initiate the primary immune responses in our body. This central role in cell-mediated immunity has made them an attractive target for immunotherapy [42–44]. To generate strongly effective dendritic cells, technologies are needed that produce high antigen expression as a result of delivering DNA- encoding antigen into the nucleus of dendritic cells, which are nondividing cells. Recently, a cellular internalization method using short peptides derived from protein-transduction domains has attracted much attention. Several cell-penetrating peptides, such as HIV-1 Tat fragments, less than 30 amino acid residues in length are capable of crossing a plasma membrane [45–47]. In addition, they can deliver their associated molecules into cells. The Tat peptide has been reported to be capable of delivering β-galactosidase to various organs when administered intraperitoneally to mice.

In order to promote the gene expression of pDNA in dendritic cells in vaginal subepithelial layer, the numerous barriers to gene delivery must be overcome. These barriers include (1) cellular adhesion and uptake, (2) escape from endosomes to the cytoplasm prior to delivery to fusion by lysosomes, and (3) decondensing the pDNA from carrier complex [48].

STR-$CH_2R_4H_2C$ was an effective multifunctional cell-penetrating peptide carrier developed in our study [49, 50]. STR-$CH_2R_4H_2C$ consists of stearic acid (STR),

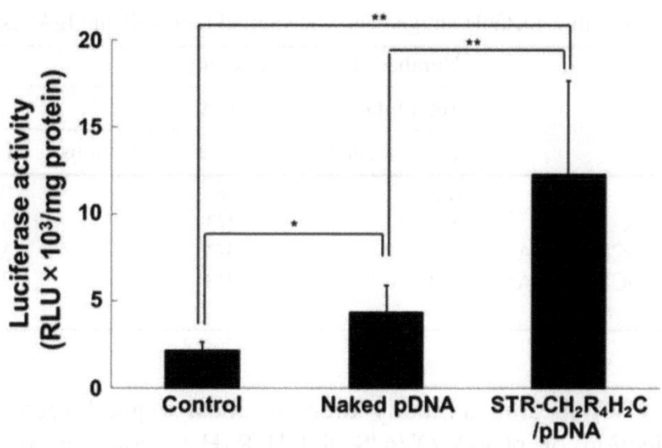

Fig. 16.3 Local luciferase activity in rabbits' vagina after intravaginal needle-free injection of pCMV-Luc/STR-$CH_2R_4H_2C$. Luciferase activity was determined after intravaginal administration of naked pDNA (pCMV-Luc) (12.5 μg), or pDNA (pCMV-Luc)/STR-$CH_2R_4H_2C$ (weight ratio 1:5) by a needle-free injector. Each *bar* represents the mean ± S.E. ($n = 3$). $^*p < 0.05$, $^{**}p < 0.01$. (Reprinted from [51], copyright 2013, with permission from Elsevier)

cysteine (C), histidine (H), and arginine (R). STR-$CH_2R_4H_2C$ strikingly enhanced in vitro transfection efficiency because of forming the stability complex and cellular uptake by STR and R, proton sponge effect by H, and the ability to release pDNA from carrier in cytoplasm by cleavage of disulfide cross linkage of C [49]. Furthermore, STR-$CH_2R_4H_2C$ carrier greatly promoted the gene expression in dendritic cells [51], which is generally difficult to achieve in nondividing cells. Therefore, the combination of needle-free injector and STR-$CH_2R_4H_2C$ carrier can deliver pDNA widely into vaginal tissue increasing the likelihood of delivery to a number of APCs, such as dendritic cells, which induce immune responses.

16.4.2 Vaginal DNA Vaccination

In order to improve the local vaginal vaccination responses generated by needle-free jet injection, a nonneedle jet injector combined with an effective peptide carrier (STR-$CH_2R_4H_2C$) were used. The local vaginal luciferase activity in rabbits is shown in Fig. 16.3. Naked pCMV-Luc or pCMV-Lus/STR-$CH_2R_4H_2C$ complex intravaginally were injected using the needle-free injector. The local luciferase activity in the pCMV-Luc administration groups was significantly greater than in the nontreated group. The pCMV-Luc solution has been shown to be more widely dispersed into the local tissue through the needle-free jet injector, likely due to the high pressure of the fluid stream. This wide distribution may account for the markedly higher luciferase activity in the local tissue. Furthermore, the luciferase activity in the rabbit's vagina was clearly higher with the STR-$CH_2R_4H_2C$ carrier than without carrier, indicating that the STR-$CH_2R_4H_2C$ carrier was able to enhance the effective delivery of pDNA into local tissue and cells [51].

Table 16.1 Rabbits with detectable antigen-specific vaginal total IgG and IgA responses [51]

Group	Number of responders/total		
	Total IgG	IgA	
	Five vaccinations	Four vaccinations	Five vaccinations
PBS	0/3	0/3	0/3
Naked pCMV-OVA	1/3	1/3	3/3
STR-$CH_2R_4H_2C$/pCMV-OVA	2/3	3/3	3/3
STR-$CH_2R_4H_2C$/pCMV-OVA/ CpG-ODN	3/3	3/3	3/3

Next, the OVA-specific antibody titers in local vaginal secretions after intravaginal injection of pCMV-OVA/STR-$CH_2R_4H_2C$ complex with or without CpG-oliogodeoxynucleotide (CpG-ODN) injection by needle-free injector were also examined. As shown in Table 16.1, there was no vaginal total IgG response in the PBS group, whereas one of the three rabbits immunized with naked pCMV-OVA and two of the three rabbits immunized with pCMV-OVA/STR-$CH_2R_4H_2C$ using the needle-free injector showed a strong increase. The vaginal IgG titer in rabbits immunized with pCMV-OVA/STR-$CH_2R_4H_2C$ tended to be higher than that with naked pCMV-OVA. Furthermore, the vaginal IgG response in rabbits immunized five times with pCMV-OVA/STR-$CH_2R_4H_2C$/CpG-ODN by the needle-free injector appeared in almost all rabbits (Table 16.1) and the vaginal IgG titer was strongly higher than that with naked pCMV-OVA. The secretory vaginal IgA titer in rabbits immunized four times with pCMV-OVA combined with STR-$CH_2R_4H_2C$ and CpG-ODN was significantly higher than that of rabbits immunized with naked pCMV-OVA only. The number of local vaginal IgA responders was one of three rabbits immunized by naked pCMV-OVA and two of three rabbits immunized by pCMV-OVA complexed with STR-$CH_2R_4H_2C$ carrier. Surprisingly, a local vaginal IgA response in rabbits vaccinated four times with STR-$CH_2R_4H_2C$ carrier and CpG-ODN adjuvant appeared in all rabbits (Table 16.1). In addition, the secretory vaginal IgA titer in rabbits immunized five times was clearly higher than that in rabbits immunized four times. Furthermore, a local IgA response in rabbits intravaginally immunized five times appeared in all rabbits.

In order to investigate the systemic immune responses following five intravaginal DNA vaccinations with STR-$CH_2R_4H_2C$ carrier and CpG-ODN adjuvant, total serum IgG in rabbits vaccinated with pCMV-OVA using the needle-free injector were also determined [51]. A serum total IgG response in the PBS groups was not present, whereas the IgG response in one rabbit out of the three immunized with naked pCMV-OVA and pCMV-OVA/STR-$CH_2R_4H_2C$ using the needle-free injector strongly increased. The serum IgG titer in rabbits immunized with pCMV-OVA/STR-$CH_2R_4H_2C$ tended to be higher than that with naked pCMV-OVA. Furthermore, the IgG response in rabbits immunized five times with pCMV-OVA/STR-$CH_2R_4H_2C$/CpG-ODN by the needle-free injector appeared in almost all rabbits, and the serum IgG titer in rabbits immunized with pCMV-OVA/STR-$CH_2R_4H_2C$/CpG-ODN was significantly higher than that with naked pCMV-OVA.

These results indicate that needle-free vaginal DNA vaccination with a combination of STR-$CH_2R_4H_2C$ and CpG-ODN adjuvant could increase the induction of not only vaginal IgG and IgA but also serum IgG secretion.

16.5 Conclusions and Future Perspectives

In this chapter, we introduced our studies into efficient vaginal DNA vaccination methods, focusing on the effects of the menstrual cycle, utilization of the needle-free injector and the novel cell-penetrating peptide carrier. We first described in this chapter that the transfection efficiency of plasmid DNA into vaginal mucosa is strongly influenced by the estrous cycle with higher luciferase gene expression observed during diestrus. We next introduced the local vaginal DNA vaccination using a needle-free jet injector. Needle-free jet injector is potentially a useful, safe, easy, and potent method for the prevention and treatment of mucosal infectious diseases. Importantly, needle-free vaccine delivery can avoid the risk of transmission of infectious disease between patients or between patients and healthcare providers [52]. Furthermore, we introduced a new peptide-based gene carrier using arginine, histidine, and cysteine that performs multiple functions, including cellular uptake, endosomal escape, and nucleic acids condensation and decondensation by disulfide cross linkage in the cytosol for more effective DNA vaccination. In addition, this multifunctional cell-penetrating peptide carrier promotes pDNA expression in not only dividing cells but also dendritic cells [49, 51]. Furthermore, intravaginal vaccination using the needle-free injector and STR-$CH_2R_4H_2C$ carrier significantly promoted secretion of vaginal IgA, in particular, and IgG as well as serum IgG compared to the control group. This probably resulted from the wider distribution of pDNA solution in the vaginal tissue injected through the needle-free jet injector, resulting in a higher contact incidence between the pDNA and APCs in vaginal tissue. An advantage of the needle-free vaccine delivery is that the risk of transmission of infectious disease between patients or between patients and healthcare providers is avoided [52]. Furthermore, the effective cell-penetrating peptides could be an attractive tool to enhance the needle-free mucosal DNA vaccination.

Currently, several treatment methods using protein or DNA vaccination, neutralizing human monoclonal antibody for HIV-1 [53], or stem cells transplantation [54] exist. Of these, a vaginal DNA vaccine could be the most attractive therapy against HIV-1 infection. A successful vaccine against HIV and HPV requires strong local cytotoxic T lymphocyte (CTL) activity at the site of viral entry, the vaginal mucosa and draining lymph node, as well as systemically. Even the most promising vaccine formulations may fail to establish protective immunity, if the route of vaccine administration is not optimal for induction of local immune responses in the local mucosa, such as the rectum or vagina. We described here that a strong vaginal immune response will be obtained as a result of strong gene expression of an antigen-coding DNA vaccine in APCs, including dendritic cells and macrophages, in vaginal tissue, and in order to promote gene expression in the vagina, the menstrual cycle phase, utilization of needle-free jet injection and cell-penetrating peptide should be considered. In the near future, it is expected that DNA vaccination using needle-free direct vaginal immunization combined with multifunctional cell-penetrating peptide carriers may be developed.

References

1. Fuller DH, Loudon P, Schmaljohn C. Preclinical and clinical progress of particle-mediated DNA vaccines for infectious diseases. Methods. 2006;40:86–97.
2. Liu MA, Ulmer JB. Human clinical trials of plasmid DNA vaccines. Adv Genet. 2005;55:25–40.
3. Kutzler MA, Weiner DB. DNA vaccines: ready for prime time? Nat Rev Genet. 2008;9:776–88.
4. Dunham SP. The application of nucleic acid vaccines in veterinary medicine. Res Vet Sci. 2002;73:9–16.
5. Dhama K, Mahendran M, Gupta PK, Rai A. DNA vaccines and their applications in veterinary practice: current perspectives. Vet Res Commun. 2008;32:341–56.
6. Tang DC, Devit M, Johnston SA. Genetic immunization is a simple method for eliciting an immune response. Nature. 1992;356:152–4.
7. Ulmer JB, Donnelly JJ, Parker SE, Rhodes GH, Felgner PL, Dwarki VJ, Gromkowski SH, Deck RR, Dewitt CM. Heterologous protection against influenza by injection of DNA encoding a viral protein. Science. 1993;259:1745–9.
8. Fynan EF, Webster RG, Fuller DH, Haynes JR, Santoro JC, Robinson HL. DNA vaccines: protective immunizations by parenteral, mucosal, and gene-gun inoculations. Proc Natl Acad Sci U S A. 1993;90:11578–82.
9. Wang B, Ugen KE, Srikantan V, Agadjanyan MG, Dang K, Refaeli Y, Sato AI, Boyer J, Williams WV, Weiner DB. Gene inoculation generates immune responses against human immunodeficiency virus type 1. Proc Natl Acad Sci U S A. 1993;90:4156–60.
10. MacGregor RR, Boyer JD, Ugen KE, Lacy KE, Gluckman SJ, Bagarazzi ML, Chattergoon MA, Baine Y, Higgins TJ, Ciccarelli RB, Coney LR, Ginsberg RS, Weiner DB. First human trial of a DNA-based vaccine for treatment of human immunodeficiency virus type 1 infection. safety and host response. J Infect Dis. 1998;178:92–100.
11. Minchess M, Tchakarov S, Zoubak S, Loukinov D, Botev C, Altankova I, Georgiev G, Petrov S, Meryman HT. Naked DNA and adenoviral immunizations for immunotherapy of prostate cancer: a Phase I/II clinical trial. Eur Urol. 2000;38:208–17.
12. Tacket CO, Roy MJ, Widera G, Swain WF, Broome S, Edelman R. Phase 1 safety and immune response studies of a DNA vaccine encoding hepatitis B surface antigen delivered by a gene delivery device. Vaccine. 1999;17:2826–9.
13. Le TP, Coonan KM, Hedstrom RC, Charoenvit Y, Sedegah M, Epstein JE, Kumar S, Wang R, Doolan DL, Maguire JD, Parker SE, Hobart P, Norman J, Hoffman SL. Safety, tolerability and humoral immune responses after intramuscular administration of a malaria DNA vaccine to healthy adult volunteers. Vaccine. 2000;18:1893–901.
14. Ulmer JB, Wahren B, Liu MA. Gene-based vaccines: recent technical and clinical advances. Trends Mol Med. 2006;12:216–22.
15. Davidson AH, Traub-Dargatz JL, Rodeheaver RM, Ostlund EN, Pedersen DD, Moorhead RG, Stricklin JB, Dewell RD, Roach SD, Long RE, Albers SJ, Callan RJ, Salman MD. Immunologic responses to West Nile virus in vaccinated and clinically affected horses. J Am Vet Med Assoc. 2005;226:240–5.
16. Garver KA, LaPatra SE, Kurath G. Efficacy of an infectious hemoatopoietic necrosis (IHN) virus DNA vaccine in Chinook *Oncorhynchus tshawytscha* and sockeye *O. nerka* salmon. Dis Aquat Organ. 2005;64:13–22.
17. Bergman PJ, Camps-Palau MA, McKnight JA, Leibman NF, Craft DM, Leung C, Liao J, Riviere I, Sadelain M, Hohenhaus AE, Gregor P, Houghton AN, Perales MA, Wolchok JD.

Development of a xenogeneic DNA vaccine program for canine malignant melanoma at the Animal Medical Center. Vaccine. 2006;24:4582–5.

18. Thacker EL, Holtkamp DJ, Khan AS, Brown PA, Draghia-Akli R. Plasmid-mediated growth hormone-releasing hormone efficacy in reducing disease associated with *Mycoplasma hyopneumoniae* and porcine reproductive and respiratory syndrome virus infection. J Anim Sci. 2006;84:733–42.
19. Azizi A, Ghunaim H, Diaz-Mitoma F, Mestecky J. Mucosal HIV vaccines: a holy grail or dud? Vaccine. 2010;28:4015–26.
20. Stevceva L, Strober W. Mucosal HIV vaccines: where are we now? Curr HIV Res. 2004;2:1–10.
21. Demberg T, Robert-Guroff M. Mucosal immunity and protection against HIV/SIV infection: strategies and challenges for vaccine design. Int Rev Immunol. 2009;28:20–48.
22. Singh M, Vajdy M, Gardner J, Briones M, O'Hagan D. Mucosal immunization with HIV-1 gag DNA on cationic microparticles prolongs gene expression and enhances local and systemic immunity. Vaccine. 2001;20:594–602.
23. White HD, Yeaman GR, Givan AL, Wira CR. Mucosal immunity in the human female reproductive tract: cytotoxic T-lymphocyte function in the cervix and vagina of premenopausal and postmenopausal women. Am J Reprod Immunol. 1997;37:30–8.
24. Kanazawa T, Takashima Y, Shibata Y, Tsuchiya M, Tamura T, Okada H. Effective vaginal DNA delivery with high transfection efficiency is a good system for induction of higher local vaginal immune responses. J Pharm Pharmacol. 2009;61:1457–63.
25. Kanazawa T, Takashima Y, Okada H. Vaginal DNA vaccination against infectious diseases transmitted through the vagina. Front Biosci (Elite Ed). 2012;4:2340–53.
26. Zhao X, Deak E, Soderberg K, Linehan M, Spezzano D, Zhu J, Knipe DM, Iwasaki A. Vaginal submucosal dendritic cells, but not Langerhans cells, induce protective Th1 responses to herpes simplex virus-2. J Exp Med. 2003;197:153–62.
27. Baker DA, Plotkin SA. Enhancement of vaginal infection in mice by herpes simplex virus type II with progesterone. Proc Soc Exp Biol Med. 1978;158:131–4.
28. Gallichan WS, Rosenthal KL. Effects of the estrous cycle on local humoral immune responses and protection of intranasally immunized female genital tract. Virolgy. 1996;224:487–97.
29. Overall JC, Kern ER, Schlitzer RL, Friedman SB, Glasgow LA. Genital herpesvirus hominis infection in mice. I. Development of an experimental model. Infect Immun. 1975;11:476–80.
30. Parr MB, Kepple L, Mcdermott MR, Drew MD, Bozzola JJ, Parr EL. Amouse model for studies of mucosal immunity to vaginal infection by herpes simplex virus type 2. Lab Invest. 1994;70:369–80.
31. Kanazawa T, Takashima Y, Hirayama S, Okada H. Effects of menstrual cycle on gene transfection through mouse vagina for DNA vaccine. Int J Pharm. 2008;360:164–70.
32. Wira CR, Rossoll RM. Antigen-presenting cells in the female reproductive tract: influence of sex hormones on antigen presentation in the vagina. Immunology. 1995;84:505–8.
33. Parr MB, Parr EL. Antigen recognition in the female reproductive tract. 1. Uptake of intraluminal protein tracers in the mouse vagina. J Reprod Immunol. 1990;17:101–14.
34. Cui Z, Baizer L, Mumper RJ. Intradermal immunization with novel plasmid DNA-coated nanoparticles via a needle-free injection device. J Biotechnol. 2003;102:105–15.
35. Haensler J, Verdelet C, Sanchez V, Girerd-Chambaz Y, Bonnin A, Trannoy E, Krishnan S, Meulien P. Intradermal DNA immunization by using jet-injectors in mice and monkeys. Vaccine. 1999;17:628–38.
36. Ren S, Li M, Smith JM, DeTolla LJ, Furth PA. Low-volume jet injection for intradermal immunization in rabbits. BMC Biotech. 2002;2:10.
37. Aquiar JC, Hedstrom RC, Rogers WO, Charoenvit Y, Sacci JB, Lanar DE, Majam VF, Stout RR, Hoffman SL. Enhancement of the immune response in rabbits to a malaria DNA vaccine by immunization with a needle-free jet device. Vaccine. 2002;20:275–80.
38. Anwer K, Earle KA, Shi M, Wang J, Munper RJ, Proctor B, Jansa K, Ledebur HC, Davis S, Eaglstein W, Rolland AP. Synergistic effect of formulated plasmid and needle-free injection for genetic vaccines. Pharm Res. 1999;16:889–95.

39. Jackson LA, Austin G, Chen RT, Stout R, DeStefano F, Gorse GJ, Newman FK, Yu O, Weniger BG. Safety and immunogenicity of varying dosages of trivalent inactivated influenza vaccine administered by needle-free jet injectors. Vaccine. 2001;19:4703–9.
40. Mumper RJ, Cui Z. Genetic immunization by jet injection of targeted pDNA-coated nanoparticles. Methods. 2003;31:255–62.
41. Kanazawa T, Takashima Y, Tamura T, Tsuchiya M, Shibata Y, Udagawa H, Okada H. Local gene expression and immune responses of vaginal DNA vaccination using needle-free injector. Int J Pharm. 2010;396:11–6.
42. Koido S, Kashiwaba M, Chen D, Gendler S, Kufe D, Gong J. Induction of antitumor immunity by vaccination of dendritic cells transfected with MUCI RNA. J Immunol. 2000;165:5713–9.
43. Rughetti A, Biffoni M, Sabbatucci M, Rahimi H, Pellicciotta I, Fattorossi A, Pierelli L, Scambia G, Lavitrano M, Frati L, Nuti M. Transfected human dendritic cells to induce antitumor immunity. Gene Ther. 2000;7:1458–66.
44. Landi A, Babiuk LA, van Drunen Little-van den Hurk S. High transfection efficiency, gene expression, and viability of monocyte-derived human dendritic cells after nonviral gene transfer. J Leukoc Biol. 2007;82:849–60.
45. Futaki S, Suzuki T, Ohashi W, Yagami T, Tanaka S, Ueda K, Sugiura Y. Arginine-rich peptides. An abundant source of membrane-permeable peptides having potential as carriers for intracellular protein delivery. J Biol Chem. 2001;276:5836–40.
46. Morris MC, Depollier J, Mery J, Heitz F, Divita G. A peptide carrier for the delivery of biologically active proteins into mammalian cells. Nat Biotechnol. 2001;19:1173–6.
47. Schewarze SR, Ho A, Vocero-Akbani A, Dowdy SF. In vivo protein transduction: delivery of a biologically active protein into the mouse. Science. 1999;285:1569–72.
48. Yang Z, Sahay G, Sriadibhatla S, Kabanov AV. Amphiphilic block copolymers enhance cellular uptake and nuclear entry of polyplex-delivered DNA. Bioconjug Chem. 2008;19:1987–94.
49. Tanaka K, Kanazawa T, Ogawa T, Suda Y, Takashima Y, Fukuda T, Okada H. A novel, bio-reducible gene vector containing arginine and histidine enhances gene transfection and expression of plasmid DNA. Chem Pharm Bull (Tokyo). 2011;59:202–6.
50. Tanaka K, Kanazawa T, Ogawa T, Takashima Y, Fukuda T, Okada H. Disulfide crosslinked stearoyl carrier peptides containing arginine and histidine enhance siRNA uptake and gene silencing. Int J Pharm. 2010;398:229–38.
51. Kanazawa T, Tamura T, Yamazaki M, Takashima Y, Okada H. Needle-free intravaginal DNA vaccination using a stearoyl oligopeptide carrier promotes local gene expression and immune responses. Int J Pharm. 2013;447:70–4.
52. Giudice EL, Campbell JD. Needle-free vaccine delivery. Adv Drug Deliv Rev. 2006;58:68–89.
53. Wu X, Yang ZY, Li Y, Hogerkorp CM, Schief WR, Seaman MS, Zhou T, Schmidt SD, Wu L, Xu L, Longo NS, McKee K, O'Dell S, Louder MK, Wycuff DL, Feng Y, Nason M, Doria-Rose N, Connors M, Kwong PD, Roederer M, Wyatt RT, Nabel GJ, Mascola JR. Rational design of envelope identifies broadly neutralizing human monoclonal antibodies to HIV-1. Science. 2010;13:856–61.
54. Allers K, Hütter G, Hofmann J, Loddenkemper C, Rieger K, Thiel E, Schneider T. Evidence for the cure of HIV infection by CCR5 Δ32/Δ32 stem cell transplantation. Blood. 2011; 117:2791–9.

Chapter 17
Vaccine Delivery Systems for Veterinary Immunization

Juan M. Irache, Ana I. Camacho and Carlos Gamazo

17.1 Introduction

This chapter offers a view about how novel strategies based on particulate delivery systems, such as nanoparticles and microparticles, may help in veterinary vaccination. As we will discuss in more depth later, these antigen delivery systems may play a critical role for the effectiveness of vaccine strategies, particularly in case of mucosal immunization.

Apart from improving animal health and productivity, veterinary vaccines have a significant impact on public health through reductions in the use of veterinary pharmaceuticals and hormones and their subsequent residues in the human food chain [1]. In addition, vaccines contribute to the well-being of livestock and companion animals, and their use is favoured by the growing animal welfare lobby.

However, despite that, veterinary vaccines comprise approximately a quarter of the global market for animal health products, the sector has not integrated new available technological advances in vaccine development as soon as they arrive.

Although the generation of veterinary vaccines has many common issues with vaccines developed for humans (to confer protective immunity, minimizing side effects, ease of handling and administration, etc.), there are specific challenges to the veterinary ones that require specific consideration. Exceptionally, vaccines for pets resemble more similar requirements to those of human vaccines, in which the costs are usually less important but the absence of side effects is more critical.

J. M. Irache (✉)
Department of Pharmacy and Pharmaceutical Technology, University of Navarra, 31080 Pamplona, Spain
e-mail: jmirache@unav.es

A. I. Camacho · C. Gamazo
Department of Microbiology, University of Navarra, Pamplona, Spain

J. das Neves, B. Sarmento (eds.), *Mucosal Delivery of Biopharmaceuticals,* DOI 10.1007/978-1-4614-9524-6_17,

17.2 Foundations of Vaccinology

The first recorded widely used deliberate active immunization was achieved during the fifteenth century in China by inhalation of powders made from the skin lesions of patients recovering from smallpox [2]. Oral route was also used in Turkey, where healthy people swallowed smallpox scabs from the infected ones. Later the skin was used to inoculate the scabs. This new practice, named "variolation", was introduced in England and western Europe in the eighteenth century. However, this was not a safe method as live viruses were just slightly attenuated after drying the scabs, and, in some cases, the host became highly infected and even died. Jenner (1796), introduced the use of the virus vaccinia from cows (lat. *vacca*), a virus antigenically related with smallpox virus, but, by far, less virulent for humans. In the late nineteenth century, Pasteur bet for a similar strategy and claimed the possibility to modify the virulence of an infectious agent for vaccination purposes, developing the first laboratory-created vaccine, a remarkable revolution in medicine. With all veneration, he proposed the term vaccine (lat. *vacca*, cow), in honour of Jenner's work.

The term immunity reflects the condition of being able to resist a particular disease (lat. *immunitas,* freedom from public service). Thus, the ultimate goal of a vaccine is to develop long-lived immunological protection by preventing the growth of a pathogenic microorganism or by counteracting the effects of its products. In this particular case, protection and immunity are analogous terms for the same meaning: freedom from the pathogen. However, vaccination is not always related with infectious diseases. In immunological terms, vaccination makes reference to the process of inducing protection against a disorder by the activation of the right branch of the immune system. In accordance, we can vaccinate against tumours, allergic and autoimmune disorders or even against fertility.

Broad-scale vaccination programmes have been successfully used in humans (smallpox has been eradicated, and others, like polio, measles or diphtheria, have been drastically reduced). But, what about zoonotic pathogens with multiple natural hosts (i.e. *Influenza, Salmonella,* or *Brucella*)? Is it possible to control these pathogens by vaccination? If by "control" we mean "eradication", the answer is no. By contrast, if control means "reduction" of the level of incidence-prevalence, we can say: yes, we can. Some pathogens have several species as natural hosts. *Brucella* species, for instance, affect sheep, goats, cattle, deer, elk, pigs, dogs, and several other vertebrates. This adaptation of some microorganisms to different hosts makes them insusceptible to vaccines. Therefore, although we cannot completely prevent infection, we can limit replication of the organisms in some hosts through vaccination.

17.3 Animal Vaccination

Prior to developing a vaccine for veterinary use, some differential features have to be considered comparing to humans. Thus, several aspects can be pointed out: (1) cost of production and delivery, particularly for species where a large number of animals with a relatively low commercial value are utilized (e.g. chickens), (2) to be effective with a single dose, especially for wildlife and animals that are handled infrequently in remote, extensive farming systems or (3) to deliver the vaccine with minimal disturbance to the animal and maximal safety to the operator. As a "contextualized" example, ballistic vaccine delivery systems have been developed which can be shot into the muscle of large wildlife species (e.g. bison) from distances of about 20 m using air-powered rifles. These biodegradable "biobullets" made of photopolymerized poly(ethylene glycol)-based hydrogels can serve as devices to deliver volumes of about 90 μL [3], while protecting the cargo from the impact upon penetration into the muscle to a depth of 10 cm. Using this method Olsen et al. demonstrated that the bison can be effectively vaccinated against *Brucella abortus* from a safe distances of 20 m [4].

17.3.1 Vaccine Types

Live-attenuated vaccines are still, by far, the most utilized ones with respect to inactivated and subunit ones, but it remains clear that the cost/benefit ratio is in favour of the subunit vaccines. Let us summarize briefly the most significant advantages and drawbacks for these elections.

The aim of attenuation is to diminish the virulence of the pathogen, but preserving its immunogenicity. As said above, Pasteur and Koch developed the methods of empirical attenuation at the end of the nineteenth century. The knowledge we have nowadays about genomics allows us to selectively knock out specific virulence genes. Advantage of this strategy is that some important antigenic determinants can be preserved by attenuated strains, able to elicit the right antibody and/or cellular immunity. Besides, the growing capacity of these attenuated vaccines provides prolonged exposure of antigens to the immune system, resulting in the production of long-lasting memory cells. However, several risks are associated with live vaccines, especially in immunocompromised individuals. Attenuated strains may still have some residual virulence due to an incomplete inactivation. Furthermore, some attenuated strains, after being released into the environment, can recover their virulence in other hosts, or can acquire genes from other microorganisms by natural genetic transfer.

Under these concerns, despite their extended use, the restrictions for the use of attenuated modified organisms in vaccination are becoming more and more stringent. To avoid the risk of live vaccines, the use of killed organisms was introduced as a safer alternative. These vaccines are made from the entire organism but inactivated (killed) by physical or chemical agents. The limitations of these kinds of vaccines are that their immunogenicity usually has to be enhanced by co-administration with adjuvants, and, in any case, multiple doses are necessary for obtaining long-term protective immunity; besides, as live vaccines, they may contain immunosuppressive antigens.

Subunit vaccines consist of crude extracts to purified antigens of the microorganism. They may be obtained by the use of recombinant DNA technology, and also include synthetic peptides or pure DNA or RNA. The primary goal of this approach is to identify the individual antigens of the pathogen that are involved in inducing protection, avoiding the immunosuppressive ones. Combining genomics with our understanding of pathogenesis, it is possible to identify specific proteins from most pathogens that are critical in inducing the right protective immune responses. The potential advantages are safety, the potential abilities to target vaccines to the site where immunity is required and also critical, to differentiate vaccinated animals from the infected ones, through the right selection of the components. However, insufficient duration of the induced immune responses remains a major difficulty, and strong adjuvants are required.

In addition, it is well known that mucosal routes are optimal for the administration of vaccines due to their capacity for triggering both systemic and mucosal immune response. Thus, putting together these two issues, subunit vaccines and mucosal administration, is a big challenge in modern vaccine development.

17.4 Immunoadjuvants

Adjuvants (lat, *adjuvare,* aid) are defined as a group of structurally heterogeneous compounds that enhance or modulate the immunogenicity of the associated antigens. The concept of adjuvant comes from the 1920s from observations such as those of Ramon [5] who noted that horses that developed an abscess at the inoculation site of diphtheria toxoid generated higher specific antibody titers. They subsequently found that an abscess generated by the injection of unrelated substances, along with the diphtheria toxoid, increased the antibody response against the toxoid [6]. Despite the recognition of many different types of adjuvant, however, little is known about their mode of action. Janeway [7] called adjuvants "the immunologists dirty little secret", because their mode of action was poorly understood. The events triggered by these immunomodulators appear to come from one or the combination of several of the following effects: formation of a slow-release depot of the antigen, direct interaction and stimulation of antigen presenting cells (APC) and/or non-specific immunostimulating effect.

It is well known that antigens in solution are mostly quickly removed by macrophages, but subsequently, they are unable to prime naive T cells. Therefore, following the antigen's disappearance, the immune response is hardly detectable. The most used adjuvants, such as oil-emulsions and antigen-absorbing aluminium salts, may retain antigen at the injection site, from where it is released in minute quantities over a prolonged period of time. These compounds mainly stimulate the production of antibodies by the induction of Th2-lymphocytes. In case of alum use, the mechanism of action seems to be due, at least in part, to the formation of a depot of free alum that would induce the recruitment and activation of immune cells to the site of inoculation [8]. However, this "favourable" local inflammation may derive in a granuloma, or even eosinophilia [9]. Besides, these adjuvants may produce allergic reactions after a re-immunization. Particulate delivery systems, as will be discussed below, may also induce the depot effect from where antigens are physically retained.

Further, the adjuvant-induced enhancement of an immune response may be ascribed to the improved delivery of antigens into the draining lymph nodes. This may be achieved by facilitating the antigen uptake by APCs, or by increasing the influx of APCs into the injection site. Whichever is the case, the result is the same: an effective priming of specific T cells derived from an increase in the provision of antigen-loaded APCs, promoting the activation state of APCs by upregulating co-stimulatory signals or the major histocompatibility complex (MHC) expression. This results in the corresponding cytokine release, enhancing the speed, magnitude, and duration of the specific immune response. Some vectors are able to target associated antigens into APCs, including particulate delivery systems (Pluronic micelles, liposomes, ISCOMs, and polymeric particles).

Finally, some adjuvants can stimulate the non-specific component of the immune system. Numerous microorganisms contain "alert signals", the so called "microbial or pathogen associated molecular patterns" (MAMPs or PAMPs, respectively), not present in mammalian cells. These structures activate immune cells through interaction with specific receptors (Toll-like receptors, TLRs). Some examples are: lipopolysaccharide (LPS), monophosphoryl lipid A (MPL), flagellin, lipoproteins, muramyl dipeptide (MDP); trehalose dimycolate (TDM), or CpG DNA motifs, among others [10, 11]. Besides, the special chemical nature of some polymers used in the formulation of vaccine delivery systems may also be recognized as scavenger ligands for the APCs [12, 13].

17.5 Particulate Delivery Systems

Upon the described situation, the strategies based on particulate delivery systems have merged for mucosal delivery since they link immunological properties as well as technical and practical features. These particulate carriers not only act as inert transport of antigens but are also capable of interacting with the immune system, exhibiting real adjuvant properties.

Biodegradable and biocompatible polymeric particles are highly useful and many antigens, regardless of their structure and water solubility, can be loaded into these systems by the use of different manufacturing techniques. As a consequence, the use of these polymeric particulates offer a number of advantages as antigen delivery systems: (1) increase the stability of the antigens incorporated; (2) protection against chemical and enzymatic inactivation in the environmental conditions of the organism; (3) improve the antigen transport to areas of the body in which produce its beneficial action, including the ability to interact with the APCs (i.e. dendritic cells, DCs, and macrophages) [14, 15]; and (4) prolong time of residence of the drug in the organism.

In other words, microparticles and nanoparticles may allow the progressive release of the antigens, delay their clearance and improve their exposure to the immune system. As a result, these devices act as adjuvants by increasing the provision of antigen-loaded APCs for cognate naive T cells, inducing cytokine release and, thus, enhancing the magnitude and duration of the immune response [16–18].

Antigens associated with particles mimic the particulate nature of pathogens. Indeed, particulate vaccines are typically a few hundred nanometres to a few microns in size; these dimensions are comparable to those of common pathogens, against which the immune system has evolved to react, and they are readily taken up by APCs. Thus, particles larger than 0.5 μm are internalized by APCs via phagocytosis, whereas uptake of soluble antigen or smaller particles is primarily mediated by endocytosis [19]. Internalization of particulate vaccines through phagocytosis into phagosomes has important consequences because phagosomes are known to be competent organelles for antigen cross presentation, thereby allowing the induction of cytotoxic T-cell responses, a feature hardly achievable when using soluble antigens [20, 21]. These immune responses are often critical for immunity to key veterinary pathogens including most viral infections. Moreover, during transport to the lymph nodes, soluble antigens are susceptible to premature degradation by proteolytic enzymes. By contrast, association of antigens with carrier particles can protect against such degradation. Other interesting features of particulate antigens include the possibility to deliver relatively large quantities of particle-associated antigen inside the APCs and, therefore, to prolong antigen release leading to extended antigen presentation compared with soluble antigen. In fact some studies have described that particulate vaccines are able to induce long-term immune protection [22–24].

Furthermore, particularly interesting is to highlight the possibility of the use of particulate systems to co-deliver antigens and immunostimulatory components [25, 26], known as "multivalent strategy" in order to achieve broad protection against different pathogens [27–29].

From a general point of view, microparticles and nanoparticles are versatile devices for vaccine design due to the availability of a number of materials (i.e. polymers, macromolecules, lipids) and methods for their synthesis. Common polymer compositions of microparticles and nanoparticles include biodegradable or bioeliminable synthetic polymers (e.g., poly(esters), poly(anhydrides), poly(amino acids), poly(ethylene glycol)s) and natural polymers (chitosan, alginate, albumin, hyaluronic acid), copolymers, and polymer blends [30, 31]. On the other hand, the formulation process and the surface properties of the resulting nanoparticles play important roles on their efficacy as adjuvants for vaccination [32, 33].

These systems protect the antigen from degradation and impact during penetration through the skin and muscle, but also, the use of particulate adjuvants also allows for additional routes of immunization, which are better suited to veterinary and wildlife species including oral delivery and long-distance ballistic intramuscular delivery. Thus, after oral administration, particulates can increase retention of antigens on mucosal surface, protect them from proteases in the mucus gels, and then, increase antigen uptake and immune responses.

Oral delivery of particulate vaccines has been shown to be successful and this approach is feasible in remote farming communities. Slow release systems have also been found to be effective as they provide a continued supply of antigen over weeks or months, which is able to boost the immune response. Encapsulation of antigen has been widely used as it is easy to deliver, provides protection of the antigen from degradation and has been found to be effective with a single dose.

Table 17.1 Classification of *Brucella* species and indication of their preferred hosts and pathogenicity

Species	Host	Comments	Reference
B. melitensis	Goats, sheep, rams, cattle, camels, cats	Smooth lipopolysaccharide (LPS); highly pathogenic for humans (70 % of cases)	[129, 130]
B. abortus	Cattle, buffalos, cows, bison, horses	Smooth LPS; pathogenic for humans (25 % of cases)	[131]
B. canis	Dogs, canids	Rough LPS; zoonotic transmission is rare	[132]
B. suis	Swine, hares, reindeers, caribous	Smooth LPS; pathogenic for humans (5 % of cases)	[133]
B. ovis	Sheep, ram	Rough LPS; zoonotic transmission has not been reported	[134]
B. neotomae	Desert wood rats	Smooth LPS; pathogenic for humans	[135]
B. pinnipedialis	Seals	Smooth LPS; potentially pathogenic for humans	[136]
B. ceti	Dolphins, porpoises, whales	Smooth LPS; potentially pathogenic for humans	[136]
B. microti	Common voles, wild red foxes	Smooth LPS; no reported infections	[137]
B. inopinata	Humans	Isolated from a breast implant wound of a woman with clinical signs of brucellosis	[138]
Brucella sp. NVSL 07-2006	Baboon		[139]

Several studies using micro- or nanoparticles for the encapsulation of antigens have been performed in different animal species including cattle, calves, elks, pigs, cows, sheep, chickens or salmon [34].

In the following section, the case of *Brucella* and the control of brucellosis by vaccination will be used as a model in the development of particulate delivery systems as adjuvants.

17.6 Brucellosis: Control Through Vaccination

Brucellosis is a zoonotic disease, caused by *Brucella* bacteria, occurring in humans and various species of domesticated and wild animals. *Brucella* spp. are considered as facultative intracellular bacteria that have the ability to avoid the killing mechanism and proliferate within the macrophages, similar to other intracellular pathogens [35].

The genus *Brucella* comprises a group of Gram-negative bacteria loosely related. Table 17.1 summarizes *Brucella* species. The species *B. melitensis* (which infects sheep and goats), *B. suis* (swine), and *B. abortus* (cattle) cause significant economic losses for animal owners and severe human disease. In most host species, the disease primarily affects the reproductive system with concomitant loss in productivity of animals affected. Thus, animals suffer from testicular alterations (in males), reduced fertility and abortions. The infected females rarely clear the pathogen from their system and tend to shed through their next parturition. It may be venereally transmitted, and shedding of the organism can be greater than 4 years in rams. Semen quality deteriorates rapidly and inflammatory cells are often present [36, 37].

From public health view point, brucellosis is considered to be an occupational disease that mainly affects slaughter-house workers, butchers, and veterinarians. *B. melitensis* is the most infectious to man (infective dose, 1–10 colony forming units, CFUs) followed by *B. suis* (10,000 CFUs) and *B. abortus* (100,000 CFUs) [38]. In addition, *Brucella* spp. are categorized as biological agents due to their high contagiousness and their impact on human and animal health [39]. Transmission typically occurs through contact with infected animals or materials with skin abrasions, by ingestion of infected food (i.e. unpasteurized milk or dairy products), inhalation of aerosols or through the conjunctiva [40, 41]. It has been estimated by the World Health Organization (WHO) that 500,000 new cases of brucellosis occur annually [42], making it one of the most frequently encountered zoonosis worldwide. In many areas of the world, brucellosis shows a high prevalence, including Central and South America, the Middle East, Mediterranean countries, northern Africa, South America, and countries of the Caucasus and Central Asia [43].

Symptoms in human brucellosis can be highly variable, ranging from non–specific, flu-like symptoms (acute form) to undulant fever, arthritis, orchitis and epididymitis [44, 45]. The disease is severely debilitating and protracted with several documented cases with signs associated with the disease lasting for over 30 years [46, 47].

Vaccination of animals is considered the best strategy to control brucellosis. It practically eliminates the clinical signs of brucellosis and reduces the likelihood that exposure to the infectious agent will cause disease in humans. For last decades, vaccination against *Brucella* infections in animals has been usually performed by administration of live attenuated smooth *Brucella* strains such as *B. melitensis* Rev.1 (efficacious against *B. melitensis* and *B. ovis* in small ruminants) and *B. abortus* S19 (against *B. abortus* in cattle). Despite the availability of these live vaccine strains, the search for improved vaccines has continued. This is in part through their remaining virulence in human hosts [48], their residual abortifacient potential in pregnant animals [49] and their interference with conventional serological assays which employ "smooth" LPS as antigen [50].

More recently, the rough strain *B. abortus* RB51 has been introduced in some countries for cattle vaccination; although vaccination of ovine or wildlife have given less encouraging results with failure to protect species including sheep, bison, reindeer or elk [51, 52]. Despite the variable success of the RB51 vaccine, it may indeed have a valued role for booster immunization of livestock immunized during calf hood with smooth S19. Use in this manner would provide immunological stimulation, but without concomitant elevation of antibodies specific for the LPS diagnostic antigen [52].

In the past, some of these live animal vaccines were also used in human beings, although were unsuccessful mainly because of the lack of attenuation of the vaccine. Other variant strains, such as *B. abortus* strain 19BA or *B. melitensis* 104M, have been used at some time in the former USSR and China, but were reactogenic and of limited efficacy [53].

17.6.1 Acellular Vaccines

In order to solve the drawbacks associated with the use of live attenuated vaccines, one possible strategy is the use of acellular vaccines. In the past, a phenol-insoluble fraction of *B. abortus* or *B. melitensis* (composed mainly of peptidoglycan, proteins, and smooth LPS) was proposed [54]. However this approach displayed a mediocre effectiveness in animal models and induced important adverse side effects [55]. A non-covalent complex between *Brucella* LPS and outer membrane proteins of *N. meningitidis* group B demonstrated interesting protective effect but, in a same way, some problems related with residual virulence arose [56].

Another interesting approach would be the use of specific subcellular fractions of the bacteria, containing highly conserved immunogenic antigens, capable to stimulate an adequate *Brucella* spp. immune response [57, 58]. So, as demonstrated by several studies, the ideal antigenic extract should contain components of the outer membrane including LPS and surface proteins (Omp 31, Omp 25, Omp 3a, Omp 3b) [59–62].

Brucella ovis is a stable rough form which lacks the O-polysaccharide side chains characteristic of the smooth strains of *Brucella* (i.e. *B. melitensis*), but contains an outer membrane composition similar to other members of the genus [63–65]. Assuming that the smooth-type *B. melitensis* Rev 1 vaccine protects sheep against rough-type *B. ovis*, a subcellular vaccine containing an outer membrane complex of *B. ovis* might be effective in protecting against infections by both rough *B. ovis* and smooth *Brucella*. It is noteworthy that this approach offers a supplementary advantage. In fact, the use of antigenic extracts based on rough microorganisms should avoid the interferences induced with the current serodiagnostic tests for the detection of animals infected with *B. melitensis*.

This strategy was validated in sheep by using an antigenic extract isolated from the strain *B. ovis* REO 198 (HS antigenic complex) [63, 66, 67]. The Haemorrhagic septicaemia (HS) vaccine induced immunity against experimental infection but protective schedule needs from booster doses.

As discussed above, the main limitation of subunit vaccines is its low immunogenicity which hampers the induction of the adequate degree of protection with only one dose. Moreover, immunity against *Brucella* requires cell-mediated mechanisms, in particular Th1 immune responses, characterized by interferon gamma (INF-γ) production [58, 68]. In addition, due to the fact that *Brucella* spp. are pathogens that initiate infection and colonization at mucosal surfaces, the mucosal delivery of the vaccine appears to be preferred, since administration by injection generally stimulate poor mucosal immune responses.

Under these premises, the association of the subcellular components to suitable adjuvants is mandatory. Among the different compounds and strategies that have been proposed as adjuvants, polymeric particulates (i.e. microparticles and nanoparticles) may be adequate in this particular case.

17.6.2 Microparticles as Vaccine Delivery Systems for Brucellosis

Poly(lactic acid) (PLA) and poly(lactic-co-glycolic acid) (PLGA) are, so far, the major synthetic polymers for the encapsulation of antigens in microparticles. The well-documented biocompatibility and safety of these materials, together with their biodegradability and controlled release capacity, has already led to their FDA approval for a number of applications [69–71]. However, their hydrolytic degradation and, as consequence, the generation of acid products during storage may negatively affect the immunogenicity of the entrapped antigens and, thus, limit the use of these polymers [72]. Another interesting biodegradable poly(ester) that has also been proposed for the formulation of microparticles is poly(ε-caprolactone) (PCL) [73]. The degradation of PCL is slower than that of PLGA, making it more appropriate for long-term delivery systems [74, 75]. Furthermore, PCL particles do not generate an acidic environment that could negatively affect the antigenicity of the encapsulated antigens [72, 76].

The antigenic extract from *Brucella ovis* (HS) was encapsulated in either PEC or PLGA 75:25 microparticles following a three step procedure. Firstly, the antigenic extract HS was homogenised with β-cyclodextrin and Pluronic® F-68 in order to improve its "hydrophilicity". Secondly, microparticles were prepared after the formation of a multiple emulsion and subsequent evaporation of the organic solvent [77–79]. Finally, the resulting microparticles were purified and lyophilized. Figure 17.1 summarizes the procedure of preparation of these microparticles.

In the first step, the HS extract was mixed with β-cyclodextrin by simple agitation. Then, the mixture was dispersed in an aqueous solution of Pluronic® F-68. In the second step, microparticles were obtained by the solvent extraction/evaporation method after the formation of a multiple emulsion $W_1/O/W_2$ either by a standard protocol involving ultrasounds and Ultraturrax ("standard method") [78, 80] or by "Total Recirculation One Machine System" (TROMS procedure) [58, 81, 82]. By the standard method, the inner aqueous phase (HS-cyclodextrin complex dispersed in an aqueous solution of Pluronic F68) was emulsified in an organic phase containing the polymer (PEC or PLGA) in methylene chloride by sonication. Then, this primary W_1/O emulsion was dispersed into a second aqueous phase containing polyvinylalcohol as stabilizer and homogenized with an Ultraturrax®. The resulting $W_1/O/W_2$ emulsion was stirred in order to allow the evaporation of the organic solvent [78]. By TROMS, the organic phase of the polymer was injected through a needle into the first vessel containing the inner aqueous phase (HS-cyclodextrin extract dispersed in an aqueous solution of Pluronic F68). The resulting W_1/O emulsion was injected into the second vessel containing the outer water solution of polyvinyl alcohol. The turbulent injection through the needle resulted in the formation of a multiple emulsion ($W_1/O/W_2$). Finally, after elimination of the organic solvents by agitation, the resulting microparticles were purified and lyophilized.

Interestingly, irrespective of the preparative method, both types of microparticles were well re-dispersed in suspension in an aqueous medium, displayed a similar mean size and the same appearance with a spherical shape and smooth surface (Fig. 17.1b). In all cases, the mean size of microparticles displayed a mean diameter around 2 μm,

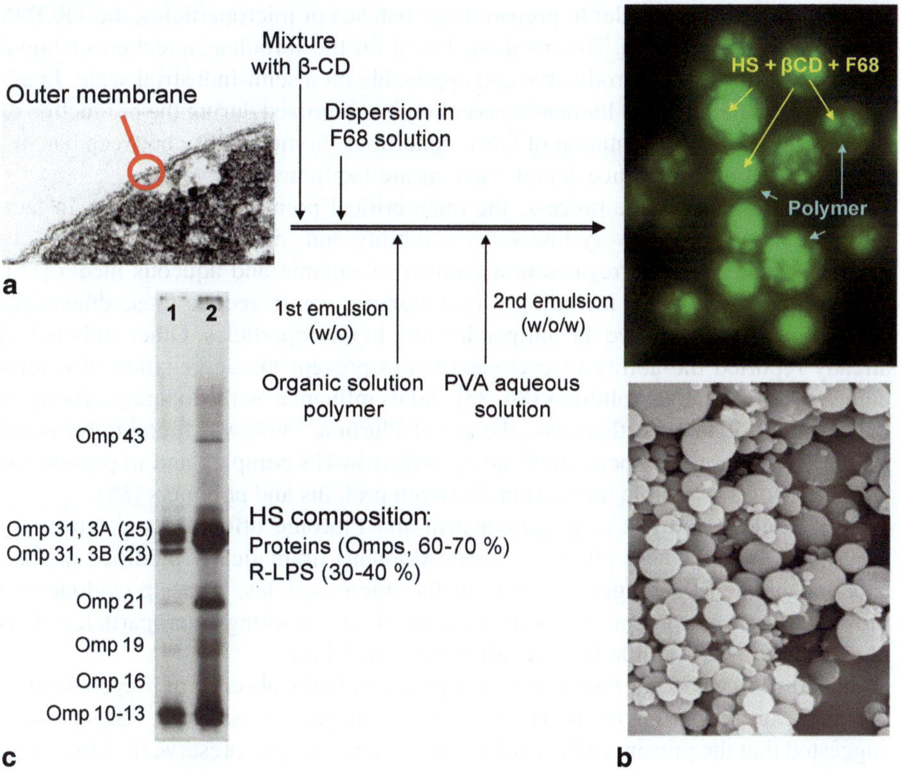

Fig. 17.1 **a** Preparation of HS-loaded microparticles by the multiple emulsion process. **b** The resulting micro-particles were characterized by SEM. **c** The integrity of encapsulated HS was evaluated by SDS-PAGE

which is considered as an optimal size to facilitate their interaction and capture by either monocyte–macrophage, DCs, or even Peyer's patches cells in the case of an oral vaccination [18, 83]. However, microparticles prepared by TROMS were more homogeneous in size than those prepared by the standard $W_1/O/W_2$ method [84]. Concerning the HS loading, no significant differences were found for microparticles prepared by either TROMS or the standard method. On the other hand, the effect of preparative process on structural integrity and antigenicity of HS were evaluated by SDS-PAGE and immunoblotting analysis, respectively. In all cases, the microencapsulation processes were safe and adequate to preserve the major HS protein constituents and their antigenicity [78, 84] (Fig. 17.1c). Regarding the in vitro release characteristics of HS from microparticles, it was confirmed that both preparative methods released the HS in a biphasic way, characterized by an initial and short release period (burst effect), followed by a longer period in which the antigen was released in a sustained way. However, TROMS microparticles displayed a lower initial release rate than "standard" microparticles. Thus, for PEC microparticles, after 4 days, only 30 % of the encapsulated HS was released from microparticles prepared by TROMS, in contrast to the 45 % released from "standard" microparticles.

In any case, and in order to prepare large batches of microparticles, the TROMS technique was preferred. This method, based on the turbulent injection of liquid phases [81], is easily reproducible and applicable on a semi-industrial scale. In addition, due to the fact that human intervention is minimized during the production of microparticles, implementation of GMP conditions, homogenicity between batches and the possibility to reduce contaminations are facilitated.

Within the preparative process, the main critical point is the first step. In fact, the HS extract shows a very low water solubility and, more importantly, tends to generate irreversible aggregates in a number of organic and aqueous media [85]. So, the main reason for including β-cyclodextrin was to reduce these drawbacks and, indirectly, to improve the antigen loading in microparticles. Other authors had already reported the ability of cyclodextrins to prevent the aggregation of several compounds in aqueous solutions [86–88] and its influence on the loading capacity of the carriers [89, 90]. Furthermore, the use of Pluronic® F68 was found to be useful to both facilitate the dispersion of the cyclodextrin-HS complex and to prevent the possibility of irreversible interactions between proteins and polymers [85].

Nevertheless, the use of β-cyclodextrin and Pluronic offered a supplementary advantage. In fact, these pharmaceutical excipients dramatically modified the distribution of the HS antigenic extract in the microparticles. Thus, β-cyclodextrin promoted the entrapment of HS in the core of the resulting microparticles. This phenomenon was amplified by including Pluronic® F68.

In addition, when microparticles were prepared in the absence of β-cyclodextrin, the antigenic properties of the HS extract were negatively affected. These results suggested that the presence of β-cyclodextrin is necessary to preserve the HS extract antigenic properties. This fact can be related to the stabilizing capacity of these oligosaccharides, based on their ability to shield hydrophobic domains of proteins [86, 91].

Evaluation of Microparticles The capability of HS-loaded microparticles (HS-PEC and HS-PLGA) to interact and activate immunocompetent cells was performed in vitro using J774.2 murine monocyte-macrophages. Phagocytosis was determined by optical microscopy, counting the number of phagocytic cells capable of taking up one or more microparticles (Fig. 17.2). In these studies, neither the antigen loading nor the composition of the internal aqueous phase of microparticles affected their internalization by cells [85, 92]. In contrast, significant differences were found depending on the polymer used, with a higher uptake of PEC-microparticles with respect to PLGA (Fig. 17.2a). Interestingly, no microparticles were observed on the surface of macrophages when the assay was performed at 4°C, validating a real phagocytosis and not just an adsorption.

These results of phagocytosis by monocytes would be explained by the higher hydrophobicity of PEC compared with PLGA. In general, it is well establish than an increase in particle surface hydrophobicity leads to an enhanced uptake by phagocytic cells [93–95].

The ingestion of microparticles may result in the activation of macrophages and, subsequently, enhance its properties as APCs [96, 97]. In order to compare the degree of activation of J744 cells after incubation with microparticles, the hydrogen

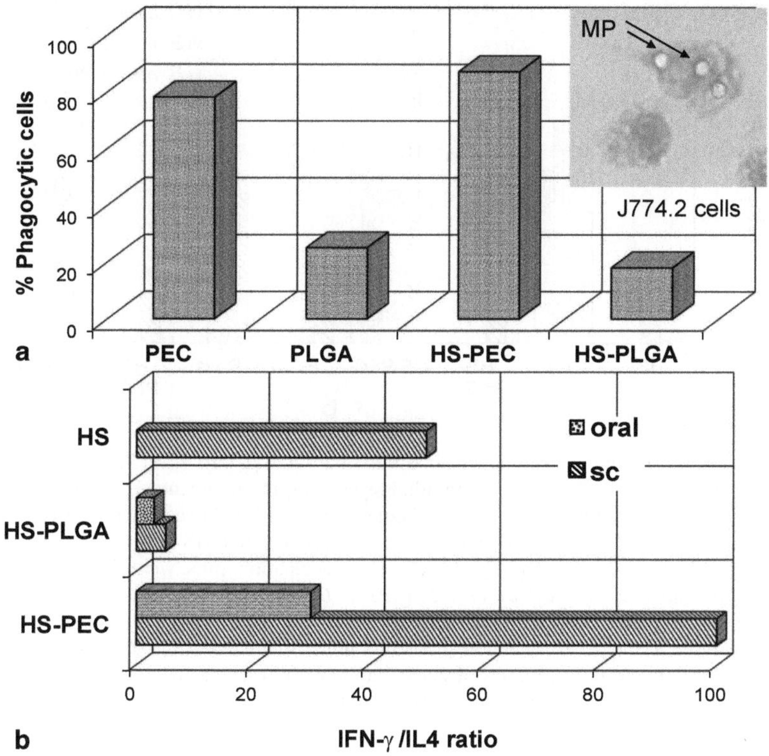

Fig. 17.2 Evaluation of HS-loaded microparticles. **a** Phagocytosis of PEC and PLGA microparticles by J774 cells. Phagocytosis was calculated as the percentage of cells capable of uptaking one of more microparticles. **b** In vitro secretion of cytokines from spleen cells elicited in mice immunized with 20 μg free HS or HS-loaded microparticles. Data are expressed as the ratio between IFN-γ and IL-4 levels

peroxide (H_2O_2) and nitric oxide (NO) production was evaluated. Regarding H_2O_2 production, both types of microparticles (PEC or PLGA) induced similar levels of hydrogen peroxide. On the contrary and as expected from phagocytosis studies, PEC microparticles were the most active NO inducers. This fact is particularly important because NO would play a major role in the intracellular killing of *Brucella* [98, 99].

The serological responses induced by the HS-loaded microparticles as well as the production of cytokines were determined in Balb/c mice after a single administration of the vaccines by either the subcutaneous or the oral routes. By the subcutaneous route, HS-loaded microparticles elicited similar and important production of both IgG1 and IgG2a antibodies; although, no significant serological responses against HS were obtained in mice immunized orally [78]. On the other hand, the administration of HS-PEC was able to activate the Th1 pathway eliciting a high IFN-γ and IL-2 release (Fig. 17.2b). In contrast, HS-PLGA elicited a Th2 response (Fig. 17.2b). These results indicate a different cytokine pattern depending on the polymer used in the formulations, suggesting again that not only the antigen intrinsic nature but also the context of its presentation by the APCs may alter the cytokine profile after a

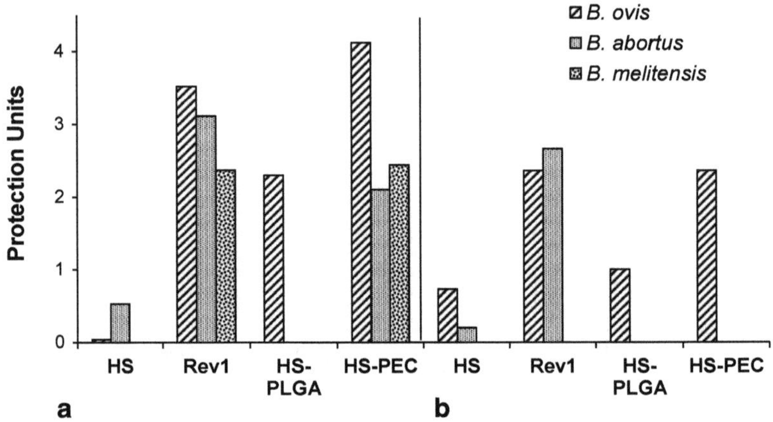

Fig. 17.3 Protection conferred by HS-loaded microparticles against *B. ovis, B. abortus,* and *B. melitensis* infection in Balb/c mice when administered by the **a** subcutaneous or **b** oral routes. Groups of mice were immunized orally or subcutaneously with HS encapsulated in either poly(ε-caprolactone) (HS-PEC) or PLGA (HS-PLGA) microparticles, free HS or 5 × 10^4 CFU/mouse of the Rev 1 reference vaccine strain. Eight weeks after vaccination, mice were challenged intraperitoneally with either 5 × 10^4 of *B. ovis* PA, 5 × 10^4 of *B. abortus* 2308 or 5 × 10^4 of *B. melitensis* H38. After the killing of animals, the spleens were aseptically removed and submitted to bacteriological analyses. Data are expressed as protection units calculated as the difference between the CFU of the virulent strain in samples of control animals (unvaccinated group), after logarithmic conversion, and the CFU of samples from immunized group

primary immunization and, in the case of HS-PLGA, induce a Th2 rather than a Th1 response. Finally, and interestingly, the seroagglutination test (Rose Bengal test, RB test) of animals immunized with HS-loaded microparticles was always negative.

For protection studies, Balb/c mice were firstly vaccinated with Rev 1 (positive control) or microparticles [78, 82]. In all cases, the animals received only one dose. Eight weeks later, the animals were experimentally infected with virulent *B. ovis* PA, *B. abortus* 2308 or *B. melitensis* H38 strains. Two or three weeks later, depending on the strain, the animals were killed and the number of viable counts (colony forming units, CFU) from spleen samples determined. Figure 17.3 summarizes these results. For animals subcutaneously vaccinated and challenged with *B. ovis*, HS-PEC offered the highest protection (PU of 4.1 logs) followed by Rev 1 (PU of 3.5 logs) and HS-PLGA (PU of 2.3 logs). When animals were vaccinated by the oral route, HS-PEC offered a similar degree of protection than the commercial vaccine Rev 1 (PU of about 2.5 logs).

On the other hand, in mice experimentally infected with *B. abortus*, the subcutaneous administration of a single dose of HS-PEC offered a significant degree of protection (Fig. 17.3), in spite of slightly lower to that conferred by Rev 1. In contrast, oral vaccination with HS-PEC did not protect animals against a challenge with *B. abortus*. Finally, concerning the challenge with *B. melitensis*, the subcutaneous administration of a single dose of HS-PEC produced significant protection with respect to unvaccinated control mice (approximately a PU of 2.5 logs). Moreover, this protection was similar to that conferred by the *B. melitensis* Rev 1 reference vaccine.

However, the relevance of mice as models for ruminant brucellosis is uncertain. In consequence, additional research was conducted on rams in order to establish the protective value of this HS-PEC vaccine [47, 58]. For this purpose, 3–4-month-old rams were used to compare the efficacy of the test vaccine (HS-PEC) against the reference Rev 1 vaccine. Each ram was only immunized subcutaneously once with either HS-PEC or Rev 1. Six months later all rams were experimentally infected with *B. ovis* PA. Then, 9 weeks after challenge, all rams were slaughtered and submitted to individual necropsy for bacteriological and pathological examinations.

Importantly, all the sera taken from the animals immunized with Rev 1 were seropositive in both RB and complement fixation (CF) tests since the first week after vaccination, and remaining seropositive all along the experiment. In contrast, no positive reactions in these tests were recorded, at any postvaccination time, when testing the sera from the rams vaccinated with the HS-PEC vaccine.

Regarding the evolution of the IgG-specific antibody response against the outer membrane proteins included in the HS antigenic extract, all immunized animals (including those vaccinated with Rev 1) developed an early and strong positive serologic response. In the 8th week post vaccination, the percentage of reactor animals decreased, respectively to 45 % and 85 %, for the Rev 1 and HS-PEC vaccinated rams. At the time of challenge (week 24 post immunization), only few of these animals remained positive. The challenge with *B. ovis* PA induced a quick anamnestic antibody response to HS antigens in all vaccinated animals that was maintained until the slaughtering. Concerning the production of IFN-γ, high levels of this cytokine were found in the majority of the vaccinated rams, independent of the type of vaccine used. Moreover, these high levels of IFN-γ were maintained at least for 6 months after immunization.

The bacteriological results obtained after the necropsy are summarized in Fig. 17.4. HS-PEC protected 54 % of vaccinated rams, this being similar to the protection induced by Rev 1 (49 %). Interestingly, the percentage of infected samples was significantly lower in the animals vaccinated with HS-PEC than in the animals vaccinated with Rev 1. The subcutaneous inoculation of both HS-PEC and Rev 1 induced variable degrees of local reactivity in the tissues surrounding the injection area. In general, this local reactivity was moderate to low and was resolved in most cases in a few weeks after vaccination. In what concerns to the pathological analysis of the animals found infected at necropsy, the animals vaccinated with the microparticle formulation were in general less affected than Rev 1 vaccinated rams. In fact, three rams vaccinated with Rev 1 were found infected at necropsy, presented important macroscopic and microscopic lesions, mainly located in the epididymides and vaginal layers. In contrast, only one of the rams immunized with HS-PEC presented severe lesions (epididymitis and fibrinous vaginalitis).

17.6.3 Nanoparticles as Vaccine Delivery Systems for Brucellosis

In case of *Brucella*, mucosal immunization may be of particular interest, since it can mimic the bacteria behaviour and generate immunity at the major portals of entry

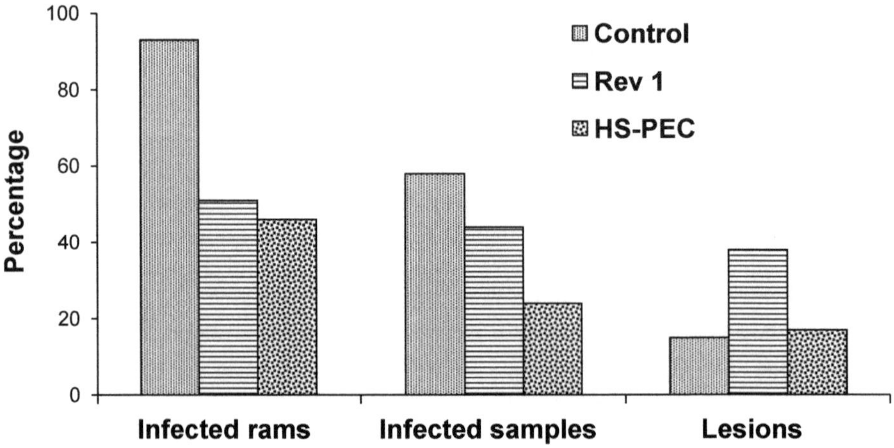

Fig. 17.4 Protective efficacy of the experimental vaccine formulations after challenge with the virulent *B. ovis* PA strain in rams. Rams were vaccinated with HS-loaded poly(ε-caprolactone) microparticles (HS-PEC) or *B. melitensis* Rev 1 vaccine. Each ram was immunized with 1 mL of the different vaccines given in a single dose by subcutaneous route. The dose of HS entrapped in microparticle was 3 mg per individual ram. The individual dose of Rev 1 vaccine was 1.6×10^9 CFU. Six months after vaccination all animals were challenged conjunctivally and preputially with 1.16×10^9 CFU of virulent *B. ovis* PA strain. Eight weeks after challenge, all rams were necropsied and selected organs and lymph nodes were submitted to bacteriological analyses. Infected samples represent the percentage of samples that were found infected during the bacteriological study. Lesions represent the percentage of infected animals at necropsy with important macroscopic and microscopic lesions in the epididymides and vaginal layers

for this pathogen. In addition, this type of administration can also be safer with less adverse effects and facilitates its dispensation and application [21, 100, 101].

The subepithelial regions of mucosal surfaces contain an abundance of immunocompetent cells such as B and T lymphocytes , as well as plasma cells and macrophages [102]. These cells are organized into the mucosal associated lymphoid tissue (MALT), which are the main components of the mucosal immune system [103–105]. Induction of mucosal responses leads to production of secretory IgA antibodies, which are not usually produced by systemic immunization, and represents the major effector mechanism of this lymphoid tissue [106, 107]. Evidence from many studies has confirmed that stimulation of the mucosal immune system at one mucosal site can lead to sIgA production in the local as well as distal mucosal surfaces [102]. This inter-connected mucosal system of sIgA induction and production has been given the name common mucosal immune system [104].

In this context, one interesting mucosal adjuvant would be the nanoparticles from the copolymer of methyl vinyl ether and maleic anhydride (Gantrez® AN). In previous studies, it has been demonstrated that these poly(anhydride) nanoparticles can effectively enhance the immune response when administered by oral route [108–110]. In fact, these nanoparticulate systems exhibit a strong capability to develop bioadhesive interactions within the mucosa and enhance and prolong the delivery of the antigen to the lymphoid cells due to their capture and internalization by the MALT [111–113]. Nevertheless, conventional poly(anhydride) nanoparticles still display a low capacity to target-specific sites [114, 115] and, although constructed by a mucoadhesive polymer, they can be eliminated to some extent by mucus shed off and physiological clearance mechanisms (i.e. peristaltism, lachrymal fluids, etc.). In order to overcome these drawbacks, the association of the nanoparticles with specific ligands, able to specifically bind within MALT components, has been proposed [116–119]. Among other ligands, poly(anhydride) nanoparticles can be easily "surface decorated" with mannosamine [110, 120]. The effectiveness of these nano-mannosylated devices in vaccination can be due to lectins with mannose-binding activity target [121], and to mannose receptors activation highly expressed in cells of the immune systems (i.e. macrophages and dendritic cells) [122–123].

HS-loaded poly(anhydride) nanoparticles were prepared by a solvent displacement method. For this purpose, the HS antigenic extract was mixed in acetone with the copolymer of methyl vinyl ether and maleic anhydride. The nanoparticles (NP-HS) were formed by addition of a hydroalcoholic mixture. Then, the resulting nanoparticles were purified by, first, elimination of the organic solvents and, secondly, by centrifugal filtration. Finally, nanoparticles were freeze-dried [109, 120].

For the preparation of mannosylated nanoparticles (MAN-NP-HS), the first step was the incubation between mannosamine and poly(anhydride) in acetone overnight at room temperature. Then, the HS was added to the mixture and after incubation, the nanoparticles were formed, purified and dried as described before [120, 124].

Overall all the nanoparticles displayed sizes in the 200–300 nm range with a quite narrow distribution. Nevertheless, mannosylated nanoparticles displayed a significantly higher size than conventional nanoparticles. Electron microscopy observations revealed nanoparticles with a spherical shape and seemingly a smooth surface, without significant differences between mannosylated and naked HS-loaded nanoparticles (Fig. 17.5a). The amount of mannosamine associated to the poly(anhydride) nanoparticles was estimated to be about 30 μg/mg nanoparticle and its localization at the surface of nanoparticles was revealed by an agglutination assay with concanavalin A (Fig. 17.5b).

The amount of HS associated to nanoparticles was found to be independent on the nanoparticle type. In fact, for conventional and mannosylated nanoparticles, HS loading was calculated to be, respectively, 28 μg and 35 μg HS per mg of nanoparticle. Interestingly, SDS-PAGE and western blot analysis revealed that the protein profile, structural integrity and antigenicity of the entrapped antigenic HS proteins were maintained. For both nanoparticle formulations, the protein profile of the HS-loaded in nanoparticles was similar to the composition of free HS. In addition, the encapsulated proteins maintained the same reactivity against a pool of sera from experimentally infected rabbits with *B. ovis*. Regarding the in vitro release kinetics of HS from nanoparticles, both releasing profiles followed a similar tendency, characterized by a biphasic release pattern in two steps: a burst effect followed by a continuous HS release for 30 days [120]. Nevertheless, mannosylated nanoparticles displayed an initial more rapid release than conventional ones (at 24 h, 26 % vs. 18 %). However, NP-HS induced a higher releasing 3 days after the beginning of the study. Then both profiles were found to be similar and sustained along time,

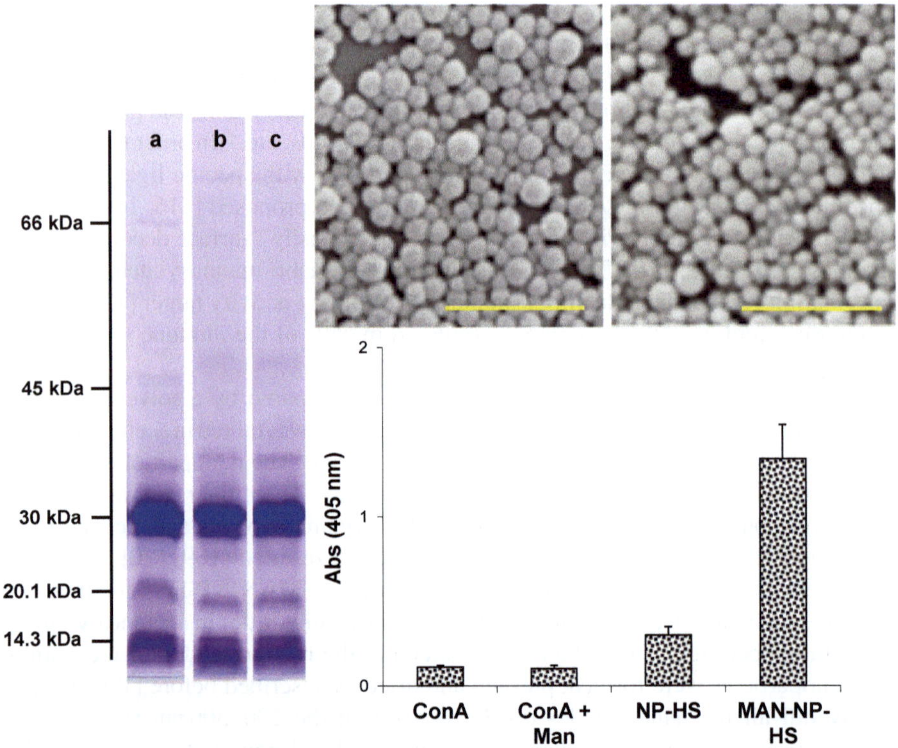

Fig. 17.5 Characterization of HS-loaded poly(anhydride) nanoparticles: **a** morphology by SEM, **b** agglutination assay of mannosylated nanoparticles with concanavalin A in order to confirm the localization of mannose residues at the surface, and **c** integrity study of HS released from nanoparticles by SDS-PAGE. **a** SEM microphotographs of *a* NP-HS and *b* MAN-NP-HS. **b** Turbidity change after incubation of nanoparticles with 50 μg Con A. A mixture of mannosamine and concanavalin A was used as a control. **c** SDS-PAGE of released HS during the in vitro release study

and at the end of the study conventional nanoparticles released about 60 % of the loaded HS whereas mannosylated nanoparticles about 65 %. In any case, SDS-PAGE (Fig. 17.5c) and western blot revealed that the HS released maintained its stability and antigenicity during the study.

The stability of poly(anhydride) nanoparticles in both lachrymal and nasal ovine fluids from rams was also evaluated [120]. After 2 h of incubation in these fluids, all nanoparticle formulations demonstrated high stability, since more than 75 % of nanoparticles maintained their integrity and size. However, both types of nanoparticles (conventional and mannosylated nanoparticles) displayed a slightly higher stability in nasal secretions than in lachrymal fluids. The slightly lower stability in the lachrymal than in the nasal fluids would be probably due to the fact that nasal secretions present higher viscosity and acidity, which hindered water access to the surface of the nanoparticles and further hydrolysis of the polymer or surface erosion of the nanoparticles.

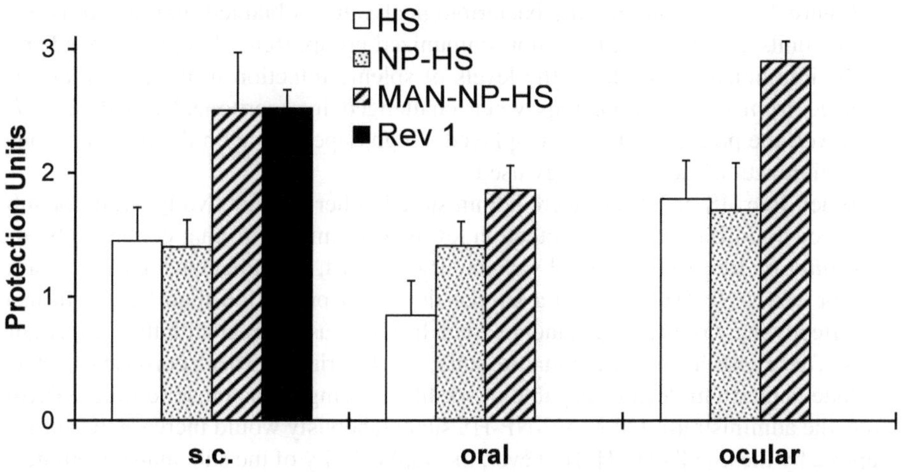

Fig. 17.6 Protective efficacy of HS-based vaccines in mice as a function of the route of administration. Mice were vaccinated with free hot saline antigenic extract (HS), HS-loaded conventional nanoparticles (NP-HS), HS-loaded mannosylated nanoparticles (MAN-NP-HS) in accordance with the following schedule: conjunctival (3 μg of HS in each eye administered with a total dose of 12 μg HS) (*right*), oral (one single dose 100 μg HS) (*middle*), or subcutaneous (one single dose of 20 μg HS) (*left*). As positive control, one group of animals was subcutaneously vaccinated with the *B. melitensis* Rev 1 vaccine (5×10^5 CFU/mouse in 100 μL buffered saline solution). Eight weeks after vaccination, mice were challenged intraperitoneally with 5×10^4 CFU/mouse of the virulent *B. ovis* PA reference strain. Three weeks after infection, animals were killed and the spleens were aseptically removed and submitted to bacteriological analysis. Data are expressed as protection units calculated as the difference between the CFU of the virulent strain in samples of control animals (unvaccinated group), after logarithmic conversion, and the CFU of samples from immunized group

Evaluation of Nanoparticles A study of the protective efficacy against the *B. ovis* PA strain was conducted in mice. Animals were vaccinated with HS, NP-HS or MAN-NP-HS, administered by one of the following routes: (1) conjunctival as eye drops, (2) oral or (3) subcutaneous. In the experiment, a group of animals immunized subcutaneously with the commercial Rev 1 vaccine was also included as positive control. Eight weeks after vaccination, mice were challenged intraperitoneally with the virulent *B. ovis* PA reference strain. Three weeks after infection, animals were killed by cervical dislocation, and the spleens were aseptically removed in order to determine the number of viable CFU of the pathogen.

Surprisingly, only strong antibody responses against HS in serum were found when nanoparticles were administered subcutaneously. When nanoparticles or HS were administered by either the oral or the conjunctival route of administration, the anti-HS IgG1 and IgG2a HS levels were of the same order, but low. Nevertheless, for MAN-NP-HS, the elicited antibody levels were significantly higher than for NP-HS. Interestingly, the strongest IgA response was observed when mannosylated nanoparticles were conjunctivally administered. Concerning the levels of cytokines secreted from splenic cells, animals vaccinated with Rev 1 or mannosylated nanoparticles expressed higher levels of IFN-γ and IL-4.

Figure 17.6 summarizes the bacteriological values obtained, expressed as protection units compared with the non-immunized group. Remarkably, independently of the immunization strategy, the levels of splenic infection in mice were always reduced from one to three logs when challenged intraperitoneally with *B. ovis*. However, the protection results displayed a high dependence on the binomial route of immunization/vaccine strategy used.

When free HS or NP-HS were administered either conjunctivally, orally or subcutaneously in Balb/c mice, protection levels were inferior to that conferred by the *B. melitensis* Rev 1 commercial vaccine. By contrast, the subcutaneous administration of MAN-NP-HS provided a similar degree of protection than Rev 1. In fact, the effectiveness of the subcutaneous route has been associated with the Langerhans cells. These cells have, in their mature state, characteristic Birbeck corpuscles, which include langerin molecules capable of specific binding to mannose residues. Therefore, the administration of MAN-NP-HS subcutaneously would increase the antigen capture by these cells [114]. However, the applicability of the subcutaneous route in veterinary mass vaccination campaigns is limited due to difficulties associated with its administration (which also increases the costs). WHO also warns about an "ideal vaccine" giving guidelines which include stark terms such as security, stability and efficiency, but also mass vaccination without needles [125].

Concerning the conjunctival administration of HS formulations, both HS and NP-HS immunized animals displayed a higher degree of protection than when vaccinated by the oral route and similar efficacy than when administered by the subcutaneous route. Nevertheless, the conjunctival administration of MAN-NP-HS yielded the highest significant protection against infection when compared to that conferred by the reference vaccine.

This response was consistent with a biodistribution study in which mannosylated nanoparticles were labelled with either technetium or rhodamine B isothiocyanate. This biodistribution study revealed that, after instillation, nanoparticles adhered to the ocular mucosa and were further distributed to the nasal mucosa, stomach and gut [124]. Thus, it is possible to hypothesize that the effective and protective value mannosylated nanoparticles when administered conjunctivally would be due to a simultaneous activation of the APCs localized at the conjunctiva associated lymphoid tissue (CALT), nasal lymphoid tissue (NALT) and gut associated lymphoid tissue (GALT). All of these epithelial surfaces contain specialized antigen-sampling cells (i.e. M cells) capable to transport antigens from the mucosal surfaces into the underlying lymphoid tissues [106] which, are rapidly internalized and processed by APCs and presented to B cells and T cells [126]. This idea is supported by the fact that MAN-NP-HS elicited the highest IgA raising and triggered strong local and distant mucosal immune responses. In agreement, Chentoufi and collaborators reported that the ocular administration of a mixture of peptides from the glycoprotein D of herpes simplex virus type 1 and CpG motifs were capable of inducing local (in conjunctiva) as well as systemic (in spleen) specific immune responses [127]. Interestingly, the surgical closure of the nasolacrimal duct did not significantly alter local ocular mucosal responses but did significantly enhance systemic responses.

Again, Hu et al. also developed an ocular mucosal administrated nanoparticulate vaccine containing DNA that conferred strong specific immune responses and effective inhibition of HSK in a HSV-1 infected murine model [128].

17.7 Conclusions and Future Perspectives

Vaccines represent one of the most successful prophylactic strategies in medical sciences. From a mechanistic perspective, vaccination manipulates the immune response through the selection, activation and expansion the memory of B and T cells. In order to determine the magnitude and quality of immune response, suitable vaccine adjuvants are required and, therefore, much effort is now focused into the finding of new, effective and non-toxic adjuvant formulations centred on the activation of key immune targets for inducing a long-term, potent and safe immune response.

Currently, vaccine delivery aims to develop innovative delivery systems, such as polymeric nanoparticles, for the delivery of acellular vaccines through the needle-free administration routes (such as ocular, oral or intranasal delivery). Furthermore, a successful mucosal immunization for antigen delivery requires a rational design of nanoparticles to facilitate their travel under physiological conditions and barriers.

To design nanoparticle vaccines that accomplish the already-assigned characteristics, several key elements are required, such as: (1) the prepared systems should be appropriately characterized, since properties (i.e. size, degradation rate and antigen release profile, surface charge) may influence the immune response, (2) the preparative process can affect the integrity and antigenicity of the loaded antigen; (3) the presence of ligands might modify the behaviour/biodistribution in the body affecting the elicited immune response after administration.

Focusing on animal health, particulate vaccine delivery systems are well suited for veterinary and wildlife vaccine strategies. Indeed, they are often applicable to a large range of species as they do not rely on specific ways of activating the immune system but rather on basic characteristics of the mostly innate immune responses. Notably, in some cases adaptations may be required due to anatomical differences. The use of particulate adjuvants also allows for additional routes of immunization, which are better suited to veterinary and wildlife species including oral delivery and long-distance ballistic intramuscular delivery. Thus, after oral administration, particulate delivery systems can increase retention of antigens on mucosal surface, protect them from proteases in the mucus gels, and then, increase antigen uptake and immune responses. In addition, these systems protect the antigen from degradation and impact during penetration through the skin and muscle.

Another important aspect to consider is that vaccine targets are also changing, with non-infectious disease targets representing a considerable emergent area. These include control of fertility, behaviour and production, by immunization against hormone or hormone receptors.

As a result of this current situation, innovative solutions have been developed to meet many of these challenges leading to new approaches to veterinary vaccine development, which may have significant advantages over more traditional approaches.

There are still many unsolved questions when the subjects vaccine and veterinary merge together.

Acknowledgements We deeply acknowledge the support received from "Instituto de Salud Carlos III" (PI12/01358), from Spain.

References

1. Shryock TR. The future of anti-infective products in animal health. Nat Rev Microbiol. 2004;2(5):425–30.
2. Hopkins DR. The greatest killer: smallpox in history. Chicago: University of Chicago Press; 2002.
3. Christie RJ, Findley DJ, Dunfee M, Hansen RD, Olsen SC, Grainger DW. Photopolymerized hydrogel carriers for live vaccine ballistic delivery. Vaccine. 2006;24(9):1462–9.
4. Olsen SC, Christie RJ, Grainger DW, Stoffregen WS. Immunologic responses of bison to vaccination with *Brucella abortus* strain RB51: comparison of parenteral to ballistic delivery via compressed pellets or photopolymerized hydrogels. Vaccine. 2006;24(9):1346–53.
5. Ramon G. Sur l'augmentation anormale de l'antitoxine chez les chevaux producteurs de serum antidiphterique. Bull Soc Centr Med Vet. 1925;101:227–34.
6. Ramon G. Procédures pour accroître la production des antitoxines. Ann Inst Pasteur. 1926;40:1–10.
7. Janeway CA, Jr. Approaching the asymptote? Evolution and revolution in immunology. Cold Spring Harb Symp Quant Biol. 1989;54(1):1–13.
8. Li H, Willingham SB, Ting JP, Re F. Cutting edge: inflammasome activation by alum and alum's adjuvant effect are mediated by NLRP3. J Immunol. 2008;181(1):17–21.
9. Gupta RK, Rost BE, Relyveld E, Siber GR. Adjuvant properties of aluminum and calcium compounds. Pharm Biotechnol. 1995;6:229–48.
10. Kawai T, Akira S. The roles of TLRs, RLRs and NLRs in pathogen recognition. Int Immunol. 2009;21(4):317–37.
11. Takeuchi O, Akira S. Pattern recognition receptors and inflammation. Cell. 2010;140(6): 805–20.
12. Tamayo I, Irache JM, Mansilla C, Ochoa-Repáraz J, Lasarte JJ, Gamazo C. Poly(anhydride) nanoparticles act as active Th1 adjuvants through Toll-like receptor exploitation. Clin Vaccine Immunol. 2010;17(9):1356–62.
13. Camacho AI, Da Costa Martins R, Tamayo I, de Souza J, Lasarte JJ, Mansilla C, Esparza I, Irache JM, Gamazo C. Poly(methyl vinyl ether-co-maleic anhydride) nanoparticles as innate immune system activators. Vaccine. 2011;29(41):7130–5.
14. Thiele L, Rothen-Rutishauser B, Jilek S, Wunderli-Allenspach H, Merkle HP, Walter E. Evaluation of particle uptake in human blood monocyte-derived cells in vitro. Does phagocytosis activity of dendritic cells measure up with macrophages? J Control Release. 2001;76(1–2):59–71.
15. Audran R, Peter K, Dannull J, Men Y, Scandella E, Groettrup M, Gander B, Corradin G. Encapsulation of peptides in biodegradable microspheres prolongs their MHC class-I presentation by dendritic cells and macrophages in vitro. Vaccine. 2003;21(11–12):1250–5.
16. De Koker S, Lambrecht BN, Willart MA, van Kooyk Y, Grooten J, Vervaet C, Remon JP, De Geest BG. Designing polymeric particles for antigen delivery. Chem Soc Rev. 2011;40(1):320–39.
17. De Temmerman ML, Rejman J, Demeester J, Irvine DJ, Gander B, De Smedt SC. Particulate vaccines: on the quest for optimal delivery and immune response. Drug Discov Today. 2011;16(13–14):569–82.

18. Jain S, O'Hagan DT, Singh M. The long-term potential of biodegradable poly(lactide-co-glycolide) microparticles as the next-generation vaccine adjuvant. Expert Rev Vaccines. 2011;10(12):1731–42.
19. Burgdorf S, Kurts C. Endocytosis mechanisms and the cell biology of antigen presentation. Curr Opin Immunol. 2008;20(1):89–95.
20. Shen H, Ackerman AL, Cody V, Giodini A, Hinson ER, Cresswell P, Edelson RL, Saltzman WM, Hanlon DJ. Enhanced and prolonged cross-presentation following endosomal escape of exogenous antigens encapsulated in biodegradable nanoparticles. Immunology. 2006;117(1):78–88.
21. Chadwick S, Kriegel C, Amiji M. Nanotechnology solutions for mucosal immunization. Adv Drug Deliv Rev. 2010;62(4–5):394–407.
22. Ahmed R, Gray D. Immunological memory and protective immunity: understanding their relation. Science. 1996;272(5258):54–60.
23. Cox E, Verdonck F, Vanrompay D, Goddeeris B. Adjuvants modulating mucosal immune responses or directing systemic responses towards the mucosa. Vet Res. 2006;37(3):511–39.
24. Demento SL, Cui W, Criscione JM, Stern E, Tulipan J, Kaech SM, Fahmy TM. Role of sustained antigen release from nanoparticle vaccines in shaping the T cell memory phenotype. Biomaterials. 2012;33(19):4957–64.
25. Malyala P, Chesko J, Ugozzoli M, Goodsell A, Zhou F, Vajdy M, O'Hagan DT, Singh M. The potency of the adjuvant, CpG oligos, is enhanced by encapsulation in PLG microparticles. J Pharm Sci. 2008;97(3):1155–64.
26. Bal SM, Slutter B, Verheul R, Bouwstra JA, Jiskoot W. Adjuvanted, antigen loaded N-trimethyl chitosan nanoparticles for nasal and intradermal vaccination: adjuvant- and site-dependent immunogenicity in mice. Eur J Pharm Sci. 2012;45(4):475–81.
27. Afrin F, Rajesh R, Anam K, Gopinath M, Pal S, Ali N. Characterization of *Leishmania donovani* antigens encapsulated in liposomes that induce protective immunity in BALB/c mice. Infect Immun. 2002;70(12):6697–706.
28. Hall MA, Stroop SD, Hu MC, Walls MA, Reddish MA, Burt DS, Lowell GH, Dale JB. Intranasal immunization with multivalent group A streptococcal vaccines protects mice against intranasal challenge infections. Infect Immun. 2004;72(5):2507–12.
29. Prasad S, Cody V, Saucier-Sawyer JK, Fadel TR, Edelson RL, Birchall MA, Hanlon DJ. Optimization of stability, encapsulation, release, and cross-priming of tumor antigen-containing PLGA nanoparticles. Pharm Res. 2012;29(9):2565–77.
30. des Rieux A, Fievez V, Garinot M, Schneider YJ, Preat V. Nanoparticles as potential oral delivery systems of proteins and vaccines: a mechanistic approach. J Control Release. 2006;116(1):1–27.
31. Mundargi RC, Babu VR, Rangaswamy V, Patel P, Aminabhavi TM. Nano/micro technologies for delivering macromolecular therapeutics using poly(D, L-lactide-co-glycolide) and its derivatives. J Control Release. 2008;125(3):193–209.
32. Katare YK, Muthukumaran T, Panda AK. Influence of particle size, antigen load, dose and additional adjuvant on the immune response from antigen loaded PLA microparticles. Int J Pharm. 2005;301(1–2):149–60.
33. Wendorf J, Singh M, Chesko J, Kazzaz J, Soewanan E, Ugozzoli M, O'Hagan D. A practical approach to the use of nanoparticles for vaccine delivery. J Pharm Sci. 2006;95(12):2738–50.
34. Scheerlinck JP, Greenwood DL. Particulate delivery systems for animal vaccines. Methods. 2006;40(1):118–24.
35. Moreno E, Cloeckaert A, Moriyon I. *Brucella* evolution and taxonomy. Vet Microbiol. 2002;90(1–4):209–27.
36. Garin-Bastuji B, Blasco JM, Grayon M, Verger JM. *Brucella melitensis* infection in sheep: present and future. Vet Res. 1998;29(3–4):255–74.
37. Seleem MN, Boyle SM, Sriranganathan N. Brucellosis: a re-emerging zoonosis. Vet Microbiol. 2010;140(3–4):392–8.
38. Young EJ. An overview of human brucellosis. Clin Infect Dis. 1995;21(2):283–9, quiz 290.

39. Doganay GD, Doganay M. *Brucella* as a potential agent of bioterrorism. Recent Pat Antiinfect Drug Discov. 2013;8(1):27–33.
40. Zúñiga Estrada A, Mota de la Garza L, Sánchez Mendoza M, Santos López EM, Filardo Kerstupp S, López Merino A. Survival of *Brucella abortus* in milk fermented with a yoghurt starter culture. Rev Latinoam Microbiol. 2005;47(3–4):88–91.
41. Magwedere K, Bishi A, Tjipura-Zaire G, Eberle G, Hemberger Y, Hoffman LC, Dziva F. Brucellae through the food chain: the role of sheep, goats and springbok (*Antidorcus marsupialis*) as sources of human infections in Namibia. J S Afr Vet Assoc. 2011;82(4):205–12.
42. Pappas G, Papadimitriou P, Akritidis N, Christou L, Tsianos EV. The new global map of human brucellosis. Lancet Infect Dis. 2006;6(2):91–9.
43. Donev DM. Brucellosis as priority public health challenge in South Eastern European countries. Croat Med J. 2010;51(4):283–4.
44. Buzgan T, Karahocagil MK, Irmak H, Baran AI, Karsen H, Evirgen O, Akdeniz H. Clinical manifestations and complications in 1028 cases of brucellosis: a retrospective evaluation and review of the literature. Int J Infect Dis. 2010;14(6):e469–78.
45. Megid J, Mathias LA, Robles CA. Clinical manifestations of brucellosis in domestic animals and humans. Open Vet Sci J. 2010;4:119–26.
46. Andriopoulos P, Tsironi M, Deftereos S, Aessopos A, Assimakopoulos G. Acute brucellosis: presentation, diagnosis, and treatment of 144 cases. Int J Infect Dis. 2007;11(1):52–7.
47. Muñoz PM, de Miguel MJ, Grilló MJ, Marín CM, Barberán M, Blasco JM. Immunopathological responses and kinetics of Brucella melitensis Rev 1 infection after subcutaneous or conjunctival vaccination in rams. Vaccine. 2008;26(21):2562–9.
48. Hoover DL, Nikolich MP, Izadjoo MJ, Borschel RH, Bhattacharjee AK. Development of new *Brucella* vaccines by molecular methods. In: Lopez-Goi I, Moriyón I, editors. Brucella: molecular and cellular biology. Norfolk: Horizon Bioscience; 2004. pp. 362–92.
49. Blasco JM. A review of the use of *B. melitensis* Rev 1 vaccine in adult sheep and goats. Prev Vet Med. 1997;31(3–4):275–83.
50. Schurig GG, Sriranganathan N, Corbel MJ. Brucellosis vaccines: past, present and future. Vet Microbiol. 2002;90(1–4):479–96.
51. Cutler S, Whatmore A. Progress in understanding brucellosis. Vet Rec. 2003;153(21):641–2.
52. Moriyón I, Grillo MJ, Monreal D, Gonzalez D, Marin C, Lopez-Goni I, Mainar-Jaime RC, Moreno E, Blasco JM. Rough vaccines in animal brucellosis: structural and genetic basis and present status. Vet Res. 2004;35(1):1–38.
53. Corbell MJ. Brucellosis in humans and animals. Geneva: WHO Press; 2006.
54. Bascoul S, Cannat A, Huguet MF, Serre A. Studies on the immune protection to murine experimental brucellosis conferred by *Brucella* fractions. I. Positive role of immune serum. Immunology. 1978;35(2):213–21.
55. Escande A, Serre A. IgE anti-brucella antibodies in the course of human brucellosis and after specific vaccination. Int Arch Allergy Appl Immunol. 1982;68(2):172–5.
56. Van De Verg LL, Hartman AB, Bhattacharjee AK, Tall BD, Yuan L, Sasala K, Hadfield TL, Zollinger WD, Hoover DL, Warren RL. Outer membrane protein of *Neisseria meningitidis* as a mucosal adjuvant for lipopolysaccharide of *Brucella melitensis* in mouse and guinea pig intranasal immunization models. Infect Immun. 1996;64(12):5263–8.
57. He Y, Xiang Z. Bioinformatics analysis of Brucella vaccines and vaccine targets using VIOLIN. Immunome Res. 2010;6 Suppl 1:S5.
58. Da Costa Martins R, Irache JM, Blasco JM, Munoz MP, Marin CM, Jesus Grillo M, Jesus De Miguel M, Barberan M, Gamazo C. Evaluation of particulate acellular vaccines against Brucella ovis infection in rams. Vaccine. 2010;28(17):3038–46.
59. Edmonds MD, Cloeckaert A, Elzer PH. *Brucella* species lacking the major outer membrane protein Omp25 are attenuated in mice and protect against *Brucella melitensis* and *Brucella ovis*. Vet Microbiol. 2002;88(3):205–21.
60. Lopez-Goni I, Guzman-Verri C, Manterola L, Sola-Landa A, Moriyon I, Moreno E. Regulation of *Brucella* virulence by the two-component system BvrR/BvrS. Vet Microbiol. 2002;90(1–4):329–39.

61. Estein SM, Cassataro J, Vizcaino N, Zygmunt MS, Cloeckaert A, Bowden RA. The recombinant Omp31 from *Brucella melitensis* alone or associated with rough lipopolysaccharide induces protection against *Brucella ovis* infection in BALB/c mice. Microbes Infect. 2003;5(2):85–93.
62. Lapaque N, Moriyon I, Moreno E, Gorvel JP. *Brucella* lipopolysaccharide acts as a virulence factor. Curr Opin Microbiol. 2005;8(1):60–6.
63. Gamazo C, Winter AJ, Moriyon I, Riezu-Boj JI, Blasco JM, Diaz R. Comparative analyses of proteins extracted by hot saline or released spontaneously into outer membrane blebs from field strains of *Brucella ovis* and *Brucella melitensis*. Infect Immun. 1989;57(5):1419–26.
64. Vizcaíno N, Cloeckaert A, Zygmunt MS, Dubray G. Cloning, nucleotide sequence, and expression of the *Brucella melitensis* omp31 gene coding for an immunogenic major outer membrane protein. Infect Immun. 1996;64(9):3744–51.
65. Tibor A, Decelle B, Letesson JJ. Outer membrane proteins Omp10, Omp16, and Omp19 of *Brucella* spp. are lipoproteins. Infect Immun. 1999;67(9):4960–2.
66. Riezu-Boj JI, Moriyón I, Blasco JM, Gamazo C, Díaz R. Antibody response to *Brucella ovis* outer membrane proteins in ovine. Infect Immun. 1990;58(2):489–94.
67. Blasco JM, Gamazo C, Winter AJ, Jimenez de Bagues MP, Marin C, Barberan M, Moriyon I, Alonso-Urmeneta B, Diaz R. Evaluation of whole cell and subcellular vaccines against *Brucella ovis* in rams. Vet Immunol Immunopathol. 1993;37(3–4):257–70.
68. Mallapragada SK, Narasimhan B. Immunomodulatory biomaterials. Int J Pharm. 2008;364(2):265–71.
69. Ungaro F, d'Angelo I, Miro A, La Rotonda MI, Quaglia F. Engineered PLGA nano- and micro-carriers for pulmonary delivery: challenges and promises. J Pharm Pharmacol. 2012;64(9):1217–35.
70. Csaba N, Garcia-Fuentes M, Alonso MJ. Nanoparticles for nasal vaccination. Adv Drug Deliv Rev. 2009;61(2):140–57.
71. Danhier F, Ansorena E, Silva JM, Coco R, Le Breton A, Preat V. PLGA-based nanoparticles: an overview of biomedical applications. J Control Release. 2012;161(2):505–22.
72. Jameela SR, Suma N, Misra A, Raghuvanshi R, Ganga S, Jayakrishnan A. Poly(epsilon-caprolactone) microspheres as a vaccine carrier. Curr Sci. 1996;70(7):669–71.
73. Irache JM, Esparza I, Gamazo C, Agueros M, Espuelas S. Nanomedicine: novel approaches in human and veterinary therapeutics. Vet Parasitol. 2011;180(1–2):47–71.
74. Kumari A, Yadav SK, Yadav SC. Biodegradable polymeric nanoparticles based drug delivery systems. Colloids Surf B Biointerfaces. 2010;75(1):1–18.
75. Sinha VR, Bansal K, Kaushik R, Kumria R, Trehan A. Poly-epsilon-caprolactone microspheres and nanospheres: an overview. Int J Pharm. 2004;278(1):1–23.
76. Florindo HF, Pandit S, Lacerda L, Goncalves LM, Alpar HO, Almeida AJ. The enhancement of the immune response against *S. equi* antigens through the intranasal administration of poly-epsilon-caprolactone-based nanoparticles. Biomaterials. 2009;30(5):879–91.
77. Ogawa Y, Yamamoto M, Okada H, Yashiki T, Shimamoto T. A new technique to efficiently entrap leuprolide acetate into microcapsules of polylactic acid or copoly(lactic/glycolic) acid. Chem Pharm Bull (Tokyo). 1988;36(3):1095–103.
78. Murillo M, Grilló MJ, Reñé J, Marín CM, Barberán M, Goñi MM, Blasco JM, Irache JM, Gamazo C. A *Brucella ovis* antigenic complex bearing poly-epsilon-caprolactone microparticles confer protection against experimental brucellosis in mice. Vaccine. 2001;19(30):4099–106.
79. Xu FH, Zhang Q. Recent advances in the preparation progress of protein/peptide drug loaded PLA/PLGA microspheres. Yao Xue Xue Bao. 2007;42(1):1–7.
80. Murillo M, Gamazo C, Irache JM, Goi MM. Polyester microparticles as a vaccine delivery system for brucellosis: influence of the polymer on release, phagocytosis and toxicity. J Drug Target. 2002;10(3):211–9.
81. del Barrio GG, Novo FJ, Irache JM. Loading of plasmid DNA into PLGA microparticles using TROMS (Total Recirculation One-Machine System): evaluation of its integrity and controlled release properties. J Control Release. 2003;86:123–30.

82. Estevan M, Gamazo C, Grillo MJ, Del Barrio GG, Blasco JM, Irache JM. Experiments on a sub-unit vaccine encapsulated in microparticles and its efficacy against *Brucella melitensis* in mice. Vaccine. 2006;24(19):4179–87.
83. McNeela EA, Lavelle EC. Recent advances in microparticle and nanoparticle delivery vehicles for mucosal vaccination. Curr Top Microbiol Immunol. 2012;354:75–99.
84. Estevan M, Gamazo C, Gonzalez-Gaitano G, Irache JM. Optimization of the entrapment of bacterial cell envelope extracts into microparticles for vaccine delivery. J Microencapsul. 2006;23(2):169–81.
85. Murillo M, Goñi MM, Irache JM, Arangoa MA, Blasco JM, Gamazo C. Modulation of the cellular immune response after oral or subcutaneous immunization with microparticles containing *Brucella ovis* antigens. J Control Release. 2002;85(1–3):237–46.
86. Sah H. Stabilization of proteins against methylene chloride/water interface-induced denaturation and aggregation. J Control Release. 1999;58(2):143–51.
87. Varca GH, Andréo-Filho N, Lopes PS, Ferraz HG. Cyclodextrins: an overview of the complexation of pharmaceutical proteins. Curr Protein Pept Sci. 2010;11(4):255–63.
88. Serno T, Geidobler R, Winter G. Protein stabilization by cyclodextrins in the liquid and dried state. Adv Drug Deliv Rev. 2011;63(13):1086–106.
89. Duchêne D, Ponchel G, Wouessidjewe D. Cyclodextrins in targeting. Application to nanoparticles. Adv Drug Deliv Rev. 1999;36(1):29–40.
90. Calleja P, Huarte J, Agueros M, Ruiz-Gaton L, Espuelas S, Irache JM. Molecular buckets: cyclodextrins for oral cancer therapy. Ther Deliv. 2012;3(1):43–57.
91. Duchêne D, Bochot A, Yu SC, Pepin C, Seiller M. Cyclodextrins and emulsions. Int J Pharm. 2003;266(1–2):85–90.
92. Murillo M, Irache JM, Estevan M, Goi MM, Blasco JM, Gamazo C. Influence of the co-encapsulation of different excipients on the properties of polyester microparticle-based vaccine against brucellosis. Int J Pharm. 2004;271(1–2):125–35.
93. Ahsan F, Rivas IP, Khan MA, Torres Suarez AI. Targeting to macrophages: role of physicochemical properties of particulate carriers–liposomes and microspheres–on the phagocytosis by macrophages. J Control Release. 2002;79(1–3):29–40.
94. Yoshida M, Babensee JE. Molecular aspects of microparticle phagocytosis by dendritic cells. J Biomater Sci Polym Ed. 2006;17(8):893–907.
95. Champion JA, Walker A, Mitragotri S. Role of particle size in phagocytosis of polymeric microspheres. Pharm Res. 2008;25(8):1815–21.
96. Artursson P, Arro E, Edman P, Ericsson JL, Sjoholm I. Biodegradable microspheres. V: Stimulation of macrophages with microparticles made of various polysaccharides. J Pharm Sci. 1987;76(2):127–33.
97. Yadav AB, Muttil P, Singh AK, Verma RK, Mohan M, Agrawal AK, Verma AS, Sinha SK, Misra A. Microparticles induce variable levels of activation in macrophages infected with *Mycobacterium tuberculosis*. Tuberculosis (Edinb). 2010;90(3):188–96.
98. Lopez-Urrutia L, Alonso A, Nieto ML, Bayon Y, Orduna A, Sanchez Crespo M. Lipopolysaccharides of *Brucella abortus* and *Brucella melitensis* induce nitric oxide synthesis in rat peritoneal macrophages. Infect Immun. 2000;68(3):1740–5.
99. Chakravortty D, Hensel M. Inducible nitric oxide synthase and control of intracellular bacterial pathogens. Microbes Infect. 2003;5(7):621–7.
100. Chen H. Recent advances in mucosal vaccine development. J Control Release. 2000;67(2–3):117–28.
101. Dietrich G, Griot-Wenk M, Metcalfe IC, Lang AB, Viret JF. Experience with registered mucosal vaccines. Vaccine. 2003;21(7–8):678–83.
102. Kaul D, Ogra PL. Mucosal responses to parenteral and mucosal vaccines. Dev Biol Stand. 1998;95:141–6.
103. Croitoru K, Bienenstock J. Characteristics and functions of mucosa-associated lymphoid tissue. In: Ogra PL, et al., editors. Handbook of mucosal immunology. San Diego: Academic Press; 1994. pp. 141–9.

104. Mestecky J, Michalek SM, Moldoveanu Z, Russell MW. Routes of immunization and antigen delivery systems for optimal mucosal immune responses in humans. Behring Inst Mitt. 1997;98:33–43.
105. Cesta MF. Normal structure, function, and histology of mucosa-associated lymphoid tissue. Toxicol Pathol. 2006;34(5):599–608.
106. Nugent J, Po AL, Scott EM. Design and delivery of non-parenteral vaccines. J Clin Pharm Ther. 1998;23(4):257–85.
107. Corthésy B. Roundtrip ticket for secretory IgA: role in mucosal homeostasis? J Immunol. 2007;178(1):27–32.
108. Salman HH, Gamazo C, Campanero MA, Irache JM. Salmonella-like bioadhesive nanoparticles. J Control Release. 2005;106(1–2):1–13.
109. Estevan M, Irache JM, Grillo MJ, Blasco JM, Gamazo C. Encapsulation of antigenic extracts of *Salmonella enterica* serovar. Abortusovis into polymeric systems and efficacy as vaccines in mice. Vet Microbiol. 2006;118(1–2):124–32.
110. Salman HH, Gamazo C, Campanero MA, Irache JM. Bioadhesive mannosylated nanoparticles for oral drug delivery. J Nanosci Nanotechnol. 2006;6(9–10):3203–9.
111. Almeida AJ, Alpar HO. Nasal delivery of vaccines. J Drug Target. 1996;3(6)455–67.
112. Gomez S, Gamazo C, Roman BS, Ferrer M, Sanz ML, Irache JM. Gantrez AN nanoparticles as an adjuvant for oral immunotherapy with allergens. Vaccine. 2007;25(29):5263–71.
113. Motwani SK, Chopra S, Talegaonkar S, Kohli K, Ahmad FJ, Khar RK. Chitosan-sodium alginate nanoparticles as submicroscopic reservoirs for ocular delivery: formulation, optimisation and in vitro characterisation. Eur J Pharm Biopharm. 2008;68(3):513–25.
114. Irache JM, Salman HH, Gamazo C, Espuelas S. Mannose-targeted systems for the delivery of therapeutics. Expert Opin Drug Deliv. 2008;5(6):703–24.
115. Irache JM, Salman HH, Gomez S, Espuelas S, Gamazo C. Poly(anhydride) nanoparticles as adjuvants for mucosal vaccination. Front Biosci (Schol Ed). 2010;2:876–90.
116. Frey A, Neutra MR. Targeting of mucosal vaccines to Peyer's patch M cells. Behring Inst Mitt. 1997;98:376–89.
117. Salman HH, Gamazo C, Agueros M, Irache JM. Bioadhesive capacity and immunoadjuvant properties of thiamine-coated nanoparticles. Vaccine. 2007;25(48):8123–32.
118. Salman HH, Gamazo C, de Smidt PC, Russell-Jones G, Irache JM. Evaluation of bioadhesive capacity and immunoadjuvant properties of vitamin B(12)-Gantrez nanoparticles. Pharm Res. 2008;25(12):2859–68.
119. Rieger J, Freichels H, Imberty A, Putaux JL, Delair T, Jerome C, Auzely-Velty R. Polyester nanoparticles presenting mannose residues: toward the development of new vaccine delivery systems combining biodegradability and targeting properties. Biomacromolecules. 2009;10(3):651–7.
120. Da Costa Martins R, Gamazo C, Irache JM. Design and influence of gamma-irradiation on the biopharmaceutical properties of nanoparticles containing an antigenic complex from *Brucella ovis*. Eur J Pharm Sci. 2009;37(5):563–72.
121. Kerrigan AM, Brown GD. C-type lectins and phagocytosis. Immunobiology. 2009;214(7): 562–75.
122. Gordon S. Alternative activation of macrophages Nat Rev Immunol. 2003;3(1):23–35.
123. McGreal EP, Martinez-Pomares L, Gordon S. Divergent roles for C-type lectins expressed by cells of the innate immune system. Mol Immunol. 2004;41(11):1109–21.
124. Da Costa Martins R, Gamazo C, Sanchez-Martinez M, Barberan M, Penuelas I, Irache JM. Conjunctival vaccination against *Brucella ovis* in mice with mannosylated nanoparticles. J Control Release. 2012;162(3):553–60.
125. World Health Organization. The development of new/improved brucellosis vaccines (WHO/EMCD//ZDI/98.14). 1997. http://whqlibdoc.who.int/hq/1998/WHO_EMC_ZDI_98.14.pdf. Accessed 15 Apr 2013.
126. Neutra MR, Kozlowski PA. Mucosal vaccines: the promise and the challenge. Nat Rev Immunol. 2006;6(2):148–58.

127. Chentoufi AA, Dasgupta G, Nesburn AB, Bettahi I, Binder NR, Choudhury ZS, Chamberlain WD, Wechsler SL, BenMohamed L. Nasolacrimal duct closure modulates ocular mucosal and systemic CD4(+) T-cell responses induced following topical ocular or intranasal immunization. Clin Vaccine Immunol. 2010;17(3):342–53.
128. Hu K, Dou J, Yu F, He X, Yuan X, Wang Y, Liu C, Gu N. An ocular mucosal administration of nanoparticles containing DNA vaccine pRSC-gD-IL-21 confers protection against mucosal challenge with herpes simplex virus type 1 in mice. Vaccine. 2011;29(7):1455–62.
129. Blasco JM, Molina-Flores B. Control and eradication of *Brucella melitensis* infection in sheep and goats. Vet Clin North Am Food Anim Pract. 2011;27(1):95–104.
130. Da Costa Martins R, Irache JM, Gamazo C. Acellular vaccines for ovine brucellosis: a safer alternative against a worldwide disease. Expert Rev Vaccines. 2012;11(1):87–95.
131. Oliveira SC, de Almeida LA, Carvalho NB, Oliveira FS, Lacerda TL. Update on the role of innate immune receptors during *Brucella abortus* infection. Vet Immunol Immunopathol. 2012;148(1–2):129–35.
132. Makloski CL. Canine brucellosis management. Vet Clin North Am Small Anim Pract. 2011;41(6):1209–19.
133. Christopher S, Umapathy BL, Ravikumar KL. Brucellosis: review on the recent trends in pathogenicity and laboratory diagnosis. J Lab Physicians. 2010;2(2):55–60.
134. Ridler AL, West DM. Control of *Brucella ovis* infection in sheep. Vet Clin North Am Food Anim Pract. 2011;27(1):61–6.
135. Hinic V, Brodard I, Thomann A, Cvetnic Z, Makaya PV, Frey J, Abril C. Novel identification and differentiation of *Brucella melitensis*, *B. abortus*, *B. suis*, *B. ovis*, *B. canis*, and *B. neotomae* suitable for both conventional and real-time PCR systems. J Microbiol Methods. 2008;75(2):375–8.
136. Foster G, Osterman BS, Godfroid J, Jacques I, Cloeckaert A. *Brucella ceti* sp. nov. and *Brucella pinnipedialis* sp. nov. for *Brucella* strains with cetaceans and seals as their preferred hosts. Int J Syst Evol Microbiol. 2007;57(Pt 11):2688–93.
137. Scholz HC, Hofer E, Vergnaud G, Le Fleche P, Whatmore AM, Al Dahouk S, Pfeffer M, Kruger M, Cloeckaert A, Tomaso H. Isolation of *Brucella microti* from mandibular lymph nodes of red foxes, *Vulpes vulpes*, in lower Austria. Vector Borne Zoonotic Dis. 2009;9(2):153–6.
138. Scholz HC, Nockler K, Gollner C, Bahn P, Vergnaud G, Tomaso H, Al Dahouk S, Kampfer P, Cloeckaert A, Maquart M, Zygmunt MS, Whatmore AM, Pfeffer M, Huber B, Busse HJ, De BK. *Brucella inopinata* sp. nov., isolated from a breast implant infection. Int J Syst Evol Microbiol. 2010;60(Pt 4):801–8.
139. Tiller RV, Gee JE, Frace MA, Taylor TK, Setubal JC, Hoffmaster AR, De BK. Characterization of novel *Brucella* strains originating from wild native rodent species in North Queensland, Australia. Appl Environ Microbiol. 2010;76(17):5837–45.

Chapter 18
Eligen® Technology for Oral Delivery of Proteins and Peptides

Sunita Prem Victor, Willi Paul and Chandra P. Sharma

18.1 Introduction

Recent advances in areas of biotechnology have resulted in the development of a large number of commercially available macromolecular drugs that have great potential for a vast range of therapeutic indications. However, the challenges of utilizing these drugs for noninvasive delivery have been daunting. Common problems that are continually faced in developing oral dosage forms include low aqueous stability, lack of permeability, rapid metabolism, intracellular trafficking, biological and chemical instability of the macromolecules, among others [1, 2]. The low permeability or absorption of biopharmaceuticals in the GI epithelium is mainly due to large molecular size and low lipophilicity [3, 4]. Over the years, research on certain formulations is ongoing on development of viable oral delivery products to overcome these issues [5–11]. On the contrary, many molecules have been limited to parental dosing because their drug properties do not facilitate oral absorption. Despite the various hurdles; oral dosing is generally considered to be the most patient friendly and convenient route of drug administration [12]. There are various companies dedicated on improving the oral delivery of existing drugs by GI absorption enhancement, which, if successful, could have the greatest impact on oral drug therapy.

Successful protein drug absorption and efficacy from the GI tract requires certain specific physicochemical properties [3, 11, 13, 14]. Moreover, it has to withstand the harsh chemical and biological milieu within the GI tract. The physiochemical properties include suitable molecular weight (typically below 500–1,000 Da), pKa (a measure of the degree of acidity or alkalinity), degree of lipophilicity (log D), as well as proper solubility [15, 16]. The hydrophilic property of these macromolecules makes it difficult for them to penetrate through the epithelial cells (via the transcellular route) because of low permeability. The absorption of large hydrophilic macromolecules is mainly limited to the paracellular pathway, which consists of

C. P. Sharma (✉) · S. P. Victor · W. Paul
Biomedical Technology Wing, Sree Chitra Tirunal Institute for Medical Sciences and Technology, Poojappura, Thiruvananthapuram 695012, India
e-mail: sharmacp@sctimst.ac.in

J. das Neves, B. Sarmento (eds.), *Mucosal Delivery of Biopharmaceuticals,* DOI 10.1007/978-1-4614-9524-6_18,

Table 18.1 Available technologies for oral protein delivery under development by pharmaceutical companies

Company	Product name	Biopharmaceuticals
Emisphere Technologies, Inc.	Eligen®	Calcitonin, insulin, growth hormone, parathyroid hormone, heparin
Altus Biologics	CLEC (cross-linked enzyme crystal)	Calcitonin, lipases, esterases, and proteases
BioSante Pharmaceuticals Inc.	BioOral™	Insulin and vaccines
Generex Biotechnology Corp.	Oral-lyn™	Insulin, macrotonin
Apollo Life Science	Oradel™	Insulin and TNF blocker
Endorex Corp.	Orasome™	Insulin and growth hormone, vaccines
NOBEX Corp. and Biocon (presently with Bristol-Myers Squibb)	Hexyl-insulin monoconjugate 2 (HIM2)/IN-105	Insulin

aqueous pores created by the cellular tight junctions. Most available drugs are either weak acids or weak bases, and under normal conditions only the nonionized fraction (the most lipophilic) crosses biological membranes, except where active transport is involved. The new technology suggested to overcome these limitations is based on carrier molecules, of amino acids having a molecular weight of 250–300 Da that are structurally diverse with different physiochemical properties [17]. These carriers possess hydrophobic moieties that can associate with the drug molecules to create a more lipophilic drug or carrier complex, enabling transport across the epithelial membrane [18, 19]. Because of the weak association between carrier and drug, the interaction is reversible, and occurs spontaneously by simple dilution on entering the blood circulation. Studies have shown that the carriers enable the systemic absorption of the drug via transcellular absorption, a common drug absorption pathway, without compromising the integrity of the intestinal epithelium.

Numerous delivery systems for oral protein delivery have been actively developed, especially by pharmaceutical companies, in hope to make them clinically viable. Although most of the works still remain in the development stage, many of them have progressed beyond the proof-of-concept stage to the clinical trials. Some of the available technologies for oral protein delivery under development by pharmaceutical companies are listed in Table 18.1.

18.2 Eligen® Technology

Emisphere Technologies, Inc. is a biopharmaceutical company pioneering the oral delivery of otherwise injectable drugs and the Eligen® technology is a broad-based platform technology developed and patented by them [20]. This oral delivery technology is founded on the design and synthesis of proprietary delivery agents, known as Emisphere® delivery agents or carriers. Emisphere's business strategy is to develop oral forms of drugs that are not currently available or have poor bioavailability in oral form, either alone or with corporate partners, by applying its proprietary

Eligen® technology to those drugs or licensing its Eligen® technology to partners who, typically, apply it directly to their marketed drugs. This technology has enabled the oral delivery of proteins, peptides, and some macromolecules. At present, Emisphere maintains a library of more than 1,800 structurally diverse carriers with different physicochemical properties. Most of these delivery agents are small organic molecules with a molecular weight of 250–350 Da, and almost all of them are amino acids.

Eligen® technology is a macromolecule-delivering platform technology where a macromolecule is used as an absorption enhancer. The macromolecule interacts with the drug molecules to create a weak, noncovalent association, the drug remaining chemically unmodified. Among the existing library of absorption-enhancing compounds, sodium N-[8-(2-hydroxybenzoyl) amino] caprylate (SNAC) or salcaprozate sodium, also called sodium N-[8-(2-hydroxybenzoyl) amino] octanoate is used by Emisphere. Emisphere contends that SNAC enhances absorption by forming a noncovalent complex with the active drug that enables transcellular absorption, without altering tight junctions [19, 21, 22]. For proteins, the mechanism may involve a reversible change in protein conformation and protection against degradation, prior to absorption. Unlike the traditional penetration enhancers, Emisphere delivery agents are believed to cause minimal histological damages to the intestinal epithelium and are applicable to diverse group of drug molecules ranging in size from 500 to $>$ 150,000 Da. The formed lipophilic drug or SNAC complex is claimed to be capable of transport across the epithelial membrane as shown in Fig. 18.1. Eligen® technology has been used to develop various types of oral formulations including solutions, tablets, and capsules. They have advanced oral formulations or prototypes of salmon calcitonin (sCT) [23–30], heparin [31–35], insulin [36–40], parathyroid hormone [18], human growth hormone [19, 21, 22, 41], Vitamin B12 [42], and cromolyn sodium. These delivery agents have been evaluated in various animal models as well as in humans for their ability to enhance the delivery of a wide array of therapeutic macromolecules that are in various stages of clinical development. To date, six oral products have undergone clinical testing with Emisphere® delivery agents. Their clinical trials have been discussed in the subsequent sections. Although the exact mechanism of absorption enhancement has not been elucidated, it has been hypothesized that the delivery agent-mediated absorption involves a sequence of advanced and novel features (Fig. 18.1) which may be generalized as follows [2, 43, 44].

A noncovalent interaction occurs between the macromolecule and the delivery agent. This interaction transiently alters the physicochemical properties of the macromolecules (e.g., hydrophobicity, conformation, stability, etc.). The complex formed has a conformation of the macromolecule that has a higher transport rate compared to its physiological conformation. This complex mimics the body's natural biomolecular transport mechanisms and the GI absorption is facilitated. Once the complexes along with the drug are transported across the epithelial cells to the circulation, the delivery agent dissociates from the drug and the drug reestablishes its native conformations, ensuring its therapeutically active state. No histological damage to the intestinal epithelium represents a significant advantage over traditional penetration enhancers that have been reported to be associated with significant disruption of the tight junctions, change in membrane fluidity, and toxicity, among others.

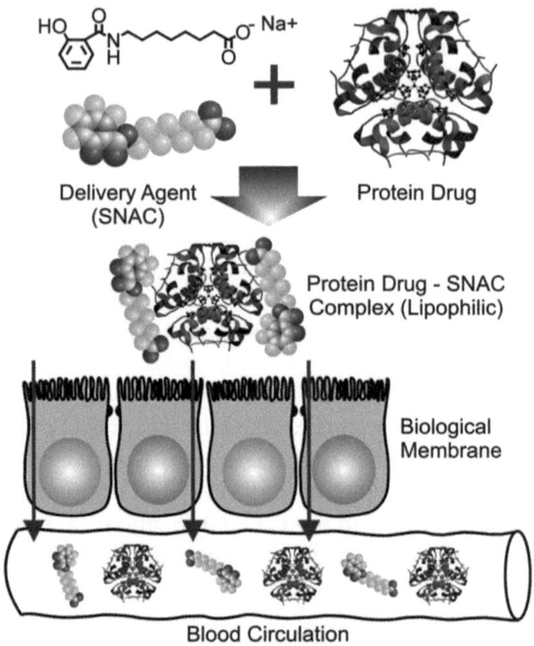

Fig. 18.1 Emisphere Eligen® oral protein delivery technology. Absorption of drug-delivery agent complex via the intestine

18.2.1 Calcitonin

Calcitonin discovered by Copp and Cameron has been available as a therapeutic agent for metabolic bone disease for more than 30 years [45]. It is a naturally occurring 32-amino acid polypeptide produced by the parafollicular C cells of the thyroid. Synthetic or recombinant calcitonin has been derived from a number of different species including human, porcine, and salmon. However, sCT, believed to be 100 times more potent than human calcitonin, has been widely preferred in clinical practice [45]. It has been approved for the treatment of postmenopausal osteoporosis, Paget's disease, bone associated pain conditions, and hypercalcemia. A unique advantage of sCT, unshared by any other antiresorptive agents, is its analgesic effect on bone pain previously demonstrated in clinical studies in patients with osteoporotic vertebral fractures and musculoskeletal disorders. sCT is commercially available as an injectable form and as a nasal spray. The short half-life of calcitonin in serum has led to several attempts to increase plasma concentrations. One very recent approach of oral calcitonin formulation is the use of the Eligen® technology. In this new formulation the carrier studied extensively is disodium salt of N-(5-chlorosalicyloyl)-8-aminocaprylic acid (5-CNAC oral calcitonin) salt [29, 46]. This carrier has been found to bind with calcitonin without changing its biological properties, thereby reducing the compound's susceptibility to degradation. It creates an insoluble entity at low pH which later dissolves at higher pH and facilitates passive absorption by the transcellular pathway by enhancing peptide transport over nonpolar biological membrane. After passage through the intestine, the delivery agent disassociates from the peptide, and the peptide is absorbed into the hepatic vein with subsequent systemic absorption. The metabolism and disposition of 5-CNAC have been investigated and they were found safe and well tolerated [47].

The pharmacokinetic profile of 5-CNAC oral calcitonin has also been evaluated in a randomized, crossover double-blinded Phase I trial, controlled by both a placebo and a parenteral verum [24]. This study demonstrated that oral delivery of sCT is feasible with reproducible absorption and systemic biological efficacy. Eight healthy volunteers received single doses of 400, 800, and 1,200 μg of sCT orally, a placebo, and a 50 IU sCT intravenous infusion. sCT was readily and reliably absorbed from the oral formulation, with an absolute bioavailability of around 1.2 % depending on the applied dose. It also demonstrated a marked, dose-dependent drop in blood and urine C-terminal telopeptide of type I collagen, a sensitive and specific bone resorption marker, with the effects of 1,200 μg exceeding those of the other formulations. It also decreased blood calcium and phosphate, and increased the circulating levels of parathyroid hormone and, transiently, the urinary excretion of calcium. It was well tolerated, with some subjects presenting mild and transient nausea, abdominal cramps, diarrheic stools, and headaches.

The efficacy and safety of another Eligen® technology-based oral formulation to deliver sCT to the circulation was assessed on postmenopausal women. A multicentric randomized, double-blind, placebo-controlled, dose-ranging clinical trial has been carried out that included 277 healthy postmenopausal women in the age group of 55–85 [30]. The women were treated with doses of sCT in combination with 200 mg of delivery agent 5-CNAC or placebo for 3 months. Subjects received treatment with daily doses of 0.15, 0.4, 1 and 2.5 mg or with 1 mg every intermittent day. They were also given 1,000 mg calcium supplements and 400 IU vitamin D daily throughout the study. Acute changes in serum urinary C-terminal telopeptide of type I collagen (CTx), N-mid osteocalcin, bone-specific alkaline phosphatase, calcium, and parathyroid hormone measured by immunoassays, were the studied efficacy parameters. The first dose of sCT brought about dose-dependent decrease in serum CTx compared with placebo and reached lowest value within 3 h after drug intake, after which gradual increase had been noticed. The results depicted effective enteral absorption, a pronounced inhibition of bone resorption with minimal alteration of formation, and reproducibility of responses over 3 months. At month 3, the placebo-corrected changes in the predose value of serum and urinary CTx were significant only in the 1.0 mg dose group (– 18.9 % and – 20.5 %, respectively, $p < 0.05$). The results thus suggested that the oral formulation was well tolerated, with mild to moderate GI and skin manifestations apparent mainly in the high-dose groups. The above 3-month trial shows that the novel Eligen® technology-based oral formulation of sCT has potential to become a safe and effective treatment for postmenopausal bone loss. Future trials are, however, necessary to assess the impact of long-term administration on changes in bone mineral density (BMD) and fracture risk.

A subsequent study was carried out to induce significant dose-dependent reductions in the biochemical marker of cartilage degradation envisaging potential chondroprotective effects [48]. This was a randomized double-blind; placebo-controlled clinical study which included 152 Danish postmenopausal women aged 55–85. The subjects received treatment with different doses of sCT coupled with Eligen® technology-based carrier molecule, or placebo for 3 months. The efficacy parameter was evaluated with the changes in the 24-h excretion of urinary CTx-I/CTx-II biomarkers of bone resorption and cartilage degradation respectively. The results depict that the 3 month treatment with oral sCT induced significant dose-dependent decreases in both urinary CTx-I and CTx-II. The maximum responses in both biomarkers were associated with treatment using 1.0 mg daily dose of sCT and it has also been noticed that women with accelerated cartilage degradation at baseline (high CTx-II) seemed to be more responsive to clinically effective dose of oral sCT. Another important finding was that women with elevated baseline urinary CTx-II were more likely to manifest with joint-related symptoms and respond with the largest decreases in the degradation product of collagen type II. So subsequently, women with high cartilage turnover are more likely to benefit from potential chondroprotective therapy. The above-mentioned study is a post hoc analysis of a completed clinical trial, investigating the efficacy and safety of sCT for the inhibition of bone turnover in postmenopausal women. However, sCT is not yet an established drug for treatment of patients with osteoarthritis. Further studies are on the anvil to assess the impact of optimal doses of sCT on cartilage mass using MRI.

A study further reports that the bioavailability and efficacy of orally administered calcitonin SMC021, is heavily influenced by meal time, amount of water used to take the tablet, and proximity to intake of a meal [29]. SMC021 is an oral formulation of sCT consisting of the peptide hormone and 5-CNAC, a unimolecular enhancer of GI peptide absorption, licensed to Novartis. The study clearly suggested that drug uptake of SMC021 is influenced by the amount of water given with the tablet. A water volume of 50 ml resulted in a two- to three-fold higher absorption of sCT in comparison with a volume of 200 ml of water. This doubling of absorption was obtained irrespective of the timing of the meal suggesting that the volume of water strongly impacts digestion and absorption. Further the biochemical marker of bone resorption demonstrated improved efficacy. These data were the first to demonstrate that water intake has an important effect on oral peptide uptake with the Eligen® technology, improving bioavailability as much as 400 %, and even more if placebo corrected. A similar study carried out was a randomized, partially blind, placebo-controlled, single dose, exploratory crossover Phase I study involving 56 healthy postmenopausal women. sCT of 0.8 mg with 50 ml of water taken 30 and 60 min prior to meal time resulted in optimal pharmacodynamic and pharmacokinetic parameters. The data suggest that this novel oral formulation may have improved absorption and reduction of bone resorption compared to that of nasal calcitonin.

Similar studies have also been reported previously with a 14-day clinical trial of twice daily oral calcitonin with 5-CNAC suggesting potentially useful reductions in biomarkers of bone resorption and cartilage degradation [26]. An abstract presented at the recent 2011 American College of Rheumatology meeting reported that an oral formulation of sCT with the Eligen® delivery system has entered Phase III clinical trials for the treatment of osteoarthritis. Oral sCT at a dose of 0.8 mg twice daily for 2 years, significantly reduced pain and stiffness, improved physical function, and slowed cartilage loss in a placebo-controlled clinical trial involving 1,169 patients with painful knee osteoarthritis. The patients had Kellgren-Lawrence grade 2 disease with mean age of trial subjects being 64 years, and mean body mass index (BMI) of 28.9 kg/m^2. Sixty-eight percent of patients were women. At month 24, oral calcitonin was also superior to placebo on 24-h visual analogue scale pain scores (P00.018), patient global assessment (P00.008), and physician global assessment (P00.014). However, by month 24, oral calcitonin subjects demonstrated a 4.5 % loss in cartilage volume on MRI in both the signal and nonsignal knee; placebo subjects demonstrated a 7 % loss in both knees. The differences were statistically significant. The most common adverse events in the oral calcitonin group versus placebo were hot flushes, nausea, dyspepsia, and diarrhea. However, despite the mentioned drawbacks the potential for a compound with improved bioavailability and efficacy in an oral preparation, combined with the established safety profile of sCT, hold suitable promise in the future.

18.2.2 Insulin

Oral insulin is an exciting area of development in the treatment of diabetes because of its potential benefit in patient compliance, rapid insulinization of liver, adequate insulin delivery while potentially avoiding adverse effects of weight gain and hypoglycemia. Insulin consists of two polypeptide chains (A and B) of 21 amino acids and 30 amino acids, respectively. Its molecular weight in monomeric form is 6,000 Da. The mechanism of the GI absorption of insulin has been studied using an Emisphere delivery agent. The molecule appears to be absorbed throughout the GI tract following oral administration, but the best site of absorption following coadministration with an Emisphere delivery agent appears to be the colon [49, 50]. A study with insulin also revealed that Emisphere delivery agents facilitate drug transport via transcellular pathways without permealization of the plasma membrane or tight junction disruption. Another study went on to investigate the mechanism of insulin absorption across Caco-2 cell monolayers with one of these drug delivery agents, SNAC. The results showed that SNAC increases insulin permeability approximately ten fold across cell monolayers and does so without affecting mannitol permeability or disrupting cell membranes. Confocal microscopy and immunocytochemistry revealed that insulin is transported transcellularly without detectable alteration of the tight junctions between adjacent cells. SNAC also appears to play some role in protecting insulin from proteolytic degradation, potentially allowing for more intact insulin to be available at the site of absorption [40].

The activity of the absorbed insulin from the GI tract was evaluated using SNAC in combination with insulin [36]. The capsules containing insulin and SNAC, in various combinations, were administered orally, as a single dose, to 12 nondiabetic subjects and four control subjects (receiving SNAC or insulin only) in order to assess its biological effect and safety. Plasma glucose levels, insulin and C-peptide concentrations, as well as SNAC levels, were determined, at timed intervals up to 4 h. In all cases, a glucose-lowering effect was demonstrated, preceded by an increase in plasma insulin levels. The nadir of plasma glucose levels appeared after 30–50 min, following the ingestion of the mixture. The plasma insulin levels were found to parallel the blood SNAC levels. Plasma C-peptide levels were suppressed by the lowered glucose levels achieved concurrent with the increasing amount of exogenous insulin absorbed, indicating that the secretion of endogenous hormone was partially abolished. There were no biological effects regarding blood glucose levels upon administration of SNAC or insulin when given alone. No adverse effects were detected during the trial or several weeks after the trial. So it was concluded that the insulin in combination with a novel delivery agent, SNAC, given orally, is absorbed through the GI tract in a biologically active form. This was also demonstrated by a glucose-lowering effect of the mixture as well as a suppression of an endogenous insulin secretion.

The oral delivery of insulin has been investigated as a representative example in a clinical trial with 10 fasted healthy volunteers following oral administration of insulin in combination with the delivery agent [38]. The results indicate that insulin was rapidly absorbed into the systemic circulation and peak plasma concentration occurred within 25 min. The corresponding maximum reductions in both plasma glucose and C-peptide (a marker of endogenous insulin production) concentrations occurred within 1 h. The results were clinically significant because insulin alone or the delivery agent alone dosed orally did not affect plasma levels of insulin or glucose. In another recent clinical study in patients with type II diabetes, a capsule preparation of insulin containing 10 mg of insulin and 200 mg of the delivery agent was evaluated. The data demonstrated that oral administration of this unformulated insulin, when administered 30 min prior to the standardized meal, reduced postprandial excursion, produced a marked increase in systemic insulin levels, and a concomitant reduction in C-peptide. In addition, plasma insulin concentrations peaked faster using Emisphere's oral unformulated dosage as compared to fast-acting injectable insulin (30 min for oral versus approximately 45 min for injectable formulations) [37].

Emisphere's oral insulin uses a proprietary permeation enhancer which helps in the absorption of insulin. The pharmacokinetics studies indicated a rapid absorption time of around 20 min from the time of administration and the plasma insulin levels return to baseline within 2 h. Based on the rapidity of absorption, it was proposed that the absorption is from the upper GI tract. In another 2-week clinical trial on patients, well controlled under dietary conditions, Emisphere's oral insulin was shown to improve both glycemic control and insulin sensitivity. Emisphere also reported completion of a placebo-controlled four treatment arm; 90 days Phase II study in 2004 [51]. The dose of insulin was fixed for the entire duration of the trial, with insulin dose ranging from 20 mg per day to 40 mg day. However, the highest studied dose showed statistically significant reduction in HbA1c over placebo, that too in patients with baseline HbA1c of 8 % and above. HbA1c decrease of 0.74 % from baseline was observed in patients on highest dose of oral insulin, while no change (0.00 %) was observed in patients on placebo ($n = 17, p = 0.03$). Emisphere's oral insulin product demonstrated a good safety profile as there were no significant differences in hypoglycemic events, serious adverse events or insulin antibody formation in comparison to placebo (additional data reported as press release). In this study, only the highest dose showed a clinically meaningful drop in HbA1c after 3 months of therapy. The high dose increases the cost of therapy and points to researchers having to pay significant attention to ensure the commercial viability of such an oral insulin drug in the marketplace.

18.2.3 Heparin

Heparin an anionic pentasaccharide is one of the most important anticoagulant drugs in current clinical use. Heparin is composed of glucosamine and L-iduronic acid or D-glucuronic acid in chains of variable length, having a molecular weight range of 5,000–30,000 [52]. It is widely used for the prevention and treatment of deep venous thrombosis, pulmonary embolism in patients undergoing orthopedic surgery and for patients with renal failure. Its hydrophilic nature, anionic structure due to presence of SO^{3-} groups and large molecular weight prevents absorption through the GI tract.

SNAC is an acetylated amino acid molecule that has been shown to facilitate the GI absorption of codelivered heparin. SNAC-mediated GI absorption of heparin occurs in a passive transcellular process without causing apparent damage to the intestinal epithelium. The pathway of oral absorption of heparin was evaluated using fluorescence microscopy to follow the transport of heparin across Caco-2 cell monolayers [32]. The localization of fluorescently labeled heparin was determined using epifluorescence and confocal microscopy. DNA dyes were used to determine the effect of SNAC on the plasma membrane integrity. F-actin was labeled with fluorescent phalloidin to investigate the stability of perijunctional actin rings in the presence of SNAC. Heparin was detected in the cytoplasm only after incubation of the cells with heparin and SNAC. No DNA staining was observed in cells incubated with a DNA dye in the presence of SNAC concentrations at which heparin transport occurred. In addition, no signs of actin redistribution or perijunctional ring disbandment were observed during the transport of heparin. The results indicate that SNAC enables heparin transport across Caco-2 monolayers via the transcellular pathway. Heparin transport in the presence of SNAC is selective and does not involve permeabilization of the plasma membrane or tight junction disruption.

SNAC was evaluated with escalating oral heparin doses in a randomized, double-blind, controlled clinical study for safety, tolerability, and effects on indexes of anticoagulation [35]. Investigations, both *in vitro* [33] and *in vivo* [34], revealed that the n-acylated nonalpha amino acid SNAC has no pharmacological activity. When dosed with 10.5 g SNAC/20,000 IU heparin, an increase in concentration of activated partial thromboplastin time (aPTT), and tissue factor pathway inhibitor (TFPI) concentrations were detected. For the entire group, 30,000 IU SNAC and heparin elevated TFPI from 74.967.6 to 254.2612.3 mg/ml (P, 0.001) 1 h after dosing (P, 0.001). Similar changes occurred in antifactor IIa and antifactor Xa. aPTT rose from 2860.5

to 42.266.3 s 2 h after dosing (P, 0.01). No significant changes in vital signs, physical examination, ECGs, or clinical laboratory values were observed. Neither 30,000 IU heparin alone nor 10.5 g SNAC alone altered the haemostatic parameters. Emesis was associated with 10.5 g SNAC. A taste-masked preparation of SNAC 2.25 g was administered orally with heparin 30,000–150,000 IU. Both aPTT and antifactor Xa increased with escalating doses of heparin. This preparation was well tolerated. These results established the feasibility of oral delivery of anticoagulant doses of heparin in humans and were believed to have broader implications for the absorption of macromolecules. Phase II clinical studies on hip replacement patients have also been promising with heparin/SNAC being comparable to subcutaneous heparin for the prevention of deep venous thrombosis [31].

18.2.4 Recombinant Human Growth Hormone (rhGH)

Human growth hormone is a protein drug (22 kDa) and has been used by patients with growth failure due to inadequate secretion of endogenous growth hormone, Turner syndrome, chronic renal insufficiency in children, and as replacement therapy for adults. The possibility of using the Emisphere delivery agent to deliver rhGH orally was first tested in rodents. A series of N-acetylated, nonalpha, aromatic amino acids were prepared and shown to promote the absorption of (rhGH) from the GI tract. Seventy compounds in this family were tested *in vivo* in rats [41]. Of the compounds tested, 4-[4-[(2-hydroxybenzoyl) amino] phenyl butyric acid was identified as a preclinical candidate and was used to demonstrate the oral delivery of rhGH in primates. A significant positive correlation was found between the relative log k′ of the delivery agents, as determined by HPLC on an immobilized artificial membrane (IAM) column, and serum rhGH concentrations following oral or colonic dosing in rats. Structure-activity relationships have also been developed on the basis of electronic effects and hydrogen-bonding characteristics of the aromatic amide substituent. Subsequently, the macromolecule was delivered in cynomolgus monkeys ($n = 4$). A mean peak serum concentration of 55 ng/ml rhGH was obtained following administration of a single oral dose of rhGH in combination with the delivery agent. Oral administration of either rhGH or delivery agent alone to these monkeys did not result in measurable circulating levels of rhGH [53].

Studies have been conducted to investigate the mechanisms of GI absorption of rhGH in the presence of an Emisphere delivery agent. The results of these mechanistic studies suggest that the oral delivery of rhGH is dependent on both the dose of the delivery agent and rhGH, and that P-glycoprotein may be involved in the hGH absorption mechanism in the presence of these delivery agents. An early phase clinical trial has been conducted in collaboration with Eli Lilly and Company to evaluate an oral formulation of rhGH in combination with an Emisphere delivery agent [54]. In another small proof of concept study, 8 GH deficient patients were given oral rhGH [19]. Novartis investigated pharmacokinetics as well as the pharmacodynamic properties of the orally delivered rhGH. The study showed that growth hormone peaks were recorded in all patients at some time points, although with considerable variability and minor endogenous growth hormone interference. An increase in IGF-I was seen in some patients, leading to a statistically significant increase in mean serum IGF-I at day 7 compared with end of wash-out. Phase I data indicated that rhGH can be absorbed when given to growth hormone-deficient (GHD) patients in a prototype oral formulation using Emisphere's Eligen® delivery technology.

18.2.5 Vitamin B12

Vitamin B12 is important for the normal functioning of the brain and nervous system and for the formation of blood. Cyanocobalamin is the stable and most widely used form of B12. Present work has been dedicated to improve available cyanocobalamin (B12) formulations directed toward achieving repletion of active B12 in B12-deficient individuals. Low levels of B12 can be the result of a lack of the vitamin in the diet, but are most likely to occur because of deficiencies in an individual's ability to absorb B12 through the natural intricate mechanism. Conditions resulting in reduced stomach acidity (such as long-term use of proton pump inhibitors or age related stomach atrophy) or GI disturbances (such as bariatric surgery, Crohn's disease, or celiac disease) can lead to B12 deficiency. In a study [42] completed in 2011, B12 deficient patients were given either a typical B12 injection regimen for 12 weeks (5 injections of 1,000 μg over the first 15 days and then one each at 21, 30, 60, and 90 days) or 1,000 μg of Eligen® B12 as a daily pill. All individuals in the study achieved rapid repletion of active B12 within 15 days (the first time point in the study) whether on injection or the oral tablet. Furthermore, all participants in the study had B12 levels that continued to be at normal levels till the end of the study. This performance placed the Eligen® B12 formulation on par with the standard regimen of frequent B12 injections, without the extra cost and inconvenience of drug injections.

Another study [55] compared the efficacy and safety profile of a new proprietary oral vitamin B12 formulation (oral B12) with intramuscular (IM) vitamin B12 (TM B12) in restoring normal serum B12 concentrations in patients with low cobalamin levels. Patients were recruited from five centers and randomly assigned to receive oral B12 1,000 μg, taken daily for 90 days, or IM B12 1,000 μg, given on study days 1, 3, 7, 10, 14, 21, 30, 60, and 90. The patients were aged between 18 and 60 years and had GI abnormalities or were on a restricted diet. The primary efficacy outcome compared the proportion of patients in each treatment arm in whom cobalamin levels were normalized ($\geq$ 350 ng/ml) following 60 days of treatment. Secondary objectives included comparing the efficacy of the two formulations after 90 days of treatment, assessing time to normalization of B12 levels, and evaluating the changes in the levels of biomarkers methylmalonic acid (MMA) and homocysteine (HC). The effect on holotranscobalamin II (active B12) levels was assessed as an exploratory end point and correlated to serum cobalamin levels in both treatment groups. Blood samples were collected at baseline (day 1) and on days 15, 31, 61, and 91. Fifty patients were recruited. Forty-eight patients (96.0 %) completed the study (22 patients [91.7 %] in the oral B12 group and 26 patients [100 %] in the IM B12 group). All patients (100 %) in both treatment groups and in both populations had a cobalamin level ≥ 350 pg/ml on day 61 and maintained it on day 91. The difference between the IM and oral treatment groups did not reach the planned level of statistical significance ($p < 0.05$) for mean percent change from baseline (PCFB) in serum cobalamin levels on day 61 and day 91. The difference between the IM and oral treatment groups did not reach the planned level of statistical significance for mean PCFB in serum MMA levels on day 61. There was a statistical difference between the IM and oral treatment groups for mean PCFB in serum MMA levels on day 91 ($p = 0.033$), with lower values in the oral B12 group. The difference between the IM and oral treatment groups did not reach the planned level of statistical significance for mean PCFB in plasma HC levels on day 61 and day 91. All patients in each treatment group achieved normalization of serum cobalamin levels by day 15. All patients in both treatment groups and in both populations had plasma holotranscobalamin levels ≥ 40 pmol/L on day 61 and on day 91. No statistical analysis was planned or performed for safety end points, which were reported only descriptively. Most observed adverse effects were considered mild or moderate in intensity. Adverse effects that were considered severe in intensity were also considered to be not related to the studied drug by the investigator. The treatment regime in this selected study population consisted of individuals with low cobalamin levels who received oral B12 (1,000 mu g/d) or IM B12 (1,000 mu g in nine injections over 3 months) for a total of 3 months. Both the oral and IM formulations were effective in restoring normal levels of serum cobalamin in all patients studied (100 %). Both formulations used in this study were well tolerated at the dose studied.

18.2.6 Parathyroid Hormone (PTH)

Parathyroid hormone (PTH), the only drug known to stimulate bone formation, is a peptide therapeutic indicated in the treatment of osteoporosis [18, 56]. It is an 84-amino acid protein and is used to regulate calcium homeostasis. Unfortunately, PTH is only effective when dosed by injection because it has no oral bioavailability. PTH is produced by the parathyroid glands to regulate the amount of calcium and phosphorus in the body. When used therapeutically, it increases bone density and bone strength to help prevent fractures. It is approved to treat osteoporosis, a disease associated with a gradual thinning and weakening of the bones that occurs most frequently in women after menopause. Untreated postmenopausal osteoporosis can lead to chronic back pain, disabling fractures, and lost mobility.

In July 2008, Emisphere announced that its partner, Novartis Pharma AG, launched a Phase I study in postmenopausal women to determine the safety and tolerability of an oral formulation PTH1–34, a combination of human PTH1–34 and the absorption enhancer 5-CNAC using Eligen® technology, for the treatment of postmenopausal osteoporosis. The study was designed to assess the bioavailability profile of increasing doses of PTH1–34 combined with different amounts of 5-CNAC administered orally. On October 19, 2009, Novartis reported results of this study which showed potentially relevant therapeutic exposure and safety profiles similar to those of the currently available injectable dosage form. These were presented at the 73rd Annual Scientific Meeting of the American College of Rheumatology in Philadelphia, PA, USA.

In April 2010, Novartis initiated a second Phase I trial for an oral PTH1–34 for the treatment of postmenopausal osteoporosis. The study was a partially blinded, placebo-controlled, active comparator study to explore the safety, tolerability, pharmacokinetics and pharmacodynamics in postmenopausal women after daily doses of PTH1–34. The study was divided into two parts (A and B) and enrolled approximately 120 women. In Part A ascending doses of oral PTH1–34 were tested for safety, tolerability, and pharmacokinetics and compared to Forsteo®. In Part B, in addition to safety and tolerability of oral PTH, pharmacodynamic responses were measured by bone biomarker levels and bone mineral density and compared to Forsteo®. On June 17, 2011, Novartis informed the results of its recently completed study for an oral PTH1–34 using Emisphere's Eligen® technology in postmenopausal women with osteoporosis or osteopenia. Novartis stated that although the study confirmed that oral PTH1–34 was both safe and well tolerated, several clinical endpoints were not met. Based on the data analyzed, Novartis has terminated the study and anticipates no further work on oral formulation of PTH1–34.

Another study demonstrated that a single dose of the novel oral PTHPTH1–34, which utilizes Eligen® technology and absorption-enhancer carrier molecule 5-CNAC, achieved potentially therapeutically relevant exposure and safety profiles to those of the currently available injectable formulation in healthy postmenopausal women. These results were from a single-center, partially blinded, incomplete crossover study conducted by Emisphere's partner Novartis Pharma AG and were presented on October 19, 2009, in a poster session at the 73rd Annual Scientific Meeting of the American College of Rheumatology in Philadelphia.

This Phase I single-center partially blinded incomplete crossover study that was designed to assess the exposure and safety of orally administered doses of PTH1–34 and different amounts of the absorption enhancer 5-CNAC was conducted in 32 healthy postmenopausal women. The subjects were randomized to receive a single dose of placebo, 20 μg of subcutaneously injected parathyroid hormone PTH1–34 (Forteo®), or one of several orally administered doses of PTH1–34 formulated with either 100 or 200 mg of Emisphere's absorption-enhancer 5-CNAC. While all doses of oral PTH1–34 were rapidly absorbed and showed appreciable blood concentrations in a dose-dependent manner, the 2.5 and 5 mg doses of oral PTH1–34 containing 200 mg 5-CNAC achieved exposure levels closest to those of 20 μg injectable PTH1–34, with a comparable incidence of adverse events. Ionized calcium remained within normal limits in all treatment groups. There were no serious adverse events in the study. Nine participants withdrew from the study. Of these, five (one on placebo, one on Forteo® and three on either 2.5 or 5 mg PTH1–34) withdrew because of symptomatic hypotension. Three patients on either 2.5 or 5 mg PTH1–34 withdrew because of delayed vomiting. One patient on 2.5 mg PTH1-34 (100 mg 5-CNAC) withdrew because of symptomatic, but unconfirmed hypercalcemia.

18.3 Conclusions and Future Perspectives

Oral administration of drugs is regarded as the most preferred route of administration, because of the convenience to large number of patient population and its cost effectiveness. However, macromolecular drugs cannot be administered orally because of the inherent properties of these drugs. Emisphere's Eligen® technology makes it possible to orally deliver a therapeutic molecule without altering its chemical form or biological integrity. Eligen® delivery agents, or "carriers," such as the absorption-enhancer 5-CNAC that facilitate or enable the transport of therapeutic molecules across the mucous membranes of the GI tract, to reach the tissues of the body where they can exert their intended pharmacological effect. Eligen® technology has been shown to enhance oral delivery of many different therapeutic molecules. This technology works especially well with water-soluble drugs, both positively and negatively charged. Enhanced oral delivery has been demonstrated in the clinic with large molecular weight drugs (such as unfractionated heparin and growth hormone), medium size biomolecules (such as peptides like calcitonin and insulin, as well as low-molecular weight heparin) and small molecules (such as cromolyn and cyanocobalamin). For drugs with low aqueous solubility, Eligen® technology has been less successful but in certain cases it can enhance oral bioavailability. The commercial success of these products certainly will depend on its increased stability, bioavailability and tolerability or high patient compliance. We can expect that more products based on Eligen® technology will be available to the patients in near future, particularly an oral insulin product which the 347 million diabetic population is expecting for the last two decades.

References

1. Donovan MD, Flynn GL, Amidon GL. Absorption of polyethylene glycols 600 through 2000: the molecular weight dependence of gastrointestinal and nasal absorption. Pharm Res. 1990;7(8):863–8.
2. Aungst BJ. Absorption enhancers: applications and advances. AAPS J. 2012;14(1):10–8.
3. Woodley JF. Enzymatic barriers for GI peptide and protein delivery. Crit Rev Ther Drug Carrier Syst. 1994;11(2–3):61–95.
4. Lipka E, Crison J, Amidon GL. Transmembrane transport of peptide type compounds: prospects for oral delivery. J Control Release. 1996;39(2–3):121–9.
5. Yun Y, Cho YW, Park K. Nanoparticles for oral delivery: targeted nanoparticles with peptidic ligands for oral protein delivery. Adv Drug Deliv Rev. 2013;65(6):822–32.
6. He C, et al. Size-dependent absorption mechanism of polymeric nanoparticles for oral delivery of protein drugs. Biomaterials. 2012;33(33):8569–78.
7. Griffin BT, O'Driscoll CM. Opportunities and challenges for oral delivery of hydrophobic versus hydrophilic peptide and protein-like drugs using lipid-based technologies. Ther Deliv. 2011;2(12):1633–53.
8. Rekha MR, Sharma CP. Oral delivery of therapeutic protein/peptide for diabetes—future perspectives. Int J Pharm. 2013;440(1):48–62.
9. Muller G. Oral delivery of protein drugs: driver for personalized medicine. Curr Issues Mol Biol. 2011;13(1):13–24.

10. Werle M, Makhlof A, Takeuchi H. Oral protein delivery: a patent review of academic and industrial approaches. Recent Pat Drug Deliv Formul. 2009;3(2):94–104.
11. Shaji J, Patole V. Protein and peptide drug delivery: oral approaches. Indian J Pharm Sci. 2008;70(3):269–77.
12. Patel VF, Liu F, Brown MB. Advances in oral transmucosal drug delivery. J Control Release. 2011;153(2):106–16.
13. Malik DK, et al. Recent advances in protein and peptide drug delivery systems. Curr Drug Deliv. 2007;4(2):141–51.
14. Park K, Kwon IC, Park K. Oral protein delivery: current status and future prospect. React Funct Polym. 2011;71(3):280–7.
15. Fasano A. Novel approaches for oral delivery of macromolecules. J Pharm Sci. 1998;87(11):1351–6.
16. Goldberg M, Gomez-Orellana I. Challenges for the oral delivery of macromolecules. Nat Rev Drug Discov. 2003;2(4):289–95.
17. Henry CM. Special delivery. Chem Eng News. 2000;78(38):49–65.
18. Leone-Bay A, et al. Oral delivery of biologically active parathyroid hormone. Pharm Res. 2001;18(7):964–70.
19. Wu SJ, Robinson JR. Transport of human growth hormone across Caco-2 cells with novel delivery agents: evidence for P-glycoprotein involvement. J Control Release. 1999;62(1–2):171–7.
20. Leone-Bay A, et al. Compounds and compositions for delivering active agents, U.S. Patent, Editor. 2003, Emisphere Technologies Inc.: US.
21. Mlynek GM, Calvo LJ, Robinson JR. Carrier-enhanced human growth hormone absorption across isolated rabbit intestinal tissue. Int J Pharm. 2000;197(1–2):13–21.
22. Wu SJ, Robinson JR. Transcellular and lipophilic complex-enhanced intestinal absorption of human growth hormone. Pharm Res. 1999;16(8):1266–72.
23. Bagger YZ, et al. Oral salmon calcitonin induced suppression of urinary collagen type II degradation in postmenopausal women: a new potential treatment of osteoarthritis. Bone. 2005;37(3):425–30.
24. Buclin T, et al. Bioavailability and biological efficacy of a new oral formulation of salmon calcitonin in healthy volunteers. J Bone Miner Res. 2002;17(8):1478–85.
25. Hamdy RC, Daley DN. Oral calcitonin. Int J Womens Health. 2012;4:471–9.
26. Karsdal MA, et al. The effect of oral salmon calcitonin delivered with 5-CNAC on bone and cartilage degradation in osteoarthritic patients: a 14-day randomized study. Osteoarthr Cartilage. 2010;18(2):150–9.
27. Karsdal MA, et al. The effects of oral calcitonin on bone collagen maturation: implications for bone turnover and quality. Osteoporos Int. 2008;19(9):1355–61.
28. Karsdal MA, et al. Investigation of the diurnal variation in bone resorption for optimal drug delivery and efficacy in osteoporosis with oral calcitonin. BMC Clin Pharmacol. 2008;8:12.
29. Karsdal MA, et al. Optimizing bioavailability of oral administration of small peptides through pharmacokinetic and pharmacodynamic parameters: the effect of water and timing of meal intake on oral delivery of Salmon calcitonin. BMC Clin Pharmacol. 2008;8:5.
30. Tanko LB, et al. Safety and efficacy of a novel salmon calcitonin (sCT) technology-based oral formulation in healthy postmenopausal women: acute and 3-month effects on biomarkers of bone turnover. J Bone Miner Res. 2004;19(9):1531–8.
31. Baughman RA, et al. Oral delivery of anticoagulant doses of heparin. A randomized, double-blind, controlled study in humans. Circulation. 1998;98(16):1610–5.
32. Brayden D, et al. Heparin absorption across the intestine: effects of sodium N-[8-(2-hydroxybenzoyl)amino]caprylate in rat in situ intestinal instillations and in Caco-2 monolayers. Pharm Res. 1997;14(12):1772–9.
33. Gonze MD, et al. Orally administered heparin for preventing deep venous thrombosis. Am J Surg. 1998;176(2):176–8.
34. Malkov D, et al. Pathway of oral absorption of heparin with sodium N-[8-(2-hydroxybenzoyl)amino] caprylate. Pharm Res. 2002;19(8):1180–4.

35. Rivera TM, et al. Oral delivery of heparin in combination with sodium N-[8-(2-hydroxybenzoyl)amino]caprylate: pharmacological considerations. Pharm Res. 1997;14(12):1830–4.
36. Kidron M, et al. A novel per-oral insulin formulation: proof of concept study in non-diabetic subjects. Diabet Med. 2004;21(4):354–7.
37. Abbas R, et al. Oral insulin: pharmacokinetics and pharmacodynamics of human insulin following oral administration of an insulin/delivery agent capsule in healthy volunteers. Diabetes. 2002;51:A48.
38. Hoffman A, Qadri B. Eligen insulin—a system for the oral delivery of insulin for diabetes. IDrugs. 2008;11(6):433–41.
39. Kapitza C, et al. Oral insulin: a comparison with subcutaneous regular human insulin in patients with type 2 diabetes. Diabetes Care. 2010;33(6):1288–90.
40. Malkov D, et al. Oral delivery of insulin with the eligen technology: mechanistic studies. Curr Drug Deliv. 2005;2(2):191–7.
41. Leone-Bay A, et al. 4-[4-[(2-Hydroxybenzoyl)amino]phenyl]butyric acid as a novel oral delivery agent for recombinant human growth hormone. J Med Chem. 1996;39(13):2571–8.
42. Castelli MC, et al. SNAC co-formulation produces significant enhancement of oral vitamin B12 bioavailability in rats. FASEB J. 2008;22:795.
43. Milstein SJ, et al. Partially unfolded proteins efficiently penetrate cell membranes—implications for oral drug delivery. J Control Release. 1998;53(1–3):259–67.
44. Alani AW, Robinson JR. Mechanistic understanding of oral drug absorption enhancement of cromolyn sodium by an amino acid derivative. Pharm Res. 2008;25(1):48–54.
45. Azria M, Copp DH, Zanelli JM. 25 years of salmon calcitonin: from synthesis to therapeutic use. Calcif Tissue Int. 1995;57(6):405–8.
46. Lee YH, Sinko PJ. Oral delivery of salmon calcitonin. Adv Drug Deliv Rev. 2000;42(3):225–38.
47. Gschwind HP, et al. Metabolism and disposition of the oral absorption enhancer 14C-radiolabeled 8-(N-2-hydroxy-5-chlorobenzoyl)-amino-caprylic acid (5-CNAC) in healthy postmenopausal women and supplementary investigations *in vitro*. Eur J Pharm Sci. 2012;47(1):44–55.
48. Bagger YZ, et al. Oral salmon calcitonin induced suppression of urinary collagen type II degradation in postmenopausal women: A new potential treatment of osteoarthritis. Bone. 2005;37(3):425–30.
49. Dinh S, Liu P, Wong V. Gastric absorption of oral insulin in rats, in Annual Meeting of American Association of Pharmaceutical Scientists. Toronto: American Association of Pharmaceutical Scientists; 2002.
50. Malkov D, Wang H-Z, Angelo R. Mechanism of oral absorption of insulin with Emisphere drug delivery agents, in The 29th Annual Meeting of Controlled Release Society. Seoul: Controlled Release Society; 2002.
51. Arbit E, et al. Oral insulin as first-line therapy in type 2 diabetes: a randomized-controlled pilot study. Diabetologia. 2004;47:A5.
52. Hirsh J, et al. Guide to anticoagulant therapy: Heparin a statement for healthcare professionals from the American Heart Association. Circulation. 2001;103(24):2994–3018.
53. Leone-Bay A, Paton DR, Weidner JJ. The development of delivery agents that facilitate the oral absorption of macromolecular drugs. Med Res Rev. 2000;20(2):169–86.
54. Leone-Bay A, et al. Oral delivery of rhGH: preliminary mechanistic considerations. Drug News Perspect. 1996;9(10):586–91.
55. Castelli MC, et al. Comparing the efficacy and tolerability of a new daily oral vitamin B-12 formulation and intermittent intramuscular vitamin B-12 in normalizing low cobalamin levels: a randomized, open-label, parallel-group study. Clin Ther. 2011;33(3):358–71.
56. Molina E, et al. The effect of parathyroid hormone (hPTH 1–34) on oxytocin-induced contractions of pregnant human myometrium "*in vitro*". FASEB J. 1996;10(3):3800.

Chapter 19
The RapidMist™ System for Buccal Delivery of Insulin

Meena Bansal, Sanjay Bansal and Rachna Kumria

19.1 Introduction

Diabetes mellitus is a progressive, chronic, systemic disease characterized by dysfunction of the metabolism of fat, carbohydrate, protein, and insulin. It also causes a change in the structure and functioning of the blood vessels and nerves. Diabetes is affecting a large population worldwide and has now attained epidemic proportions [1]. Clinically, diabetes can be classified into four categories: type 1, type 2, gestational diabetes, and other type of diabetes (secondary diabetes). Type 1 diabetes, also referred to as insulin-dependent diabetes mellitus (IDDM), results from the destruction of β-cells and this in turn leads to a complete deficiency of insulin secretion. IDDM is a juvenile onset diabetes. Patients in this case must essentially take exogenous insulin to maintain blood glucose levels.

Type 2 diabetes, described as non-insulin dependent diabetes mellitus (NIDDM) or the adult-onset diabetes, results from progressive defect in insulin secretion as well as development of insulin resistance. Generally, type 2 diabetes is the most prevalent diabetes and accounts for nearly 90 % of all cases of diabetes. NIDDM is characterized by a decreased sensitivity of glucoreceptors towards insulin. Obesity, overeating, age and lack of physical activity are the key factors responsible for type 2 diabetes. Once considered as a disease of adults, it has been found to occur at all ages and more so in children. Gestational diabetes is the glucose intolerance developed during pregnancy, while secondary diabetes is a term used to describe diabetes attributed to diseases of the pancreas, genetic diseases, drugs, among other causes. The common outcome of all types of diabetes is hyperglycemia, which is associated with several diabetic complications. The management of this disease

R. Kumria (✉)
Swift School of Pharmacy, Village Ghaggar Sarai, Rajpura, Panjab, India
e-mail: kumria_r@yahoo.com

M. Bansal · S. Bansal
Mehr Chand Polytechnic College, Jalandhar, Panjab, India

J. das Neves, B. Sarmento (eds.), *Mucosal Delivery of Biopharmaceuticals,* DOI 10.1007/978-1-4614-9524-6_19,

requires lifestyle changes and weight control. Treatment focuses on regulating blood glucose levels akin to those observed in nondiabetics.

Normally, glucose regulation needs a balance between insulin, counter-regulatory hormones, intestinal incretins, and amylin. Insulin is responsible for stimulating glucose transport across the cell membrane and also for promoting storage of glucose. It further prevents the mobilization of fat for energy by inhibiting glucose production from liver and muscle glycogen, promotes incorporation of amino acids into protein, and decreases the breakdown of fatty acids. Diabetes is associated with abnormal glucose regulation and hyperglycemic emergencies. In uncontrolled diabetes, the disease progresses from an initial abnormality in glucose metabolism to life-threatening diabetes ketoacidosis or hyperglycemic hyperosmolar nonketotic syndrome.

In healthy non-diabetic humans, insulin is secreted by β-cells of islets of Langerhans in response to rising blood glucose levels (e.g., postprandial state). Insulin increases the glucose uptake by the cells of liver, muscles and adipose tissue. This reduces glucose levels to normal and, as a consequence, insulin secretion recedes. Secretion of insulin is thus a closely monitored process required to maintain normal blood glucose levels [2].

Type 1 and type 2 diabetic patients respond very well to intravenous (IV) insulin and as such it has been adopted as the most effective measure of controlling blood sugar levels mainly due to its quick onset, predictable hypoglycemia, and safety profile. But IV route has its own shortcomings and needle-free drug delivery would be highly preferred [3]. As such a number of noninvasive methods for insulin delivery are being intensively pursued. Success of insulin administration will depend upon the ability of delivery systems to elicit effective and predictable hypoglycemia and reduce the risk of diabetic complications. Additionally, delivery systems should be safe, simple, and patient friendly.

Transdermal, buccal, oral, pulmonary, nasal, ocular, and rectal routes are some of the recently explored routes for insulin delivery. Vaginal, rectal and colonic delivery of insulin have exhibited a certain degree of success but are associated with social limitations. Nasal delivery of insulin is not acceptable due to the fragility of the nasal mucosa. Oral route is not preferred for insulin administration as it degrades in the gastro-intestinal tract. Pulmonary insulin, once proposed to be a promising approach, has been found to pose risk of lung cancer and all pulmonary insulin products have been withdrawn from the market [4–5].

Mucosal delivery of insulin was found to be less effective than parenteral administration owing to poor membrane permeability and drug metabolism in the mucosa [6]. However, buccal mucosa has been used for thousands of years and is known to absorb a variety of drugs and natural substances. In recent years, it has emerged as a promising site for insulin delivery. The buccal mucosa offers numerous advantages as a drug delivery site [7] and these include (1) by-pass of pre-systemic metabolism, (2) relatively large surface area for absorption (100–200 cm^2), (3) ease of accessibility, (4) presence of high vasculature particularly in some regions, (5) possibility to place the drug delivery system according to the permeability features of the target area, (6) low enzymatic activity, (7) presence of a robust mucosa, (8) protection of insulin from destructive acidic environment of the stomach, and (9) low variability in pH. Also, it is patient friendly.

The drawbacks of the buccal route are (1) Unlike intestinal mucosa, which is a prime absorptive site, buccal mucosa is not an absorptive organ. Its histological features are those of a lining membrane, posing barrier to drugs [5, 9–10]. Thus, absorption of drug through buccal mucosa is challenging, (2) the relatively small area available for absorption (as compared to small intestine), (3) the presence of scavenging effect of saliva, (4) the challenge in delivering molecules with high molecular weight, and (5) the possibility of accidental swallowing of delivery systems (not relevant to spray-based delivery systems). All these limitations need to be taken into account while exploring the buccal mucosa as a delivery site for insulin [8, 11]. The main challenge for buccal insulin delivery is the size of the active molecule. Buccal delivery systems need to be designed in such a way that they promote the penetration of the macromolecule insulin across the mucosa, and additionally protect it from environmental degradation. To address the above listed limitations, a number of strategies have been developed, namely the use of (1) absorption enhancers, (2) fine-mist forming spray formulations capable of utilizing maximum available surface area, (3) physical shielding of the drug molecules, (4) enzyme inhibitors and solubilizers /surfactants, and (5) use of bioadhesive systems, among others. The use of such functional excipients is essential for successful buccal insulin delivery.

19.2 Insulin

The discussion of detailed information about insulin is beyond the scope of this chapter. However, a brief introduction about insulin, as relevant to the present case study, is outlined. Insulin and its analogs are categorized into different types depending upon their onset of action, T_{max}, and the duration of action: (1) rapid acting or ultra-short acting insulin analogs (lispro, aspart, and glulisine), (2) short-acting insulin (regular), (3) intermediate-acting insulin (isophane, detemir), (4) long-acting (glargine, levemir), and (5) premixed insulin products.

Insulin is required for all patients with type 1 diabetes and also in patients with type 2 diabetes where other injectable or oral therapies are not able to manage or achieve the required diabetic control. Insulin may be used in combination with other injectable antidiabetic agents to achieve glycemic control. In rapid-acting insulin, onset of action is less than 0.5 h, peaks around 0.5–1.5 h and is effective for 3–4 h. This type of insulin is administered immediately before meals (Table 19.1). Oral-lyn™ utilizes regular recombinant human insulin in its formulation. This is an insulin analog synthesized by making use of recombinant DNA (r DNA) technology. Regular insulin shows onset of action within 0.5–1 h and peaks between 2–4 h when administered subcutaneously (SC).

Table 19.1 Different types of insulin and their key pharmacokinetic properties when administered by SC injection

Insulin type	Onset of activity (h)	Time to peak (h)	Duration of action (h)	Comments
Rapid acting analogues (lispro, aspart, glulisine)	0.5	0.5–1.5	3–4	Administered immediately before meals
Short acting (regular, neutral)	0.5–1	2–4	5–8	Administered within 30 min before meals
Intermediate acting (NPH)	1–2.5	4–12	16–24	Administered once or twice daily
Long acting analogues (glargine)	1–2	–	24	Administered once daily

19.3 Buccal Insulin

The buccal mucosa is the inner lining of the cheek comprising of epithelium, lamina propria and submucosal area with a rich vasculature (Fig. 19.1). The buccal formulations are placed in the mouth between the upper gingivae (gums) and the cheeks, both for local and systemic delivery. Insulin reaching the systemic circulation by absorption through buccal mucosa is known as buccal insulin. Transport through the buccal mucosa is either transcellular or paracellular. Insulin traverses the epithelium via paracellular route [5].

The absorption across the buccal mucosa is influenced by molecular weight, charge, hydrophilicity/lipophilicity , stereospecificity, solubility, and partition coefficient. Small molecules such as nitroglycerine cross easily, whereas large molecules like insulin require permeability facilitators. After penetration, molecules rapidly enter the bloodstream, in contrast to the lag time seen with SC injection. This allows great flexibility in pin-pointing the time of action in the case of buccal insulin.

Oh and Ritschel studied the biopharmaceutical aspects of insulin absorption through buccal mucosa using rabbits [12]. They concluded that insulin penetration across the buccal mucosa was enhanced by using absorption enhancers. In another study, it was found that a minimum contact time is required for insulin molecule to be taken up by the buccal mucosa and the absorption profile of insulin was dependent on the time of application [13]. A number of similar studies have been conducted in small animals in order to evaluate the absorption enhancing effect of a number of compounds. The results of these studies have shown high variability and the absorption across the buccal mucosa could not be effectively increased. This is probably attributable to the fact that rodents have a highly keratinized mucosa, thus resulting in negligible insulin permeation. However, the buccal mucosa of humans has high permeability, especially in the sublingual non-keratinized region. The hard palate, on the other hand, is poorly permeable [7].

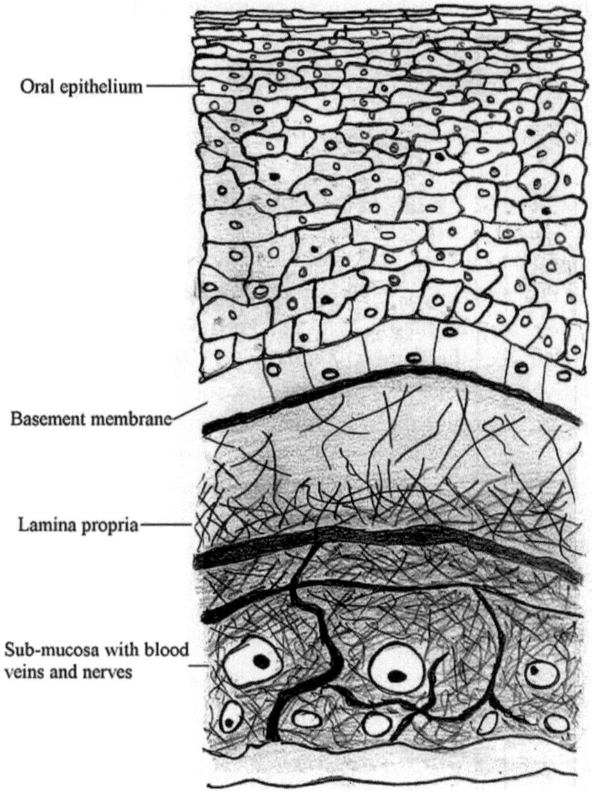

Fig. 19.1 Transverse section of buccal mucosa

19.4 RapidMist^{TM} Technology for Insulin Delivery

Generex Biotechnology developed a platform delivery device, RapidMist^{TM}, that delivers formulations such as Oral-lyn^{TM} directly into the buccal mucosa [14]. RapidMist^{TM} is similar to an inhaler device used by asthma patients. Oral-lyn^{TM} is the liquid formulation containing insulin for buccal delivery. It is the only buccal insulin formulation available till date. It is a tasteless, odorless liquid aerosol formulation of regular recombinant human insulin, with a spray propellant that forms a fine mist when activated. This insulin delivery system has been proposed for achieving glycemic control of prandial insulin.

The RapidMist^{TM} offers the following advantages: (1) simple and patient-friendly procedure, (2) rapid delivery into the blood circulation, (3) accurate dosing, and (4) bolus drug delivery. Each canister (28 ml) holds 400 IU of regular human insulin. The insulin formulation is stable at room temperature (North America) for 6 months. Once activated, the liquid insulin from the device is sprayed as micelles. The insulin-containing micelles so formed are relatively large in size ($> 7\ \mu m$, with 85 % having mean size $> 10\ \mu m$). Therefore, they cannot enter the deep lungs regardless of effort [5, 14]. Some subjects have reported a feeling of coldness upon initial use of the product probably caused by the propellant. However, this tends to resolve with regular use.

The manually actuated mechanism incorporated in RapidMist^{TM} introduces a metered dose into the oral cavity. Each spray of Generex Oral-lyn^{TM} is claimed to deliver 10 IU of insulin. The absorption efficiency of Oral-lyn^{TM} is around 10 %; consequently, 1 IU of insulin is delivered into the blood stream with each spray [15]. RapidMist^{TM} is known to deliver same amount of insulin each time and, as such, the first and last spray are identical. This allows exact dosing rather than approximate and makes this delivery device superior to other alternative forms of insulin delivery.

This technology utilizes the formation of microfine, thin membrane micelles, comprising a combination of insulin and specific absorption enhancers which encapsulate and protect the insulin molecules for safe and effective delivery. The RapidMist^{TM} delivery device introduces the aerosol at high velocity ($\approx$ 100 mph or 160 km/h) into the oropharyngeal cavity for absorption through local buccal mucosa [15]. A study was conducted to trace the absorption of insulin following an oral spray (reviewed in [5]). Radio-labelled insulin was administered to adults and the tracer was followed. It was observed that the radio-labelled insulin was located in the mouth, oropharynx, and gastrointestinal area up to esophagus only and no drug appeared in the lungs. This may be attributed to the large size of the droplets (7–10 μm) which prevents their entry into the lungs. This study also suggests that most of the insulin absorption occurs via buccal mucosa.

19.4.1 Composition and Mechanism of Action of Oral-lyn^{TM}

The major hurdle in devising a buccal insulin formulation is the large size of insulin molecule which is difficult to get across the inner lining of the oral mucosa. Generex Biotechnology has overcome this obstacle by incorporating an adequate combination of ingredients. Oral-lyn^{TM} contains a surfactant, a solubilizer, a micelle-creating agent, and emulsifying agents, all "generally regarded as safe" (GRAS) excipients, and prepared in a manner so as to allow insulin to penetrate in a predictable way [5]. These substances modify the lipophilicity of insulin and help the transport of micellized insulin across lipoidal cell membrane or across the lipid components of the intercellular permeability barrier. Surfactants added in the Oral-lyn^{TM} increase the effects of absorption enhancers. Once the RapidMist^{TM} device is activated, a fine mist of insulin micelles (with all functional excipients) deposits in the buccal cavity at high velocity (Fig. 19.2). The impelled micelles traverse the superficial layers of buccal mucosa and, with the aid of absorption enhancers, insulin molecules get rapidly absorbed into the bloodstream and appear in circulation within 5–10 min after administration [5, 14].

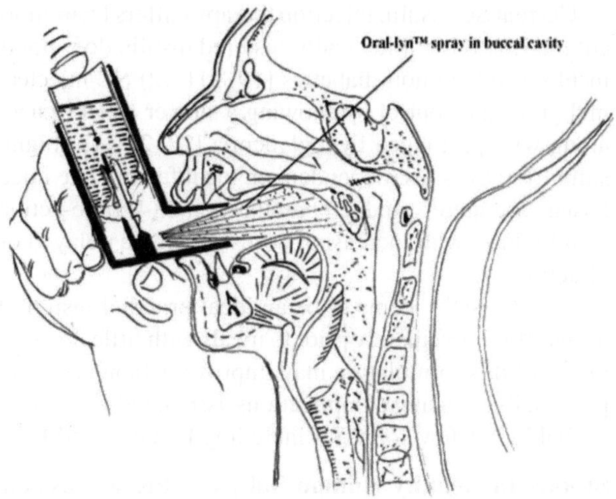

Fig. 19.2 A proposed mechanism of spray pattern of oral insulin mist from RapidMist^{TM}. (Adapted from [5])

19.4.2 The RapidMist^{TM} System

The RapidMist^{TM} drug delivery system uniquely combines pharmaceutical agents in a proprietary aerosolized liquid formulation with a compact, easy-to-use device. This patented system is capable of putting 150 or more compounds present as a solution, into fine mist form as in Oral-lyn^{TM}. Till date, the concept has been tested with morphine, fentanyl, and low molecular weight heparin [5]. It is proposed to address the needs of both type 1 and type 2 diabetic patients utilizing a patented spray device, similar to asthma inhalers. Insulin dose may be split into before-meal dose and after-meal dose, thereby reducing the risk of hypoglycemia. This flexibility provides both type 1 and type 2 patients, an opportunity to have a better control over glucose levels. The use of a familiar spray system (similar to asthma inhaler) allows both patients and physicians to readily accept it. The spray delivers accurately metered doses (equivalent to 1 IU of systemic insulin) with each spray. Thus, the patient has to take multiple sprays as per his requirement or as suggested by the physician.

19.4.3 Oral Insulin Spray: Pharmacokinetics

Various clinical studies conducted by the Diabetes Control and Complications Trial (DCCT) and the UK Prospective Diabetes Study (UKPDS) Group have proven that a proper management of glycemic levels remarkably decreases the risk of long-term diabetic complications in both type 1 and type 2 diabetic patients [16–17]. Guidelines from the International Diabetes Federation have strongly recommended the optimal management of postprandial hyperglycemia as a critical aim of diabetes therapy [18].

Current SC insulin injection therapy suffers from shortfalls such as (1) the pharmacokinetic (PK) profile of administered insulin does not match with the physiological insulin levels in non-diabetics [19–21], (2) SC injected insulin provides a suboptimal glycemic control by showing a slower onset of action and a lower insulin peak along with prolonged hypoglycemia [22–23], (3) highly variable absorption of insulin from SC sites (dependent on local factors like injection site, depth of injection, tissue vasculature, and temperature), and (4) time-action profile is dependent on the insulin dose, with increasing doses causing a delayed onset and prolonged duration of action.

RapidMist^{TM} seems to be an ideal prandial insulin that can be conveniently administered immediately prior to meals with little prospect of hypoglycemia. The ease of use of this formulation may improve patient acceptance and compliance, thereby potentially reducing complications associated with diabetes. The product is currently available in a few countries including Ecuador and India.

Studies in Healthy Human Subjects PK studies conducted on RapidMist^{TM} revealed that the absorption of insulin delivered via oral insulin spray was directly related to the administered dose. There was a quick onset and a shorter duration of action as compared to SC regular insulin. This PK feature makes oral insulin spray an ideal candidate for prandial glycemic control.

Effect of single dose Cernea et al. studied the PK and pharmacodynamic (PD) properties of Oral-lyn^{TM} on healthy subjects to study its dose-response profile and metabolic effects [24–25]. Insulin absorption and glucose uptake properties of a single-dose oral insulin spray were compared with SC insulin injection in six healthy individuals using euglycemic clamp technique. On the first visit, subjects were given 0.1 IU/kg SC regular insulin injection. On a subsequent visit 7–14 days later, the subjects received 15 oral sprays (150 IU, 10 IU/spray) of oral insulin. Insulin absorption and glucose uptake was studied. Results showed that oral insulin spray had a higher peak serum level (C_{max}) and a shorter time to achieve this (T_{max}). Oral-lyn^{TM} showed shorter time to peak glucose uptake in comparison to SC insulin. The total amount of glucose infused over 6 h was higher with SC insulin but the initial glucose consumption in the first 2 h was found to be similar in both cases.

Dose-response relationship Cernea et al. compared the dose-response profile of oral insulin spray with SC regular insulin in seven healthy individuals utilizing the glucose clamp technique [25]. This was a randomized, five-way crossover study. Subjects were given three different doses of oral insulin spray: five sprays (50U), ten sprays (100 IU), 20 sprays (200 IU) and a single dose of 0.10 IU/kg of SC regular insulin. Each spray of oral insulin delivered 10 IU. Serum insulin levels were recorded over 6 h. It was observed that with oral insulin spray, the onset of action was quicker (31.7 ± 12 vs. 77.8 ± 3 min), peak insulin levels were attained earlier (44.2 ± 10 min vs. 159.2 ± 68 min), and duration of action was shorter (85.1 ± 25 min vs. 319.2 ± 45 min) as compared to SC insulin. It was also observed that with an increase of the dose in insulin (use of multiple sprays) there was an increased response as shown by C_{max} and the area under-the-curve (AUC).

Studies in Type 1 Diabetics A number of different studies reported the PD and PK properties of oral insulin spray to evaluate the effect of this new delivery system under various conditions in type 1 diabetic subjects (e.g., biokinetic analysis, mealtime analysis, lunch time study, study in subjects taking NPH dosage, and the dose-response relationship). The results of these studies have been found to be similar to those obtained in healthy subjects. Important details and outcomes of these studies are given in the following sections.

Biokinetic analysis A biokinetic study was conducted in 18 subjects with type 1 diabetes. Comparison of plasma glucose, insulin, and C-peptide levels with SC regular insulin regimen at breakfast, to an equivalent amount of oral insulin spray was conducted on two consecutive days. During this period, patients did not receive intermediate or long-acting insulin [26]. No significant differences were observed in levels of insulin, C-peptide or hypoglycemic potential during meals between SC and oral insulin spray regimen. However, the duration of hypoglycemic effect was prolonged with regular SC insulin in comparison to oral insulin spray and the effect was more pronounced at 180 and 240 min. The study demonstrated that Oral-lyn™ is as effective as SC regular insulin in reducing blood glucose levels in type 1 diabetics, since plasma glucose levels between the insulin regimens showed no significant variations.

Comparison at mealtime in subjects on twice-daily insulin injection A biphasic study has been conducted to compare the PD of glucose at mealtime after oral insulin spray and SC injection [27]. The basal insulin treatment regimen of twice-daily insulin injection was continued throughout the study. Blood glucose levels were monitored by subjects. Levels of fructosamine and glycated hemoglobin (HbA1C) were determined at baseline and at the end of 12 days. In the first phase (3 days) subjects were given pre-meal injection of SC regular insulin before each major meal. In the second phase (9 days) this pre-meal injection was replaced with 2 doses (8–12 sprays/dose) of oral insulin spray. No significant differences were observed in the AUC values, levels of fructosamine, and HbA1C between the two treatment regimens.

Oral insulin spray administered at lunchtime in juvenile subjects Guevara-Aguirre evaluated the safety and efficacy of oral insulin administered at lunchtime in juvenile type 1 subjects receiving basal glargine insulin and pre-breakfast and pre-dinner SC regular insulin [28]. For a variety of social and behavioral reasons, lunch time is the most common meal for skipping insulin. The pre-lunch insulin injection was replaced with buccal insulin for more than 3 months and a continued fall in HbA1C value was observed with the buccal insulin at lunch in comparison to other meal times.

Oral insulin spray in subjects maintained on twice-daily NPH insulin Guevara-Aguirre also studied a group of well controlled type 1 diabetes mellitus subjects maintained on twice daily NPH insulin [29]. They were given NPH twice daily and before meals, either regular insulin injection or split doses of buccal insulin for 99 days. HbA1C dropped from a mean of 9 to 8.1 % at the stability point. After 99 days of therapy with oral insulin spray, HbA1C reduced to a value between 6–7 %.

Dose-ranging effects Dose-ranging studies of oral insulin spray and SC injected regular insulin were carried out by Cernea et al. in a randomized, single-blinded, open-label, five-way crossover trial [30]. Six male subjects received three different doses of oral insulin spray, i.e., five sprays, ten sprays, 20 sprays, one dose of placebo, and one dose of SC regular insulin injection on five separate days. All the other parameters were kept the same. The absorption of insulin and glucose uptake was assessed using euglycemic clamp technique. There was no significant difference in C_{max} and incremental insulin AUC from 0 to 2 h between 200 IU of oral insulin spray and SC insulin. However, serum insulin concentrations declined after 2 h of oral insulin spray. Oral insulin spray showed an earlier onset of action, a shorter time to reach C_{max}, and a shorter duration of action when compared to SC regular insulin. A dose-dependent relationship was observed for peak serum insulin levels with the three doses of oral insulin spray.

Studies in Type 2 Diabetics Only a few studies have been carried out on type 2 diabetic subjects using the oral insulin spray. The metabolic effects and PK of oral insulin spray in these subjects is briefly mentioned in this section.

Oral insulin spray in subjects failing on oral hypoglycemic agents (OHAs) Guevera-Aguirre et al. conducted a randomized, crossover, open-label study in type 2 subjects with suboptimal glucose control [31]. This study was conducted on two different days. All the subjects were given two treatment regimens separated by 3–10 days. While one treatment regimen included metformin, glyburide, and placebo sprays, the other regimen included metformin, glyburide, and 100 IU of oral insulin spray (four sprays; 25 IU/spray) equivalent to 7–8 IU of SC insulin. Baseline values of serum insulin, C-peptide, and glucose levels were recorded. Ten minutes after treatment, a standard breakfast was provided. The results indicated that the postprandial glucose levels were significantly reduced with oral insulin spray than with OHAs alone. At the end of 4 h period glucose levels were reduced by 38 %. C-peptide levels were also significantly reduced with oral insulin spray. Serum insulin levels were significantly higher and a quick onset of action was observed with oral insulin treatment. Thus, oral insulin spray can be used as a combination therapy with failing OHAs in type 2 diabetic patients to control postprandial glucose levels.

Effect of oral insulin spray on subjects managed with multiple insulin injections Guevara-Aguirre et al. performed another investigation to study the metabolic effect of oral insulin spray in type 2 diabetics who were maintained on multiple insulin injections daily [32]. Evening dose of regular insulin was continued with evening meal. An open-label, crossover, randomized study was conducted 3–7 days apart in 23 subjects. In one treatment, a daily dose of SC regular human insulin injection (0.1 IU/kg) was given; as for the other treatment, 100 IU of oral insulin spray (four sprays; 25 IU/spray) was administered. Blood samples were taken at baseline and at regular intervals to determine blood glucose, insulin, and C-peptide levels. There was a significant rise in serum insulin levels at 30 min with oral insulin spray as compared to SC insulin. Serum glucose levels and C-peptide levels were significantly lowered with oral insulin treatment during the first 60 min. The study concluded that oral insulin spray can be used safely instead of meal time insulin injections in type 2 diabetic patients to control postprandial hyperglycemia.

Oral insulin spray in subjects failing diet and exercise Modi et al. studied the response of Oral-lyn^{TM} formulation in 15 subjects failing diet and exercise [15]. Before a standard breakfast, subjects were given oral insulin spray or placebo spray and their blood glucose, insulin and C-peptide levels were recorded at regular intervals during the 4-h study period. The postprandial glucose levels were significantly lower with oral insulin spray in comparison to placebo. This suggests that instead of OHAs, oral insulin spray can be introduced early in the treatment of newly diagnosed type 2 patients not responding to diet and exercise alone. The early introduction of Oral-lyn^{TM} can be helpful to manage effectively the postprandial glucose levels and preserve β-cell function, thus avoiding long-term complications of diabetes. Though the outcome of all these studies is favorable to Oral-lyn^{TM}, more confirmatory data are required.

19.5 Safety Profile

Oral insulin spray was found to be well tolerated by almost all subjects, without any significant adverse effects. No major episodes of hypoglycemia or hyperglycemia were reported in type 1 and type 2 diabetic patients [29–30]. There were no abnormalities in vital functions or physicochemical parameters in subjects, though mild dizziness of limited duration was experienced by some individuals [24–25, 30]. A long-term (2 years) toxicological study was performed on 40 beagle dogs. Biopsies and scrapings of the buccal cavity, taken at regular intervals were examined by oral pathologists and no evidence of any change in buccal epithelium was found by observation or cytopathology [5, 14].

A proof-of-concept study was conducted by Palermo et al. to investigate the safety and efficacy of treatment with Oral-lyn^{TM} [33]. It was demonstrated that the buccal insulin spray is a simple and valuable therapy for reducing postprandial hyperglycemia in obese subjects with impaired glucose tolerance. Importantly, this treatment was safe and none of the study subjects experienced hypoglycemia. Also, no acute side effects were reported in the clinical studies using the buccal mucosa as a delivery site. This may probably be attributed to the pluristratified buccal epithelium which is more resistant than a monolayer of enterocytes in the gut towards insulin and also the excipients used in the buccal insulin spray. Though no acute side effects have been observed, there is a need for evaluation in a larger population. Only after large-scale safety studies are satisfactorily conducted, can the long-term safety of this buccal insulin formulation be confirmed.

19.6 Conclusions and Future Perspectives

Findings from clinical trials indicate that oral insulin spray seems to be an ideal candidate for postprandial glycemic control. The absorption of insulin has been found to be directly related to the administered dose. Oral insulin spray has a rapid onset of action and a relatively shorter duration of action as compared to SC insulin. A dose response relationship clearly depicts an increased PK/PD response with increasing dose. This insulin formulation can be used effectively as an add-on therapy in type 1 and type 2 diabetics. In type 2 diabetics on multiple daily injections, Oral-lyn™ can be used safely to replace meal time injections. Also oral insulin spray may be beneficial in patients who are non-compliant and refuse to take injections as the former is easy to use, self-administrable, needle-free, and pain-free.

However, a few more studies need to be conducted in order to evaluate the following:

1. The effect of type and concentration of surfactants used in the formulation and the harmful effects of these on long-term use. Such information is of utmost consideration since buccal insulin will have to be administrated daily throughout life [14]. Also inclusion of absorption enhancers over a long period of time may lead to potential cell damage;
2. Variation in response of the patients using the device at different times, i.e., intra-individual variability. The reproducibility of the metabolic effect induced has not been studied appropriately so far with oral insulin spray. A considerable difference in the intra-individual metabolic effect has been documented after SC injection of prandial or basal insulin in the same dose of the same insulin [34];
3. The distribution of the formulation into the oral cavity after spray. As different regions of the oral cavity have different permeability, the best way and the best region where the device may be placed during spray needs to be optimized and subjects need to be trained in this technique;
4. Performance of Oral-lyn™ with respect to rapid acting analogs. As till date, no appropriate comparative study with a rapid-acting insulin analog has been performed;
5. Relative biopotency of oral insulin spray, as this has not been much studied. Reference to this issue has been made only in one of the euglycemic clamp studies. Biopotency seems to be quite low, i.e., in the range of 1–2 %, and more than 95 % of the administered insulin is ingested by patients;
6. Also, large scale clinical trials need to be conducted as available studies are of limited scale.

The Oral-lyn™ system has a few drawbacks as well. These include: (1) administration of a dose equivalent to 10 IU systemic insulin requires ten sprays (this procedure may be quite time consuming and not user friendly); (2) though it is claimed that the device delivers a metered dose, identical from first spray to last, appropriate dosing of the buccal insulin requires some sort of training [14]; and (3) the low potency and heavily patented technology may lead to exuberant price of the product, but this may still be acceptable, as it provides the patient a needle free, rather safe approach to tackle diabetes.

The clinical implications of these findings are quite significant and oral insulin spray seems to be effective in managing diabetes. However, it may be too early to extrapolate its actions and uses in large diabetic populations, since the studies were performed on a very small number of subjects. Generex Biotechnology Corporation is on its way to undergo a large, randomized clinical trial to evaluate the safety, efficacy and acceptability of Oral-lyn™ formulation and its delivery device RapidMist™ [35]. The study will also assess the effectiveness of buccal insulin in comparison with SC regular human insulin. Apart from blood glucose measurements, other parameters such as insulin antibodies, buccal mucosa cytopathology, and the usual health parameters will also be studied. Moreover, better understanding of the mechanism behind this novel route of insulin administration, transport of insulin through buccal mucosa, and more detailed in vitro and in vivo absorption studies may lead to widespread acceptance of this novel approach in the scientific and medical community. There is a good chance that the buccal insulin formulation currently in the end phase of its clinical development will come to the market in the USA and Europe.

References

1. Wild S, Roglic G, Green A, Sicree R, King H. Global prevalence of diabetes: estimates for the year 2000 and projections for 2030. Diabetes Care. 2004;27(5):1047–53.
2. Varshosaz J. Insulin delivery system for controlling diabetes. Recent Pat Endocr Metab Immune Drug Discov. 2007;1(1):25–40.
3. Kumria R, Goomber G. Emerging trends in insulin delivery: buccal route. J Diabetol. 2011;2(1):1–9.
4. Turner R, Cull C, Holman R. United Kingdom Prospective Diabetes Study 17: a 9-year update of a randomized, controlled trial on the effect of improved metabolic control on complications in non-insulin-dependent diabetes mellitus. Ann Intern Med. 1996;124(1 Pt 2):136–45.
5. Bernstein G. Delivery of insulin to the buccal mucosa utilizing the RapidMist system. Expert Opin Drug Deliv. 2008;5(9):1047–55.
6. Lee VH. Enzymatic barriers to peptide and protein absorption. Crit Rev Ther Drug Carrier Syst. 1988;5(2):69–97.
7. Heinemann L, Jacques Y. Oral insulin and buccal insulin: a critical reappraisal. J Diabetes Sci Technol. 2009;3(3):568–84.
8. Chidambaram N, Srivatsava AK. Buccal drug delivery systems. Drug Dev Ind Pharm. 1995;21(9):1009–36.
9. al-Waili NS. Sublingual human insulin for hyperglycaemia in type I diabetes. J Pak Med Assoc. 1999;49(7):167–9.
10. Veuillez F, Kalia YN, Jacques Y, Deshusses J, Buri P. Factors and strategies for improving buccal absorption of peptides. Eur J Pharm Biopharm. 2001;51(2):93–109.
11. Rossi S, Sandri G, Caramella CM. Buccal drug delivery: A challenge already won? Drug Discov Today Technol. 2005;2(1):59–65.
12. Oh CK, Ritschel WA. Biopharmaceutic aspects of buccal absorption of insulin. Methods Find Exp Clin Pharmacol. 1990;12(3):205–12.
13. Oh CK, Ritschel WA. Absorption characteristics of insulin through the buccal mucosa. Methods Find Exp Clin Pharmacol. 1990;12(4):275–9.
14. Generex Biotechnology Corp. Data on File, 2008.
15. Modi P, Mihic M, Lewin A. The evolving role of oral insulin in the treatment of diabetes using a novel RapidMist System. Diabetes Metab Res Rev. 2002;18(Suppl 1):S38–42.

16. The Diabetes Control and Complications Trial Research Group. The effect of intensive treatment of diabetes on the development and progression of long-term complications in insulin-dependent diabetes mellitus. The Diabetes Control and Complications Trial Research Group. N Engl J Med. 1993;329(14):977–86.
17. UK Prospective Diabetes Study (UKPDS) Group. Intensive blood-glucose control with sulphonylureas or insulin compared with conventional treatment and risk of complications in patients with type 2 diabetes (UKPDS 33). UK Prospective Diabetes Study (UKPDS) Group. Lancet. 1998;352(9131):837–53.
18. International Diabetes Federation Guideline Development Committee. Guideline for Management of Postmeal Glucose. Brussels: International Diabetes Federation; 2007.
19. Berger M, Cuppers HJ, Hegner H, Jorgens V, Berchtold P. Absorption kinetics and biologic effects of subcutaneously injected insulin preparations. Diabetes Care. 1982;5(2):77–91.
20. Binder C, Lauritzen T, Faber O, Pramming S. Insulin pharmacokinetics. Diabetes Care. 1984;7(2):188–99.
21. Patton JS, Bukar J, Nagarajan S. Inhaled insulin. Adv Drug Deliv Rev. 1999;35(2–3):235–47.
22. Galloway JA, Spradlin CT, Nelson RL, Wentworth SM, Davidson JA, Swarner JL. Factors influencing the absorption, serum insulin concentration, and blood glucose responses after injections of regular insulin and various insulin mixtures. Diabetes Care. 1981;4(3):366–76.
23. Almér LO, Wollmer P, Jonson B, Troedsson Almér A. Insulin inhalation with absorption enhancer at meal-times results in almost normal postprandial insulin profiles. Clin Physiol Funct Imaging. 2002;22(3):218–21.
24. Cernea S, Kidron M, Wohlgelernter J, Modi P, Raz I. Comparison of pharmacokinetic and pharmacodynamic properties of single-dose oral insulin spray and subcutaneous insulin injection in healthy subjects using the euglycemic clamp technique. Clin Ther. 2004;26(12):2084–91.
25. Cernea S, Kidron M, Wohlgelernter J, Modi P, Raz I. Dose-response relationship of oral insulin spray in healthy subjects. Diabetes Care. 2005;28(6):1353–7.
26. Pozzilli P, Manfrini S, Costanza F, Coppolino G, Cavallo MG, Fioriti E, Modi P. Biokinetics of buccal spray insulin in patients with type 1 diabetes. Metabolism. 2005;54(7):930–4.
27. Guevara-Aguirre J, Guevara-Aguirre M, Saavedra J, Bernstein G, Rosenbloom AL. Comparison of oral insulin spray and subcutaneous regular insulin at mealtime in type 1 diabetes. Diabetes Technol Ther. 2007;9(4):372–6.
28. Guevara-Aguirre J, Guevara-Aguirre M, Saavedra J, Bernstein G. 6-month safety and efficacy of lunch-time oral insulin in juvenile type-1 DM subjects receiving basal glargine insulin and pre-breakfast and pre-dinner S.C. regular insulin (abstract), Endocrine Society 89th Annual Meeting, 2007.
29. Guevara-Aguirre J, Guevara-Aguirre M, Saavedra J. A comparison of metabolic control of preprandial S.C. regular insulin versus prandial split doses of an oral insulin (Generex Oral-lyn) in well controlled type 1 diabetes mellitus subjects maintained on twice daily NPH insulin (abstract), Endocrine Society 89th Annual Meeting, 2007.
30. Cernea S, Kidron M, Wohlgelernter J, Raz I. Dose-response relationship of an oral insulin spray in six patients with type 1 diabetes: a single-center, randomized, single-blind, 5-way crossover study. Clin Ther. 2005;27(10):1562–70.
31. Guevara-Aguirre J, Guevara M, Saavedra J, Mihic M, Modi P. Beneficial effects of addition of oral spray insulin (Oralin) on insulin secretion and metabolic control in subjects with type 2 diabetes mellitus suboptimally controlled on oral hypoglycemic agents. Diabetes Technol Ther. 2004;6(1):1–8.
32. Guevara-Aguirre J, Guevara M, Saavedra J, Mihic M, Modi P. Oral spray insulin in treatment of type 2 diabetes: a comparison of efficacy of the oral spray insulin (Oralin) with subcutaneous (SC) insulin injection, a proof of concept study. Diabetes Metab Res Rev. 2004;20(6):472–8.
33. Palermo A, Napoli N, Manfrini S, Lauria A, Strollo R, Pozzilli P. Buccal spray insulin in subjects with impaired glucose tolerance: the prevoral study. Diabetes Obes Metab. 2011;13(1):42–6.
34. Heinemann L. Variability of insulin absorption and insulin action. Diabetes Technol Ther. 2002;4(5):673–82.
35. Generex Biotechnology Corp. Active Comparator Study of Generex Oral-lyn™ Spray and Injected Human Insulin, ClinicalTrials.gov Identifier: NCT00668850, 2008.

Chapter 20
The Pharmaceutical Development of rhDNase (Dornase Alpha) for the Treatment of Cystic Fibrosis

Steven J. Shire and Thomas M. Scherer

20.1 Introduction

20.1.1 Cystic Fibrosis and Treatment with DNase

Cystic fibrosis (CF), which occurs at a rate of one in 2,500 births, is considered the most common lethal inherited genetic disease in Caucasians [1]. The major cause of morbidity and mortality in CF is due to chronic obstruction of the airways by thick mucosal secretions. It has been shown that the major genetic defect responsible for this pathology is in the cystic fibrosis transmembrane conductance regulatory protein (CFTR), which regulates chloride ion transport across respiratory epithelial cell membranes. There have been over 1,300 mutations identified [2], of which 20 occur most commonly in CF patients [3, 4]. The most prevalent mutation that makes up 70 % of all observed CF mutations is a deletion of phenylalanine 508 that results in improper protein transport and inadequate incorporation of the CFTR into the membrane [3]. It is believed that lack of this important ion regulatory protein results in an abnormal ion transport, which leads to dehydration and thickened viscous mucosal secretions in the airways of CF patients [5, 6]. This thickened mucus may contribute to a decreased mucociliary clearance, which promotes persistent bacterial infection. These persistent infections elicit an immune response whereby neutrophils are mobilized to the site of the infection, which results in a large concentration (3–14 mg/g of sputum) of extracellular DNA from the lysed neutrophils [5–6]. The presence of a high concentration of DNA contributes further to an increase in sputum viscosity, which in turn may result in further decrease in mucocilliary clearance resulting in persistence of bacterial infection. Thus, this vicious cycle continues making it very difficult for the patient to mount an effective immune response. Since increased concentrations of DNA exacerbate the problem, it was reasoned that use of an enzyme that could cleave the DNA should result in lower sputum viscosity and increase

T. M. Scherer (✉) · S. J. Shire
Genentech, Late Stage Pharmaceutical Development, San Francisco, CA, USA
e-mail: tscherer@gene.com

J. das Neves, B. Sarmento (eds.), *Mucosal Delivery of Biopharmaceuticals,* DOI 10.1007/978-1-4614-9524-6_20,

mucociliary clearance. In fact, it was shown that incubation of purulent sputum with bovine pancreatic DNase I (bDNase) resulted in a large decrease in sputum viscosity [7, 8], and that treatment of patients with this enzyme delivered by the pulmonary route resulted in a decrease of viscosity of purulent sputum [9]. Although there was no clear clinical response in the initial studies with bDNase there were no adverse events, and thus it was concluded that the doses used were inadequate since the observed decrease of sputum viscosity lasted less than 12 h. In a later study where bDNase was used to treat various bronchopulmonary diseases, it was concluded that patients had beneficial effects within 0.5–1 h after treatment [10]. The reported beneficial effects included increased amounts of expectorated sputum compared to that before treatment and the change in the physical appearance of the sputum, but this study lacked appropriate placebo controls, or measurements of improved lung function. Eventually, on the basis of the observed effects of purulent sputum from patients, bDNase was approved for human use in 1958 under the generic name dornase or the trade name Dornavac (Merck, Sharp and Dome Research Laboratories). Later studies using dornase to treat patients with pneumonia [11] and CF [12] led to the conclusion that there were benefits to the patients due to noticeable thinning and measured decrease in sputum viscosity as determined by cone-and-plate viscometry, as well as an increased volume of expectorated sputum. Altogether the studies with bDNase did suggest that the drug had a reasonable safety profile and was effective in reducing the viscosity of purulent lung secretions. However, occasional adverse reactions that could be attributed to an allergic reaction as a consequence of administration of a foreign protein or irritation due to contaminating proteases in the bDNase preparations, [12, 13] led to the withdrawal of Dornavac from the market. In order to mitigate the problems that resulted from using a foreign protein, human DNase (rhDNase) was cloned from a human pancreatic cDNA library and expressed in human embryonic kidney 293 cells (HEK 293) [14]. The rhDNase used in clinical trials was expressed in Chinese hamster ovary cells (CHO), purified and formulated for delivery as an aerosol by jet nebulizers.

Unlike previous clinical trials of bDNase, the stability of the rhDNase formulation before and after nebulization was characterized, along with the physical properties of the aerosol, further discussed in this chapter. In two Phase I dose-escalation studies, [15, 16] no acute adverse events occurred and most significantly no antibody titers to rhDNase were detected after a single dose. In addition, the studies showed an indication of efficacy as determined by improvements in the subject's mean forced expiratory volume in 1 s (FEV1). Additional safety and efficacy data from two randomized placebo-controlled Phase II clinical trials [17, 18] and an extensive Phase III trial using 968 patients [15] showed no life-threatening adverse events and demonstrated efficacy as assessed by increases in pulmonary function determined by changes in forced vital capacity (FVC) and FEV1. Moreover, patients dosed with aerosolized rhDNase spent fewer days in the hospital and had fewer respiratory exacerbations requiring antibiotic treatment. These studies demonstrated conclusively that rhDNase was both an effective and safe drug for treatment of CF and resulted in FDA approval of rhDNase under the trade name Pulmozyme® on December 30,

1993. Since its introduction, Pulmozyme® has been approved in approximately 70 countries and has been shown to be a safe and effective treatment for CF patients [19].

20.2 Pharmaceutical Development of Pulmozyme®

When launched in 1993, Pulmozyme® was the first new therapy in 30 years for the treatment of CF. The rapid development time, 5 years from cloning and expression to commercial introduction, was all the more remarkable because of the unique challenges faced with the development of a protein, therapeutic for specific local delivery to the lung. In particular, formulations had to be developed that ensured stability over a 2-year shelf life, as well as during delivery by devices that generate aerosols. Since patients were already using jet nebulizers to dispense CF medications such as antibiotics, the developed formulations needed to be compatible with these devices and also provide stability of the protein during the aerosolization process. In addition to these requirements, the formulation coupled with the aerosol delivery device should result in a robust delivery of the protein drug to the airways. Thus, studies on both protein stability and aerosol delivery of the formulated protein needed to be done using the specific delivery device. Here we summarize the formulation and the aerosol characterization studies that were used to register Pulmozyme® as a licensed pharmaceutical. Many of the jet nebulizer and air compressor systems that were approved for use with Pulmozyme® are now considered obsolete, and thus it is highly desired to enable use of new delivery technology that may also be more convenient for patients. In particular, CF patients use many inhaled medications that are delivered by jet nebulizers and bulky air compressors, which require substantial time for aerosol delivery, and may restrict travel due to portability issues. Shortening the time for delivery of the drug or providing smaller portable devices would likely be welcomed by CF patients who already have a high daily treatment burden. Recent studies investigating new nebulizer technology are also summarized in this chapter.

20.3 Formulation Development of rhDNase

20.3.1 Major Route of Degradation of rhDNase Solutions

The formulation for a protein must stabilize the protein drug over its intended shelf life, and requires investigation into the major routes of degradation. A variety of assays including size-exclusion chromatography, SDS polyacrylamide gel electrophoresis (SDS-PAGE), tentacle ion-exchange chromatography, and an activity assay based on release of an intercalated dye from DNA during enzymatic digestion were used to monitor product stability. The tentacle ion-exchange chromatography assay detects a specific deamidation at Asparagine (Asn) 74 [20], which

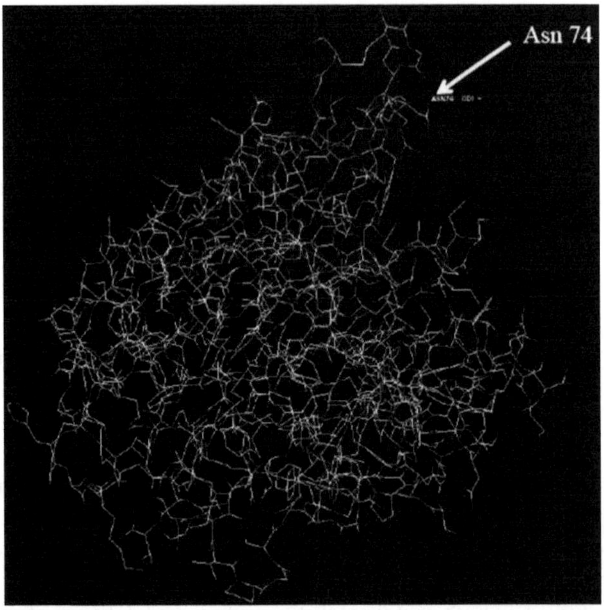

Fig. 20.1 rhDNase X-ray crystal structure showing exposed Asn 74 in DNA binding pocket. (Reproduced with permission from Protein Data Bank (www.rcsb.org))

is in an exposed loop involved with the binding of the DNA substrate (Fig. 20.1). The deamidation at Asn 74 is a major route of degradation of rhDNase in solution that results in a decrease in activity as assessed by the methyl-green activity assay. This degradation route, however, does not lead to a completely inactive molecule but rather a protein with 50–60 % of the initial activity of the nondeamidated protein. The correlation between deamidation and activity is shown in Fig. 20.2 for rhDNase formulated in 5 mM Tris, 150 mM NaCl at pH 8 and stored at 37 °C. This decrease in activity appears to be related to a decrease in binding of the substrate in the binding pocket occupied by Asn 74. The conversion of this Asn residue to either a negatively charged Aspartic acid (Asp) or isoaspartic acid (isoasp) residue [21, 22] appears to result in an electrostatic repulsion of the negatively charged phosphate backbone of the DNA substrate, thus decreasing the binding of the substrate. The rates of deamidation as a function of pH at 2–8, 15, 25, and 37 °C of rhDNase were determined using tentacle ion-exchange chromatography. As expected, the pseudo-first-order rate constants for deamidation are highly pH dependent and decrease with a reduction in pH as well as temperature (Table 20.1) have been previously observed for deamidation of peptides (Robinson and Rudd 1974; Wright, 1991; Cleland et al. 1993). Although the rate constant for deamidation at pH 5 is smaller than at pH 8 ($\approx$0.004 versus 0.1 day^{-1} at 37 °C storage), precipitation occurs at 37 °C and pH 5. These data suggested that the formulation pH should be kept low enough to effectively control the rate of deamidation but not too low since precipitation could occur upon storage. Storage in the pH range of 6–8 resulted in minimal detection of protein aggregation ($<$ 1 %) as determined by SDS PAGE and native size-exclusion chromatography (data not shown).

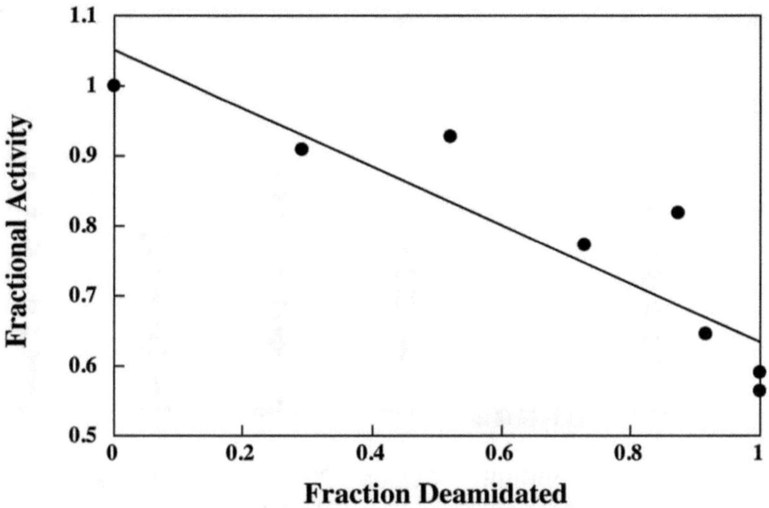

Fig. 20.2 Correlation of fractional activity of rhDNase with fractional deamidation at pH 8 and 37°C. The fractional activity is computed as the active concentration as determined by the methyl green-DNA activity assay divided by the concentration determined using the UV absorbance at 280 nm (absorptivity for rhDNase for a 1 cm pathlength = 1.7 mg/mL cm^{-1}). Fraction deamidated is computed as $C_{deam}/C^0{}_{deam}$ where C_{deam} and $C^0{}_{deam}$ are concentrations of deamidated rhDNase at time t and 0, respectively. (Reproduced from reference [34], with kind permission from Springer Science and Business Media. Copyright 2002)

Table 20.1 First order rate constants ($days^{-1}$) for deamidation of rhDNase as assessed by tentacle ion exchange chromatography[a]. (Data adapted from reference [34])

Formulation[b]	2–8 °C	15 °C	25 °C	37 °C
Acetate, pH 5	+	≈0	$7 \pm 1 \times 10^{-4}$	$3.7 \pm 0.15 \times 10^{-3}$
Succinate, pH 5	$3.6 \pm 10 \times 10^{-4}$	$7 \pm 7 \times 10^{-5}$	$5.6 \pm 1.5 \times 10^{-4}$	$3.8 \pm 1.5 \times 10^{-3}$
Citrate, pH 5	+	+	$7.3 \pm 1.4 \times 10^{-4}$	$4.0 \pm 1.3 \times 10^{-3}$
Histidine, pH 6	$3 \pm 3 \times 10^{-4}$	$1.3 \pm 0.2 \times 10^{-3}$	$3.6 \pm 0.15 \times 10^{-3}$	$1.4 \pm 0.03 \times 10^{-2}$
Succinate, pH 6	+	$7 \pm 7 \times 10^{-5}$	$1.6 \pm 0.1 \times 10^{-3}$	$9.96 \pm 0.16 \times 10^{-3}$
Maleate, pH 6	+	$2.4 \pm 1 \times 10^{-4}$	$1.3 \pm 0.08 \times 10^{-3}$	$9.66 \pm 0.65 \times 10^{-3}$
Tris, pH 7	$1.9 \pm 0.5 \times 10^{-3}$	$5.7 \pm 0.8 \times 10^{-3}$	$1.3 \pm 0.08 \times 10^{-2}$	$3 \pm 0.15 \times d10^{-2}$
Tris, pH 8	$1.1 \pm 0.1 \times 10^{-2}$	$2.3 \pm 0.13 \times 10^{-2}$	$4.8 \pm 0.3 \times 10^{-2}$	$1.3 \pm 0.03 \times 10^{-1}$

[a](+) indicates that slope was slightly positive. The errors in the determined rate constants of deamidation at 2–8 and 15 °C at pH values below seven were large because of the small degree of deamidation. In some cases the experimental error was larger than the observed change in deamidation resulting in apparent positive slopes
[b]Buffers consist of 5 mM buffer salt, 150 mM NaCl and 1 mM $CaCl_2$

20.3.2 Biological Compatibility of rhDNase Formulations

In addition to providing stability, the formulation must be compatible with the lung airway tissues. Critical variables that impact the bronchoconstriction and potential adverse reactions during pulmonary delivery by nebulization include the osmolality and pH of the drug solution [23–28]. Moreover, it was recommended that nebulizer

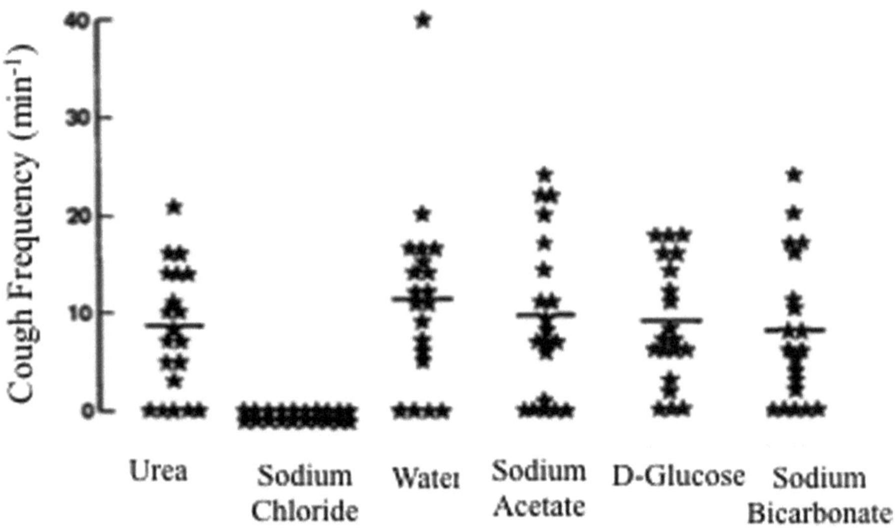

Fig. 20.3 Cough frequency during challenge with isotonic urea at pH 8.2, isotonic NaCl at pH 6.0, isotonic sodium acetate at pH 7.5, isotonic D-glucose at pH 3.6, isotonic sodium bicarbonate at pH 8.9, and water. Cough did not occur with NaCl, but occurred with similar frequency for all the other solutions. (Reproduced with permission from reference [31]. Copyright (1986) the Biochemical Society)

solutions should be isotonic at pH > 5 to prevent adverse reactions [24]. It has also been shown that formulations that are not isotonic can alter the droplet size distribution of the aerosol as a result of uptake of moisture from the airways [29, 30]. As will be discussed later, the droplet size distribution can affect the amount of drug that gets delivered to the airways.

In addition to the impact of tonicity, it has been shown that buffer components can cause adverse reactions such as involuntary cough responses as shown in Fig. 20.3 [31]. Thus, Pulmozyme® was formulated as an unbuffered isotonic NaCl solution at 150 mM at pH 6.3 ± 0.7. Although the control of the pH in an unbuffered formulation may be a major concern, especially since the major degradation route of rhDNase is deamidation and therefore highly pH dependent, rhDNase formulated at 1 mg/ml provides sufficient buffering capacity so that the pH of the formulated drug product is quite stable over the recommended storage life of the drug [32].

20.4 Addition of Calcium for Stabilization of rhDNase

Since calcium and other bivalent cations regulate bDNase activity, substrate specificity, and conformation [33–35], it was anticipated that calcium may be required for an rhDNase formulation. The binding of calcium to rhDNase at pH 5–6 was studied to determine if the calcium binding properties of the bovine and human protein were similar. Over the pH range of 5–6, there are four weak binding sites and one strong binding site for Ca^{++} in rhDNase [34], which is comparable to what has been

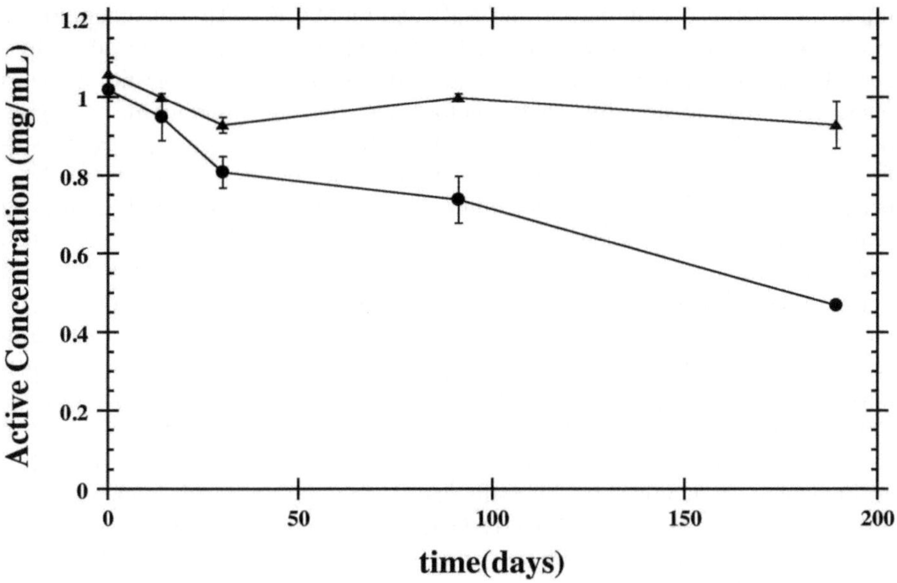

Fig. 20.4 Effect of calcium on activity of rhDNase stored at 25°C at ≈ pH 6. rhDNase was either formulated in 150 mM NaCl and 1 mM $CaCl_2$ (solid triangles) or treated with EDTA and formulated in isotonic 10 mM PO4 (solid circles). The active concentration was determined by the DNA-methyl green activity assay. (Reproduced with permission from reference [32]. Copyright (1994) American Chemical Society)

observed for bDNase at pH 5 [36]. After treatment of rhDNase with EDTA, there are 1–1.5 calcium ions that remain bound to rhDNase, and subsequent formulation into a phosphate buffer results in a loss of activity when stored at 25 °C (Fig. 20.4). Although it has been shown that phosphate catalyzes deamidation in many peptides [37], the rate of deamidation was similar with and without the phosphate buffer (data not shown). Altogether these data suggest that the phosphate buffer effectively competes with any remaining calcium bound tightly to the protein, and that removal of this essential and tightly bound calcium results in loss of activity. Additional studies of rhDNase stored at 40 °C showed that after treatment with EDTA the rate of deamidation did increase compared to an rhDNase solution with calcium. All of these observations suggested that addition of calcium would be required to maintain stability. The addition of 1 mM calcium to the 150 mM NaCl formulation (≈ 33-fold molar excess compared to rhDNase) was sufficient to maintain stability at the recommended storage temperature of 2–8 °C [32].

20.4.1 Formulation and Stability of rhDNase in Final Primary Packaging

All of the preformulation studies described above and the initial dose response studies in clinical trials [17] led us to formulate rhDNase at 1 mg/ml in an unbuffered isotonic liquid formulation containing 150 mM NaCl and 1 mM $CaCl_2$. The container-closure system for this formulation was originally a 5-cc glass vial with siliconized, Teflon-coated, gray, butyl rubber stoppers. An alternative to glass vials are plastic ampoules manufactured using blow-fill-seal technology, which is often used for the packaging of aerosol products such as normal saline and the bronchodilator, metaproterenol sulfate (Alupent®). This process uses thermoplastic low-density polyethylene (LDPE) pellets that in one self-contained manufacturing line extrudes, blow-molds, fills, and seals plastic vials or ampoules in one continuous operation. The advantages of such a system include, contact of one material with drug since stopper enclosures are not required, reduction of breakage of vials, elimination of vial preparation, and more convenience for the patient. However, because of the need to ship bulk drug substance (DS) to an outside contractor specializing in this technology and using a new fill manufacturing process for rhDNase, additional studies needed to be done. This included freeze-thaw studies and validation of a process whereby DS would be shipped to another site and thawed, sterile filtered, and transferred to the blow-fill-seal machine. In addition, the blow-fill-seal technology can result in the exposure of product to temperatures as high as 37°C for several minutes. Thus, before any large-scale fills were done at the contractor site, studies were performed to test the quality of rhDNase after a 15-min exposure to 37 °C in plastic ampoules made from resins from two potential suppliers [38]. Empty ampoules in a research and development study were filled with rhDNase at 4 mg/ml, sealed, and placed in an incubator at 37 °C for 15 min. The rhDNase was then analyzed by assessment of color and clarity, UV absorption spectroscopy, ELISA, DNA-methyl green activity assay, and gel-sizing chromatography. The stability of rhDNase in plastic ampoules before and after incubation at 37 °C for 15 min was shown to be fully active without any aggregate formation. The near- and far-UV circular dichroism of rhDNase at 4 mg/ml before and after incubation of the protein in the plastic ampoules at 37 °C for 15 min as compared to the rhDNase in glass vials was unaltered, showing that rhDNase conformation is not affected by a 15-min incubation at 37 °C in plastic ampoules [34].

rhDNase was filled in glass vials for the first clinical studies, and therefore a direct stability comparison was made between rhDNase in glass vials and rhDNase filled into plastic ampoules using blow-fill-seal technology. Since this manufacturing process uses LDPE resins that result in containers that are gas permeable, it is conceivable that the protein concentration may change as a result of water vapor transmission through the walls of the ampoules. It is also possible that permeation of oxygen into the ampoules could result in oxidation of the protein potentially leading to another major product degradation route. This problem can be mitigated by packaging the plastic ampoules in a gas-impermeable foil-laminate pouch, which may also be filled with nitrogen. The stability study was therefore also designed to determine if the stability in plastic ampoules is different when the ampoules are foiled with or without a nitrogen atmosphere [38]. The presence of a foil barrier with a nitrogen atmosphere did not have any appreciable protective effect on the activity of rhDNase (Fig. 20.5) or aggregate formation as assessed by SDS-PAGE and gel sizing chromatography or color and clarity of product (data not shown). Although

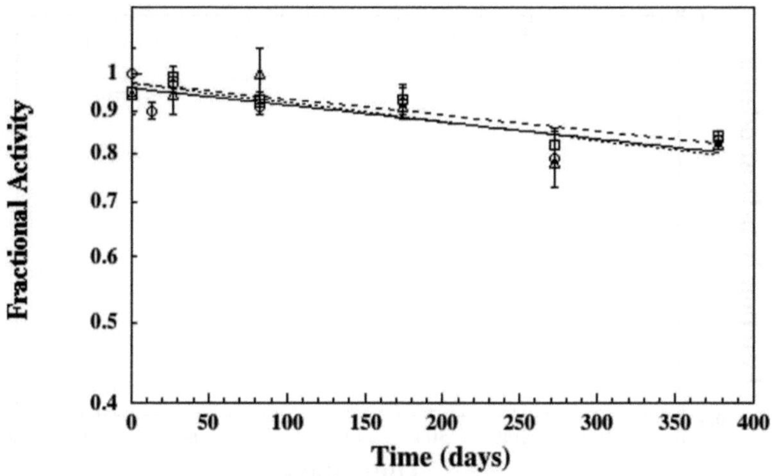

Fig. 20.5 Stability of rhDNase at 25 °C in plastic ampoules made with Dupont 20 resin. The fractional activity is computed as the active concentration as determined by the DNA-methyl green activity assay divided by the concentration determined using the UV absorbance at 280 nm (absorptivity = 1.7 cm^{-1} mL/mg). The data were analyzed by pseudo first order kinetics for unfoiled ampoules (open circles, solid line, $k = 4.7 \times 10^{-5}$ day^{-1}), foiled ampoules (open squares, dashed line, $k = -4.5 \times 10^{-5}$ day^{-1}), and ampoules foiled in the presence of a nitrogen atmosphere (open triangles, dotted line, $k = -5.2 \times 10^{-5}$ day^{-1}). (Reproduced from reference [34], with kind permission from Springer Science and Business Media. Copyright 2002)

these data suggest that a foil barrier is not required to maintain stability and quality of rhDNase, there was a noticeable increase in protein concentration during storage at 37 °C due to water vapor transmission through the low-density polyethylene ampoule walls [34]. Moreover, exposure to intense fluorescent light (≈ 1,600 foot-candles) results in aggregate formation as detected by SDS-PAGE (data not shown). Altogether, these data supported the decision to include a foil overlay for packaged rhDNase ampoules.

During the various stability investigations, it became evident that the rates of deamidation were similar for rhDNase stored at 2–8 °C in foiled and unfoiled plastic ampoules, but substantially different from rhDNase stored in glass vials (Fig. 20.6). The pseudo-first-order rate constants at 2–8, 25, and 37 °C were two- to threefold greater for rhDNase stored in glass vials than in plastic ampoules (Table 20.2). Additional investigative work showed that the pH of unbuffered, formulated rhDNase in a plastic ampoule was found to increase by 0.5 pH unit in less than 2 h after direct transfer into a glass vial and remained constant after the initial increase (data not shown). Thus, it is likely that due to the leaching of ions from the glass surface and the subsequent replacement of these ions by protons from the water leads to a significant increase in pH, especially considering the low buffer capacity of the formulation, resulting in faster rates of deamidation.

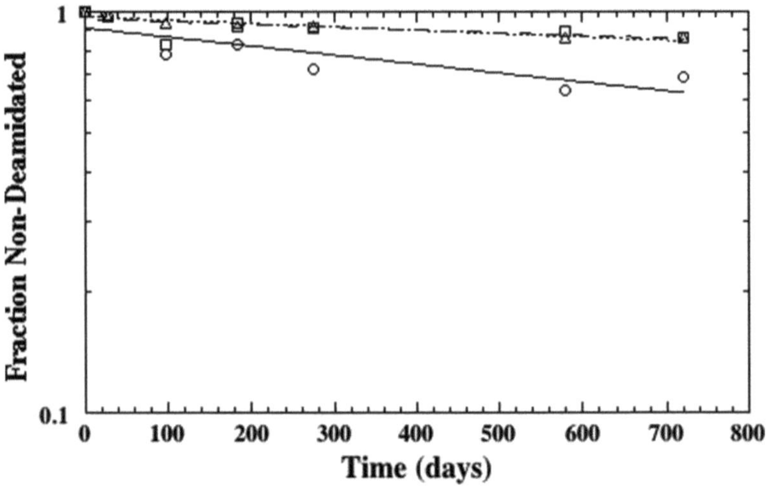

Fig. 20.6 Kinetics at 2–8 °C of deamidation of rhDNase formulated in 150 mM NaCl, 1 mM $CaCl_2$ stored in glass vials (open circles), foiled Dupont 20 plastic ampoules (open squares), and unfoiled Dupont 20 plastic ampoules (open triangles). Fraction Deamidated is computed as $Cd_{eam}/C^0{}_{deam}$ where C_{deam} and $C^0{}_{deam}$ are concentrations of deamidated rhDNase at time t and 0, respectively. (Reproduced from reference [34], with kind permission from Springer Science and Business Media. Copyright 2002)

Table 20.2 First order rate constants ($days^{-1}$) for deamidation of rhDNase as assessed by tentacle ion-exchange chromatography. (Data adapted from reference [34])

Container	2–8 °C	25 °C	37 °C
Glass	$5.7 \pm 1.6 \times 10^{-4}$	$7.2 \pm 0.8 \times 10^{-3}$	$2.6 \pm 1.2 \times 10^{-2}$
Unfoiled plastic	$2.0 \pm 0.3 \times 10^{-4}$	$2.2 \pm 0.1 \times 10^{-3}$	$8.2 \pm 0.9 \times 10^{-3}$
Foiled plastic	$1.7 \pm 1.1 \times 10^{-4}$	$1.7 \pm 0.3 \times 10^{-3}$	$7.4 \pm 0.8 \times 10^{-3}$

20.4.2 Stability of Nebulized rhDNase

The requirement of using a device to generate an aerosol of rhDNase for delivery to the lung poses additional challenges since the protein may be subjected to a variety of stresses as a result of the aerosol generation process. In particular, the aerosol is generated by entraining the solution into a rapidly accelerating air stream whereby droplets are formed and further reduced in size after impacting onto an internal baffle system at the speed of sound before exiting the device for delivery to the patient's airways. Thus, the formulation must not only provide long-term stability throughout the storage shelf life but also maintain an unaltered product in the aerosol for pulmonary delivery. In particular, the activity of the protein needs to be retained to ensure product efficacy and degradation products, especially aggregates, may need to be avoided to prevent safety issues. Apparent links between protein aggregation and immunogenicity have been noted [39–41].

Table 20.3 rhDNase before and after delivery by jet nebulization. (Data adapted from reference [42])

rhDNase sample	1 mg/mL		4 mg/mL	
	Activity[a]	% monomer	Activity[a]	% monomer
Before nebulization	1.00 ± 0.10	100 ± 0.1	1.00 ± 0.10	99.9 ± 0.1
Marquest Acorn II				
Residua after nebulization	0.94 ± 0.10	100 ± 0.1	ND	ND
Collected aerosol	0.93 ± 0.10	98.5 ± 0.7	ND	ND
Marquest Customized Respirgard II				
Residua after nebulization	1.01 ± 0.10	100 ± 0.1	ND	ND
Collected aerosol	0.96 ± 0.10	100 ± 0.1	ND	ND
Hudson T Up-Draft II				
Residua after nebulization	1.00 ± 0.10	100 ± 0.1	1.01 ± 0.10	99.9 ± 0.1
Collected aerosol	0.90 ± 0.10	100 ± 0.1	1.00 ± 0.10	99.6 ± 0.1
Baxter Airlife Misty				
Residua after nebulization	1.01 ± 0.10	100 ± 0.1	0.97 ± 0.10	99.9 ± 0.1
Collected aerosol	0.96 ± 0.10	100 ± 0.1	0.96 ± 0.10	99.9 ± 0.1

[a] Normalized activity expressed as ratio of active concentration of rhDNase as determined by methyl green activity assay to concentration determined by UV absorption spectroscopy relative to a value of 1.00 for the control rhDNase sample. The error in the value for the normalized activity is about 10 %. The activity concentration is an average of six replicates assayed in the methyl green assay. The concentration determined by UV spectroscopy is based on one measurement only

The properties of 1 and 4 mg/mL rhDNase solutions formulated in 150 mM NaCl, 1 mM $CaCl_2$ at pH ≈ 7 were determined before and after nebulization using four different jet nebulizers [42]. The generated aerosol was collected via a connecting tube from the mouthpiece of the nebulizer directly by impaction into a test tube that was partially immersed in an ice bath to minimize protein degradation during the collection process. The activity, pH of the solution, protein conformation, amount of monomer, and aggregate of rhDNase pre and post nebulization were determined by the methyl green activity assay, pH meter equipped with glass combination electrode, circular dichroism, and size exclusion chromatography, respectively. The results showed that at both concentrations, the rhDNase as formulated was unaltered after nebulization using the four jet nebulizers as determined by activity and molecule size distribution (Table 20.3). In addition, the overall rhDNase tertiary and secondary structures were unaltered after nebulization as determined by near- and far-UV circular dichroism spectrophotometry (data not shown). Although there was a ≈ 0.5 pH decrease in the 1 mg/mL solutions after nebulization, it was shown that this was likely the result of exposure of the aerosol to CO_2 (resulting in the generation of carbonic acid) since repeat experiments under a nitrogen atmosphere resulted in no change in the pH. The ability of the rhDNase to serve as a buffering agent was demonstrated by much smaller decreases in pH for the 4 mg/mL solutions and the large decreases seen for nebulized formulation solution without rhDNase. Additional studies designed to characterize rhDNase aerosols generated by seven additional jet nebulizers coupled with different air compressor systems showed that the structural integrity and activity of rhDNase in aerosols produced by the jet nebulizers remained unaltered [43]. This contrasted with the observation that some ultrasonic nebulizers caused denaturation of the rhDNase, probably due to the elevated temperature during ultrasonic nebulization.

20.4.3 Characterization of rhDNase Aerosols

In order to treat CF patients, rhDNase needs to be delivered at sufficient concentrations locally into the lung. Many of the early clinical studies with bDNase used jet [9–11, 13] and ultrasonic nebulizers [12] to deliver bDNase aerosols. However, little was done to characterize the physical properties of the aerosols generated by these nebulizers.

As stated earlier, stability of rhDNase during nebulization was demonstrated using four different jet nebulizers in support of the human clinical trials conducted in the USA [32, 42]. In addition to the stability, the aerosols were also characterized for size distribution and most importantly the amount of rhDNase delivered to the airways. In these early studies the droplet size distribution was obtained using a cascade impactor to collect the emitted aerosol from the nebulizer. The droplet size distribution in an aerosol is critical to determining deposition and appropriate delivery of an inhaled aerosol drug. Droplets larger than 6 μm will deposit mainly in the oropharynx, whereas droplets less than 1 μm are likely to be exhaled during normal tidal breathing. Accordingly, the mass percent respirable fraction was defined in these early studies as the percent of the total droplet size from 1–6 μm. Later studies further suggested that increasing the fraction of small droplet sizes by lowering the median mass aerodynamic diameter to around 2 μm could also be efficient for drug deposition due to potential deposition in the lower airways [44]. The droplet size distribution of nebulized rhDNase formulated at 1 mg/ml in unbuffered isotonic saline with 1 mM $CaCl_2$ was similar for the four jet nebulizers initially tested (Fig. 20.7) and was between 46 and 51 %. The nebulizer efficiency, defined as the percent of the total amount of drug loaded into the nebulizer that is actually delivered as an aerosol to the mouthpiece of the nebulizer, ranged from 44–55 %, and was typical for other drugs delivered by jet nebulization [45]. The delivery efficiency, defined as the percent of the initial rhDNase dose delivered to the mouthpiece in the respirable range of 1–6 μm was between 20 and 28 %. Thus, since the patients are exhaling $\approx$50 % of the time during nebulization using a nebulizer with no shut-off options during dosage, it would result in an overall efficiency of drug delivered by the four jet nebulizers tested between 10–15 %. rhDNase sold as Pulmozyme® is filled at 2.5 mL at 1 mg/mL in plastic ampoules, and of this total dose only 0.3–0.4 mg of drug is delivered to the airways of CF patients through the use of jet nebulizer delivery systems. Later studies that characterized seven additional jet nebulizers coupled with different air compressor systems also showed that large differences exist in the droplet size distribution and delivery efficiency of aerosols produced by the different delivery systems [43]).

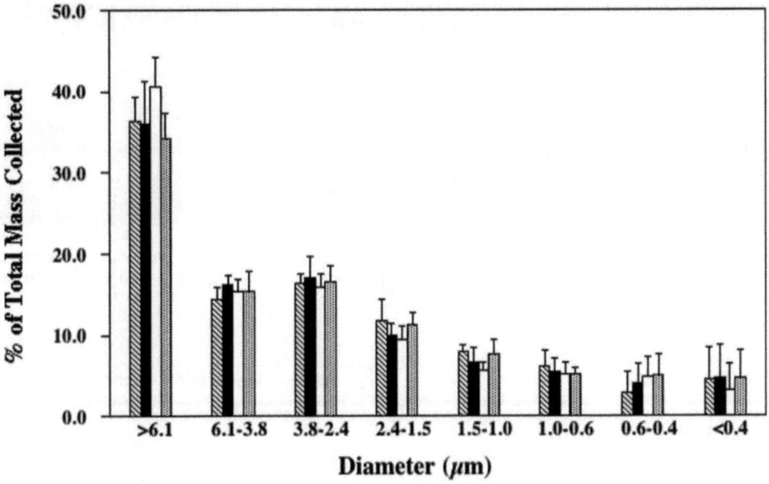

Fig. 20.7 Droplet size distribution of rhDNase aerosols. Aerosols were generated over 10 min using four different jet nebulizers loaded with 2.5 mL of rhDNase at 1 mg/mL. The droplet size distribution was determined with a seven-stage cascade impactor (In-Tox Products, Albuquerque, NM, USA). Nebulizers used were (diagonal stripes): Respirgard II model #124030 (Marquest, Englewood, CA, USA) modified by removal of the expiratory one-way valve, (solid black): Acorn II model 124014 (Marquest, Englewood, CA, USA), (solid white): Airlife Misty with Tee Adapter model # 0020308 (Baxter-American Pharmaseal Co., Valencia, CA, USA) and (solid greay): T Up-Draft II model #1734 (Hudson RCI, Temecula, CA, USA). The error bars are the standard deviations that result from 7–8 independent determinations. (Reproduced from reference [42], with kind permission from Springer Science and Business Media. Copyright 2002)

20.4.4 Pulmozyme® Delivery by Small Portable Nebulizers

When approved in 1993, it was stipulated that Pulmozyme® should be used only with the jet nebulizer compressor combinations that had been used in the clinical trials. Unfortunately many of the nebulizers and compressors are increasingly outdated and some are no longer being manufactured. Many of the jet nebulizer compressor combinations require long nebulization times, are bulky, and require an AC power source, which restricts their portability. Currently there are several newer technologies that may shorten dosage times, and due to the smaller size, allow for easier transport. Devices that generate an aerosol by forcing pressurized liquids through nozzles are promising [46] but may require reformulation of Pulmozyme® at higher concentrations to deliver the same doses as a jet nebulizer. Other devices that use a vibrating stainless steel mesh or porous membrane are portable, silent, do not require a compressed air source, and can operate with either battery or AC power.

The PARI eFlow® platform consists of electronic jet nebulizers that use a vibrating stainless steel porous membrane/mesh technology (Fig. 20.8). These nebulizers allow for adjustment of droplet size distribution by appropriate selection of membranes with different pore sizes [47]. The PARI eFlow® nebulizer has been approved in the USA as a general-purpose nebulizer for use by adult and pediatric patients for whom doctors have prescribed medication for nebulization. Currently three eFlow®

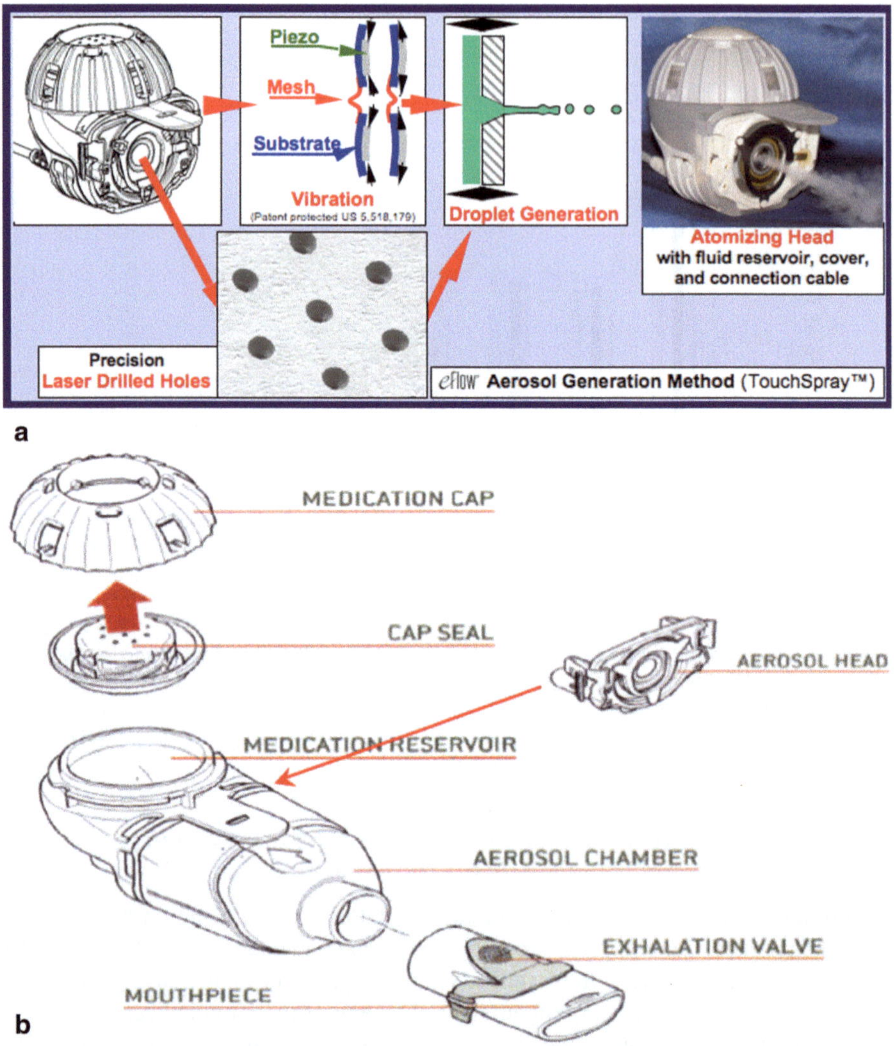

Fig. 20.8 **a** Aerosol generation principle of the eFlow® electronic nebulizer platform technology. **b** Main components of the eFlow® electronic nebulizer. (Reproduced from reference [50], with kind permission from John Wiley and Sons. Copyright 2010)

nebulizers are approved for use in the USA; the eFlow® (registered as the Trio®) is available from a limited number of compounding pharmacies for CF therapies, the Altera® is an eFlow® nebulizer that has been customized for use with the antibiotic, Cayston® (astreonam lysine for inhalation), and the eFlow® Rapid that has been used in Europe and approved for use in the USA as a general purpose nebulizer. Clinical experience with the eFlow® Rapid device to deliver Pulmozyme® and other CF drugs has been reported [48]. The essential difference between the eFlow® and eFlow® Rapid nebulizers is that the eFlow® Rapid has a larger reservoir for drug loading as well as larger aerosolization chamber. The larger reservoir bowl results in ≈ 1 mL of loaded drug that will not be in contact with the vibrating mesh [49], which in the case of Pulmozyme delivery results in approximately half the contents being converted to an aerosol.

In a recent in vitro study [50], aerosols of Pulmozyme® generated by Pari eFlow® and eFlow® Rapid devices were compared at two different laboratories (Nemours Children's Clinic in Orlando (NCC) and PARI Pharma GmbH (PARI) in Germany) with those generated by LC Plus®, and LC Star® jet nebulizers driven by a PARI ProNeb Ultra® compressor. The Pari eFlow® and eFlow® Rapid nebulizers were configured to match the aerosol size distribution generated by the LCPlus®. As previously discussed, the earlier characterization of rhDNase aerosols was done using cascade impactor technology [42, 43]. Laser diffraction has been used to determine droplet size distribution where instead of mass median aerodynamic diameter a volume median aerodynamic diameter (VMD) is measured [51, 52]. Since this technique is more rapid and easier to use than cascade impaction, the assessment of the aerosol droplet size distribution emitted by these nebulizers was conducted by laser diffraction. Results obtained with Pulmozyme® show that the VMD of the PARI LC PLUS® (4.1–4.4 mm) is closely matched by the eFlow® (4.0–4.3 mm) and the eFlow® Rapid (3.9 mm) nebulizers. The droplet size distributions of both the eFlow® and eFlow® Rapid devices are narrower than the droplet size distributions of the three jet nebulizers, as reflected in the Geometric Standard Deviation (GSD), a parameter that indicates the spread of the size distribution curves (lower GSD indicates more uniform particle sizes). Overall the eFlow® electronic nebulizers have a GSD of 1.57 compared to values of ≈ 2 for the jet nebulizers and had fewer droplets above 5 mm.

Determination of delivered doses was done by using a COMPASS (PARI, GmbH, Germany) breath-simulator by collecting and determining the amount of Pulmozyme® deposited on the inhalation filter. The eFlow® nebulizer had significantly better nebulization efficiency over all jet nebulizers tested, with a respirable dose that was ≈ twofold that of the LC Plus® (≈ 0.9 vs. 0.45 mg). The respirable dose of the eFlow® Rapid was comparable to the LC Plus®, and was able to deliver the equivalent dose in half the time (2.3 vs. 5.7 min). This difference compared to the eFlow® nebulizer is due to the fact that ≈ 50 % of the loaded dose is not nebulized in the eFlow® Rapid. This design modification provides an optimal delivery time while maintaining delivered doses equivalent to those from jet nebulizer devices, doses previously found safe and efficacious in clinical evaluations.

As discussed earlier, the efficacy of Pulmozyme® administered by nebulization depends not only on the respirable dose, but also on the ability of this therapeutic protein to retain activity and integrity throughout the nebulization and delivery process. The actual process used to generate the aerosol will subject the protein drug to different stresses.

Vibrating membrane technology potentially imparts energy at high frequency of vibration of a stainless steel perforated membrane, subjecting the protein to forces during extrusion through the pores. Thus, the process of aerosol generation and delivery has the potential to alter the protein as a result of different stresses resulting in conformational changes or generation of aggregates that may impact the safety and efficacy of this drug. Some recent in vitro studies of Pulmozyme® using devices with perforated vibrating membrane technology are incomplete since the use of assays that only measure activity may not detect these alterations and other potential impacts to product quality and rhDNase integrity [53, 54]. Thus, assessments of Pulmozyme® quality following delivery by a nebulizer should incorporate additional appropriate assays to assess the stability and product quality during aerosol delivery.

A much more rigorous and thorough evaluation of the stability of Pulmozyme® after nebulization using the eFlow® was performed (Tables 20.4 and 20.5) with the assays that are used for Pulmozyme® drug release as well as long-term stability studies to support drug shelf life [50]. The release tests that were used to assess Pulmozyme® quality after nebulization using the eFlow® technology include tentacle ion exchange chromatography (IEC), to assess % deamidation, size exclusion chromatography (SEC), to assess % monomer, pH, protein concentration by spectrophotometric scan, color, appearance and clarity (CAC), turbidity, osmolality, and the methyl green specific activity assay (MG assay). Overall, the data (Tables 20.4 and 20.5) show no changes in the quality of Pulmozyme® (results of all assays within the precision of the assay) after nebulization using an eFlow® nebulizer, and thus no detrimental effects due to the aerosol generation process used in the eFlow® system. The first 45 nebulizations were done with product loaded at 22 °C and done at 22 °C (Table 20.4) and the additional 15 nebulizations were done by loading product at 5 °C and immediately starting the nebulization at 22 °C (Table 20.5). Although these analytical studies of product quality after nebulization were not done with the eFlow® Rapid nebulizer, it is likely that similar results would be obtained since both electronic nebulizers use the same mechanism and highly similar components to generate an aerosol, so that Pulmozyme® would be subjected to the same stresses and conditions.

20.4.5 Robustness of Nebulization by the eFlow® Technology

Jet nebulizer handsets are generally sold as single use disposable devices, but often are used several times with cleaning. It has been shown that with proper care and washing between uses that the performance of the disposable jet nebulizers tested does not deteriorate as assessed by determination of the particle size distribution and output in mL/min [55]. Unlike a jet nebulizer, the eFlow® platform technology is intended to be a reusable system with a replaceable stainless steel membrane which may be subject to changes in performance with use over time. It has been reported that changes over time in performance of the Pari eFlow® Rapid delivering the antibiotic Tobramycin occur and that these changes appear to be due to clogging of the pores of the vibrating mesh [56]. In a guidance note on the web Geller and Kesser also discuss change in aerosol performance due to clogging of the pores when using the antibiotic Cayston [57]. Thus, in order to ascertain if the eFlow® nebulizer could maintain performance, average nebulization times to deliver Pulmozyme® were determined for 60 actuations [50] of four different devices and membranes. A linear regression analysis of a plot of average nebulization time vs. the number of actuations of the device was extrapolated to show the expected nebulization times over a period of 3 months for an eFlow® membrane with 95 % confidence limits and

Table 20.4 rhDNase before (controls 1 and 2) loading at 22 °C and after delivery by eFlow® nebulization ($n = 4$). All nebulizations performed at 22 °C. (Reproduced from reference [50], with kind permission from John Wiley and Sons. Copyright 2010)

Nebulization no.	Nebulization time	IEC % deamidated (min:sec)	SEC % monomer	pH	Protein conc. (mg/mL)	COC			Turbidity neat	Osmolality mOsm/kg	MG assay specific activity x 10^3 U/mL
						Appearance	Clarity	Color			
1	2:26	66.05	99.94	6.24	0.97	Liquid	Clear	None	0.009	283	1.01
2	2:39	65.94	99.92	6.26	0.97	Liquid	Clear	None	0.007	283	1.00
3	2:44	65.99	99.93	6.26	1.03	Liquid	Clear	None	0.005	285	0.96
Control 1	0	65.71	99.93	6.12	0.97	Liquid	Clear	None	0.002	282	1.00
General SD	0:18	0.16	0.02	0.09	0.04	–	–	–	0.00	1.49	0.05
15	2:54	65.88	99.91	6.35	0.98	Liquid	Clear	None	0.009	285	1.02
30	3:00	65.81	99.92	6.29	0.98	Liquid	Clear	None	0.010	283	0.94
45	2:52	65.92	99.91	6.38	0.97	Liquid	Clear	None	0.008	283	1.00
Control 2	0	65.82	99.93	6.34	0.99	Liquid	Clear	None	0.012	284	1.01
General SD	0:33	0.07	0.01	0.06	0.01	–	–	–	0.01	1.41	0.09

Table 20.5 rhDNase before (Controls 1 and 2) loading at 5 °C and after delivery by eFlow® nebulization ($n = 4$). All nebulizations performed at 22 °C. (Reproduced from reference [50], with kind permission from John Wiley and Sons. Copyright 2010)

Nebulization no.	Nebulization time	IEC % deamidated (min:sec)	SEC % monomer	pH	Protein conc. (mg/mL)	COC			Turbidity neat	Osmolality mOsm/kg	MG assay specific activity x 10^3 U/mL
						Appearance	Clarity	Color			
46	3:02	65.64	99.91	6.53	0.96	Liquid	Clear	None	0.01	282	0.96
47	3:16	65.67	99.91	6.45	0.95	Liquid	Clear	None	0.00	282	0.98
48	3:18	65.73	99.91	6.45	0.96	Liquid	Clear	None	0.00	282	0.98
60	3:16	65.61	99.91	6.46	0.96	Liquid	Clear	None	0.00	282	0.96
Control 3	0	65.58	99.95	6.48	0.96	Liquid	Clear	None	0.002	281	1.00
General SD	0:35	0.09	0.03	0.07	0.01	–	–	–	0.00	0.92	0.08

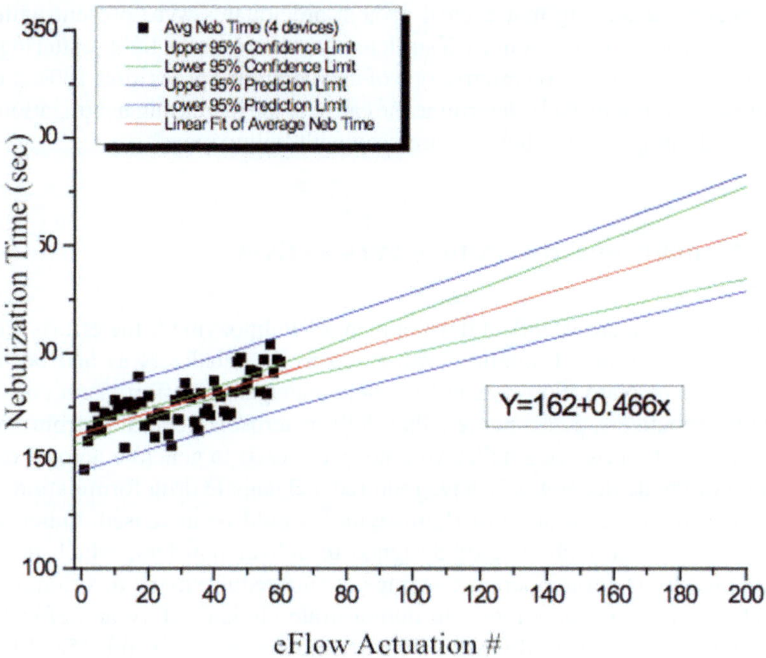

Fig. 20.9 Plot of average nebulization time vs. actuation # with linear fit of data for actuations 1–60 for four investigational eFlow® devices. A linear fit to data extended to show expected nebulization times over PARI's recommended use lifetime of the investigated eFlow® membrane. 180 actuations correspond to 2-times daily Pulmozyme use by the patient for 3 months. (Reproduced from reference [50], with kind permission from John Wiley and Sons. Copyright 2010)

95 % prediction limits (Fig. 20.9). The analysis extended out to 180 nebulizations (twice daily Pulmozyme® dosing for 3 months) provides from the calculated lower 95 % prediction limit (upper line) an estimate of between 4 and 4.5 min for the average duration of the 180th nebulization. Thus, if the device performance continues to decline in a linear fashion, then in 3 months the delivery time will still be less than 5 min, close to the shortest time for delivery by most jet nebulizers [43]. Since these results are based on a linear regression analysis of limited data, they should not be used to predict long-term device performance with Pulmozyme®, as actual device performance trends may be nonlinear and dependent on the care/handling of the nebulizer membrane components.

The observed increase of time for dosing may be due to several factors including gradual change in the vibrating membrane performance as a result of membrane assembly fatigue or potential clogging of the membrane pores as seen with the use of antibiotics such as Tobramycin and Cayston [56, 57]. The clogging of the pores may occur either from a small amount of foreign particulate or from aggregated protein. However, the data in Tables 20.3 and 20.4 show that at least after 60 actuations there is no change in protein quality, and in particular, no evidence of increase protein aggregation. It should also be noted that the turbidity of recovered nebulized solutions is unchanged, suggesting that even if protein precipitates were present within the pores, this material did not slough off sufficiently to increase the light scattering of the solution. An investigation by microscopy of the membrane pores after 180 actuations is anticipated to conclusively determine the cause of the increasing nebulization times when the vibrating mesh nebulizers are used with Pulmozyme®.

20.5 Conclusions and Future Perspectives

During the early pharmaceutical development of Pulmozyme®, the efforts were directed to deliver 1 mL of solution with standard jet nebulizers as had been done previously for bDNase. Subsequent life-cycle management efforts later considered alternative nebulizers, i. e., nonjet, that deliver aerosols generated from solution were considered, but since smaller volumes were used to generate aerosol on each activation of the device it would have required a change in drug formulation, where in particular the concentration of Pulmozyme® would be increased. Other aerosol delivery devices and technologies designed to deliver powders, which may have an advantage for room-temperature storage, would require many additional studies to develop powder formulations and demonstrate clinical safety and efficacy. Although initial efforts using rhDNase showed some potential for this [58] life-cycle development strategy, the uncertainties and need for extensive clinical trials in conjunction with their associated costs, discouraged further development. Thus, much of the pharmaceutical development for Pulmozyme® centered on the use of jet nebulizers capable of delivering ≈ 1 mL in ≈ 5 min of therapy. Initially only the jet nebulizer/compressor combinations that were used in the clinic were placed on the Pulmozyme® label. Additional in vitro and some limited in vivo clinical trials were done to evaluate additional jet nebulizers. This effort was important since many of the original jet nebulizers on the label have become either out-of-date or replaced by newer models with similar performance characteristics, or are no longer manufactured. Thus, physicians may be compelled to use nebulizers that are available but they may not have been tested previously with Pulmozyme®. In the case of nebulizers which deliver a continuous aerosol spray consisting of liquid droplets, it may be argued that if in vitro characterization of aerosols demonstrate similarity of droplet size distribution and dosage delivery compared to those generated by the older jet nebulizers, combined with a thorough analysis of protein drug quality (such as described in our studies), shows no alterations, that the delivery systems should be deemed equivalent. Acceptance of such a data-driven scientific rationale could limit evaluations to user studies without the requirement for repetitious clinical trials designed to explore safety and efficacy.

In this review we summarized the development of rhDNase as a pharmaceutical including development and stability assessment of a formulation, compatibility with jet nebulization, characterization of the aerosols produced by jet nebulization, and compatibility and storage of the final drug product configuration in the final blow-fill seal plastic ampoules, which was eventually registered under the brand name

Pulmozyme®. Recent life-cycle development efforts have involved performing in vitro technical feasibility studies to determine if Pulmozyme® can be delivered by a portable electronic mesh nebulizer without alteration of the product and with similar dosage to what was used to originally register the drug. As to whether such in vitro assessments are sufficient to license Pulmozyme® with new nebulizer systems, enabling future modernization of patient aerosol delivery options without extensive clinical data remains a question for regulatory health agencies to consider.

References

1. Boat T, Welsh MJ, Beaudet A, Cystic Fibrosis. In: Scriver C, et al., editor. The metabolic basis of inherited disease. New York: McGraw Hill;1989. pp. 2649–2860.
2. MacDonald KD, McKenzie KR, Zeitlin PL. Cystic fibrosis transmembrane regulator protein mutations: 'class' opportunity for novel drug innovation. Paediatr. Drugs. 2007;9(1):1–10.
3. Tsui LC, Buchwald M. Biochemical and molecular genetics of cystic fibrosis. Adv Hum Genet. 1991;20:153–266.
4. Abeliovich D, Lavon IP, Lerer I, Cohen T, Springer C, Avital A, Cutting G. Screening for five mutations detects 97 % of cystic fibrosis (cf) chromosomes and predicts a carrier frequency of 1:29 in the Jewish Ashkenazi population. Am J Hum Genet. 1992;51:951–6.
5. Chernick WS, Barbero GJ. Composition of tracheobronchial secretions in cystic fibrosis of the pancreas and bronchiectasis. Pediatrics. 1959;24:739–45.
6. Potter JL, Matthews LW, Lemm J, Spector S. Human pulmonary secretions in health and disease. Ann N Y Acad Sci. 1963;106:692–7.
7. Armstrong JB, White JC. Liquefaction of viscous purulent exudates by deoxyribonuclease. Lancet. 1950:739–40.
8. Chernick WS, Barbero GJ, Eichel HJ. In-vitro evaluation of effect of enzymes on tracheobronchial secretions from patients with cystic fibrosis. Pediatrics. 1961;27:589–96.
9. Elmes PC, White JC. Deoxyribonuclease in the treatment of purulent bronchitis. Thorax. 1953;8:295–300.
10. Salomon A, Herchfus JA, Segal MS. Aerosols of pancreatic dornase in bronchopulmonary disease. Anna Allergy. 1954;12:71–9.
11. Spier R, Witebsky E, Paine JR. Aerosolized pancreatic dornase and antibiotics in pulmonary infections. JAMA. 1961;178:878–86.
12. Lieberman J. Enzymatic dissolution of pulmonary secretions. Am J Dis Chil. 1962;104:342–8.
13. Raskin P. Bronchospasm after inhalation of pancreatic dornase. Am Rev Respir Dis. 1968;98:597–8.
14. Shak S, Capon DJ, Hellmiss R, Marsters SA, Baker CL. Recombinant human DNase I reduces the viscosity of cystic fibrosis sputum. Proc Natl Acad Sci U S A. 1990;87(23):9188–92.
15. Aitken ML. Clinical trials of recombinant human DNase in cystic fibrosis patients. Monaldi Arch Chest Dis. 1993;48(6):653–6.
16. Hubbard RC, McElvaney NG, Birrer P, Shak S, Robinson WW, Jolley C, Wu M, Chernick MS, Crystal RG. A preliminary study of aerosolized recombinant human deoxyribonuclease I in the treatment of cystic fibrosis. N Engl J Med. 1992;326(12):812–5.
17. Ramsey BW, Astley SJ, Aitken ML, Burke W, Colin AA, Dorkin HL, Eisenberg JD, Gibson RL, Harwood IR, Schidlow DV, et al. Efficacy and safety of short-term administration of aerosolized recombinant human deoxyribonuclease in patients with cystic fibrosis. Am Rev Respir Dis. 1993;148(1):145–51.
18. Ranasinha C, Assoufi B, Shak S, Christiansen D, Fuchs H, Empey D, Geddes D, Hodson M. Efficacy and safety of short-term administration of aerosolised recombinant human DNase I in adults with stable stage cystic fibrosis. Lancet. 1993;342(8865):199–202.

19. Pressler T. Review of recombinant human deoxyribonuclease (rhDNase) in the management of patients with cystic fibrosis. Biologics. 2008;2(4):611–7.
20. Cacia J, Quan CP, Vasser M, Sliwkowski MB, Frenz J. Protein sorting by high-performance liquid chromatography. I. Biomimetic interaction chromatography of recombinant human deoxyribonuclease I on polyionic stationary phases. J Chromatogr. 1993;634(2):229–39.
21. Cleland JL, Powell MF, Shire SJ. The development of stable protein formulations: a close look at protein aggregation, deamidation and oxidation. Crit Rev Ther Drug Carrier Syst. 1993;10(4):307–77.
22. Wright HT. Deamidation of asparaginyl and glutaminyl residues in proteins. CRC Crit Rev Biochem Mol Biol. 1991;26:1–52.
23. Balmes JR, Fine JM, Christian D, Gordon T, Sheppard D. Acidity potentiates bronchoconstriction induced by hypoosmolar aerosols. Am Rev Respir Dis. 1988;138:35–9.
24. Beasley R, Rafferty P, Holgate ST. Adverse reactions to the non-drug constituents of nebulizer solutions. Br J Clin Pharmac. 1988;25:283–7.
25. Desager KN, Van Bever HP, Stevens WJ. Osmolality and pH of antiasthmatic drug solutions. Agents Actions. 1990;31:225–8.
26. Fine JM, Gordon T, Thompson JE, Sheppard D. The role of titratable acidity in acid aerosol-induced bronchoconstriction. Am Rev Respir Dis. 1987;135:826–30.
27. Sant'Ambrogio G, Anderson JW, Sant'Ambrogio FB, Mathew OP. Response to laryngeal receptors to water solutions of different osmolality and ionic composition. Respir Med. 1991;85(Supplement A):57–60.
28. Snell NJC. Adverse reactions to inhaled drugs. Respir Med. 1990;84:345–8.
29. Gonda I, Kayes JB, Groom CV, Fildes FJT. Characterization of hygroscopic inhalation aerosols. In: Stanley-Wood NG, editor. Particle size analysis. New York: Wiley; 1982. pp. 31–43.
30. Gonda I, Phipps PR. Some consequences of instability of aqueous aerosols produced by jet and ultrasonic nebulizers. In: Masuda S, Takahashi K, editor. Aerosols. Vol. 1. New York: Pergamon Press; 1991. pp. 227–30.
31. Godden DJ, Borland C, Lowry R, Higenbottam TW. Chemical specificity of coughing in man. Clin Sci. 1986;70:301–06.
32. Cipolla D, Gonda I, Meserve K, Weck S, Shire SJ. Formulation and aerosol delivery of recombinant deoxyribonucleic acid derived human deoxyribonuclease I. In: Cleland JL, Langer R, editor. ACS Symposium Series 567, formulation and delivery of proteins and peptides. Washington: American Chemical Society; 1994. pp. 322–42.
33. Poulos TL, Price PA. Some effects of calcium ions on the structure of bovine pancreatic deoxyribonuclease A. J Biol Chem. 1972;247:2900–4.
34. Shire SJ. Stability characterization and formulation development of recombinant human deoxyribonuclease I [Pulmozyme®, (Dornase Alpha)]. In: Pearlman R, Wang YJ, editor. Formulation, characterization and stability of protein drugs. New York: Plenum; 1996. pp. 393–426.
35. Wiberg JS. On the mechanism of metal activation of deoxyribonuclease I. Arch Biochem Biophys. 1958;73:337–58.
36. Price PA. Characterization of Ca^{++} and Mg^{++} binding to bovine pancreatic deoxyribonuclease A. J Biol Chem. 1972;247:2895–9.
37. Capasso S, Mazzarella L, Zagari A. Deamidation via cyclic imide of asparaginyl peptides: Dependence on salts, buffers and organic solvents. Pept Res. 1991;4(4):234–8.
38. Meserve K, Weck S, Shire SJ. Stability of recombinant deoxyribonuclease I (rhDNase) in plastic vials manufactured by the automatic liquid packaging (ALP) system. Pharm Res. 1994;11(10):S–74.
39. Rosenberg AS. Effects of protein aggregates: an immunologic perspective. AAPS J. 2006;8(3):E501–7.
40. Wang W, Singh SK, Li N, Toler MR, King KR, Nema S. Immunogenicity of protein aggregates—concerns and realities. Int J Pharm. 2012;431(1–2):1–11.
41. Kumar S, Singh SK, Wang X, Rup B, Gill D. Coupling of aggregation and immunogenicity in biotherapeutics: T- and B-cell immune epitopes may contain aggregation-prone regions. Pharm Res. 2011;28(5):949–61.

42. Cipolla D, Gonda I, Shire SJ. Characterization of aerosols of human recombinant deoxyribonuclease I (rhDNase) generated by jet nebulizers. Pharm Res. 1994;11(4):491–8.
43. Cipolla DC, Clark AR, Chan H-K, Gonda I, Shire SJ. Assessment of aerosol delivery systems for recombinant human deoxyribonuclease. STP Pharma Sci. 1994;4(1):50–62.
44. Geller DE, Eigen H, Fiel SB, Clark A, Lamarre AP, Johnson CA, Konstan MW. Effect of smaller droplet size of dornase alfa on lung function in mild cystic fibrosis. Pediatr Pulmonol. 1998;25(2):83–7.
45. Byron PR. Aerosol formulation, generation, and delivery using nonmetered systems. In: Byron PR, editor. Respiratory drug delivery. Vol. Chapter 6, CRC Press; 1990. pp. 143–65.
46. Geller DE. New liquid aerosol generation devices: systems that force pressurized liquids through nozzles. Resp Care. 2002;47(12):1392–404; discussion 1404–1395.
47. Knoch M, Keller M. The customised electronic nebuliser: a new category of liquid aerosol drug delivery systems. Expert Opin Drug Deliv. 2005;2(2):377–90.
48. Naehrig S, Lang S. Lung function in adult patients with cystic fibrosis after using the eFlow® rapid for one year. Eur J Med Res. 2011;16(2):63–6.
49. Vecellio L. The mesh nebulizer: a recent technical innovation for aerosol delivery. Breathe. 2006;2(3):253–60.
50. Scherer T, Geller DE, Owyang L, Tservistas M, Keller M, Boden N, Kesser KC, Shire SJ. A technical feasibility study of dornase alfa delivery with eFlow (R) vibrating membrane nebulizers: aerosol characteristics and physicochemical stability. J Pharm Sci. 2011;100(1):98–109.
51. Clark AR. The use of laser diffraction for the evaluation of the aerosol clouds generated by medical nebulizers. Int J Pharm. 1995;115(1):69–78.
52. Keller M, Tservistas M, Bucholski A, Hug M, Knoch M. Correlation of laser diffraction and cascade impaction data for aqueous solutions aerosolized by the eFlow® electronic nebulizer. Proc Respir Drug Deliv. 2006:745–748.
53. Johnson JC, Waldrep JC, Guo J, Dhand R. Aerosol delivery of recombinant human DNase I: in vitro comparison of a vibrating-mesh nebulizer with a jet nebulizer. Respir Care. 2008;53(12):1703–8.
54. Potter RW, Hurren TJ, Nickerson C, Hatley RH. Comparison of the delivery characteristics of dornase alpha from the I-NEB1 AAD® system and the Sidestream® jet nebulizer. Pediatic Pulm Supp. 2008;31:483.
55. Standaert TA, Morlin GL, Williams-Warren J, Joy P, Pepe MS, Weber A, Ramsey BW. Effects of repetitive use and cleaning techniques of disposable jet nebulizers on aerosol generation. Chest. 1998;114(2):577–86.
56. Rottier BL, van Erp CJP, Sluyter TS, Heijerman HGM, Frijlink HW, de Boer AH. Changes in performance of the pari eFlow (R) rapid and pari LC Plus™ during 6 months use by CF patients. J Aerosol Med Pulm D. 2009;22(3):263–9.
57. Geller DE, Kesser KC. Guidance on the use of eFlow nebulizers (Altera® and Trio®). 2010.
58. Maa YF, Nguyen PA, Andya JD, Dasovich N, Sweeney TD, Shire SJ, Hsu CC. Effect of spray drying and subsequent processing conditions on residual moisture content and physical/biochemical stability of protein inhalation powders. Pharm Res. 1998;15(5):768–75.

Chapter 21
Development of the Exubera® Insulin Pulmonary Delivery System

Cynthia L. Stevenson and David B. Bennett

21.1 Introduction

Insulin is the primary therapy for Type 1 diabetes mellitus (10 % of diabetics), and is also an integral component for the treatment of Type 2 diabetes (90 % of diabetics). The lifetime risk for developing Type 2 diabetes, if diagnosed at age 40, is 11.6 years for men and 14.3 years for women [1]. The lifetime risk for developing diabetes, if born in the year 2000, increases to 32.8 % for men and 38.5 % for women [1]. Insulin was once considered the therapy of last resort for patients with Type 2 diabetes, but physicians are now utilizing it earlier in the treatment of the disease state. Furthermore, insulin is an antihyperglycemic agent with a proven long-term safety profile [2].

Many patients with Type 2 diabetes have poor glucose control, where their glycosylated hemoglobin (HbA1c) level is greater than 7 % [3]. Patients on insulin, or insulin with an oral therapy regimen, can restore normal levels of glycemia [4]. Currently, the use of insulin regimens is limited by their lack of flexibility, needle phobia, and patient motivation. Specifically, anxiety associated with insulin needles may affect patient compliance [5].

The pulmonary route of administration was chosen in an effort to increase patient compliance and quality of life for those patients who found multiple daily subcutaneous injections of insulin burdensome [6, 7]. The advantages of pulmonary delivery of macromolecules include a well-established delivery route, large alveolar surface for absorption ($\approx 100\,m^2$), rapid absorption across alveolar membranes without absorption enhancers, and needle-free administration [8–11]. Absorption of peptides and proteins into the systemic circulation is higher where the cell layers between air space and pulmonary capillaries are thin: in the alveoli, alveolar ducts, and respiratory bronchioles, collectively termed the deep lung [12]. Pulmonary delivery of insulin was designed to optimize alveolar deposition of the drug while minimizing deposition in the mouth, throat, and upper airways.

C. L. Stevenson (✉) · D. B. Bennett
Pharmaceutical Consultants, 100 W. El Camino Real #48, Mountain View, CA 94040, USA
e-mail: cynthialstevenson@gmail.com

J. das Neves, B. Sarmento (eds.), *Mucosal Delivery of Biopharmaceuticals,* DOI 10.1007/978-1-4614-9524-6_21,

Development of Exubera® for pulmonary delivery of insulin presented significant challenges: (a) a room temperature stable formulation using excipients suitable for safe administration to the lungs, (b) a powder manufacturing process providing particles suitable for efficient aerosolization, (c) a reproducible, high throughput powder filling and packaging process for low mass doses, (d) a rugged, reusable mechanical device for powder dispersal and reliable dosing to the patient independent of inhalation flow rate, and (e) an extensive clinical program to demonstrate the safe and effective treatment of a chronic metabolic disease involving extensive pulmonary-function assessments [13, 14]. Successfully surmounting these challenges led to the development and the US/EU approval of Exubera®, representing the first of its kind pharmaceutical product for the pulmonary delivery of recombinant human insulin to treat Type 1 and Type 2 diabetes [4, 15–20].

21.2 Formulation Development

21.2.1 Physical Form and Composition

The Exubera® insulin powder for inhalation was designed as a dry, amorphous solid. Optimization of the formulation and particle engineering factors are required for physicochemical stability and enhanced powder dispersibility [21]. Microbial growth can be prevented or minimized in low-water content solids, thus avoiding the need for preservatives and minimizing the number of excipients. Insulin stability improves in amorphous solids compared to crystalline solids and solutions [22]. In addition, a greater insulin payload per unit inhaled mass could be achieved using a powder compared to a solution, which enabled a wide range of doses in only one to three breaths.

The amorphous insulin powder was composed of 60.0 % insulin with 27.1 % sodium citrate (dihydrate), 10.0 % mannitol, 2.6 % glycine, and 0.3 % sodium hydroxide. Sodium citrate buffered the aqueous bulk solution prior to spray drying, and imparted a high glass transition temperature (T_g) when dry. Similarly, glycine increased buffer capacity and pH control of the aqueous solution prior to spray drying. Mannitol served as a lyoprotectant to stabilize insulin as water was flash evaporated during spray drying.

21.2.2 Physical Characterization

Amorphous pharmaceutical solids have been shown to be more soluble than their crystalline counterparts [23]. Therefore, the insulin powder would dissolve rapidly in epithelial lining fluid once delivered to the deep lung. Amorphous solids (e.g., solid solutions) do not exhibit a defined melting temperature, but a T_g that depends on the formation conditions, processing history, chemical composition, and the time

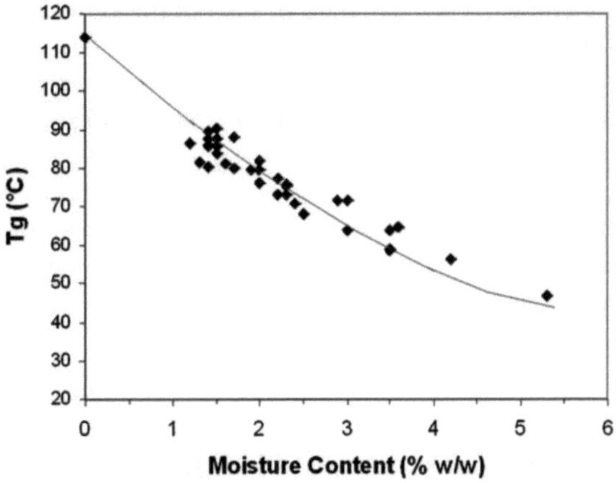

Fig. 21.1 Glass transition temperature (T_g) of the amorphous spray-dried insulin powder as a function of moisture content. (Reprinted from [13])

scale of observation. The T_g onset is the temperature at which a transition begins from a mechanical solid phase to a viscoelastic, super-cooled rubber phase.

When amorphous solids are stored well below their T_g, molecular motions and chemical reactivity are slowed. Molecular mobility becomes insignificant when the storage temperature is significantly (i.e., >50 °C) below the T_g [24]; therefore, T_g of the insulin powder was targeted to be at least 75 °C to enable long-term stability at room temperature [25]. The T_g for insulin powder was assessed at 2° min^{-1} by modulated differential scanning calorimetry (MDSC), as the extrapolated onset temperature of the C_p transition. The T_g (onset) for spray-dried powder was ≈ 80 °C and water content was ≈ 2 % (w/w) following manufacture at standard conditions. The T_g was dependent on moisture content (Fig. 21.1), but the powder exhibited a single T_g, regardless of moisture content, indicative of a single amorphous phase [26]. No evidence of component crystallization or melting was detected in the heat flow signal from 0–150 °C. Thus, even with exposure to temperature conditions favorable for crystallization ($T \gg T_g$), insulin powder exhibited no signs of crystallization.

The spray-dried insulin formulation was characterized as a strong glass, with a temperature range of 20 °C between the extrapolated onset and the extrapolated endpoint of the broad glass transition [27]. A strong glass exhibits changes in mobility and chemical reactivity that may be approximated by an Arrhenius relationship over a wide temperature range. The mobility (viscosity) of a strong glass also has weaker-temperature dependency relative to fragile glasses. Fragile glasses deviate from Arrhenius behavior and exhibit rapid increases in molecular mobility (solid softening and chemical reactivity) above the T_g [28, 29]. The insulin powder was a strong glass expected to demonstrate Arrhenius-like chemical kinetics for insulin degradation.

Water acted as a plasticizer, increased insulin mobility and thus, its reactivity [30]. Therefore, the primary determinant of stability (physical and chemical) was ensuring control of moisture on storage. In order to maintain an appropriate difference between T_g and storage temperature, moisture content was controlled to no more than 5 % throughout product shelf life.

The amorphous nature of insulin powder was characterized by the complete absence of sharp diffraction peaks using X-ray powder diffraction (XRPD). Further, the XRPD and the particle morphology by scanning electron microscope (SEM) did not change upon exposure to increased moisture or temperature. Moisture sorption curves at 25 °C also showed no sign of water-induced crystallization [26]. Each of the individual components (insulin, sodium citrate, mannitol, and glycine) was potentially capable of crystallization at specific conditions of concentration, temperature, and moisture content; however, the formulated insulin powder remained amorphous at all pharmaceutically relevant conditions due to mutual dilution of the components within the amorphous matrix.

21.2.3 Chemical and Conformational Characterization

Circular Dichroism (CD) and Fourier Transform Infrared (FTIR) Spectroscopy of drug substance and drug product were essentially unchanged after spray drying and temperature-accelerated storage [31]. CD spectra obtained before and after spray drying indicated similar α-helical conformations [32]. FTIR confirmed that the structure of insulin was primarily α-helical [33]. Furthermore, no aggregation, precipitation, or fibrillation was observed by SDS-PAGE and DLS. The spray-drying process did not affect the secondary structure of insulin, assuring biological activity. Bioactivity was confirmed in vivo according to USP <121>.

Insulin was monitored for chemical changes following storage at accelerated and long-term conditions by reversed-phase high-performance liquid chromatography (RP-HPLC) and size exclusion chromatography (SEC) [34, 35]. The major degradation pathways for insulin were deamidation of Asn^{A21} (A21) and aggregation into high-molecular weight products (HMWP). The sum of all other degradation products was termed insulin-related substances (IRS). Formation of A21, HMWP, and IRS were found to fit an Arrhenius kinetic model over a wide temperature range, above and below the T_g, (Fig. 21.2) [34, 35]. These results confirmed that the spray-dried insulin powder was a strong glass with low molecular mobility [27].

The Exubera® formulation met all compendial requirements for insulin impurities ($A21 \leq 5.0\,\%$, $HMWP \leq 2.0\,\%$, and $IRS \leq 6.0\,\%$) throughout its shelf life. Proper formulation design, a nondestructive manufacturing process, and stringent control of moisture enabled the first room temperature-stable insulin formulation.

21.3 Manufacturing

21.3.1 Spray Drying and Particle Characterization

The application of spray drying technology to produce particles for pulmonary use required process conditions outside those normally used for pharmaceutical products.

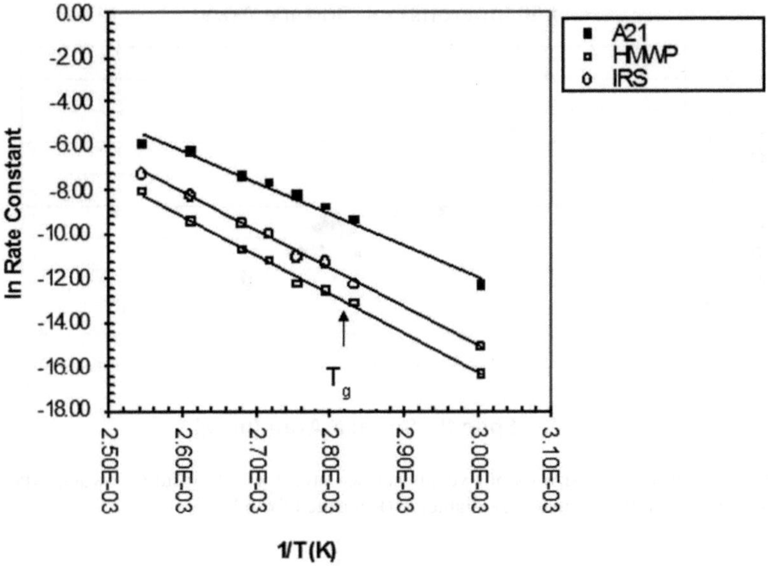

Fig. 21.2 Arrhenius plot of the rate of formation of major degradation products: deamidated Asn^{A21} insulin (A21), insulin-related substances (IRS), and high molecular weight protein (HMWP). (Modified from [13])

Generally, spray drying is used to produce particles much larger than that required for aerosol delivery: Food applications generally range from 50–100 μm and typical pharmaceutical applications range from 10–50 μm. Spray drying was selected over other particle-processing technologies for its ability to produce homogenous particles within the desired small particle size range (<5 μ m), low moisture content, and high drug purity. Solution spray drying ensured compositional homogeneity of the insulin powder, since the insulin and the excipients are dissolved prior to spray drying. Furthermore, spray drying is a continuous production method, scalable for commercial production volumes.

The formulation was compounded by dissolving the excipients in purified water, adding the insulin, and adjusting the pH to 7.3 with a small amount of sodium hydroxide. The compounded solution was chilled and filtered (0.45 and 0.2 μm) before spray drying. Spray drying involved several stages occurring in milliseconds. The compounded bulk solution was first atomized through a 2-fluid jet assist nozzle into a stream of heated air. The droplets rapidly cooled to their wet bulb temperature due to flash evaporation of water. Moisture loss continued and once solute concentration was exceeded, a "skin" formed at the droplet surface. Diffusion of the remaining water from the particle interior through the skin resulted in rugose, raisin-shaped particles [36, 37]. The particles approached the spray-drying outlet temperature during the final drying and were collected using a high efficiency cyclone. Final moisture content was determined by control of the relative humidity within the cyclone [38]. Predetermined collector changed out intervals limited powder exposure to the outlet temperature.

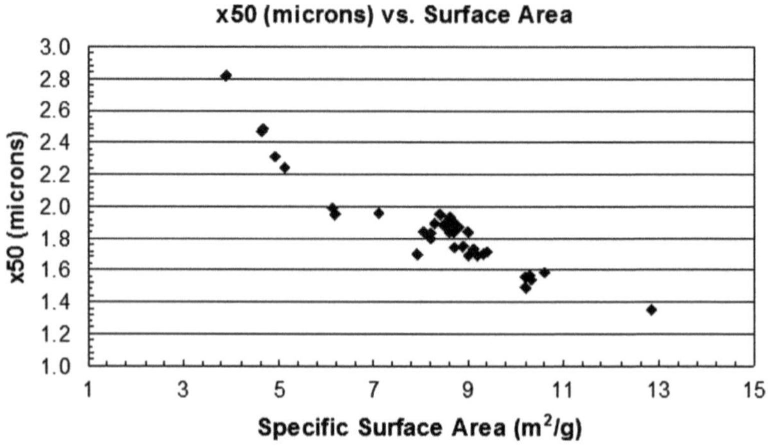

Fig. 21.3 Median particle size (× 50) versus surface area for bulk insulin powder, where × 50 represents 50 % cumulative undersize diameter. (Reprinted from [13])

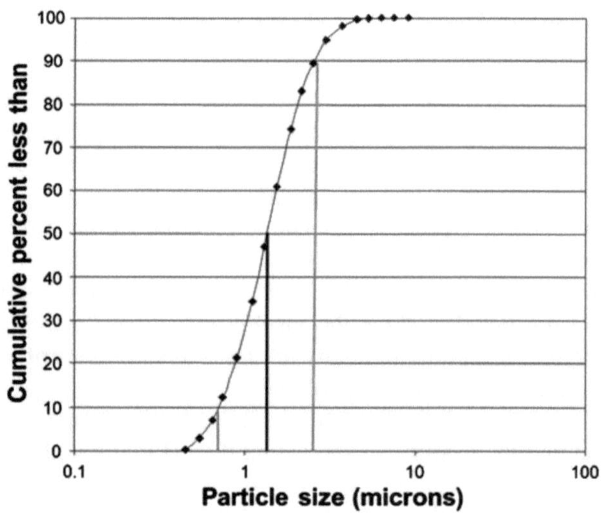

Fig. 21.4 Primary particle size distribution of spray-dried insulin powder

Process robustness was confirmed by correlating specific surface area (nitrogen adsorption/BET) and primary particle size (Sympatec Helos) (Fig. 21.3) [28]. The particles were a narrow, log-normal distribution ($X_{10} = 0.7$ μm, $X_{90} = 2.7$ μm) with a volumetric median (mass median) diameter $X_{50} = 1.5$–2.1 μm (Fig. 21.4). Specific surface area varied ($\approx$ 4–13 m^2/g) as a function of processing conditions (varied outside standard conditions), where specific surface area showed a strong correlation with particle size distribution and solid content of the bulk solution [28]. At standard processing conditions, powder-specific surface area was consistently $\approx$ 9 m^2/g. The Exubera® spray-dried bulk insulin powder was characterized for true density

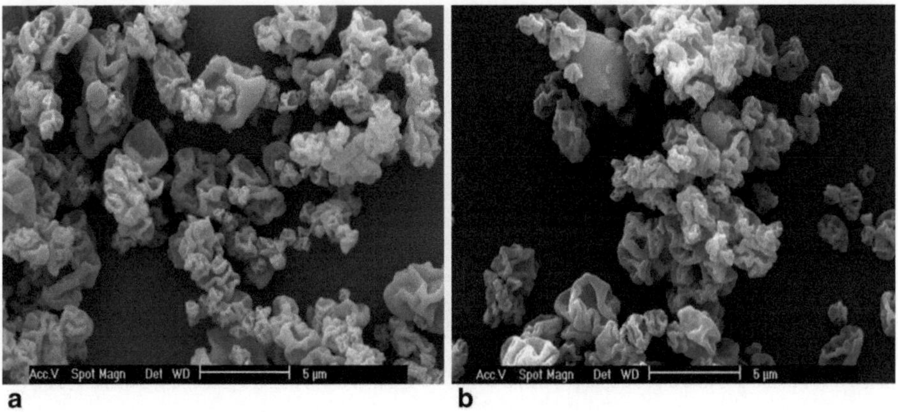

Fig. 21.5 Scanning electron micrographs of spray-dried insulin powder **a** before and **b** after unprotected exposure to 25 °C/75 %RH for 36 h. (Modified from [13])

$\approx 1.48\ g/cm^3$ (helium pycnometry) and bulk density ($\approx 0.2\ g/cm^3$) [28]. These powder characteristics have been shown to be typical for fine powders of a size suitable for inhalation applications.

Spray-dried insulin solution resulted in uniform particle morphology with a raisin-like appearance, as shown by SEM (Fig. 21.5a). The resulting insulin powder for inhalation proved physically stable when exposed to 75 % RH for 36 h (at 25 °C) (Fig. 21.5b). No solid–liquid transitions, material flow, or inter-particle bridges were observed, and the particle morphology (characteristic wrinkled surface) did not change.

Insulin powder for inhalation was spray dried from a uniform aqueous solution which provided an advantage over conventional dry powder pulmonary products that are typically blends of crystalline and/or amorphous materials that may undergo particle segregation and powder separation. The spray-drying process maintained insulin integrity, yielded a homogenous amorphous powder of low moisture content, and provided precise control of the particle size distribution.

21.3.2 Powder Filling and Packaging

In order to provide a pharmaceutical alternative to rapid acting subcutaneous insulin injections, dose flexibility was required to accommodate the varying insulin dose needs of diabetic patients. To enable dosage flexibility, a premetered unit dose concept was developed, utilizing two product strengths: 1 and 3 mg insulin (1.7 and 5.1 mg nominal powder fill weight, respectively). Combinations of these product strengths enabled individual dose titration.

Due to the low fill powder masses required to achieve the insulin unit dose blisters, novel filling technology was developed. The equipment was designed to induce controllable powder flow and volumetric dose metering of loose powder compacts into blister packages at low relative humidity. Specific design challenges for

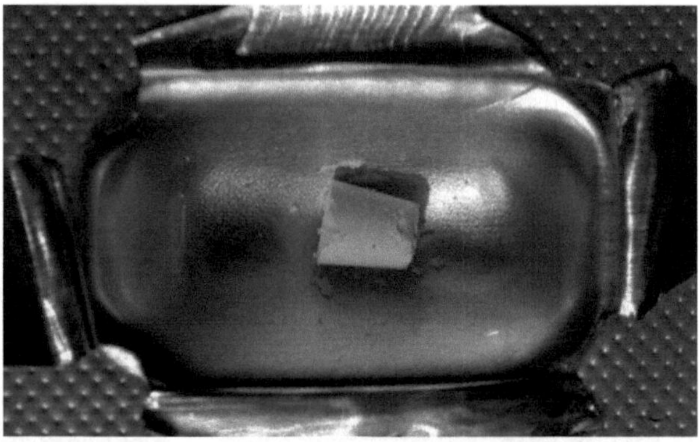

Fig. 21.6 Insulin powder "puck" in an opened unit-dose blister. (Reprinted from [13])

filling included low powder density, powder cohesivity, extremely low fill weights (<5 mg), need for high throughput at commercial scale (>1,500 fills/min), and stringent controls on accuracy, precision, and consistency. The powder-filling dose metering system was integrated with packaging equipment that formed aluminum foil blisters on-line and heat-sealed the filled blisters. Dose metering at the powder-filling head was accomplished by ultrasonically inducing powder flow from a trough into a truncated cone-shaped filling cavity located on a rotating cylinder. A specified negative pressure was drawn from within the rotating cylinder at the base of each filling cavity to allow precise control of the filled powder "puck" density. The cylinder of multiple filling cavities was then rotated 180° about its long axis to invert the powder pucks above a line of formed lower blister packaging laminate. The pucks were expelled into the open blister cavities by controlled reversal of the airflow within the cylinder. The geometry of a representative powder puck within the cavity of an opened blister package is shown in Fig. 21.6; the puck was extremely friable to slight vibration and its shape was not retained upon further processing. Control of the unit dose filling process was demonstrated through exquisite fill weight control with RSD <2 % (Fig. 21.7).

The primary packaging design was comprised of a compact blister as the unit dose to facilitate patient handling and ensure compatibility at the device interface [21]. The blister package dimensions (cavity depth) and registration in the device were critical to performance. Single blisters were produced for clinical studies and multiblister cards (six removable blisters per card) were designed as the commercial presentation. Virtually impermeable materials were selected for the primary package to minimize moisture transfer to the powder on storage: bottom web of cold-formable PVC-foil laminate and top web of heat-sealable foil laminate lidstock. The only possible path of (limited) moisture or oxygen transfer would be between the heat-sealed laminate layers originating from the blister edge inward the interior cavity. Optimizing the path length from blister edge to interior cavity was important not only for proper registration with the pulmonary inhaler, but also to enable prolonged storage (>1 month) at accelerated RH conditions after the protective secondary package overwrap was removed. The blister-sealing process was optimized for temperature (top and bottom plates) and dwell time to ensure complete and continuous thermal

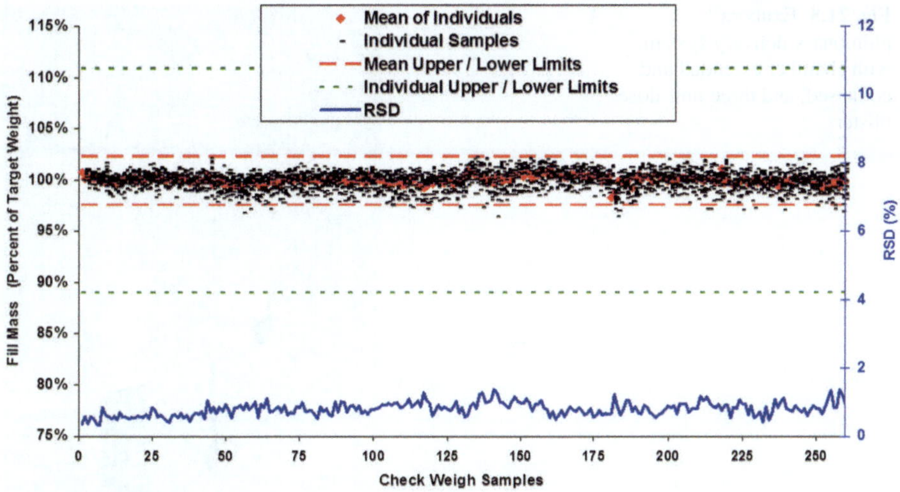

Fig. 21.7 Process-control chart for the powder filling process. (Reprinted from [13])

welding of the laminate layers and to minimize material temperature change at the walls of the blister cavity. Powder pucks sampled immediately following blister filling and heat sealing showed no increase in powder crystallinity (amorphous by XRPD) or increase in insulin degradation products.

Technological innovations in the production of stable and respirable insulin particles, precise metering of ultra-fine powders, and design of a functional and protective blister were successfully brought to commercial-scale as part of Exubera® development.

21.4 Pulmonary Inhaler Development

21.4.1 Inhaler Design

Typical limitations of traditional inhalation devices include low efficiency, variable dosing, poor moisture barriers, low drug content per inhalation, inapplicability to macromolecules, and sensitivity to the breathing maneuver [21]. In comparison, the Exubera® inhaler solved many of these challenges.

The PDS is a reusable dry powder inhaler that has been designed to deliver insulin to the small airways and alveoli for systemic insulin absorption [39, 40]. The inhaler was designed to provide reproducible powder extraction, deagglomeration and dispersion, capable of aerosolizing relatively small amounts of cohesive powder (1–10 mg). The design was solely mechanical, and required only modest patient effort to operate. Unlike typical high resistance inhaler systems, aerosolization of insulin powder by the Exubera® inhaler was independent of patient inspiratory effort [21, 41].

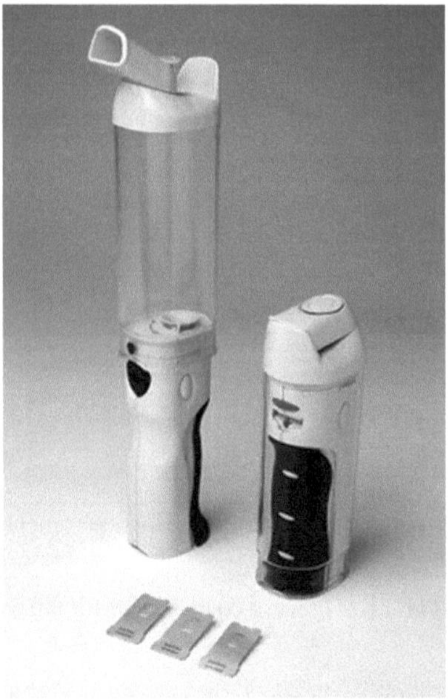

Fig. 21.8 Exubera® pulmonary delivery system, with chamber extended and collapsed, and three unit-dose blisters

The Exubera® inhaler consisted of three subsystems: base, transJector, and chamber (with mouthpiece) (Fig. 21.8). The base and chamber were washable, and each designed to last for at least 1 year under accelerated use conditions. The transJector was not washable, since the interior was not easily dried, and was intended to be replaced after 1–2 weeks of normal use. The Exubera® inhaler performed four key functions: puncture of the insulin powder-containing blister after loading into the device, extraction of the powder from the blister, dispersal of the powder as an aerosol into the chamber, and facilitation of inhalation delivery of the aerosol cloud to the patient. The patient opens and closes the pump handle once before actuation to draw in and compress ambient air ($\approx$ 8 mL).

Upon actuation, an individual blister is raised to contact the bottom of the transJector where it is punctured and the compressed air is released through jets within the core of the transJector. The sudden ($\approx$ 70 msec) sonic speed discharge of this mass of air through the narrow transJector jets creates a negative pressure within the blister cavity at the base of the transJector core, causing "chase air" to flow through the blister cavity scrubbing the insulin powder from the package and expelling it into the chamber as a standing aerosol cloud (Fig. 21.9). The energy generated by the transJector jets ($\approx$ 40 Watts) was sufficient to disperse the powder into particles 3–5 μm in aerodynamic diameter, which were appropriately sized for alveolar depositions [42].

The patient, visually queued by the aerosol cloud within the chamber, rotated the mouthpiece 180° into the open position and slowly inhaled the aerosol through the mouthpiece with a slow, deep breath. The aerosol was evacuated from the chamber ($\approx$ 200 mL) within the first 300–400 mL of the inhalation maneuver (Fig. 21.10). Thereafter, the following 500 mL of inspired volume moved the airborne dose

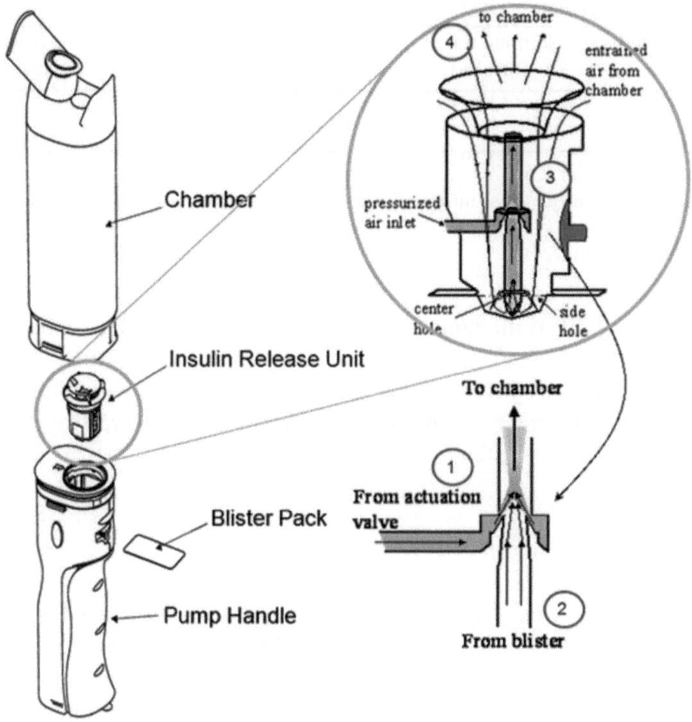

Fig. 21.9 Exubera® pulmonary delivery system showing airflow pathways within the transJector. *1* Pathway of compressed air to jets. *2* "Chase" air flow drawn from blister. *3* Powder dispersion occurs within core of transJector. *4* Aerosol ejected to chamber. (Modified from [13])

Fig. 21.10 Schematic representation of inhaled dose. The aerosol dose is administered in the first seconds of inhalation and the remaining inhaled volume carries the dose to the deep lung. (Modified from [13])

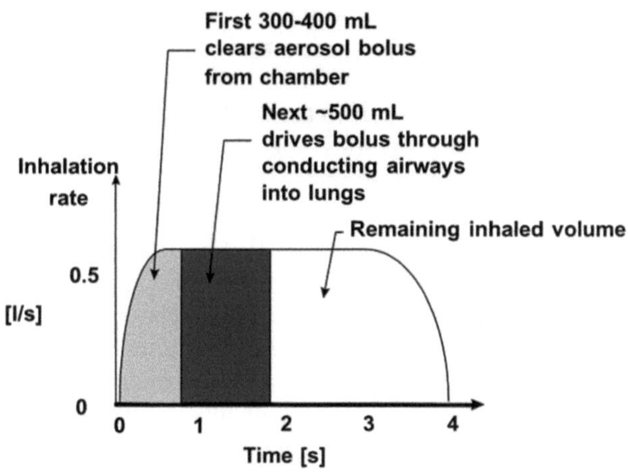

through the conducting airways into the deep lung. By comparison, the average tidal volume is ≈500 mL and the volume of a deep breath may range from 1 to 3 L above tidal volume in healthy adults. After completing the inhalation maneuver, the

Table 21.1 Dose nomenclature

Fill mass (mg powder)	Nominal dose (mg insulin)	Emitted dose[a,b] (mg insulin)	Fine particle dose[b,c] (mg insulin)
1.7	1.0	0.53	0.4
5.1	3.0	2.03	1.0

[a]Flow rate of 30 L/min for 2.5 s
[b]Emitted dose and fine particle dose information are not intended to predict actual pharmacodynamic response
[c]Flow rate of 28.3 L/min for 3 s

patient was instructed to maintain a brief breath hold after inhalation. The design and function of the Exubera® inhaler did not require exact coordination of inhaler actuation and inhalation by the patient since powder dispersal was controlled by the inhaler device. Insulin powder could be delivered into the absorptive regions of the deep lung independent of patient inspiratory flow rate [14].

21.4.2 Inhaler Performance

Product performance was characterized by emitted dose (ED) per USP <601> at 30 L/min. The aerosol particle size distribution (PSD) and Fine Particle Dose (FPD, $< 3.3\ \mu$ m) were determined using the Anderson Mark II cascade impactor operated at 28.3 L/min. ED, PSD, and FPD were also tested under a wider range of simulated patient use and environmental conditions.

Exubera® delivered specific ED and FPD of insulin; however, the doses delivered were not linear between the 1 and 3 mg insulin blisters (Table 21.1) [43]. The lack of dose linearity between the low and high strength insulin blisters was due to powder retained within blisters: $\approx$ 40 and $\approx$ 25 % of the contents of 1 and 3 mg blisters, respectively, remained adhered to the interior surfaces of the blister packages after actuation. In vitro ED was consistent among all flow rates tested above 10 L/min. The aerosol particle size distribution (mass median aerodynamic diameter, MMAD $= \approx 3\ \mu$m) was predicted by properties of the spray dried powder due to the consistent dispersive energy imparted by the inhaler (Fig. 21.11).

Aerosol delivery across a range of flow rates (5–60 L/min), flow volumes (400–1,400 mL), inhaler orientations (0–270° from vertical), standing cloud dwell times), repeated actuations, temperature (5–40 °C), humidity 20–90 % RH, and reduced atmospheric pressure (up to the equivalent of 10,000 ft altitude) did not materially affect aerosol performance metrics for either the 1 or the 3 mg blisters [17, 41].

ED and PSD showed negligible dependence on inhaler priming (Fig. 21.12), relative humidity (Fig. 21.13), or inhaler orientation for both 1 and 3 mg blisters (Fig. 21.14). All inhaler orientations (0°, 24°, 90°, 180°, and 270° from vertical) gave mean ED and FPD values within 2–3 % of the controls.

Long-term in-use performance consistency (aerosol and mechanical) was demonstrated through in vitro use life simulations and through testing of inhalers retrieved from clinical trials following up to 2 years of inhaler (base) usage (Fig. 21.15). These studies resulted in the demonstration of improved performance and durability over standard dry powder inhalers.

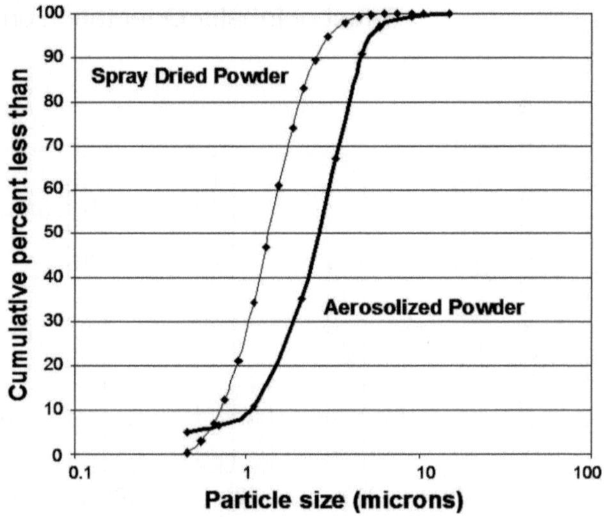

Fig. 21.11 Spray-dried primary particle size distribution compared with aerosol particle size distribution. (Reprinted from [13])

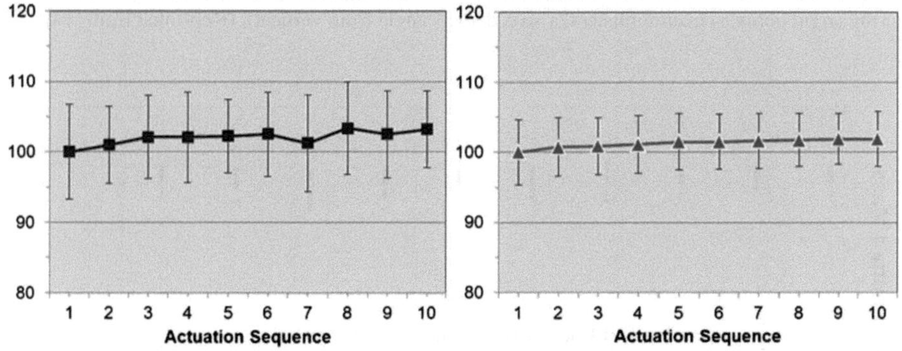

Fig. 21.12 *Vertical axis* shows emitted dose normalized to first actuation (mean ±1 SD) as a function of actuation sequence, depicting absence of priming effect (*left*: 1 mg $n = 81$; *right*: 3 mg $n = 269$)

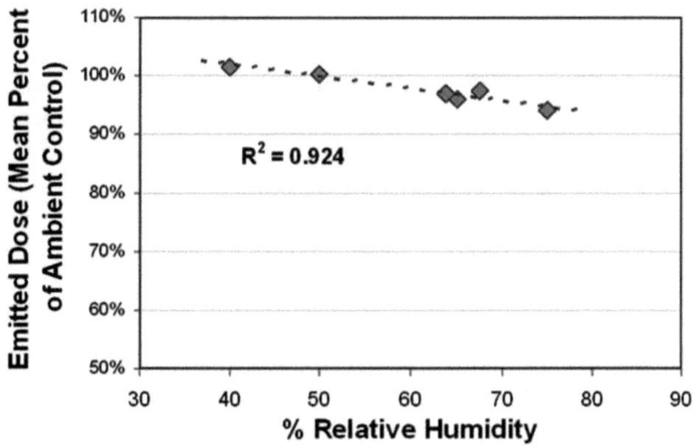

Fig. 21.13 Minimal effect of increasing relative humidity on emitted dose normalized to ambient control (50 % RH). (Reprinted from [14])

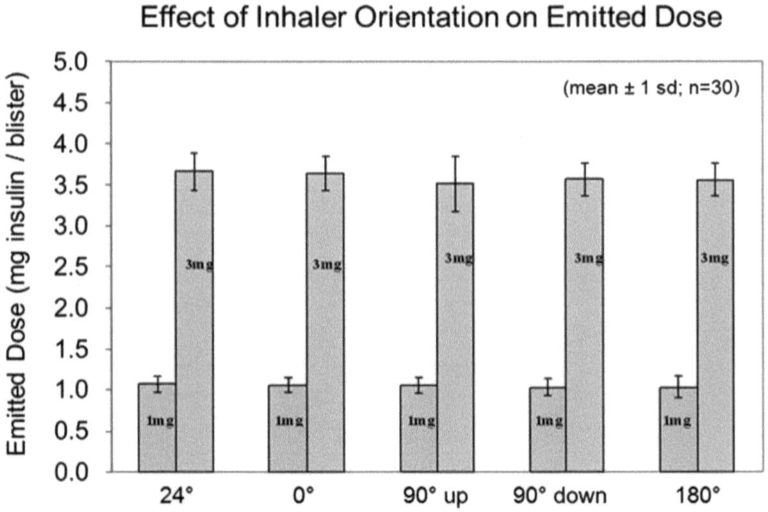

Fig. 21.14 Effect of Exubera® inhaler orientation on emitted dose from 1 mg (*left column*) and 3 mg (*right column*) insulin blisters (*x-axis* depicts angle from vertical). (Reprinted from [14])

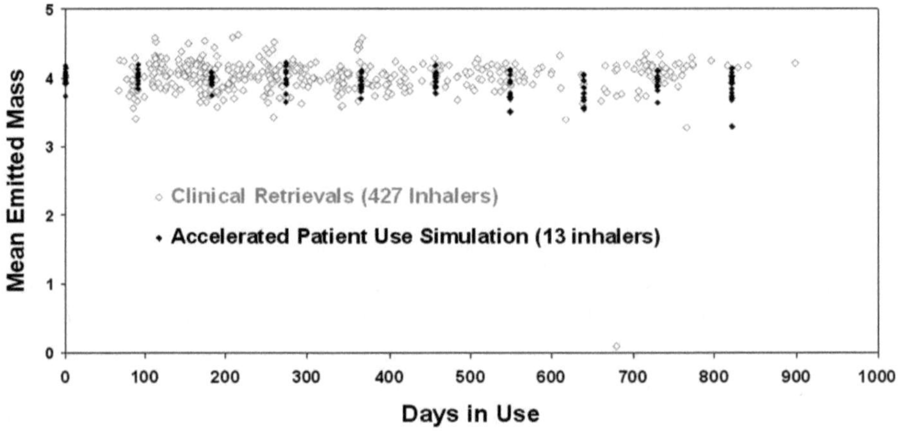

Fig. 21.15 Exubera® inhaler robustness represented by consistent long-term use (1 year shelf life + 1 year use life). *Vertical axis* represents mean emitted mass (mg) from 3 mg insulin blisters (5.1 mg fill mass). (Reprinted from [14])

21.5 Clinical Studies

Exubera® was approved in the USA in 2006 for treatment of Type 1 and Type 2 diabetes, with or without concomitant oral agents and/or long-acting insulin, and was contraindicated in smokers and patients with underlying lung diseases [44].

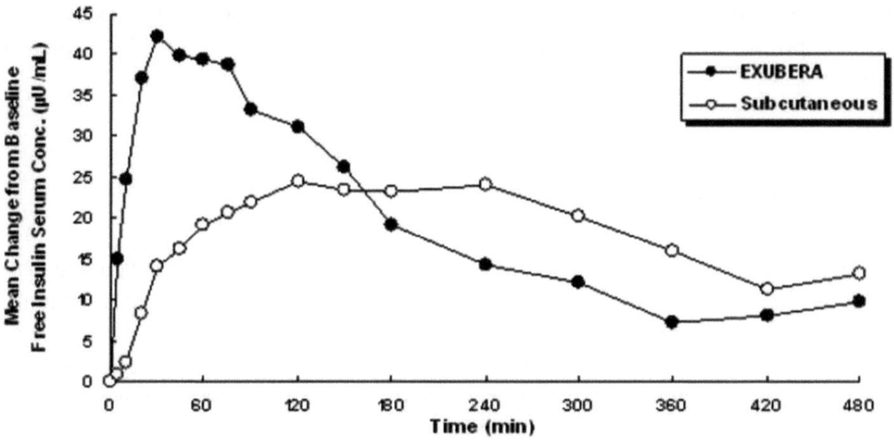

Fig. 21.16 Mean changes in free insulin serum concentrations (μU/mL) in patients with Type 2 diabetes following administration of single doses of inhaled insulin from Exubera® (6 mg) and subcutaneous regular human insulin (18 IU)

In clinical trials over 2,500 patients received insulin administered using Exubera®, some for as long as 8 years.

Exubera® efficacy showed equivalent HbA1c control and bioavailability (≈ 10 %) to subcutaneous injection [44, 45]. The safety profile showed that the incidence of hypoglycemia was the same as subcutaneous injection, with a mild cough being the most common side effect [44, 45].

21.5.1 Pharmacokinetics

Insulin bioavailability from Exubera® was found to be approximately 10 % relative to subcutaneous injection. Absorption following inhalation was rapid and similar to subcutaneously administered rapid-acting insulin analogs, and was more rapid than subcutaneously administered regular insulin (Fig. 21.16) [43].

In Type 1 and Type 2 diabetic patients serum peak concentration reached peak concentrations more quickly after inhalation of Exubera® than after subcutaneous injection of regular human insulin, 49 min (30–90 min) compared to 105 min (60–240 min), respectively [43]. In healthy subjects, AUC and C_{max} increased with increasing dose following administration of Exubera® from 1 to 6 mg when administered as combinations of 1 and 3 mg insulin blisters [43]. Apparent from the in vitro FPD results listed in Table 21.1, three 1 mg blisters were not bioequivalent to one 3 mg blister. No safety and efficacy concerns arose from the lack of interchangeability of both blisters and the lack of dose equivalence between the two blister strengths was addressed by clear labeling. The C_{max} and AUC determined following administration of three 1 mg blisters were approximately 30 and 40 % greater, respectively, than after administration of three 1 mg blisters [43]. In smokers, the systemic insulin exposure for Exubera® was estimated to be two to five times higher than in nonsmokers.

21.5.2 Pharmacodynamics

Exubera® has a more rapid onset of glucose-lowering activity than subcutaneously administered regular human insulin, and was similar to subcutaneously administered rapid-acting insulin analogs. The onset of glucose-lowering activity occurs within 10–20 min following inhalation. In Type 1 and Type 2 diabetics, Exubera® exhibited a greater glucose-lowering effect within the first 2 h after dosing when compared with subcutaneously administered regular human insulin. Following inhalation from Exubera®, smokers experienced a more rapid onset of glucose-lowering action compared to nonsmokers.

The duration of glucose-lowering activity following inhalation was comparable to subcutaneously administered regular human insulin and longer than subcutaneously administered rapid-acting insulin analogs. The maximum effect on glucose lowering occurs approximately 2 h after inhalation with duration of approximately 6 h. The intra-subject variability of glucose-lowering activity following inhalation was generally comparable to that of subcutaneously administered regular human insulin in patients with Type 1 and Type 2 diabetes. Following inhalation from Exubera®, smokers experienced a greater total glucose-lowering effect (particularly during the first 2–3 h after dosing) compared to nonsmokers.

21.5.3 Safety and Efficacy

Clinical noninferiority of Exubera® was demonstrated through equivalent control of HbA1c levels compared with subcutaneous injected insulin. Inter- and intra-patient variability of HbA1c levels was the same as subcutaneous injection; and the incidence of hypoglycemia was also similar [43, 46]. Exubera® was studied in Type 1 and Type 2 patients across a variety of treatment regimens. In Type 1 and Type 2 diabetics, a daily regimen of three premeal inhaled insulin administrations plus subcutaneous basal insulin resulted in similar reductions in HbA1c from baseline compared to a regimen of three premeal subcutaneous short-acting insulin injections plus subcutaneous basal insulin. The percentages of patients achieving HbA1c $< 8\,\%$ or $< 7\,\%$ were comparable to or better than those achieved with the subcutaneous insulin regimen. Patients also preferred using Exubera®, were more satisfied with their overall treatment and showed greater improvements in symptoms and cognitive function [46].

The frequency and nature of adverse events were similar for Exubera® and control groups. The most common side effect unique to Exubera® was mild cough which decreased with increased exposure [43, 46, 47]. A small nonprogressive difference in pulmonary function tests, but without clinical manifestation, was also observed between a limited group of Exubera® and control patients. Patients treated with Exubera® exhibited small (<2 %), early, nonprogressive declines in forced expiratory volume in 1 s (FEV1) and carbon monoxide diffusing capacity (DLCO) compared with similar declines (<1 %) in patients treated with subcutaneous insulin [47].

Treatment with inhaled insulin was associated with increased insulin antibody levels compared with patients treated with subcutaneous short-acting insulin or oral therapies. For subjects receiving Exubera®, antibody levels were higher in Type 1 patients than Type 2 patients. Among Type 2 diabetics, insulin antibody levels were higher in patients using insulin rather than patients not using insulin at baseline. Mean insulin antibody levels increased during the first 6 months and reached a plateau after approximately 6–12 months in Exubera®- treated patients. Insulin antibody levels decreased upon discontinuation of inhaled insulin. Antibodies associated with inhaled-insulin exposure were qualitatively similar to those induced by subcutaneous-insulin exposure. No associations between antibody levels and insulin dose, glycemic control, pulmonary function, or adverse clinical outcomes were identified. No consistent safety signals were identified among patients with the highest insulin antibody levels [46, 48].

In more recent clinical trials of Exubera®, there were six diagnosed cases of primary lung malignancies among Exubera®-treated patients, and one diagnosed case among comparator treated patients. There was one postmarketing report of a primary lung malignancy in an Exubera®-treated patient. The incidence of new primary lung cancers per 100 patient-years of study drug exposure was 0.13 (5 cases over 3,900 patient-years) for Exubera®-treated patients and 0.02 (1 case over 4,100 patient-years) for comparator-treated patients. There were too few cases to determine whether the emergence of these events is related to Exubera®. All patients who were diagnosed with lung cancer had a prior history of cigarette smoking [49].

21.6 Market Withdrawal

Despite regulatory approval in the USA and Europe, Exubera® was pulled from the market. The decision was not based on safety or efficacy, but for the lack of sales. Several factors influencing Exubera®'s market opportunity after launch can be grouped into the following categories: product, marketing, price/reimbursement, physicians, and patient feedback [50]. These factors are briefly discussed below.

Overall, the product itself took longer to develop than anticipated (>10 years), so that by the time it was launched, other competitor products (e.g., pen injectors, rapid-acting insulin analogs) had filled various medical niches intended to be met by Exubera®. The inhaler was deemed too big to use discretely by some users. Another limitation of the product was that it required patients and physicians to convert their previous injected insulin dose from "IU" to an inhaled dose in "mg insulin." Moreover, the ability to finely titrate dose was limited with only two blister strengths of insulin powder. And importantly, one 3 mg blister was not equivalent to three 1 mg blisters, raising concern about dose targeting and reproducibility as patient use moved between the two blister strengths.

The primary marketing issue may have been that Pfizer lacked a major presence in the Endocrinology therapeutic area, unlike its more established competitors Eli Lilly and Novo Nordisk. To compound matters, Pfizer's strategically switched from one nonendocrinology sales force to a second nonendocrinology sales force shortly after launch, resulting in a major force reeducation and lost momentum. Furthermore, the sales force that inherited the product was from their cardiovascular therapeutic area and those professionals were primarily focused on maintaining the eroding Lipitor® sales. Exubera® direct-to-consumer advertising campaign was begun 18 months after launch, and was too late to create sufficient demand among patients before Pfizer began their significant cost-reductions related to the anticipate loss of Lipitor® revenues.

The retail price of Exubera®, approved as a Tier 3 product, was approximately $5/dose, while injectable insulin products were priced at about $3/dose. Pfizer's shared royalty arrangement with Nektar (and earlier with Aventis) and the significant cost of goods further reduced their already thin product margins to the point that they were unsustainable with a prolonged ramp in sales volume.

Physician feedback indicated that potential patient aversion to needles was not their primary concern. Generally, healthcare providers were more comfortable with the established route of administration given its familiarity and relative certainty in dosing. The added requirement for patient education and spirometry equipment for the office put additional burdens on the physician's ability to see the maximum number of patients efficiently (per billable hour). Physicians were also concerned about overall lung safety.

Patients provided very positive feedback to inhaled insulin during clinical development; however, the pain/phobia associated with needles had been overstated once the general population of diabetics was introduced to Exubera®. Increasingly, the use of thinner needles and pen/auto injectors resulted in less pain upon administration and greater compliance with injectable insulin. Neither could Exubera® completely erase all the need for injections since many patients required injections of long-acting insulin for basal control. Furthermore, although diabetics on insulin are sometimes averse to handling a syringe in public for fear of the potential negative perceptions, the size of the Exubera® inhaler was not small enough to provide sufficient discretion. And significantly, many Type 2 patients were reluctant to switch from oral therapies to insulin, even if they did not have good HbA1c control, due to the perception that they had personally "failed" their therapeutic regimen if insulin was required.

In summary, the launch of a novel drug/device combination product requires not only commercializable solutions to scientific and technical issues, but it also requires state-of-the-art pharmacoeconomic and marketing preparation years before product launch to enable commercial success.

21.7 Conclusions and Future Perspectives

A primary strategy for improving insulin compliance was to provide a noninvasive delivery system; and pulmonary delivery was selected because of the large absorptive area of the deep lung. The challenge taken on by the many pharmaceutical scientists and engineers in the development of Exubera® was to create and enable an inhalable insulin product for commercialization. Each of the individual technical challenges was inherently difficult, and each was further complicated by the need to integrate with the other novel technologies and the requirement to provide equal glucose control compared to insulin injection. Exubera® was the first room temperature stable insulin formulation, the first durable powder inhaler, and the first noninvasive insulin product to gain approval in the USA and the EU.

Withdrawal of Exubera® from the market had repercussions beyond Pfizer's bottom line (Pfizer wrote off $2.8 billion in 2007) and the hundreds of lost jobs suffered by Pfizer and Nektar staff. Shortly thereafter Novo Nordisk and Eli Lilly also discontinued their inhaled insulin programs at a one-time cost of DKK 1.3 billion and $90–120 million, respectively. And while MannKind continues to pursue approval of Afrezza® after two complete response letters from the FDA, it remains unclear how Afrezza® will overcome the challenges that Pfizer and Exubera® could not. Beyond the economic figures, the negative connotation imparted to pulmonary delivery of proteins has resulted in a hiatus for follow-on biopharmaceuticals. Nevertheless, the promise of exploiting the large, absorptive surface area of the deep lung for systemic therapies will remain for future entrepreneurs.

The most obvious lesson learned from Exubera® is that successful technological innovation by itself does not guarantee commercial success.

Acknowledgments The authors would like to thank the Exubera® team members within Pfizer, Nektar, and Sanofi-Aventis for their technical inspiration and teamwork.

References

1. Narayan KMW, Boyle JP, Thompson TJ, Sorensen SW, Williamson DF. Lifetime risk for developing diabetes mellitus in the United States. JAMA. 2003;290:1884–90.
2. Home PD, Boulton AJM, Jimenez J, Landgraf R, Osterbrink B, Christiansen JS. Issues relating to the early or earlier use of insulin in type 2 diabetes. Pract Diabetes Int. 2003;20:63–71.
3. American Diabetes Association (ADA). Test of glycemia in diabetes. Diabetes Care. 2002;25:S97–S9.
4. Rosenstock J, for the Exubera® Phase III Study Group. Mealtime rapid-acting inhale insulin (Exubera®) improves glycemic control in patients with type 2 diabetes failing combination oral agents: a 3-month, randomized, comparative trial. Diabetes. 2002;51:A132.
5. Zambanini A, Newson RB, Maisey M, Feher MD. Injection related anxiety in insulin-treated diabetes. Diabetes Res Clin Pract. 1999;46:239–46.
6. Home PD. Intensive insulin therapy in clinical practice. Diabetologia. 1997;40:S83–S7.
7. Cefalu WT. Rationale for and strategies to achieve glycemic control. In: Leahy JL, Cefalu WT, editors. Insulin therapy. New York: Marcel Dekker; 2002. pp. 1–11.

8. Patton JS. Mechanisms of macromolecule absorption by the lungs. Adv Drug Del Rev. 1996;19:3–36.
9. Smith SJ, Bernstein JA. In: Hickey AJ, editor. Inhalation aerosols: physical and biological basis for therapy. New York: Marcel Dekker; 1996. pp. 233–69.
10. Patton JS, Bukar J, Nagarajan S. Inhaled insulin. Adv Drug Del Rev. 1999;35:235–47.
11. Patton JS, Bukar JG, Eldon MA. Clinical pharmacokinetics and pharmacodynamics of inhaled insulin. Clin Pharmacokinet. 2004;43:781–801.
12. Patton JS, Platz RM. Routes of delivery: case studies. Pulmonary delivery of peptides and proteins for systemic action. Adv Drug Del Rev. 1992;8:176–96.
13. White S, Bennett DB, Cheu S, Conley PW, Guzek DB, Gray S, Howard J, Malcolmson R, Parker JM, Roberts P, Schumacher JD, Sadrzadeh N, Seshadri S, Sluggett GW, Stevenson CL, Harper NJ. EXUBERA®: pharmaceutical development of a novel product for pulmonary delivery of insulin. Diabetes Technol Ther. 2005;7:896–906.
14. Harper NJ, Gray S, de Groot J, Parker JM, Sadrzadeh N, Schuler C, Schumacher JD, Seshadri S, Smith AE, Steeno GS, Stevenson CL, Taniere R, Wang M, Bennett DB. Design and performance of the exubera pulmonary insulin delivery system. Diabetes Technol Ther. 2007;9:16–27.
15. Skyler JS, for the Exubera® Phase III Study Group. Efficacy and safety of inhaled insulin (Exubera®) compared to subcutaneous insulin therapy in an intensive insulin regimen in patients with type 1 diabetes: results of a 6-month, randomized, comparative trial. Diabetes. 2002;51:A134.
16. Gelfand RA, Schwartz S, Horton M, Law CG, Pun EF. Pharmacological reproducibility of inhaled human insulin pre-meal dosing in patients with type 2 diabetes mellitus (NIDDM). Diabetes. 1998;47:0388.
17. Balanger A, for the Exubera® Phase III Study Group. Efficacy and safety of inhaled insulin (Exubera®) compared to subcutaneous insulin therapy in an intensive insulin regimen in patients with type 2 diabetes: results of a 6-month, randomized, comparative trial. Diabetologia. 2002;45:A260.
18. Gelfand RA, Schwartz SL, Horton M, Law CG, Pun EF. Pharmacological reproducibility of inhaled human insulin dosed pre-meal in patients with type 2 diabetes mellitus. Diabetes. 2002;51:A202.
19. Quattrin T, Belanger A, Bohannon NJV, Schwartz SL. Efficacy and safety of inhaled insulin (Exubera) compared with subcutaneous insulin therapy in patients with type 1 diabetes. Diabetes Care. 2004;27:2622–27.
20. Hollander PA, Blonde L, Rowe R, Mehta AE, Milburn JL, Hershon KS, Chaisson JL, Levin SR. Efficacy and safety of inhaled insulin in patients with type 2 diabetes: a 6-month, randomized, comparative trial. Diabetes Care. 2004;27:2356–62.
21. Ashurst I, Malton A, Prime D, Sumby S. Latest advances in the development of dry powder inhalers. Pharm Sci Technol Today. 2000;3:246–56.
22. Pikal MJ, Rigsbee DR. Dynamics of pharmaceutical amorphous solids: the study of enthalpy relaxation by isothermal microcalorimetry. Pharm Res. 1997;14:1379–87.
23. Hancock BC, Parks M. What is the true solubility advantage for amorphous pharmaceuticals? Pharm Res. 2000;17:397–404.
24. Hancock BC, Shamblin SL, Zografi G. Molecular mobility of amorphous pharmaceutical solids below their glass transition temperatures. Pharm Res. 1995;12:799–806.
25. Franks F, Hatley RHM, Mathias SF. Materials science and the production of shelf-stable biologicals. Pharm Technol Int. 1991;3:24–34.
26. Lechuga-Ballesteros D, Kuo MC, Liang Y, Malcolmson R, Miller DP, Sekulic S, Seshadri S, Stults CLM, Tan T, Joshi V, Zhen C, Williams L, Bennett DB. The physical stability of insulin powder for inhalation. AAPS J. 2004;6:R6137.
27. Kajiwara K, Franks F, Echlin P, Greer AL. Structural and dynamic properties of crystalline and amorphous phases in raffinose-water mixtures. Pharm Res. 1999;16:1441–8.
28. Lechuga-Ballesteros D, Miller DP, Zhang J. Residual water in amorphous solids: measurements and effects on stability. In: Levine H, ed. Progression amorphous food and pharmaceutical systems. London: The Royal Society of Chemistry; 2002.

29. Duddu SP, Zhang G, Dal Monte PR. The relationship between protein aggregation and molecular mobility below the glass transition temperature of lyophilized formulations containing a monoclonal antibody. Pharm Res. 1997;14:596–600.
30. Hancock BC, Zografi G. The relationship between glass transition temperature and the water content of amorphous pharmaceutical solids. Pharm Res. 1994;11:471–7.
31. Suen C, Bennett DB, Sadrzadeh N, Seshadri S, Stevenson CL, Tan MM, Wang ML, Kelly KE. Solution and solid state structural stability of insulin. AAPS J. 2004;6:M1261.
32. Pittman I, Tager HS. A spectroscopic investigation of the conformational dynamics of insulin in solution. Biochemistry. 1995;34:10578–90.
33. Dong A, Caughey WS. Infrared methods for study of hemoglobin reactions and structures. Methods Enzymol. 1994;232:139–75.
34. Sadrzadeh N, Wang ML, Yu M, Antonino L, Stevenson CL, Bennett DB, Kelly ME. The chemical stability of spray dried insulin powder for inhalation. AAPS J. 2004;6:M1261.
35. Sadrzadeh N, Miller DP, Lechuga-Ballesteros D, Harper N, Stevenson CL, Bennett DB. Solid-state stability of spray-dried insulin powder for inhalation: chemical kinetics and structural relaxation modeling of exubera above and below the glass transition temperature. J Pharm Sci. 2010;99:3698–710.
36. Malcolmson R, De Moor CP, Miller DP, Liang Y, Zhen C, Kim Y, Merchant J, Bennett D, Mazumder MK, Saracovan I, Sekulic S. Physical properties of bulk insulin powder for inhalation. AAPS J. 2004;6:R6169.
37. Stahl K, Claesson M, Lilliehorn P, Linden H, Backstrom K. The effect of process variables on the degradation and physical properties of spray dried insulin intended for inhalation. Int J Pharm. 2002;233:227–37.
38. Vehring R, Tep V, Foss WR. Novel experimental method indicates proteins and peptides are protected from high gas temperatures during spray drying. AAPS PharmSci. 2003;5:M1247.
39. Schuler C, Mao Z, Cameron J. Nektar's dry-powder durable pulmonary drug delivery system: principles of operation. Proceedings of the 31st International Symposium on Controlled Release of Bioactive Materials, Hawaii; 2004.
40. Bakshi A, Paboojian A, Rasmussen D, Tuttle D, Snyder H, Clark A, Smith A, Schuler C. Inhale's dry-powder pulmonary drug delivery system: challenges to current modeling of gas-solid flows. 3rd ASME/JSME Fluids Engineering Conference, San Francisco, CA; 1999.
41. Gray S, Parker J, deGroot J, Ozawa L, Chwa T, Harper N. Performance robustness characteristics of insulin powder for inhalation using a pneumatic inhalation device. Drug Delivery to the Lungs 15, London, UK; 2004.
42. Byron PB. Prediction of drug residence times in regions of the human respiratory tract following aerosol inhalation. J Pharm Sci. 1986;75:433–8.
43. Exubera Product Label, www.accessdata.fda.gov/drugsatfda_docs/label/2008/021868S016S017lbl.pdf.
44. Kaptiza C, Heise T, Pfutzner A, Steiner S, Heiniemann L, Kave K. Dose-response characteristics for a new pulmonary insulin formulation and inhaler. Diabetologia. 2000;40(Supp 1):A46.
45. FDA Advisory Committee Meeting, September 5, 2005 slides, http://www.fda.gove/ohrms/dockets/ac/o5/slides/2004-4169S1_00_slide-Index.htm.
46. Exubera Briefing Document, http://www.fda.gov/ohrms/dockets/ac/05/briefing/2005-4169B1_01_01-Pfizer-Exubera.pdf
47. Skyler JS, Jovanovic L, Klioze S, Reis j, Dugan W. Two-year safety and efficacy of inhaled human insulin (Exubera) in adult patients with type 1 diabetes. Diabetes Care. 2007;30:579–85.
48. Quattrin T, Bélanger A, Bohannon NJ, Schwartz SL. Efficacy and safety of inhaled insulin (Exubera) compared with subcutaneous insulin therapy in patients with type 1 diabetes results of a 6-month, randomized, comparative trial. Diabetes Care. 2004;27:2622–27.
49. http://clinicaltrials.gov/show/NCT00734591.
50. Heinemann L. The failure of exubera: are we beating a dead horse? J Diabetes Sci Technol. 2008;2:518–29.

Chapter 22
Technosphere®: An Inhalation System for Pulmonary Delivery of Biopharmaceuticals

António J. Almeida and Ana Grenha

22.1 Introduction

The lung has been explored for a long time for biopharmaceutical administration, one of the oldest reports dating back to 1924 and referring to inhaled insulin [1]. The advantages of this route for systemic drug delivery are well known, including the large alveolar surface available for absorption, the very thin diffusion path to the bloodstream, the extensive vascularization, the relatively low metabolic activity and the possibility to avoid hepatic first-pass effect [2–4]. However, several limitations have to be considered as well, mainly related with the sinuous architecture of the lung tree that impairs particle flowing, the reduced amount of liquid for drug dissolution and diffusion, and specific defense mechanisms like the mucociliary clearance [3, 5]. A more detailed description of these considerations can be found in Chap. 7, which is fully dedicated to pulmonary delivery of biopharmaceuticals.

Notwithstanding its promising characteristics, systemic pulmonary delivery of drugs is not an established approach. This is due not only to the difficulty in designing adequate drug carriers that overcome the referred limitations, reaching the alveoli successfully and in sufficient amount, but also to the safety concerns raised by the alveolar deposition of drugs, particularly when chronic administration is considered.

In summary, designing adequate carriers for systemic pulmonary delivery demands addressing key considerations related with aerodynamic properties, the ability to provide complete drug release in reduced amount of liquid and, above all, the safety of the drug carrier. In the following sections, the details on how Technosphere® technology met these requirements will be described.

A. Grenha (✉)
CBME—Centre for Molecular and Structural Biomedicine/IBB—Institute for Biotechnology and Bioengineering, Faculty of Sciences and Technology, University of Algarve, Campus de Gambelas, 8005–139 Faro, Portugal
e-mail: amgrenha@ualg.pt

A. J. Almeida
iMed.UL, Faculty of Pharmacy, University of Lisbon, Lisbon, Portugal

J. das Neves, B. Sarmento (eds.), *Mucosal Delivery of Biopharmaceuticals,* DOI 10.1007/978-1-4614-9524-6_22,

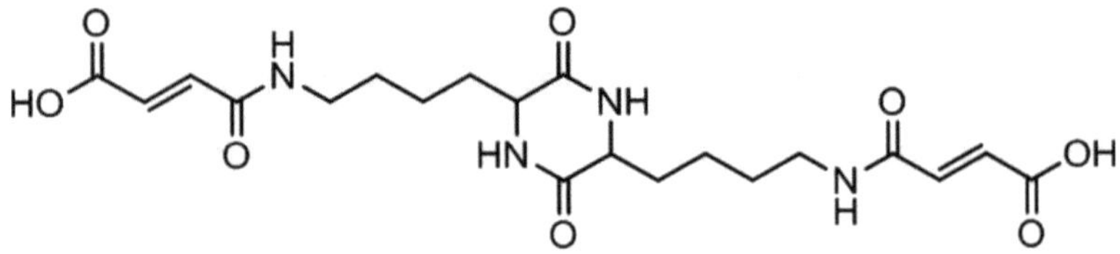

Fig. 22.1 Chemical structure of fumaryl diketopiperazine

22.2 The Technosphere® System

Developing inhalable dry powders demands meeting several requirements, not only regarding the difficulties posed by the lung defense mechanisms and airway structure, but also addressing issues related with drug release and stability. An ideal particle engineering technology should permit working on a wide size range to accommodate several therapies which, in turn, also requires an application to different molecules, such as small drugs and large biopharmaceuticals. Additionally, drug pharmacokinetics should be adequate, the involved excipients eliminated and, finally, the technology should be scalable and cost effective.

The Technosphere® technology is a registered trademark of MannKind Corporation and was developed to meet all these requirements. It comprises microparticles mainly composed of fumaryl diketopiperazine (FDKP), an excipient that is also property of that company, while residual amounts of Tween 80® are further included [6–9]. FDKP is a derivative of diketopiperazines, a group of small cyclic dipeptides commonly found in natural products [10]. They are advantageous in comparison with linear peptides, namely regarding the stability to proteolysis, conformational rigidity, and the promotion of interactions with biological targets by hydrogen bonding mediated by donor and acceptor groups [11]. The latter characteristic is of major importance, because hydrogen bonds are the driving force for the preparation of Technosphere® particles. In fact, FDKP (bis-3,6(4-fumarylaminobutyl)-2,5-diketopiperazine) is a fumaramide derivative of diketopiperazine (Fig. 22.1) that self-assembles into larger constructs by means of intermolecular hydrogen bonding, as described for diketopiperazine-based molecules [12–13]. It was actually identified as the derivative providing the optimal properties for the self-assembly into microspheres, justifying its selection [14]. The self-assembly occurs at acidic pH (< 5.2) [7, 15] in a process that is thought to be mediated either by the carboxylic acid or the amide groups [14]. A deep analysis of the molecular events governing the self-assembly of FDKP is available in [14].

Depending on the method used to process FDKP for obtaining the microparticles, the final carriers can be either crystalline or amorphous. The morphological differences between the two types of microparticles are intense (Fig. 22.2). In a publication authored by MannKind scientists, a curious morphological description of the crystalline microparticles was provided, referring that "the particle can be envisioned as a three-dimensional sphere constructed from a deck of playing cards. Each card represents a FDKP nanocrystal and the sphere constructed from the cards represents

Fig. 22.2 Technosphere® crystalline particle (*left*) and amorphous particle (*right*). Average particle size is about 2 μm in diameter. (Images gently provided by MannKind Corporation)

a Technosphere particle" [16]. The preparation of these crystalline microparticles occurs in solution and involves the formation of FDKP nanocrystals in a controlled, pH-dependent crystallization, followed by the self-assembly of the nanocrystals into microparticles [6, 16]. The incorporation of the drugs occurs by mixing with the excipient solution during the precipitation process, resulting in the adsorption to the nanocrystal surface [6, 17]. Amorphous particles are, in turn, obtained by spray-drying. A salt of FDKP is mixed with the drug and the resultant microparticles are a homogeneous composite of both components [16].

In both cases the particles are monodisperse and exhibit high porosity, thus resulting in low density and suitable aerodynamic properties for deep lung deposition [16]. This aspect appears as a real advantage of Technosphere®, as most standard inhalable dry powder formulations consist of saccharides blended with micronized drug powders, frequently resulting in final heterodisperse particles [18]. The aerodynamic diameter of microparticles ranges between 2 and 2.5 μm [6, 19], more than 90 % being in the respirable size range (0.5–5.8 μm) [6–7].

The Technosphere® engineering technology has been reported as highly versatile, permitting the production of small particles for deep lung inhalation or larger particles for deposition in the upper airways [20]. An optimization of the process parameters is reported as sufficient to endow the microparticles with a preselected size [16]. Therefore, upon formation of either type of microparticles (crystalline or amorphous), no further processing is necessary for size modulation. A final process of freeze-drying is applied to crystalline particles to endow the dry powders with the most suitable properties for inhalation via a small inhaler device [15, 21].

Bearing in mind that this technology was developed to provide systemic lung delivery, aerodynamic suitability provides the guarantee that most of the emitted dose will reach the alveolar zone. However, a problem remains to be solved regarding the release of the drug. As referred before, the amount of lung lining fluid is very small and ensuring drug release and dissolution might be a challenge. Technosphere® has

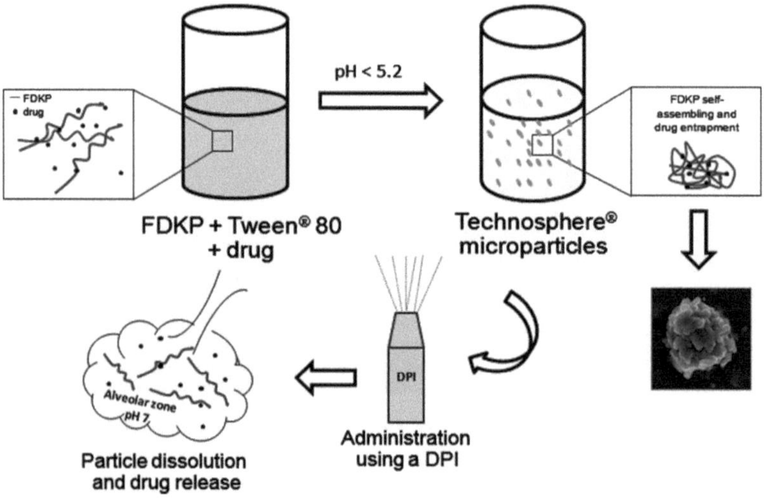

Fig. 22.3 Technosphere® drug carrier: mechanism of particle formation and behavior upon administration. (Microparticle image gently provided by MannKind Corporation)

been described to meet this challenge in a very positive manner. In fact, FDKP is highly soluble at pH values above 6 [7, 16]. Therefore, as the pH of alveolar lung lining fluid is approximately 7 [22–23], Technosphere® particles dissolve rapidly upon reaching the alveolar zone, releasing the drug in a complete manner [17]. A complete scheme on Technosphere® drug carrier is available in Fig. 22.3.

The dissolution profile of inhaled Technosphere® drugs is reported to be very fast, so that their pharmacokinetic profile closely resembles that of intravenous (IV) injection [16, 24], depending on the proper capacity of the drug to dissolve in the lung environment. Absorption begins almost immediately after inhalation and circulating drug concentrations peak within minutes of administration [7]. In vitro studies performed in Calu-3 cells (bronchial cell line) [19] and in vivo in rats [6] revealed that FDKP does not act as permeation enhancer, the rapid drug absorption being attributed to both FDKP and drug dissolution profile.

This characteristic of FDKP also contributes for a rapid elimination. FDKP has a plasmatic half-life of 190 min in diabetic patients with normal renal function [6], being cleared from the lung lining fluid with a half-life of approximately 1 h [7]. Similar results were reported in other studies, with FDKP t_{max} determined 10–15 min after dosing [18, 25], as depicted in Fig. 22.4. Altogether, these results describe a very important feature of FDKP, largely contributing for its safety.

Although developing inhalation dry powders is known as a challenging task, MannKind appears to have addressed most of the key considerations to be successful with the use of FDKP to produce Technosphere® microparticles. In summary, Technosphere® is a drug carrier technology with simple assembly, suitable aerodynamic properties for deep lung inhalation, permitting a rapid release and, thus, absorption of the carried drugs.

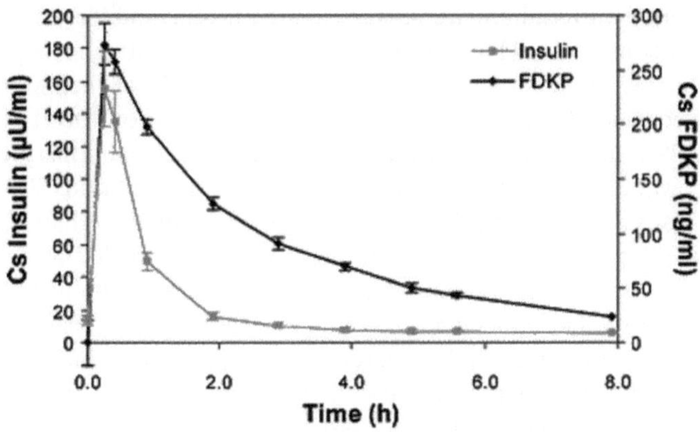

Fig. 22.4 Serum insulin and fumaryl diketopiperazine concentrations vs. time (mean ± SE; $n = 12$). (Reprinted from [7] with kind permission from Springer Science+Business Media)

22.3 Different Drugs for the Same Carrier

Technosphere® is a versatile technology, permitting the association of molecules with distinct properties, namely concerning their molecular size. Small molecules such as felbamate (238 Da) [26] or biopharmaceutical drugs, such as insulin [7], glucagon-like peptide 1 [18], or parathyroid hormone [27] (molecular weight 4–6 kDa), were successfully associated to the microparticles. Felbamate-loaded Technosphere® microparticles [26] were only reported for IV administration and, therefore, will not be further analyzed.

Insulin was the first biopharmaceutical drug to be approached in the context of inhalation and is the most studied. Technosphere® insulin is currently in Phase III clinical trials and Sect. 4 provides a complete overview of the formulation. Technosphere® glucagon-like peptide 1 (GLP-1) is also in Phase I clinical trials for the treatment of diabetes [6], with several reports providing the related data. Parathyroid hormone, oxyntomodulin, peptide YY, atropine, monoclonal antibodies, and bacterial antigens are other molecules that have been formulated in Technosphere® microparticles. However, while data on the former can be found in the literature [27], the others have been mentioned to be tested nonclinically [6] or are only mentioned in a patent [28], no data being available.

Parathyroid hormone (PTH) is a regulator of calcium homeostasis, with a function in maintaining serum calcium levels [29]. When calcium concentration decreases, PTH is secreted and calcium is mobilized from skeletal stores [30]. This makes PTH an unlikely agent for treating osteoporosis, but its intermittent administration has demonstrated to induce bone formation more than bone reabsorption, at least in the first 12 months of therapy [31]. PTH-mediated bone rebuilding in osteoporosis needs rapid absorption and elimination, as prolonged exposure can induce bone loss. Therefore, the pulmonary administration of PTH using Technosphere® is an appealing alternative to the current subcutaneous (SC) injection, as it theoretically meets that important requirement. A study on ten healthy subjects demonstrated a faster and higher increase in PTH (1–34) concentrations resulting from pulmonary as compared with SC administration. The inhalation of 1,600 IU of Technosphere®/PTH (1–34) elicited much lower t_{max} compared to SC injection of 400 IU PTH (1–34) (10 min vs. 29 min), with 48 % relative bioavailability of pulmonary PTH (1–34) for 6 h. C_{max} was also much higher upon pulmonary administration (309 pM vs. 102 pM) [27]. This is the only published study on the formulation, perhaps the peak-like pharmacokinetic profile of Technosphere®/PTH (1–34) revealed to have different effects on bone metabolism as compared with the parenteral administration.

GLP-1 is the second molecule integrating a Technosphere®-based formulation undergoing Phase I clinical trials. Secreted at the gastrointestinal tract, it stimulates insulin secretion by pancreatic β-cells after meals and is further involved in the regulation of glucagon release and gastrointestinal motility [32–33]. Very importantly, studies performed in rodents evidenced the ability of GLP-1 to stimulate β-cell growth and replication [34]. Altogether, these effects suggest a broad role of GLP-1 as mediator of postprandial glucose homeostasis and as a potential enhancer of β-cell functioning, with important therapeutic potential in type 2 diabetes. Its utility is somewhat hindered by both a very short circulating half-life ($\approx$ 2 min) and the need for injection [35]. The adequate GLP-1 therapy has been described as one in which the drug is administered at mealtime with exposure limited to the postprandial period [18], thus mimicking its physiological pattern, reaching peak levels soon after a meal and rapidly rising insulin concentration [36]. Inhaled Technosphere® GLP-1 might comply with this requirement. A preliminary study in rats, administered a single inhaled dose, revealed a 5–10 min t_{max}, 10–15 min $t_{1/2}$ and reduced food intake [18]. A single Phase I clinical trial is available, enrolling 26 healthy normal subjects and 20 type 2 diabetic subjects. Inhaled GLP-1 produced peak plasmatic concentrations within 5 min in both groups, leading to insulin peak at 10–15 min. GLP-1 returned to baseline within 30 min. Interestingly, subjects in the fasting state, whether healthy or diabetic, registered a decrease of insulin levels to baseline at 30–40 min, whereas diabetic subjects who had eaten a meal showed meal-stimulated insulin levels for several hours. Low bioavailability was estimated (0.5–1.6 %) due to both a low fraction of drug entering the lungs, as the inhaler was not specifically adapted to the formulation, and to the activity of dipeptidyl peptidase-4 present in the lungs and lung vasculature [36]. Generally, it was concluded that the inhalation of GLP-1 Technosphere® in type 2 diabetes patients produced a rapid and transient secretion of insulin that mimics the first-phase response of healthy subjects, restraining typical early glucose postprandial excursions.

As previously mentioned, GLP-1 also has a function on the regulation of gastrointestinal motility. A synthetic GLP-1 analog (ROSE-010) modified to protect against dipeptidyl peptidase-4 proteolytic action [37], evidenced the ability to relax intestinal smooth muscle and relieve pain associated with irritable bowel syndrome in preclinical and clinical assays [38]. A study of ROSE-010 Technosphere® inhaled by a rat model of the disease showed comparable reduction of intestinal motility as IV/SC injections. The effect was found similar to that of normal endogenous GLP-1 [37], evidencing the potential of ROSE-010 Technosphere® to be used as an alternative treatment of irritable bowel syndrome patients.

22.4 Technosphere® Insulin: An Effective Case-Study

Insulin therapy is a broad commercial hit as type 1 diabetes mellitus patients are totally dependent on the protein and, in parallel, the increasing prevalence of type 2 diabetes mellitus and the fact that people are being diagnosed at an early age, indicates that many of these patients will also develop severe insulin deficiency owing to pancreatic β-cell loss over time. Technosphere® insulin is, therefore, the most advanced of the Technosphere® drugs, not only because of the relevant market, but also because it is expected to attain good patient acceptance, eliminating/decreasing the need for regular uncomfortable injections.

Technosphere® insulin is a dry powder formulation of recombinant human insulin, predominantly composed of insulin and FDKP (1:9, w/w) under the form of crystalline microparticles [7], which is proposed for type 1 and type 2 diabetes treatment under the commercial name of Afrezza®. Available studies were mostly performed in type 2 diabetes patients, as the first Phase III clinical trial on type 1 diabetes patients has finished very recently (May 2013) [39]. The main characteristic of the formulation is perhaps to provide short insulin t_{max} (10–15 min, Fig. 22.4) [21, 24–25, 40–41], closely resembling the normal physiological profile of prandial insulin [42]. Moreover, it provides about 60 % of its glucose lowering effect within 3 h post administration [25]. This pharmacokinetic profile, which is also exhibited in patients with chronic obstructive pulmonary disease [43], is consistent with 2-compartment disposition, as reported for IV/SC administration [44], suggested an application as ultra-rapid-acting prandial insulin, addressing the synchronization between the post-prandial action of prandial insulin and the postprandial glucose, thus reducing the incidence of hyperglycemia and hypoglycemia [21]. This allows patients to inhale a dose 5–10 min before a meal, instead of injecting insulin 20–30 min ahead. In comparison, Technosphere® insulin has better pharmacokinetic profile than other inhaled insulin formulations, which provide 50–65 min t_{max} [42, 45–49], while SC insulin has worst performance ($t_{max} \approx 120$–140 min) [25, 40, 42], leading to less than 30 % of the total hypoglycemic effect occurring in the first 3 h [25]. This endows the SC formulation with several drawbacks. Particularly, as peak levels usually appear after the meal is already digested, there is a high risk of hypoglycemia, frequently demanding the intake of a preventive snack, apart from very strict treatment regimens [25, 50]. Nevertheless, inhaled insulin is probably less efficient than SC insulin, involving insulin loss within the inhaler and mouth during inhalation [42]. In this context, several studies have reported the relative bioavailability of inhaled insulin to be around 20–30 % [24, 41, 51]. Furthermore, it could also be assumed that pulmonary absorption is more regular than SC, as many parameters affecting the latter are not to be considered in the lung, such as the variable amount of fat, the injection site containing or not connective tissue, and the depth. However, comparing the administration of Technosphere® insulin and SC, regular human insulin in 13 type 2 diabetes patients, revealed a lower (but not statistically significant) intra-subject variability in Technosphere® insulin pharmacokinetic parameters during the first 3 h and significantly higher between subject variability. No significant differences were observed at pharmacodynamic parameters [25].

The first Phase I clinical study with Technosphere® insulin involved 5 nonsmoking healthy volunteers and compared the administration of 100 IU inhaled regular human insulin with that of 5 IU given by IV route and 10 IU by SC route, using euglycemic glucose clamp procedure. It was demonstrated for the first time that this inhaled insulin formulation elicited a very rapid rise of insulin levels (t_{max} at 13 min as compared with 120 min for SC), similar to that observed upon IV administration. The corresponding maximal metabolic effect on glucose occurred more than 2 h earlier with inhaled insulin, with a return to baseline within 3 h. Importantly, the insulin area under-the-curve (AUC) for this period was more than twice as high as those for IV and SC injection [24]. This first study used a commercially available inhaler, but it was soon identified that an improved performance would require a specifically designed device. MannKind then developed a specific inhaler for the Technosphere® technology (MedTone®), which provides a mean emitted dose of 65 % [7], being used in subsequent studies. The ability to improve glycemic control at mealtime was demonstrated in a meal-challenge study involving 16 nonsmoking type 2 diabetes subjects (administered 48 IU Technosphere® insulin or 14 IU SC regular human insulin), where blood glucose AUC_{0-240} of inhaled insulin was $\approx$52 % that of SC insulin. Total serum insulin exposure was almost identical in both treatments [40]. Importantly, it was verified that, when selecting an optimized dose of inhaled insulin, patients could ingest meals with variable carbohydrate content or skip meals without severe hypoglycemia. This is the main result of a study evaluating eight type 2 diabetes patients taking Technosphere® insulin (dose of 15 UI or 30 UI, optimized for each subject) with a meal of different carbohydrate content (0, 50, 200 %). Postprandial glucose excursions were determined to be minimal for 0 and 50 % carbohydrate content meals. The meal adjusted to 200 % registered moderate increase of glucose, in any case below American Diabetes Association targets. Additionally, a general decrease of HbA1c around 1.6 % was observed [52].

A first study opposing Technosphere® insulin to placebo Technosphere® demonstrated that the active formulation strongly decreased glycated hemoglobin (HbA1c), reduced postprandial glucose excursions by 56 % as compared with baseline and maximal postprandial glucose levels by 43 % as compared with placebo. This study enrolled 126 insulin-naïve type 2 diabetes patients suboptimally controlled with oral agents for a total of 12 weeks [53]. It was further demonstrated that the metabolic activity induced by Technosphere® insulin is dose-dependent. Using the euglycemic glucose clamp technique, 11 healthy subjects received three different doses (25, 50, and 100 IU) on three different days. The doses resulted in insulin peaks at 42, 50, and 58 min, respectively, which occurred 2 h earlier than upon SC injection (10 IU). Over the first 3 h, the relative bioavailability was 36, 35, and 32 %, respectively [41]. Comparable results were reported in a similar study [54]. A more prolonged study (11 weeks) tested the dose-response of four Technosphere® insulin doses (14, 28, 42, and 56 IU) on 227 type 2 diabetes patients with inadequate glycemic control. Inhaled insulin was administered daily before each of three meals, in combination with insulin glargine. Technosphere® insulin demonstrated to induce statistically significant dose-dependent reduction of HbA1c, whether versus baseline or placebo (Technosphere® powder alone). Inhaled insulin generally decreased the postprandial maximum glucose concentration (except for the lowest dose) and reduced significantly the postprandial glucose AUC for the two higher doses [55].

Considering the fact that most patients will require the administration of Technosphere® insulin along with other insulin formulations, testing combined effects is important. A one-year study involving 485 type 2 diabetes patients established a group administering prandial inhaled insulin plus SC insulin glargine at bedtime, and another group consisting twice daily SC premixed biaspart insulin. Changes in HbA1c were similar in both groups, but patients from the group of Technosphere® insulin registered a significantly lower weight gain and fewer mild-to-moderate and severe hypoglycemic events [51].

Although the pharmacological efficacy of Technosphere® insulin has been demonstrated, many scientists and physicians are still skeptic about its use. This is mainly due to the issue of insulin accumulation in lung tissue, and the continuous inhalation of a powder, which might compromise common lung functions. Several clinical studies addressed these concerns, evaluating the clearance of the formulation and testing distinct parameters of pulmonary function. The administration of ^{99m} Tc-radiolabelled particles to five healthy volunteers revealed that 60 % of the emitted dose reaches the lung (remainder is swallowed), distributing homogeneously to both lung sides. A bronchoalveolar lavage (BAL) study demonstrated that inhaled insulin is rapidly cleared from the lung, with undetectable concentrations after 12 h. Clearance half-life was determined to be ≈ 1 h [7].

All studies on pulmonary function reported acceptable results, with only minimal changes observed upon treatment with inhaled insulin, which were not statistically different from those observed in other test groups. The most usual adverse effect was cough, in which intensity decreases with treatment continuation [7, 24, 25]. One particular study tested the effects on specific parameters of pulmonary function (forced expiratory volume for 1 s, forced vital capacity, total lung capacity and lung diffusion capacity for carbon monoxide) over 2 years. Results from 910 subjects indicated that the evolution of pulmonary function was similar for patients (either with type 1 or type 2 diabetes) using prandial inhaled insulin or usual care (oral antidiabetics or SC insulin). The lung function actually declined in both groups through the study period, showing a tendency for lung function decline associated with diabetes, in which underlying mechanism remains unclear [56].

One important aspect of using inhaler devices is being trained on their use and having the necessary physiological ability to use with maximum benefit. As referred above, after using a commercially available inhaler in the first tests, MannKind developed its proper inhaler MedTone® to be used with the Technosphere® technology. It was demonstrated to work properly in a study with 56 type 1 and type 2 diabetes subjects who have shown to provide the necessary inspiratory effort for Technosphere® insulin inhalation [57]. However, MannKind has developed a second-generation inhaler for Affreza®, called DreamBoat® (Fig. 22.5), which is smaller and uses a lower dose compared with MedTone® [58].

In summary, it was evidenced by the described assays that Technosphere® insulin generally provides a better glycemic control than other formulations, resulting in lower weight gain and less hypoglycemic events. Therefore, it is taking the forefront for commercialization.

Fig. 22.5 Dreamboat® dry powder inhaler. (Images gently provided by MannKind Corporation)

22.5 Technosphere® Insulin from the Patient Perspective

Improving quality of life is the ultimate goal of any therapeutic regimen. Therefore, perceiving patient reactions to medication is essential, as the patient must be truly committed to enable therapeutic success. After the failure of Exubera® it became very important to address these concerns cautiously, as several reports indicated the inability of the patients to use the inhaler or their discomfort in handling a cumbersome device [59–60]. Some works evaluate type 2 diabetes patients' perceptions of inhaled insulin therapy effect on health-related quality of life and treatment satisfaction, using a measure of health-related quality of life (SF-36) and the Inhaled Insulin Treatment Questionnaire [61], which were filled before and after treatment. The latter measure assesses diabetes worries, perceptions of insulin therapy, and inhaler performance. One study compared insulin-naïve subjects (119) receiving inhaled Technosphere® insulin or placebo Technosphere® formulation, revealing that perceptions of insulin therapy were significantly more positive after using Technosphere® insulin. Participants on this arm reported no negative impacts on quality of life or worries about diabetes, attitude improvement toward insulin therapy, and satisfaction associated with the inhaler device. The only negative aspect regarding the inhaler was related with the difficulty in evaluating cartridge emptiness [62]. Another study compared patients (618) taking Technosphere® insulin in combination with basal insulin glargine with those taking premixed biphasic rapid-acting insulin 70/30. The decrease in HbA1c was similar in both groups but, with similar overall glucose levels among groups, patients in the Technosphere®/glargine arm had significantly lower weight gain. The same was observed for the incidence of hypoglycemia. Diabetes worries declined significantly in the group taking Technosphere® insulin and glargine, whose participants also reported no negative impacts on quality of life, while perceptions of insulin therapy, treatment satisfaction, and treatment preference improved in both arms of the study [63]. The same authors also addressed the perceptions of both diabetic patients and physicians regarding the use of inhaled insulin. An internet survey, in which 1,094 American patients participated, suggested they evaluate diabetes medications primarily in terms of the ability to control postprandial hyperglycemia and reduce discomfort and inconvenience [64]. The same methodology was used to register the opinion of 602 physicians who treat diabetic adults. Physicians who self-identified as medical innovators or who reported high levels of involvement with patients, tended to rate the inhaled insulin higher. In contrast, those who self-identified as diabetes experts or who avoided using insulin, tended to attribute a lower rate. Interestingly, family physicians were most likely and endocrinologists least likely to say they would recommend inhaled insulin for a variety of patient profiles [60]. The conditions for a patient appealing therapeutic are apparently guaranteed.

22.6 Safety Concerns of Technosphere®

The withdrawal of Exubera® (Pfizer) and the possibility that it may have been associated with safety reasons, namely lung cancer, has hampered the development and licensing of protein-containing formulations intended for inhalation. Exubera® was only available for a short period (August 2006 to October 2007) and was withdrawn because of cost and bulky device [58]. Nevertheless, this formulation was reported to cause cough, dyspnea, increased sputum, and epistaxis [65]. Although patient satisfaction was higher with inhaled insulin compared with SC injections, regular lung function tests are needed because long-term safety has not been established [66]. The safety reasons that may have been involved pertain mainly to the chronic exposure of the alveolar epithelium and underlying connective tissue to biologically active insulin [67]. Insulin acts as a weak growth factor by binding to the IGF-1 receptor, which raises serious safety concerns [1]. The association of inhaled insulin with small, consistent reductions in lung function has been shown [68], while it was also reported that inhaled insulin rapidly aggregates at the lung air-tissue interface, forming amyloid structures causing a significant reduction in pulmonary air flow [69]. In addition, the role of the lung as immunological organ and the consequences of a chronic deposition of inhaled proteins in the lung epithelium are not fully understood. Therefore, we tend to agree when Bailey and Barnett [67] state "proceed with caution".

The Technosphere® system has been proposed for inhaled insulin and was submitted to the FDA in 2009 as a New Drug Application, under the trade name Afrezza®. It has shown an acceptable safety profile in the clinical development, overcoming some drawbacks that contributed to the demise of Exubera® [51, 53, 56]. From a technological perspective, the FDKP self-assembly process for microsphere assembly has the advantage of avoiding the use of organic solvents while using mild formulation conditions compatible with protein stability. As aforementioned, after administration the particles dissolve rapidly in the alveolar pH-neutral environment and readily liberate insulin for systemic absorption [70–72]. FDKP is not metabolized, being excreted in urine as ammonium salts within hours of administration [6–7].

The in vivo fate of pulmonary delivered particulate systems depends on their composition and physicochemical characteristics. The uptake by lung macrophages and translocation across respiratory epithelia, either to the systemic circulation or lymphatic circulation, depend on particle size, charge, and hydrophobicity [73–74].

Moreover, inhaled micro- and nanoparticles may influence drug absorption by controlling the release and retaining the drug within the lungs [3, 74–75]. The rapid dissolution of FDKP microspheres is most probably the key for the success of Technosphere®, including its safety. After particle dissolution and protein release, the in vivo fate of the latter depend on its physicochemical properties rather than formulation parameters. The same applies to protein toxicity within the lung compartment. However, as a novel excipient, FDKP microspheres will have to comply with harsh regulatory demands, including full details of manufacture, characterization and controls with cross references to supporting safety, including data concerning toxicology according to the dosage form and the route of administration of the medicinal product. The information should provide the same level of detail as that provided for a drug substance [76–78]. The literature concerning the efficacy and safety assessment of FDKP microspheres is scarce and mostly based on studies performed by MannKind, some of which were already analyzed in this chapter [6–7, 19, 70]. In vitro studies carried out using the Calu-3 human cell line demonstrate it is not cytotoxic [19] and the absence of a pharmacological effect of FDKP was further demonstrated both in vitro and in vivo [6]. A pharmacokinetic study performed on healthy nonsmoking subjects receiving IV FDKP revealed the elimination of more than 90 % of the excipient in 8 h. It was predominantly cleared unchanged by the kidney with negligible oral bioavailability. In another study, diabetic subjects with mild-to-moderate nephropathy were exposed to FDKP only 18–25 % of the time determined for diabetic subjects with normal renal function, which is consistent with the predominant renal clearance of FDKP. As insulin and FDKP cleared from the lungs with a half-life of $\approx$ 1 h by systemic absorption, authors concluded that the potential for accumulation on chronic administration is minimal because there is an overnight washout period [6]. Nonetheless, only limited data are available on long-term effects of inhaled proteins/excipients of Technosphere® system. This is particularly relevant for chronic diseases and treatments such as diabetes, where any inhalation delivery system must demonstrate long-term pulmonary and systemic safety before it can be approved.

22.7 Conclusions and Future Perspectives

Counting on many years of testing, Technosphere® appears as one of the most promising formulations designed for the pulmonary delivery of biopharmaceuticals. Technosphere® insulin is taking the forefront of the approach and will soon receive a final decision from FDA. If positive, it might initiate a new meaning for lung drug delivery. Being mostly composed of an excipient highly soluble in the lung fluid, FDKP, Technosphere® usually permits a strong metabolic effect shortly after administration. The rapid lung clearance of FDKP and its renal elimination in nonmetabolized form strongly contribute for its safety profile. The question remaining to be answered is clearly related with the long-term safety of the formulations, which can only be clarified upon the performance of long-term assays, preferably before marketing authorization.

Declaration of Interest This chapter was written by the authors with no involvement of MannKind Corporation. The authors report no financial disclosures as related to products discussed in this article.

Acknowledgments The authors acknowledge the Portuguese government (Fundação para a Ciência e Tecnologia) and FEDER: research project PTDC/SAU-FCF/100291/2008, PTDC/DTP-FTO/0094/2012 and strategic projects PEst-OE/SAU/UI4013/2011 and PEst-OE/EQB/LA0023/2013.

References

1. Siekmeier R, Scheuch G. Inhaled insulin—does it become reality? J Physiol Pharmacol. 2008;59(Supp 6):81–113.
2. Pilcer G, Amighi K. Formulation strategy and use of excipients in pulmonary drug delivery. Int J Pharm. 2010;392(1–2):1–19.
3. Grenha A, Al-Qadi S, Seijo B, Remuán-López C. The C potential of chitosan for pulmonary drug delivery. J Drug Deliv Sci Technol. 2010;20(1):33–43.
4. Almeida AJ, Souto E. Solid lipid nanoparticles as a drug delivery system for peptides and proteins. Adv Drug Deliv Rev. 2007;59(6):478–90.
5. Ungaro F, d'Angelo I, Miro A, La Rotonda MI, Quaglia F. Engineered PLGA nano- and micro-carriers for pulmonary delivery: challenges and promises. J Pharm Pharmacol. 2012;64(9):1217–35.
6. Potocka E, Cassidy J, Haworth P, Heuman D, van Marle S, Baughman JR. Pharmacokinetic characterization of the novel pulmonary delivery excipient fumaryl diketopiperazine. J Diabetes Sci Technol. 2010;4(5):1164–73.
7. Cassidy J, Amin N, Marino M, Gotfried M, Meyer T, Sommerer K, Baughman R. Insulin lung deposition and clearance following Technosphere® insulin inhalation powder administration. Pharm Res. 2011;28(9):2157–64.
8. Feldstein R, Glass J, Steiner S, Self assembling diketopiperazine drug delivery system. United States Patent 5352461 A, 1994.
9. Steiner S, Feldstein R, Lian H, Rhodes C, Shen G, Microparticles for lung delivery comprising diketopiperazine. United States Patent 6071497 A, 2000.
10. Cornacchia C, Cacciatore I, Baldassarre L, Mollica A, Feliciani F, Pinnen F. 2,5-Diketopiperazines as neuroprotective agents. Mini Rev Med Chem. 2012;12:2–12.
11. Martins MB, Carvalho I. Diketopiperazines: biological activity and synthesis. Tetrahedron. 2007;63(40):9923–32.
12. Bergeron RJ, Phanstiel O, Yao GW, Milstein S, Weimar WR. Macromolecular self-assembly of diketopiperazine tetrapeptides. J Am Chem Soc. 1994;116(19):8479–84.
13. Luo T-JM, Palmore GTR. Influence of structure on the kinetics of assembly of cyclic dipeptides into supramolecular tapes. J Phys Org Chem. 2000;13(12):870–9.
14. Kaur N, Zhou B, Breitbeil F, Hardy K, Kraft KS, Trantcheva I, Phanstiel O IV. A delineation of diketopiperazine self-assembly processes: Understanding the molecular events involved in N$\in$-(fumaroyl)diketopiperazine of l-Lys (FDKP) interactions. Mol Pharm. 2008;5(2):294–315.
15. Neumiller J, Campbell R. Technosphere insulin: an inhaled prandial insulin product. Biodrugs. 2010;24(3);165–72.
16. Leone-Bay A, Baughman R, Smutney C, Kocinsky J. Innovation in drug delivery by inhalation. OnDrugDelivery Magazine. 2010;4–8. Available at www.ondrugdelivery.com.
17. Neumiller J, Campbell R, Wood L. A review of inhaled Technosphere insulin. Ann Pharmacother. 2010;44:1231–9.

18. Leone-Bay A, Grant M, Greene S, Stowell G, Daniels S, Smithson A, Villanueva S, Cope S, Carrera K, Reyes S, Richardson P. Evaluation of novel particles as an inhalation system for GLP-1. Diabetes Obes Metab. 2009;11(11):1050–9.
19. Angelo R, Rousseau K, Grant M, Leone-Bay A, Richardson P. Technosphere® insulin: defining the role of technosphere particles at the cellular level. J Diabetes Sci Technol. 2009;3(3): 545–54.
20. Leone-Bay A, Smutney C, Kocinsky J. Pulmonary drug delivery—simplified. OnDrugDelivery Magazine. 2011;18–21. Available at www.ondrugdelivery.com.
21. Boss AH, Petrucci R, Lorber D. Coverage of prandial insulin requirements by means of an ultra-rapid acting inhaled insulin. J Diabetes Sci Technol. 2012;6(4):773–9.
22. Walters DV. Lung lining liquid—the hidden depths. Neonatology. 2002;81(Suppl. 1):2–5.
23. Kyle H, Ward J, Widdicombe J. Control of pH of airway surface liquid of the ferret trachea in vitro. J Appl Physiol. 1990;68:135–40.
24. Steiner S, Pfützner A, Wilson B, Harzer O, Heinemann L, Rave K. Technosphere™ /insulin—proof of concept study with a new insulin formulation for pulmonary delivery. Exp Clin Endocrinol Diabetes. 2002;110(1):17–21.
25. Rave K, Heise T, Heinemann L, Boss A. Inhaled Technosphere® insulin in comparison to subcutaneous regular human insulin: time action profile and variability in subjects with type 2 diabetes. J Diabetes Sci Technol. 2008;2(2):205–12.
26. Lian H, Steiner SS, Sofia RD, Woodhead JH, Wolf HH, White HS, Shen GS, Rhodes CA, McCabe RT. A self-complementary, self-assembling microsphere system: application for intravenous delivery of the antiepileptic and neuroprotectant compound felbamate. J Pharm Sci. 2000;89(7):867–75.
27. Pfützner A, Flacke F, Pohl R, Linkie D, Engelbach M, Woods R, Forst T, Beyer J, Steiner S. Pilot study with Technosphere/PTH(1–34)—a new approach for effective pulmonary delivery of parathyroid hormone (1–34). Horm Metab Res. 2003;35:319–23
28. Steiner S, Gelber C, Feldstein R, Pohl R, Compositions for treatment or prevention of bioterrorism, United States Patent 20040018152 A1, 2006.
29. Hoyer H, Perera G, Bernkop-Schnürch A. Noninvasive delivery systems for peptides and proteins in osteoporosis therapy: a retroperspective. Drug Dev Ind Pharm. 2010;36(1):31–44.
30. Toulis K, Anastasilakis A, Polyzos S, Makras P. Targeting the osteoblast: approved and experimental anabolic agents for the treatment of osteoporosis. Hormones. 2011;10(3):174–95.
31. Lim V, Clarke BL. New therapeutic targets for osteoporosis: beyond denosumab. Maturitas. 2012;73(3):269–72.
32. Ahrén B, Holst JJ, Mari A. Characterization of GLP-1 effects on β-cell function after meal ingestion in humans. Diabetes Care. 2003;26(10):2860–4.
33. Yusta B, Baggio LL, Estall JL, Koehler JA, Holland DP, Li H, Pipeleers D, Ling Z, Drucker DJ. GLP-1 receptor activation improves β cell function and survival following induction of endoplasmic reticulum stress. Cell Metab. 2006;4:391–406.
34. D'Alessio DA, Vahl TP. Glucagon-like peptide 1: evolution of an incretin into a treatment for diabetes. Am J Physiol Endocrinol Metab. 2004;286(6):E882–90.
35. Vilsbøll T, AgersøH, Krarup T, Holst JJ. Similar elimination rates of glucagon-like peptide-1 in obese type 2 diabetic patients and healthy subjects. J Clin Endocrinol Metab. 2003;88(1):220–4.
36. Marino M, Costello D, Baughman R, Boss A, Cassidy J, Damico C, van Marle S, van Vliet A, Richardson P. Pharmacokinetics and pharmacodynamics of inhaled GLP-1 (MKC253): proof-of-concept studies in healthy normal volunteers and in patients with type 2 diabetes. Clin Pharmacol Ther. 2010;88(2):243–50.
37. Hellström PM, Smithson A, Stowell G, Greene S, Kenny E, Damico C, Leone-Bay A, Baughman R, Grant M, Richardson P. Receptor-mediated inhibition of small bowel migrating complex by GLP-1 analog ROSE-010 delivered via pulmonary and systemic routes in the conscious rat. Regul Pept. 2012;179(1–3):71–6.
38. Hellström PM, Hein J, Bytzer P, Björnssön E, Kristensen J, Schambye H. Clinical trial: the glucagon-like peptide-1 analogue ROSE-010 for management of acute pain in patients with

irritable bowel syndrome: a randomized, placebo-controlled, double-blind study. Aliment Pharmacol Ther. 2009;29(2):198–206.

39. MannKind. 2013-06-20; http://www.mannkindcorp.com/.
40. Rave K, Heise T, Pfützner A, Boss AH. Coverage of postprandial blood glucose excursions with inhaled Technosphere insulin in comparison to subcutaneously injected regular human insulin in subjects with type 2 diabetes. Diabetes Care. 2007;30(9):2307–8.
41. Rave K, Potocka E, Boss AH, Marino M, Costello D, Chen R. Pharmacokinetics and linear exposure of AFRESA™ compared with the subcutaneous injection of regular human insulin. Diabetes Obes Metab. 2009;11(7):715–20.
42. Mandal T. Inhaled insulin for diabetes mellitus. Am J Health Syst Pharm. 2005;62:1359–64.
43. Potocka E, Amin N, Cassidy J, Schwartz S, Gray M, Richardson P, Baughman R. Insulin pharmacokinetics following dosing with Technosphere insulin in subjects with chronic obstructive pulmonary disease. Curr Med Res Opin. 2010;26(10):2347–53.
44. Potocka E, Baughman RA, Derendorf H. Population pharmacokinetic model of human insulin following different routes of administration. J Clin Pharmacol. 2011;51(7):1015–24.
45. Heise T, Brugger A, Cook C, Eckers U, Hutchcraft A, Nosek L, Rave K, Troeger J, Valaitis P, White S, Heinemann L. PROMAXX® inhaled insulin: safe and efficacious administration with a commercially available dry powder inhaler. Diabetes Obes Metab. 2009;11(5):455–9.
46. Perera AD, Kapitza C, Nosek L, Fishman RS, Shapiro DA, Heise T, Heinemann L. Absorption and metabolic effect of inhaled insulin: intrapatient variability after inhalation via the Aerodose insulin inhaler in patients with type 2 diabetes. Diabetes Care. 2002;25(12):2276–81.
47. Kim D, Mudaliar S, Chinnapongse S, Chu N, Boies SM, Davis T, Perera AD, Fishman RS, Shapiro DA, Henry R. Dose-response relationships of inhaled insulin delivered via the Aerodose insulin inhaler and subcutaneously injected insulin in patients with type 2 diabetes. Diabetes Care. 2003;26(10):2842–7.
48. Rave K, Bott S, Heinemann L, Sha S, Becker RHA, Willavize SA, Heise T. Time-action profile of inhaled insulin in comparison with subcutaneously injected insulin lispro and regular human insulin. Diabetes Care. 2005;28(5):1077–82.
49. Heinemann L. New ways of insulin delivery. Int J Clin Pract. 2010;64(Suppl 166):29–40.
50. Pfützner A, Mann A, Steiner SS. Technosphere™/insulin—a new approach for effective delivery of human insulin via the pulmonary route. Diabetes Technol Ther. 2002;4(5):589–94.
51. Rosenstock J, Lorber DL, Gnudi L, Howard CP, Bilheimer DW, Chang PC, Petrucci RE, Boss AH, Richardson PC. Prandial inhaled insulin plus basal insulin glargine versus twice daily biaspart insulin for type 2 diabetes: a multicentre randomised trial. Lancet. 2010;375(9733):2244–53.
52. Zisser H, Jovanovic L, Markova K, Petrucci R, Boss A, Richardson P, Mann A. Technosphere insulin effectively controls postprandial glycemia in patients with type 2 diabetes mellitus. Diabetes Technol Ther. 2012;14(11):997–1001.
53. Rosenstock J, Bergenstal R, DeFronzo RA, Hirsch IB, Klonoff D, Boss AH, Kramer D, Petrucci R, Yu W, Levy B. Efficacy and safety of Technosphere inhaled insulin compared with Technosphere powder placebo in insulin-naive type 2 diabetes suboptimally controlled with oral agents. Diabetes Care. 2008;31(11):2177–82.
54. Pfützner A, Forst T. Pulmonary insulin delivery by means of the Technosphere™ drug carrier mechanism. Expert Opin Drug Deliv. 2005;2(6):1097–106.
55. Tack C, Christov V, de Galan B, Derwahl K-M, Klausmann G, Pelikánová T, Perušičová J, Boss A, Amin N, Kramer D, Petrucci R, Yu W. Randomized forced titration to different doses of Technosphere® insulin demonstrates reduction in postprandial glucose excursions and hemoglobin A1c in patients with type 2 diabetes. J Diabetes Sci Technol. 2008;2(1):47–57.
56. Raskin P, Heller S, Honka M, Chang PC, Boss AH, Richardson PC, Amin N. Pulmonary function over 2 years in diabetic patients treated with prandial inhaled Technosphere insulin or usual antidiabetes treatment: a randomized trial. Diabetes Obes Metab. 2012;4(2):163–73.
57. Smutney C, Friedman E, Polidoro J, Amin N. Inspiratory efforts achieved in use of the Technosphere® insulin inhalation system. J Diabetes Sci Technol. 2009;3(5):1175–82.

58. Kling J. Dreamboat sinks prospects for fast approval of inhaled insulin. Nat Biotech. 2011;29(3):175–6.
59. Heinemann L. The failure of Exubera: are we beating a dead horse? J Diabetes Sci Technol. 2008;2(3):518–29.
60. Rubin R, Peyrot M. Factors associated with physician perceptions of and willingness to recommend inhaled insulin. Curr Med Res Opin. 2011;27(2):285–94.
61. Rubin R, Peyrot M. Psychometric properties of an instrument for assessing the experience of patients treated with inhaled insulin: the Inhaled Insulin Treatment Questionnaire (IITQ). Health Qual Life Outcomes. 2010;8(1):32.
62. Peyrot M, Rubin R. Effect of Technosphere inhaled insulin on quality of life and treatment satisfaction. Diabetes Technol Ther. 2010;12(1):49–55.
63. Peyrot M, Rubin R. Patient-reported outcomes in adults with type 2 diabetes using mealtime inhaled Technosphere insulin and basal insulin versus premixed insulin. Diabetes Technol Ther. 2011;13(12):1201–6.
64. Peyrot M, Rubin R. Perceived medication benefits and their association with interest in using insulin type 2 diabetes: a model of patients' cognitive framework. Patient Prefer Adherence. 2011;5:255–65.
65. Setter SM, Levien TL, Iltz JL, Odegard PS, Neumiller JJ, Baker DE, Campbell RK. Inhaled dry powder insulin for the treatment of diabetes mellitus. Clin Ther. 2007;29(5):795–813.
66. Brunton S. Insulin delivery systems: reducing barriers to insulin therapy and advancing diabetes mellitus treatment. Am Med. 2008;121(Suppl 6):35–41.
67. Bailey CJ, Barnett AH. Inhaled insulin: new formulation, new trial. Lancet. 2010;375(9733):2199–201.
68. Hegewald M, Crapo R, Jensen R. Pulmonary function changes related to acute and chronic administration of inhaled insulin. Diabetes Technol Ther. 2007;9(Suppl 1):S. 93–101.
69. Lasagna-Reeves CA, Clos AL, Midoro-Hiriuti T, Goldblum RM, Jackson GR, Kayed R. Inhaled insulin forms toxic pulmonary amyloid aggregates. Endocrinology. 2010;151(10):4717–24.
70. Richardson P, Boss A. Technosphere insulin technology. Diabetes Technol Ther. 2007;9(Suppl 1):65–72.
71. Sarala N, Bengalorkas G, Bhuvana K. Technosphere: new drug delivery system for inhaled insulin. Pract Diabetes. 2012;29:23–4.
72. Depreter F, Pilcer G, Amighi K. Inhaled proteins: challenges and perspectives. Int J Pharm. 2013;447(1–2):251–80.
73. Videira M, Botelho MF, Santos AC, Gouveia LF, Pedroso de Lima JJ, Almeida AJ. Lymphatic uptake of pulmonary delivered radiolabelled solid lipid nanoparticles. J Drug Target. 2002;10:607–13.
74. Al-Qadi S, Grenha A, Carrión-Recio D, Seijo B, Remuán-López C. Microencapsulated chitosan nanoparticles for pulmonary protein delivery: in vivo evaluation of insulin-loaded formulations. J Control Release. 2012;157:383–90.
75. Poyner EA, Alpar HO, Almeida AJ, Gamble MD, Brown MRW. A comparative study on the pulmonary delivery of tobramycin encapsulated into liposomes and PLA microspheres following intravenous and endotracheal delivery. J Control Release. 1995;(31):41–8.
76. International Conference on Harmonisation, Guideline ICH S7A, Safety pharmacology studies for human pharmaceuticals, 2000.
77. European Medicines Agency, EMEA/CHMP/QWP/396951/2006, Guideline on excipients in the dossier for application for marketing authorisation of a medicinal product; 2006.
78. Food and Drug Administration, Guidance for industry: nonclinical studies for the safety evaluation of pharmaceutical excipients; 2005.

Chapter 23
ChiSys® as a Chitosan-Based Delivery Platform for Nasal Vaccination

Peter Watts, Alan Smith and Michael Hinchcliffe

23.1 Introduction

Many pathogenic organisms infect the body via the mucosal membranes, especially those lining the respiratory and gastrointestinal tracts. The nasal mucosa provides an important immunological defensive barrier to inhaled antigens and for this reason makes an excellent target site for delivering vaccines. Intranasal vaccination offers many attractions, including the ability to induce both local and systemic immune responses, the potential for effecting immunity at mucosal sites distant to the site of vaccination, convenience and ease of administration, the avoidance of syringes and needles for dosing, and the presence of a relatively benign environment to maintain antigen integrity [1–3].

A number of authors discuss the complex process by which a mucosally administered antigen generates an antibody response [1–7]. The primary site for the activation of immune responses is the lymphoid tissues. The nasal-associated lymphoid tissue (NALT) is principally found in the nasopharyngeal tonsil (adenoids) and palatine and lingual tonsils. M cells are found on the surface of these tissues which are able to bind and transport particulate antigens for processing in the lymph nodes, where antibodies are generated. Small soluble antigens may be able to penetrate the nasal epithelium, assisted by appropriate formulation additives, where they can be captured by dendritic cells and macrophages, and carried to the lymph nodes. The nasal epithelium also contains receptors (e.g., Toll-like receptors) which can identify foreign organisms and trigger an immune response; these receptors may form specific targets for vaccine adjuvants.

P. Watts (✉) · A. Smith
Archimedes Development Ltd, Albert Einstein Centre,
Nottingham Science Park, Nottingham NG7 2TN, UK
e-mail: peterwatts@archimedespharma.com

M. Hinchcliffe
Paracelsis Ltd, Nottingham, UK

J. das Neves, B. Sarmento (eds.), *Mucosal Delivery of Biopharmaceuticals,* DOI 10.1007/978-1-4614-9524-6_23,

Currently there are two approved intranasally administered human vaccines. FluMist® (AstraZeneca) is a seasonal influenza vaccine [8]. Nasovac® (Serum Institute) is licensed in India as a vaccine for H1N1 pandemic influenza ("swine flu") [9]. Both of these vaccines contain live-attenuated organisms and neither uses an adjuvant.

The use of intranasal vaccines is more widespread in companion animals and livestock, with licensed products available to immunize against a range of infections in a number of different species. As with the aforementioned human nasal influenza vaccines, intranasally administered animal vaccines primarily contain live-attenuated organisms [10].

Antigens which are live-attenuated organisms are typically able to induce adequate mucosal responses when administered alone (i.e., without adjuvant) in a relatively simple vehicle. However, antigens which are subunits or purified components of an organism will generally be unable by themselves to generate a sufficient immune response and require the coadministration of adjuvants. The ideal adjuvant needs to be safe and well tolerated, both locally and systemically, effective with a wide range of antigens, and available in a quality suitable for pharmaceutical use.

ChiSys® is the proprietary name given by Archimedes Development Limited to its suite of intellectual property relating to the mucosal delivery of therapeutic agents using chitosan and certain derivatives, and includes the use of chitosan as a nasally administered vaccine adjuvant.

Chitosan is a copolymer of glucosamine and *N*-acetyl glucosamine and has been widely investigated as an ingredient to facilitate the delivery of drugs and vaccines by the nasal route. It has been demonstrated to enhance the transmucosal uptake, and thus increase bioavailability, of a wide range of small molecule, peptide, and protein drug compounds [11, 12]. Chitosan also has a wide range of other potential biomedical applications, including wound healing [13], tissue engineering (scaffolds) [14, 15], and as a dietary supplement for weight loss [16].

This chapter focuses on the use of ChiSys® as a delivery platform for nasal vaccination; the data presented primarily relate to chitosan in glutamate salt form although the use of chitosan derivatives is also described.

23.2 Chemistry of Chitosan

23.2.1 Chemical Structure

Chitosan is a linear copolymer of β1-4 linked monomers of *D*-glucosamine and *N*-acetyl-*D*-glucosamine (see Fig. 23.1). Many properties of chitosan are defined by the number and distribution of acetyl groups. The degree of deacetylation (DD) is typically at least 40 % [17]. The molecular weight of chitosan will be dependent on the raw material source and the processes used for extraction and purification. A range of molecular weight grades is available commercially, typically in the range 30–600 kDa. Chitosan oligomers are also available for specialist applications (e.g., NOVAFECT O produced by NovaMatrix).

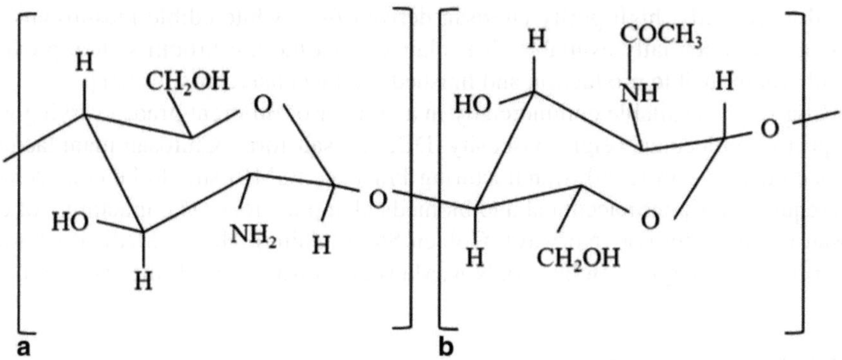

Fig. 23.1 Chemical structure of chitosan, illustrating the a *D*-glucosamine and b *N*-acetyl-*D*-glucosamine monomer groups

23.2.2 Physicochemical Properties

A defining feature of chitosan is its cationic nature, the positive charge being provided by the primary amine groups. Chitosan is insoluble in water, dissolves in most organic acids (e.g., acetic, glutamic, lactic, succinic) but, with the exception of hydrochloric acid, is generally less soluble in inorganic acids. It is essentially insoluble in organic solvents [18]. The aqueous solubility of chitosan depends on both the DD and the distribution of acetyl groups along the polymer chain. A DD of at least 50 % is reportedly required in order for chitosan to dissolve in acid [19].The pH of a 1 % aqueous chitosan solution will lie in the range 4–6 [18]. The pKa of the primary amine groups on chitosan is approximately 6.5 and the polymer rapidly becomes insoluble as the pH rises above 6. To maintain chitosan in solution long term, the pH should be maintained below 6. High concentrations of salts may suppress chitosan solubility [17, 20].

23.2.3 Sources and Production

Chitosan is derived from the partial deacetylation of chitin, a naturally occurring biopolymer comprising repeating groups of *N*-acetyl-*D*-glucosamine. The most common industrial sources of chitin are the exoskeletons of crustaceans, especially shrimps and fungi (e.g., mushrooms).

Extraction of chitin from crustacean shells involves the removal of proteins using alkali and the removal of minerals by treatment with acid. The extracted chitin is N-deacetylated to produce chitosan by treating with sodium hydroxide at elevated temperature. The resulting precipitate is washed with water, dissolved in acid, and reprecipitated by adding alkali. The final precipitate may be washed and dried to produce chitosan base or dissolved in acid and dried to form a water-soluble salt. Very high purity chitosan grades are available, which have undergone additional purification steps such as ultrafiltration [20].

More recently, high purity chitosan derived from white edible mushrooms has become commercially available. It is claimed that the mushroom source provides highly reproducible production and finished product characteristics [21].

Chitosan is available commercially in a number of different grades varying with respect to molecular weight, viscosity, DD, and salt form. Chitosan manufactured in accordance with Good Manufacturing Practices (GMP) should be considered a prerequisite for pharmaceutical and biomedical applications. Manufacturers of chitosan include Chitinor (Norway), Golden Shell (China), Heppe Medical Chitosan (Germany), Kitozyme (Belgium), NovaMatrix (Norway), and Primex (Iceland).

23.2.4 Chitosan Derivatives

The amino, and to lesser extent hydroxyl, groups on the chitosan molecule may be modified to produce derivatives with different physicochemical or biological properties. The primary motivation for modifying chitosan is in order to change its solubility characteristics, specifically to increase solubility at neutral and alkaline pH. This can be of particular importance when formulating chitosan with therapeutic agents which are not soluble or stable below neutral pH. A wide range of chitosan derivatives has been described, the most common ones being carboxymethyl chitosan and trimethyl chitosan [22, 23]. Quaternary piperazine derivatives of chitosan have been prepared by coupling a quaternary piperazinium acetic acid onto chitosan or by attaching a tertiary 1,4 dimethylpiperazineonto N-chloroacetyl-6-O-triphenyl-methylchitosan [24, 25].

There is also the possibility to modify the biological properties of chitosan by chemical derivatization. For example, chitosan has been thiolated as a means to increase mucoadhesiveness. It has also been suggested that this modification can improve oral drug absorption by inhibiting intestinal efflux pumps [26]. Pegylation has been reported to enhance the stability of nanoparticles in biological fluids [27], and amino acids and sugars have been joined to chitosan as targeting ligands [28, 29].

23.3 Biological Properties of Chitosan

The mechanism by which chitosan may act as a vaccine adjuvant when administered intranasally is not fully established. However, chitosan has a number of in vivo effects which are described below; these may contribute to varying degrees to its adjuvant activity.

23.3.1 Mucoadhesion

A prominent feature of chitosan is its ability to interact with mucus and provide bioadhesion. Mucin is the primary protein component of mucus and contains sialic acid moieties which are negatively charged at physiological pH. The binding of chitosan to mucin is primarily by electrostatic binding between amino groups and sialic acid residues, although hydrogen bonding reportedly plays a role too [30–32].

In vivo studies have been performed to evaluate the bioadhesive properties of chitosan when administered intranasally. Radiolabeled chitosan solution and powder formulations were administered to human volunteers with a chitosan-free solution of radiolabel being used as a control. Clearance half-lives of the nasal doses were determined by measuring radioactive counts using a gamma camera. Mean clearance half-lives were 21, 41, and 84 min for the control, chitosan solution and chitosan powder, respectively, indicating that chitosan has bioadhesive properties and is able to increase retention in the nasal cavity [33]. A study was also conducted in sheep and a similar clearance pattern was demonstrated [34].

23.3.2 Modulation of Tight Junctions

Although bioadhesion undoubtedly contributes to chitosan's absorption-enhancing properties by prolonging the contact of therapeutic agent with the mucosal surface, it has also been demonstrated that chitosan is able to modulate epithelial intercellular tight junctions, thus enabling increased paracellular transport of drug molecules. In particular, it has been reported that chitosan interacts with proteins which regulate tight junctions, such as ZO-1 and JAM-1 and is able to initiate signalling pathways involving protein kinase C which in turn interrupts the integrity of tight junctions. Such effects appear to be transient, with the integrity of the tight junctions rapidly returning [35–37].

Both solution and nanoparticle forms of chitosan have been demonstrated to have a comparable modulatory effect on tight junctions [38]. For four alkylated chitosan derivatives, tight junction opening was found to be less efficient when the polymers were in nanoparticle form compared to soluble form; this effect was attributed to reduced surface charge [39].

23.3.3 Immunostimulant Properties

It is notable that chitosan can exert adjuvant effects when administered subcutaneously. Given by this route, chitosan was shown to induce humoral and cell-mediated immune responses and was equipotent with incomplete Freund's adjuvant and superior to alum. It was postulated that chitosan may be exerting an immune adjuvant effect by retaining antigen in a depot at the injection site and via stimulation of NK cells and macrophages; chitosan resulted in a large cellular expansion in local lymph nodes [40].

Elsewhere, it has been reported that chitosan exhibits a range of immunological effects, including macrophage activation and stimulating the production of a range of inflammatory cytokines [41]. Similar effects have been demonstrated in intestinal cells following oral administration of chitosan to rats [42].

23.4 Safety Aspects of Chitosan

For any excipient, it is necessary to demonstrate that the intended mode of use is safe. In the case of utilizing chitosan as a nasally administered adjuvant, local and systemic safety needed to be considered.

Chitosan is generally considered to be a relatively nontoxic and biocompatible material [43, 44]. The absorption of nasally administered chitosan is limited as a consequence of its high molecular weight and the majority is likely to be cleared from the nasal cavity into the gastrointestinal tract. In some species chitosan undergoes enzymatic degradation in the intestines by endogenous enzymes and those arising from the microflora, although intestinal chitosan digestion in humans is understood to be limited. In the event of systemic absorption, chitosan is likely to be degraded to lower molecular weight fractions (including glucosamine and *N*-acetylglucosamine monomers) and excreted. In producing chitosan derivatives, the possibility for reducing biodegradability and systemic elimination need to be taken into consideration [44].

For any material derived from a natural source, purity is the key safety consideration. Although chitosan is commonly extracted from shellfish, the authors are not aware of any allergic reactions arising from its nasal administration during clinical trials. Tropomyosin (a protein) is identified as the main shellfish allergen although other proteins such as arginine kinase myosin light chain may also play a role [45]. These proteins are found in the flesh of the shellfish as opposed to the shell from which chitosan is produced. Additionally, chitosan suitable for pharmaceutical and medical applications is a highly purified material and contains little or no residual protein. When chitosan-containing bandages (HemCon) were applied to patients with reported shellfish allergies, no allergic reactions were seen [46].

In terms of local safety, Archimedes has conducted a number of GLP-compliant studies examining nasally administered chitosan glutamate, including studies of up to 28 days duration in rats, 3 months in dogs and 6 months in rabbits. Data from these studies have successfully supported the conduct of Phase I–III clinical trials on intranasal ChiSys®-based formulations in Europe and the USA. Details of these unpublished studies are summarized in Table 23.1.

There is a significant body of human data for intranasal ChiSys® administration. In excess of 2,900 nasal chitosan doses (powder and aqueous solution forms as glutamate salt) have been administered to a total of more than 1,000 human subjects in clinical trials; the majority of the doses have contained drugs or vaccines. Details of a number of the published studies are provided in Table 23.2. Collectively, data from these studies have demonstrated that overall intranasally administered chitosan glutamate solutions and powders are safe and well tolerated.

Table 23.1 Summary of ChiSys® preclinical safety studies

Study	Species/strain (no. animals dosed, male/female)	Dosing regimen	Chitosan formulation
1	Rat/Sprague Dawley (10M/10F)	Once daily (both nostrils) for 14 days	5 mg/ml chitosan glutamate solution
2	Rat/Sprague Dawley (10M/10F)	Once daily (both nostrils) for 14 days	5 mg/ml chitosan glutamate solution
3	Rat/Sprague Dawley (10M/10F + Recovery: 5M/5F	Twice daily (alternate nostrils) for 28 days (with 28-day recovery phase)	5 mg/ml chitosan glutamate solution
4	Rabbit/NZW (5M/5F)	Once daily (both nostrils) for 10 days	20 mg/ml chitosan glutamate solution
5	Rabbit/NZW (5M/5F)	Once daily (both nostrils) for 14 days	Chitosan glutamate powder
6	Rabbit/NZW (Interim: 3M/3F Main: 6M/6F)	Twice daily (both nostrils) for 3 (interim) or 6 months	Saline solution (control), 5, 10 and 20 mg/ml chitosan glutamate solution
7	Dog/Beagle (3M/3F)	Once daily (both nostrils) for 14 days	5 mg/ml chitosan glutamate solution
8	Dog/Beagle (3M/3F)	Once daily (both nostrils) for 14 days	5 mg/ml chitosan glutamate solution
9	Dog/Beagle (Main: 3M/3F + Recovery: 2M/2F)	Twice daily (both nostrils) for 3 months (with 28-day recovery phase in selected animals)	5 mg/ml chitosan glutamate solution

23.5 Regulatory Status of Chitosan

A review of the regulatory status of chitosan has recently been published [56]. To date, no pharmaceutical products containing chitosan have been licensed, although regulatory agencies have approved a number of clinical trials for ChiSys® formulations (vaccines and other therapeutic agents). As noted earlier, chitosan is consumed orally in large quantities as a health supplement purported to aid in weight loss but such products are not regulated as pharmaceuticals.

Chitosan is a component of a number of medical devices (wound dressings) approved in the USA, e.g., HemCon® and Celox™ [56].

Monographs for chitosan (base) and chitosan hydrochloride appear in the United States Pharmacopoeia-National Formulary [57] and European Pharmacopoeia [58], respectively.

Chitosan manufacturers which currently have active Drug Master Files lodged with the US FDA are FMC Biopolymer and Golden Shell [59].

23.6 Formulating Antigens with Chitosan

An overview of approaches for coformulating ChiSys® and vaccine antigens is provided in Fig. 23.2. Factors which need to be considered when developing a formulation are highlighted below.

Table 23.2 Summary of selected human clinical trials using ChiSys®

Short title [reference]	ChiSys® formulation	Dosing regimen	No. subjects	Comments on nasal safety
Clearance characteristics of chitosan [33]	Aqueous solution	Two sprays to one nostril	8	No information provided on local tolerability/adverse events
	Powder	One powder spray to one nostril	8	
Mucociliary transport study [47]	Aqueous solution	One spray to one nostril for 7 days	10	Chitosan was well tolerated and the nasal membrane appeared healthy and normal in all volunteers following endoscopic examination
Mucociliary clearance of chitosan [48]	Aqueous solution	Multiple doses (one spray to one nostril once daily for 5 days)	14	The formulation appeared to be generally well tolerated
Immunology studies on nasal diphtheria vaccine [49]	Powder	One powder spray to one nostril at 0 and 28 days	10	Well tolerated with only transient and mild-to-moderate symptoms
Immunology studies on nasal influenza antigen [50]	Aqueous solution	One spray per nostril; two doses, four weeks apart	46 (23 at each of two antigen dose levels)	Treatment related adverse events were all mild in nature and mostly the symptoms of rhinorrhea
Pharmacokinetic profile of alniditan nasal spray [51]	Aqueous solution	One spray to one nostril on each of two occasions	27	Formulation generally well tolerated; shortly after drug administration, patients reported some nose and throat irritation as well as taste disturbance
Efficacy and safety of intranasal morphine formulation in a postsurgical dental pain model [52]	Aqueous solution	One spray to one nostril	225: 135 received active spray and 90 received placebo	Well tolerated; local effects consistent with application of any nasal spray (including placebo)

Table 23.2 (continued)

Short title [reference]	ChiSys® formulation	Dosing regimen	No. subjects	Comments on nasal safety
Efficacy and safety of morphine-chitosan nasal solution in patients following orthopedic surgery [53]	Aqueous solution	Ranged from one spray in one nostril to two sprays in both nostrils	90 completed a single-dose treatment phase) 177 completed a multiple dose treatment phase	Local effects at morphine doses of 15 mg or lower reported to be "usually mild and transient"
Immunology studies on dry powder norovirus vaccine formulations [54]	Powder	Two intranasal doses, three weeks apart	78: 60 received vaccine and 18 received adjuvants (including chitosan) only	Intranasal doses were generally well tolerated
Immunology/challenge study on dry powder norovirus vaccine formulation [55]	Powder	Two intranasal doses, three weeks apart	50: 47 received two doses of vaccine and 3 received one dose	Intranasal doses were generally well tolerated

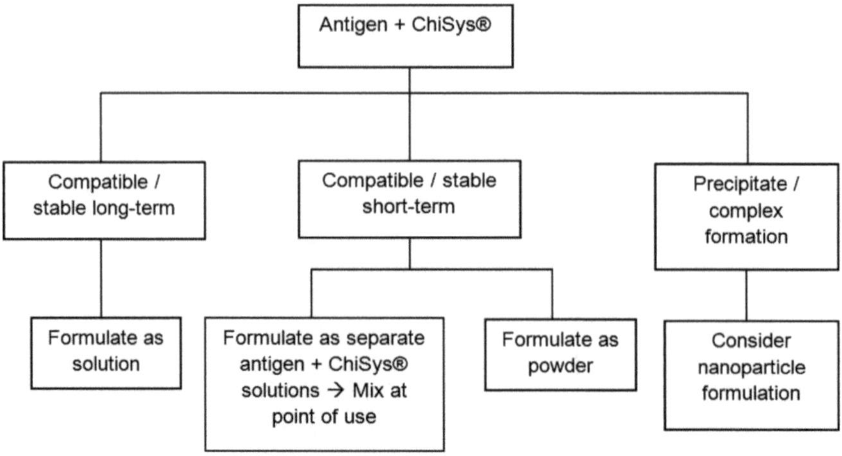

Fig. 23.2 Overview of formulation strategies for ChiSys®-based vaccines

23.6.1 *Physicochemical Considerations*

As discussed in Sect. 2.2, the key feature of chitosan which determines its formulation characteristics is the positive charge when dissolved in aqueous solution. If an antigen is negatively charged, this can potentially lead to an electrostatic interaction with the chitosan sufficient to cause precipitation of polymer and antigen. In many instances this will be an undesirable outcome, although it can be exploited, notably for the delivery of nucleic acids (DNA, siRNA) which have a strong negative charge. The interaction of nucleic acids with chitosan to form particulate entities, commonly termed "polyplexes" has been widely investigated and is a potential approach for vaccination [60–63]. The surface charge of such polyplexes should ideally be positive in order to retain the functional characteristics of chitosan. A potential advantage of formulating a vaccine as a chitosan nanoparticle is that extra protection may be provided to the antigen.

23.6.2 *Liquid Formulations*

The preferred presentation for a nasal formulation is often an aqueous solution due to ease of manufacture and the availability of multiple device options for intranasal delivery. However, the stability of antigen in the presence of chitosan is a key consideration. For example, conventional (underivatized) chitosan must be maintained below pH 6 to ensure it remains in solution whereas influenza antigen is unstable at acidic pH [64]. One option for resolving such a conflict is to keep the chitosan and antigen components separate until the point of administration. This could simply be achieved by mixing together the contents of two containers and then transferring it into a nasal delivery device, or more conveniently, by using a device in which the liquid components are mixed in situ just prior to or during the administration process.

It is generally preferable for nasal solutions to be administered as a spray rather than drops in order to provide more uniform distribution and clearance [65]. Appropriate consideration must be given to factors such as viscosity and concentration of chitosan to ensure that the solution can be atomized to form an acceptable and reproducible spray plume (shape and droplet size).

23.6.3 Powder Formulations

Formulating an intranasal vaccine as a powder increases the complexity of manufacture and dose delivery. However, a ChiSys® powder may offer advantages afforded by longer antigen retention in the nasal cavity (see Sect. 3.1), and the potential for improving antigen stability which in turn may allow avoidance of cold-chain distribution. Powder vaccine could optionally be reconstituted to produce an aqueous solution prior to use.

Ensuring uniformity of antigen content, controlling particle size, and achieving good flow properties are all challenges to be met when developing a powder formulation. The antigen may be isolated in solid form by freeze-drying or spray drying, and the nature of stabilizing and bulking agents used in such processes is an important consideration when developing a powder-vaccine formulation.

23.7 ChiSys® In Vivo Case Studies

23.7.1 Pertussis

A study was conducted in which mice received intranasally an aqueous solution containing pertussis toxin and filamentous hemagglutinin, with and without chitosan glutamate. A negative control group received chitosan solution alone intranasally and the positive control received the antigens in combination with alum intraperitoneally. The immunizations were repeated in appropriate animals at 28 days. Samples of blood, nasal washes, and lung lavage were collected from separate subgroups of animals at 28 and 42 days. Results indicated that systemic and mucosal immune responses were induced by the intranasally administered antigens and these were significantly augmented when chitosan was present [66].

23.7.2 Diphtheria

Based on the positive outcome of preclinical investigations in mice and guinea pigs [67], a clinical study has been conducted in 20 human subjects whereby single nasal immunizations were administered of two powder formulations comprising either diphtheria antigen, chitosan glutamate, and mannitol, or diphtheria antigen and mannitol [49]. A further group of five subjects received a standard alum-adsorbed diphtheria vaccine by intramuscular (IM) injection. A single nasal immunization induced serum antitoxin IgA and IgG and protective levels of toxin-neutralizing antibody. The second immunization boosted neutralizing activity. The response was highest in the subjects receiving the chitosan-containing nasal doses. Levels of secretory IgA in nasal wash samples were greater than tenfold higher in the chitosan nasal group compared to chitosan free.

23.7.3 Seasonal Influenza

Sixty-eight subjects participated in a clinical trial in which they received either an IM injection of conventional trivalent subunit vaccine or the same vaccine at doses of 7.5 or 15 μg as an intranasal solution containing chitosan glutamate [50]. Two doses of the nasal vaccines were administered, 4 weeks apart. Results were compared to the European Committee for Proprietary Medicinal Products (CPMP) requirements for registration of injectable influenza vaccines, namely that at least one of the three criteria need to be fulfilled for the antigen strains contained within the vaccine: (1) serum HI antibody titer of ≥ 40 in 70 % of subjects; (2) $\geq$ fourfold increase in HI antibody titers in 40 % of subjects and; (3) ≥ 2.5-fold increase over the preimmunization HI antibody titer level. Subjects receiving the 15 μg dose of intranasal vaccine met at least two of the three CPMP criteria after one or two doses. At a 7.5 μg dose, one or two of the criteria were met after one or two doses. Results were not statistically different from those obtained with the IM injection thus demonstrating the potential for providing effective influenza vaccination via intranasal immunization with a ChiSys® formulation.

23.7.4 Avian Influenza (H5N1)

A challenge study using a highly pathogenic avian influenza (HPAI) H5N1 virus has been conducted in the preferred animal model (ferret) in order to evaluate the immunogenicity and protective efficacy of intranasally administered vaccine candidates containing H5N1 subunit NIBRG-14 antigen. The efficacy of chitosan glutamate or trimethyl chitosan as vaccine adjuvants was evaluated by comparison to unadjuvanted (chitosan-free) vaccine and placebo (phosphate buffered saline) treatments [68].

Significant seroconversion occurred in vitro against clade 1 H5N1 A/Vietnam/1194/2004 following immunization with the chitosan-adjuvanted vaccines; the strongest response was seen with trimethyl chitosan. In contrast, seroconversion was not significant in animals treated with unadjuvanted vaccine or placebo. The two chitosan-containing vaccines were also shown to induce cross protective antibodies in vitro against representative clade 2.1 and 2.2 H5N1 viruses.

All animals immunized with the vaccines containing chitosan glutamate and trimethyl chitosan survived and were protected against H5N1-related infection (fever, body weight loss, virus replication, and pathological findings) resulting from intratracheal or intranasal challenge with HPAI A/Vietnam/1194/2004 (H5N1) virus whereas immunization with unadjuvanted vaccine provided little or no protection. Again, trimethyl chitosan provided the best response.

23.8 Anthrax

Intranasal dry powder formulations have been developed comprising recombinant anthrax protective antigen (rPA), monophosphoryl lipid A (MPL; a Toll-like receptor agonist) and chitosan. In a first study, single doses of the formulations were administered intranasally to rabbits [69]. Positive control groups received a solution of rPA in alum by IM injection and negative control groups received a nasal dose comprising MPL and chitosan only. Intranasal rPA doses were 50, 100, and 150 μg and the IM dose was 100 μg. Three weeks after immunization, the intranasal dose of 150 μg rPA had significantly increased serum anti-rPAIgG compared to the negative control. Although a stronger antibody response was seen with the IM injection at 3 weeks, by 6 weeks it was not significantly different to the negative control whereas antibody levels remained elevated for the 150 μg nasal dose.

The second experiment was a challenge study performed in rabbits. Nine weeks following a single intranasal immunization of 150 μg rPA in the MPL/chitosan formulation, the rabbits received an anthrax aerosol challenge. Eight out of ten animals receiving the nasal vaccine were protected whereas no animals survived in the negative control group [69].

23.9 Norovirus

A nasal vaccine containing chitosan and MPL as adjuvants has also been developed for norovirus. Two Phase I double-blind, controlled, clinical studies on candidate intranasal Norwalk virus-like particle (VLP) vaccines have been conducted in healthy subjects [54]. Two doses, given 3 weeks apart, of 5, 15, or 50 μg VLP antigen were evaluated in a preliminary dose-escalation study. In the second study a comparison was made between VLP antigen (doses of 50 and 100 μg were evaluated, $n = 20$/group), adjuvant control (VLP-free powder, $n = 10$), and true placebo (puff of air, $n = 11$). Subjects recorded symptoms for 7 days postdosing and safety was followed for 180 days. Blood samples were collected for serology, antibody-secreting cells (ASC) and analysis of ASC homing receptors. Norovirus VLP-specific IgG and IgA antibodies increased around fivefold and ninefold respectively at the 100 μg dose level (Fig. 23.3). Hemagglutination inhibition antibody titer, a measure of functional antibodies which may contribute to protection against the virus, increased

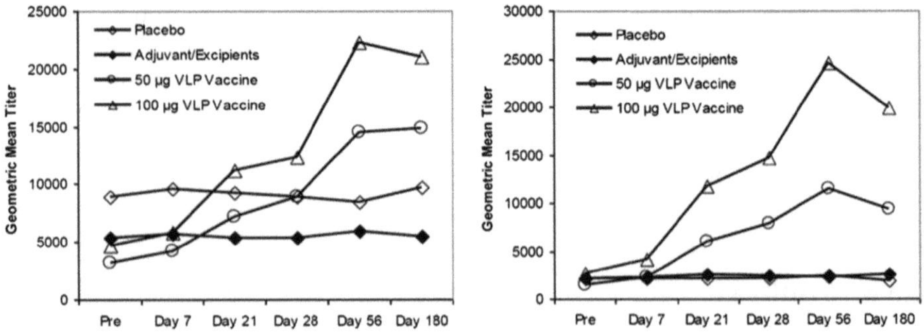

Fig. 23.3 Norwalk VLP-specific serum IgG and IgA geometric mean antibody titers following intranasal immunizations. 61 healthy adult subjects were randomized to receive either two doses separated by 21 days of *1* 50 μg of Norwalk VLP vaccine (20 subjects, *circles*), *2* 100 μg of Norwalk VLP vaccine (20 subjects, *open triangles*); *3* adjuvant control (10 subjects, *filled diamonds*), or *4* true placebo consisting of a puff of air (11 subjects, *open diamonds*). Serum IgG and IgA values shown in *left* and *right* graphs, respectively. (Reprinted from reference [54], Copyright (2010), with permission from Infectious Diseases Society of America)

ninefold at the 100 μg dose level. All subjects who received the 50 or 100 μg vaccine dose developed IgA ASC in peripheral blood. In an ancillary study, all of the subjects receiving 100 μg and 90 % receiving 50 μg vaccine had significant IgA and IgG B memory responses [70].

The efficacy of the intranasal norovirus VLP vaccine in preventing acute gastroenteritis following live norovirus challenge was demonstrated in a Phase II study [55]. In the study, healthy adults completed the challenge after receiving two doses, 3 weeks apart of the vaccine (100 μg norovirus VLPs) or placebo. Seventy-seven of the subjects completed the trial in accordance with the protocol. Norovirus-associated gastroenteritis occurred in 69 % of subjects receiving placebo and 37 % of those receiving active vaccine. Norovirus infection occurred in 82 % and 61 % of subjects receiving placebo and active vaccine respectively. These differences between active and placebo response were statistically significant. The vaccine was generally well tolerated.

23.10 Conclusions and Future Perspectives

Although nasal vaccination offers a number of attractions, typically only antigens in the form of live-attenuated organisms are able, by themselves, to generate a sufficient immune response. Subunit or purified antigens generally need to be administered with an adjuvant (or coadjuvants) in order to elicit adequate immunity.

ChiSys® (chitosan) is an attractive nasal adjuvant: It is safe and well tolerated, with a significant database of human exposure, has demonstrable efficacy, and is available in a quality suitable for pharmaceutical use. The solubility characteristics of chitosan may present incompatibilities with some antigens but there are a number of strategies which can be used to resolve such issues.

There is the potential of further enhancing the immune response of nasal vaccines by combining chitosan with another adjuvant; this approach is especially attractive for other adjuvants which are expensive and/or reactogenic or toxic since the addition of chitosan may allow a reduction in doses.

For the future, chitosan also holds promise as a delivery agent for intranasal administration of nucleic acid-based vaccines.

References

1. Davis SS. Nasal vaccines. Adv Drug Del Rev. 2001;51:21–42.
2. Jabbal-Gill I. Nasal vaccine innovation. J Drug Target. 2010;18:771–86.
3. Pavot V, Rochereau N, Genin C, Verrier B, Paul S. New insights in mucosal vaccine development. Vaccine. 2012;30:142–54.
4. Brandtzaeg P. Potential of nasopharynx-associated lymphoid tissue for vaccine responses in the airways. Am J Respir Crit Care Med. 2011;183:1595–604.
5. Papadaki HA, Velegraki M The immunology of the respiratory system. Pneumon. 2007; 20:384–94.
6. Neutra MR, Kozlowski PA. Mucosal vaccines: the promise and the challenge. Nat Rev Immunol. 2006;6:148–58.
7. Casaba N, Garcia-Fuentes M, Alonso MJ. Nanoparticles for nasal vaccination. Adv Drug Del Rev. 2009;61:140–57.
8. FluMist®. http://www.flumist.com/. Accessed: 6 Jan 2013.
9. Nasovac®. http://www.seruminstitute.com/content/products/product_nasovac.htm. Accessed: 6 Jan 2013.
10. Gerdts V, Mutwiri GK, Tikoo SK, Babiuk LA. Mucosal delivery of vaccines in domestic animals. Vet Res. 2006;37:487–510.
11. Illum L. Nasal drug delivery—recent developments and future prospects. J Control Release. 2012;161:254–63.
12. Amidi M, Mastrobattista E, Jiskoot W, Hennink WE. Chitosan-based delivery systems for protein therapeutics and antigens. Adv Drug Del Rev. 2010;62:59–82.
13. Dai T, Tanaka M, Huang YY, Hamblin MR. Chitosan preparations for wounds and burns: antimicrobial and wound-healing effects. Expert Rev Anti Infect Ther. 2011;9:857–79.
14. Riva R, Ragelle H, des Rieux A, Duhem N, Jérome C, Préat V. Chitosan and chitosan derivatives in drug delivery and tissue engineering. Adv Polym Sci. 2011;244:19–44.
15. Jiang T, Deng M, Abdel-Fattah WI, Laurencin CT. Chitosan-based biopharmaceutical scaffolds in tissue engineering and regenerative medicine. In: Sarmento B, Das Neves J, editors. Chitosan-based systems for biopharmaceuticals: delivery, targeting and polymer therapeutics. Chichester: Wiley; 2012.
16. Egras AM, Hamilton WR, Lenz TL, Monaghan MS. An evidence-based review of fat modifying supplemental weight loss products. J Obes. 2011. doi: 10.1155/2011/297315.
17. Illum L. Chitosan and its use as a pharmaceutical excipient. Pharm Res. 1998;15:1326–31.
18. Chitosan. In: Rowe RC, Sheskey PJ, Cook WG, Fenton ME, editors. Handbook of pharmaceutical excipients seventh edition. London/Washington: Pharmaceutical Press/American Pharmaceutical Association; 2012. pp. 172–5.
19. Rinaudo M. Physical properties of chitosan and derivatives in sol and gel states. In: Sarmento B, das Neves J, editors. Chitosan-based systems for biopharmaceuticals: delivery, targeting and polymer therapeutics. Chichester: Wiley; 2012.
20. Skaugrud O, Hagen A, Borgersen B, Dornish M. Biomedical and pharmaceutical applications of alginate and chitosan. Biotechnol Genet Eng. 1999;16:23–40.
21. KiOmedine-CsU®, the world's first GMP ultra-pure chitosan of non-animal origin for medical and pharmaceutical applications. http://www.drugdevcompare.com/1488-White-Papers/

37547-KiOmedine-CsU-The-World-s-First-GMP-Ultra-pure-Chitosan-of-Non-animal-Origin-for-Medical-and-Pharmaceutical-Applications/. Accessed: 6 Jan 2013.

22. Upadhyaya L, Singh J, Agarwal V, Tewari RP. Biomedical applications of carboxymethylchitosans. Carbohyd Polym. 2013;91:452–66.
23. Sahni JK, Chopra S, Ahmad FJ, Khar RK. Potential prospects of chitosan derivative trimethyl chitosan chloride (TMC) as a polymeric absorption enhancer: synthesis, characterization and application. J Pharm Pharmacol. 2008;60:1111–9.
24. Holappa J, Nevalainen T, Safin R, Soininen P, Asplund T, Luttikhedde T, Masson M, Jarvinen T. Novel water-soluble quaternary piperazine derivatives of chitosan: synthesis and characterization. Macromol Biosci. 2006;6:139–44.
25. Novel quaternary polymers, International Publication Number WO 2007/034032.
26. Werle M, Bernkop-Schnürch A. Thiolated chitosans: useful excipients for oral drug delivery. J Pharm Pharmacol. 2008;60:273–81.
27. Prego C, Torres D, Fernandez-Megia E, Novoa-Carballal R, Quiñoá E, Alonso MJ. Chitosan–PEG nanocapsules as new carriers for oral peptide delivery: effect of chitosan pegylation degree. J Control Release. 2006;111:299–308.
28. Casettari L, Vllasaliu D, Lam JK, Soliman M, Illum L. Biomedical applications of amino acid-modified chitosans: a review. Biomaterials. 2012;33:7565–83.
29. Kim TH, Jiang HL, Dhananjay J, Park IK, Cho MH, Nah JW, Akaike T, Cho CS. Chemical modification of chitosan as a gene carrier in vitro and in vivo. Prog Polym Sci. 2007;32:726–53.
30. Sogias IA, Williams AC, Khutoryanskiy VV. Why is chitosan mucoadhesive? Biomacromolecules. 2008;9:1837–42.
31. Silva CA, Nobre TM, Pavinatto FJ, Oliveira ON. Interaction of chitosan and mucin in a biomembrane model environment. J Coll Interf Sci. 2012;376:289–95.
32. Leithner K, Bernkop-Schnürch A. Chitosan and derivatives for biopharmaceutical use: mucoadhesive properties. In: Sarmento B, das Neves J, editors. Chitosan-based systems for biopharmaceuticals: delivery, targeting and polymer therapeutics. Chichester: Wiley; 2012.
33. Soane RJ, Frier M, Perkins AC, Jones NS, Davis SS, Illum L. Evaluation of the clearance characteristics of bioadhesive systems in humans. Int J Pharm. 1999;178:55–65.
34. Soane RJ, Hinchcliffe M, Davis SS, Illum L. Clearance characteristics of chitosan based formulations in the sheep nasal cavity. Int J Pharm. 2001;217:183–91.
35. Sonaje K, Chuang EY, Lin KJ, Yen TC, Su FY, Tseng MT, Sung HW. Opening of epithelial tight junctions and enhancement of paracellular permeation by chitosan: microscopic, ultrastructural and computed-tomographic observations. Mol Pharmaceutics. 2012;9:1271–9.
36. Smith J, Wood E, Dornish M. Effect of chitosan on epithelial cell tight junctions. Pharm Res. 2004;21:43–9.
37. Smith JM, Dornish M, Wood EJ. Involvement of protein kinase C in chitosan glutamate-mediated tight junction disruption. Biomaterials. 2005;26:3269–76.
38. Vllasaliu D, Exposito-Harris R, Heras A, Casettari L, Garnett M, Illum L, Stolnik S. Tight junction modulation by chitosan nanoparticles: comparison with chitosan solution. Int J Pharm. 2010;400:183–93.
39. Sadeghi AMM, Dorkoosh FA, Avadi MR, Weinhold M, Bayat A, Delie F, Gurny R, Larijani B, Rafiee-Tehrani M, Junginger HE. Permeation enhancer effect of chitosan and chitosan derivatives: comparison of formulations as soluble polymers and nanoparticulate systems on insulin absorption in Caco-2 cells. Eur J Pharm Biopharm. 2008;70:270–8.
40. Zaharoff DA, Rogers CJ, Hance KW, Schlom J, Greiner JW. Chitosan solution enhances both humoral and cell-mediated responses to subcutaneous vaccination. Vaccine. 2007;25:2085–94.
41. Petrovsky N, Cooper PD. Carbohydrate-based immune adjuvants. Expert Rev Vaccines. 2011;10:523–37.
42. Canali MM, Porporatto C, Aoki MP, Biancol D, Correa SG. Signals elicited at the intestinal epithelium upon chitosan feeding contribute to immunomodulatory activity and biocompatibility of the polysaccharide. Vaccine. 2010;28:5718–24.
43. Baldrick P. The safety of chitosan as a pharmaceutical excipient. Reg Toxicol Pharmacol. 2010;56:290–9.

44. Kean T, Thanou M. Biodegradation, biodistribution and toxicity of chitosan. Adv Drug Del Rev. 2010;62:3–11.
45. Lopata AL, O'Hehir RE, Lehrer SB. Shellfish allergy. Clin Exp Allergy. 2010;40:850–8.
46. Waibel KH, Haney B, Moore M, Whisman B, Gomez R. Safety of chitosan bandages in shellfish allergic patients. Mil Med. 2011;176:1153–6.
47. Aspden TJ, Mason JD, Jones NS, Lowe J, Skaugrad O, Illum L. Chitosan as a nasal delivery system: the effect of chitosan solutions on in vitro and in vivo mucociliary transport rates in human turbinates and volunteers. J Pharm Sci. 1997;86:509–13.
48. Newman SP, Simpson M, Fisher T, Iqbal K. Quantification of lung deposition and nasal mucociliary clearance for a nasally administered drug formulation containing chitosan. In: Dalby RN, Byron PR, Peart J, Suman JD, Farr SJ, editors. Respiratory drug delivery IX. Virginia Commonwealth University; 2004.
49. Mills KHG, Cosgrove C, McNeela EA, Sexton A, Giemza R, Jabbal-Gill I, Church A, Lin W, Illum L, Podda A, Rappuoli R, Pizza M, Griffin GE, Lewis DJM. Protective levels of diphtheria-neutralising antibody induced in healthy volunteers by unilateral priming-boosting intranasal immunization associated with restricted ipsilateral mucosal secretory immunoglobulin A. Infect Immun. 2003;71:726–32.
50. Read RC, Naylor SC, Potter CW, Bond J, Jabbal-Gill I, Fisher A, Illum L, Jennings R. Effective nasal influenza vaccine delivery using chitosan. Vaccine. 2005;23:4367–74.
51. Roon KI, Soons PA, Uitendaal MP, de Beukelaar F, Ferrari MD. Pharmacokinetic profile of alniditan nasal spray during and outside migraine attacks. Br J Clin Pharmacol. 1999; 47:285–90.
52. Christensen KS, Cohen AE, Mermelstein FH, Hamilton DA, McNicol E, Babul N, Carr DB. The analgesic efficacy and safety of a novel intranasal morphine formulation (morphine plus chitosan), immediate release oral morphine, intravenous morphine, and placebo in a postsurgical dental pain model. Anesth Analg. 2008;107:2018–24.
53. Stoker DG, Reber KR, Waltzman LS, Ernst C, Hamilton D, Gawarecki D, Mermelstein F, McNicol E, Wright C, Carr DB. Analgesic efficacy and safety of morphine-chitosan nasal solution in patients with moderate to severe pain following orthopaedic surgery. Pain Med. 2008;9:3–12.
54. El-Kamary SS, Pasetti MF, Mendelman PM, Frey SE, Bernstein DI, Treanor JJ, Ferreira J, Chen WH, Sublett R, Richardson C, Bargatze RF, Sztein MB, Tacket CO. Adjuvanted intranasal Norwalk virus-like particle vaccine elicits antibodies and antibody-secreting cells that express homing receptors for mucosal and peripheral lymphoid tissues. J Infect Dis. 2010;202:1649–58.
55. Atmar RL, Bernstein DI, Harro CD, Al-Ibrahim MS, Chen WH, Ferreira J, Estes MK, Graham DY, Opekun AR, Richardson C, Mendelman PM. Norovirus vaccine against experimental human Norwalk virus illness. N Engl J Med. 2011;365:2178–87.
56. Dornish M, Kaplan DS, Arepalli SR. Regulatory status of chitosan and derivatives. In: Sarmento B, das Neves J, editors. Chitosan-based systems for biopharmaceuticals: delivery, targeting and polymer therapeutics. Chichester: Wiley; 2012.
57. Chitosan. In: United States Pharmacopeia 35-National Formulary 30 United States Pharmacopeial Convention, Rockville; 2011.
58. Chitosan hydrochloride. In: European Pharmacopoeia 7.0. EDQM, Strasbourg; 2010.
59. Drug Master Files. http://www.fda.gov/Drugs/DevelopmentApprovalProcess/FormsSubmissionRequirements/DrugMasterFilesDMFs/default.htm. Accessed 6 Jan 2013.
60. Xu W, Shen Y, Jiang Z, Wang Y, Chu Y, Xiong S. Intranasal delivery of chitosan-DNA vaccine generates mucosal SIgA and anti-CVB3 protection. Vaccine. 2004;22:3603–12.
61. Yang X, Yuan X, Cai D, Wang S, Zong L. Low molecular weight chitosan in DNA vaccine delivery via mucosa. Int J Pharm. 2009;375:123–32.
62. Mao S, Sun W, Kissel T. Chitosan-based formulations for delivery of DNA and siRNA. Adv Drug Del Rev. 2010;62:12–27.
63. Holzerney P, Ajdini B, Heusermann W, Bruno K, Schuleit M, Meinel L, Keller M. Biophysical properties of chitosan/siRNA polyplexes: profiling the polymer/siRNA interactions and bioactivity. J Control Release. 2012;157:297–304.

64. Quan FS, Li ZN, Kim MC, Yang D, Compans RW, Steinhauer DA, Kang SM. Immunogenicity of low-pH treated whole viral influenza vaccine. Virology. 2011;417:196–202.
65. Bryant ML, Brown P, Gurevich N, McDougall IR. Comparison of the clearance of radiolabelled nose drops and nasal spray as mucosally delivered vaccine. Nucl Med Commun. 1999;20: 171–4.
66. Jabbal-Gill I, Fisher AN, Rappuoli R, Davis SS, Illum L. Stimulation of mucosal and systemic antibody responses against *Bordetella pertussis* filamentous haemagglutinin and recombinant pertussis toxin after nasal administration with chitosan in mice. Vaccine. 1998;16:2039–46.
67. McNeela EA, O'Connor D, Jabbal-Gill I, Illum L, Davis SS, Pizza M, Peppoloni S, Rappuoli R, Mills KH. A mucosal vaccine against diphtheria: formulation of cross reacting material (CRM (197)) of diphtheria toxin with chitosan enhances local and systemic antibody and Th2 responses following nasal delivery. Vaccine. 2000;19:1188–98.
68. Mann AJ et al. Intranasal H5N1 vaccination, adjuvanted with novel chitosan derivatives, protects against highly pathogenic Avian Influenza intranasal and intratracheal challenge, in unprimed ferrets. PLOS ONE (in press).
69. Klas SD, Petrie CR, Warwood SJ, Williams MS, Olds CL, Stenz JP, Cheff AM, Hinchcliffe M, Richardson C, Wimer S. A single immunization with a dry powder anthrax vaccine protects rabbits against lethal aerosol challenge. Vaccine. 2008;26:5494–502.
70. Ramirez K, Wahid R, Richardson C, Bargatze RF, El-Kamary SS, Sztein MB, Pasetti MF. Intranasal vaccination with an adjuvanted Norwalk virus-like particle vaccine elicits antigen-specific B memory responses in human adult volunteers. Clin Immunol. 2012;144:98–108.

Chapter 24
Development of a Cationic Nanoemulsion Platform (Novasorb®) for Ocular Delivery

Frédéric Lallemand, Philippe Daull and Jean-Sébastien Garrigue

24.1 Introduction

Drug delivery across mucosal barriers has always been a challenge, and crossing the eye mucosa is no exception. The eye surface is a unique and complex mucosa with its own physiology and mechanisms of protection. This chapter illustrates how, from a clear understanding of the eye mucosal barrier structure, a new delivery system was designed to better treat ocular surface diseases. This case study describes the development of a new drug delivery system, the Novasorb® platform, designed to overcome ocular barriers to improve ophthalmic drug delivery. This technology is based on cationic emulsions primarily developed in the late 1990s by University of Jerusalem professor Simon Benita. Several years later, and after the creation of a spin-off company, Novagali Pharma (Evry, France) in 2001, the Novasorb® technology was successfully transferred to clinical use. The main steps of the development are briefly presented; from concept formulations to preclinical pharmacokinetics (PK) and toxicity studies to the clinic.

24.1.1 Eye Protection Systems

Of the sensory organs, the eye is probably the most precious and the organ upon which our daily activities depend most. As an extension of the central nervous system, the eye needs to be well protected although continuously exposed to and threatened by an external and aggressive environment. As a consequence, the body has developed several effective protection mechanisms to preserve the eye's structure and function. However, these protection mechanisms make it particularly difficult to access the eye's inner tissues, even the different corneal layers, when treating the eye becomes necessary.

J.-S. Garrigue (✉) · F. Lallemand · P. Daull
Novagali Pharma, 1 rue Pierre Fontaine, 91058 Evry Cedex, France
e-mail: jean-sebastien.garrigue@santen.fr

J. das Neves, B. Sarmento (eds.), *Mucosal Delivery of Biopharmaceuticals,* DOI 10.1007/978-1-4614-9524-6_24,

For other mucosal tissues (pulmonary, nasal, intestine, etc.), mucus production and excretion is one of the main protection mechanisms, primarily by preventing pathogen adherence and tissue infection through their entrapment in a thick external mucus layer. For example, the thickness of the mucus layer varies from 100 μm in the large intestine to only 0.5 μm on the ocular surface. Hence, the membrane-tethered mucins (the mucus layer) of the ocular surface cannot be the only mechanism of protection.

The ocular mucosa is protected by both physical and dynamic mechanisms. The eye is firstly protected by the eyelid, which blinks every 4–5 s. The eyelids possess several important ocular functions, with the primary objective of protecting both the anterior globe (cornea) from injury and the retina from excessive incoming light. Eyelid blinking also helps spread and maintain the ocular tear film. Eyelid behaviors achieving these functions include blinking (voluntary, spontaneous, or reflexive) and voluntary eye closure (gentle or forced) [1]. Blinking swipes away the overloads of tears as well as xenobiotic or solid particles present on the ocular surface, including any active ingredients administered topically.

The precorneal tear film is also of major importance in the protection and health of the eye. The tear film is a nourishing, lubricating, and protecting layer that bathes the ocular surface. It is continuously replenished through cycles of production and elimination via evaporation, absorption, and drainage. These processes are often referred to as tear-film dynamics [2]. Tears fight desiccation, microbial contamination, and the effects of xenobiotic and solid particles. This film also maintains surface humidity to provide transparency and optical quality of the cornea as a refracting surface [3]. The tear film comprises three layers: a thin superficial layer of meibomian lipid; an intermediate aqueous layer containing dissolved mucins, salts and proteins; and an internal layer of mucus network, secreted mainly from conjunctival goblet cells whose chemical structure has now been fully described [4]. These ocular mucins are highly negatively charged, with the majority terminated by sialic acid, while those from rabbits are mainly neutral and terminated by alpha 1-2 fucose and/or alpha 1-3 *N*-acetylgalactosamine [3]. In addition, these O-glycan mucins prevent bacterial adhesion and endocytic activity and maintain epithelial barrier function through interactions with galectins [5]. However, in addition to their traditional protective functions (selective barrier to the penetration of xenobiotics, antiadhesive that prevent pathogen adherence, and lubrication), the membrane mucins are also signaling molecules via their cytoplasmic tails and epidermal growth factor (EGF) domains [6]. While providing lubrication and protection of the ocular surface, this layer reduces the efficacy of pharmaceutical treatment by limiting tissue penetration. However, as we will see below, this negatively charged mucosa may also be diverted from its original function and may be a critical player in the improvement of ocular drugs formulated in cationic emulsions.

The presence of a constant tear flow of approximately 1.2 μL/min (0.5–2.2 μL/min) is also an important protection system. This constant flow results in a tear turnover rate of 16 % per minute during waking hours, and the reflex lachrymation may increase this rate up to 100-fold, to 300 μL/min. The high turnover rate of this precorneal tear film contributes to the protection of the eye, but also to the low availability of topically administered drugs [7].

The inner part of the eye is protected by the cornea, which has a complex structure of three different layers with varying physiological properties alternating between lipophilicity and hydrophilicity [8]. This layered and alternated construction makes corneal crossing of most drugs very difficult. Optimal permeant molecules should have a log D of 2–3 [9]. Examining in greater detail the corneal structure, first the outermost layer of the cornea is distinguished, i.e., lipophilic epithelium, which is formed by epithelial cells linked by tight junctions providing a strong barrier to the molecules present in the tear fluid. The corneal epithelium is consequently almost impermeable to any substance larger than 500 Da [10]. The next layer is made up of the stroma, a hydrophilic layer composed of fibrous tissue made of large collagen fibers and proteoglycans that form the major part of the cornea. Finally, the endothelium is a monolayer of hexagonal cell interfaces, which is also quite lipophilic [11].

In addition, it should be noted that the cornea is innervated by sensory nerve terminals of the trigeminal ganglion. Physical and chemical agents acting on the ocular surface (extreme environmental temperatures, wind, foreign bodies, and chemicals) induce conscious sensations and reflex motor and autonomic responses (blinking, lacrimation, conjunctival vasodilation) aimed at protecting the eye from further injury and drug penetration [12].

These successive physical and biological protections result in less than10 % of an instilled drug being absorbed by ocular tissues [13], leading to poor efficacy and the need for repeated instillations for the vast majority of eye drops. As a result of extending and maintaining the efficacy of eye drop solutions, the drug concentration needs to be increased but with potential exacerbation of local and/or systemic side effects with potent drugs such as timolol (a beta-blocker). Consequently, there is a need for new formulations that will improve efficacy while also limiting the risk of local side effects.

24.1.2 Options to Overcome Physiological Barriers

Ocular drug absorption from the lacrimal fluid to the anterior ocular tissues via transcorneal absorption is determined by two major factors: drug permeability through the cornea and contact time of the product with ocular tissues. Based on these two principles, scientists have created several valuable approaches to overcome the barriers.

24.1.3 Enhancing Penetration

Firstly, to promote drug permeability, penetration enhancers have been added to aqueous eye drops with some success [14]. These excipients, based on their surface-active property, are able to open corneal epithelium tight junctions and desmosomes leading to penetration toward the anterior chamber. The literature is very rich in examples of studies testing penetration enhancers, their effect toward hydrophilic and lipophilic drugs, the size and charge of molecules, etc. [15]. However, all authors agree on one fact: Although penetration enhancers can be good adjuvants for ocular penetration, as regards their surfactant nature they are intrinsically deleterious for the ocular surface [16]. Long-term use of penetration enhancers might result in poor patient compliance due to chronic discomfort, thus limiting the use of penetration enhancers in ophthalmology.

Another way to increase corneal penetration is to increase the specific uptake of the drug through the use of vectors, such as nanoparticles or liposomes, which have been described as being specifically taken up by corneal cells, as shown and discussed by Calvo et al. [17] and Diebold et al. [18]. However, the exact mechanism is not yet fully elucidated, therefore, limiting the use of this promising approach.

Corneal penetration can also be influenced by the specific contact surface created by the colloidal system with the cornea. Emulsions are a typical example of such systems. An oil-in-water emulsion is a dispersion of oil into a water phase. This dosage form has been used in pharmacy for decades but only recently in ophthalmology as eye drops. Ophthalmic emulsions have shown promise in topical ocular delivery as they are nontoxic systems, easy and inexpensive to manufacture, and able to deliver a lipophilic active agent with enhanced corneal penetration [19]. In addition, ophthalmic emulsions are able to protect unstable active agents from chemical degradation (such as latanoprost) and to mask the irritation potential of some drugs. The first ophthalmic emulsion to be approved in the USA was Restasis® in 2002. This dosage form is now routinely used to treat dry eye conditions. A few years later, Durezol® was marketed to treat ocular inflammation. The exact mechanistic processes regarding enhanced corneal penetration has still not been fully elucidated [20] but they most likely involve several physicochemical and biological mechanisms whose increased specific surface of exchange with the cornea has major importance.

24.1.4 Increasing Retention Time

Increased retention time by enhanced viscosity or bioadhesion has been widely used and described during the past 20 years [8], leading to a number of ophthalmic products. Mucoadhesion is based on noncovalent bonds (hydrogen or electrostatic bonds) between polymers and mucus or physical entanglement. In spite of their current wide use to decrease the frequency of administration and/or concentration in the solution, hydrogels present only a limited value because such aqueous formulations are eliminated by the usual routes in the ocular domain and cause blurred vision and patient discomfort.

24.1.5 Enhancing Penetration and Increasing Retention Time

Improving the eye surface's drug retention alone is inadequate in bringing a significant improvement of drug bioavailability. It should be combined with penetration

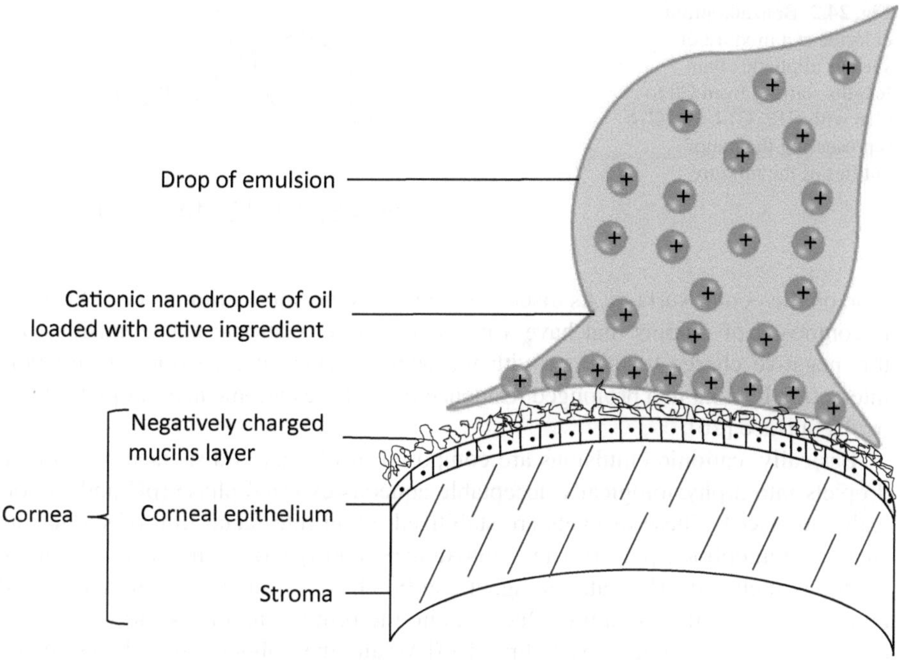

Fig. 24.1 Novasorb® enhanced the spread of a drop emulsion on the corneal surface and increased the contact surface of oily drops on the ocular surface

enhancement. Several examples of combined systems (penetration and retention) have been described in the literature, such as liposomes combined with hydrogels [21], liposomes combined with a collagen shield [22], and even more solid lipid nanoparticles associated with a bioadhesive hydrogel such as chitosan [23]. Although promising, these examples all remain today at the prototype stage. Nevertheless, a new and innovative system has come to the market combining the effect of a very large specific surface of colloidal systems with improved bioadhesion properties. This drug-delivery system is the cationic nanoemulsion registered under the trademark Novasorb® technology (Fig. 24.1).

24.2 Cationic Emulsions

24.2.1 Cationic Agent

While ophthalmic emulsions are becoming an essential tool for topical delivery, cationic emulsions combining intrinsic advantages of emulsions with a bioadhesive mechanism have provided enhanced efficacy of emulsions [24, 25]. This innovative approach uses the physiological barrier of mucus as a tool to increase retention

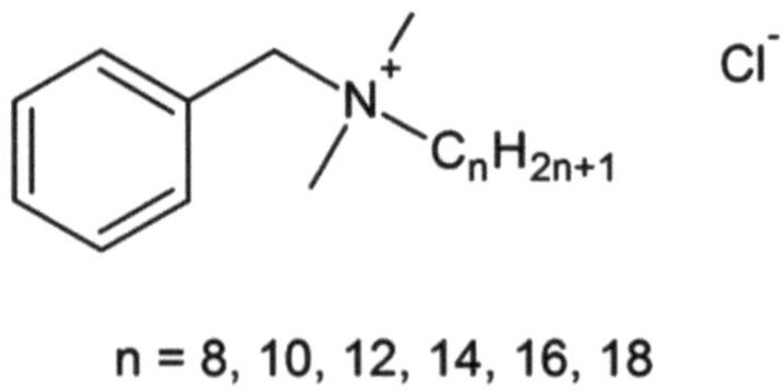

Fig. 24.2 Benzalkonium chloride is a mixture of various aliphatic chain lengths starting from C8 to C18 with C12, C14, and C16 representing the major entities of the mixture

time on the ocular surface. As discussed above, the last protection layer of the tears is composed of mucins that have a negative charge. Cationic nanoemulsions use this negative charge to interact with the ocular surface via a strong electrostatic interaction leading to a prolonged residence time of the cationic nanodroplets on the ocular surface [26].

Basically, cationic emulsions are composed of oil that is dispersed in ultrafine droplets into a physiologically acceptable aqueous external phase (pH and osmotically adjusted). These droplets are stabilized by an interfacial film of surfactants such as cremophors, polysorbates, poloxamers, and tyloxapol in which a cationic charge is included. The cationic agent can be chosen over those commonly used and described in the literature. One can cite the primary amines stearylamine and oleylamine, the cationic phospholipid DOTAP and the polymers poly-L-lysine and polyethylenimine. Yet, these cationic agents are not suitable for use in pharmaceutical products in terms of regulatory requirements, stability, or toxicity issues. The alternative is to use registered excipients such as quaternary amines usually used as preservative agents in ophthalmic aqueous solutions. The most widely used are cetylpyridinium chloride, benzalkonium chloride (BAK), and benzethonium chloride. However, the current tendency is to withdraw these preservatives from eye drops, due to long-term intolerance to them [27, 28], and to replace them either with soft preservatives (sodium perborate, Sofzia™) or single-use containers or preservative-free multidose containers. Nonetheless, BAK has several significant advantages over the other cationic agents: it is listed in all pharmacopeias, used in more than 80 % of eye drops at a concentration of 0.02 %, and has excellent interfacial properties, making this excipient a potential cationic agent candidate.

Ten years ago, Sznitowskaat et al. [29] noted that when used in combination with oil-in-water emulsions, the preservative efficacy of BAK was drastically decreased. This observation was explained by the inclusion of part of BAK in the oily phase of the emulsion, leading to a lower molecular concentration available in the aqueous phase. Only the freely soluble molecules of BAK present in the aqueous phase can exert their antimicrobial effects on bacteria. Concretely, BAK is a mixture of several quaternary ammoniums (Fig. 24.2) with varying lipophilicity. According to US and European pharmacopeias, three main entities should be present in the mixture: benzododecinium chloride (C12-substituted alkyl chain), myristalkonium chloride (C14), and cetalkonium chloride (C16), as presented in Fig. 24.3. In presence of oil droplets, the most lipophilic entity of BAK, cetalkonium chloride, is rearranged at the oil or water interface, providing a cationic charge on the droplets while hydrophilic

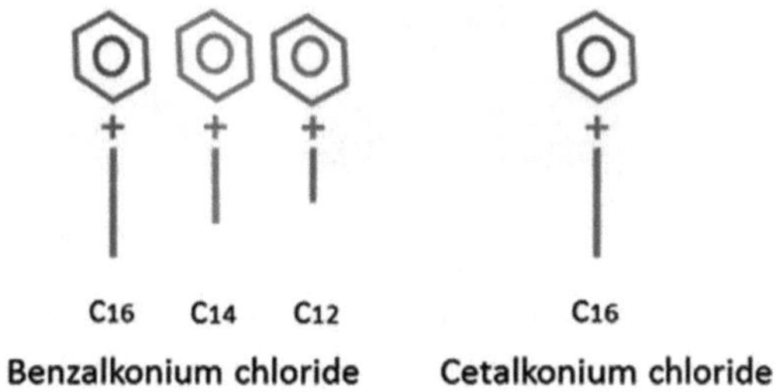

Fig. 24.3 Benzalkonium chloride made of three different entities of increasing lipophilicity (cetalkonium chloride, myristalkonium chloride, and benzododecinium chloride) and cetalkonium chloride schematic developed formulae

entities remain in the aqueous phase. Based on this observation, it was decided to use only cetalkonium chloride at a very low concentration (0.002–0.005 %) as a cationic agent to make cationic nanoemulsions [30].

24.2.2 Physicochemical Considerations Regarding Novasorb®

Cationic nanoemulsions are characterized by two main physicochemical properties. First, the zeta potential, which is defined as the electrical potential difference between the dispersion medium (i.e., water) and the stationary layer of fluid attached to the dispersed oil nanodroplets [31]. This surface charge will provide the system with its bioadhesion property on the ocular surface as well as the stability of the emulsion by providing electric repulsion between droplets. For optimal electrostatic interaction between cornea and product, a zeta potential of about +20 mV is sufficient [30].

The second main property is the oil droplet size. As described by the Stokes law (Fig. 24.4), the smaller the droplet size, the slower the dispersed system will separate, thus providing greater stability to the system. Even more importantly, the size of the droplets significantly participates in the penetration rate of the drug in the ocular tissue. With the active ingredient being solubilized in the lipid nanodroplets, a smaller particle size should provide a greater contact surface, hence a higher tissue concentration. In the case of the Novasorb® technology, a size between 100 and 200 nm was demonstrated to be small enough to provide a stable emulsion.

Surface bioadhesion was demonstrated by measurement of the spreading properties of the emulsion in contact angle studies. On excised rabbit eyes, a drop of cationic emulsion is applied and compared to anionic emulsion and hyaluronic gel in terms of contact angle (Fig. 24.5). Immediately after drop deposit, the cationic

$$Vs = \frac{2}{9}\frac{(\rho_p - \rho_f)}{\mu} g R^2$$

Fig. 24.4 Stokes law equation where v_s is the particle creaming velocity, g is the gravitational acceleration, ρ_p is the mass density of the particles, ρ_f is the mass density of the fluid and μ is the viscosity of the continuous phase

Fig. 24.5 Dynamic contact angle measurements on rabbit eyes confirm optimal and immediate spreading of cationic emulsions compared to anionic emulsions and hyaluronic acid-based product (adapted from Lallemand et al. [30])

emulsion based on Novasorb® spread all over the cornea while other tested products remained with high contact angles.

Other physicochemical parameters are important for the development and use of Novasorb®. It should be noted that to avoid reflex tearing and blinking the emulsions should be compatible with biological parameters (pH and osmolality). pH is adjusted to physiological pH (about 7–7.2). Osmolality should be adjusted to avoid osmotic stress to the epithelial cells. Neutral molecules should be used to avoid charge masking and emulsion destabilization such as mannitol, sorbitol or, glycerol. Glycerol is favored because this compound possesses an intrinsic beneficial demulcent property on the eye surface that is particularly useful in dry eye disease (DED) [32].

Finally, several lipophilic active ingredients were added to the cationic emulsion either to address unmet medical needs or to improve existing products: cyclosporin A (CsA), latanoprost, antihistaminics, anti-inflammatories, antibiotics, and antifungals. For example, latanoprost, the active ingredient of the antiglaucoma blockbuster product Xalatan®, was included in the emulsion providing new properties. Latanoprost is an unstable molecule in presence of water due to ester hydrolysis. When encapsulated in oil, the concerned ester function is masked in the oily phase, thus avoiding hydrolysis and providing an advantage over Xalatan®. In addition, as discussed below, the beneficial effect of Novasorb® on the ocular surface should be beneficial to patients with combined glaucoma and ocular surface disease.

Another example is CsA. This molecule has always been a huge challenge to formulate in an adequate and efficient product [33] due to its high lipophilicity. Novasorb® seems the most appropriate technology to administer this molecule to patients.

24.3 Nonclinical Evaluation

24.3.1 Pharmacokinetics and Efficacy of Cationic Emulsions

The ocular delivery of Novasorb® cationic emulsions loaded CsA and latanoprost was evaluated in rabbits. The first indications that cationic emulsions were effective ocular drug-delivery vehicles with prolonged precorneal residence time come from the work of Benita et al., who used delta 8-tetrahydrocannabinol and pilocarpine as model drugs [34, 35]. These results were confirmed later with CsA cationic emulsions, which have an increased ocular absorption, especially in the conjunctiva and cornea, when compared to anionic emulsions [20, 36]. However, these first cationic emulsions all used noncompendial excipients, or primary amines, such as stearylamine and oleylamine as the cationic agent, which are not devoid of toxicity. Based upon these results, the Novasorb® technology was developed to create cationic emulsions of CsA and latanoprost with improved ocular tolerance safety profiles. CsA cationic emulsions were compared to Restasis®, a commercially available unpreserved BAK-free anionic emulsion of 0.05 % CsA, in single- and multiple-dose PK studies [37]. The corneal absorption of CsA following a single instillation of a 0.05 % CsA cationic emulsion was approximately twice that observed with Restasis® (Fig. 24.6a), and a 0.025 % CsA cationic emulsion was as effective as Restasis® at delivering CsA to the cornea. The area under the curve (AUC) in the cornea following a single instillation was 14,210, 14,476, and 26,476 ng h/g for Restasis®, 0.025 % CsA cationic emulsion, and 0.05 % CsA cationic emulsion, respectively (Fig. 24.6b). Interestingly, a second peak is observed 4 h post instillation with the 0.05 % CsA cationic emulsion, which is not present with the Restasis® anionic emulsion. This suggests that the cationic emulsion does indeed possess a prolonged precorneal residence time. With the 0.1 % CsA cationic emulsion, this second peak was observed 12 h post instillation [37], confirming that for such a peak to be present, a prolonged residence time in the precorneal space was necessary. This prolonged residence time can be explained by the presence of the positive charge on the oil droplets of the emulsion, which can interact with the negatively charged corneal epithelium. In the multiple-dose PK studies, bis in die (BID) (twice daily) instillations for 7 days of the 0.05 % CsA cationic emulsion or once daily instillation for 7 days of the 0.1 % CsA cationic emulsion were not accompanied by an increased systemic absorption of CsA.

Confirmation of the good absorption following instillations of the CsA cationic emulsions was obtained with a 0.005 % latanoprost cationic emulsion. In a monkey model of laser-induced elevated intraocular pressure (IOP), the 0.005 % latanoprost cationic emulsion was as effective as Xalatan® (0.005 % latanoprost) at reducing elevated IOP [38], suggesting an equivalence in the latanoprost-delivered dose between the two 0.005 % latanoprost formulations. Xalatan® contains 0.02 % BAK, a quaternary ammonium that at this high concentration acts both as a preservative for the eye drop solution and a permeation enhancer for latanoprost. For example, Allergan increased the BAK concentration, from 0.005 to 0.02 %, while decreasing the concentration of bimatoprost (a prostaglandin analog, PGA) from 0.03 to 0.01 % in order for the new Lumigan® 0.01 % bimatoprost eye drop solution to have the same efficacy as the original 0.03 % bimatoprost formulation. In this regard, increasing the BAK concentration potentiated the absorption of bimatoprost and helped in maintaining its efficacy. This formulation change was motivated by the side effect of the PGA class: hyperemia, i.e., conjunctival redness, which results from PGA-induced conjunctival vein vasodilation. Hyperemia is the major cause of antiglaucoma PGA treatment cessation and poor compliance.

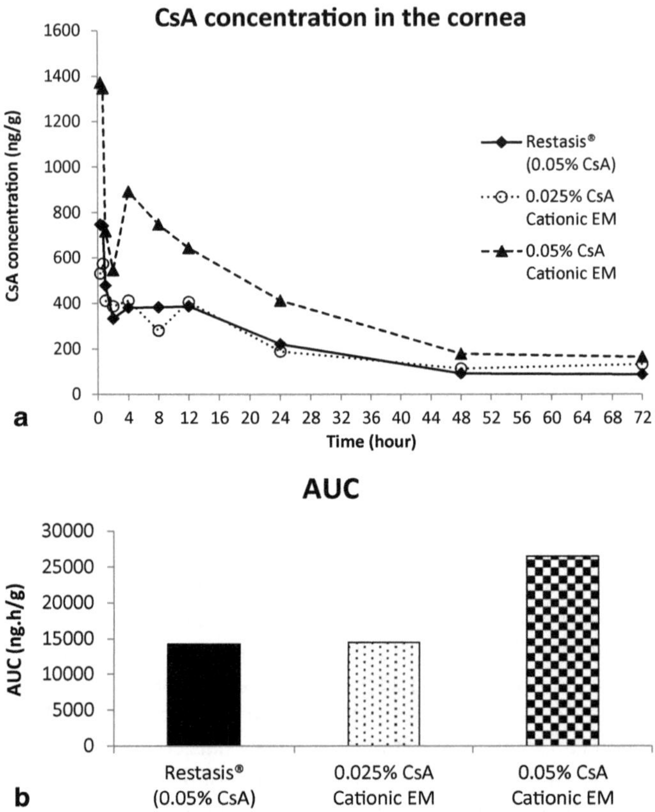

Fig. 24.6 **a** CsA concentrations in the cornea following a single instillation in the rabbit eye of Restasis®, 0.025 % CsA, and 0.05 % CsA cationic emulsion. **b** CsA area under the curve (*AUC*) in the cornea of rabbits instilled with one drop of the various CsA-containing eye drop preparations. (Adapted from references [30] and [37], with permission from Lippincott Williams & Wilkins, Inc., Copyright 2013)

It is interesting to note that the 0.005 % latanoprost cationic emulsion, which is as effective as Xalatan® at reducing elevated IOP, has an incidence of hyperemia that is twice as low following repeated instillations over 28 days in the rabbit [38]. This can be explained by a lower latanoprost free acid conjunctival AUC; 197 vs. 121 pg h/mg, or peak concentration (C_{max}); 90 ± 34 vs. 64 ± 44 pg/mg for Xalatan® and the 0.005 % latanoprost cationic emulsion, respectively [38]. Four hours after instillations with either Xalatan® or the 0.005 % latanoprost cationic emulsion, latanoprost free acid concentrations were equivalent in the different ocular tissues, even though the C_{max} was higher with Xalatan®. This suggests that the elimination rates were also different with the two 0.005 % latanoprost formulations. Indeed, the improved residence time of the 0.005 % latanoprost cationic emulsion accounts for this slower apparent elimination rate because absorption and elimination take place concomitantly for a longer period of time following instillation of the 0.005 % latanoprost cationic emulsion.

24.3.2 Safety and Ocular Tolerance of Cationic Emulsions

Cationic emulsions are a new type of ocular drug-delivery vehicles. They comprise an oil phase constituted of 200nm oil droplets loaded with lipophilic drugs (e.g., CsA, latanoprost, etc.), stabilized by surfactants and harboring a positive charge at their surface stemming from a lipophilic cationic agent and an aqueous phase that accounts for 90–95 % of the cationic emulsion. As with all new vehicles or devices, a thorough evaluation of the cationic emulsions was undertaken to evaluate its toxicity. Classic 28-day local tolerance studies in the rabbit demonstrated that cationic emulsions, with or without any loaded drug, were very well tolerated. No signs of ocular irritation or histology findings were recorded following as many as six daily instillations with the 0.1 % CsA cationic emulsion [30, 37]. Twice daily instillations of the 0.005 % latanoprost cationic emulsion resulted in an incidence of hyperemia approximately twice as low as BID instillations of Xalatan® [38].

In an acute in vivo toxicity rabbit model of repeated instillations (15 instillations over 75 min, one instillation every 5 min), the integrity of the cornea was evaluated through in vivo confocal microscopy (IVCM) 4 h and 1 day after the last instillation. Empty cationic emulsions were demonstrated to be very well tolerated by the ocular surface [39], even though they used BAK (0.02 %) or cetalkonium chloride (CKC, 0.002 %) as the cationic agent. Cationic emulsions of CsA (0.05 %) and latanoprost (0.005 %) were demonstrated to be very well tolerated as well. Both cationic emulsions have low IVCM scores, close to those observed following repeated instillations of PBS, and performed much better than their references, i.e., Restasis® and Xalatan® (Fig. 24.7a) [40, 41]. Interestingly, the repeated instillations of the CsA cationic emulsion resulted in a lower inflammatory cell count in the conjunctival-associated lymphoid tissue (CALT) when compared to Restasis®, a BAK-free unpreserved anionic emulsion known to be very well tolerated by the ocular surface (Fig. 24.7b).

Moreover, when the cationic emulsions were applied onto debrided rat corneas, i.e., on diseased corneas, BID instillations for 5 days of the 0.005 % latanoprost cationic emulsions resulted in safe and almost scar-free healing of the corneal epithelium [42]. In addition, treatment with the 0.005 % latanoprost cationic emulsion was also accompanied by a reduced number of infiltrated inflammatory cells in the cornea. The same observations were made with empty cationic emulsions (Cationorm®), whose BID instillations over 5 days were able to manage the corneal inflammation of the debrided cornea better than classic eye drops [43].

Altogether these data confirm that the Novasorb® cationic emulsions; Cationorm® (empty vehicle) as well as the CsA and latanoprost-loaded cationic emulsions, are very well tolerated by the ocular surface in animal models. This good safety

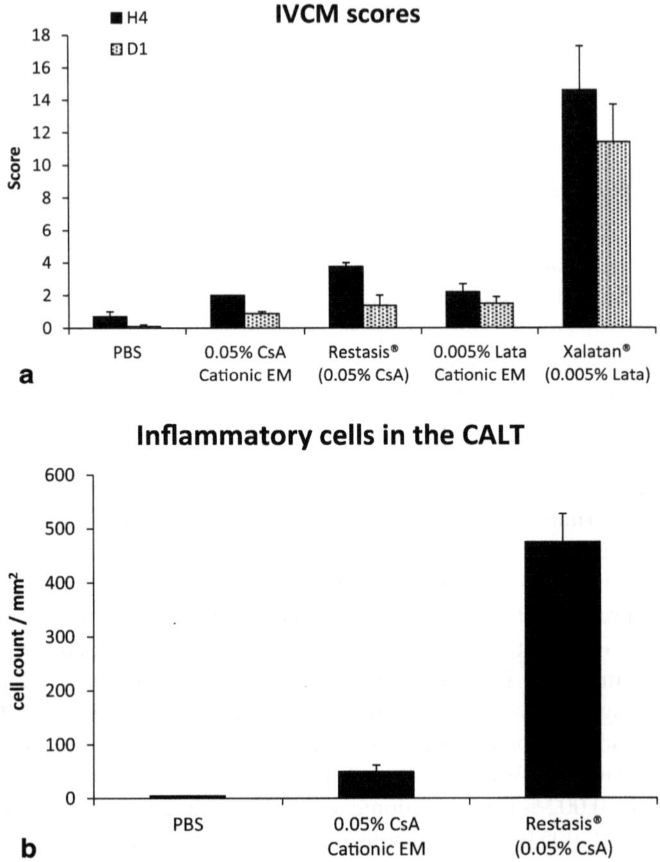

Fig. 24.7 **a** In vivo confocal microscopy (*IVCM*) scores of the ocular surface of rabbits receiving 15 instillations of the various test items over 75 min, (one instillation every 5 min) 4 h (*H4*), and 1 day (*D1*) after the last instillation of CsA and latanoprost-loaded cationic emulsions compared to Restasis® and Xalatan®, respectively. PBS served as negative control. (Adapted from [40] and [41].) **b** Inflammatory cell count in the conjunctiva-associated lymphoid tissue (*CALT*) of rabbits treated with Restasis® and 0.05 % CsA cationic emulsion. (Adapted from [41])

profile of the cationic emulsions is corroborated by the numerous clinical trials conducted by Novagali alongside the development of its pipeline of ocular surface and antiglaucoma therapies.

24.3.3 Nonclinical Discussion

It was clearly demonstrated that cationic emulsion could enhance corneal penetration of lipophilic molecules compared to anionic emulsions. Nevertheless, the exact mechanism underlying these observations remained to be explored to precisely determine the fate of the oil nanodroplets in the precorneal space and their interaction with the eye mucosa and mucins. Based on the current data, if the oil nanodroplets are immediately disrupted in contact with tears, the oil and the solubilized active principles, such as CsA, will float above the aqueous layer of the tear film as a consequence of the lower oil density. In this case, reduction to no contact with the cornea epithelial cells is likely to occur, resulting in poor CsA penetration. This is contradicted by the improved CsA PK profile and the excellent physical stability of the emulsion.

On the contrary, if the oil nanodroplets remain intact and are attracted through electrostatic interactions via their cationic charge, they spread on the mucin layer and remain in close contact with the corneal epithelial cells. At this point, in contrast to free drugs, which can cross the cell membrane by active or passive diffusion, nanocarriers, generally have combined mechanisms of penetration. In this configuration, the nanocarrier can either release the active ingredient outside the cells, which will then diffuse into the cell membranes, a lipophilic environment, or be directly taken up as a whole by the targeted cells. In the latter case, depending on their size, the nanoparticles may enter the cells through phagocytosis (as in specialized immune cells), pinocytosis or, endocytosis via the multiple subtypes of clathrin-dependent and clathrin-independent pathways. Thus far, several coexisting mechanisms have been described [9] in epithelial cells, which may be responsible for active ingredient uptake.

24.4 Clinical Evidence of the Benefits of Cationic Emulsions for the Treatment of Eye Diseases

A wide clinical program was designed in close interaction with health agencies (EMA and FDA) to demonstrate the tolerance, safety, and efficacy of the Novasorb® technology. A first-in-man clinical trial conducted in 2003 with the drug-free cationic emulsion vehicle confirmed the tolerance and safety of the cationic emulsions and paved the way for the clinical development of the cationic emulsions as a new type of ocular drug delivery vehicle for lipophilic drugs. Table 24.1 lists the different clinical trials conducted with the Novasorb® technology-derived cationic emulsions.

24.4.1 Clinical Evaluation of Cationorm®

The prolonged residence time of Novasorb® on the ocular surface due to the electrostatic attraction between the lipid nanodroplets and the ocular surface led to the assumption that cationic emulsions could be beneficial for the ocular surface even in the absence of active ingredient. A total of five trials were conducted with the cationic emulsion vehicle that led to the commercialization of Cationorm®. In the

Table 24.1 Clinical trial conducted with Novasorb. (Updated table from Lallemand et al. [30])

Year	Phase type	Product	Objectives	Indication	Patients
2003	Phase I	Vehicle #1	Tolerance and safety	None	16
2004	Phase II		Tolerance and safety, Exploratory efficacy	Dry eye	50
2007	Phase II	Cationorm® (Vehicle #2)	Efficacy, tolerance and safety vs. Refresh®	Dry eye	79
2008	Phase II		Efficacy, tolerance and safety vs. Optive® or Emustil®	Dry eye	71
2011	Phase IIIb		Efficacy, tolerance and Safety vs. Vismed®	Dry eye	81
2005	Phase IIa	CsA emulsion	Tolerance and safety; exploratory efficacy	Dry eye disease	48
2008	Phase IIb		Exploratory efficacy, tolerance and safety	Dry eye disease	132
2007	Phase III (Sic-canove)		Efficacy, tolerance and safety	Dry eye disease	496
2011	Phase III (Sansika)		Efficacy, tolerance and safety	Dry eye disease	246
2006	Phase IIb/III	Vekacia®	Efficacy, tolerance and safety	Active VKC	118
2009	Phase IIb		Efficacy, tolerance and safety	Nonactive VKC	34
2011	Phase II	Catioprost®	Exploratory efficacy, tolerance and safety	Glaucoma	22
2011	Phase IIb		Exploratory efficacy, tolerance and safety vs. Travatan Z®	Glaucoma and ocular surface disease	105

VKC vernal keratoconjunctivitis

three trials conducted in 2005, 2007 [44], and 2011, Cationorm® was compared to referenced products such as Refresh Tears® (carboxymethylcellulose sodium solution), Optive® (carboxymethylcellulose and glycerol solution), Emustil® (anionic emulsion), and Vismed® (hyaluronic acid solution) to assess the ocular tolerance (safety) and efficacy in patients with signs and symptoms of mild to moderate DED. While the local tolerance was globally good with all the products, DED symptoms improved significantly better with Cationorm®. Cationorm® was then launched in the market in 2008 as an artificial tear for the relief of patients with mild to moderate signs and symptoms of dry eye.

24.4.2 Clinical Evaluation of Cyclosporine A Emulsion

Cationic emulsion loaded with CsA was developed concomitantly for severe DED and vernal keratoconjunctivitis. Only the DED results will be presented here. DED is defined as a multifactorial disease of the tears and ocular surface that results in symptoms of discomfort, visual disturbance, and tear film instability with potential damage to the ocular surface (Dry Eye Workshop report 2007 [DEWS]). The disease is accompanied by an increased osmolality of the tear film and inflammation of the ocular surface, potentially necessitating an anti-inflammatory treatment for patients not responding to artificial tears. CsA is widely used in this indication in the USA (Restasis®, Allergan), whereas in the EU no treatment has been approved yet. Two Phase III clinical trials were conducted by Novagali Pharma in 2009 and 2011, the Siccanove and Sansika studies, respectively. The Siccanove study was a 6-month Phase III, multicenter, randomized, controlled, double-masked trial of cationic emulsion at 0.1 % CsA administered once daily vs. its emulsion vehicle in 492 patients with moderate to severe DED. Patients treated with CsA showed a statistically significant improvement in corneal fluorescence staining grade compared to the vehicle and continued to improve until month 6. The benefit of treatment with CsA emulsion was greater in patients with the most severe keratitis at baseline [45].

24.4.3 Catioprost® Latanoprost Emulsion Evaluation

Glaucoma is a leading cause of blindness among elderly individuals and it is estimated that up to 60 % of glaucoma patients have ocular surface disease. Topical IOP-lowering therapy with a PGA is the first-line treatment for patients with glaucoma and ocular hypertension. Approximately three out of four topical ophthalmic preparations, including IOP-lowering drops, contain BAK as a preservative agent. BAK is known to cause ocular surface damage by disrupting the tear film and increasing conjunctival inflammation [28]. The coexistence of ocular surface disease in glaucoma patients has a negative impact on quality of life and reduces the compliance to IOP-lowering therapies in these patients, resulting in the risk of glaucoma progression. Therefore, a change from or a reduction in the use of preserved ocular hypotensive medications could prevent or ameliorate the signs and symptoms of OSD in patients with glaucoma or ocular hypertension. As such, Novagali developed Catioprost® to address the unmet medical need, an unpreserved IOP-lowering therapy that also treats the signs and symptoms of ocular surface disease at the same time.

Two proof-of-concept clinical trials evaluating the effect of Catioprost® in patients with glaucoma or ocular hypertension with signs and/or symptoms of ocular surface disease were conducted. The first study was a multicenter, Phase II investigator-masked, randomized study evaluating the safety and efficacy of Catioprost® compared to Travatan Z® (Alcon), a commercially marketed PGA that uses the "soft preservative" Sofzia® system in place of BAK, in subjects with glaucoma or ocular hypertension and ocular surface disease (Fig. 24.8). The second study was an open label, multicenter Phase II pilot study to assess the efficacy and safety of Catioprost® in patients with ocular hypertension or open-angle glaucoma with ocular surface disease initially treated with Xalatan®. The aim of these 3-month studies was to determine if the positive effect of Catioprost® on reducing elevated IOP and on improving and protecting previously injured ocular surface suggested by preclinical studies can be demonstrated in patients. The results showed that while the IOP-lowering effect of Catioprost® in patients was similar to that of Xalatan® and

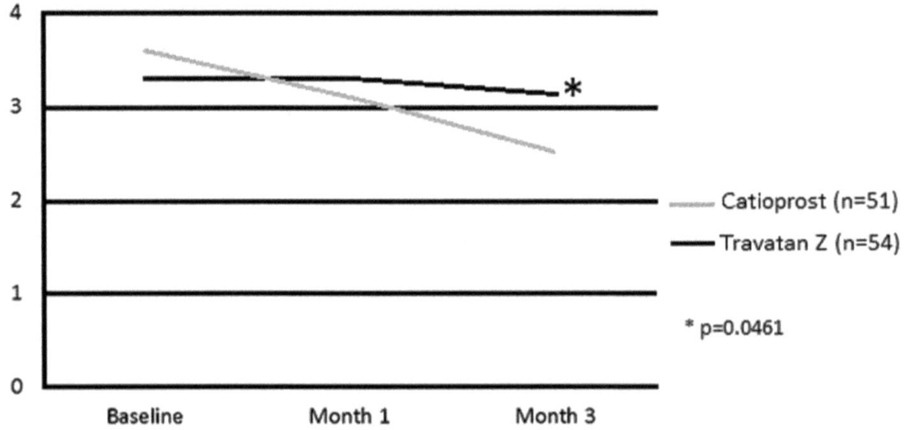

Fig. 24.8 Corneal fluorescein staining (ORA Scale—corneal sum) of Catioprost® vs. Travatan Z® administered once daily for 3 months

Travatan Z®, the effect on the ocular surface was enhanced with Catioprost® due to the absence of preservative and the inherent effect of the cationic emulsion on the ocular surface signs and symptoms as evidenced in patients treated with Cationorm®. For example, Fig. 24.6 shows a statistically significant improvement of the ocular surface damage at month 3 with 30 % vs. 5 % reduction in corneal fluorescence staining compared to baseline (using the ORA Scale) for Catioprost® and Travatan Z®, respectively. It is therefore clear that in addition to providing at least an equivalent efficacy, Catioprost® has a beneficial effect on the cornea that was not observed with the soft preserved Travatan Z®.

24.5 Conclusions and Future Perspectives

In the case of other body mucosae, the goal is to cross the very thick mucus layer by disrupting sulfur bridges of mucins to reach epithelial cells. In ophthalmology, the strategy is reversed. It is of great importance not to disturb the tears' equilibrium and mucus layer to avoid an outbreak of all reflex protections, which would decrease the efficacy of the treatment. The key of Novasorb® was to use the specific biological properties of mucin barriers (negative charge) to obtain enhanced efficacy of ocular treatment without deleterious effects on the eye. Based on this fact, careful attention was paid to ocular tolerance of the formulation and preservation of eye surface integrity during the development of Novasorb®. About 10 years of development were necessary to prove the suitability of Novasorb® in patient treatment. From early-stage formulation work to animal studies and human clinical trials, the same strategy was adopted to bring to patients with unmet medical needs an innovative, high-quality and efficient pharmaceutical product. As of today, more than 1,000 patients have been treated with Novasorb® technology-derived products within clinical trials and thousands of people over the world use Cationorm® on a daily basis for the relief of dry eye symptoms with no signs of local intolerance, thus confirming its place in the ophthalmologist's armamentarium.

Disclosure All authors have a direct financial relation with the company Santen SAS and the products described herein.

References

1. Rucker JC. Normal and abnormal lid function. Handb Clin Neurol. 2011;102:403–24.
2. Stahl U, Willcox M, Stapleton F. Osmolality and tear film dynamics. Clin Exp Optom. 2012;95(1):3–11.
3. Royle L, Matthews E, Corfield A, Berry M, Rudd PM, Dwek RA, Carrington SD. Glycan structures of ocular surface mucins in man, rabbit and dog display species differences. Glycoconj J. 2008;25(8):763–73.
4. Mantelli F, Argueso P. Functions of ocular surface mucins in health and disease. Curr Opin Allergy Clin Immunol. 2008;8(5):477–83.
5. Argueso P. Glycobiology of the ocular surface: mucins and lectins. Jpn J Ophthalmol. 2013;57(2):150–5.
6. Govindarajan B, Gipson IK. Membrane-tethered mucins have multiple functions on the ocular surface. Exp Eye Res. 2010;90(6):655–63.
7. Gan L, Wang J, Jiang M, Bartlett H, Ouyang D, Eperjesi F, Liu J, Gan Y. Recent advances in topical ophthalmic drug delivery with lipid-based nanocarriers. Drug Discov Today. 2013; 18(5–6):290–7.
8. Ludwig A. The use of mucoadhesive polymers in ocular drug delivery. Adv Drug Deliv Rev. 2005;57(11):1595–639.
9. Mannermaa E, Vellonen KS, Urtti A. Drug transport in corneal epithelium and blood-retina barrier: emerging role of transporters in ocular pharmacokinetics. Adv Drug Deliv Rev. 2006;58(11):1136–63.
10. Hamalainen KM, Kananen K, Auriola S, Kontturi K, Urtti A. Characterization of paracellular and aqueous penetration routes in cornea, conjunctiva, and sclera. Invest Ophthalmol Vis Sci. 1997;38(3):627–34.
11. Edward A, Prausnitz MR. Predicted permeability of the cornea to topical drugs. Pharm Res. 2001;18(11):1497–508.
12. Belmonte C, Gallar J. Cold thermoreceptors, unexpected players in tear production and ocular dryness sensations. Invest Ophthalmol Vis Sci. 2011;52(6):3888–92.
13. Keister JC, Cooper ER, Missel PJ, Lang JC, Hager DF. Limits on optimizing ocular drug delivery. J Pharm Sci. 1991;80(1):50–3.
14. Kaur IP, Smitha R. Penetration enhancers and ocular bioadhesives: two new avenues for ophthalmic drug delivery. Drug Dev Ind Pharm. 2002;28(4):353–69.
15. Tang-Liu DD, Richman JB, Weinkam RJ, Takruri H. Effects of four penetration enhancers on corneal permeability of drugs in vitro. J Pharm Sci. 1994;83(1):85–90.
16. Chetoni P, Burgalassi S, Monti D, Saettone MF. Ocular toxicity of some corneal penetration enhancers evaluated by electrophysiology measurements on isolated rabbit corneas. Toxicol In Vitro. 2003;17(4):497–504.
17. Calvo P, Alonso MJ, Vila-Jato JL, Robinson JR. Improved ocular bioavailability of indomethacin by novel ocular drug carriers. J Pharm Pharmacol. 1996;48(11):1147–52.

18. Diebold Y, Enriquez de Salamanca A, Jarrin M, Callejo S, Vila A, Alonso MJ. Uptake of bioadhesive nanoparticles in a normal human conjunctiva (NHC) epithelial cell line. ARVO Meet Abstr 2003;44:3785.
19. Vandamme TF. Microemulsions as ocular drug delivery systems: recent developments and future challenges. Prog Retin Eye Res. 2002;21(1):15–34.
20. Tamilvanan S, Benita S. The potential of lipid emulsion for ocular delivery of lipophilic drugs. Eur J Pharm Biopharm. 2004;58(2):357–68.
21. Zhang HH, Luo QH, Yang ZJ, Pan WS, Nie SF. Novel ophthalmic timolol meleate liposomal-hydrogel and its improved local glaucomatous therapeutic effect in vivo. Drug Deliv. 2011;18(7):502–10.
22. Grammer JB, Kortum FA, Wolburg H, Ludtke R, Schmidt KH, Thiel HJ, Pleyer U. Impregnation of collagen corneal shields with liposomes: uptake and release of hydrophilic and lipophilic marker substances. Curr Eye Res. 1996;15(8):815–23.
23. Sandri G, Bonferoni MC, Gokce EH, Ferrari F, Rossi S, Patrini M, Caramella C. Chitosan-associated SLN: in vitro and ex vivo characterization of cyclosporine A loaded ophthalmic systems. J Microencapsul. 2010;27(8):735–46.
24. Klang SH, Siganos CS, Benita S, Frucht-Pery J. Evaluation of a positively charged submicron emulsion of piroxicam on the rabbit corneum healing process following alkali burn. J Control Release. 1999;57(1):19–27.
25. Benita S. Prevention of topical and ocular oxidative stress by positively charged submicron emulsion. Biomed Pharmacother. 1999;53(4):193–206.
26. Rabinovich-Guilatt L, Couvreur P, Lambert G, Dubernet C. Cationic vectors in ocular drug delivery. J Drug Target. 2004;12(9–10):623–33.
27. Paimela T, Ryhanen T, Kauppinen A, Marttila L, Salminen A, Kaarniranta K. The preservative polyquaternium-1 increases cytoxicity and NF-kappaB linked inflammation in human corneal epithelial cells. Mol Vis. 2012;18:1189–96.
28. Baudouin C, Labbe A, Liang H, Pauly A, Brignole-Baudouin F. Preservatives in eyedrops: the good, the bad and the ugly. Prog Retin Eye Res. 2010;29(4):312–34.
29. Sznitowska M, Janicki S, Dabrowska EA, Gajewska M. Physicochemical screening of antimicrobial agents as potential preservatives for submicron emulsions. Eur J Pharm Sci. 2002;15(5):489–95.
30. Lallemand F, Daull P, Benita S, Buggage R, Garrigue JS. Successfully improving ocular drug delivery using the cationic nanoemulsion Novasorb. J Drug Deliv. 2012;2012:604204. doi:604210.601155/602012/604204.
31. Rabinovich-Guilatt L, Couvreur P, Lambert G, Goldstein D, Benita S, Dubernet C. Extensive surface studies help to analyse zeta potential data: the case of cationic emulsions. Chem Phys Lipids. 2004;131(1):1–13.
32. Gensheimer WG, Kleinman DM, Gonzalez MO, Sobti D, Cooper ER, Smits G, Loxley A, Mitchnick M, Aquavella JV. Novel formulation of glycerin 1 % artificial tears extends tear film break-up time compared with systane lubricant eye drops. J Ocul Pharmacol Ther. 2012;28(5):473–8.
33. Lallemand F, Felt-Baeyens O, Besseghir K, Behar-Cohen F, Gurny R. Cyclosporine A delivery to the eye: a pharmaceutical challenge. Eur J Pharm Biopharm. 2003;56(3):307–18.
34. Naveh N, Muchtar S, Benita S. Pilocarpine incorporated into a submicron emulsion vehicle causes an unexpectedly prolonged ocular hypotensive effect in rabbits. J Ocul Pharmacol. 1994;10(3):509–20.
35. Muchtar S, Almog S, Torracca MT, Saettone MF, Benita S. A submicron emulsion as ocular vehicle for delta-8-tetrahydrocannabinol: effect on intraocular pressure in rabbits. Ophthalmic Res. 1992;24(3):142–9.
36. Abdulrazik M, Tamilvanan S, Khoury K, Benita S. Ocular delivery of cyclosporin A. II. Effect of submicron emulsion's surface charge on ocular distribution of topical cyclosporin A. STP Pharma Sci. 2001;11(6):427–32.
37. Daull P, Lallemand F, Philips B, Lambert G, Buggage R, Garrigue JS. Distribution of cyclosporine A in ocular tissues after topical administration of cyclosporine A cationic emulsions to pigmented rabbits. Cornea. 2013;32(3):345–54.

38. Daull P, Buggage R, Lambert G, Faure MO, Serle J, Wang RF, Garrigue JS. A comparative study of a preservative-free Latanoprost cationic emulsion (Catioprost) and a BAK-preserved Latanoprost solution in animal models. J Ocul Pharmacol Ther. 2012;28(5):515–23.
39. Liang H, Brignole-Baudouin F, Rabinovich-Guilatt L, Mao Z, Riancho L, Faure MO, Warnet JM, Lambert G, Baudouin C. Reduction of quaternary ammonium-induced ocular surface toxicity by emulsions: an in vivo study in rabbits. Mol Vis. 2008;14:204–16.
40. Liang H, Baudouin C, Faure MO, Lambert G, Brignole-Baudouin F. Comparison of the ocular tolerability of a latanoprost cationic emulsion versus conventional formulations of prostaglandins: an in vivo toxicity assay. Mol Vis. 2009;15:1690–9.
41. Liang H, Baudouin C, Daull P, Garrigue JS, Brignole-Baudouin F. Ocular safety of cationic emulsion of cyclosporine in an in vitro corneal wound healing model and an acute in vivo rabbit model. Mol Vis. 2012;18:2195–204.
42. Liang H, Baudouin C, Daull P, Garrigue JS, Brignole-Baudouin F. In vitro and in vivo evaluation of a preservative-free cationic emulsion of latanoprost in corneal wound healing models. Cornea. 2012;31(11):1319–29.
43. Daull P, Feraille L, Elena PP, Baudouin C, Garrigue JS. Comparison of the anti-inflammatory effects of artificial tears in a rat model of corneal scraping. The 2012 European Association for Vision and Eye Research Conference (EVER), Nice.
44. Aragona P, Spinella R, Rania L, Postorino E, Roszkowska A, Versura P, Profazio V, Rolando M. Assessment of the efficacy of Cationorm® in patients with moderate dry eye compared with Optive® and Emustil® eye drops. Acta Ophthalmologica. 2011;89:S246.
45. Buggage RR, Amrane M, Ismail D, Lemp MA, Leonardi A, Baudouin C. The effect of Cyclokat® (unpreserved 0.1 % cyclosporine cationic emulsion) on corneal involvement in patients with moderate to severe dry eye disease participating in a phase III, multicenter, randomized, controlled, double-masked, clinical trial. Eur J Ophthalmol. doi:10.5301/EJO.2011.7544.

Part IV
Regulatory, Toxicological and Market Issues

Chapter 25
Regulatory Aspects and Approval of Biopharmaceuticals for Mucosal Delivery: Quality, Toxicology, and Clinical Aspects

Karen Brigitta Goetz, Yuansheng Sun, Katrin Féchir, Evelyne Kretzschmar and Isabel Buettel

Abbreviations

BA	Bioavailability
BE	Bioequivalence
EMA	European Medical Agency
ERA	Environmental risk assessment
FDA/CBER	Food and Drug Agency/Center for Biologics Evaluation and Research
FBS	Fetal bovine serum
GMO	Genetically modified organism
ICH	International Conference of Harmonization
IMP	Investigational medicinal product
MABEL	Minimal anticipated biological effect level
NOAEL	No observed adverse effect level
PCR	Polymerase Chain Reaction
PD	Pharmacodynamics
PDCO	Pediatric Committee
Ph. Eur.	European Pharmacopoeia
PIP	Pediatric investigation plan
PK	Pharmacokinetics
USP	United States Pharmacopoeia

25.1 Introduction

The past decades have seen increasing interest in mucosal delivery of vaccines and other biopharmaceuticals [1]. Beyond traditional polio oral vaccines, several registered vaccines, including inactivated influenza vaccines and live measles-containing

K. B. Goetz (✉) · Y. Sun · K. Féchir · E. Kretzschmar · I. Buettel
Paul-Ehrlich-Institut, Federal Institute for Vaccines and Biomedicines,
Paul-Ehrlich-Straße 51–59, 63225 Langen, Germany
e-mail: goeka@pei.de

J. das Neves, B. Sarmento (eds.), *Mucosal Delivery of Biopharmaceuticals,* DOI 10.1007/978-1-4614-9524-6_25,

vaccines, have been explored for needle-free delivery, e.g., via an intranasal or aerosol route. In addition, there have been a number of vaccines under development for mucosal delivery, such as DNA and live attenuated or recombinant vectored vaccines against HIV and other challenging diseases [2]. Protein and naked DNA vaccines for mucosal delivery often require novel formulations including mucosal adjuvants, and use of delivery systems, such as biodegradable microparticles, liposomes and bioadhesive polymers, in order to enhance their efficacy. Similarly, several human peptide/protein hormones and growth factors produced by recombinant DNA technology have been registered for aerosol or intranasal use, and the number of such products may increase with improvement of the delivery system. However, the development and evaluation of mucosally delivered biopharmaceuticals present regulatory challenges as criteria to evaluate their safety profile may not exist. In fact, the safety of intranasally delivered vaccines is largely unknown, and the same holds for the long-term safety profile of aerosolized biologicals based on human peptide/protein hormones and growth factors. Manufacturers of biopharmaceuticals often have questions about the type of information and extent of data that would be required to support proceeding to clinical studies with mucosal formulations. Existing regulatory guidelines issued by International Conference of Harmonization (ICH) [3], WHO [4, 5], EU [6–10], and US/FDA [11–13] provide valuable guidance, however, they are general in nature and frequently lack sufficient details for recommendations on specific testing programs. Given the importance and the complexity of the issues, clear guidance on nonclinical and clinical programs required for a particular mucosal biopharmaceutical would help the manufacturers proceed more efficiently on their path towards development of such products.

In this chapter, we discuss regulatory requirements concerning the quality of future mucosal products, and review the requirements for nonclinical testing and types of toxicology studies with mucosal biological products and formulations, as well as regulatory expectations from the clinical development perspective. Published regulatory guidelines and recommendations from ICH, WHO, European Medical Agency (EMA), and US/FDA have been taken into account, and the key biological/immunological and technical aspects of mucosal formulations and delivery systems are also considered. The mucosal routes include oral, nasal, pulmonary, rectal and vaginal mucosa, which are all considered as potential sites for immunization.

25.2 Quality

25.2.1 General Considerations

There is a wide diversity of different types of mucosal applications of pharmaceutical products, including oral, buccal, ocular, vaginal, pulmonary, and nasal delivery. Drugs applied to the nasal mucosa or to the lungs may exert either a local or a systemic effect. This diversity accounts for, to some extent, different quality requirements depending on the particular drug product, the delivery route, and the delivery device.

For example, in ocular delivery, the sterility of the pharmaceutical product is vitally important, while for nasal delivery sterility may not be essential.

Mucosal delivery of biopharmaceuticals can offer some advantages over the classical invasive delivery routes. For example, administration of vaccinations using a noninvasive application form, such as nasal delivery, is of great interest due to the possibility of easy and needle-free application, fewer problems with contamination, cold-chain independence and reduction of the need for trained personnel to administer the vaccination. While oral administration generally remains the most popular route for drug administration, delivery of biopharmaceutical products across the nasal mucosa is becoming more and more important as a therapeutically relevant route for both localized and systemic drug delivery. Some reasons for this trend are the easy accessibility of the mucosal tissue, the potential for rapid systemic drug absorption, and the enhanced bioavailability (BA) for substances that are presystemically metabolized after gastrointestinal absorption [14, 15]. In comparison to injected formulations, nasal application may also improve patient compliance, a factor not to be underestimated, especially for chronic therapies. However, there are also limitations to nasal drug application. Due to the anatomical properties of the nasal cavity, drug administration is restricted to a relatively low volume (100–150 μl). Furthermore, the absorption of larger biomolecules with high molecular weight and a polar structure may be reduced in this delivery method. However, improved absorption may be obtained by application of a nasal absorption promoter system [16]. During pharmaceutical development, these and other factors influencing nasal drug absorption should be carefully taken into account. A thorough characterization of the drug product as well as sufficient generation of nonclinical and clinical data is essential in this context.

Most of the European guidance documents relevant to mucosal delivery of pharmaceutical products deal with nasal and inhalation delivery. This is mainly due to the long-term usage of nasal and inhalation formulations for the treatment of localized diseases and for the development of pulmonary-applied small molecular weight drugs for systemic delivery. However, some biopharmaceuticals for nasal application and inhalation have already been granted marketing authorization in Europe. Examples include the nasal influenza vaccine Fluenz® and the inhalation product Pulmozyme® (Dornasealfa) for the treatment of cystic fibrosis. Further biomolecules for nasal and pulmonary application are under development. Therefore, the main focus of the following sections will be on quality requirements for nasal and inhalation drug formulations. General quality considerations applying to all biopharmaceutical drugs will also be addressed. However, it should be kept in mind that due to their variety and individual characteristics, the regulatory quality requirements for biopharmaceuticals may be at least partly product specific and must finally always be determined on a case-by-case basis.

25.2.2 Characterization

Generally, a comprehensive characterization of the biological pharmaceutical product constitutes a prerequisite to ensure product quality and consistency and forms the basis for establishment of drug substance and drug product specifications. A product characterization should be performed during pharmaceutical development as well as after significant process changes. The characterization program includes the determination of physicochemical properties, biological activity, purity and impurities. For the definition of physicochemical properties, the composition, physical properties and primary and higher order structures of the biological substances should be analyzed. Methods used for this purpose include electrophoresis, chromatography, or spectroscopy profiles. Some biological compounds may not have a uniform structure due to posttranslational modifications, e.g., glycosylation. In these cases, the manufacturer should define the pattern of heterogeneity and show consistency of the compounds between different lots. The structure of proteins and large biomolecules is often closely linked to its biological function and activity. Therefore, for biologically active compounds such as vaccines that are combined with other pharmaceutical ingredients or adjuvants, the structural and functional integrity of the antigen should be controlled during manufacture and storage. This is typically achieved using different spectroscopy methods [17]. Biological activity or potency of a product can be measured by animal- or cell culture-based biological assays as well as by biochemical assays. Product-related impurities such as degradation products or molecular variants with different properties and process-related impurities such as host cell proteins or growth media must be identified and addressed quantitatively. As determination of absolute purity is difficult and highly method dependent, the use of a combination of different analytical methods to identify impurities should be considered. Comparison to an appropriate reference standard, which may be an in-house established reference material, should be performed to the extent possible. The pharmaceutical properties of the drug product should also be thoroughly investigated, particularly for device-assisted applications.

For biopharmaceuticals, where the process defines the product, the quality of the final product is highly dependent upon the quality of the starting material. Therefore, both raw material testing and in-process controls are important tools to guarantee adequate product quality. Critical process parameters should be identified for each step of the manufacturing process and suitable in-process control limits should be established to monitor ongoing production. Performance of adequate in-process controls may also justify the reduction of drug substance and drug product testing. Finally, for all biotechnology products it should be clear that the pharmaceutical quality system has to be in compliance with cGMP standards and the production process should be well defined and sufficiently validated [18].

Specifications In contrast to a full characterization, specifications are intended to confirm the quality of the drug substance and the drug product by focusing on molecular and biological characteristics considered to be suitable to guarantee safety and efficacy of the product [18]. For each specification test, adequate acceptance criteria must be defined. The manufacturer should justify the choice of specifications, taking into account that they are linked to the manufacturing process. Furthermore, it is desirable that the lots used for establishment of specifications are also used in preclinical and clinical studies. Degradation of drug substance and drug product during storage should also be considered. The set of specifications is usually product specific. Drug substance specifications generally include appearance and description, identity testing, purity and impurities, potency, and the quantification of the active substance. If possible, potency testing of drug substance and drug product should be performed with different methods. Pharmacopoeial tests should be used where applicable. Drug product specifications should include identity, purity, and potency testing. Ideally, the potency test should be linked to the biological activity of the product. If drug product impurities are identical to those of drug substance, repeated testing may be omitted. However, impurities that appear during the drug production process and storage also have to be considered. Further testing of unique dosage forms may be needed, depending on the particular drug product. Analytical methods performed in specification assays should be fully validated and their suitability should be shown [18]. Analytical tests prescribed in a Pharmacopoeia (Ph. Eur., USP) have to be taken into account. Sterility and endotoxin testing are mandatory for all products to be delivered via the ocular route and should also be performed according to an accepted pharmacopoeial test. Further pharmacopoeial monographs are available for microbial limits, volume in container, particulate matter, and uniformity of dosage. For lyophilized products, such as dry powder vaccines for mucosal application, testing of moisture content is recommended. The Ph. Eur. also provides different monographs for vaccines for human use. For novel products, pharmacopoeial monographs for other product groups also may be applicable. For example, the monograph for typhoid vaccine (live, oral, strain Ty21a) may be used as a relevant guide for oral gene therapy products using the *S. typhi* Ty21a strain as a vector.

If the manufacturing process of biological products such as vaccines or gene therapy products involves the use of cell substrates and raw materials of animal origin contamination of these materials with adventitious agents must be considered. Adventitious agents include viruses, bacteria, mycoplasma, or TSE agents. Testing for adventitious agents should be performed on raw materials, virus seeds, master and working cell banks, as well as on the final product. Test methods for foreign viruses in viral vaccines include in vivo and in vitro testing, cell culture safety tests, transmission electron microscopy, or biochemical techniques such as PCR [19]. If raw materials of animal origin (e.g., fetal bovine serum) are used in the production process, certificates of suitability for TSE safety must be provided according to the "Note for guidance on minimizing the risk of transmitting animal spongiform encephalopathy agents via human and veterinary medicinal products" (EMA/410/01 rev.3). FBS used in the manufacturing process should be virus inactivated (e.g., Gamma-irradiated).

Specific Considerations for Nasal and Inhalation Products Biopharmaceutical products for nasal administration or inhalation require particular tests to examine the specific drug formulation or delivery device. Characterization of the pharmaceutical properties is especially important for formulations that are applied using sprays or inhalers.

Several points should be considered during the pharmaceutical development of inhalation and nasal products. The physical characterization of the drug substance should involve assessment of solubility, size, shape, density, rugosity, charge, and crystallinity, since these parameters may influence the functionality of the final product. For all nasal and inhalation products in powder form, the drug substance and drug product specifications should include a particle size test. Currently, the most commonly used particle sizing technique is laser diffraction [17]. Acceptance criteria should be set to assure a consistent particle size distribution. For inhalation products the manufacturer should also assess the fine particle mass, preferably by using the minimum recommended dose. Further, it has to be demonstrated that the delivered dose is consistent over the lifetime of the container. At least ten different doses should be examined at the beginning, middle, and end of the container to be tested. Priming requirements should be defined to ensure that all doses to be applied meet the specification limits for delivered dose uniformity. Concerning the mean delivered dose, tolerance limits of $\pm$ 15 % are in principle regarded as acceptable, if compatible with the specific product. The drug delivery rate and the total drug delivered should be determined for the batches to be used in clinical studies. For example, this may be achieved through the use of a breath simulator. For metered dose inhalers, it should be shown that the container minimum fill complies with the intended numbers of actuations. For plastic or rubber container closure components that are not described in a Pharmacopoeia and that are in contact with the drug substance, an extractable profile should be determined with different solvents. Leachables should be identified for all plastic parts and a safety assessment should be performed [20].

For all products that have to be shaken before use, shaking instructions should be provided. Foam formation due to excessive shaking should also be considered and examined by testing of delivered dose uniformity. Furthermore, product performance at different temperatures, including low temperatures (below 0 °C) and temperature cycling, as well as cleaning requirements should be tested to establish respective user instructions. Finally, the robustness of the product should be verified by tests which simulate actual patient use, such as frequent activation, carrying or dropping of the delivery device [20].

A detailed description of all specification tests for nasal and inhalation products depending on the particular delivery device can be found in the "Guideline on the pharmaceutical quality of inhalation and nasal products" (EMEA/CHMP/QWP/49313/2005Corr).

25.2.3 Comparability

For biotechnology products in particular, changes in the manufacturing process are commonly implemented throughout the lifecycle of the product to improve product quality, production yield, or process economics. These modifications may concern different aspects of the production process, such as changes of raw materials, excipients, container/closure system, storage and shipping conditions, manufacturing site, scale, cell culture conditions etc. The changes may be introduced in the development phase as well as after marketing authorization. In both situations, and especially when preclinical or clinical studies have been performed, comparability between the two products must be shown [21]. Generally it should be clear that in the later phases of development and after approval criteria for comparability studies are more stringent than in the early stages. In this regard, establishment of reference standards in early stages might be helpful [22]. The aim of the comparability exercise is to ensure the quality, safety, and efficacy of the drug product produced by the changed manufacturing process. Regarding the approach for demonstrating comparability, the European and US perceptions slightly differ. While the EMA states that "the comparability study should demonstrate that the quality, safety and efficacy profiles are the same before and after a change," the FDA focuses on the "lack of adverse effect of the change on the identity, strength, quality, purity or potency of the product as it relates to the safety and effectiveness." However, in Europe the product quality attributes need not necessarily be identical before and after the production change as well, but they should be highly similar. In addition, it must be shown that the change of attributes does not exert any adverse impact upon safety or efficacy of the drug product.

Depending on the expected impact of the change, a step-wise comparability approach may include characterization studies, validation of the manufacturing process, release criteria, stability data and, if applicable, nonclinical and clinical data. In the most cases, comparison of only the product specifications will not be sufficient to assess comparability because the specifications are not intended to provide a full product characterization. However, the manufacturer should demonstrate that the specifications are still valid for the product after the process change. The suitability of the chosen analytical tests must be carefully evaluated. If comparability of the quality attributes cannot be sufficiently shown by analytical testing or biological assays, the manufacturer may consider introduction of further tests, as well as performance of nonclinical and/or clinical studies [23]. Generally physicochemical properties, biological activity (potency), impurities and stability of a product should be considered in comparability testing. Establishment of a correlation between physicochemical properties and biological activity would be favorable. Due to the often complex nature of many biopharmaceutical products, it may be necessary to apply different analytical tests to define certain quality parameters. In cases where the end product is a complex mixture of molecules, tests should address all these product-related substances to show batch-to-batch consistency. With regard to impurities, a combination of analytical methods might be used to adequately evaluate the purity profile of the changed product. If a difference in the purity profile is detected, the new impurities should be identified, characterized, and their impact on the safety and efficacy of the product determined. However, it should be kept in mind that impurities do not always have a negative impact on product quality. In rare cases, they might act as stabilizers or prevent aggregation [22]. There may be further product-specific aspects affecting product quality, efficacy, or safety that should also be considered. As an example, for an oral gene therapy product containing a bacteria-based vector, the viability of the bacteria would be a major concern in comparability studies. Since even slight modifications leading to a changed protein structure or impurity profile of a product may influence its degradation over time, performance of real-time stability studies should also be considered in any case. Stability studies under accelerated and stress conditions may further provide useful information concerning this matter.

For comparability exercises carried out for a product claimed to be highly similar to a previously authorized product, it is anticipated that not all necessary information may be accessible to the manufacturer. In these cases reference to bibliographical data or pharmacopoeial monographs can be made. However, to sufficiently show comparability, an extensive comparability exercise based on a clearly identified reference product will also be required. The extent of nonclinical and/or clinical bridging studies will depend on the product characteristics as well as on potential differences in comparison to the reference product [21, 24]. If the dosage form or the formulation of a product is changed during clinical development further problems may arise that need to be addressed in comparability studies. One example would be protein aggregation after change of an excipient leading to altered product stability and BA. Another example would be the change of a delivery device resulting in container-closure interactions caused by leachates [22].

With regard to locally acting nasal products, the FDA provides guidance for product quality studies to determine BA and bioequivalence. BA depends on several factors, most importantly the release of the active substance from the drug product and the delivery to the mucosa. Release of the drug substance may be demonstrated using in vitro methods, while drug delivery to the site of action may be assessed using nonclinical and clinical studies. The in vitro BA studies should be conducted on at least three relevant batches used in production, stability studies, and clinical trials. Detailed testing should comprise single actuation content through container life, droplet size distribution, drug particle size distribution, spray pattern, plume geometry, and priming/repriming. Droplet size distribution may be analyzed by laser diffraction or cascade impactor. For suspension products, the drug particle size and extent of agglomerates, important for the rate of dissolution, should be assessed by microscopic methods. Shape and size of spray patterns can be characterized by either manual or automated image analysis. Priming and repriming data should be provided in multiple orientations of the delivery device. Finally, all in vitro tests should be validated for accuracy and precision [25].

25.2.4 Stability

Many biopharmaceutical products are particularly susceptible to different environmental conditions, e.g., high temperature, light, oxidation, and humidity. Oftentimes they have shorter shelf lives compared to chemical drugs. Therefore, performance of stability studies is very important to support the claimed shelf life of the final product, to define adequate storage conditions, and to guarantee satisfactory functionality of the product in this period. In the stability studies, the different external conditions affecting the identity, purity, and potency of a product should be considered. Assessment of stability should always be based on long-term, real-time, real-condition stability studies. However, studies under accelerated conditions (e.g., high temperature, light exposure, vibration, or pH changes) may also provide useful supportive data for the definition of the expiration date and storage conditions, for identification of degradation products, and to elucidate product characteristics under unusual conditions that may occur in atypical transport situations. For products that are packaged in containers protecting them against humidity, stability studies at different degrees of humidity may be omitted if the protective character of the container can be confirmed. In case the drug substance is stored before formulation of the final product, stability data should be provided for at least three batches that are representative of the material used in preclinical and clinical studies. Stability data for the final product should be provided for at least three batches produced under conditions used in manufacturing scale as well. It is desirable that testing be performed on drug products that are derived from different batches of bulk material. If a shelf life of more than 6 months is pursued, a minimum of 6 months stability data should be available at the time of submission. For products with proposed storage periods of more than 1 year, the studies should be performed every 3 months during the first year of storage, every 6 months during the second year, and annually thereafter. Many biopharmaceutical products such as nasally delivered vaccines may have a much shorter shelf life. In these cases, the real-time stability studies should be conducted monthly for the first 3 months and at 3-month intervals thereafter. The samples that are subject to stability testing may be selected so as to be derived from different batches, different strengths, different sizes, and different container/closure systems, insofar as they behave similarly under the respective storage conditions. In this case, data should be provided showing that the approach represents the stability of all samples. The stability-testing program should at least include analysis of the drug product specifications such as visual appearance, biological activity, purity, and sterility testing. Assays for container/closure integrity and degradation of additives or excipients may be implemented where required. Liquid products should also be stored in an upside down position to exclude deleterious effects from contact with the closure. The stability of freeze-dried products after reconstitution has to be in accordance with the labeling on the container and package inserts. Potency testing should be performed by a quantitative method and compared to a reference standard. Since determination of purity is highly method dependent, application of different methods is regarded as indispensable for detecting all impurities and degradation products. Typical methods used for this purpose include electrophoresis, high-resolution chromatography, and peptide mapping. Acceptable limits for degradation products should be established and justified. Generally, the specifications of all products should be maintained throughout the products' shelf life. In exceptional cases, release specifications and specifications at the end of shelf life may differ. However, this must be adequately justified and supported by respective data [26].

Recently, a novel stability model named "estimation model" has been established for vaccines by the WHO. It is a statistical model that is based on regression analysis of potency loss over storage time. The estimation model is supposed to provide a more comprehensive description for stability of vaccines than the current "compliance model." It may be applied as soon as at least three stability time points have been examined for a product. However, the larger the analyzed lots and the later the time points of stability analysis, the more reliable data will be obtained [27].

Specific Considerations for Different Mucosal Application Forms Dry powders for mucosal application (e.g., vaccines) exhibit a greater stability than liquid formulations. Products are often freeze-dried as in this form they tend to undergo less degradation, and storage periods can be extended. However, physical and chemical degradation may also take place in solid formulations. An important factor for the stability of proteins and peptides in the solid state is the glass transition temperature (T_g) of the matrix. T_g is defined as the temperature at which an amorphous material transits from a hard state into a rubber-like state. The higher the glass transition temperature, the higher is the stability of the product. Thus, the storage temperature of proteins should be far underneath T_g. The presence and concentrations of cryoprotectants such as sugars or polyols, other excipients and moisture content may have critical impact on the glass transition temperature of the matrix. Attention should also be paid to a possible crystallization of excipients, since this may lead to phase separation and protein degradation [17].

For products to be applied by an inhaler, the storage orientation may be a relevant issue for product performance. Accordingly, in stability studies containers should be stored in various positions to determine the effect of orientation. If the product has a secondary packaging that serves to protect it from light and/or humidity (e.g., dry powder inhaler inside a foil overwrap), the product should be removed from the packaging and stored according to the instructions for use prior to the performance of the stability studies. Different tests and limits may be established for the time of release and the end of shelf life; however, this should be thoroughly justified [20].

Nanoemulsion drug delivery systems have proven to offer some advantages for intranasal and ocular application forms, including increased BA and efficient target delivery. In contrast to liposomes and other vesicular delivery forms, nanoemulsions show a much higher stability. However, stability of nanoemulsions and their long-term storage still poses a great challenge to manufacturers. The main reason for this problem is the potential of nanoemulsions for Ostwald ripening, which means the diffusion of smaller particles in solution and deposition on larger ones. This leads to an altered droplet size distribution resulting in turbidity and changed delivery of the formulation. Different additives, such as polymeric surfactants, have the ability to prevent the effect of Ostwald ripening. In either case, stability studies for nanoemulsions should include analysis of droplet size as well as the determination of viscosity and refractive index [28].

25.2.5 Environmental Risk Assessment for GMO-Containing Medicinal Products

Legal Basis In view of a marketing authorization, every new medicinal product has to be evaluated for its potential risk to the environment by performance of an environmental risk assessment (ERA). To this effect, biopharmaceuticals like gene therapy products or live viral vaccines consisting of or containing a genetically modified organism (GMO) have to comply with pharmaceutical as well as environmental legislation. Regulation (EC) No 726/2004 laying down "Community procedures for the authorization and supervision of medicinal products for human and veterinary use and establishing a European Medicines Agency" requires performance of an ERA for GMO-containing products that is based on the information described in Directive 2001/18/EC on the deliberate release into the environment of GMOs.

Methodology There are several possibilities of how GMOs may be disseminated into the environment during clinical use. Among these is the unintended dispersal of products during normal handling and use, incomplete or inappropriate decontamination, and disposal of waste and distribution through patient excreta. Consequences of an unintentional release of a GMO into the environment may include spread, genetic or phenotypic changes of the GMO, competition with other species, and transfer of genetic material from the GMO to viruses, microorganisms, animals, plants, or humans.

For proper identification of possible harmful effects of a GMO and the associated risks for the environment or public health, the ERA should follow the methodology described in the "Guideline on environmental risk assessment for medicinal products consisting of, or containing GMOs" (EMEA/CHMP/BWP/473191/2006corr) [74] and in the "Guideline on scientific requirements for the environmental risk assessment of gene therapy medicinal products" (EMEA/CHMP/GTWP/125491/2006). The evaluation of the ERA includes six steps.

First, the characteristics of the GMO which may cause potential adverse effects should be identified. Points to consider in this regard include the pathogenicity, virulence, infectivity, host range, tissue tropism, replication mechanism, latency/reactivation, survival, and stability of the GMO. It should also be kept in mind that attenuating modifications of the GMO can be reversed over time. Therefore, analysis of the pathogenicity of the parental organism should also be taken into account. For replication-incompetent viral vectors, contaminating replication-competent vectors emerging from recombination events may pose another challenge. The risk of generating replication-competent vectors during manufacture can be reduced by the use of appropriate cell lines and vectors with minimal sequence homology. Furthermore, confirmation of the absence of replication-competent vectors should be implemented in the routine analysis. Another point to be addressed is the genetic stability of the GMO, since rearrangement of the genome or insertion of foreign sequences could alter its pathogenicity or other characteristics. Finally, every induced genetic modification should be checked for its ability to alter the pathogenicity or other characteristics of the GMO, including tissue tropism or susceptibility to the immune system or to medical therapies.

In the second step of the ERA, the potential consequences of each adverse effect have to be evaluated. Thus, the severity is either ranked as high, moderate, low or negligible, depending on the consequences on the environment and human health. Vulnerable individuals, such as immune-compromised persons, pregnant women, and young children, are usually considered separately. Moreover, the application conditions, the exposed environment and available measures for reducing adverse effects are also taken into consideration when estimating the potential consequences of the adverse effects.

The third step of the ERA consists of the evaluation of the likelihood of the occurrence of each identified potential adverse effect. Points to consider in this step include the probability of the GMO to establish an infection, to reverse its attenuation, to disseminate into the environment, and to be transmitted to thirds. As it is often difficult to make quantitative predictions, a worst-case scenario can be defined instead, in order to guarantee that all hazards will be adequately estimated. Generally, the release of the GMO into the environment after application of the medicinal product should be investigated in nonclinical and clinical studies.

In the fourth step, the risk posed by each identified characteristic of the GMO will be estimated by combination of the likelihood of occurrence (Step 3) and the magnitude of consequences (Step 2).

Step 5 details the application of management strategies for lowering the risks associated with the deliberate release of the GMO. Such management strategies may include hygienic measures, decontamination of patient's excreta, and avoidance of contact with vulnerable individuals. These directions may further be complemented by a monitoring plan to assess the effectiveness of the risk management strategy.

Finally, in Step 6, the ERA results in the determination of the overall risks of the GMO based on the evaluation of the previous five steps. This part should also include a summary and conclusion of the overall environmental risk of the GMO-containing medicinal product [29].

Considerations for Mucosal Application of GMOs To date, GMOs in medicinal products are mainly derived from viruses or bacteria. Depending on the particular product, GMOs in biopharmaceuticals for mucosal application will presumably differ from those applied intravenously with regard to infectivity and the route of excretion and therefore might also have a different impact on the environment. The route of application largely influences the biodistribution and excretion of GMOs present in medicinal products. Intravenous (IV) administration of a GMO results in a maximal systemic exposure and the GMO may be shed into various patients' fluids and excreta. In contrast, GMOs in medicinal products for ocular application might only be present in the tears in relevant concentrations. Orally administered gene therapy products based on bacterial strains, e.g., *S. typhi*, are expected to be primarily shed into saliva and stool. Compared to IV application, mucosal application will result in a higher concentration of the GMO in the respective body fluid. For orally administered gene therapy products, the GMO concentration will be particularly high in the saliva. Since pronounced shedding leads to an increased incidence of negative environmental effects (ERA step 3), products with such properties should not exhibit any characteristics with risk potential for the environment or other organisms (ERA step1 and 2).

Another point to consider is the ability of the GMO or its parental organism to colonize in the gastrointestinal tract. The survival of the bacteria after gastrointestinal excretion should also be investigated. Due to their sensitivity, NAT tests are often performed to analyze bacterial DNA. However, these tests do not give any evidence about viability and infectivity of the excreted bacteria and further testing is necessary to evaluate these characteristics. Finally, it should be kept in mind that oral administration of bacteria could also lead to systemic exposure by absorption via damaged mucosa. In these cases, signs of infection as well as survival of bacteria in the systemic circulation or in peripheral tissues should be addressed in nonclinical studies.

25.2.6 Medical Devices, Formulation, and Route Requirements

Requirements for Delivery Devices Many biopharmaceutical products need a specific formulation or have to be combined with a medical device in order to be applied via the different mucosal routes. In particular, products for nasal application or inhalation purposes need to be applied in combination with a delivery device such as a spray or an inhaler. There are several types of inhalers that may be used depending on the type of product, e.g., pressurized metered dose inhalers, dry powder inhalers that may be device-metered or premetered, nebulizers or nonpressurized metered dose inhalers. Nasal application can be achieved by drops intended for single or multiple use, pressurized and nonpressurized metered sprays or nasal powders. Depending on the type of inhaler or spray, different characterization tests have to be performed in pharmaceutical development studies. As an example, shaking requirements would be relevant for nebulization products but not for dry powder inhalers. Detailed guidance for these specific tests can be found in the "Guideline on The Pharmaceutical Quality of Inhalation and Nasal Products" (EMEA/CHMP/QWP/49313/2005). If the delivery device is breath activated, it should be shown that all patient groups that are intended to be treated with the product are able to activate the device. For patient convenience, equipment of device-metered dry powder inhalers with a counter or fill indicator is desirable. For mucosally applied medicinal products, the most important factors to achieve a high BA of the active substance are the release of the drug substance from the drug product and the delivery to the mucosa. Therefore, products such as sprays or inhalers should always be tested for the performance of the actuation release mechanism. Further, particular attention should be paid to the confirmation of reproducible drug delivery. The description of the container closure system should include general information, such as composition and material supplier, as well as specifications for the selected plastic material. The manufacturer has to demonstrate that the composition of the container-closure system complies with pharmacopoeial or other relevant standards. Furthermore, additives like antioxidants or plasticizers and colorants should be identified. Another important issue is the compatibility of the primary packaging material with the medicinal product, which is especially important for nonsolid active substances. Compatibility can be shown by performance of extraction and interaction studies. Extraction studies exposing a material sample to a suitable solvent should preferably be carried out under stress conditions to increase the rate of extraction. All identified substances should be listed in the specification of the packaging material. Interaction studies may be performed with the respective material or the container itself and may also include migration studies to analyze leaching of substances from the plastic material and/or sorption studies to exclude adsorption effects of the drug. Toxicological information for extractables and leachables may also be provided, where applicable [30]. For liquid dosage forms to be applied with a measuring device, dosing accuracy is of particular importance. An adequate graduation of the device is important to ensure accurate and precise dosing from release until the end of the shelf life of the product. Since a glued label may detach, the graduation should preferentially be embossed in or printed on the material. The suitability of the dosing device for the respective medicinal product must be shown considering dosing accuracy and precision as well as the risk of overdosing and physical properties of the liquid that may influence dosing performance [31].

Finally, all medical devices used must have undergone a conformity assessment resulting in a CE mark for its intended use according to the Directive 93/42/EEC concerning medical devices.

Specific Formulation Requirements For many different reasons, in mucosal drug application, active substances are frequently combined with pharmaceutical excipients such as preservatives, adjuvants, stabilizers, solubilizers, flavoring substances, etc. Preservatives like benzalkonium chloride or organic mercurials are particularly added to multi-dose ophthalmic preparations to maintain product sterility during use. However, preservatives should not be used in products to be applied during eye surgery since they could damage the corneal endothelium under these conditions. Nasal formulations are often complemented with absorption enhancers due to the low nasal BA of many drug substances. Absorption enhancers may act by changing the permeability of the epithelial cell layer or by opening of tight junctions. Among the substances used for this purpose are: surfactants, bile salts, fatty acids, or polymeric enhancers. One advantage of the use of polymeric enhancers is that they are not absorbed due to their high molecular weight and thus, they exhibit a very low toxicity. An example of a polymeric absorption enhancer is the biopolymer chitosan that is often used in nasal formulations because of its biodegradability, biocompatibility, and bioadhesive characteristics. Cyclodextrins are also used as complexing agents to improve nasal drug absorption by increasing drug solubility and stability [14]. Beside the use of excipients, factors as viscosity or pH of a solution may influence nasal absorption and have to be adjusted accordingly.

Especially in the field of nasal vaccination, nanoparticulate delivery systems are emerging because they provide improved protection and facilitated transport of the antigen and may also improve antigen recognition by the immune cells. Nanoparticles are sized between 1 and 100 nm and consist of macromolecular material composed of lipids or polymeric substances, such as lectins or polysaccharides. There are several methods to produce nanoemulsions, including high pressure homogenization, microfluidization, phase inversion, or solvent displacement. In addition to standard parameters, characterization of nanoemulsions should include determination of morphology, nanoemulsion droplet size, viscosity, thermodynamic stability, and surface characteristics [28].

If possible, all excipients used should be of pharmaceutical grade. Excipients that are described in the Ph. Eur. or another accepted pharmacopoeia have to comply with the respective monograph. If an excipient is not described in a pharmacopoeia, appropriate specifications should be established based on physical characteristics, identification and purity tests, in addition to other relevant assays. For novel excipients, a detailed characterization including supporting safety data must be provided. In these cases, performance of nonclinical studies may also be appropriate. Furthermore, compatibility of excipients with the active substance has to be shown [32].

25.3 Nonclinical Aspects

25.3.1 General Considerations

Nonclinical studies, in vitro and in vivo, are essential to support initiation of early phase of clinical trials and marketing authorization for mucosally administered biologicals [33, 34]. The testing program required may vary, depending upon product characteristics and existing knowledge about its mode of action and safety profile. For a new product, of which neither the drug substance and formulation nor the route of administration is approved, a complete program for full characterization of its toxicity is required. Whereas for previously licensed products which are newly formulated for mucosal delivery or proposed for a new mucosal route, some evaluations in pharmacology, pharmacokinetics (PK), and toxicology may have been performed with original formulation and route. In this case, focus of nonclinical evaluation should be to address any preexisting deficiencies associated with the proposed particular formulation or route [11]. The greater the change made in formulation composition and way of product delivery, the more likely it is to justify additional nonclinical studies.

As for any pharmaceutical, safety evaluation of mucosally administered biologicals is a step-wise process. Normally, the acute and repeat-dose toxicity studies with safety pharmacology information as well as in vitro genotoxicity data (if applicable) are considered adequate for support of first administration in humans. Whereas reproductive and developmental toxicity data, which is necessary for a biological-intended use in women of child bearing potential, can be submitted at late clinical development stage to support large-scale Phase III studies in the USA, or even at the time of marketing authorization application in the EU. The pivotal nonclinical safety data used to support clinical trials and marketing authorization should be produced from definitive studies, which need to be conducted according to good laboratory practices (GLP) using the final product/formulation. The study design should consider several important aspects including animal model selection and the way of product delivery, as discussed below.

Animal Model Selection Animal models should mimic the physiological/disease state in humans and be able to develop responses anticipated in humans. For vaccine products, the ideal model is that the species is susceptible to human pathogen infection under study. However, such a model is not always available, as is the case for smallpox, HPV, human CMV and HBV, among others. For recombinant human protein products, a more relevant species should be used for chronic toxicology studies, if appropriate, e.g., based on pharmacological and short-term toxicological studies. In addition, anatomy and physiology of proposed mucosal site of administration as well as its reception to the particular way of delivery should be considered. For example, mice and rats are useful models for droplet but not for spray for intranasally administered products. Whereas rabbits and dogs are useful for use of spray devices, their olfactory bulbs are highly protected which may allow to underestimate the neurotoxicity risk, and if this is a potential issue, specific techniques would be required to ensure the test article to reach this organ [4]. Nonhuman primates may be considered for pharmacology and/or toxicology studies, when test article is pharmacologically inactive in other species [3, 4]. The choice of animal model should be justified.

Normally, one relevant species may be sufficient for short-term general toxicity studies if the pharmacological activity of the product is well known. In some instances, two or more species may be required, e.g., if there is a species-specific or strain-specific difference, novel adjuvants, limited knowledge about mechanism of protection (e.g., intranasally administered influenza and measles vaccines, among others), or when a safety concern is raised.

Dose and Dosing Regimen For any mucosal route, dose and dosing regimen should generally reflect intended human exposure. However, consideration should be given to the total volume of the administered test article, which may affect the outcome of safety study. For example, intranasal administration of more than 5 μl of the test volume per nostril to a mouse would result in the test article being swallowed, rather than being adsorbed by the nasal mucosa [4].

Endpoints and Route-specific Considerations In addition to parameters prescribed for general toxicity studies, additional outcome measures may be needed for addressing specific concerns associated with a particular route and target organ. For example, if concerns exist for potential passage of product components (drug substance, vaccine adjuvant, live virus, etc.) to the brain following intranasal administration, immunohistology, and neurotixicity studies and examinations need to be performed. For products administered by inhalation, pulmonary function tests and data on histopathology of lungs need to be provided [11].

Immunogenicity Assessment For various vaccine products being developed for mucosal delivery, mucosal immune responses to drug substances or immunogens should be measured, in addition to measuring serological responses. Development of appropriate assays for vaccine antigen-specific T-cell responses, antibody responses, and cytokine product should be considered. For human protein products manufactured via recombinant DNA technology, immunogenicity in terms of antidrug antibodies (ADA) should be carefully evaluated for safety and efficacy reasons [3].

25.3.2 Pharmacodynamic (PD) Studies

For any route of administration, the proof-of-concept information should be generated in a relevant species or in in vitro studies when no relevant species is available. Type of proof-of-concept studies is determined by the product characteristics and proposed indication. If a licensed parenteral formulation is proposed to be administered via a particular mucosal route as alternative, comparative design should be considered.

Challenge studies should preferably be expected for prophylactic vaccines, if an appropriate model is available that can reflect infections and diseases in humans [4, 10]. Attempts should be made to identify any correlation between the level of protection from infection or disease and an immune response measured, including antibodies, $CD8^+$ CTL (cytotoxic T lymphocytes), or local cytokine responses. At minimum, immunogenicity data are expected from a relevant species in order to support initiation of clinical trials. The studies should assess relevant immune responses to each vaccine antigen, including mucosal and systemic immunity (e.g., secreted IgA, serological total IgG, functional immunity such as neutralizing antibodies, opsonophagocytic activity) that contributes to protection. The level, class, and subclass of antibodies produced and duration of immune responses should be characterized, according to dose and dosing intervals. Information on cell-mediated immunity including $CD8^+$ CTL response is also often relevant for mucosal vaccines [1], especially live vaccines and inactivated vaccines formulated with mucosal adjuvants should be collected as far as possible. Activation of innate immune system is often part of the mode of action of vaccine adjuvants, DNA vaccines, and live-vectored vaccines that exert inherent adjuvant activity, and such information should be considered as well. For DNA and live-vectored vaccines, expression and production of the vaccine antigens (heterologous antigen) in appropriate targeted site must be demonstrated [5, 8, 9, 12]. For adjuvanted vaccines or those containing new delivery systems such as biodegradable microparticles, liposomes, and bioadhesive polymers, studies should be designed to generate comparable data so that their use/inclusion in final formulation can be justified [6]. It should be recognized that frequently, animal models may not predict immunogenicity and efficacy in humans, depending upon how closely the animal model resembles the human disease and human immune response, and how comparable the potential of live vector replication is in human versus in animal model. Thus, interpretation of these nonclinical pharmacodynamic (PD) data must be with caution.

In addition, for vectored vaccines, preexisting and vaccine-induced immune responses to the vector should be evaluated, to estimate the possibility of repeated vaccination or reuse of the same vector virus in another vaccine [9].

For recombinant human protein hormones and growth factors, the in vitro studies, like species-specific receptor-binding affinity studies and cell-based assays, are often undertaken to provide evidence about functional activity of the products. In these situations, use of homologous molecules in diseased animal models to explore potential clinical effect is encouraged [3].

25.3.3 Pharmacokinetics (PK), Biodistribution, and Integration Studies

Mucosal formulations of human protein products often require stabilizer and absorption enhancer, which can sometimes have profound effects on product's immunogenicity potential and safety profile. Thus, absorption, disposition, and clearance of new formulations need to be evaluated adequately in relevant animal models, prior to clinical studies [11]. PK studies should reflect clinical route of administration and, whenever possible, utilize preparations that are adequately representative of that intended for toxicity testing and clinical use. Single and multiple dose PK and tissue distribution studies in relevant species are useful, whereas routine studies attempting to assess mass balance and classic biotransformation studies are not needed. It should be noted that, when comparing PK of a new formulation with a licensed formulation is applicable, both the shape of concentration/time curve and the total area under curve (AUC) should be examined [11]. If PK data for new formulation are not available, an assumption of 100 % BA from proposed clinical dose might be used to judge adequacy of available systemic toxicity information.

Some intrinsic attributes of this type of biologicals may deserve careful considerations during interpretation of study results: (1) altered PK profile due to immune-mediated clearance mechanisms; (2) inter-species difference in PK profile; and (3) significantly delayed or prolonged PD effects relative to PK profile/plasma level of some products, like cytokines. These aspects could significantly impact predictability of animal studies including dose-response relationship assessment.

Mucosal vaccines containing new adjuvants and/or new delivery systems also require PK investigations, at least local disposition studies to assess retention of vaccine components at the administration site and draining lymph nodes [4, 6]. However, determination of serum or tissue concentrations of vaccine antigen(s) is normally not needed.

For naked DNA vaccines, plasmid biodistribution, persistence and integration studies should be conducted on a case-by-case basis [5, 7, 8, 12]. Although there is no need for such studies for a common plasmid vector with documented biodistribution/integration profile, biodistribution and integration studies are necessary for those who utilize new vectors, new formulations, new delivery methods, new routes of administration, or any other modifications suspected or expected to significantly enhance the cellular uptake and biodistribution, and/or to increase the capacity of plasmid DNAs to enter the nucleus. A typical biodistribution and persistence study should assess presence of plasmid collected from a panel of tissues at multiple time points ranging from a few days to several months post administration. The panel of tissues typically includes blood, heart, brain, liver, kidney, bone marrow, ovaries/testes, lung, draining lymph nodes, spleen, muscle at administration site and its subcutis. Tissue distribution/persistence and plasmid levels are typically evaluated using a quantitative real time polymerase chain reaction (qRT-PCR) assay validated for sensitivity, specificity, and the absence of inhibitors. The sensitivity of this assay should be sufficient to quantify < 100 copies of plasmid per microgram of host DNA [12].

The potential integration of plasmid DNAs into the host's chromosomes should be studied, in case of instances discussed above, or when a common plasmid is detected to persist in any tissue of any animal at levels exceeding 30,000 copies/μg of host DNAs by study termination [12]. A typical integration study will assess all tissue(s) containing persisting DNA plasmid. It is recommended that at least four independent DNA samples be analyzed. Each sample may include DNA pooled from several different donors.

Similarly, biodistribution studies and an investigation into the potential of integration into host cell genome should be conducted for all of live attenuated or recombinant vectored vaccines proposed for mucosal delivery [9]. Biodistribution studies should be performed in a full set of tissues and organs including the brain, especially for the intranasal route of administration. One species is considered sufficient if scientifically justified. Endpoints of distribution studies may include recovery of infectious virus, detection of viral antigens, or detection of viral genetic material. Crossing of blood-brain barrier might be an indication of potential neurovirulence. Published in vitro and/or in vivo studies and data on detection of integration of the same vector may provide useful additional information. Specific integration studies are warranted, if applicable.

The need for testing of inadvertent germ line transmission, for both DNAs and live-vectored vaccines, should be assessed, especially when nucleic acid signal is detected in the gonads [35].

25.3.4 *Toxicology*

For all mucosal routes and each biopharmaceutical, the core package of the non-clinical safety assessment as constituted by general toxicity studies and safety pharmacology should be submitted to support first administration in humans. Definitive toxicity studies should be conducted under GLP conditions and using final product. In addition, additional studies may be required on a case-by-case basis, according to the product category, target population, and published safety concerns, as described below.

Safety Pharmacology Studies To support the first human administration, the product should be evaluated for the potential undesirable core effects on pivotal physiological functions, i.e., effects on central nervous system, cardiovascular system, and respiratory system. Aerosolized product may have direct effect on lung functions. Inflammatory responses induced by vaccines and biotechnology-derived proteins may exert adverse effects on blood pressure and cardiac responses. However, separate studies are generally not needed, if safety pharmacology parameters, such as body temperature, electrocardiogram, and central nervous system evaluations could be incorporated into acute or repeated dose toxicity studies [3, 4].

Single-Dose Toxicity Studies The purpose of these studies is to determine acute adverse effects by assessing mortality, clinical signs, body weight, and food consumption. Such parameters can be incorporated into repeated-dose study design thus obviating the need for single-dose toxicity studies [36], especially when repeated dose studies are pivotal to support a clinical trial.

A single-dose toxicity study might still be useful to generate dose-response data to support dose selection and safety assessment in some circumstances, e.g., significant intrinsic toxicity exists, marked pharmacological action of the product substance, or immune responses induced by the first administration significantly alters reactions to a second administration, e.g., neutralization of live vectors, neutralization of human cytokine used as adjuvant. In case that a single-dose study is pivotal to support a clinical trial, its design should be extended to include more parameters, i.e., hematology, clinical pathology, necropsy, and microscopic examination of a limited set of tissues and organs. Such extended design should also include further evaluations 2 weeks later after a single administration, to allow assessment of delayed toxicity and/or recovery.

Repeated Dose Toxicity Studies Repeated dose toxicity studies are regularly the pivotal studies that support the first human administration. The fundamental selection of animal model was mentioned above. The design of these studies is broadly modeled on the repeat dose toxicity study design for medicinal products [37], but considerations must be given to vaccine-specific issues as discussed in WHO Guidelines [4].

Briefly, the route of administration and treatment regime in nonclinical studies should mimic the proposed clinical route and regime. Vaccine doses should be given episodically rather than daily, and an interval of 2–3 weeks between successive administrations is generally considered sufficient. However, in certain circumstances where the kinetic of immune responses is poorly understood, preliminary studies over an extended period of time may need to be conducted. Since repeated vaccinations may result in an increasingly pronounced immune response, it is generally expected that the number of vaccine doses in toxicity studies should exceed the number proposed for human administration [4, 6], commonly, at least one more administration in animals than in humans. When the product is to be administered in humans using a particular device, the same device should be used in animal studies, where feasible (e.g., aerosolized measles vaccine in monkeys).

Regarding the level of vaccine dose, the full human dose should be tested, without scaling for body weight or surface area, whenever feasible. Where this is not possible due to formulation constrains, the maximum feasible dose should be administered and this dose should exceed the human dose on an mg/kg basis and induce an immune response in the selected animal species. Alternatively, it may be possible to administer the total volume to more than one site using the same route of administration. On occasions small multiples of full human dose (e.g., 3-fold, 5-fold) may be tested to evaluate the dose response of a finding or to allow some flexibility in choice of clinical dose.

PD monitoring, i.e., immunogenicity data of vaccines are a valuable element of toxicity studies, as it can not only confirm relevance of the selected animal model and their exposure to the product, but also help identify responder/nonresponder animals and thus establish a correlation between toxic effects and measured immune responses.

The observation period must cover both immediate events and any delayed toxicities that arise from the immune response and persistent effects [4]. The early in vivo and postmortem investigations need to be made when these events and responses are expected to be at peak levels, normally 2–7 days after the last dose (1–3 days for clinical pathology), whereas the late phase investigations are focused upon the persistence and exacerbation of and/or recovery from any adverse effects, which normally require a treatment-free period of 2–4 weeks, but may be longer in case of pronounced persistent effects.

Histopathological examination is usually performed on a full range of tissues, as defined in Annex of repeat-dose toxicity study [4, 37]. However, a reduced list may be appropriate, in some circumstances, but must include pivotal organs such as brain, kidney, liver, heart, lung, gonads, and special attention to immune organs (local and remote lymph nodes, thymus, spleen, bone marrow, Peyer's patches, and others).

By comparison, recombinant human protein products should be assessed using regime-reflecting proposed clinical and treatment duration as recommended in ICH M3 R2 [36]. The high dose should be included in repeat-dose toxicity studies, which can be selected based on PK/PD information [3]. The high dose refers to a dose that gives the maximum intended PD effect in the species or a dose which gives up to 10 × exposure multiple over the maximum exposure to be achieved in the clinic, whichever the higher one. If PD data are not available, a dose up to 10 × multiple over the highest anticipated clinical exposure should be sufficient. Note that the 10 × multiple refers to multi-dose, steady state AUC exposure as compared to the maximal clinical dose currently under investigation (i.e., not to a lower clinical dose anticipated for marketing). Comparative receptor binding/activation kinetics of the product for human and selected species targeted receptors will also need to be considered and doses adjusted (if needed) to achieve an appropriate 10 × exposure multiple for chronic studies. Evaluation of systemic exposure, i.e., toxicokinetics (TK) is generally required for biotechnology-derived human protein products [3], although it is not necessary for vaccine products. The study design should include appropriate concurrent control groups to enable (1) detection of toxicity and reactogenicity of the product compared to a placebo; (2) evaluation of the recovery of observed toxic effects; and (3) screening for any delayed adverse effects. Recovery may take considerably longer time, which needs to be taken into account in study design. Noteworthy is that demonstration of complete reversibility is normally not required.

Local Tolerance Studies The potential irritation at administration site(s) inadvertently coming into contact with the product (e.g., eye exposure for intranasal spray formulation) should be evaluated, both by gross observation and by histopathology [3, 4, 10]. A grading system, e.g., Draize scoring, should be used. If marked reactions are observed, follow-up studies may need to examine the persistence of material at administration site and in draining lymph nodes. This evaluation can usually be made during single- or repeated-dose toxicity study when route of administration used is the same. In some cases, however, a stand-alone study may be preferable for more detailed investigations.

Local tolerance studies should be conducted with formulations for marketing or clinical studies, or at least be comparable to these materials.

In addition to above-mentioned safety studies, there are several additional studies that need to be considered, according to intended clinical population, product category, or published safety concerns.

Reproductive and Developmental Toxicity Studies Fertility studies in male and female animals are generally not required, if general toxicity studies do not show positive finding in histopathology of reproductive organs [10]. When cause of concern exists, further evaluation is needed, e.g., on the menstrual cyclicity, sperm count, morphology, motility, testicular volume, and male/female reproductive hormone levels. In addition, fertility and general reproductive function should also be considered, if biodistribution studies with live vectors and DNA vaccines show localization in gonads with resultant germ-line alterations, e.g., foreign genetic material within germline cells and genomic DNAs [5, 8, 9, 12].

The developmental toxicity studies are necessary when the product will be indicated for use in women of childbearing potential. To date, only one regulatory guideline [13] has been issued on the design of such study for vaccine products, and in the USA, the results of embryo-fetal toxicity study should normally be made available for regulatory review prior to large-scale (Phase III) clinical studies. Overall, the design is broadly modeled on ICH S5a [38], however, vaccine-specific attributes must be taken into account, with respect to the dosing regimen and the number of species. One relevant species is generally sufficient for vaccine products. No daily dosing is employed. Female animals are usually dosed twice or more: one at a few weeks before mating and others during the period from the day of embryonic implantation through closure of the hard palate and to the end of pregnancy, defined as stages C, D, and E in ICH S5a. By doing so, the maximum exposure of the embryo and the fetus to the various components of vaccine formulation during organogenesis and to the induced immune responses, including IgG antibodies after the second half of pregnancy, respectively, can be ensured. Notably, transfer of maternal antibodies to the embryo-fetus is very low during organogenesis/first trimester both in animals (e.g., rabbits) and in humans; whereas in rats, antibody transfer is high only during lactation which is not a model of human exposure.

The adverse effects on development of embryo/fetus and offspring can be evaluated in a single experiment [38]: The caesarean subgroup is used for fetal examination and routine teratology investigation on gestation day (GD) 20 for rats, GD 18 for mice, GD 29 for rabbits. The other subgroup of pregnant animals is allowed to litter and the development of pups is monitored until weaning. Each subgroup typically includes 20 pregnant females to ensure availability of at least 16–18 litters per group for outcome analysis. Endpoints for fetal examinations include, but are not limited to, viability, resorptions, abortions, fetal body weight, and morphology of fetuses. Endpoints for pups from birth to weaning include viability, normal growth, body weight gain, nursing activity, physical development (incisor eruption, fur growth, eye opening) and reflexes (surface righting, auditory reflex, pupil response), and necropsy at weaning (21 days of age for rodents, 35 days of age for rabbits). The study may be extended to evaluate the postweaning development of offspring if equivocal effects are seen.

For human protein-based products [3], if pharmacologically active in two species, one rodent, one nonrodent (i.e., rabbits), both species should be used for embryo-fetal developmental studies, unless embryo-fetal lethality or teratogenicity has already been identified in one of these species. Where the nonhuman primates are the only relevant species, the developmental toxicity studies should be conducted in this species, usually cynomolgus monkeys. In this case, a single enhanced pre- and postnatal developmental study can be considered: Only one single cohort of gestationally exposed females is used to assess pregnancy outcome at natural delivery, without caesarian section. A group size of 12–14 female cynomolgus monkeys is recommended. The dosing period should start on GD 20 throughout gestation until natural birth (GD160–165). Offspring should be evaluated for viability and survival, external malformations, skeletal effects (e.g., by X-ray) or visceral morphology (at necropsy), and others if relevant (immune function, neurobehavior). The duration of postnatal phase monitoring ranges from 1 to 12 months, dependent on the relevant endpoints, and is generally 6 months (PND160). The lack of caesarean section for fetal examinations is not expected to impact on the capacity of the study to detect dysmorphogenesis of the offspring, since in utero X-ray examinations are performed, the delivered babies are given a morphological examination at birth and any aborted or stillborn fetuses are recovered for examination.

It is noteworthy that the study design discussed above is intended to apically screen for any adverse effects on fetus/offspring development [38], irrespective of the causality and the mechanism (e.g. direct toxicity of a component of the formulation, induction of an immune response, or immune imbalance of the mother). If the toxicity is identified, further studies will most likely be necessary to identify the causative agent and elucidate the mechanism.

Juvenile Animal Studies Juveniles differ from adults with respect to vulnerability to insults. Adult human and animal data are considered poor predictive of pediatric population below 2 years of age [39]. Juvenile animal studies may provide the opportunity to address whether immature animals have increased susceptibility to systemic toxicity of the product compared with adults, whether adverse influences exist on growth and development and whether these influences can be reversible. Currently, such study is not routinely needed for vaccine products, although many vaccines are given to children below 2 years of age. However, The necessity of juvenile studies should be examined, with considerations being given to clinical trial results and postmarketing surveillance in adults, information on the usage in the pediatric population, the age of pediatric population to which the product is being applied, comparison of the pediatric population to which the product is applied with juvenile animals (toxicity target, the development of organs/functions, and difference in PKs), content of the package of existing nonclinical studies, and the data of drugs which fall into the same pharmacological category. Juvenile safety studies are encouraged in some circumstances, e.g., inclusion of novel adjuvants and new delivery systems in mucosal formulations intended for use in neonates, observed adverse effects of any product on developing systems in mature animals, or expectation of significant alteration in PK profile of mucosal formulation in juveniles versus adults.

The design of juvenile animal studies should include appropriate endpoints to be specifically addressed. The age of animals should be chosen with their relative stage of maturity at the start of dosing equivalent to that of the youngest child included in pediatric study. So far, the database of biologicals in juveniles is very limited, but this may change as more studies are submitted.

Genotoxicity Studies Genotoxicity studies are generally not required for biopharmaceuticals [3, 4, 6]. However, this type of studies may be required in some circumstances, such as novel organic linker in a conjugated protein, newly synthesized chemical used as DNA complexing material, adjuvant, excipient, or a component of the delivery system. Testing with these novel compounds alone should be sufficient. Synthetic peptides or oligonucleotides used as adjuvants can be exempted from genotoxicity studies, as accumulating evidence has revealed that such classes of compounds do not carry a genotoxic risk.

Carcinogenicity/Tumorigenicity Studies Carcinogenicity studies are generally not needed for adjuvants and vaccines [4, 6]. For DNA and live-vectored vaccines, the product-specific assessment of carcinogenicity should be conducted [5, 7–9, 12], e.g., investigation into the presence of oncogene sequences, oncogene protein, or mode of action of the product in genome (i.e., integration and insertional mutagenesis). If oncogenic potential is detected, e.g. an expressed gene product has very prolonged expression of a growth factor or growth factor receptors or immunosuppressive molecules, a study on tumourigenicity may be needed in appropriate in vitro/in vivo models, e.g., by analyzing proliferative capacity, dependence on exogenous stimuli, response to apoptosis stimuli and genomic modification, using a variety of cell lines to investigate changes in cell morphology, functions, and behavior.

Similarly, product-specific assessment of carcinogenic potential applies to recombinant human protein products [3], and the need for this assessment should also be determined with regard to the duration of clinical dosing (short-term vs. chronic use), patient population (young vs. old), and biological activity. A weight-of-evidence approach should be used, including a review of relevant animal and human data from a variety of sources and class effects. If existing knowledge can clearly support the presence/absence of concern, no additional studies are required. Otherwise, a strategy should be designed to address the issue, e.g., understanding of target biology related to potential carcinogenic concern or inclusion of additional endpoints in toxicity studies. At the end if this more extensive assessment supports a concern, additional carcinogenicity study should be considered to mitigate the concern, or, the product labeling should reflect this concern. In addition to this labeling proposal, clinical monitoring, postmarketing surveillance, or a combination of these approaches should be incorporated into the risk management plan.

Notably, standard lifetime rodent bioassays (or short-term carcinogenicity studies) with homologous products are generally of limited value and not required [3].

Neurotoxicity Studies Neurovirulence testing is required for live vaccines, attenuated or recombinant vectored, which have theoretical or established potential for reversion of attenuation or for neurotropic activity, either intrinsic or acquired via selection/passages on neural tissues for attenuation [4, 9]. Beyond neurotropism consideration, clinical experience is another determinant factor. Examples include viruses and vectors derived from yellow fever, polio, dengue, influenza, smallpox, and JE [39–43], as well as new strains of some live vaccines against measles, rubella, varicella, and some strains of mumps [44]. Note that the current vaccine strains of measles, rubella, and varicella viruses have a good safety record and minimal change to seed lots or to manufacture, and do not require reperformance of neurovirulence tests. Similar is true for live rotavirus vaccines. However, this is specifically referred to the s.c. route of administration. Neurovirulence testing should be considered for these current strains of measles, rubella, and varicella vaccines, if they are administered through intranasal or aerosol route.

In addition, there is scientific reason to perform neurovirulence testing for inactivated intranasal vaccines that contain mucosal adjuvants [45]. Both vaccine antigen and adjuvant can reach the olfactory bulb of the brain after i.n. administration, and neurotoxicity effects have been reported for seasonal intranasal influenza vaccines containing LT.

Immunotoxicity Vaccines are a class of biologicals likely to cause immunotoxicity or undesired consequences such as persistent immune depression or a skewed Th1/Th2 balance, associated with increased susceptibility to infection, infestation, allergy (e.g., asthma), and autoimmunity. Note that current methods available are only to detect immune depression in the context of nonclinical safety testing, whereas no acceptable animal models exist for prediction of human allergy and autoimmune diseases.

Nonetheless, a basic battery of investigations could be incorporated into repeat-dose toxicity or juvenile-study design. An initial assessment of the integrity of the immune system and function can be provided by routine parameters, i.e., clinical pathology (e.g., white blood cell counts), organ weights, and histopathology of immune organs (e.g., spleen, thymus, lymph nodes) and bone marrow cellularity. An evaluation of lymphocyte subsets may be a useful addition. In the case of specific concerns, functional tests may be performed, such as NK activity, macrophage function, mitogen-/antigen-stimulated lymphocyte proliferative response, a primary antibody test to T-cell-dependent antigen like sheep red blood cells or a humoral response to keyhole limpet hemocyanin. One evaluation of immune function is already inbuilt in repeat-dose or juvenile study, i.e., the ability of animal to raise an immune response to administered vaccine antigen. However, extensive functional tests are not routinely recommended. In certain circumstances, e.g., autoimmunity signal detected in clinic, underlying mechanisms may need to be investigated, in vitro using human cells or even in animal models. Additional measures including clinical monitoring and surveillance, refinement of trial inclusion/exclusion criteria and target population, and product labeling proposal should be incorporated into risk mitigation plan.

For plasmid/naked DNA vaccines, induction of immune tolerance in neonatal mice is reported and this concern cannot be neglected, especially when a DNA vaccine is intended for neonates [5, 12]. In addition, bacterial plasmid DNA can induce IgG anti-DNA autoantibodies and repeated DNA vaccination can stimulate a $\leq$ 5-fold increase in anti-DNAs autoantibody levels if a sensitive ELISA is used. However, such increase may not be detected by less sensitive clinical antinuclear antibody screening and this level observed is well below that associated with development of autoimmune disease. Routine analysis of anti-DNA antibodies in nonclinical program is not generally warranted [5, 12], unless significant improvement in DNA delivery efficiency and frequency are made.

Antibodies raised in animals against human proteins may also have severe consequences on efficacy and/or on safety. However, such immunogenicity is unwanted and irrelevant to predict product immunogenicity in humans, but rather useful for interpretation of nonclinical study results [3]. ADA responses to some protein products can result in greatly altered plasma drug levels over time, with both increased and decreased levels being observed, which may complicate study results interpretation. Therefore, appropriate serum samples during study should be preserved. The ADA measurement should be conducted, when (1) altered PD/loss or enhancement of efficacy is evident, (2) unexpected changes in PK, or (3) immune-mediated reactions are seen (acute effects like infusion reaction, hypersensitivity, anaphylaxis; delayed effects; immune complex-mediated like arthralgia, rash, myalgia, or cross reactivity with endogenous protein). Neutralizing potential of antibodies should be characterized in case of ADAs and lack of PD marker to demonstrate sustained activity in animals.

Preexisting or induced immunity, humoral or cell-mediated, to live vectors may impact on reuse or repeated use of the vector as well as on study result interpretation, and should be assessed [9].

Environmental Risk/Virus Shedding Assessment For live attenuated or live recombinant restored viral vaccines, shedding of infectious virus and environment risk should be assessed.

Inhalation Toxicity Studies Such studies are necessary, when aerosol route of administration gives rise to altered PK, qualitatively or quantitatively, compared to other routes [11]. In fact, inhaled product may have a local effect in airways, either short-term (effect on ciliary function or other signs of local irritation) or long term (emphysema, bronchitis, malignancy), which may not be observed for other routes of administration. Similarly, inhalation toxicity studies should be considered for new formulation that has not been tested by inhalation.

Two species should be used for short-term studies, followed by up to a 6-month study in the most appropriate species with new formulation for a chronically indicated product. Animals chosen should be free of pulmonary infection and have a low incidence of other pulmonary pathology. The method of administration should ensure the substance to reach the desired site, e.g., directly into airways via a nasotracheal tube or through a tracheotomy in acute studies, use of head-only or nose-only exposure chambers or masks for inhalation in long-term exposure studies. Normally three dose levels and one or more control groups (sham control, vehicle control) as appropriate should be used, and choice of dose levels should be justified. The duration of study should be related to intended human exposure. Observations should include any local effects including ciliary function and microflora, lung weights of all animals and histopathological examination of tissues taken from all exposed levels of respiratory tract and from associated lymphoid tissues.

Others Delayed hypersensitivity needs to be evaluated for new formulation that is administered through the vaginal route. Whereas no additional studies are recommended in addition to the acute and repeat dose toxicity studies, for the new formulation proposed for the rectal route of administration [11].

25.3.5 Common Issues and Current Challenges

Different animal species, including rodents (mice, rats, hamster, guinea pig), rabbits, ferrets, dogs, mini-pigs, and nonhuman primates have been used for nonclinical safety evaluation of biopharmaceuticals. However, significant differences in anatomical characteristics as well as physiology and immune regulatory pathways (including innate immunity system) exist between laboratory animals and humans. Currently, regulatory acceptable animal model is usually selected by an in vivo demonstration of pharmacological activity, for example, an induction of antibody in the animal for vaccine products. However, the animal model chosen for the nonclinical safety assessment should, ideally, be susceptible to the pathogen against which the vaccine antigen is directed. Whereas many currently used animal toxicology species are not permissive to the human pathogens. Therefore, to date, there are no animal models that can reliably predict the risk of an active substance, or an ingredient in final formulation of biopharmaceuticals, that may cause specific adverse events in the clinic. This could be due to inter-species differences, to varying degrees, in innate immune receptor distribution, cytokine repertoire, and expression and recognition or binding of the target of interest (for products such as recombinant human protein, vaccine antigen, adjuvant, nucleic acid, plasmid DNA), and for live vaccines (attenuated or recombinant vectored), the different replicating potential, thereby resulting in different physiopathological and immunopathological reactions between the laboratory animal models and humans.

For recombinant human proteins, a human cytokine used as the vaccine adjuvant, and some of other novel adjuvants that exert species-specific effects, the use of species-specific homologues of these biologicals (instead of clinical candidates) represents an optional or alternative approach for nonclinical safety assessment, and may be considered. However, the difference in species-specific receptor/target protein distribution, the lack of historical data, as well as the fact of difference between the formulation used in animals and the clinical candidate being used in the clinic, will limit the interpretation of the data derived from such studies.

Nonclinical safety evaluations are usually conducted using healthy adult animal models, e.g., for all kinds of preventative vaccine products and for biotechnology-derived protein products. For the latter intended for use in the pediatric population, an assessment of need/no need for juvenile toxicity studies is needed, and in some circumstances (poor predictability of adult data, target organs identified in immature/developmental systems from general toxicity studies, altered PK profile), such studies should be considered, to detect any increased vulnerability of immature animals to the systemic toxicity of the product compared with the adult, and to assess adverse effects on growth and development. However, there are currently no such requirements for vaccine products, even though most preventative vaccines are given to children. This situation may change, especially for novel adjuvants and adjuvanted vaccines with an neonate indication, which seems reasonable, considering the immunotoxicity potential of such products specifically intended to alter the immune system, the prolonged period of postnatal development of the immune system (reaching maturity at around the time of adulthood), and that adjuvants are often developed for use in several pediatric vaccines. However, while some of the juvenile models are in early development, they are often not available for nonclinical safety assessments, and unavailability of reagents and invalidation of the animal models can hamper immune evaluation. In addition, interpretation of findings derived from such models presents with challenges, especially since it may be very difficult, if not impossible, to extrapolate immune system developmental stages from the animal model to humans. In addition, since vaccines may be developed for specific subpopulations (e.g., elderly and immunosuppressed subjects), the question has been raised whether the nonclinical safety assessment should be studied in specific animal models.

Furthermore, for most, if not all, classical inactivated vaccines (with or without adjuvants) and other relevant biopharmaceuticals, immunogenicity is currently being evaluated as a part of nonclinical safety assessments, generally restricted to vaccine (antigen)-induced immune responses, e.g., antibody levels. However, this immunological parameter does not allow the full understanding of the safety of these products. Therefore, it may be considered to incorporate additional immune markers into general toxicity studies, such as cytokine profiles (e.g., IL-4, IL-13, IL-10; IFN-g, IL-12, IL-17, IL-23) , T-cell subsets (Th1, Th2, Th17, CD8 + CTL), IgE, and other biomarkers (e.g., IL-6, C-reactive protein and fibrinogen). Such supplementary information may be helpful for understanding and evaluating the potential of a product's immunotoxicity, including the risk of autoimmunity, allergy (hypersensitivity), and chronic inflammation, in particular for new adjuvanted vaccines. The choice of individual markers depends upon the product's PD, an understanding of the mechanism of action of the adjuvant used, and an assessment of the risks about the possibility to trigger or exacerbate potential immune disease events. However, efforts are still to be made to validate the relevant assays and/or to make reagents available for many animal species. Although there is some suggestion to assess the autoimmune risk in autoimmune-diseased animal models, the use of such diseased models in nonclinical safety assessments of prophylactic vaccines is, to date, exploratory. Indeed, there are no models that are able to predict autoimmunity in humans. Interpretation of data from currently available autoimmune diseased models presents with a real challenge, and the lack of historical data and the potential for confounders due to the disease itself presents some of the concerns regarding use of such models in nonclinical safety assessments. Similarly, we are still unable to predict respiratory hypersensitivity (e.g., immediate symptoms include rhinitis, bronchoconstriction, and asthma), and we do not have an acceptable model for prediction of systemic anaphylaxis, both are adverse reactions involving immunological mechanism and of relevance to biopharmaceutical products or aerosolized formulations. Therefore, these signals cannot be neglected.

The aforementioned limitations of current animal models and studies also apply to developmental toxicity studies in animal models for vaccine products and other biopharmaceuticals. For example, the primary purpose of developmental toxicity studies is to serve as a signal for detection of potential developmental hazards in humans. However, factors including the species specificity of induced immune response, species-specific differences in developmental time lines, species-specific differences in anatomy and physiology of reproductive organs, and differences in the dosing regimen between various species, etc., do exist which may complicate risk prediction. In addition, current endpoints used in regulatory developmental toxicity studies may not sufficiently address potential adverse effects on the physiology, immune system, and development of the offspring. However, the lack of validated assays and lack of animal model(s) that resemble human pregnancy present challenges with respect to what can be assessed to date.

So far, the safety profiles of mucosal adjuvants delivered by the nasal route are largely unknown, and the potential of antigen or adjuvant transfer to neuronal tissue via olfactory bulb in animal models must be fully explored. In addition, nasal delivery may include various absorption enhancers which may result in adverse effects that only become apparent after many months, whereas current animal studies usually only last a few weeks. The same is true for aerosol route administration of human protein hormones and growth factors, for which the long-term safety is very poorly understood. Many protein biopharmaceuticals, even those with fully human amino acid sequences, are known to cause an immune response. Pulmonary delivery of insulin product Exubera® resulted in a higher level of antibodies than subcutaneously injected insulin. However, such immune response to the product, either immediate (e.g., anaphylactic shock) or delayed, is poorly predicted by animal models.

Prediction of serious adverse events (SAEs) in humans from animal model data is a real challenge since these adverse events (AEs) occur at such low rates that it would be nearly impossible to detect them in animal models. Similarly, there are no nonclinical studies that can be used to predict adverse events such as vasculitis or autoimmunity. These limitations must be taken into account for study result interpretation. Even if no SAEs are observed in an extensive nonclinical toxicological study, it cannot be guaranteed that the adjuvanted vaccine formulation presents no risks to vaccinees, and unexpected events may occur.

In summary, characterization of pharmacological and toxicological profile of a mucosally delivered biopharmaceutical in animal models is of limitations which should be acknowledged. Predictability of human pharmacological effect or activity of biopharmaceuticals based on proof-of-concept studies in an animal may be limited due to the species-specificity of the response. Similarly, local and systemic adverse effects observed in a nonclinical safety study may not be directly translatable to the clinic. Moreover, currently designed safety studies in animal models are not expected to detect rare and/or late onset adverse events that may occur in human subjects. Nevertheless, nonclinical safety studies represent the best currently available tools to help maximize the benefit-risk assessment by evaluation of preclinical safety and pharmacology of mucosally delivered biopharmaceuticals.

25.3.6 Concluding Remarks on Nonclinical Aspects

Regulations of the EU and other ICH regions require that nonclinical evaluation of mucosally administered biopharmaceuticals including vaccine products is needed to provide scientific rationale for support of their initial testing in proposed clinical studies. In addition, nonclinical safety studies should be conducted to recommend a safe dose, schedule, and route to be investigated in human subjects. Therefore, nonclinical evaluation is an essential part of the mucosal biopharmaceutical development, as it not only provides proof-of-concept information but also helps establish the safety of the product to allow entry into clinical studies. The regulatory guidelines published by the EU, WHO, and the US/FDA allow flexibility in testing requirements, which are based on evidence and level of information available. The use of most relevant test models and careful design of the testing program will enable the most predictive nonclinical safety assessment. However, it should be recognized that gaps remain between currently available tools to assess pharmacology and toxicology of mucosal biopharmaceutical products and the fully relevant preclinical evaluations. With further experience being gained and new methods being developed, approaches to nonclinical evaluations of mucosal biopharmaceuticals will continue to evolve, aiming at optimization of evaluating product safety, prevention of unnecessary use of animals, and the ultimate support of product development.

25.4 Clinical Aspects

25.4.1 General Considerations

Generating clinical data is not only necessary for a later marketing authorization but also often a mine field for developers regarding the own versus the regulators' expectations.

Clinical data are the keystones, and each type of study phase provides a different view on one's own products as well as gives opportunity to optimize and streamline the development. On the other hand, regulators prefer to have—at least in the developer's view—a huge amount of data which simply means that a lot of money needs to be spent. To balance both sides' wishes and not unnecessarily delay a product's way to the market, this chapter tries to explain what regulators need for a decision on the risk-benefit profile of a product and at which points in the clinical-development program hurdles might be encountered.

25.4.2 Size of Database

Safety The size of the available safety database is always a key point at the time of applying for marketing authorization. For some products the regulators have decided on specific numbers whereby balancing the information versus the uncertainties on defined risks. The ICH Guideline E1 states that approximately 1,500 subjects should have been exposed to the IMP to attempt a license; this includes also "short-term" exposure. For novel vaccines for example, there is an EMA Guideline [46] stating that the safety database should at least include 3,000 persons that have received the final formulation of the vaccine in the intended dose and scheme (e.g., three initial doses plus one booster dose). This number was picked to be relatively sure that at least uncommon adverse events ($\geq 1/1{,}000$) are noted prior to marketing the vaccine. The extent of the database is rather easily achieved for vaccines as they are commonly used in various age-groups (all adding to the total number) and in healthy persons as a prophylactic agent.

Other products for human use, especially if used only for small indications, will never reach that number of subjects prior to marketing. Here, proving safety will mainly be done postlicensure and the extent of the database needed before licensure can and should be discussed with (national) regulatory agencies [47]. This also helps in defining which and how expected risks should be closely observed, and if this can be even done prior to licensure. Safety post the grant of marketing authorization is covered in the risk management plan and by experiences with similar products already providing safety data in the real-life application. Postlicensure usually also covers safety in special populations that are usually not the main target of an indication but are also concerned: immunocompromised (e.g., HIV or due to drugs), preterm newborns or pregnant women. For these groups (if not the main target of the product) usually special studies are done in the postmarketing phase but these studies and rough timelines are already agreed on during the authorization procedure.

Efficacy or Correlates Even though the European guideline for new vaccines states a specific limit of safety data, there is no such clear guidance for the number of subjects needed when it comes to efficacy/immunogenicity or similar proof of a product's intended effect. But as the recording of safety data is always an endpoint (even if only secondary) the extent of database will be very similar although probably slightly smaller for efficacy due to safety information coming from early studies.

All products will need an adequate size of data to achieve the right dose and scheme in the target population. For vaccines there is also the need to proof persistence of immunological protection, booster ability, and that vaccines concomitantly given do not affect the new product and are themselves not negatively affected. These prerequisites alone will increase the size of database. To not overly do so, there is, for vaccines, always the possibility to bridge results, e.g., from one age-group to another but this bridging also requires either well-proven correlates of protection or at least the existence of a certain cut-off level. For other products bridging might be easier as those affect certain clinical laboratory parameters (e.g., plasma glucose, liver enzymes, etc.) or have a measurable plasma level. Those parameters are affected by age and organ functionality but are often easier defined and comparable to other schemes, doses, populations, etc. than immunological parameters necessary to describe vaccines' effects.

Extrapolation is commonly used to bridge results from adults to children. Here the FDA uses an algorithm based on similarities (disease progression, response to intervention, etc.) [48], and ICH and EMA have established similar guidance[49, 50]. The EMA has now published a draft guidance document that takes these "similarity"-principles used for pediatrics to a more general and extended use to avoid replication of studies for ethical and resource reasons [51].

If the product is intended for an orphan disease, huge databases will not be reached and the applicability of extrapolation of data becomes a paramount goal.

Here, even a combined evaluation of single case studies can lead to licensure if a systematic review is made possible by careful planning of the studies beforehand: Treatment conditions and data collection need to be standardized, data should be of high quality and of course adhere to GCP standards. A combination of analyzing individual case reports and observational studies can also be considered. Detailed knowledge of the PK and PD of the drug (added by preclinical data) as well as detailed knowledge of the diseases' pathophysiology are vital to design those clinical studies and calculate the amount of data needed. There are also a number of size minimization measures, both design related and statistical, that are discussed in a European guideline for small sample sizes [52]. An FDA guideline covering the proof of effectiveness per se also discusses different database sizes and their possible implications or hurdles [53].

For mucosal delivery of a biopharmaceutical there might already exist a licensed "conventional" formulation that can be used for bridging of effectiveness and justification for an overall smaller database. Also if a product has shown its effect in one disease stage a single study might be sufficient to show that it is also effective, even to a different extent, in another disease stage.

Design, amount of data, control, and the choice of clinical or surrogate endpoints is best discussed with regulatory authorities either on a national or international basis to avoid later problems in the marketing authorization procedure. Overall it can be said that even one single pivotal study can be enough for licensure—if the study is of sufficient quality, data are compelling and the circumstances (disease frequency or severity, among others) justify this approach.

25.4.3 Controlling Safety

Controlling and assessing the safety of a new product is a central point in all drug development. The necessary extent of surveillance usually decreases with experience (e.g., later study phases) but increases if specific issues are seen in either the new product or related products or product-classes and becomes more focused. With the intramuscular (IM) application of vaccines, for example, one will nearly always see a certain extent of the same local reactions (redness, swelling, induration, or pain) and systemic reactions (fever, fatigue, and in infants, often, irritability). Those reactions are more or less class related: due to the local reaction of the immune system at the area the vaccine was injected and the systemic reaction of the immune system towards antigens, which involves certain cytokine cascades resulting in fever and general unwell feeling. Apart from those solicited reactions, there might be adverse events that only occur very rarely but can be attributed to the vaccination (if no other natural or artificial agent—a disease or another vaccine—can be held responsible): Guillan–Barre syndrome and Kawasaki syndrom are sometimes discussed in this respect. To evaluate if these syndromes are really caused by sometimes long time past vaccinations, licensed products all have a risk management plan that clearly defines risks and special medical conditions that should be closely monitored in the postauthorization phase. This document is regularly updated and results from various safety databases assessed to incorporate new issues and decide if the benefit of a given product still outweighs its risks.

This is all regardless of the application route but for the different application routes special aspects of safety should be kept in mind.

Specifics Concerning the Oral Route of Delivery Risks resulting from the oral application of a biological product occur usually beyond the gastric acid barrier either in later intestinal organs during replication of attenuated bacteria, transport-agent-related intestinal difficulties, or with the product during its pathway through various tissues and fluids exposed to enzymes, cytokines, and other cascades. The first-pass effect, for example, can not only diminish a product's effect but the product itself can damage all tissues involved here as well. So a clear knowledge about the products and its vehicles (capsules to overcome the gastric acidity, etc.) is needed to be able to closely scrutinize eventual vulnerable areas in the clinical use: liver parameters, blood parameters, and others.

Specifics Concerning the Buccal Route of Delivery In this route there will be no first-pass effect but a rapid systemic effect. The rapidity of the positive effect is even higher than for the rectal route in antiseizure benzodiazepines [54]. Nevertheless, locally high doses of very concentrated product might lead to irritations and even stones. Lesions anywhere in the mouth cavity might cause additional local pain and irritation.

Specifics Concerning the Pulmonary Route of Delivery Inhaling a product is in itself so difficult that it always requires the training of patients. Here, adverse events firstly result from wrong handling of devices resulting in incomplete effect in the target organ and possibly also irritation of the esophagus and intestinal apparatus due to inadvertent swallowing of the product. Also, concomitant infections that take place in the upper respiratory tract can reduce a product's uptake or the product might—due to its formulation—worsen an already present condition like swelling of the throat, coughing, and even respiratory distress in asthmatics.

Specifics Concerning the Nasal Route of Delivery Application of a drug to the nose will usually lead to a stuffy or runny nose as solicited AEs. Also if a low grade cold or an allergic predisposition already exists their symptoms will most likely worsen. In some clinical trials even asthma was seen as SAE following the use of a nasal vaccine [55]. Again, the correct use of the medical device needed to apply nasally will have to be trained and be nevertheless the first source of safety issues. Swallowing of wrongly applied product or too large quantities might occur as easily as with the pulmonary route. Long-term usage might lead to irritation, lesions, and scars of the nasal mucosa, especially if the medical device releases very high pressure "mist" or if the product's formulation is too concentrated. Apart from the nasal mucosa even the nervus olfactorius might be irritated and loss of smell and taste might occur. Additionally, due to only the ethmoid bone (perforated with nerve endings) separating the exposed nose cavity from brain tissue sterility of the supplied drug must be assured to avoid ascending infection.

Specifics Concerning the Ocular Route of Delivery Sterility, sterility, and sterility are the watchwords for the ocular route. No immune system prevents infection here, so sterile solution of the drug as well as sterile application are mandatory. Common AEs at this site are dryness of the eye or increased flow of tears along with reddening of the conjunctiva. Again, if an underlying condition like cold or allergic predisposition exists, the product might worsen the symptoms including clotted nose as (excess) solution is drained via the nasolacrimal ducts. Also, due to this draining path product might be swallowed analogue to the nasal route application.

Specifics Concerning the Vaginal Route of Delivery Using the vaginal route one needs to keep in mind the very balanced flora of the external and internal vaginal barrier as well as the menstrual-related pathway to the uterus. Ascending (bacterial) infection of the uterus can lead to infertility. Rising of the vaginal pH can very easily increase the danger of fungal and bacterial infections of vagina, uterus, and urine bladder. Specific applicators are usually needed for this route so training of use might again be a bottleneck for both safety and efficacy.

Specifics Concerning the Rectal Route of Delivery The rectal route is rather unproblematic and is frequently being used in infants or when oral uptake is difficult due to vomiting or other medical conditions. A very dense capillary network allows rapid uptake but also rapid elimination. Rectal diazepam for instance is well absorbed and has a rapid onset of action but is rapidly redistributed and can accumulate with repeated doses [56]. Elimination is increased if diarrhea is present and the drug might end up in the wrong places if, due to underlying diseases, ulcers or even fistulas are present. If applicators are used to supply the drug the risk of contaminating the vaginal area is given (see above).

25.4.4 Controlling Effectiveness

The choice of control is affected by the availability of a "similar" licensed product or in case there is none placebo or even "best standard of care" should be considered to guarantee the best possible unbiased estimate of effect. In case of very rare diseases even historical controls might be acceptable. For those groups, patient registers may supply important information for the assessment of effectiveness and safety either due to the natural course of disease or historical study data.

Vaccines There are two distinct ways in which the effect of a vaccine can be proven: either by efficacy studies which are always rather large and sometimes depending on the epidemiology long and difficult to achieve, or by finding and establishing an (functional) immunological correlate or surrogate of protection. The latter involves specific, validated, and reliable assays which in itself might proof very difficult if, for example, there is a difference between using animal or human sera. For meningococcal vaccines, there exist two distinct assays one uses baby-rabbit and one uses human sera; the results of these two assays are very different and not in the least correlated. Which of them now measures the "real" functional immune protection is unclear.

Nevertheless, when it comes to transferring the application route of a vaccine from IM or ID (intradermal) to any mucosal route, mucosal and nonmucosal antibody titers are the easiest comparability endpoint. The same applies to mucosal vaccines newly developed against diseases for which other vaccines already exist. For an entirely new vaccine against a disease for which so far no effective vaccine is licensed, the proof of effect must be done in the same way as for any other new vaccine: choosing between efficacy and immunogenicity.

Whether and how bridging other products, historical experiences, or on a genetic level is possible should best be discussed with regulatory authorities on an international basis. There is even the possibility to ask for a combined scientific advice by EMA and FDA.

Other Biopharmaceutical Products When proving the effect of a mucosal formulation of an already licensed product the comparison of PK and PD (especially the consequence of losing/ adding of the first-pass effect) of the new versus the established product are essential. Here, noninferiority or even superiority should be shown. For entirely new products, the disease's, pathophysiology needs to be well characterized to generate correlate(s) of clinical effect and PK and PD need to be described as detailed as with all other IM or IV biological products. Doses-response studies with other formulations or similar products can help to estimate the effective dose of the mucosal product by comparing the PK and PD of the substances[8]. For endpoints either clinical endpoints (e.g., improvement organ function parameters, complete remission, overall survival, progression-free survival, time to progression, etc.) can be chosen or the improvement/normalization of laboratory values as correlates of effect. Here, the comparison should be made with either standard of care, placebo, or similar products depending on the nature of disease, population, and availability of a treatment.

As always the careful design and adequate power of the study are vital to prove the product's benefit versus risks.

Food, Drugs, and Rock 'n Roll—Interference with Real Life? As medicinal products are seldom given on their own, drug–drug interaction as well as the influence of other life-style agents (food, drink, etc.) has to be evaluated and described for prescribers and users of the new medicinal product. The interaction can hereby be split into the effect of the other medicinal product/food/drink/others on the PK/PD of the new drug and the effect of the new drug on the PK/PD of the other medicinal product.

The effect of food, for example, on the delivery and uptake of an oral product needs to be measured during clinical trials, BA and BE are essential parameters to describe this effect in relation to fasted drug delivery and/or other (established) routes or products. Both FDA and EMA guidelines also describe methods for these studies that take into account interaction with other medicinal products, gastrointestinal function, diurnal rhythm, etc. [57, 58].

Medicinal product interaction studies also take into account tissue-specific enzyme and transporter alteration as well as the nature and possible inhibition of metabolites to the final elimination of the new and the concomitantly given/taken/present established drug or other agent. The European Guideline on drug interaction offers extensive discussion on various points of interaction as well as examples and strategies for clinical evaluation of the same [59].

Local or Systemic? When using the mucosal route of application, controlling the effect also needs to take into account whether the drug is intended to work locally or if a systemic effect is attempted. An intended systemic effect is easy to measure (see above). The measurement of the systemic absorption (e.g., due to inadvertent swallowing or due to blood transfer of the mucosal substance) of a locally effective drug often requires very sensitive assays ranging from the low ng/mL to pg/mL. Proving equivalence of locally intended drugs to a licensed product relies on both in vitro (quality) and in vivo data [60]:

- Qualitative and quantitative sameness of formulation
- Equivalence of in vitro tests (e.g., particle/droplet size, spray pattern, etc.)
- Equivalence of systemic exposure or systemic absorption
- Equivalence of the local delivery study (comparative design with clinical endpoints)

For new products that cannot be compared with a licensed drug only the BA is described with in vitro studies complemented by in vivo studies:

- Release of the substance from the drug product
- Availability to local sites of action

An example of the interacting factors influencing release and absorption is given for the nasal route in Fig. 25.1.

The addition of in vivo studies is made because based on the delivery method it might be impossible to adequately characterize particle size distribution and thus, the

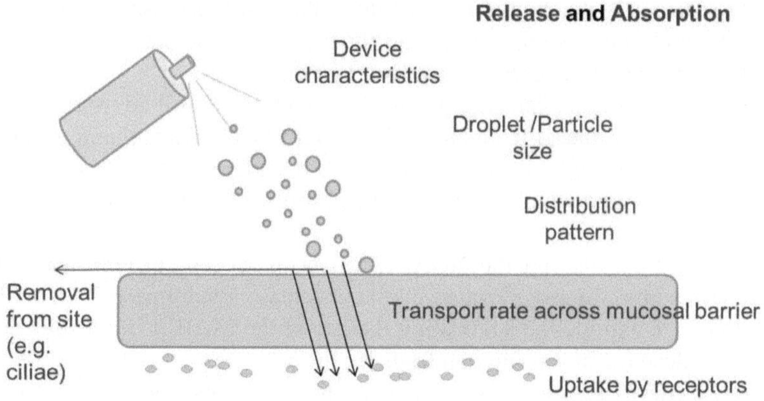

Fig. 25.1 Release and absorption factors

estimate of a local clinical effect so that the intended clinical effect should directly be tested.

Additionally to the single dose usage the PK and PD effects of repeated dosing need to be determined if the disease to be treated requires repeated doses.

If an existing product is changed either from a nonmucosal to mucosal use, formulation, dose, or other relevant, BA studies are used to determine:

- The rate and extent of absorption
- Fluctuations in drug concentrations
- Variability in PKs arising from of the drug formulation
- Dose proportionality
- Factors influencing the performance of the modified drug formulation
- The risk of unexpected release characteristics (e.g., dose dumping)

The extent and applicability of in vitro versus in vivo studies is best discussed with the national regulatory agencies.

25.4.5 First-in-Human Studies

At the start of clinical studies there are some milestones to be achieved for a new product: a generally good idea or even the proof of the product's mode of action and safety, both coming from animal models as well as a defined, if not yet completely validated, manufacturing process. Ideally for safety, more than one species has been exposed to the new drug with a dose that at least should equal the planned starting dose for the first clinical trial. To pick that starting dose it usually helps if a closely related product is already licensed, then this product's dose can be chosen. If none is yet licensed the best animal dose could be selected plus a dose one or two power of ten lower and one dose one or two power of ten higher. Two generally different approaches are used to select a "safe" human starting dose: the no-observed-adverse-effect level (NOAEL) which focuses on toxicity and the minimal-anticipated-biological-effect level (MABEL) which focuses on any observed effect.

The classical approach is using the NOAEL and this is still advised by the FDA guideline for FIM studies. First a NOAEL is determined, then the human equivalent dose is calculated using body surface and/or body weight conversions and finally, a safety factor is applied which represents the power to ten increase/decrease already mentioned [61].

In the TGN1412 disaster the NOAEL had been chosen but was insufficient to predict the resulting highly elevated cytokine-release syndrome [62]. This led for Europe to create a guideline advocating the use of the MABEL for a selection of a safe starting dose for first-in-human trials [63].

Using the MABEL or even NOAEL with the "adjustors" mentioned in the FDA guideline above is appropriate for most biopharmaceutical products. For vaccines on the other hand, depending on the antigen, both approaches can be completely inappropriate and result in a false feeling of safety concerning the transfer of nonclinical knowledge to the first clinical use. Here, the toxicity seen with various human doses in the animal models is usually a good indicator for the human safety but efficacy might be only seen in extensionally higher doses. The first use of a vaccine should be based on a differentiated risk evaluation taking into account the adjuvant (if any), vaccine antigen, schedule and immune response elicited. These four determinants allow for a testing strategy of potential risks that covers tissue cross-reactivity with human tissue, sequence database searches, comparison with "related" vaccines, and with natural infections and its sequelae in addition to the responsiveness to toxicity and to the natural pathogen in the relevant animal model [64]. These factors combined will lead to an individual risk assessment of the product (see Fig. 25.2).

From the patient's or trial subject's point of view another risk assessment has to be made, irrespective of whether the product is a vaccine or any other biopharmaceutical product. Risk identification includes vulnerability of the target population, availability of a relevant species for proof of concept and toxicity, and host factors. Risk assessment weighs identified risks versus anticipated and unexpected findings in animal studies or seen in similar concepts for both acute and chronic use. Details can be found in Fig. 25.3.

25.4.6 Age-Related Specialties

Age affects many points in medication: starting with the use and handling of a product to how the premature, developing, or ageing organ system reacts to a given drug.

Adults and the Elderly The FDA has available a guideline for the study of drugs likely to be used in the elderly which describes special concerns related to the age or diseases more likely occurring at older age [65]. For the EU, such guidance is planned to be drafted in the future, so far a very general ICH guideline exists [66].

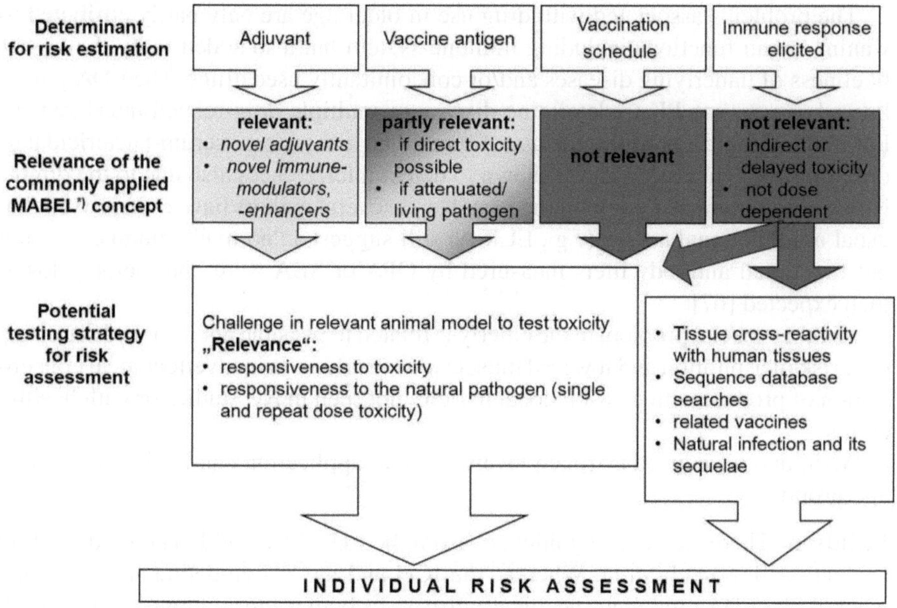

Fig. 25.2 Factors to be considered for the starting dose of a vaccine for first-in-human administration

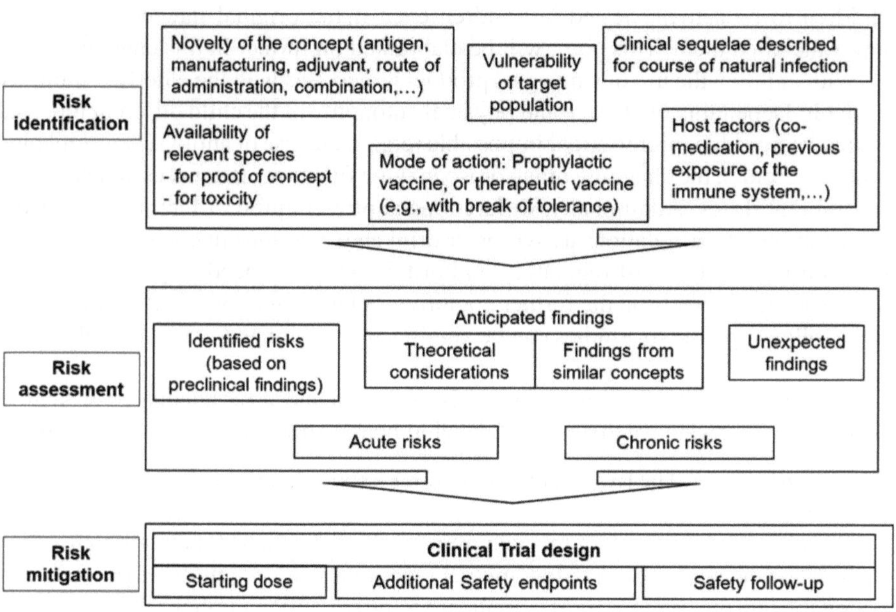

Fig. 25.3 Risk assessment for a vaccine intended for first-in-human administration

The problems associated with drug use in older age are only partly attributed to waning organ functions including immune system but also added to by the higher likeliness of underlying diseases and/or concomitantly used drugs. The FDA guideline advises to use PK to determine disease or multiple drug-related interferences. For immunological products the use of functional assays (e.g., serum-bactericidal or opsonophagocytic assays) have shown to detect interferences also due to the diminishing aging immune system in the past. Here, several authors have even proven that usual nonfunctional assays (e.g., ELISA) still suggest a "normal" immune reaction but functional antibody titers measured by OPA or SBA were considerably lower than expected [67].

Another special problem in the elderly is related to possible neurological sequalae, e.g., risk of aspiration and lowered muscle tone resulting in inadvertent local concentration of product with possible local toxicity not seen in NC studies or with healthy adults.

Also, due to reduced estrogen levels, vaginal application can be difficult in that age-group.

Children There are several guidelines from the EU, FDA, and ICH regarding drug use and testing in children. Whereas the ICH and the EU guideline are somewhat general, the FDA guideline specifically lists risks both acute and long term for each age-group and development status from prenatal to adolescents [68–70].

Some regions (e.g., the EU) require all products that can possibly be used in children to be actually tested in children even if the original intended indication never included them. This point will be elaborated on further in this chapter.

With children the use of a mucosal product is also—as with the elderly—multiply difficult: Depending on the age and physical autonomy of the child different mucosal routes are possible (preferred) or impossible (e.g., rectal use of antipyretics or inhaled corticosteroids), the changing organ or isoenzyme function often necessitates the use of different drugs. Additionally, different ages often require different doses or the use of different formulations as syrups or drinkable solutions in contrast to a tablet. Considering safety, most regulatory authorities stress the need to prove that there are no negative effects on the further cognitive, behavioral or physical development, thus, follow-up time for studies can be significantly prolonged if a "vulnerable" age is studied.

All the guidelines mentioned above also stress the need for adequate NC studies focusing on reproductive toxicity and developmental effects seen in juvenile animals.

PIP Requirements The EU legislation requires explicitly testing drugs for a possible pediatric use [71, 72].

To assure the fulfillment of this requirement all applications for authorization filed in the EU, whether for centralized, mutual recognition or decentralized procedure or national licensure, require for the applicant to prove "PIP compliance". That means that during the later stages of development a so called pediatric investigation plan (PIP) is drafted by the company responsible for the product development describing to what extent the said product will potentially be usable in any pediatric clientele. The PIP then needs to be filled with all studies per age-group, starting with nonclinical as necessitated by the human age-group's needs, that are considered necessary to reliably test a later use in children up to the age of 18 years. The Pediatric Committee (PDCO) of the EMA then decides on the applicability and adequacy of that test program and those "agreed" studies have to be undertaken (or agreed to be delayed after licensure) before a marketing authorization is applied for. By now, this system has been in place in the EU for more than 5 years and experience shows that it is advantageous to supply the PIP at a stage of development when the first data of human use in adults are already available (to allow for a risk-benefit analysis) but the clinical development program is not more or less done.

If the company or persons responsible for a product are of the opinion that their product is not of any use in children this has to be justified in the PIP and agreed on by the PDCO. The intended indication (e.g., Alzheimer's disease) is not valid for a waiver but the mode of action must be described in detail and it has to be shown that this mode of action is of no use in even very different diseases seen in children. For instance, monoclonal antibodies against "adult tumors" (e.g., malign melanoma) might use the same mode of action although in different effector cells in common children's tumors.

A Question of Formulation: Adjusting Adult's Product or Start a new? When it comes to actually using a product in children there might be the necessity to have different formulations for different age-groups. Here, it is of importance whether the originally developed product can be easily transformed into more suitable presentations for children or if the formulation change also requires a dose adjustment or conjugation/mixing with other agents. Vaccines that are available for both IM and oral use, for instance, use very different formulations with completely distinct product development programs. The EMA reflection paper "Formulations of choice for the pediatric population" [73] lists preferred formulations per age-group as well as risks and benefits due to different ways of application. It is stressed that especially excipients might be handled differently by the juvenile organism than is known from adults. This might also facilitate the need for a separate product development.

25.4.7 Implications of Genetically Modified Organisms

As discussed before a special guideline covers the necessary steps for the ERA [74]. For mucosal delivery, especially, shedding is of importance be it occurring naturally (after oral, vaginal, or rectal use) or by accidentally using too much product ocular or nasally.

The GMO is usually compared to the same "non-GMO" under corresponding conditions (e.g., survival of mycobacterium tuberculosis outside the human body) and the nature and technique of the genetic modification is also taken into account (whether or not viral vectors are used or if similar organisms were combined).

From a clinical point of view, it has to be taken into account that the patient is not an entity himself but will have contact with other persons who then might be at risk of exposure from the GMO. Thus, in clinical studies that do not completely take place in hospitals the health status of household contacts is of concern as well. Usually, double measures are taken to prevent pregnancies and even special antimicrobial soaps or cleaners need to be used.

25.5 Advanced Therapy Medicinal Products and their Delivery by Mucosal Routes

25.5.1 Introduction and Legal Background

In recent years the continuous progress in the areas of biology, biotechnology, and medicine originated promising therapies for prevention and treatment of diseases and dysfunctions of the human body. Especially the field of biopharmaceuticals benefitted a lot from technical innovations and a more substantiated scientific knowledge. ATMP reflect the most complex and innovative class of biopharmaceuticals as these products are highly research-driven and heterogeneous with regard to their origin, type, and complexity. Furthermore, ATMP usually combine various aspects of medicine, cell, and molecular biology and material science and are mostly characterized by innovative manufacturing processes. The class of ATMP integrates gene therapy medicinal products (GTMP), somatic cell therapy medicinal products (sCTMP), and tissue engineering products (TEP). The fact that many ATMP represent individualized and patient-specific therapies is a further challenge with which developers and regulators have to deal. Moreover, advanced therapies are often developed by small and medium sized enterprises, university and academia, for whom regulatory requirements could present a major hurdle on a successful way from science to market.

The entry into force of Regulation (EC) 1394/2007—the so called "ATMP Regulation"—represents a milestone in providing a clear regulatory framework for the development of advanced therapies. In 2009, a new European committee was established: the Committee for Advanced Therapies (CAT). This committee gathers the best available expertise on this type of products in the European Community whereas its composition ensures appropriate coverage of all scientific areas relevant to advanced therapies, i.e., gene therapy, cell therapy, tissue engineering, but also medical devices, pharmacovigilance, and ethics. Moreover, representatives of patient associations and clinicians are also present in the CAT to get more insight in ATMP-related matters from a distinct point of view. In its "Work Program 2010–2015" the committee pointed out its main future activities as, for example, to foster innovation and to facilitate development of ATMPs and access to marketing authorization procedure [75].

The CAT is responsible for all regulatory procedures concerning ATMP in the EU, inter alia the classification, certification, and scientific evaluation of ATMP in centralized marketing authorizations. It collaborates with other EMA scientific committees, working parties, and others, e.g., the Pediatric Committee, the Scientific

Advice Working Party, and EMA's Innovation Task Force. One important task of the committee is its scientific recommendation of classification of ATMP according to Art. 17 of the "ATMP-Regulation." Article 17 states that any applicant developing a product based on genes, cells, or tissues may request a scientific recommendation on classification to find out whether the referred product falls, on scientific grounds, within the definition of an ATMP. Thus, the CAT delivers free of charge scientific recommendations on ATMP classification after consultation with the European Commission within 60 days after receipt of the request. The outcome of all scientific recommendations on classification can be obtained from the EMA website whereas all summary reports are listed from July 2011 onwards, certainly, after deletion of all information of commercial confidential nature [76]. More guidance and experience on the ATMP classification procedure is provided in the "Reflection paper on classification of advanced therapy medicinal products" that was adopted by the CAT at the end of 2012 [77]. From a legal perspective classifications given by the CAT are not legally binding as the classification of medicinal products lies within the responsibility of the respective National competent authority (NCA) where the product is manufactured. However, the CAT classification procedure is one important step forward to harmonization of regulatory thinking and requirements in the European scientific and regulatory community. Furthermore, it is a valuable tool and also a good opportunity for developers to get in early touch with the EU regulatory body in preliminary stages of product development but also for regulators to get an insight in the product pipeline of ATMP.

25.5.2 Classifications of ATMP Applied by Mucosal Routes

Among other ATMP that are currently in development in the EU and elsewhere this chapter will focus on four ATMPs which have been subject to the CAT's classification procedure and which are all applied by mucosal routes: Two are for drinking purposes, one for mouth rinsing, and one is formulated as enteric-coated capsules or enema. All of them were classified by the CAT as gene therapy medicinal products and interestingly, they are all based upon two genetically modified bacteria strains: *Lactococcuslactis* and *Salmonella typhi* Ty21a. Two of these products are intended for the treatment of inflammatory diseases, the other two represent cancer immunotherapies. Three of the products already reached the clinical stage; one is currently in the nonclinical stage of development.

During their classification procedure some general discussions came up regarding the fulfillment of the GTMP definition as written in Directive 2009/120/EC amending Directive 2001/83/EC Annex I Part IV. Here it is stated under 2.1 that "a gene therapy medicinal product means a biological medicinal product which has the following characteristics: (a) it contains an active substance which contains or consists of a recombinant nucleic acid used in or administered to human beings with a view to regulating, repairing, adding or deleting a genetic sequence; and (b) its therapeutic, prophylactic or diagnostic effect relates directly to the recombinant nucleic acid sequence it contains, or to the product of genetic expression of this sequence." It was discussed whether the definition implies that the recombinant nucleic acid has to be transferred into the human-recipient cell or genome and has to be expressed by the eukaryotic machinery. As this is not the case for this kind of products, it was talked over if they rather fulfill solely the definition of GMOs as the definition of GTMP. However, reading the GTMP definition thoroughly word by word—without adding or deleting anything—the majority of the CAT was of the opinion that these products fulfill the definition of GTMP—clearly also the definition of GMO—as the therapeutic effect is directly related to the genetic expression of the recombinant nucleic acid sequence they contain.

"Genetically Modified *Lactococcus lactis* Secreting Human Interleukin-10 (hIL-10)" This product is intended for the treatment of inflammatory bowel disease. *Lactococcus lactis* is one of the most common microorganisms in the dairy industry, used for manufacture of dairy products like buttermilk and cheese. In this product, the recombinant bacteria are either applied orally in capsules or by topical rectal application in form of enemas. Introduced into the patient's gastrointestinal (GI) tract or the colon, they reside for a limited period of time secreting hIL-10. Interleukin-10 (IL-10) is an anti-inflammatory cytokine that has pleiotropic effects in immunoregulation and inflammation. Furthermore, it has been shown to play a role in epithelial integrity and modulation of the mucosal immune system [78]. As stated by the company by delivering hIL-10 locally at inflamed tissue in the intestine, it is believed that, compared to hIL-10 given by injection, the effectiveness may be increased, with fewer adverse effects. A Phase I trial with the product in patients with Cohn's disease reinforces this assumption [79]. For the evaluation of the safety, tolerability, PD, and efficacy of the transgenic bacteria expressing hIL-10 a Phase IIa clinical study was initiated in subjects with moderately active ulcerative colitis. This study has been completed in September 2009 and the company claimed that all primary endpoints of the study have been successfully met.

"Medicinal Product Composed of Living, Genetically Modified *Lactococcus lactis* Bacteria, Containing the Human Trefoil Factor 1 (hTFF1) Gene" Another medicinal product that is also composed of living, genetically modified *Lactococcus lactis* bacteria is intended for the prevention and treatment of chemotherapy-induced and/or radiotherapy-induced oral mucositis in patients with cancer of the head and the. Oral mucositis is a painful inflammation and ulceration of the mouth, affecting nearly every patient that receives radiotherapy of the head/neck region and also many patients suffering from solid tumors treated with chemotherapy or radiation therapy. Thus, it is a common and often debilitating site effect of cancer treatment, which can be accompanied by local infections by viruses, bacteria, or fungi. The *Lactococcus lactis* bacteria containing the human trefoil factor 1 (hTFF1) gene is formulated as an oral mouth rinse formulation with which the recombinant bacteria are introduced in the subject's oral cavity. As in the other example the expression construct is under the control of a prokaryotic promoter and it is not integrated into the human cell or genome. Members of the trefoil (TFF) family play an essential role in epithelial restitution and repair. They are nonmitogenic, protease-resistant, stable secretory proteins expressed in GI mucosa and postulated to protect the mucosa from insults, stabilize the mucus layer, and affect healing of the epithelium. It could be shown in a hamster model for radiation-induced oral mucositis that the topical administration of the hTFF1 protein producing engineered bacteria delivers therapeutic concentrations of the protein significantly reducing the severity and course of the inflammatory process [80]. Twenty-four hours after dosing, the bacteria and also the protein were still detectable at the administration site, indicating that the bacteria adhere to the buccal mucosa and actively secrete protein locally, resulting in homogeneous exposure to the entire mucosal surface. However, even if the company claims that the bacteria cannot survive in the systemic circulation of the animals, there is some risk for potential infections caused by the recombinant bacteria staying at the gut and its lumen. A Phase Ib clinical study in cancer patients at risk of developing oral mucositis has already been started to test different dose levels and dosing frequencies but also the safety and tolerability of the topically applied GTMP. As stated by the company the study provided also initial efficacy data, indicating a 35 % reduction in the duration of ulcerative oral mucositis in recombinant bacteria-treated patients versus the placebo-treated patients and close to 30 % response rate in the active group versus no response in the placebo group.

"*Salmonella typhi*-Based Oral Cancer Immunotherapeutics" As mentioned before, the next two GTMP classified by the CAT are both oral cancer immunotherapeutics consisting of an engineered, virulence-attenuated strain of *Salmonella typhi Ty21a* bacteria, a strain that has also been used for licensed oral typhoid vaccines. Whereas the GTMP "*Salmonella typhi* strain genetically modified to secrete a fusion protein of the prostate specific antigen (PSA) and a protein leading to an increased antigenicity" is intended for the treatment of prostate cancer by directly attacking the PSA-expressing tumor cells, the other product "DNA vaccine targeting vascular endothelial growth factor receptor 2 (VEGFR-2) using *Salmonella typhi* Ty21a as a vector" elicits an immune response against VEGFR 2 that leads to CTL-mediated killing of proliferating endothelial cells. Thus, this product is not directed against the tumor cells themselves but it is breaking the peripheral immune tolerance against the VEGF receptor 2 thereby preventing the angiogenesis of the tumor and inhibiting its growth [81]. Both medicinal products are provided for drinking purposes and as in the examples before, all elements are controlled by prokaryotic promoters.

The principle of "*Salmonella typhi*" strain genetically modified to secrete a fusion protein of the prostate specific antigen (PSA) and a protein leading to an "increased antigenicity" is based on the recombinant expression of prostate-specific antigen fused to the B subunit of cholera toxin and a secretion signal in the presence of the *Escherichia coli* type I hemolysin secretion system. Consequently, the expression and secretion of PSA is accompanied by an intrinsic, immunological adjuvant leading to an enhanced induction of CD8 T-cell responses. Studies in a mouse tumor challenge model seem to demonstrate proof-of-concept with a mouse homologue of the product.

"DNA vaccine targeting vascular endothelial growth factor receptor 2 (VEGFR-2) using *Salmonella typhi* Ty21a as a vector" is intended for the treatment of solid malignancies with or without metastases. It is thought to have a broad spectrum of anticancer activity as the target, i.e., VEGFR 2, is abundantly present on the tumor vasculature. After oral vaccination VEGFR 2-specific cytotoxic T cells, so called killer cells, mediate a systemic antiangiogenic effect by targeting and destroying the neovascular endothelial cells. Consequently, tumor growth is inhibited, vessels start to leak and dissolve, and the tumor shrinks. Furthermore, due to missing vessels the risk of metastasis formation is reduced. As the product is thought to cause inflammation in proximity to the tumor, an immune response against the tumor itself is initiated, too. Such an antitumor activity was demonstrated in various animal studies for different tumor types [82]. Here, no impairment of fertility, neuromuscular performance or hematopoiesis could be observed. However, some delay in wound healing was detected and should be adequately addressed as a potential risk during further product development. Currently, human clinical trials enrolling pancreatic cancer patients have already been initiated, further clinical studies (Phase II) are planned to be conducted in other cancer indications [83].

25.5.3 Mucosal ATMPs: Specific Aspects

Taken together, several GTMP applied by mucosal routes, i.e., the oral mucosa, the GI tract, or the colon, are currently under research and development and a part of the ATMP pipeline. The strategy of topical delivery of genetically modified bacteria is thought to circumvent systemic site effects thereby leading to safer medicinal products. Furthermore, such genetically modified bacterial carrier systems could be seen as platform technologies which can be applied to a wide range of diseases. Thus, this technology could build up the basis for different types of GTMP depending on the gene/protein of interest with several treatment opportunities. Accordingly, these products share the same principles but they also bare similar risks that should be taken into account during drug development. Administering genetically modified bacteria, i.e., GMOs could also have negative implications for the patient and also for its environment. In general, an appropriate data package with a comprehensive risk management plan is one important criterion when the medicinal product enters clinical trials or marketing authorization. According to the nature of the product different questions have to be addressed in the risk management. Concerning the patient's health such questions could be for example:

- Does treatment with genetically modified bacteria lead to infections of the human body?
- Are treatment effects on body weight gain or food consumption detectable?
- Does the drug induce systemic or long-term side effects?

Considering environmental risks the deliberate release of GMOs into the environment, its shedding and recombination-caused GMO formation should be properly considered. An ERA needs to be carried out based on a scientifically sound premise, empirically derived data and/or clinical use. The ERA generally needs to consider potential adverse effects for nonpatients, e.g., staff in the clinic involved in administering the product or in patient care, family members (including infants or immunosuppressed persons), the "general public," but also for animals, plants, and microorganisms. There are several guidelines addressing such risks and providing support during product development, e.g., ICH Considerations: General Principles to Address Virus and Vector Shedding [84], Guideline on Scientific Requirements for the Environmental Risk Assessment of Gene Therapy Medicinal Products [85], and Guideline on Follow-up of patients administered with GTMP [86]. According to these guidance documents and the identified product-specific risks information on proposed monitoring and risk management strategies needs to be provided for entering clinical trials or as part of the marketing authorization application.

However, all these GTMP are promising innovative therapeutics combining cutting-edge biotechnology, microbiology, oncology, and immunology. The application of GTMP via the mucosal route may open a new treatment opportunity for potentially safer innovative advanced therapies.

25.6 Conclusions and Future Perspectives

Developing a biopharmaceutical mucosal product has its own hurdles and the available regulatory information is diverse and might seem confusing for developers. It nevertheless pays to generally know common pitfalls and that locally, regionally, and even globally regulators are available to give advice and help product development with their own regulatory input. Regulatory and legal requirements are under constant revision to keep up with scientific progress and patients' needs; also there are sometimes pronounced differences in those requirements between the different regions of the world (e.g., Europe and the USA). Thus, it is only feasible to leave the knowledge of details to the regulatory experts as scientific and product details are left to you as developers. Bringing both sides of knowledge together is the all-fitting key to success in this respect.

Contact with regulators can be made from very early points of development and during the whole cycle of a product's life. Even after licensure when it might be required or useful to change things about the product itself or other circumstances concerning the product regulators are always happy to assist with guidance to help smoothing the obligatory legal procedures.

References

1. Holmgren J, Czerkinsky C. Mucosal immunity and vaccines. Nat Med. 2005;11(4):45–53.
2. Ryan EJ, Daly LM, Mills KHG. Immunomodulators and delivery systems for vaccination by mucosal routes. Ternds Biotech. 2001;19(8):293–304.
3. ICH Topic S6 (R1) Preclinical safety evaluation of biotechnology-derived pharmaceuticals. 1997. Addendum to the Parent Guideline (2009).
4. WHO guidelines on nonclinical evaluation of vaccines. 2003. www.who.int/biologicals/publications/nonclinical_evaluation_vaccines_nov_2003.pdf.

5. Guidelines for assuring the quality and nonclinical safety evaluation of DNA vaccines. WHO Technical Report Series No 941. 2007.
6. Guideline on adjuvants in vaccines for human use. 2005. EMEA. EMEA/CHMP/VEG/134716/2004.
7. Guideline on the non-clinical studies required before first clinical use of gene therapy medicinal products. 2008. EMEA. EMEA/CHMP/GTWP/125459/2006.
8. Note for guidance on the quality, preclinical and clinical aspects of gene transfer medicinal products. 2001. EMEA. CPMP/BWP/3088/99.
9. Guideline on quality, non-clinical and clinical aspects of live recombinant viral vectored vaccines. 2010. EMA. EMA/CHMP/VWP/141697/2009.
10. Note for guidance on preclinical pharmacological and toxicological testing of vaccines. 1997. EMEA. CPMP/SWP/465/95.
11. Guidance for industry and review staff. "Nonclinical safety evaluation of reformulated drug products and products intended for administration by an alternative route". 2008.
12. Guidance for Industry:"Considerations for plasmid DNA vaccines for infectious disease Indications". 2007. http://www.fda.gov/biologicsbloodvaccines/guidancecomplianceregulatory information/guidances/vaccines/ucm074770.htm.
13. Guidance for Industry: " Consideration for developmental toxicity studies for preventive and therapeutic vaccines for infectious disease indications". 2006. http://www.fda.gov/Biologics BloodVaccines/GuidanceComplianceRegulatoryInformation/Guidances/Vaccines/ucm074827. htm.
14. Pires A, Fortuna A, Alves G, et al. Intranasal drug delivery: how, why and what for? J Pharm Pharm Sci. 2009;12(3):288–311.
15. Makidon PE, Nigavekar SS, Bielinska AU, et al. Characterization of stability and nasal delivery systems for immunization with nanoemulsion-based vaccines. J Aerosol Med Pulm Drug Deliv. 2010;23(2):77–89.
16. Illum L. Nasal drug delivery—recent developments and future pro-spects. J Control Release. 2012;161(2):254–63.
17. Wang SH, Thompson AL, Hickey AJ, et al. Dry powder vaccines for mucosal administration: critical factors in manufacture and delivery. Curr Top Microbiol Immunol. 2012;354:121–56.
18. ICH Topic Q6B: specifications: test procedures and acceptance criteria for biotechnological/biological products.
19. FDA guidance for industry: characterization and qualifications of cell sub-strates and other biological materials used in the production of viral vaccines for infectious disease indications.
20. Guideline on the Pharmaceutical quality of inhalation and nasal products; EMEA/CHMP/QWP/49313/2005.
21. Guideline on comparability of medicinal products containing biotechnology-derived proteins as active substance: quality issues; EMEA/CPMP/BWP/3207/00/Rev1.
22. Towns J, Webber K. Demonstrating comparability for well-characterized biotechnology products. Bioprocess Int. 2008;2(6):32–43.
23. ICH Topic Q5E Comparability of biotechnological/biological products.
24. Guideline on similar biological medicinal products containing biotechnology-derived proteins as active substance: quality issues; EMEA/CHMP/BWP/49348/2005.
25. US Food and Drug Administration (FDA) Guidance for industry: bioavailability and bioequivalence studies for nasal aerosols and nasal sprays for local action; FDA Washington, DC; 2003.
26. ICH topic Q5C quality of biotechnological products: stability testing of bio-technological/biological products.
27. Guidelines on stability evaluation of vaccines; WHO/BS/06.2049.
28. Lovelyn C, Attama A. Current state of nanoemulsions in drug delivery. J Biomater Nanobiotechnol. 2011;2(5A):626–39.
29. Anliker B, Longhurst S, Buchholz CJ. Bundesgesundheitsbl. 2010;53:52–7.
30. Guideline on plastic immediate packaging materials; CPMP/QWP/4359/03.

31. Guideline on the suitability of the graduation of delivery devices for liquid dosage forms; EMEA/CHMP/QWP/178621/2004.
32. Guideline on Excipients in the dossier for application for marketing authorization of a medicinal product; EMEA/CHMP/QWP/396951/2006.
33. Directive 2001/83/EC of the European parliament and of the council of 6 November 2001 on the Community code relating to medicinal products for human use. Official J L. 2001;311:67-128
34. Code of Federal Regulation, Title 21, Part 312, Washington, DC, US Government Printing Office. 2010. http://www.accessdata.fda.gov/scripts/cdrh/cfdocs/cfcfr/CFRSearch.cfm?CFRPart=210&showFR=1.
35. Guideline on non-clinical testing for inadvertent germline transmission of gene transmission of gene transfer vectors. 2005. EMEA. EMEA/273974/05.
36. ICH Guideline M3 (R2) on non-clinical safety studies for the conduct of human clinical trials and marketing authorization for pharmaceuticals. Step 5. 2009. EMA/CPMP/ICH/286/1995.
37. Guideline on repeated dose toxicity. 2010. CPMP/SWP/1042/99 Rev 1 Corr.
38. ICH—S5A Guideline on detection of toxicity to reproduction for medicinal products. 1994.
39. Guideline on the need for non-clinical testing in juvenile animals on human pharmaceuticals for paediatric indications. 2005. EMEA. Doc. Ref. EMEA/CHMP/SWP/169215/2005.
40. WHO Guidelines on the quality, safety and efficacy of dengue tetravalent vaccines (live, attenuated). 2011.
41. WHO recommendations to assure the quality, safety and efficacy of live attenuated poliomyelitis vaccine (oral). 2012.
42. WHO Guidelines on the quality, safety and efficacy of japanese encephalitis vaccine (live, attenuated) for human use. 2012.
43. Note for guidance on the development of vaccinia virus-based vaccines against smallpox. 2002. EMEA. CPMP/1100/02.
44. IABS scientific workshop on neurovirulence tests for live attenuated viral vaccines. 2005. WHO.
45. Mutsch M, Zhou W, Rhodes P, Bopp M, Chen RT, Linder T, Spyr C, Steffen R. Use of the inactivated intranasal influenza vaccine and the risk of Bell's palsy in Switzerland. N Engl J Med. 2004;350(9):896–903.
46. Guideline on clinical evaluation of new vaccines, EMEA/CHMP/VWP/164653/2005; www.ema.europa.eu.
47. ICH Topic E, 1 Population Exposure: the extent of population exposure to assess clinical safety, www.ich.org.
48. Pediatric Decision Tree. US Food and Drug Administration. Specific requirements on content and format of labelling for human prescription drugs: revision of "pediatric use" subsection in the labeling: final rule. FedRegist. 1994;59. www.fda.gov.
49. ICH E11 Clinical Investigation of medicinal products in the paediatric population, www.ich.org.
50. Role of Pharmacokinetics in the development of medicinal products in the Paediatric Population, CHMP/EWP/147013/2004, www.ema.europa.eu.
51. Concept paper on extrapolation of efficacy and safety in medicine development, EMA/129698/2012, www.ema.europa.eu.
52. Guideline on clinical trials in small populations, CHMP/EWP/83561/2005, www.ema.europa.eu.
53. Providing clinical evidence of effectiveness for human drugs and biological products, http://www.fda.gov/cder/guidance/.
54. Ellis SJ, Baddely. Buccal midazolam and rectal diazepam for epilepsy. Lancet. 1999; 353(9166):1796–7.
55. Clinical AR of FluMist® www.fda.gov/downloads/BiologicsBloodVaccines/Vaccines/ApprovedProducts.
56. Rey E, Treluyer JM, Pons G. Pharmacokinetic optimization of benzodiazepine therapy for acute seizures. Focus on deliveryroutes, Clinpharmacokinet. 1999;36:409-24.
57. Food-effect bioavailability and fed bioequivalence studies, FDA guidance for industry, www.fda.gov/cder/guidance/.

58. Note for guidance on modified release oral and transdermal dosage forms: SECTION II (Pharmacokinetic and clinical evaluation), CPMP/EWP/280/96, www.ema.europa.eu.
59. Guideline on the investigation of drug interactions; CPMP/EWP/560/95/Rev. 1, www.ema.europa.eu.
60. Bioavailability and bioequivalence studies for nasal aerosols and nasal sprays for local Action, FDA Guidance for Industry, http://www.fda.gov/cder/guidance/.
61. US Food and Drug Administration (FDA). Guidance for industry and reviewers. estimating the safe starting dose in clinical trials for therapeutics in adult healthy volunteers (FDA, Washington, DC; 2002). www.fda.gov/downloads/Drugs/GuidanceComplianceRegulatoryInformation/Guidances/.
62. Suntharalingam G, et al. N Engl J Med. 2006;355:1018–28.
63. Guideline on Strategies to identify and mitigate risks for first-in-human clinical trials with investigational medicinal products, CHMP/SWP/28367/07, www.ema.europa.eu.
64. Goetz KB, Pfleiderer M, Schneider CK. First-in-human clinical trials with vaccines–what regulators want. Nat Biotechnol. 2010;28(9):910–6.
65. Guideline for the study of drugs likely to be used in the elderly, US Department of Health and Human Services, Food and Drug Administration, 1989. www.fda.gov/cder/guidance/.
66. Studies in support of special populations: geriatrics, ICH Topic E7, www.ich.org.
67. Romero-Steiner S, et al. Reduction in functional antibody activity against streptococcus pneumoniae in vaccinated elderly individuals highly correlates with decreased IgG antibody avidity. Clin Infect Dis. 1999;29(2):281–8.
68. General considerations for the clinical evaluation of drugs in infants and children, US Department of Health and Human Services, Food and Drug Administration, 1977. www.fda.gov/cder/guidance.
69. Guideline on the role of pharmacokinetics in the development of medicinal products in the paediatric population, EMEA/CHMP/EWP/147013/2004, www.ema.europa.eu.
70. Clinical investigation of medicinal products in the paediatric population, ICH Topic E11, www.ich.org.
71. Regulation (EC) No 1901/2006 of the European Parliament and of the Council of 12 December 2006 on medicinal products for paediatric use and amending Regulation (EEC) No 1768/92, Directive 2001/20/EC, Directive 2001/83/EC and Regulation (EC) No 726/2004, http://ec.europa.eu/health/documents/eudralex/.
72. Regulation (EC) No 1902/2006 of the European Parliament and of the Council of 20 December 2006 amending Regulation 1901/2006 on medicinal products for paediatric use. http://ec.europa.eu/health/documents/eudralex/.
73. Reflection paper: Formulations of choice for the paediatric population, EMEA/CHMP/PEG/194810/2005, www.ema.europa.eu.
74. Guideline on environmental risk assessments for medicinal products consisting of, or containing, genetically modified organisms (GMOs), EMEA/CHMP/473191/06 Corr, www.ema.europa.eu.
75. Committee for Advanced Therapies (CAT) Work Programme 2010–2015 (EMA/CAT/235374/2010).
76. European Medicines Agency, advanced-therapy-medicinal-product classification. Summaries of scientific recommendations.
77. Reflection paper on classification of advanced therapy medicinal products (EMA/CAT/600280/2010).
78. Emami CN, Chokshi N, Wang J, Hunter C, Guner Y, Goth K, Wang L, Grishin A, Ford HR. Role of interleukin-10 in the pathogenesis of necrotizing enterocolitis. Am J Surg. 2012 Apr;203(4):428–35.
79. Braat H, Rottiers P, Hommes DW, Huyghebaert N, Remaut E, Remon JP, van Deventer SJ, Neirynck S, Peppelenbosch MP, Steidler L. A phase I trial with transgenic bacteria expressing interleukin-10 in Crohn's disease. Clin Gastroenterol Hepatol. 2006 Jun;4(6):754–9.
80. Caluwaerts S, Vandenbroucke K, Steidler L, Neirynck S, Vanhoenacker P, Corveleyn S, Watkins B, Sonis S, Coulie B, Rottiers P. AG013, a mouth rinse formulation of Lactococcuslactis

secreting human Trefoil Factor 1, provides a safe and efficacious therapeutic tool for treating oral mucositis. Oral Oncol. 2010 Jul;46(7):564–70.
81. ActoGeniX announces positive results from a Phase 1 PK study of AG013 for the prevention of oral mucositis in cancer patients. Ghent, Belgium, August 22, 2012.
82. Niethammer AG, Xiang R, Becker JC, Wodrich H, Pertl U, Karsten G, Eliceiri BP, Reisfeld RA. A DNA vaccine against VEGF receptor 2 prevents effective angiogenesis and inhibits tumor growth. Nat Med. 2002 Dec;8(12):1369–75.
83. Niethammer AG, Lubenau H, Mikus G, Knebel P, Hohmann N, Leowardi C, Beckhove P, Akhisaroglu M, Ge Y, Springer M, Grenacher L, Buchler MW, Koch M, Weitz J, Haefeli WE, Schmitz-Winnenthal FH. Double-blind, placebo-controlled first in human study to investigate an oral vaccine aimed to elicit an immune reaction against the VEGF-Receptor 2 in patients with stage IV and locally advanced pancreatic cancer. BMC Cancer. 2012;12:361.
84. ICH considerations—general principles to address virus and vector shedding (EMEA/CHMP/ICH/449035/2009).
85. Guideline on scientific requirements for the environmental risk assessment of gene therapy medicinal products (EMEA/CHMP/GTWP/125491/2006).
86. Guideline on follow-up of patients administered with gene therapy medicinal products (CHMP/GTWP/60436/07).

Index

J. das Neves, B. Sarmento (eds.), *Mucosal Delivery of Biopharmaceuticals*, DOI 10.1007/978-1-4614-9524-6,

MIX
Papier aus verantwortungsvollen Quellen
Paper from responsible sources
FSC® C105338
FSC
www.fsc.org

If you have any concerns about our products,
you can contact us on
ProductSafety@springernature.com

In case Publisher is established outside the EU,
the EU authorized representative is:
Springer Nature Customer Service Center GmbH
Europaplatz 3, 69115 Heidelberg, Germany

Printed by Libri Plureos GmbH
in Hamburg, Germany